Magnetic Stimulation in Clinical Neurophysiology
Second Edition

Magnetic Stimulation in Clinical Neurophysiology
Second Edition

Edited by:

Mark Hallett, M.D.
Chief, Human Motor Control Section, National Institute of Neurological Disorders and Stroke, National Institutes of Health, Bethesda, Maryland

Sudhansu Chokroverty, M.D., F.R.C.P., F.A.C.P.
Professor and Co-Chair of Neurology, Clinical Neurophysiology and Sleep Medicine, New Jersey Neuroscience Institute at JFK and Seton Hall University, Edison, New Jersey

ELSEVIER
BUTTERWORTH
HEINEMANN

ELSEVIER
BUTTERWORTH
HEINEMANN

The Curtis Center
170 S Independence Mall W 300E
Philadelphia, Pennsylvania 19106

MAGNETIC STIMULATION IN CLINICAL NEUROPHYSIOLOGY   ISBN 0-7506-7373-7
Copyright 2005, Elsevier, Inc. All rights reserved.

No part of this publication may be reproduced or transmitted in any form or by any means, electronic or mechanical, including photocopy, recording, or any information storage and retrieval system, without permission in writing from the publisher. Permissions may be sought directly from Elsevier's Health Sciences Rights Department in Philadelphia, PA, USA: phone: (+1) 215 238 7869, fax: (+1) 215 238 2239, e-mail: healthpermissions@elsevier.com. You may also complete your request on-line via the Elsevier homepage (http://www.elsevier.com), by selecting 'Customer Support' and then 'Obtaining Permissions'.

Notice

Medicine is an ever-changing field. Standard safety precautions must be followed but as new research and clinical experience broaden our knowledge, changes in treatment and drug therapy may become necessary or appropriate. Readers are advised to check the most current product information provided by the manufacturer of each drug to be administered to verify the recommended dose, the method and duration of administration, and contraindications. It is the responsibility of the treating physician, relying on experience and knowledge of the patient, to determine dosages and the best treatment for each individual patient. Neither the Publisher nor the editors assumes any liability for any injury and/or damage to persons or property arising from this publication.

The Publisher

First Edition 1990.   Second Edition 2005.

Library of Congress Cataloging-in-Publication Data

Magnetic stimulation in clinical neurophysiology / [edited by] Mark Hallett and
Sudhansu Chokroverty.–2nd ed.
   p. ; cm.
      Includes bibliographical references and index.
      ISBN 0-7506-7373-7
   1. Magnetic brain stimulation.   I. Hallett, Mark.   II. Chokroverty, Sudhansu.
      [DNLM: 1. Nervous System Physiology. 2. Electric Stimulation.
   3. Magnetics–therapeutic use. 4. Nervous System Diseases–therapy. WL 102 M196 2005]
      RC386.6.M32M34 2005
616.8–dc22   2004061889

Acquisitions Editor: Susan F. Pioli
Developmental Editor: Joan Ryan
Project Manger: David Saltzberg

Working together to grow
libraries in developing countries

www.elsevier.com | www.bookaid.org | www.sabre.org

Printed in the United States of America

Last digit is the print number: 9 8 7 6 5 4 3 2 1

ELSEVIER | BOOK AID International | Sabre Foundation

Contents

Contributing Authors vii
Preface xi

1 Basic Electromagnetism 1
Pedro Cavaleiro Miranda

2 Basic Physics and Design of Transcranial Magnetic Stimulation Devices and Coils 17
Jarmo Ruohonen and Risto J. Ilmoniemi

3 Activation of Peripheral Nerve and Nerve Roots 31
Marcela Panizza and Jan Nilsson

4 Transcranial Electrical and Magnetic Stimulation of the Brain: Basic Physiological Mechanisms 43
John C. Rothwell

5 The Physiology and Safety of Repetitive Transcranial Magnetic Stimulation 61
Zafiris J. Daskalakis and Robert Chen

6 Central Motor Conduction and Its Clinical Application 83
Christian W. Hess

7 Evaluation of Myelopathy, Radiculopathy, and Thoracic Nerve 105
Vincenzo Di Lazzaro and Antonio Oliviero

8 Cranial Nerves 129
Giorgio Cruccu and Mark Hallett

9 Transcranial Magnetic Stimulation and Brain Plasticity 143
Adriana B. Conforto and Leonardo G. Cohen

10 Transcranial Magnetic Stimulation in Amyotrophic Lateral Sclerosis 155
Ryuji Kaji and Nobuo Kohara

11 Motor System Physiology 165
Joseph Classen

12 Transcranial Magnetic Stimulation in Movement Disorders 181
Nicola Modugno, Antonio Currà, Francesca Gilio, Cinzia Lorenzano, Sergio Bagnato, and Alfredo Berardelli

13 Cerebellar Stimulation in Normal Subjects and Ataxic Patients 197
Yoshikazu Ugawa and Nobue K. Iwata

14 The Role of Transcranial Magnetic Stimulation in the Study of Fatigue 211
Paul Sacco, Gary W. Thickbroom, and Frank L. Mastaglia

15 Treatment of Movement Disorders 223
Hartwig R. Siebner

16 Transcranial Magnetic Stimulation in Stroke 239
Paolo M. Rossini and Flavia Pauri

17 Evaluation of Epilepsy and Anticonvulsants 253
Ulf Ziemann

18 Components of Language as Revealed by Magnetic Stimulation 271
Agnes P. Funk and Charles M. Epstein

19 Other Cognitive Functions 281
Marjan Jahanshahi

20 Individual Differences in the Response to Transcranial Magnetic Stimulation of the Motor Cortex 303
Eric M. Wassermann

21 Potential Therapeutic Uses of Transcranial Magnetic Stimulation in Psychiatric Disorders 311
Mark S. George, Ziad Nahas, F. Andrew Kozel, Xingbao Li, Kaori Yamanaka, Alexander Mishory, Sarah Hill, and Daryl E. Bohning

22 External Modulation of Visual Perception by Transcranial Magnetic and Direct Current Stimulation of the Visual Cortex 329
Andrea Antal, Michael A. Nitsche, and Walter Paulus

23 Somatosensory System 341
Massimiliano Oliveri

24 Eye Movements 349
René M. Müri, Christian W. Hess, and Charles Pierrot-Deseilligny

25 Intraoperative Monitoring of Corticospinal Function Using Transcranial Stimulation of the Motor Cortex 365
David Burke

26 Magnetic Stimulation in the Assessment of the Respiratory Muscle Pump 381
Michael I. Polkey and John Moxham

27 Clinical Applications of Functional Magnetic Stimulation in Patients with Spinal Cord Injuries 393
Vernon Lin and Ian Hsiao

28 Transcranial Magnetic Stimulation in Migraine 411
Alain Maertens de Noordhout, Anna Ambrosini, Peter S. Sándor, and Jean Schoenen

29 Transcranial Magnetic Stimulation in Sleep and Sleep-Related Disorders 419
Stefan Cohrs and Frithjof Tergau

30 Transcranial Magnetic Stimulation Studies in Children 429
Marjorie A. Garvey

Index 439

Contributing Authors

Anna Ambrosini, M.D., Ph.D.
Neurologist Consultant, Headache Clinic, INM
Neuromed, IRCCS, Pozzilli (Isernia), Italy

Andrea Antal, Ph.D.
Research Associate, Department of Clinical Neurophysiology, Georg-August University of Göttingen, Göttingen, Germany

Sergio Bagnato, M.D.
Department of Neurological Sciences, University of Rome La Sapienza, Rome; Department of Neurological Sciences, Instituto Mediterraneo Neuromed, University of Rome, Pozzilli, IS, Italy

Alfredo Berardelli, M.D.
Professor of Neurology, Department of Neurological Sciences, University of Rome La Sapienza, Rome, Italy

Daryl E. Bohning, Ph.D.
Professor of Radiology, Medical University of South Carolina, Charleston, South Carolina

David Burke, M.D., D.Sc.
Dean of Research and Development, College of Health Sciences, The University of Sydney, Sydney N.S.W., Australia

Robert Chen, M.A., M.B.BChir., M.Sc., F.R.C.P.C.
Associate Professor of Neurology, University of Toronto; Staff Neurologist, University Health Network, Toronto, Ontario, Canada

Joseph Classen, M.D.
Senior Lecturer in Neurology and Chief, Human Cortical Physiology and Motor Control Laboratory, Neurologische Klinik, Julius-Maximilians-Universität Würzburg, Würzburg, Germany

Leonardo G. Cohen, M.D.
Chief, Human Cortical Physiology Section, National Institute of Neurological Disorders and Stroke, National Institutes of Health, Bethesda, Maryland

Stefan Cohrs, M.D.
Department of Psychiatry and Psychotherapy, University of Göttingen, Göttingen, Germany

Adriana B. Conforto, M.D.
Professor of Neurology, São Paulo University; Staff Neurologist, Hospital das Clínicas, São Paulo, Brazil

Giorgio Cruccu, M.D.
Full Professor of Neurology, Department of Neurological Science, University of Rome La Sapienza; Head, Department of UOC Electromyography and Neurophysiology, Policlinico Umberto I, Rome, Italy

Antonio Currà, M.D., Ph.D.
Assistant Professor of Neurology, Department of Neurological Sciences, University of Rome La Sapienza, Rome, Italy

Zafiris J. Daskalakis, M.D., Ph.D., F.R.C.P.(C)
Assistant Professor of Psychiatry, University of Toronto; Staff Psychiatrist, Schizophrenia and Continuing Care Program, Centre for Addiction and Mental Health, Toronto, Ontario, Canada

Vincenzo Di Lazzaro, M.D.
Assistant Professor of Neurology, Institute of Neurology, Universitá Cattolica, Rome, Italy

Charles M. Epstein, M.D.
Professor of Neurology, Emory University School of Medicine and Atlanta VA Medical Center, Atlanta, Georgia

Contributing Authors

Agnes P. Funk, Ph.D.
Postdoctoral Fellow, Department of Neurology, Emory University School of Medicine, Atlanta, Georgia

Marjorie A. Garvey, M.B., B.Ch.
Chief, Pediatric Movement Disorders Unit, Human Motor Control Section, National Institute of Neurological Disorders and Stroke, National Institutes of Health, Bethesda, Maryland

Mark S. George, M.D.
Distinguished Professor of Psychiatry, Radiology, and Neurology, Medical University of South Carolina; Director, Brain Stimulation Laboratory and Director, Center for Advanced Imaging Research, Ralph H. Johnson VA Medical Center, Mental Health Service, Charleston, South Carolina

Francesca Gilio, M.D.
Department of Neurological Sciences, University of Rome La Sapienza, Rome; Department of Neurological Sciences, Instituto Neurologico Mediterraneo Neuromed IRCCS, University of Rome, Pozzilli, IS, Italy

Mark Hallett, M.D.
Chief, Human Motor Control Section, National Institute of Neurological Disorders and Stroke, National Institutes of Health, Bethesda, Maryland

Christian W. Hess, M.D.
Professor and Chairman, Department of Neurology, University of Bern; Medical Director, Department of Neurology, Inselspital (University Hospital of the Cantone of Bern), Bern, Switzerland

Sarah Hill, B.S., B.A.
Medical Student, College of Medicine, Medical University of South Carolina, Charleston, South Carolina

Ian Hsiao, Ph.D.
Department of Physical Medicine and Rehabilitation, College of Medicine, University of California, Irvine; Spinal Cord Institute, VA Long Beach Healthcare System, Long Beach, California

Risto J. Ilmoniemi, Ph.D.
Managing Director, Nexstim Ltd., Helsinki, Finland

Nobue K. Iwata, M.D., Ph.D.
Visiting Researcher, Department of Neurology, Division of Neuroscience, Graduate School of Medicine, The University of Tokyo, Tokyo, Japan

Marjan Jahanshahi, Ph.D.
Sobell Department of Motor Neuroscience and Movement Disorders, Institute of Neurology, The National Hospital for Neurology and Neurosurgery, London, United Kingdom

Ryuji Kaji, M.D., Ph.D.
Professor and Chairman, Department of Neurology, Tokushima University Graduate School of Medicine, Tokushima, Japan

Nobuo Kohara, M.D., Ph.D.
Director, Department of Neurology, Kobe City General Hospital, Kobe City, Japan

F. Andrew Kozel, M.D., M.S.
Assistant Professor of Psychiatry and Behavioral Neurosciences, Medical University of South Carolina; Psychiatry/Neuroscience Fellow, Ralph H. Johnson VA Medical Center, Charleston, South Carolina

Xingbao Li, M.D.
Research Scientist, Department of Psychiatry, Brain Stimulation Laboratory, Medical University of South Carolina, Charleston, South Carolina

Vernon Lin, M.D., Ph.D.
Associate Professor of Physical Medicine and Rehabilitation, University of California, Irvine; Associate Chief of Staff, VA Long Beach Healthcare System; Director, Spinal Cord Institute, VA Long Beach Healthcare System, Long Beach, California

Cinzia Lorenzano, M.D.
Department of Neurological Sciences, University of Rome La Sapienza, Rome; Department of Neurological Sciences, Instituto Neurologico Mediterraneo Neuromed IRCCS, University of Rome, Pozzilli, IS, Italy

Alain Maertens de Noordhout, M.D.
Professor and Chairman, University Department of Neurology, State University of Liège, Hôpital de la Citadelle, Liège, Belgium

Frank L. Mastaglia, M.D., F.R.C.P., F.R.A.C.P.
Professor and Director, Centre for Neuromuscular and Neurological Disorders, University of Western Australia; Consultant Neurologist, Sir Charles Gairdner Hospital, Perth, Western Australia

Pedro Cavaleiro Miranda, Ph.D.
Assistant Professor, Physics Department, Institute of Biophysics and Biomedical Engineering, Faculty of Science, University of Lisbon, Lisbon, Portugal

Alexander Mishory, M.D.
Department of Psychiatry, Brain Stimulation Laboratory, Medical University of South Carolina, Charleston, South Carolina

Nicola Modugno, M.D., Ph.D.
Consultant in Neurology, University of Rome La Sapienza, Rome, Italy

John Moxham, M.D., F.R.C.P.
Professor of Respiratory Medicine, Guy's Kings and St Thomas' School of Medicine, King's College; Professor of Respiratory Medicine and Consultant Physician, King's College Hospital, London, United Kingdom

René M. Müri, M.D.
Professor, Department of Neurology and Clinical Research, Perception and Eye Movement Laboratory, Inselspital, Bern, Switzerland

Ziad Nahas, M.D.
Assistant Professor of Psychiatry and Medical Director, Brain Stimulation Laboratory, Medical University of South Carolina, Charleston, South Carolina

Jan Nilsson, B.S.E.E.
Biomedical Engineer, Department of Biomedical Engineering and Medical Informatics, Salvatore Maugeri Foundation, IRCCS, Castel Goffredo (MN), Italy

Michael A. Nitsche, M.D.
Resident in Clinical Neurophysiology, Georg-August University of Göttingen, Göttingen, Germany

Massimiliano Oliveri, M.D., Ph.D.
Dipartimento di Psicologia, Universitá di Palermo; Laboratorio di Neurologia Clinica e Comportamentale, Fondazione Santa Lucia IRCCS, Rome, Italy

Antonio Oliviero, M.D.
Instituto di Neurologia, Universitá Cattolica, Rome, Italy; Fennsi Group, Hospital Nacional de Parapléjicos, Toledo, Spain

Marcela Panizza, M.D.
Clinical Neurophysiologist, Department of Electromyography, Hospital San Clemente, Mantova, Italy

Walter Paulus, M.D.
Professor of Clinical Neurophysiology and Head of Department of Clinical Neurophysiology, Georg-August University and University Hospital of Göttingen, Göttingen, Germany

Flavia Pauri, M.D.
Professor of Neurology and Otolaryngology, La Sapienza University; Department of Clinical Neuroscience, Ospedale S. Giovanni Calibita, Fatebenefratelli, Isola Tiberina, Rome, Italy

Charles Pierrot-Deseilligny, M.D.
Professor, Service de Neurologie, Hôpital de la Salpêtriére, Paris, France

Michael I. Polkey, Ph.D., F.R.C.P.
Reader in Respiratory Medicine, National Heart and Lung Institute; Consultant Physician, Department of Respiratory Medicine, Royal Brompton Hospital, London, United Kingdom

Paolo M. Rossini, M.D., Ph.D.
Professor of Neurology, University Campus Bio-Medico; Chairman, Department of Neuroscience, Ospedale S. Giovanni Calibita, Fatebenefratelli, Isola Tiberina; Scientific Director, Scientific Institute, Centro S. Giovanni di Dio, Brescia, Rome, Italy

John C. Rothwell, Ph.D.
Professor of Human Neurophysiology, Sobell Department of Neurophysiology, Institute of Neurology, London, United Kingdom

Jarmo Ruohonen, Ph.D.
Engineering Director, Nexstim Ltd., Helsinki, Finland

Paul Sacco, Ph.D.
Research Scientist, Centre for Neuromuscular and Neurological Disorders, University of Western Australia; Senior Lecturer, School of Biomedical and Sports Science, Edith Cowan University, Perth, Australia

Peter S. Sándor, M.D.
Professor of Neurology, University of Zurich, Switzerland

Jean Schoenen, M.D.
Professor of Neuroanatomy and Neurology, University of Liège; Consultant and Director, Headache Research Unit, Department of Neurology, CHR Citadelle, Liège, Belgium

Hartwig R. Siebner, M.D.
Privatdozent, M.D., Department of Neurology, Christian-Albrechts-University; Consultant, Department of Neurology, Schleswig-Holstein University Hospital, Kiel; Principal Investigator, Neuroimaging Centre, NeuroImage-Nord Hamburg-Kiel-Lübeck, Hamburg, Germany

Frithjof Tergau, M.D.
Priv.-Doz. Dr. med., Department of Clinical Neurophysiology, University of Göttingen, Göttingen, Germany

Gary W. Thickbroom, Ph.D.
Centre for Neuromuscular and Neurological Disorders, University of Western Australia, Perth, Western Australia

Yoshikazu Ugawa, M.D., Ph.D.
Associate Professor of Neurology, Division of Neuroscience, Graduate School of Medicine, The University of Tokyo, Tokyo, Japan

Eric M. Wassermann, M.D.
Adjunct Professor of Neurology, Uniformed Services University of the Health Sciences; Clinical Specialty Consultant, National Institute of Neurological Disorders and Stroke, National Institutes of Health, Bethesda, Maryland

Kaori Yamanaka, M.D.
Brain Stimulation Laboratory, Institute of Psychiatry, Medical University of South Carolina, Charleston, South Carolina; Department of Psychiatry, Gunma Hospital, Gunma, Japan

Ulf Ziemann, M.D.
Associate Professor of Neurology, J.W. Goethe University and Hospital of J.W. Goethe University, Frankfurt am Main, Germany

Preface

Every now and then in a scientific field there is a major event that changes the discipline. Such an event was witnessed in 1831 with the discovery by the English physicist Michael Faraday of electromagnetic induction (the induction of electric current in a circuit subjected to a changing magnetic field), and his introduction of the term *magnetic field* in 1845. Another milestone came in 1896 when d'Arsonval reported that the brain could be magnetically stimulated when the head was placed inside a coil carrying a high current producing magnetophosphenes caused by the induced currents. These observations were later verified by Thompson, Dunlap, Magnusson, and Stevens in the beginning of the 20th century. Just two decades ago, in 1985, Barker and associates succeeded in producing muscle contractions following transcranial magnetic stimulation using a powerful magnetic stimulator. Since then, rapid advances in the field ensued. In less than two decades, clinical neurophysiology has been energized with this new tool that has utility for diagnosis, research, and therapy. An enormous amount of new research has been enabled that has allowed a new understanding of the physiology of the brain, including even intracortical circuitry of the intact human. The literature is growing rapidly, and, for example, work with magnetic stimulation is one of the most frequent topics now for the journal *Clinical Neurophysiology*.

This book attempts to summarize the current state of the art, and should be useful for both the beginner and expert. The disciplines interested in magnetic stimulation include neurology, physical medicine, psychiatry, and neuroscience. We begin with chapters about the physics of magnetism and magnetic stimulation. The subsequent early chapters deal with methods, including methods for stimulation of peripheral nerve. The rest of the book deals with specific applications for routine clinical diagnosis, current state of the art in research, and new approaches to therapy.

So, what can magnetic stimulation do that makes it so valuable? Importantly, it is capable of stimulating nerve or brain almost painlessly. This characteristic makes it highly acceptable for clinical and research studies, much more than, for example, transcranial electrical stimulation, which is very painful. The lack of pain seems due to two features: the magnetic field penetrates tissue without loss of strength and the very brief pulse duration is not efficient in activating sensory nerves. Magnetic stimulation can activate brain or inhibit it, inhibition coming in large part from activation of inhibitory neurons. Simple activation allows measurement of central motor conduction time in the corticospinal tract, and this has given rise to clinical applications. Since much of the activation is transsynaptic, clever methods of delivering the stimulation produce techniques for examining synaptic networks. Assessment of brain excitability is extremely valuable in understanding its physiology. While a few single pulses of stimulation have no lasting effect on the brain, if stimulation is given repetitively in trains, there can be short or longer changes in excitability. These changes are being exploited for therapy.

This is the second edition of the book, first published in 1990. Comparing chapters and contents makes clear the rapid growth of the field. We are grateful to the expert authors who have taken off time from their other activities to contribute to this book.

Mark Hallett, M.D.
S. Chokroverty, M.D.

Magnetic Stimulation in Clinical Neurophysiology
Second Edition

Basic Electromagnetism

Pedro Cavaleiro Miranda

Magnetic stimulation makes use of a time-varying current flowing in a coil to stimulate excitable cells without resorting to electrodes. This phenomenon occurs because the time-varying magnetic field created by the current pulse in the coil gives rise to an induced electric field that depolarizes the cellular membrane.[1–5] Understanding the origin and properties of this electric field requires some knowledge of the basic concepts of electromagnetism. The aim of this chapter is to introduce some of these concepts, focusing on the aspects that are most relevant for magnetic stimulation.

The electric field due to static electric charges and its relationship to the electric potential are described. Such electric fields can make a significant contribution to the total electric field induced in tissues by magnetic stimulation. The concept of flux of a vector quantity is introduced, and the significance of Gauss's law, which deals with the flux of the electric field, is then examined. Moving on from electrostatics to magnetostatics, steady currents are considered next and a more general formulation of the familiar Ohm's law is presented. The magnetic flux density $\vec{B}$ generated by a steady current flowing in a coil, such as those used in magnetic stimulation, is then described, and the more useful magnetic vector $\vec{A}$ potential is introduced. Electromagnetic induction, whereby a magnetic field changing in time gives rise to an electric field, is presented, and the relationship between the magnitude of the induced electric field and the rate of change of current in the stimulation coil is established. Further details on these topics can be found in textbooks on electromagnetism, such as *Introductory Electromagnetics*[6] or *Introduction to Electrodynamics*.[7] In the last two sections of this chapter, the quasistatic approximation is brought up to justify the selective treatment of electromagnetism, and the impact of electrical conductivity boundaries on the spatial distribution of the induced electric field is highlighted.

■ Electric Charge, Coulomb's Law, and the Electric Field

There are only two types of electric charge, which are most conveniently described as positive and negative. Static electric charges exert considerable force on each other. In 1784, Coulomb determined experimentally that this force is repulsive for like charges and attractive for unlike charges, that it acts along the line joining the two particles, that it is proportional to the charge of each particle, and that it is inversely proportional to the square of the distance between the particles. In its simplest form, Coulomb's law and can be written as:

$$F = k\frac{q_1 q_2}{r^2} \qquad (1)$$

In this equation, F is the force acting on the charged particles, q_1 and q_2 are the charges on

2 Basic Electromagnetism

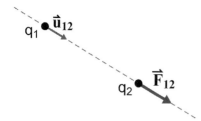

Figure 1-1 The electrostatic force exerted between charged particles at rest. This attractive or repulsive force acts along the line joining the two particles and decreases as the inverse of the square of the distance between the particles.

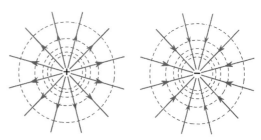

Figure 1-2 The electric field due to a positive charge *(left)* and a negative charge *(right)* is represented by means of electric field lines *(solid lines)*. The *dashed lines* represent equipotential lines, which are analogous to contour lines in topographic maps. In such maps, the direction of the electric field is analogous to the direction of steepest descent. Electric field lines and equipotential lines are always perpendicular to each other.

the two particles, r is the distance between the two pointlike particles, and k is a constant. The unit of charge is the Coulomb (C). Unlike charge, which is a scalar quantity, force is a vector* quantity defined by a magnitude and a direction. For this reason, Coulomb's law is more completely described in vector form as:

$$\vec{F}_{12} = \frac{1}{4\pi\varepsilon_0} \frac{q_1 q_2}{r_{12}^2} \vec{u}_{12} \qquad (2)$$

As shown in Figure 1-1, $\vec{F}_{12}$ is the force exerted by particle 1 on particle 2, and $\vec{u}_{12}$ is a vector of unit length pointing from particle 1 to particle 2. For practical reasons, the constant of proportionality is written as $\frac{1}{4\pi\varepsilon_0}$, in which ε_0 is the permittivity of free space and whose experimentally determined value is 8.854×10^{-12} Farad per meter (Fm^{-1}). If several charged particles are present at different positions in the vicinity of a test particle with charge q_0, the total force on that particle is calculated by finding the force exerted by each of the other particles. Summing up the individual vectors,†

$$\vec{F} = \sum_i \frac{1}{4\pi\varepsilon_0} \frac{q_0 q_i}{r_i^2} \vec{u}_i = \frac{q_0}{4\pi\varepsilon_0} \sum_i \frac{q_i}{r_i^2} \vec{u}_i \qquad (3)$$

---

*A vector $\vec{v}$ is usually written as $\vec{v} = x\vec{i} + y\vec{j} + z\vec{k}$, where $\vec{i}$, $\vec{j}$, and $\vec{k}$ are vectors of unit length along the x, y, and z axis, respectively. The length of the vector is $v = \sqrt{x^2 + y^2 + z^2}$, and its direction is determined by the relative values of the coordinates x, y, and z.
†The sum of two vectors is the vector whose coordinates are the sums of the corresponding coordinates of the individual vectors.

In Equation 3, the summation sign $\sum_i$ represents the sum over all charges q_i, and r_i is the distance between the particle with charge q_0 and the particle with charge q_i. It is assumed that the charge is small enough not to perturb the distribution of the other charges.

An electric field exists in a region of space if a force is exerted on a static charged particle placed in that region. The electric field, or electric field intensity, is defined at every point in space by a vector $\vec{E}$ whose direction is that of the force acting on the test particle and whose magnitude is the magnitude of the force divided by the charge of the test particle:

$$\vec{E} = \frac{\vec{F}}{q_0} = \frac{1}{4\pi\varepsilon_0} \sum_i \frac{q_i}{r_i^2} \vec{u}_i \qquad (4)$$

This vector field is usually represented by electric field lines, as shown in Figure 1-2. The direction of the electric field is always tangential to these lines and points in the direction of the arrow. A positive charge placed at rest in the electric field starts moving in the direction of the arrow, whereas a negative charge moves in the opposite direction. For a continuous distribution of charges described by a volume charge density ρ, which is usually a function of position, the expression in Equation 4 becomes:

$$\vec{E} = \frac{\vec{F}}{q_0} = \frac{1}{4\pi\varepsilon_0} \int_v \frac{\rho dv}{r^2} \vec{u}_r \qquad (5)$$

In Equation 5, dv is a small element of the total volume (V) where charge exists, $\vec{u}_r$ is a unit vector pointing from the position of the volume element dv to the point where $\vec{E}$ is being calculated, and r is the distance between the two points. The integral sign indicates that the sum is carried out over an infinitely large number of infinitesimally small volume elements, dv, within each of which ρ is considered constant.

The Electric Potential

The electric potential, ϕ, provides an equivalent way of describing the properties of the field surrounding static electric charges. It is a scalar field (i.e., it is defined by a magnitude at each point in space), and it is therefore often simpler to manipulate. The relationship between the electric field and the electric potential can be derived as follows. In a region where an electric field $\vec{E}$ exists, a charge q is subject to a force:

$$\vec{F} = q\vec{E} \qquad (6)$$

Consider the special case in which the electric field points along the x axis so that the previous equation simplifies to $F_x = qE_x$. Suppose that under the influence of this force, the charged particle moves by a small distance dx, along which E_x can be assumed to be constant. By definition, the work done by the electric field is given by $F_x dx$. As a result, the potential energy of the particle in the field changes by minus this amount, and the total energy is conserved. The resulting change in the electric potential, $d\phi$, is equal to the change in potential energy per unit charge:

$$d\phi = -\frac{F_x dx}{q} = -E_x dx \qquad (7)$$

The electric field E_x is related to the electric potential ϕ by:

$$E_x = -\frac{d\phi}{dx} \qquad (8)$$

E_x is minus the rate of change of ϕ with x. For infinitesimally small values of dx, the ratio $\frac{d\phi}{dx}$ is the derivative of ϕ with respect to x. The electric potential is measured in volts (V),

and the electric field is measured in volts per meter (Vm^{-1}).

If the electric potential is also a function of y and z, the partial derivatives[‡] must be used, and the following expressions apply:

$$E_x = -\frac{\partial \phi}{\partial x}, \quad E_y = -\frac{\partial \phi}{\partial y}, \quad E_z = -\frac{\partial \phi}{\partial z} \qquad (9)$$

If ϕ is known, the total electric field is then given by:

$$\vec{E} = -\left(\frac{\partial \phi}{\partial x}\vec{u}_x + \frac{\partial \phi}{\partial y}\vec{u}_y + \frac{\partial \phi}{\partial z}\vec{u}_z\right) \qquad (10)$$

In Equation 10, $\vec{u}_x$, $\vec{u}_y$, and $\vec{u}_z$ are unit vectors pointing along the direction of the x, y, and z axes, respectively. The electric field $\vec{E}$ is minus the gradient of the potential ϕ; at a given point in space, the electric field vector points in the direction along which the potential is decreasing most rapidly, and its magnitude is given by the rate of change of ϕ along that direction. Equation 10 can be written in a more compact and universal form as:

$$\vec{E} = -\nabla \phi \qquad (11)$$

In Equation 11, ∇ is the del operator, defined in cartesian coordinates by:

$$\nabla = \left(\frac{\partial}{\partial x}\vec{u}_x + \frac{\partial}{\partial y}\vec{u}_y + \frac{\partial}{\partial z}\vec{u}_z\right) \qquad (12)$$

If the electric field is known, the potential difference between any two points can be calculated as follows. Let the vector $\vec{d\ell}$ represent a small displacement along a path from A to B, whose components are dx, dy, and dz. The change in the potential energy of a test charge is equal to minus the work done by the electric field on the charge as it moves along $\vec{d\ell}$ per unit charge:

$$d\phi = -\frac{1}{q_0}(F_x dx + F_y dy + F_z dz)$$
$$= -(E_x dx + E_y dy + E_z dz) = -\vec{E} \cdot \vec{d\ell} \qquad (13)$$

[‡]If ϕ is a function of x, y, and z, then $\frac{\partial \phi}{\partial x}$ is the partial derivative of ϕ with respect to x and stands for the rate of change of ϕ with x when y and z are held constant.

in which Equation 6 was used to substitute for F in terms of E. The last expression introduces the use of the dot product.§ The potential difference, $\phi_B - \phi_A$, can be estimated by dividing the path from A to B into a large number of small displacements, $\vec{d\ell}$ and summing the potential changes, $d\phi$, due to each step. The true result is obtained in the limit of infinitesimally small displacements and is written using an integral sign rather than a summation sign:

$$\phi_B - \phi_A = \int_A^B d\phi = -\int_A^B \vec{E} \cdot \vec{d\ell} \quad (14)$$

The work done in going from B to A, going backward along the path taken from A to B, is equal but of opposite sign to the work done in going from A to B, because the direction of $\vec{d\ell}$ is reversed. Moreover, it can be shown that $\phi_B - \phi_A$ is independent of the path chosen. This means that the work done in moving a charged particle around any closed path back to its original position is zero:

$$\oint \vec{E} \cdot \vec{d\ell} = 0 \quad (15)$$

In Equation 15, the circle in the integral sign indicates that this line integral is carried over a closed path. The electrostatic field is said to be conservative.

Equation 14 defines only the electrical potential difference between two points. A potential of 0 V is arbitrarily assigned to a reference point, relative to which all potential values are specified. For example, an expression for the electric potential due to a positive charge Q at the origin can be obtained by assuming that the potential is zero at infinity, $\phi_\infty = 0$. Given the spherical symmetry of the arrangement (see Fig. 1–2), the electric potential depends only on the distance from the origin, r. Taking point A to be at infinity and point B to be at position r, the combination of Equation 14 and Equation 4 gives:

$$\phi_r = -\int_\infty^r E_r dr = \frac{Q}{4\pi\varepsilon_0 r} \quad (16)$$

The electric potential field is usually represented with the help of equipotential lines, as shown in Figure 1–2.

■ Flux of a Vector Quantity, Electric Flux, and Gauss's Law

The concept of flux through a surface applies to any vector field, but it is more easily explained in terms of the flow of a fluid. Imagine water flowing in a straight pipe, and suppose that the water is moving parallel to the pipe's axis with a velocity $\vec{v}$ (in ms^{-1}), at every point of a perpendicular cross section whose surface area is S (in m^2). The flux through this surface is the volume of water passing through the surface per unit time and is given by the product vS (in m^{3}s^{-1}). In reality, the velocity is higher in the center of the pipe and decreases closer to the wall, which means that the magnitude of the velocity depends on position. This can be taken into account by dividing the cross section into an infinite number of infinitesimal surface elements of area, dS, over which the velocity can be considered constant, and integrating the product, vdS. If $\vec{v}$ is not perpendicular to the surface element, only its perpendicular component will contribute to the flux. This can be expressed mathematically if the small surface element is represented by a vector, $\vec{dS}$ whose direction is perpendicular to its surface and whose magnitude is the area of the element. Then the flux through this element is given by $v\cos(\theta)dS = \vec{v} \cdot \vec{dS}$, where θ is the angle between $\vec{v}$ and $\vec{dS}$, as shown in Figure 1–3. The total flux through the surface is then given by:

$$\int_S \vec{v} \cdot \vec{dS} \quad (17)$$

---

§The dot or scalar product of two vectors, denoted by $\vec{A} \cdot \vec{B}$, is a scalar quantity equal to the sum of the products of the corresponding components: $x_A x_B + y_A y_B + z_A z_B$. It is also equal to the product of the lengths of the two vectors and the cosine of the angle θ between them: $AB\cos(\theta) = \sqrt{x_A^2 + y_A^2 + z_A^2}\sqrt{x_B^2 + y_B^2 + z_B^2}\cos(\theta)$. This corresponds to multiplying the length of one of the vectors by the length of the projection of the second vector along the direction of the first vector. It is equal to zero when the two vectors are perpendicular to each other.

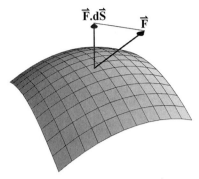

Figure 1–3 The flux of a vector field $\vec{F}$ through a surface S is given by the integral of the normal component $\vec{F} \cdot d\vec{S}$ over the whole surface. $\vec{F} \cdot d\vec{S}$ represents the product of the component of $\vec{F}$ normal to the surface element $d\vec{S}$ times the area of that element. The component of $\vec{F}$ parallel to $d\vec{S}$ does not contribute to the flux through the surface.

Flux has a magnitude and a sense. For example, the water in the pipe can flow through the surface S from left to right, but it could also flow in the opposite direction. These two situations can be distinguished by arbitrarily assigning a positive sign to the flux in one of the cases and a negative sign in the other.

A quantity of fundamental interest in electromagnetism is the flux through a closed surface. Imagine a surface enclosing a volume of water inside the pipe. By convention, the normal to a closed surface is directed outward so that flux out of the enclosed volume is positive and flux into it is negative. In the steady state, the amount of water in the volume is constant so as much water flows through the surface into the volume as flows out of it, independently of the shape of the closed surface. The total flux of water through the closed surface is therefore equal to zero:

$$\oint_S \vec{v} \cdot d\vec{S} = 0 \qquad (18)$$

The flux of the electric field through a surface S is given by:

$$\int_S \vec{E} \cdot d\vec{S} \qquad (19)$$

In a plot of electric field lines, the electric flux through a surface is large if many lines cross that surface at approximately right angles.

Gauss's law states that the electric flux through a closed surface is equal to the total charge Q enclosed by the surface, divided by ε_0:

$$\oint_S \vec{E} \cdot d\vec{S} = \frac{Q}{\varepsilon_0} \qquad (20)$$

This result is a consequence of the $1/r^2$ dependence of the electric field due to a single charged particle. It is often used to calculate the electric field when the charge distribution is known and has a high degree of symmetry. If the volume enclosed by the surface contains no charge ($Q = 0$), the electric field flux through the surface is zero:

$$\oint_S \vec{E} \cdot d\vec{S} = 0 \qquad (21)$$

In this case, the number of electric field lines entering the closed surface is equal to the number of lines leaving it. Gauss's law states that positive and negative electric charges are the source and sink of electrostatic field lines.

The combination of Equation 11 and Equation 21 written in differential form ($\nabla \cdot \vec{E} = 0$) gives Laplace's equation for the electric potential in a region where no charges are present:

$$\nabla^2 \phi = \left(\frac{\partial^2 \phi}{\partial x^2} + \frac{\partial^2 \phi}{\partial y^2} + \frac{\partial^2 \phi}{\partial z^2} \right) = 0 \qquad (22)$$

Problems in electrostatics are often tackled by solving Laplace's equation for ϕ and then calculating its gradient to get $\vec{E}$.

Conductors in an Electrostatic Field

The presence of a conducting volume has a significant impact on the electrostatic field. Conducting media have a large number of charged particles, such as electrons or ions, which are free to move. As a result, when a conductor is placed in an electrostatic field, positive charges move in the direction of the electric field (away from the positive charges that created the field), and negative charges move in the opposite direction. These displaced charges give rise to an electric field that has the opposite direction to the external electric field and therefore tends to cancel it. The movement of free charges ceases only when there is no net force acting on them (i.e., when the total electric field inside the conductor is zero). The electrostatic field inside any conductor, charged or uncharged, is therefore always zero. It follows that the electric flux through the surface of the conductor is zero, and according to Gauss's law, the net charge within the volume of the conductor must also be zero. Any excess charge in a conductor in an electrostatic field must reside strictly on its surface. The electric potential must be the same everywhere on the surface and in the bulk of the conductor; otherwise, a potential gradient (i.e., an electric field) would exist. The conductor is equipotential.

The charge redistribution that takes place when a conductor is placed in an electrostatic field is extremely rapid. The time constant associated with this process is given by the ratio of the permittivity to the electrical conductivity (ϵ/σ) of the medium and for biological tissues it is less than 1 µs. Because the electric field induced by magnetic stimulation rises and falls in a time interval that is of the order of 100 µs or more, the charge redistribution that follows it can be considered to occur instantaneously.

■ Electromotive Force, Steady Currents, Ohm's Law, and Current Continuity

The flow of steady current in a conductor cannot be achieved with an electrostatic field. If charged particles are introduced at one end of a wire, some will flow away from that end, rearranging themselves so that the electric field inside the wire is zero and the whole wire becomes equipotential. To keep a steady current flowing, a potential difference must be maintained between the two ends of the wire by means of a source of energy such as a battery, which continuously pushes charge from its low potential terminal to the high potential terminal. The source is said to have an electromotive force, emf (in volts), equal to the potential difference that it maintains between the two ends of a conductor, assuming a negligible internal resistance for the source.

Inside a conducting wire of uniform cross section and length L subject to an emf V, the electric potential falls linearly by V volts along its length. The electric field (i.e., potential gradient) inside the wire is therefore constant and given by:

$$E = \frac{V}{L} \quad (23)$$

This relationship between E and the emf V can be written as $EL = V$. More generally, if the electric field is not constant along the path, this equality must be rewritten as:

$$\int \vec{E} \cdot \vec{d\ell} = V \quad (24)$$

This equation is fundamentally different from Equation 15 in the electrostatics section because the emf V can give rise to a nonzero value for the integral of $\vec{E}$ around a closed path. Such an electric field cannot be attributed to static electric charges.

In the presence of an electric field, the free charges inside a conductor are accelerated by the force $q\vec{E}$ acting on them, giving rise to a net flow of charge in the direction of the field. The flow is hindered by collisions with the atoms and molecules that make up the conductor. In these collisions, the extra kinetic energy acquired between collisions is transferred to the conductor, heating it. As a result of these two processes, charged particles acquire a mean drift velocity, $\vec{v}$. The resulting current density, $\vec{J}$ in amperes per square meter (Am^{-2}), at any point in the conductor is the net amount of charge passing through a section perpendicular to $\vec{v}$ per second and per unit area of the section. It is given by:

$$\vec{J} = Nq\vec{v} \quad (25)$$

In Equation 25, N is the number of free charged particles per unit volume, and q is the particle's charge. For most conductors, the relationship between the applied electric field and the current density is linear:

$$\vec{J} = \sigma \vec{E} \quad (26)$$

In Equation 26, σ is the electrical conductivity of the conductor given in Siemens per meter (Sm^{-1}). Usually, the conductor is assumed to be homogeneous and isotropic (i.e., σ is a scalar quantity whose value does not depend on position or direction). If σ is known, Equation 26 can be used to calculate $\vec{J}$ when $\vec{E}$ is known and vice versa.

The current, or current intensity, I flowing through a surface S is given by the total flux of through that surface:

$$I = \int_S \vec{J} \cdot d\vec{S} \qquad (27)$$

If the current density is uniform across the wire, the current can be obtained by multiplying the current density by the cross-sectional area A of the conductor. Making use of Equations 26 and 23 gives:

$$I = AJ = A\sigma E = \sigma \frac{A}{L} V = \frac{V}{R} \qquad (28)$$

In Equation 28, R is the resistance of the wire. Equation 26 is a more general form of Ohm's law, of which Equation 28 ($V = RI$) is a particular case; whereas the latter holds for a specific conductor (e.g., a particular wire), the former holds at any point in a conductor.

In the steady state, the current going into a closed surface that does not enclose any sources or sinks of current must equal the current coming out. If this were not the case, a net current would flow into the enclosed volume, and the amount of charge within it would change with time, contrary to the steady-state assumption. The total flux of the current density vector through a closed surface that does not enclose any current sources or sinks must be zero:

$$\oint_S \vec{J} \cdot d\vec{S} = 0 \qquad (29)$$

In particular, this statement of current continuity is valid inside a volume conductor, such as the head, where current flow is induced by magnetic stimulation. It is a direct consequence of the conservation of charge. It is often written in differential form as $\nabla \cdot \vec{J} = 0$.

■ Magnetic Flux Density and the Magnetic Vector Potential

In magnetic stimulation, magnetic fields are created by large currents flowing in coils. The

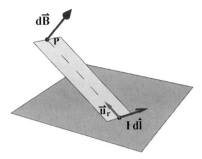

Figure 1–4 The Biot-Savart law is used to compute the contribution $d\vec{B}$ to the total magnetic flux density at point P due to a small current element $I d\vec{\ell}$ in a coil. The vector $d\vec{B}$ is perpendicular to the plane containing $I d\vec{\ell}$ and $\vec{u}_r$ with its sense is given by the right-hand rule and its magnitude is given by $\mu_0 I d\ell \sin\theta / 4\pi r^2$, where θ is the angle between $d\vec{\ell}$ and $\vec{u}_r$.

magnetic flux density $\vec{B}$ due to a current flowing in a wire of arbitrary shape can be calculated by considering first the contribution $d\vec{B}$ due to a small current element $I d\vec{\ell}$ where I is the current intensity and $d\vec{\ell}$ is a vector tangent to the wire and whose length is the length $d\ell$ of the small segment of wire under consideration, as shown in Figure 1–4. The magnetic flux density $d\vec{B}$ is given by the Biot-Savart law:

$$d\vec{B} = \frac{\mu_0 I}{4\pi} \frac{d\vec{\ell} \times \vec{u}_r}{r^2} \qquad (30)$$

In Equation 30, μ_0 is a constant called the magnetic permeability of vacuum, $\vec{u}_r$ is a unit vector pointing from the position of $d\vec{\ell}$ toward the point P where $d\vec{B}$ is being calculated, and r is the distance between these two points. By definition of the cross product,[||] $d\vec{B}$ is perpendicular to $d\vec{\ell}$ and $\vec{u}_r$.

The magnetic flux density $\vec{B}$ given in Tesla (T), due to a current I flowing in a closed loop

---

[||]The cross or vector product of two vectors, denoted by $\vec{A} \times \vec{B}$, is a vector perpendicular to the plane containing $\vec{A}$ and $\vec{B}$ and whose length is equal to the product of the lengths of the two vectors and the sine of the angle θ between them: $AB\sin(\theta)$. It is equal to zero when the two vectors are parallel. The direction of $\vec{A} \times \vec{B}$ is determined by the right-hand rule: If the fingers of the right hand twist around the raised thumb from $\vec{A}$ to $\vec{B}$, then $\vec{A} \times \vec{B}$ points along the direction of the thumb.

8 Basic Electromagnetism

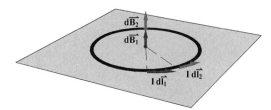

Figure 1–5 Calculation of the magnetic flux density $\vec{B}$ at the center of a round coil. The vector $\vec{u}_r$ always points radially inward, and the vector $d\vec{\ell}$ is always tangential to the circle so that the angle between them is always 90 degrees. The integral of $d\ell$ along the circular path of the coil is equal to its perimeter, $2\pi a$.

is given by:

$$\vec{B} = \frac{\mu_0 I}{4\pi} \oint \frac{d\vec{\ell} \times \vec{u}_r}{r^2} \quad (31)$$

This formula represents the sum of an infinitely large number of terms (e.g., Equation 30) as $d\vec{\ell}$ follows the path of the loop. Equation 31 can be used to calculate the magnetic flux density at the center of a round coil of radius a in the following way. In this case, $d\vec{\ell}$ and $\vec{u}_r$ are always perpendicular to each other, as shown in Figure 1–5. The cross product $d\vec{\ell} \times \vec{u}_r$ defines a vector that is perpendicular to the plane of the coil and points in the direction given by the right-hand rule (i.e., if the right thumb points along I, the curved fingers point along B). Its length is equal to $d\ell\, u_r \sin 90$ degrees, which is equal to $d\ell$. The distance r from the coil center to the wire is constant and equal to the radius a. Vector $\vec{B}$ at the center of the coil points into the sheet of paper, and its magnitude is given by:

$$B = \frac{\mu_0 I}{4\pi} \oint \frac{d\ell}{a^2} = \frac{\mu_0 I}{4\pi} \frac{2\pi a}{a^2} = \frac{\mu_0 I}{2a} \quad (32)$$

It is proportional to the current in the coil and inversely proportional to its radius. For a typical magnetic stimulation coil the magnetic flux density at the center of the coil can be as high as 2 T.

The flux density due to a current flowing in a circular wire is illustrated in Figure 1–6 by means of magnetic field lines. Where these lines are closer together, the field is more intense, as was the case with electric field lines.

Figure 1–6 The flux density due to a round coil is represented by means of magnetic field lines. The vector $\vec{B}$ is tangential to the field lines. For clarity, only the lines lying in one plane are shown; other lines can be obtained by rotation of the pattern shown about the coil axis. The flux through the coil is obtained by integrating the flux density over the area defined by the coil.

However, whereas electrostatic field lines originated at positive charges and terminated at negative charges, magnetostatic field lines form closed loops around the current carrying conductor. The magnetic flux Φ through a surface S, such as the flat circular surface defined by the coil in Figure 1–6, is the flux of vector $\vec{B}$ through that surface:

$$\Phi = \int_S \vec{B} \cdot d\vec{S} \quad (33)$$

This is the reason why $\vec{B}$ is often referred to as the *magnetic flux density* rather than simply the *magnetic field*. Because there are no sources or sinks of magnetic field lines (i.e., there are no magnetic monopoles that would play the role that electric charges play in electrostatics), the magnetic flux entering a closed surface must equal the flux leaving it (i.e., magnetic flux is conserved). This is expressed mathematically as follows (compare with Gauss's law in electrostatics, Equations 20 and 21):

$$\oint_S \vec{B} \cdot d\vec{S} = 0 \quad (34)$$

The magnetic field is most often described in terms of the magnetic flux density $\vec{B}$ but the solution of some problems is simplified by the introduction of yet another vector field, the magnetic vector potential $\vec{A}$. The relationship between $\vec{A}$ and $\vec{B}$ bears some resemblance to the relationship between the electric field $\vec{E}$ and the electric scalar potential ϕ (see Equation 11). However, $\vec{A}$ is a vector field, whereas ϕ is a scalar field, and $\vec{B}$ is derived from $\vec{A}$ by a different differential operator: $\vec{B} = \nabla \times \vec{A}$. The curl operator is denoted by $\nabla \times$, where ∇ stands for the del operator (see Equation 12) and $\times$ represents the cross product.¶ The magnetic vector potential is often easier to calculate than $\vec{B}$ because it is related to the current flowing in a closed loop by:

$$\vec{A} = \frac{\mu_0 I}{4\pi} \oint \frac{\vec{d\ell}}{r} \qquad (35)$$

In Equation 35, all the symbols have the same meaning as in Equations 30 and 31.

Even though the integral in the expression for $\vec{A}$ is considerably simpler to evaluate than the one in the Biot-Savart law, it still leads to a complicated formula for the magnetic vector potential field produced by a round coil of radius a. For more complicated coil geometries, it is often evaluated using numerical methods. A scheme for computing the magnetic vector potential produced at a point P by a round coil lying in the xy plane and centered on the origin is shown in Figure 1–7A. The perimeter of the round coil is divided into a large number of small elements, and the coordinates of the vector $\vec{d\ell}$ tangential to each element are found. Each vector is then divided by the distance r between the center of the element and point P, which amounts to dividing each one of the three coordinates by r. All the weighted vectors $\frac{\vec{d\ell}}{r}$ are then added by adding all the x (y or z) coordinates to obtain the x (y or z) coordinate of the resultant vector. By symmetry, the vector $\vec{A}$ at point P is tangential to a circle centered on the axis of the coil (the z axis) and lying in a plane parallel to the plane of the coil (the xy plane). This can be demonstrated graphically by drawing the weighted vectors with their origin at point P and verifying that the radial components add up to zero, whereas the tangential ones do not. Along this circle, $\vec{A}$ has a constant magnitude (i.e., such circles are equipotential lines for the magnetic vector potential). If P is located at the center of the coil, all the elements are divided by the same distance, the coil radius a, and the sum of all

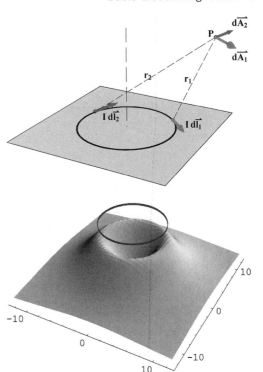

Figure 1–7 The magnetic vector potential $\vec{A}$ due to a round coil. **A (top),** The potential at point P is obtained by summing the contributions $\vec{dA}$ from the various elements $I\vec{d\ell}$. **B (bottom),** Plot of the magnetic vector potential as a function of position (x, y) in a plane parallel to a coil of radius a and a distance $z = 0.4a$ below it.

the tangential vectors is the null vector (i.e., zero length). The magnetic vector potential is equal to zero everywhere on the axis of a round coil, unlike the magnetic flux density, and is highest near the current-carrying conductor (see Figure 1–7B).

In this section, it was implicitly assumed that the current-carrying conductors were placed in vacuum. Because air and most biological tissues have a magnetic permeability that is very close to that of vacuum, no further elaboration of the equations presented here is required.

■ Time-Varying Magnetic Fields and Faraday's Law of Electromagnetic Induction

In 1831, Faraday attempted without success to produce a steady electric current in a loop of wire using static magnetic fields. Instead, he

---

¶In cartesian coordinates, $\nabla \times \vec{A} = \left(\frac{\partial A_z}{\partial y} - \frac{\partial A_y}{\partial z}\right)\vec{u}_x + \left(\frac{\partial A_x}{\partial z} - \frac{\partial A_z}{\partial x}\right)\vec{u}_y + \left(\frac{\partial A_y}{\partial x} - \frac{\partial A_x}{\partial y}\right)\vec{u}_z$.

observed that currents did flow momentarily when a magnet was brought up to the coil or moved away from it but not when it was held stationary relative to the coil. A similar effect can be achieved by replacing the moving magnet by a moving coil carrying a steady current or by a stationary coil carrying a time-varying current. In all cases, the magnetic flux density $\vec{B}$ at the pickup coil is changing with time. The current in the pickup coil indicates that an electric field exists in the coil, causing the free charges in it to flow.

The electric field associated with a time-varying magnetic field is usually referred to as the *induced* electric field. The force exerted by this field on charged particles is still given by Equation 6. However, unlike the case of an electrostatic field, the integral of the induced electric field around the closed loop of the coil is not equal to zero; otherwise, there would be no current flowing in the coil. Faraday attributed this new electromotive force (see Equation 24) to the change in magnetic flux Φ threading the coil with time:

$$V = \oint \vec{E} \cdot d\vec{\ell} = -\frac{d\Phi}{dt} \qquad (36)$$

The minus sign arises because the current flows in a direction such as to oppose the change in magnetic flux (i.e., if the flux is increasing, the current in the coil produces a magnetic field that decreases the total flux through the coil). This is known as Lenz's law and follows from the principle of the conservation of energy. The flux can be written in terms of the magnetic flux density so that:

$$V = \oint \vec{E} \cdot d\vec{\ell} = -\frac{d}{dt}\int_S \vec{B} \cdot d\vec{S} \qquad (37)$$

In Equation 37, the integration is carried out over a surface S defined by the coil. If the coil is not moving, this surface does not change with time, and the equation becomes:

$$\oint \vec{E} \cdot d\vec{\ell} = -\int_S \frac{\partial \vec{B}}{\partial t} \cdot d\vec{S} \qquad (38)$$

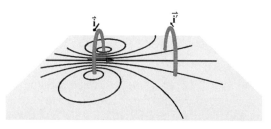

Figure 1–8 Electromagnetic induction. The emf induced in the pickup coil *(right)* is proportional to the rate of change of the flux threading it. For stationary coils, it is determined by the rate of change of the current in the stimulation coil *(left)*. The field lines represent the flux density, as explained in Figure 1–6.

If in addition to the induced electric field there is an electrostatic field due to some charge distribution, the previous equation is still valid because the static contribution to the integral on the left is zero.[#] Figure 1–8 illustrates the emf induced in a pickup coil by a current pulse flowing in a nearby coil. The magnetic flux density at the pickup coil can be calculated using Equation 31.

Equation 38 relates the integral of the induced electric field around a closed loop to the rate of change of the magnetic flux through a surface whose rim is the loop. It is Faraday's law of electromagnetic induction written in integral form. By appropriate manipulation of this equation, it is possible to derive an equivalent relationship between $\vec{E}$ and $\vec{B}$:

$$\nabla \times \vec{E} = -\frac{\partial \vec{B}}{\partial t} \qquad (39)$$

$\nabla \times$ is the differential operator curl, and Equation 39 is the differential version of Equation 38. Whereas the first equation holds for a given contour (the coil) and a surface defined by it, the differential equation must be satisfied at every point in the field. Substituting $\vec{B} = \nabla \times \vec{A}$ into Equation 39 and inverting the order of the differential

---

[#]Tofts and Branston[8] describe an important consequence of this regarding the measurement of the total induced electric field.

operators $\nabla \times$ and $\frac{\partial}{\partial t}$ leads to:

$$\nabla \times \vec{E} = -\nabla \times \left(\frac{\partial \vec{A}}{\partial t}\right)$$

$$\nabla \times \left(\vec{E} + \frac{\partial \vec{A}}{\partial t}\right) = 0 \quad (40)$$

This equality must hold at every point in space, which is possible only if the expression in parenthesis is everywhere zero:

$$\vec{E} = -\frac{\partial \vec{A}}{\partial t} \quad (41)$$

This simpler statement of Faraday's law of electromagnetic induction gives the induced electric field in terms of the rate of change in time of the magnetic vector potential. If the coil is stationary, as it is in magnetic stimulation, the direction of $\vec{A}$ at any point in space does not change with time, and Equation 41 means that the direction of $\vec{E}$ is parallel but opposite to the direction of $\vec{A}$ when $\vec{A}$ is increasing with time and that the magnitude of $\vec{E}$ is equal to the rate of change of the magnitude of $\vec{A}$.

In magnetic stimulation, the rate of change of $\vec{A}$ with time is determined by the rate of change of the current through the coil. Taking the time derivative of Equation 35 gives the following expression for the rate of change of $\vec{A}$:

$$\frac{\partial \vec{A}}{\partial t} = \frac{\mu_0}{4\pi} \frac{dI}{dt} \oint \frac{d\vec{\ell}}{r} \quad (42)$$

The induced electric field strength $\frac{\partial \vec{A}}{\partial t}$ is proportional to $\frac{dI}{dt}$. This equation also means that the time course of the induced electric field is equal to the time course of $\frac{dI}{dt}$. During the rising phase of the current pulse, the current can be increasing by as much as 100 A μs^{-1}, and electric fields of the order of 100 Vm^{-1} can be induced in the human cortex.

■ The Quasistatic Approximation

The current pulses used in magnetic stimulation have a rise time of the order of 100 μs and the time-varying induced electric field therefore has a frequency spectrum that typically peaks at less than 10 kHz. Given the relatively slow rate of change with time of the induced electric field and given the electric properties of the stimulated tissues, three simplifying assumptions are invariably made that are usually referred to collectively as the *quasistatic approximation*.[9,10] With these simplifications, the equations presented dealing with electrostatics, magnetostatics, and electromagnetic induction are sufficient to describe magnetic stimulation of biological tissues, despite the fact that magnetic stimulation relies on (slowly) time-varying electric and magnetic fields.

In air, electromagnetic waves propagate at the velocity of light c and have a wavelength λ that is related to their frequency f by $\lambda = c/f$. At any one moment, the amplitude of the electric field associated with the wave varies from maximum to minimum, passing through zero over the spatial extent corresponding to one wavelength. Propagation effects can be neglected as long as the wavelength associated with the induced electric field is much larger than the dimensions of the object of interest, such as the head. This is the case, because the wavelength of a 10 kHz electromagnetic wave is 30 km. The induced electric field can be considered to follow exactly the same time course everywhere within the region of interest without any significant propagation delays. This is what is usually meant by quasistatic approximation.

The second assumption is that the current that flows in the tissues in response to the induced electric field produces a magnetic field that is negligible compared with the magnetic field produced by the current in the coil. In conformity with Lenz's law, the magnetic field produced by the induced current opposes changes in the inducing magnetic field. As a result, the total magnetic field, the induced electric field, and the current density in the tissue decay exponentially with depth with a characteristic length known as the *skin depth*. In magnetic stimulation, the skin depth is several meters, whereas stimulation takes place only a few centimeters below the tissue surface. The extra attenuation of the induced electric field with depth due to the skin effect can therefore also be neglected. This assumption holds because of the low frequencies involved in magnetic stimulation, the relatively low conductivity of biological tissues and the small dimension of the stimulation coils compared with the skin depth.

In general, the response of a biological tissue to the induced electric field is governed by its electrical conductivity, σ, and by its dielectric permittivity, ε. These parameters determine the current density and the charge density, respectively, generated in the tissue by the electric field. The first term corresponds to a resistive current, whereas the second term gives rise to a capacitive current. The ratio of the resistive to the capacitive current is given by $\frac{2\pi f \varepsilon}{\sigma}$ and is typically less than 1% for most biological tissues for frequencies of about 10 kHz. To a good approximation, the tissues can be considered to be purely resistive (i.e., the electric properties of the tissues are characterized by a single parameter σ as in Equation 26.

■ Effect of Electrical Conductivity Boundaries on the Spatial Distribution of the Induced Electric Field

When a magnetic stimulation coil is fired in a homogeneous and isotropic medium, such as in air, and far away from any boundaries between media with different electric conductivity values, the spatial distribution of the induced electric field can be calculated using Equations 41 and 35. It is determined exclusively by the geometry of the coil windings.[11,12]

In practice, the coil will be placed on a subject's scalp or skin, and it will inevitably be close to several conductivity boundaries. A simple model is to assume that the tissue is homogeneous and isotropic so that there is only one boundary, the one that separates the tissue from the surrounding air. The induced electric field causes currents to flow in the tissue but not in the air. In the steady state, current continuity requires that the normal component of current density $\vec{J}$ at all points on the air-tissue boundary be zero inside the conductor as it is outside:

$$\vec{J} \cdot \vec{n} = 0 \quad (43)$$

In Equation 43, $\vec{n}$ is a unit vector perpendicular to the boundary. However, the magnetic vector potential $\vec{A}$ and therefore the induced electric field $\vec{E}$ generally have a component normal to the boundary, as illustrated in Figure 1–9. As a result, a normal component of the current density exists for a short period, causing charge to build up on the boundary. The charge redistribution stops as soon as the electric field generated by the surface charge and the electric field generated by the magnetic vector potential have normal components that are exactly equal but opposite. In purely resistive tissues, this cancellation of the normal component of the induced electric field would occur instantaneously. In practice, the cancellation of the normal component is sufficiently rapid for its effect to be negligible.

The surface charge contribution to the total electric field can be expressed as the gradient of a scalar potential, unlike the magnetic vector potential contribution, and is usually denoted by $-\nabla\phi$. The total electric field is therefore given by:

$$\vec{E} = -\frac{\partial \vec{A}}{\partial t} - \nabla\phi \quad (44)$$

Because the curl of the gradient of any scalar function ϕ is zero $[\nabla \times (\nabla\phi)]$, Equation 44 is the most general solution to Equation 40. In most cases, the induced electric field has two components, one whose spatial distribution is determined by the coil configuration and position, $-\frac{\partial \vec{A}}{\partial t}$, and another whose spatial distribution is also determined by the location of the conductivity boundaries, $-\nabla\phi$. Because the tissues are assumed to be purely resistive, $-\nabla\phi$ follows $-\frac{\partial \vec{A}}{\partial t}$ instantaneously so that the time course of both components is determined by the time course of $\frac{dI}{dt}$.

The presence of the surface charge at the boundary of a homogeneous isotropic medium has important consequences. It generally reduces the magnitude of the total induced electric field because it cancels its normal component. It also makes the spatial distribution of the induced electric field dependent on the shape of the boundary. The cancellation of the normal component implies that it is not possible to induce an

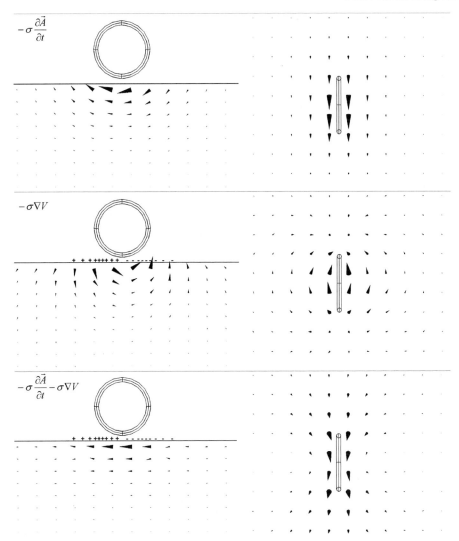

Figure 1–9 The current density is induced in a saline bath by a circular coil placed above and perpendicular to the air-bath interface. **Left column,** side view of the plane containing the coil. **Right column,** top view of the plane just below the bath's surface. **Top row,** contribution from the changing current in the coil; the arrows are tangent to circles that are concentric with the coil and lie in planes parallel to it. **Middle row,** contribution from the surface charge that builds up at the interface. **Bottom row,** the total current density is the sum of the two contributions above; it has no component perpendicular to the interface, and the current density lines form closed loops in planes parallel to the interface. Different scales were used in the different rows.

electric field component perpendicular to a planar infinite boundary or to induce a radial component in a spherical conductor.[13–15] The electric field distribution has been calculated and measured for several simple boundary geometries and coil shapes by many investigators.[16–31]

The stimulated volume is likely to be highly heterogeneous and anisotropic. At each one of the numerous boundaries within it, the perpendicular component of the current density, $\vec{J} \cdot \vec{n}$, must be the same on both sides. To satisfy this condition, charge builds up on the boundaries. For example, it can be shown that this charge accumulation at boundaries can have a significant effect on the electric field distribution within the brain, given the degree of heterogeneity and anisotropy known to exist there.[32] To calculate the electric field induced in a real brain, the conductivity must be known at all points in the head. This information may be derived from diffusion tensor magnetic resonance images by taking the three principal axes of the conductivity tensor to be those of the measured water diffusion tensor and by scaling the diffusion coefficients along these directions into conductivity values.[33,34] These values can then be used in conjunction with numerical methods, such as the finite element method,[35,36] to calculate the distribution of the induced electric field in realistic models of the head. The spatial distribution of the gradient of the induced electric field, which is relevant in peripheral nerve stimulation, can be derived analytically or numerically from that of the electric field.

REFERENCES

1. Roth BJ, Basser PJ. A model of the stimulation of a nerve fiber by electromagnetic induction. IEEE Trans Biomed Eng 1990;37:588–597.
2. Basser PJ, Roth BJ. Stimulation of a myelinated nerve axon by electromagnetic induction. Med Biol Eng Comput 1991;29:261–268.
3. Amassian VE, Eberle L, Maccabee PJ, et al. Modelling magnetic coil excitation of human cerebral cortex with a peripheral nerve immersed in a brain-shaped volume conductor: the significance of fiber bending in excitation. Electroencephalogr Clin Neurophysiol 1992;85:291–301.
4. Nagarajan SS, Durand DM, Warman EN. Effects of induced electric fields on finite neuronal structures: a simulation study. IEEE Trans Biomed Eng 1993;40:1175–1188.
5. Maccabee PJ, Amassian VE, Eberle LP, et al. Magnetic coil stimulation of straight and bent amphibian and mammalian peripheral nerve in vitro: locus of excitation. J Physiol 1993;460:201–219.
6. Popovic ZB, Popovic BD. Introductory Electromagnetics. Upper Saddle River, NJ, Prentice Hall, 2000.
7. Griffiths DJ. Introduction to Electrodynamics, 3rd ed. Upper Saddler River, NJ, Prentice Hall, 1999.
8. Tofts PS, Branston NM. The measurement of electric field, and the influence of surface charge, in magnetic stimulation. Electroencephalogr Clin Neurophysiol 1991;81:238–239.
9. Roth BJ, Cohen LG, Hallett M. The electric field induced during magnetic stimulation. Electroencephalogr Clin Neurophysiol Suppl 1991;43:268–278.
10. Plonsey R, Heppner DB. Considerations of quasi-stationarity in electrophysiological systems. Bull Math Biophys 1967;29:657–664.
11. Cohen LG, Roth BJ, Nilsson J, et al. Effects of coil design on delivery of focal magnetic stimulation. Technical considerations. Electroencephalogr Clin Neurophysiol 1990;75:350–357.
12. Grandori F, Ravazzani P. Magnetic stimulation of the motor cortex—theoretical considerations. IEEE Trans Biomed Eng 1991;38:180–191.
13. Tofts PS. The distribution of induced currents in magnetic stimulation of the nervous system. Phys Med Biol 1990;35:1119–1128.
14. Roth BJ, Saypol JM, Hallett M, et al. A theoretical calculation of the electric field induced in the cortex during magnetic stimulation. Electroencephalogr Clin Neurophysiol 1991;81:47–56.
15. Heller L, van Hulsteyn DB. Brain stimulation using electromagnetic sources: theoretical aspects. Biophys J 1992;63:129–138.
16. Ueno S, Tashiro T, Harada K. Localized stimulation of neural tissues in the brain by means of a paired configuration of time-varying magnetic fields. J Appl Phys 1988;64:5862–5864.
17. Roth BJ, Cohen LG, Hallett M, et al. A theoretical calculation of the electric field induced by magnetic stimulation of a peripheral nerve. Muscle Nerve 1990;13:734–741.
18. Maccabee PJ, Eberle L, Amassian VE, et al. Spatial distribution of the electric field induced in volume by round and figure '8' magnetic coils: relevance to activation of sensory nerve fibers. Electroencephalogr Clin Neurophysiol 1990;76:131–141.
19. Maccabee PJ, Amassian VE, Eberle LP, et al. Measurement of the electric field induced into inhomogeneous volume conductors by magnetic coils: application to human spinal neurogeometry. Electroencephalogr Clin Neurophysiol 1991;81:224–237.
20. Nyenhuis JA, Mouchawar GA, Bourland JD, et al. Energy considerations in the magnetic (eddy-current) stimulation of tissues. IEEE Trans Magn 1991;27:680–687.
21. Cohen D, Cuffin BN. Developing a more focal magnetic stimulator. Part I. Some basic principles. J Clin Neurophysiol 1991;8:102–111.
22. Yunokuchi K, Cohen D. Developing a more focal magnetic stimulator. Part II. Fabricating coils and measuring induced current distributions. J Clin Neurophysiol 1991;8:112–120.
23. Eaton H. Electric field induced in a spherical volume conductor from arbitrary

23. [continued] coils: application to magnetic stimulation and MEG. Med Biol Eng Comput 1992;30:433–440.
24. Esselle KP, Stuchly MA. Neural stimulation with magnetic fields: analysis of induced electric fields. IEEE Trans Biomed Eng 1992;39:693–700.
25. Durand D, Ferguson AS, Dalbasti T. Effect of surface boundary on neuronal magnetic stimulation. IEEE Trans Biomed Eng 1992;39:58–64.
26. Esselle KP, Stuchly MA. Quasi-static electric field in a cylindrical volume conductor induced by external coils. IEEE Trans Biomed Eng 1994;41:151–158.
27. Roth BJ, Maccabee PJ, Eberle LP, et al. In vitro evaluation of a 4-leaf coil design for magnetic stimulation of peripheral nerve. Electroencephalogr Clin Neurophysiol 1994;93:68–74.
28. Ren C, Tarjan PP, Popovic DB. A novel electric design for electromagnetic stimulation—the Slinky coil. IEEE Trans Biomed Eng 1995;42:918–925.
29. Ruohonen J, Ravazzani P, Grandori F. An analytical model to predict the electric field and excitation zones due to magnetic stimulation of peripheral nerves. IEEE Trans Biomed Eng 1995;42:158–161.
30. Ruohonen J, Ravazzani P, Nilsson J, et al. A volume-conduction analysis of magnetic stimulation of peripheral nerves. IEEE Trans Biomed Eng 1996;43:669–678.
31. Zimmermann KP, Simpson RK. "Slinky" coils for neuromagnetic stimulation. Electroencephalogr Clin Neurophysiol 1996;101:145–52.
32. Miranda PC, Hallett M, Basser PJ. The electric field induced in the brain by magnetic stimulation: a 3-D finite-element analysis of the effect of tissue heterogeneity and anisotropy. IEEE Trans Biomed Eng 2003;50:1074–1085.
33. Basser PJ, Mattiello J, LeBihan D. MR diffusion tensor spectroscopy and imaging. Biophys J 1994;66:259–267.
34. Tuch DS, Wedeen VJ, Dale AM, et al. Conductivity tensor mapping of the human brain using diffusion tensor MRI. Proc Natl Acad Sci USA 2001;98:11697–11701.
35. Wang W, Eisenberg SR. A three-dimensional finite element method for computing magnetically induced currents in tissues. IEEE Trans Magn 1994;30:5015–5023.
36. Cerri G, De Leo R, Moglie F, et al. An accurate 3-D model for magnetic stimulation of the brain cortex. J Med Eng Technol 1995;19:7–16.

2

Basic Physics and Design of Transcranial Magnetic Stimulation Devices and Coils

Jarmo Ruohonen and Risto J. Ilmoniemi

In this chapter, we outline the basic electrophysiology and physics relevant for transcranial stimulation. We first introduce the essential mechanisms of the cellular response and explain the key differences between transcranial electrical stimulation (TES) and transcranial magnetic stimulation (TMS). The focus then shifts to TMS and the design of TMS devices and coils. In the last part of the chapter, we present formulas that relate the coil shape to its inductance, resistance, warming, and the electric field induced in the brain.

■ Basic Phenomena behind Transcranial Stimulation

Electrophysiological Basis

An electric field **E** applied on neuronal tissue can excite nerve cells (Fig. 2–1). The field forces free charges into coherent motion both in the intracellular and extracellular spaces. In other words, the electric field drives an intracranial electrical current $\mathbf{J} = \sigma \mathbf{E}$ in the brain, where σ is the electrical conductivity of the brain. Cell membranes that interrupt the current flow become depolarized or hyperpolarized. If the depolarization is sufficient, a progressing depolarization front, or action potential may be initiated. This is the mechanism of action for electrical and magnetic stimulation.

In practice, the electric field in the cortex should be on the order of 100 V/m to elicit motor-cortex activation sufficiently strong to lead to hand muscle twitches. With $\sigma = 0.4$ S/m in the cortex, this field strength gives rise to a cortical current density of about $40\,\mu\mathrm{A/mm}^2$. The activation of neurons depends on the strength of the applied electric field and on its direction with respect to the neurons and their parts. However, because neurons have complex shapes and their electrical properties are generally unknown, and it is difficult to make precise predictions. Even after almost 15 years of research with TMS and TES, our understanding of the neuronal response to transcranial stimulation is mostly qualitative.

A Model of Long Axons

To understand brain stimulation, it is useful to first consider the peripheral nerves, for which the geometry is simpler (see Chapter 3). The subthreshold behavior of the transmembrane potential V of long and straight axons is described by the cable equation.[1,2] The cable equation states that the gradient of the electric field component along the axon is mostly responsible for stimulation (i.e., rate of change along the axon). There are several major implications (Fig. 2–2):

1. Straight long axons are most easily stimulated where the gradient of the electric field along the axon is the strongest (dE_x/dx for axons along the x axis).

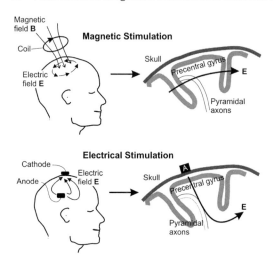

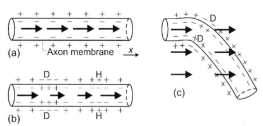

Figure 2–2 Schematic illustration of activation mechanisms for magnetic or electrical stimulation of a long, straight axon. The membrane polarization is sketched for different externally applied electric field patterns (arrows): uniform **E** along the axon, with no change from the resting state (**A**); gradient activation, with $\partial E_x/\partial x \neq 0$ (**B**); and gradient activation for a bent axon in uniform **E** (**C**). Regions of depolarization (D) and hyperpolarization (H) are indicated.

Figure 2–1 Principles of transcranial magnetic stimulation (TMS) (top) and transcranial electrical stimulation (TES) (bottom). In TMS, current $I(t)$ in the coil generates a magnetic field **B** that induces an electric field **E**. The lines of **B** go through the coil; the lines of **E** form closed circles. The drawing (upper right) schematically illustrates a lateral view of the precentral gyrus in the right hemisphere. Two pyramidal axons are shown with a typical orientation of the intracranial **E**. The electric field is parallel to the scalp, probably stimulating pyramidal axons near bends or other fibers parallel to the cortex. In TES, current flows from the anode to the cathode, with **E** in the same direction. The lines of **E** are both parallel and radial to the scalp (shown schematically here). TES probably stimulates the pyramidal axon deeper in the white matter.

2. Short axons are most easily stimulated at their ends.
3. Curved axons are most easily stimulated at their bends, where the effective electric field gradient along the axon is the strongest.

For the activation of distal nerves, the required gradient[3,4] of the electric field is about 10 mV/mm^2. The exact value depends on the length of the stimulating pulse.

Strength-Duration Relationship. The strength of transcranial stimulation depends on the strength of the externally applied electric field and on the temporal waveform of the stimulation pulse. The capacitance and resistance of the cell membrane form an electrical pathway that behaves as a leaky integrator (see "Strength-Duration Curve"). The shorter the stimulus pulses, the less stimulus energy is required.[5,6] However, shortening the pulse requires an increase in the stimulus amplitude. Figure 2–3 illustrates the strength-duration and energy-duration curves for a typical biphasic TMS device. Similar curves can be plotted for TES.

Locus of Activation in Transcranial Electrical Stimulation and Transcranial Magnetic Stimulation

The TES compound muscle action potential (CMAP) recordings consist of a short-latency D wave and several following I waves.[7] The D waves are caused by direct activation of pyramidal axons, whereas the I waves are thought to originate from the deep fiber system that is tangential to the cortex and that synaptically excites the pyramidal cells. Because the D waves are the dominant response to TES, activation originates most likely from the white matter.

In their fundamental works, Day[7] and Amassian[8] and their colleagues reported that the latencies of CMAPs with TMS are about 2 ms longer than with TES. This means that TMS probably stimulates more superficial structures than TES. It is therefore possible that the site of excitation is in the gray matter for TMS and in the white matter for TES. By appropriately orienting the TMS coil, the

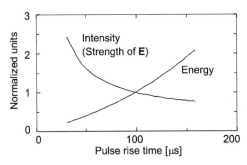

Figure 2–3 Stimulation pulse strength-duration and energy-duration relationships. The pulse duration is the zero-to-peak rise time of the TMS coil current. Current shape for the calculation was biphasic, and the membrane time constant was $\tau = 150\,\mu s$; the results are normalized to a pulse duration of $100\,\mu s$. Although not shown, the energy-duration curve levels off at a constant value for very short pulses.

CMAP latencies match, leaving the possibility that TMS can activate the pyramidal axons also directly.

Modeling. Mathematical modeling further helps to identify which cell structures are stimulated. Neuronal excitability is affected by various geometrical factors, such as axon terminals, bending, branching, nonuniformity, and tapering and volume-conductor nonuniformities.[9] In particular, axon bends are thought to play a key role in TMS.[10,11] The activation is probably strongest where the electric field changes most quickly along the path of the axon. Because the characteristic dimensions of the shapes of the cortical neurons are small compared with the distances over which **E** varies, the strongest variation of **E** along the length of the axon is achieved at locations where the axons bend. TMS activation probably takes place at the maximum of the induced **E**. In vitro experiments[3] and in vivo data support this view; there is good agreement with the data from locating the somatosensory cortex with TMS and the anatomical and functional information obtained with functional magnetic resonance imaging (fMRI), magnetoencephalography (MEG), positron-emission tomography (PET) and direct cortical stimulation.[12–17]

■ Transcranial Electrical Stimulation

Principles

In TES, electrical current is applied to the head by two or more scalp electrodes, typically one anode and one or more cathodes. The anode is the stimulating electrode and is placed approximately over the target area (see Fig. 2–1). In bifocal stimulation, the cathode electrode is placed a few centimeters away from the anode, such as over the vertex. The shorter the distance between the anode and the cathode (e.g., 2 cm), the better focused the stimulation is.[8] Focusing reduces the strength of the current reaching the brain, increases the scalp currents, and may result in pain. In unifocal stimulation, several widely spaced electrodes replace the single cathode electrode, which reduces the effects of currents near the electrodes.

The electrode current is normally of the order of 500 mA. The current pulses typically decay exponentially and last about 50 to $100\,\mu s$. Pulses can be generated for example by discharging a $1\text{-}\mu F$ capacitor from a voltage of 500 V through the electrodes. It is relatively easy engineering work to generate different electrode current pulses and repeat the stimulation pulses quickly.

Electric Field in Transcranial Electrical Stimulation

Most of the current in TES flows through the scalp, never reaching the brain. The highly resistive skull permits only small part of the current to reach the brain. Consequently, the needed strong stimulating current often causes intolerable pain due to activation of scalp muscles and pain receptors.

Figure 2–4 shows the electric field **E** in the brain; the field was calculated using a three-sphere spherical model of the head.[19,20] The field is maximal just below the electrodes, where it is directed radially or toward the center of the head. The electric field is distributed over a diffuse area. Because the current takes the path of least resistance, it is possible that the cerebral location of the strongest current does not move when the electrode position is changed. It may be difficult to predict the exact location of

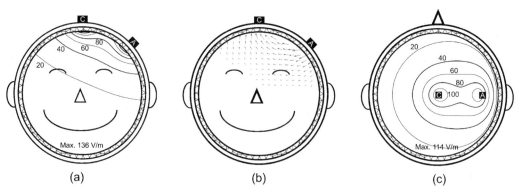

Figure 2–4 Electric field **E** generated by bifocal stimulation. The cathode is over the vertex, and the anode is 45 degrees (~7 cm) from the cathode. The electrode current is 500 mA. **A,** Electric field **E** in the plane of the electrodes. **B,** The electric field vectors in the same plane. **C,** Electric field **E** on a spherical surface 3 mm below the cortex or 15 mm below the scalp surface.

stimulation even with an accurate model of the conductivity of the skull.

Transcranial Magnetic Stimulation

Principles

TMS is based on electromagnetic induction (see Fig. 2–1), which is described by Faraday's law. A time-varying current in a primary circuit (TMS coil) induces an electric field and consequently a current flow (eddy current) in a secondary circuit (brain). The changing current in the coil generates a magnetic field **B**(t), but the field itself has no effect on neuronal activation; the magnetic field merely mediates the interaction. The electric field induced is proportional to the rate of change of **B**; no neural excitation occurs with slowly changing or stationary **B**.

Because the tissue is transparent to the magnetic field in TMS, the scalp and the skull do not resist the field. Unlike TES, TMS would not benefit from removing the skull and the scalp. However, the stimulating effect weakens quickly with distance from the coil. Only a minor part—much less than 1/100,000th—of the coil's magnetic energy is transferred into the brain. Because the skull does not shield the brain from TMS, much less current is induced in the scalp than is present in TES.

To summarize, there are two important differences between TMS and TES. TMS does not require any contact with the stimulator coil, and TMS is usually painless, because the current in the scalp is much smaller.

Electric Field in Transcranial Magnetic Stimulation

The electric field induced in the brain by TMS depends on the shape of the coil, the location and orientation of the coil with respect to the head, and the electrical conductivity structure of the scalp, skull and brain. The total electric field **E** in the tissue is the sum of *primary* and *secondary* electric fields, in which case the primary finduced directly by the changing magnetic field **B**(t) of the coil. In any object with non-zero conductivity σ, **E**$_1$ causes a flow of current **J**$_1 = \sigma$ **E**$_1$. Any conductivity changes along the path of the current **J**$_1$ cause electric charges to accumulate, giving rise to the secondary field **E**$_2 = -\nabla V$. Expressing **B** in terms of the vector potential **A** (i.e., **B** $= \nabla \times$ **A**), the total **E** is:

$$\mathbf{E} = \mathbf{E}_1 + \mathbf{E}_2 = \partial \mathbf{A}/\partial t - \nabla V$$

The electrical potential V obeys Laplace's equation, $\nabla^2 V = 0$. More extensive analysis of the TMS fields can be found from references 21 and 22 (see "Induced Electric Field" later in this chapter for calculation of **E**).

No Field in Radial Direction in Transcranial Magnetic Stimulation

The general boundary conditions for electromagnetism imply that the TMS-induced electric field near the skin-air boundary will always be tangential to the boundary. When the head is modeled as a sphere, this condition means that the induced field is always parallel to the nearest sphere surface. The field in the radial direction (i.e., perpendicular to the skull surface) is zero. Because all directions are radial in the center of a sphere, the field must always vanish there.

For TES, the boundary conditions are different. The stimulating field has both tangential and radial directions, and the field is generally non-zero in the sphere's center.

No Three-Dimensional Focusing

No TMS or TES arrangement can stimulate deep brain structures without stronger stimulation of the more superficially located cortex. It is not possible to focus TES or TMS or any combination of them in three dimensions. Summarizing the work of Heller and van Hulsteyn,[23] the field maximum is always on the surface of any homogeneous volume-conductor compartment. Many interesting studies could be performed if deep brain structures could be stimulated without interfering with the function of the superficially located cortex, and coil designs capable of three-dimensional focusing are occasionally suggested, but such trials are doomed to failure. Naturally, TMS and TES will affect indirectly deep brain structures that have neural connections with the stimulated cortical regions.

Comparison of Transcranial Magnetic Stimulation and Transcranial Electrical Stimulation Fields

Distribution and orientation of the electric fields in TES and TMS are significantly different. In TMS, the field is always in the direction along the scalp; there is no radial electric field. In TES, there are both field directions, and the relative strengths of the field components parallel and perpendicular to the scalp depend on the electrode configuration. Figure 2–5 shows that the electric field from TMS and TES decrease quickly with distance from the surface of the brain. The field from TMS decreases faster than that from TES, and this is probably one of the reasons why TMS stimulates mostly the superficial cortex. The fields from a 40-mm-diameter, 8-shaped TMS coil ($dI/dt = 120$ A/μs) and bifocal TES (7-cm electrode separation and 500 mA current) are nearly equally strong on the cortical surface (Fig. 2–6).

Three main characteristics make the effects of TMS easier to predict. The electric field can be calculated with good precision for any coil orientation; the skull affects the field less; and stimulation is most likely limited to the superficial cortex. It is also easier to focus the stimulation in TMS because focused TES requires very closely placed

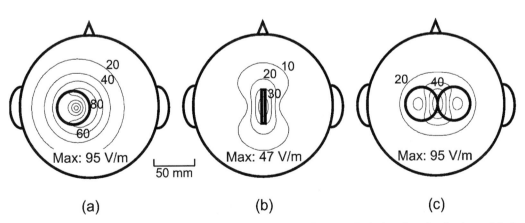

Figure 2–5 Contour maps of the strength of the electric field **E** on a spherical surface for circular and 8-shaped coils. The diameter of the coils was 40 mm, and $dI/dt = 100$ A/μs. The coils had 10 turns. The peak value of **E** is given below each plot. The depth of the spherical surface below the coil was 15 mm.

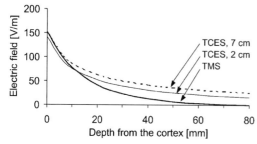

Figure 2–6 The strength of the electric field **E** as function of distance from surface of the cortex for transcranial electrical stimulation (TES) and transcranial magnetic stimulation (TMS). The TMS coil had a figure-of-eight shape and 40-mm-diameter wings; the current parameters were $dI/dt = 120$ A/μs and $I_{max} = 7$ kA, corresponding to the stimulation intensity that usually causes twitches in the hand muscles. The field strengths were computed below the center of the coil. The TES electrodes were 7 or 2 cm distant; the current strength was 500 mA. The field was calculated below one of the electrodes.

electrodes, which increases the painful scalp stimulation.

Transcranial Magnetic Stimulation Instrumentation

Stimulus Pulse Generation

The major challenge for TMS is that the submillisecond current pulse through the stimulating coil must have peak amplitude of up to 10,000 amperes.[24-26] The basic TMS power electronics circuit (Fig. 2–7A) consists of a capacitor (with capacitance C), a thyristor switch, and the stimulating coil (with inductance L). There is also resistance R in the coil, cables, thyristor, and capacitor, and the actual circuit is an RCL oscillator. The capacitor is first charged to some kilovolts and then discharged through the coil by gating the thyristor into the conducting state. The resulting damped sinusoidal current pulse has a peak value of 5 to 10 kA. Depending on the exact details of the circuit, the TMS pulse form is most commonly biphasic or monophasic (see Figure 2–7B,C). In monophasic pulses, the magnetic field rises rapidly from zero to the peak value and returns slower back to zero. The biphasic pulse consists of one damped cycle of a sinusoid. Some devices use polyphasic pulses, such as multiple-cycle damped sinusoids (see Fig. 2–7D). Because of the high power levels, it is much more difficult to modify the pulse shapes in TMS than in TES.

Discharging of the capacitor results first in flow of current in one direction through the thyristor in biphasic circuits (clockwise in Fig. 2–7A); after reaching its peak value, the current flows in the opposite direction through the diode. If the thyristor gating is terminated during the second half-cycle, the oscillation ends when the cycle is completed. Monophasic pulses are obtained (e.g., by connecting the diode forward parallel directly with the capacitor). Polyphasic current pulse shapes are obtained with a similar circuit as biphasic pulses. They may also appear at high stimulation intensities as a result of ringing of the pulse circuit due to a failure of the thyristor to stop conducting after the first sinusoid. See "Current Pulse Shape" later in this chapter for expressions for the current $I(t)$ in the coil.

Types of Stimulators

Single-Pulse Devices. Single-pulse TMS devices are the most widely used stimulator type. They are able to generate stimuli with a repetition rate lower than 1 Hz. Single-pulse devices have current pulse duration of 200 to 300 μs for biphasic and about 600 μs for monophasic pulses; the peak current is 2 to 8 kA. The operating voltage of TMS devices is typically 2 to 3 kV, and the power consumption 2 to 3 kW at maximum stimulus rate and intensity. Single-pulse TMS is most useful in the assessment of the descending motor pathways and in all basic research studies where good temporal resolution is essential.

Paired-Pulse and Dual-Pulse Devices. In paired-pulse stimulation technique, two pulses are driven through the same coil. The pulses are separated by an adjustable time interval from 1 to 1,000 ms in 0.1-ms steps. The intensity of the individual pulses can be adjusted independently. Accordingly, quadruple-pulse devices are capable of delivering four pulses. Some manufacturers have add-on modules to their single-pulse devices that can connect two

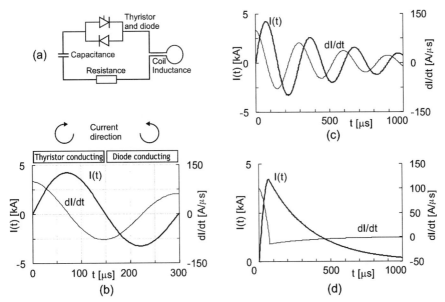

Figure 2–7 A schematic illustration of the stimulator circuit (**A**). The current *I(t)* and its rate of change *dI/dt* are shown for biphasic (**B**), polyphasic (**C**), and monophasic (**D**) current pulses. *Insets* for **B** depict the conducting periods of the thyristor (T) and diode (D) and the direction of the current in the coil. Parameters used include coil inductance of $L = 15\,\mu H$, capacitance of $100\,\mu F$, resistance of $50\,m\Omega$, and an initial capacitor voltage of 2,000 V.

stimulator units to one coil. The paired-pulse technique is an attractive way to determine the excitability of the cortex and a tool for the study of fast dynamic inhibitory and facilitatory phenomena. Two stimulator units can be used together to drive separate coils to stimulate different regions at the same time or in quick succession, a mode called dual-pulse TMS.

Rapid-Rate Devices. Rapid-rate stimulation (rTMS) devices operate at up to a 50-Hz repetition rate and at 40% to 100% of the maximum intensity of single pulses. The pulses are biphasic. The duration of sustained operation is typically limited by coil heating to 100 to 1,000 pulses at maximum power. With proper coil cooling, the duration of the stimulus train is not limited by coil heating. The rTMS devices require high power input levels and are the most challenging stimulator type to design. rTMS pulses can be used, for instance, to interfere with the performance of different tasks. However, because the pulse train may last seconds, rTMS has limited temporal resolution.

Image Guidance: Stereotaxy. The stimulator coil is normally positioned manually above the head on the basis of external landmarks or the coil location that elicits strongest motor responses. However, the coil can also be positioned stereotactically over the target location on the basis of MR images. This method is called image-guided TMS or navigated brain stimulation (NBS). Stereotactic stimulus targeting requires accurate localization of the coil with respect to the head. Frameless, three-dimensional localization systems can be used to guide the coil.

It is mandatory for stereotactic TMS to have an advanced user interface that guides the user to plan, perform, and monitor the experiments and that also documents the experiments in a controlled manner to improve the reproducibility and repeatability. An important part of stereotactic TMS is the calculation or prediction of the electric field induced in the brain. Because of the strong dependence of the field on the distance from the coil, even millimeter-level changes in position of the coil or small changes in the orientation of the coil may change the stimulating field significantly. Together with

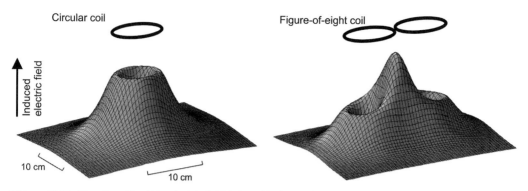

Figure 2–8 The strength of the electric field induced below a circular *(left)* and an 8-shaped coil *(right)*.

electric field calculation, stereotaxy allows the selection of the desired strength of the induced field instead of selecting the stimulus amplitude as the percentage of the maximum stimulator output.

Coils

The shape, size and orientation of the coil are the main factors that determine the size of the stimulated area as well as the direction of the induced current flow. Successful stimulus targeting to desired regions requires good knowledge of the field pattern generated by the coil. Likewise, any significant engineering modifications in the coil and stimulator design must be made with good understanding of the electromagnetic fields around the coil. In practice, however, the design changes are greatly limited by the high amount of energy that must be driven through the coil in a very short time. This energy level is up to 500 J, which suffices to throw a weight of 1 kg to a height of 50 m.

Circular Coil. The region activated by the circular coil is roughly under the circumference of the coil. The circular coil is normally placed so that its edge is tangential to the scalp. Figure 2–8 shows the shape of the field induced on a plane parallel to the coil. The electric field is diffuse, and circular coils are particularly useful when there is uncertainty about the exact location of the target area. They stimulate a relatively large area and it is easier to position the coil. Large circular coils can stimulate more effectively deeper cortical structures than small coils.

8-Shaped Coil. The region activated by the 8-shaped coil[27,28] (also known as a double, butterfly, or figure-of-eight coil) is under its center. The 8-shaped coil induces a more concentrated field than the circular coil. The induced field is also stronger than the field from a circular coil, making the 8-shaped coil more effective for TMS. Because of more concentrated field, the 8-shaped coils are normally used in all studies involving mapping. Figure 2–8 shows the magnitude of the field induced by an 8-shaped coil on a plane parallel to the coil. The field strength from the 8-shaped coil decreases about as quickly with distance as the field from a circular coil of the same size as one of the wings of the 8-shaped coil.

Caplike or Cone Coils. Sometimes the 8-shaped coils are made caplike by inclining their wings so that the coil fits the curvature of the head. The caplike or cone coil induces a slightly stronger field than the standard flat 8-shaped coil; the field is also slightly less concentrated. The caplike coils are useful for stimulation of deep cortical structures, such as the leg motor areas.[29] They are not especially suitable for examinations involving mapping.

Coils with Ferromagnetic Core. It is known from early discoveries of electromagnetism that ferromagnetic material (e.g., iron) can increase and direct the magnetic flux of coils. An iron core also can be used in TMS to strengthen the magnetic field without increasing the strength of the current pulse. This design is challenging because most of the ferromagnetic materials saturate at magnetic fields stronger than

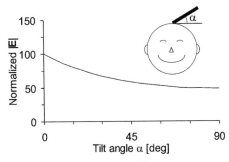

Figure 2–9 Coil tilting for sham stimulation. The normalized strength of the electric field **E** plotted as a function of the tilt angle for a circular coil. The coil diameter was 50 mm, and the computation points were 20 mm from the coil.

about 0.5 T. Once saturated, the core enhances the magnetic field less. Moreover, because the ferromagnetic materials behave nonlinearly and their characteristics vary with temperature, it may be difficult to adjust the stimulus pulse amplitude precisely. Iron-core coils have been successfully built, and they may offer benefits in selected applications.

Sham or Placebo Coils. Controlling of any possible placebo effects of TMS is required in cognitive studies and in therapeutic applications. Sham TMS aims at applying pulses that do not significantly stimulate the brain but that do cause the perception of real TMS. One method to obtain sham stimulation is to tilt or lift the coil so that the electric field induced in the brain decreases. Figure 2–9 plots the magnitude of **E** in the spherical model as a function of the tilt angle for a 50-mm-diameter circular coil. One edge of the coil was held against the head at the same point and the field point was 20 mm below the coil edge (see insert). Tilting the coil from the tangential to perpendicular orientation reduces the induced field only to about 50%. If the coil is tilted 45 degrees, the field strength is almost 60% of the maximal field. Lifting the coil is a more effective way of obtaining placebo stimulation; lifting the coil by 20 mm decreases the field strength to about one third or less (see Fig. 2–6).

Another possibility for sham stimulation is to use a dedicated sham coil (Fig. 2–10), which has two separate coils built inside.[30] The direction of the current in one of the

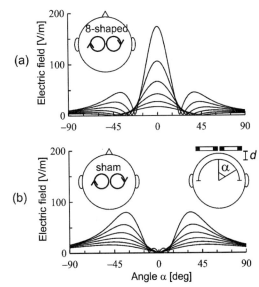

Figure 2–10 Sham coil. The electric field is shown in points of a circular arc going from the left ear to the right ear, when the arc is at different distances from the coil ($d = 10$, 20, 30, 40, 50, and 60 mm). Each of the 70-mm-diameter coils of the 8-shaped coil consisted of nine concentric turns wound of 1-mm-thick wire; the coils were driven with current that varied at a rate of 10^8 A/s. The *insets* show the geometry.

coils can be switched. If the coil is built 8-shaped, it is then possible to select whether normal or sham stimuli are delivered by simply switching the current direction. When the current directions in the wings are opposite, the coil functions as a normal 8-shaped coil. When the currents are in the same direction, the fields induced by loops cancel each other below the coil. This generates sham TMS clicks similar to the standard coil and produces similar scalp sensations, but it does not induce any field below the coil center (see Fig. 2–10).

Coil Cooling. Part of the strong current pulse through the coil is lost in the electrical resistance of the coil windings and the cables. This results in heating of the coil. With repetitive stimulation, the heating can be significant. It is possible to cool the coils by circulating air, oil, or water inside them.

Other Coil Shapes. A variety of other coil types for TMS exist, including the 8-shaped coil with drop-shaped windings. The slinky coil, a half-toroid named after the sprung toy coil with multiple wings in different angles, has also been suggested.[31] Field maps would reveal that these coils behave very similarly with the conventional 8-shaped coil with coplanar circular wings.

Stimulator Construction

Electronics. The power electronics for TMS is relatively simple, but selection of the components is not trivial. The thyristors for the pulse circuit should be fast; they must resist high forward currents and high dv/dt and di/dt without becoming conducting. Snubber circuits are needed to protect against overvoltages. Because the forward and reverse voltages over the thyristor can be up to 3 kV, it is advisable to use several thyristors connected in series. Power components also heat significantly; especially in rTMS, attention must be paid to the determination of the appropriate heat sink. Pulse transformers can be used for the triggering of the thyristors. Standard capacitors suitable for high-performance TMS equipment are usually not available but must be ordered according to specifications. Capacitors for biphasic pulse circuits should withstand reverse voltages of at least some two thirds of the operating voltage. Modern capacitors can store up to 1.5 kJ/kg, but such capacitors are expensive for commercial TMS units. One or more chopper power sources are usually the best choice for capacitor charging.

Coil Construction. The current in the coil produces strong forces, which can exceed tens of kilonewtons. Thick wire must be used and the coil's shape must be designed to withstand these forces. The forces are proportional to the square of the current.

In rTMS, tens of W/Hz of power is dissipated as heat in the coil. The standards for medical devices limit the temperature of the coil surface to 41°C. Built-in temperature sensors and effective cooling can be used to guard against excessive temperatures. Reducing the coil's resistance, determined by the wire gauge and coil geometry, can alleviate problems with power consumption and coil heating. Striped, foil, or litz wire can be used to reduce the skin and proximity effects that increase the resistance. The skin effect causes the current to flow mainly on the surface of thick wire; hence, the resistance of tubular wire is not much higher than that of solid wire. Liquid coolant can then be made to flow inside the wire.

The voltage over the coil's connectors may be 3 kV; the voltage across adjacent turns can be from 200 to 1,000 V. The wire insulation (e.g., varnish, film, mylar paper) must have the necessary dielectric strength and resist chemical solvents of the potting material (e.g., epoxy resin, polyurethane foam). The electrical and liquid coolant connectors must be tightly fastened and well insulated.

Physical Terms and Definitions

This chapter collects the most important quantities needed for extensive analysis of TMS device or coil performance and design.

Induced Electric Field

The electric field induced below a circular coil with radius a is:

$$\mathbf{E}(\mathbf{r}) = \mu_0 \frac{dI}{dt} \frac{a(y\mathbf{i} + x\mathbf{j})}{2\pi k \rho^2} [(2 - k^2)K(k) - 2E(k)],$$

where the field is computed at $\mathbf{r}$ (x, y, z). The center of the coil is in the origin and $k^2 = 4a\rho[(a + \rho)^2 + z^2]^{-1}$ and $\rho^2 = x^2 + y^2$. The functions E and K are the complete elliptic integrals,[33] and $\mu_0 = 4\text{p} \times 10^{-7}$ H/m is the permeability of free space.

This equation does not take into account the boundaries of the conductor, which may reduce the field strength by about one third. The most useful approximation of the conductivity profile of the head is the spherical conductor.[34]

Induced Current Density

The induced current density in the brain is:

$$\mathbf{J} = \sigma \mathbf{E}$$

where $\sigma \approx 0.4$ S/m is the conductivity of the tissue.

Current Pulse Shape

The coil current $I(t)$ for biphasic and polyphasic current waveforms is obtained from:

$$I(t) = (U_0/L\omega)e^{-\alpha t}\sin(\omega t)$$

where $\alpha = R/2L$ and $\omega^2 = (LC)^{-1} - \alpha^2$, and U_0 is the capacitor's initial voltage. R, C, and L are the total resistance, capacitance, and inductance of the circuit, respectively. Typical values are $R = 50$ mΩ, $C = 100$ μF, and $L = 15$ μH.

For a monophasic pulse circuit with the diode connected forward-parallel with the capacitor, (instead of anti-parallel with the thyristor in the bipolar circuit), the current pulse is:

$$I(t) = (U_0/L\omega)e^{-\alpha t}\sin(\omega t), \quad \text{when } 0 \leq t \leq T$$

$$I(t) = (U_0/L\omega)e^{-\alpha T}\sin(\omega T)e^{-R(t-T)/L},$$
$$\text{when } t > T$$

where $T = (1/\omega)\arctan(\omega/\alpha)$ is the rise time from zero to maximum, and α and ω are the same as for the biphasic pulse.

Strength-Duration Curve

The relationship between the pulse shape and the pulse's effect on the transmembrane potential is obtained from the temporal part of the cable equation, which is a differential equation[1]:

$$\tau \partial V(t)/\partial t + V(t) = -dF(t)/dt$$

where $\tau = 100$ to 200 μs is the membrane time constant,[5] $V(t)$ is the transmembrane potential measured from rest, and $F(t) \propto I(t)$ is a function proportional to the coil current. To compare different pulse shapes, the maximum of $-V(t)$ (i.e., depolarization) must be calculated for each current pulse shape.

Peak Magnetic Energy in the Coil

When the capacitor is discharged, its electrostatic energy is transformed into the coil's magnetic energy. The peak magnetic energy in the coil is:

$$W = \frac{1}{2}LI_{max}^2$$

where L is the coil's inductance and I_{max} the peak current in the coil. Coil designs with small inductance and low peak current are most desirable.

Coil's Inductance

For circular coils, a rough estimate of the inductance is[35]:

$$L \approx \mu_0 N^2 a(\ln(8r/w) - 2)$$

where a is the radius of the coil, w is the radius of the wire, and N is the number of turns. The inductance of the 8-shaped coil is approximately twice the inductance of one wing of the coil. The inductance of a caplike 8-shaped coil with inclined wings is slightly larger than the inductance of a flat 8-shaped coil.

Power Consumption

The average power dissipated as Joule heating in the coil, cables, and electronics components is

$$P_{coil} = fR\int_0^T I^2(t)dt$$

where f is the stimulus repetition rate, R the total resistance in the circuit, and T is the pulse duration. Most of the circuit's total resistance is in the coil (e.g., 15 of 20 mΩ), so that the dissipation in the coil at $f = 1$ Hz is about $P_{coil} = 40$ W for a typical biphasic pulse at maximal strength. The dissipated power grows with the square of the stimulation intensity.

Coil Warming

The temperature of the copper wire rises after one pulse roughly by:

$$\Delta T = \int_0^T I^2(t)dt = P_{coil}/mcf$$

where m is the copper mass in the coil, and $c = 387$ J/(kg °C) the specific heat. If the mass

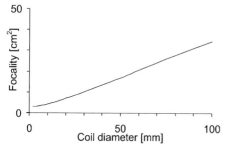

Figure 2–11 Focality, expressed as the surface area bound by the half-maximum of **E**, is plotted as a function of the coil diameter for the 8-shaped coil. The spherical 8-cm-radius computation surface was 20 mm below the coil; the sphere model was used.

of the copper wire in the coil is 200 g, and $P_{coil} = 40$ W at $f = 1$ Hz, the temperature increase may be as high as $\Delta T \approx 50°C/100$ pulses. Some of this heat is dissipated in the environment, so that the coil's temperature actually increases less. The coil heating grows with the square of the stimulation intensity.

Forces and Sound

The coil is subject to high internal forces that tend to increase or decrease its radius. The total radial force in the coil is approximately:

$$F = \partial W / \partial a \approx \tfrac{1}{2} I_{max}^2 \mu_0 N^2 (\ln(8a/w) - 1)$$

With typical values, the radial force is on the order of 10 kN.

The loud clicking sound during TMS arises from internal stresses in the coil, cables, and the capacitor, with the peak sound pressure being 120 to 300 dB 10 cm from the coil. Most of the acoustic energy is in the frequency range 2 to 7 kHz.

Focality of Stimulation

Focality of stimulation can be defined in many ways. One convenient definition is the area of the plane (in cm²), or spherical surface, where the strength of the electric field **E** is greater than 50% or 90% of the maximum of **E**.

Figure 2–11 plots the focality, defined as the area of the spherical surface bound by the half-maximum of **E**, as a function of the wing diameter of an 8-shaped coil. The computation was done in the spherical conductivity model of the head. The smaller the coil, the better focused is the field; however, the smaller the coil, the more current is needed in the coil to generate the field. In practice, it is very difficult and impractical to build coils smaller than about 20 mm in diameter for TMS.[36]

Summary

Transcranial stimulation is based on applying an electric field in the brain tissue. The electric field drives ionic currents in the tissue, charging the capacitances of neuronal membranes and thereby triggering the firing of neurons. The physiological principles at the cellular level are the same for TMS and TES.

Whereas TES drives through scalp electrodes directly a current into the brain, TMS is based on electromagnetic induction. The electromagnetic theory for TMS is well established and can be divided into three separate parts: the electrical operation of the stimulator circuit, the computation of the macroscopic electromagnetic fields being imposed on the brain due to current in the stimulator coil, and the detailed flow of current and buildup of charge on cellular membranes as a result of imposing the macroscopic electric field in the brain.

The theory of the macroscopic magnetic and electric fields in TMS is perfectly understood, and the computation is quite precise, even though many details are usually neglected because they are not available or because they complicate the computations excessively. Likewise, the electrical operation of the TMS circuit is well understood, although several details of the power electronics components and the coil cause practical difficulties. Only the principles underlying the cellular-level events leading to neuronal excitation are understood; the effects of cellular shapes, gray-white matter boundaries, local tissue anisotropy, glial cells, and the background neuronal activity remain largely unknown.

REFERENCES

1. McNeal DR. Analysis of a model for excitation of myelinated nerve. IEEE Trans Biomed Eng 1976;23:329–337.

2. Basser PJ, Roth BJ. Stimulation of myelinated nerve axon by electromagnetic induction. Med Biol Eng Comput 1991;29:261–268.
3. Maccabee PJ, Amassian VE, Eberle L, et al. Magnetic coil stimulation of straight and bent amphibian and mammalian peripheral nerves in vitro: locus of excitation. J Physiol 1993;460:201–219.
4. Ruohonen J, Panizza M, Nilsson J, et al. Transverse-field activation mechanism in magnetic stimulation of peripheral nerves. Electroencephalogr Clin Neurophysiol 1996;101:167–174.
5. Barker AT, Garnham CW, Freeston IL. Magnetic nerve stimulation: the effect of waveform on efficiency, determination of neural membrane time constants and the measurement of stimulator output. In Levy WJ, Cracco RQ, Barker AT, et al. (eds): Magnetic Motor Stimulation: Basic Principles and Clinical Experience. Amsterdam, Elsevier Science, 1991:227–237.
6. Panizza M, Nilsson J, Roth BJ, et al. Relevance of stimulus duration for activation of motor and sensory fibers: implications for the study of H-reflexes and magnetic stimulation. Electroencephalogr Clin Neurophysiol 1992;85:22–29.
7. Day B, Dressler D, de Noordhart C, et al. Electrical and magnetic stimulation of the human motor cortex: surface EMG and single motor unit responses. J Physiol (Lond) 1989;412:449–473.
8. Amassian VE, Cracco RQ, Maccabee PJ. Focal stimulation of human cerebral cortex with the magnetic coil: a comparison with electrical stimulation. Electroencephalogr Clin Neurophysiol 1989;74:401–416.
9. Roth BJ. Mechanisms for electrical stimulation of excitable tissue. Crit Rev Biomed Eng 1994;22:253–305.
10. Abdeen MA, Stuchly MA. Modeling of magnetic field stimulation of bent neurons. IEEE Trans Biomed Eng 1994;41:1092–1095.
11. Hyodo A, Ueno S. Nerve excitation model for localized magnetic stimulation of finite neuronal structures. IEEE Trans Magn 1996;32:5112–5114.
12. Morioka T, Yamamoto T, Mizushima A, et al. Comparison of magnetoencephalography, functional MRI, and motor evoked potentials in the localization of the sensory-motor cortex. Neurol Res 1995;17:361–367.
13. Ruohonen J, Ravazzani P, Ilmoniemi RJ, et al. Motor cortex mapping with combined MEG and magnetic stimulation. In Barber C, Celesia G, Comi G, et al. (eds): Functional Neuroscience. Amsterdam, Elsevier Science, 1996:317–322.
14. Wassermann EM, Wang B, Zeffiro TA, et al. Locating the motor cortex on the MRI with transcranial magnetic stimulation and PET. Neuroimage 1996;3:1–9.
15. Carter N, Zee DS. The anatomical localization of saccades using functional imaging studies and transcranial magnetic stimulation. Curr Opin Neurol 1997;10:10–17.
16. Paus T, Jech R, Thompson CJ, et al. Transcranial magnetic stimulation during positron emission tomography: a new method for studying connectivity of the human cerebral cortex. J Neurosci 1997;17:3178–3184.
17. Krings T, Buchbinder BR, Butler WE, et al. Stereotactic transcranial magnetic stimulation: correlation with direct electrical cortical stimulation. Neurosurgery 1997;41:1319–1325.
18. Cohen LG, Hallett M. Noninvasive mapping of human cortex. Neurology 1988;38:904–909.
19. Rush S, Driscoll DA: Current distribution in the brain from surface electrodes. Anesth Analg 1968;47:717–723.
20. Grandori F, Rossini P. Electrical stimulation of the motor cortex—theoretical considerations. Ann Biomed Eng 1988;16:639–652.
21. Ilmoniemi RJ, Ruohonen J, Karhu J. Transcranial magnetic stimulation—a new tool for functional imaging of the brain. Crit Rev Biomed Eng 1999;27:241–284.
22. Ruohonen J, Ilmoniemi RJ. Modeling of the stimulating field generation in TMS. Electroencephalogr Clin Neurophysiol 2003;Suppl 51:1–14.
23. Heller L, van Hulsteyn DB. Brain stimulation using electromagnetic sources: theoretical aspects. Biophys J 1992;63:129–138.
24. Barker AT. An introduction to the basic principles of magnetic nerve stimulation. J Clin Neurophysiol 1991;8:26–37.
25. Cadwell J. Principles of magnetoelectric stimulation. In Chokroverty S (ed): Magnetic Stimulation in Clinical Neurophysiology. Boston, Butterworth, 1990:13–32.
26. Jalinous R. Technical and practical aspects of magnetic stimulation. J Clin Neurophysiol 1991;8:10–25.
27. Ueno S, Tashiro T, Harada K. Localized stimulation of neural tissue in the brain by means of a paired configuration of time-varying magnetic fields. J Appl Phys 1988;64:5862–5864.
28. Cohen, D. Feasibility of a magnetic stimulator for the brain. In Weinberg H, Stroink G, Katila T (eds): Biomagnetism: Applications & Theory. New York, Pergamon Press, 1985:466–470.
29. Kraus KH, Gugino LD, Levy WJ, et al. The use of a cap-shaped coil for transcranial magnetic stimulation of the motor cortex. J Clin Neurophysiol 1993;10:353–362.
30. Ruohonen J, Ollikainen M, Nikouline V, et al. Coil design for real and sham transcranial magnetic stimulation. IEEE Trans Biomed Eng 2000;47:145–148.

31. Ren C, Tarjan PP, Popovic DB. A novel electric design for electromagnetic stimulation—the Slinky coil. IEEE Trans Biomed Eng 1995;42:918–925.
32. Arfken G. Mathematical Methods for Physicists, 3rd ed. San Diego, Academic Press, 1985:321–327.
33. Grandori F, Ravazzani P. Magnetic stimulation of the motor cortex—theoretical considerations. IEEE Trans Biomed Eng 1991;38:180–191.
34. Ravazzani P, Ruohonen J, Grandori F, et al. Magnetic stimulation of the nervous system: induced electric field in unbounded, semi-infinite, spherical, and cylindrical media. Ann Biomed Eng 1996;24:606–616.
35. Grover FW. Inductance Calculations: Working Formulas and Tables. Special edition prepared for Instrument Society of America. New York, Dover Publications, 1973.
36. Cohen D, Cuffin BN. Developing a more focal magnetic stimulator. Part I. Some basic principles. J Clin Neurophysiol 1991;8:102–111.

3 Activation of Peripheral Nerve and Nerve Roots

Marcela Panizza and Jan Nilsson

Magnetic stimulation of excitable tissue has been studied for more than a century,[1-7] and probably one of the first human studies was the observation of phosphenes by D'Arsonval in 1896. However, it was only in 1982 that the Sheffield group developed an instrument that clinically could be used for stimulation of the median nerve in humans and recording action potentials from the thumb muscles.[8,9] A real revolution in clinical neurophysiology came in 1985, when Barker and coworkers[10] demonstrated painless transcranial magnetic stimulation applied to the scalp over the motor area and recording from the abductor digiti minimi muscle in humans. Immediately after these demonstrations, stimulators began appearing commercially.

Magnetic stimulation is not a new technique; it is an improvement of an old technique. As stated by John Cadwell,[11] "It is not magic." He also said, "The principles of magnetic stimulation are not complex. The complexity and the mystery are in the organization of the nervous system."

■ Technical Considerations

To understand the mechanisms of peripheral nerve magnetic stimulation, results are often compared with those obtained with electrical stimulation. In electrical stimulation, it has been shown experimentally that electric fields oriented parallel to the nerve fibers are optimal for nerve excitation.[12] For example, when applying surface electrodes longitudinally along the nerve, current passes between the anode and cathode and causes an outward current across the nerve membrane under the cathode. When this outward current has carried sufficient charge to depolarize the membrane to a threshold level, an action potential is generated.

There are essential differences between electrical and magnetic stimulation. In electrical stimulation, the current often passes through the skin by means of surface electrodes into the body near the nerve, and it is normally only a fraction of the resulting charge that arrives to the excitable membranes and can cause depolarization. In magnetic stimulation, a changing magnetic field causes an induced current based on the scientific principles of mutual inductance described by Faraday in 1831. If a pulse of magnetic field is passed to the body, the induced electrical field will cause a current to flow. If the amplitude, duration, and spatial characteristics of the induced current are adequate, depolarization will occur.[13]

Magnetic stimulation is a useful instrument to study the human nervous system noninvasively. Magnetic stimulation has found widespread use for motor evoked potentials when stimulating the motor cortex transcranially, and it can be used to determine the conduction velocity of motor and sensory nerves. Because up-to-date magnetic stimulators still lack reproducibility, focality, and intensity for peripheral nerve stimulation, the technique has gained very little recognition. However, magnetic stimulation has several advantages over electrical stimulation: no direct contact with the skin is necessary, it causes little or no pain, it suffers only small interference from electrical and geometrical properties of the intervening tissue, and deep neural structures can easily be reached noninvasively.

Despite the obvious advantages, magnetic stimulation of peripheral nerves has found very little clinical use, and one reason may be the wide variety of coils and of the waveforms of the induced current. Almost every producer of commercially available magnetic stimulators has its own design of the magnetic coils, and the waveforms that cause the induced current are different as well. These factors make it very difficult to predict the exact point of depolarization along the nerve.

In principle, there are two kinds of waveforms used to induce currents: a monophasic or near-monophasic waveform (Fig. 3–1B) and a damped oscillatory waveform or biphasic waveform (see Fig. 3–1C). The near-monophasic waveforms induce currents that have a very rapid rise time—within $5\,\mu s$—and have an almost linear decay to zero of about $100\,\mu s$.[14] Stimulation occurs during the first rapid phase of the induced waveform, and if the direction of the current is clockwise in the stimulating coil, the resulting induced current will be counterclockwise near the nerve. By flipping the coil, the direction of the induced current is reversed. The oscillatory waveforms induce currents that have a very rapid rise time (still within $5\,\mu s$) and typically a decay of about $80\,\mu s$ to the zero crossing of the first phase. The following phases are usually smaller in amplitude and have a longer decay. Stimulation occurs normally during the initial phase of the oscillatory induced current. Specifically, for a waveform to behave monophasically, the amplitude of the second phase must be small relative to the first, and the duration of the second phase must be greater than the membrane time constant, which for large motor axons is on the order of $100\,\mu s$. If these conditions are not met, stimulation can occur on both the first and second phases of the stimulus, and the determination of the site of stimulation becomes much more difficult.

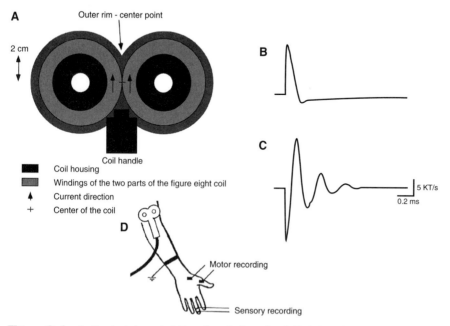

Figure 3–1 **A:** Physical characteristics of an 8-shaped coil. **B:** Induced-current waveform for a monophasic stimulator. **C:** Induced-current waveform for a biphasic stimulator. **D:** Positioning of the 8-shaped coil and recording electrodes on the arm and hand. (From Nilsson J, Panizza M, Roth BJ, et al. Determining the site of stimulation during magnetic stimulation of a peripheral nerve. Electroencephalogr Clin Neurophysiol 1992;85:253–264.)

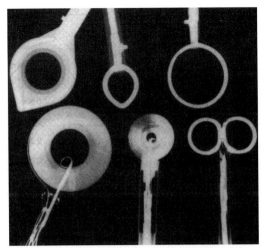

Figure 3–2 X-ray films of different magnetic coils. *Top left to right*: 9-cm-diameter Cadwell coil with an angulated extension, 5-cm-diameter Cadwell coil with an angulated extension, and a 9-cm-diameter Cadwell round coil. *Bottom left to right*: 14-cm-diameter Magstim flat-spiral round coil, 6.7-cm-diameter Magstim flat-spiral round coil, and a Cadwell 8-shaped coil with a 4-cm diameter of each wing. (From Cohen LG, Roth BJ, Nilsson J, et al. Effects of coil design on delivery of focal magnetic stimulation. Electroencephalogr Clin Neurophysiol 1990;75:350–357.)

We have used the following stimulators: for a monophasic waveform: Magstim (formerly Novametrix), Medtronic (formerly Dantec), and Digitimer; for a biphasic waveform: Cadwell. However, up-to-date versions of most magnetic stimulators are delivered with a control allowing a change between monophasic and biphasic waveforms.

We have studied the effect of varying the duration of the monophasic waveform of the magnetic stimulus.[15] To produce a similar strength-duration curve, a triangular waveform such as the first phase of the monophasic pulse must have a duration of more than three times longer than a square wave pulse.[15] In practice, this study showed that, for example, with a duration of the magnetic stimulus of 180 μs, the effective duration was less than 60 μs.

Magnetic stimulators typically can be used with different diameters of round coils, and different sizes of butterfly or 8-shaped coils (Fig. 3–2). Round coils are flat-spiral coils (see Fig. 3–2, *bottom left*) or tightly wound coils (see Fig. 3–2, *top right*).[16] Smaller round coils produce a more focal stimulus but with less intensity. The 8-shaped coil is much more focal and because the two "wings" are wound with opposite current direction, and the resulting current at the intersection of the wings is increased.[17] If the current direction for the 8-shaped coil is pointing away from the handle, the virtual cathode can be found under the handle.[18]

Apart from these standard coils, experimental coils have been developed to improve the localization of the magnetic field. These coils—the four-leaf clover design,[19] the slinky coil,[20] and a three-dimensional differential coil[21]—all improve the focality, but the clinical utility still must be proved.

■ Activation of Peripheral Nerves

Magnetic stimulation may become a valuable tool for the determination of peripheral nerve conduction velocity. However, to make it a suitable instrument, it is important to know the exact location of the virtual cathode and anode and to understand the behavior and importance of the different magnetic coils and waveforms.

We previously performed some experiments to locate and characterize the virtual cathode,[18] and these experiments described the typical behavior of magnetic stimulation for different coils and different waveforms. Briefly, in the first experiment, we stimulated the median nerve at equally spaced locations along the nerve. Under these conditions, we should experience a uniform shift in latency when changing the position. We recorded a uniform latency change using the 8-shaped coil (Fig. 3–3), and this was the case when using a stimulator with a monophasic or a biphasic waveform. This experiment was a control to ensure that there were no localized sites of low threshold or current focusing.[22] However, when we applied the round coils in the same setup, we observed sudden latency changes of 1 to 2 ms over a small distance, and we attributed this different behavior to the more diffuse electric field distribution for the round coils and perhaps to the stimulation near a region of low impedance.[23] The

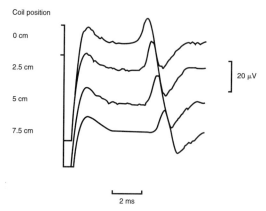

Figure 3-3 Typical recording of the response from the right median nerve to magnetic stimulation at the elbow. Traces show the sensory response (digit III) elicited when the 8-shaped coil was moved in steps of 2.5 cm proximally. The recording distance from position "0 cm" was 36 cm. Output of the magnetic stimulator was at 100%, and the stimulating waveform was monophasic. (From Nilsson J, Panizza M, Roth BJ, et al. Determining the site of stimulation during magnetic stimulation of a peripheral nerve. Electroencephalogr Clin Neurophysiol 1992;85:253–264.)

8-shaped coils were less influenced by such inhomogeneities because they have a much more focal electric field.

In a second experiment, we found the virtual cathode to be 3.0 ± 0.5 cm distal to the center of an 8-shaped coil (Fig. 3–4).

This result was in agreement with other studies.[24,25]

With the monophasic stimulators, we observed latency shifts of 0.65 ± 0.05 ms when reversing the virtual cathode and anode. These latency shifts were longer than when reversing the electrical stimulator (0.4 ± 0.14 ms), and estimated from the conduction velocity, we found a displacement of the site of stimulation of about 4.1 cm for the round and 8-shaped coils. However, for a magnetic stimulator with a biphasic waveform, we found much shorter latency shifts, 0.20 to 0.25 ms by flipping the round and 8-shaped coils. This latency difference may be caused by the oscillating pulse shape producing stimulation at two sites at slightly different times.[26]

In another experiment, we used the observation that H reflexes are preferentially activated over motor responses[15,27,28] when stimulating the tibial nerve with a round coil using a monophasic pulse. This observation showed similar anode-like and cathode-like behavior to results from electric stimulation.[28] We found no preferential activation of the H reflexes relative to motor responses when using a magnetic stimulator with an oscillating waveform (Fig. 3–5).

These four experiments characterize in detail the location and behavior of the virtual

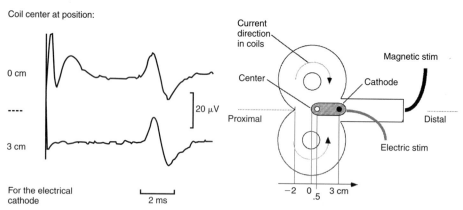

Figure 3-4 Typical recordings of magnetic and electrical stimulation of the right median nerve at the elbow. *Upper trace* shows the sensory response (digit III) elicited by the 8-shaped magnetic coil (stimulator output: 100%, monophasic waveform). *Lower trace* shows the sensory response elicited by electrical stimulation, with the cathode position 3 cm closer to the recording electrodes than the center of the magnetic coil. The recording distance from position "0 cm" was 36 cm. (From Nilsson J, Panizza M, Roth BJ, et al. Determining the site of stimulation during magnetic stimulation of a peripheral nerve. Electroencephalogr Clin Neurophysiol 1992;85:253–264.)

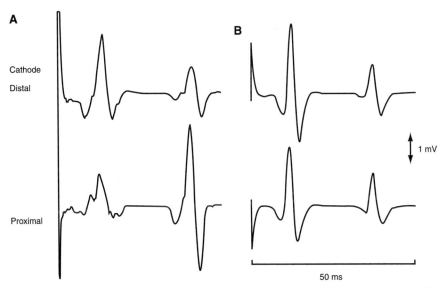

Figure 3–5 Typical responses to magnetic stimulation of the left tibial nerve at the popliteal fossa elicited using a circular coil with a monophasic waveform (**A**) and a biphasic waveform (**B**). (From Nilsson J, Panizza M, Roth BJ, et al. Determining the site of stimulation during magnetic stimulation of a peripheral nerve. Electroencephalogr Clin Neurophysiol 1992;85:253–264.)

cathode during magnetic stimulation. They are also consistent with a mathematical model of magnetic stimulation of axons.[26] The position of the virtual cathode relative to the coil and its change in position on reversal of the stimulus polarity are both explained by the negative derivative of the electric field parallel to the axon being responsible for stimulation.

There remains quantitative disagreement between theory and experiment regarding the precise location of the virtual cathode. There are several sources of this discrepancy. Perhaps the largest uncertainty is produced by the dependence of the virtual cathode on the stimulus strength and threshold intensity. Our experiments were typically carried out at about twice threshold, but for example, if the stimulus was actually 2.5 times threshold, the site of stimulation would move about 0.5 cm distal. Another complicating factor is that threshold stimulus strength depends on axon parameters such as diameter[29] and depth.[30]

Our results agree with some previous reports of cathode location during magnetic stimulation and disagree with others. Olney and colleagues[24] compared the site of activation using electric and magnetic stimulation of the median and ulnar nerves. They placed the center of a Cadwell 8-shaped coil parallel to the nerve and estimated the point of nerve activation to be 0.3 to 1 cm in front of the coil center. Ravnborg and coworkers[31] found a latency difference using electrical and magnetic stimulation, implying that the virtual cathode was 7.2 cm more distal than the electrical cathode, when placing the outer edge of a round Dantec coil (4-cm inner diameter, 12-cm outer diameter) over the brachial segment of the ulnar nerve and recording from the first dorsal interosseus. Claus and coworkers[32] found latency shifts in the response from the abductor pollicis brevis (APB) when centering round Digitimer or Novametrix coils over the median nerve at the elbow, implying a shift in the site of stimulation of 6.5 cm for the Digitimer stimulator, and 3.5 cm for the Novametrix stimulator. They found no latency shifts when flipping a Cadwell coil, a result they attributed to the oscillating stimulus waveform. Chokroverty[33] and Chokroverty and associates[34] placed a Cadwell coil tangential to the nerve and parallel to the surface of the arm for conduction velocity studies. They pointed out the importance of always using the same current direction to avoid changes of the configuration, amplitude, and latency of the compound action potential.

Contrary to these studies are the ones in which it was impossible to determine the site of stimulation or its position was nearly independent of stimulus polarity. Evans and colleagues,[25] using a Novametrix magnetic stimulator to activate the median nerve at the wrist, could not relate the geometry and position of the coil to the site of nerve depolarization. Hallett and coworkers[35] centered a round Cadwell coil over the median nerve at the wrist and the elbow and found consistent latency changes in the APB response when moving the coil near the wrist, no consistent changes when moving the coil near the elbow, and latency changes of about 0.2 ms when flipping the coil at the elbow. Maccabee and coworkers[36,37] observed little or no latency shifts when flipping a Cadwell coil held orthogonally above the median nerve at the elbow and recording from APB with a single-fiber needle electrode. They found a maximum latency shift of only 0.1 ms, implying that the virtual cathode shifted position by at most 0.5 to 0.6 cm. This small latency shift could be due the oscillating waveform creating multiple sites of stimulation. Maccabee and associates[37] used a modified Cadwell stimulator, in which the stimulus waveform did not oscillate and a hardware switch reversed the current polarity without having to flip the coil. They found no latency shifts in the thenar muscle response when reversing the current polarity.

In another study, Ruohonen and coworkers[38] used a Cadwell magnetic stimulator with a placement of the 8-shaped coil and with current directions that theoretically produced no virtual cathode. On the basis of the classic cable theory, it has been justified that long and straight fibers can be activated by an electric field that has a large enough negative spatial gradient along the axons.[39,40] This theory is in agreement with experiments with electrical stimulation. Longitudinal orientation of the stimulating electrodes with respect to the nerve causes a high negative spatial gradient near the cathode. The longitudinal placement is reportedly four or five times more efficient than a transverse orientation, which theoretically should not produce an electric field parallel with the fiber. That activation is achieved with the latter orientation may be attributed to tissue inhomogeneities and bends in the nerves causing local focusing of the induced electric field.[41,42] In contradiction with the predictions of the cable theory, however, it is known that strong muscle responses can be evoked by coil orientations that should not induce any electric field parallel with the nerve fibers.[24,35,43] These orientations were referred to as *inconsistent*. An inconsistent orientation is, for example, the transverse 8-shaped coil placed symmetrically over the nerve. A typical conventional orientation, or consistent, is the symmetrical longitudinal 8-shaped coil and the edge-tangential round coil. These orientations are illustrated in Figure 3–6, together with muscle responses that were elicited by them in one experiment. Muscle responses with similar waveforms and amplitudes are evoked by the inconsistent and consistent orientations.

Ruohonen and colleagues[38] showed that the responses elicited by inconsistent orientations were frequently nearly maximal or even

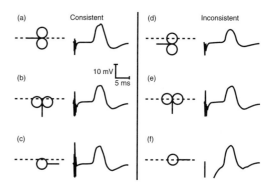

Figure 3–6 The coil orientations on the left are consistent with the classic cable theory predictions, whereas the orientations on the right should not yield excitation of peripheral nerves (i.e., they are inconsistent). Recordings are shown from the abductor pollicis brevis after stimulation at the elbow with the indicated orientation of the coils: symmetric-longitudinal 8-shaped coil (**A**); edge-tangential transverse 8-shaped coil (**B**); edge-tangential round coil (**C**); shifted longitudinal 8-shaped coil (**D**); symmetric-transverse 8-shaped coil (**E**); and symmetric round coil (**F**). The nerve is shown by a *dotted line*. The scale is consistent in all recordings. (From Ruohonen J, Panizza M, Nilsson J, et al. Transverse-field activation mechanism in magnetic stimulation of peripheral nerves. Electroencephalogr Clin Neurophysiol 1996;101:167–174.)

supramaximal, and they concluded that the responses were unlikely due only to tissue inhomogeneities or bends in the nerves. Supporting this consideration, smooth shifts in latency were recorded while moving the stimulation coil along the nerve, excluding the possibility of activation taking place at low-threshold points or so-called hot spots. Otherwise, the recordings were in good agreement with the literature, and for instance, the conventional coil orientations caused excitation to initiate near sites of high electric field parallel with the nerve (i.e., near the virtual cathode). The study showed that the core conductor model and the cable theory do not completely explain magnetic stimulation of peripheral nerves, and additional mechanisms must be involved. To find the possible sources of the discrepancy between theory and experiments, it is necessary to check the assumptions that lead to the classic theoretical predictions. Using mathematical modeling to simulate the induced electric fields in the tissue and superimpose the evoked muscle responses on simulated maps of the electric field, Ruohonen and associates[38] concluded that the unexpected excitation with the inconsistent orientations of the coil is most probably caused by a strong electric field component perpendicular to the nerve. Consequently, Ruohonen and colleagues[38] modified the activation function of a peripheral nerve to a combination of the electric field component perpendicular to the electric field of the nerve and the spatial gradient of the electric field parallel with the nerve.

The stimulation of the peripheral nerve using the consistent and inconsistent coil orientations reported by Ruohonen and co-workers[38] was confirmed in a study by Sun and colleagues,[44] using a Magstim 200 stimulator with a monophasic waveform and an 8-shaped coil. However, this study was probably much more precise due to the monophasic waveform that was used, with respect to the biphasic waveform used by Ruohonen and associates.[38]

Models of the active straight nerve fiber do not predict activation when the fiber is exposed to transverse electric fields. This is because the component of the electric field along the nerve, which is assumed to be the driving force of the activation mechanism, vanishes in this case. If an undulating nerve fiber is used in computer simulations,[45] the electric field lines are not perpendicular to the fiber at all points. Schnabel and Struijk[45] showed that even for small undulation amplitudes, the component of the electric field along the fiber is enough to induce an excitation current inside the axon. Schnabel and Struijk[45] concluded that the effect of fiber undulation could fully account for the experimental results reported by Ruohonen and colleagues.[38]

Discrepancies between the theory, the developed models and the experiments have usually been explained by inhomogeneities such as: localized sites of low threshold or current focusing[22,46] or stimulation in a region of low impedance,[23] and one possible cause for these inhomogeneities could be explained by an interference from bone and soft tissue. Kobayashi and associates[47] extended this analysis to include inhomogeneities from muscle and fat, and using the classic long straight nerve model they could explain responses from consistent and inconsistent coil orientations by applying a soft tissue model that took into consideration the inhomogeneities between muscle and fat.

Hsu and coworkers[48] presented theoretical results based on models of rabbit and human peripheral nerves. Their simulations suggested a flat coil design for optimal stimulation efficiency. They also found that the four-leaf clover design[19] achieved the highest efficiency for infinitely long fibers, whereas the 8-shaped coil was optimal for terminating or bending fibers.

Clinical application of magnetic stimulation for peripheral nerve stimulation can be summarized by the work of Bischoff and colleagues.[49] They compared the ability of round and 8-shaped coils to activate peripheral nerves in healthy subjects. No differences in motor threshold intensities were found between the coils, but the intensities necessary to elicit maximum compound muscle action potentials were different. To guarantee supramaximal stimulation, intensity usually has to be increased by 10% to 15% above the level necessary to elicit a maximum compound muscle action potential. For magnetic stimulation, this is difficult and sometimes impossible because the range between the lowest intensity necessary to elicit a maximum response and the maximum output intensity of the device is much smaller than that for electrical stimulation. Moreover, a prolongation of stimulus duration that increases the excitatory effect of the electrical stimulation is for technical reasons still impossible with magnetic stimulators. Small circular coils were superior to larger coils in terms of the lower intensities necessary to elicit maximum compound muscle action potentials, and provided a more focal stimulus. For deep nerves, amplitudes were always submaximal, and in their study, they found that one of the main drawbacks was the coactivation of nearby nerves and underlying muscles.

The optimal stimulus waveform for peripheral nerve stimulation is a monophasic pulse that will give a typical virtual cathode situation. An 8-shaped coil will result in a more focal induced current. To ensure stimulation of a body part where there are no or very few inhomogeneities, it is recommended to monitor latency shifts of the sensory or motor response when moving the coil in steps of 2.5 cm over a distance of at least 5 to 10 cm. Eventually, to be sure that the magnetic stimulus is supramaximal, the stimulus intensity can be compared with electrical stimulation by placing the electrical cathode at the same point as the virtual cathode.

■ Activation of Nerve Roots

The technique, theory, and models that have been described under the activation of peripheral nerves apply also for the activation of nerve roots; however, there are other considerations that must be applied for nerve roots. Cohen and Cuffin[50] were the first to claim that bone, because of its extremely low conductivity, was the major controller of induced current flow in the human body. For example, in the cranium, induced currents concentrate below the inner skull and fall off very fast toward the center. In the spine, a correctly oriented round or 8-shaped coil induces current parallel to the neuroforaminal axis, and the current is most likely focally concentrated there.[51] Magnetic stimulation is thought to activate spinal roots near their exit from the spinal cord.[14] Maccabee and coworkers[23] studied a segment of cervical-thoracic spine immersed in a saline volume conductor, using a technique of recording the transneuroforaminal electric field and its first spatial derivative (Fig. 3–7). These observations were consistent with human studies, and demonstrated the absence of spinal cord stimulation compared with the relative ease of nerve root stimulation. The observation was consistent with the lack of latency shift of elicited compound action potentials and motor units at threshold when moving the magnetic coil along the rostrocaudal axis of the vertebral column.[22,34,52–54]

Several investigators have listed the optimal placements of the magnetic stimulation coils

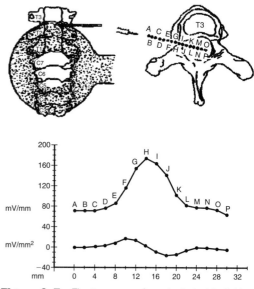

Figure 3–7 The transneuroforaminal electric field recorded proximal to, within, and distal to the intervertebral foramen between T2 and T3. A 4-mm linear coaxial cable probe was used. Each data point designated by an alphabetic letter, obtained at 2-mm intervals, corresponds to the midpoint of the probe. The *lower trace* indicates the first spatial derivative (mV/mm^2) of the electric field. The magnetic coil was oriented symmetrically and tangentially. (From Maccabee PJ, Amassian VE, Eberle LP, et al. Measurement of the electric field induced into inhomogeneous volume conductors by magnetic coil: application to human spinal nerve geometry. Electroencephalogr Clin Neurophysiol 1991;81:224–237.)

over the spinal cord.[14,34] The optimal stimulus waveform does not exist, and monophasic or biphasic stimulators are equally good. With the coil centered on the spinal cord, latencies are about the same with most stimulators.[14] For clinical evaluation, recordings should be done bilaterally, and responses elicited by both current directions in the magnetic stimulator coil usually should be recorded.

In clinical practice, the use of magnetic stimulation of peripheral nerves has found limited use, mostly because of the impossibility to obtain reliable supramaximal stimulation intensity and because of the uncertainty of the site of stimulation. These uncertainties all make it very difficult to estimate the conduction velocity. Contrary to peripheral nerves, magnetic stimulation of the cervical and lumbar roots is widely used in clinical studies to estimate the central conduction time at the level of the cervical and lumbar enlargements.[52,55–57]

In the past decade, magnetic stimulation was used to study nerves or roots that usually were difficult to study with conventional electrical stimulation. The technique has shown to be of interest, for example, for the study of the phrenic nerve, thoracic radiculopathy, and truncal neuropathies. Studying these situations using the conventional methods of electrical stimulation is very difficult, and the use of needle electrodes makes it painful and invasive. Chokroverty and colleagues[58] described a method to study thoracic nerves and roots using magnetic stimulation. In 1997, Carls and associates[59] compared electrical and magnetic stimulation of intercostal nerves. Both studies concluded that despite submaximal amplitudes of the motor potentials, the latencies were constant and reproducible, and because the method is well tolerated by the patients, it can be a useful clinical tool.

Magnetic stimulation at the base of the skull is an alternative painless technique that permits eliciting motor responses of deeply situated nerves such as the accessory nerve.[60] Pelliccioni and coworkers[61] used this technique to study patients with spinal accessory nerve palsy, recording evoked muscle responses from sternocleidomastoid and trapezius muscles. Using electrical stimulation by a needle electrode caused significant discomfort and pain, whereas magnetic stimulation was painless and easily activated muscle responses. Polkey and colleagues[62] applied magnetic stimulation to the femoral nerve and reported this as a painless supramaximal method of addressing quadriceps strength and fatigue. The method elicited reproducible results and was even used to monitor fatigue in subjects who were unable to perform a maximum voluntary contraction force properly.

REFERENCES

1. d'Arsonval A. Dispositifs pour la mesure des courants alternatifs de toutes frequencies. C R Soc Biol 1896;2:450–451.
2. Thompson SP. A physiological effect of an alternating magnetic field. Proc R Soc (Lond) 1910;B82:396–399.
3. Bickford RG, Fremming BD. Neural stimulation by pulsed magnetic fields in animals and man. Digest of the 6th International Conference of Medical Electronics and Biological Engineering, Tokyo, 1965:7–6.
4. Maass JA, Asa MM. Contactless nerve stimulation and signal detection by inductive transducer. IEEE Trans Magn 1970;6:322–326.
5. Oberg PA. Magnetic stimulation of nerve tissue. Med Biol Eng 1973;11:55–64.
6. Hallgren R. Inductive nerve stimulation. IEEE Trans Biomed Eng 1973;20:470–472.
7. Ueno S, Matsumoto S, Harada K, et al. Capacitive stimulatory effect in magnetic stimulation of nerve tissue. IEEE Trans Magn 1978;14:958–960.
8. Polson MJ, Barker AT, Freeston IL. Stimulation of nerve trunks with time-varying magnetic fields. Med Biol Eng Comput 1982;20:243–244.
9. Freeston IL, Barker AT, Jalinous R, et al. Nerve stimulation using magnetic fields. Presented at the 6th Annual Conference of the IEEE Engineering in Medicine and Biology Society 1984:557–561.
10. Barker AT, Jalinous R, Freeston IL. Non-invasive magnetic stimulation of human motor cortex. Lancet 1985;11:1106–1107.
11. Cadwell J. Principles of magnetoelectric stimulation. In Chokroverty S (ed): Magnetic Stimulation in Clinical Neurophysiology. Stoneham, MA, Butterworth, 1990:13–32.
12. Rushton WAH. Effect upon the threshold for nervous excitation of the length of the nerve exposed and the angle between current and nerve. J Physiol (Lond) 1927;63:357–377.
13. Barker AT. Magnetic nerve stimulation. Basic principles and early development. In Nilsson J, Panizza M, Grandori F (eds): Advances in

Occupational Medicine & Rehabilitation. Pavia, Italy, PI–ME Press, 1996:1–11.
14. Rossini PM, Barker AT, Berardelli A, et al. Non-invasive electrical and magnetic stimulation of the brain, spinal cord and roots: basic principles and procedures for routine clinical application. Report of an IFCN committee. Electroencephalogr Clin Neurophysiol 1994;91:79–92.
15. Panizza M, Nilsson J, Roth BJ, et al. Relevance of stimulus duration for activation of motor and sensory fibers: implications for the study of H-reflexes and magnetic stimulation. Electroencephalogr Clin Neurophysiol 1992;85:22–29.
16. Cohen LG, Roth BJ, Nilsson J, et al. Effects of coil design on delivery of focal magnetic stimulation. Electroencephalogr Clin Neurophysiol 1990;75:350–357.
17. Ueno S, Tashiro T, Harada K. Localised stimulation of neural tissues in the brain by means of a paired configuration of time-varying magnetic fields. J Appl Phys 1988;64:5862–5864.
18. Nilsson J, Panizza M, Roth BJ, et al. Determining the site of stimulation during magnetic stimulation of a peripheral nerve. Electroencephalogr Clin Neurophysiol 1992;85:253–264.
19. Roth BJ, Maccabee PJ, Eberle LP, et al. In vitro evaluation of a 4-leaf coil design for magnetic stimulation of peripheral nerve. Electroencephalogr Clin Neurophysiol 1994;93:68–74.
20. Ren C, Tarjan PP, Popovic DB. A novel electric design for electromagnetic stimulation—the slinky coil. IEEE Trans Biomed Eng 1995;42:918–925.
21. Hsu KH, Durand DM. A 3-D differential coil design for localized magnetic stimulation. IEEE Trans Biomed Eng 2001;48:1162–1168.
22. Maccabee PJ, Amassian VE, Cracco RQ, et al. Stimulation of human nervous system using the magnetic coil. J Clin Neurophysiol 1991;8:38–55.
23. Maccabee PJ, Amassian VE, Eberle LP, et al. Measurement of the electric field induced into inhomogeneous volume conductors by magnetic coil: application to human spinal nerve geometry. Electroencephalogr Clin Neurophysiol 1991;81:224–37.
24. Olney RK, So YT, Goodin DS, et al. A comparison of magnetic and electrical stimulation of peripheral nerves. Muscle Nerve 1990;13:957–963.
25. Evans BA. Magnetic stimulation of the peripheral nervous system. J Clin Neurophysiol 1991;8:77–84.
26. Roth BJ, Basser PJ. A model of the stimulation of a nerve fiber by electromagnetic induction. IEEE Trans Biomed Eng 1990;37:588–597.
27. Veale JL, Mark RF, Rees S. Differential sensitivity of motor and sensory fibres in human ulnar nerve. J Neurol Neurosurg Psychiatry 1973;36:75–86.
28. Panizza M, Nilsson J, Hallett M. Optimal stimulus duration for H-reflexes. Muscle Nerve 1989;12:576–579.
29. Basser PJ, Roth BJ. Stimulation of a myelinated nerve axon by electromagnetic induction. Med Biol Eng Comput 1991;29:261–268.
30. Basser PJ. Focal magnetic stimulation of an axon. IEEE Trans Biomed Eng 1994;41:601–606.
31. Ravnborg M, Blinkenberg M, Dahl K. Significance of magnetic coil position in peripheral motor nerve stimulation. Muscle Nerve 1990;13:681–686.
32. Claus D, Murray NMF, Spitzer A, et al. The influence of stimulus type on the magnetic excitation of nerve structures. Electroencephalogr Clin Neurophysiol 1990;75:342–349.
33. Chokroverty S. Magnetic stimulation of the human peripheral nerves. Electromyogr Clin Neurophysiol 1989;29:409–416.
34. Chokroverty S, Spire JP, DiLullo J, et al. Magnetic stimulation of the human peripheral nerves. In Chokroverty S (ed). Magnetic Stimulation in Clinical Neurophysiology. Stoneham, MA, Butterworth, 1990:249–273.
35. Hallett M, Cohen LG, Nilsson J, et al. Differences between electrical and magnetic stimulation of human peripheral nerve and motor cortex. In Chokroverty S (ed). Magnetic Stimulation in Clinical Neurophysiology. Stoneham, MA, Butterworth, 1990:275–287.
36. Maccabee PJ, Amassian VE, Cracco RQ, et al. An analysis of peripheral motor nerve stimulation in humans using magnetic coil. Electroencephalogr Clin Neurophysiol 1988;70:524–533.
37. Maccabee PJ, Amassian VE, Cracco RQ, et al. Effective anode and cathode are very close together when stimulating peripheral nerve with the magnetic coil. Abstract for the Book Society for Neuroscience, 20th Annual Meeting. St. Louis, Missouri, October 28 to November 2, 1990;516.7, p 1261.
38. Ruohonen J, Panizza M, Nilsson J, et al. Transverse-field activation mechanism in magnetic stimulation of peripheral nerves. Electroencephalogr Clin Neurophysiol 1996;101:167–174.
39. Rattay F, Aberham M. Modeling axon membranes for functional electrical stimulation. IEEE Trans Biomed Eng 1993;40:1201–1209.
40. Basser PJ, Wijesinghe R, Roth BJ. The activation function for magnetic stimulation derived from a three-dimensional volume

conductor model. IEEE Trans Biomed Eng 1992;39:1207–1210.
41. Reilly JP. Peripheral nerve stimulation by induced electric currents: exposure to time-varying magnetic fields. Med Biol Eng Comput 1989;27:101–110.
42. Nagarajan SS, Durand DM, Warman EN. Effects of the induced electric field on finite neuronal structures: a simulation study. IEEE Trans Biomed Eng 1993;40:1175–1188.
43. Cros D, Day TJ, Shahani BT. Spatial dispersion of magnetic stimulation in peripheral nerves. Muscle Nerve 1990;13:1076–1082.
44. Sun SJ, Tobimatsu S, Kato M. The effect of magnetic coil orientation on the excitation of the median nerve. Acta Neurol Scand 1998;97:328–335.
45. Schnabel V, Struijk JJ. Magnetic and electrical stimulation of undulating nerve fibres: a simulation study. Med Biol Eng Comput 1999;37:704–709.
46. Maccabee PJ, Amassian VE, Eberle LP, et al. Magnetic coil stimulation of straight and bent amphibian and mammalian peripheral nerve in vitro: locus of excitation. J Physiol (Lond.) 1993;460:201–219.
47. Kobayashi M, Ueno S, Kurokawa T. Importance of soft tissue inhomogeneity in magnetic peripheral nerve stimulation. Electroencephalogr Clin Neurophysiol 1997;105:406–413.
48. Hsu KH, Nagarajan SS, Durand DM. Analysis of efficiency of magnetic stimulation. IEEE Trans Biomed Eng 2003;50:1276–1285.
49. Bischoff C, Riescher H, Machetanz J, et al. Comparison of various coils used for magnetic stimulation of peripheral motor nerves: physiological considerations and consequences for diagnostic use. Electroencephalogr Clin Neurophysiol 1995;97:332–340.
50. Cohen D, Cuffin HN. Developing a more focal magnetic stimulator. Part 1. Some basic principles. J Clin Neurophys 1991;8:102–111.
51. Maccabee PJ, Amassian VE, Cracco RQ, et al. Mechanisms of neuromagnetic stimulation of peripheral nerve. In Nilsson J, Panizza M, Grandori F (eds): Advances in Occupational Medicine & Rehabilitation. Pavia, Italy, PI-ME Press, 1996:117–128.
52. Ugawa Y, Rothwell JC, Day BL, et al. Magnetic stimulation over the spinal enlargements. J Neurol Neurosurg Psychiatry 1989;52:1025–1032.
53. Britton TC, Meyer BU, Herdman J, et al. Clinical use of the magnetic stimulator in the investigation of peripheral conduction time. Muscle Nerve 1990;13:396–406.
54. Epstein CM, Fernandez-Beer E, Weissman JD, et al. Cervical magnetic stimulation: the role of the neural foramen. Neurology 1991;41:677–680.
55. Barker AT, Freeston IL, Jalinous R, et al. Magnetic stimulation of the human brain and peripheral nervous system: an introduction and the results of an initial clinical evaluation. Neurosurgery 1987;20:100–109.
56. Maertens de Noordhout A, Remacle JM, Pepin JL, et al. Magnetic stimulation of the motor cortex in cervical spondylitis. Neurology 1991;41:75–80.
57. Tavy DL, Wagner GL, Keunen RW, et al. Transcranial magnetic stimulation in patients with cervical spondylitic myelopathy: clinical and radiological correlations. Muscle Nerve 1994;17:235–241.
58. Chokroverty S, Deutsch A, Guha C, et al. Thoracic spinal nerve and root conduction: a magnetic stimulation study. Muscle Nerve 1995;18:987–991.
59. Carls G, Ziemann U, Kunkel M, et al. Electric and magnetic stimulation of the intercostal nerves: a comparative study. Electromyogr Clin Neurophysiol 1997;37:509–512.
60. Priori A, Berardelli A, Inghilleri M, et al. Electrical and magnetic stimulation of the accessory nerve at the base of the skull. Muscle Nerve 1991;13:477–478.
61. Pelliccioni G, Scarpino O, Guidi M. Magnetic stimulation of the spinal accessory nerve: normative date and clinical utility in an isolated stretch-induced palsy. J Neurol Sci 1995;132:84–88.
62. Polkey MI, Kyroussis D, Hamnegard CH, et al. Quadriceps strength and fatigue assessed by magnetic stimulation of the femoral nerve in man. Muscle Nerve 1996;19:549–555.

4 Transcranial Electrical and Magnetic Stimulation of the Brain: Basic Physiological Mechanisms

John C. Rothwell

Although it is well over 100 years since the electrical excitability of the brain was discovered by Fritsch and Hitzig and David Ferrier, surprisingly little work has addressed the question of which cells are stimulated and where. Apart from constructing maps of the motor cortex, stimulation in animal experiments is rarely used as a tool to investigate cortical physiology. Experiments on most brain areas, especially in recent years, focus on recording natural patterns of activity rather than provoking artificial discharges.

However, the development of noninvasive methods of stimulating the brain in humans has produced a mass of data that suggests that a half-dozen or more different varieties of neurons can be targeted by different intensities or polarities of stimulation. Our problem now is that the results in humans have seemingly outstripped basic anatomical and physiological knowledge in animals, leading investigators to postulate the existence of particular populations of cortical neurons to explain their data. In this chapter, I explain the limits of our knowledge and discuss the theories that describe the most recent data.

■ Electrical Brain Stimulation

Experiments on Animals

Electrical stimulation is most commonly used to map the organization of the motor cortex. Such maps, which plot the sites where stimulation provokes activation of particular muscles or movements, are essentially maps of the output organization of the cortex. In the 1950s and 1960s, as the detail of such maps became complex, it became important to study how the output (corticospinal) neurons were recruited by the stimulus. Given the complexity of cortical organization and the possible number of different neurons that could be activated, the pattern of recruitment is remarkably straightforward.

Studies on the exposed motor cortex of monkeys showed that a single pulse of stimulation given through an electrode resting on the surface of the brain tended to recruit corticospinal neurons in two different ways.[1,2] The stimulus activated the cell directly, probably at the proximal portion of its axon where threshold is lowest, or indirectly through synaptic connections. These modes of activation could be distinguished in recordings taken from the descending corticospinal tract on the basis of their latency. Direct activation produced the earliest descending volley (i.e., D waves), whereas synaptic activation gave rise to later volleys (i.e., I waves). The D wave was still present after cooling the cortex or even removing the gray matter entirely, consistent with direct axonal stimulation, whereas the I waves were abolished.[3] The D and I volleys traveled at the same (high) velocity and were thought to involve activity in the fast-conducting, large-diameter fibers of the tract (Fig. 4–1).

In most cases, corticospinal recordings represent the summed activity of all the fibers

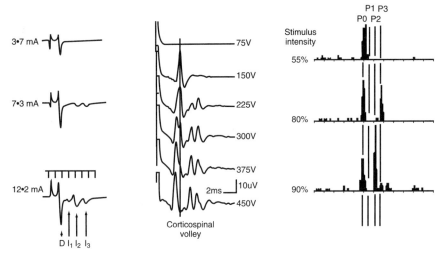

Figure 4–1 *Left panel*: Recordings of descending volleys made from the surface of the pyramidal tract of a monkey after direct electrical stimulation of the surface of the motor cortex at gradually increasing intensities. Notice the recruitment of D and I waves. *Middle panel*: Similar recordings made from the epidural space of the spinal cord of a human patient undergoing scoliosis surgery after transcranial electrical stimulation (TES) of the motor cortex. *Right panel*: Poststimulus time histograms of the time of firing of a single motor unit in an intact conscious subject after TES of the motor cortex. Notice how at low intensities of stimulation the unit fires once only at a short latency. As the intensity is increased, the unit begins to fire at other latencies, representing the arrival of I waves at the spinal motor neuron. (From Rothwell JC, Thompson PD, Day BL, et al. Stimulation of the human motor cortex through the scalp. Exp Physiol 1991;76:159–200.)

in the tract; single-unit recordings are rare. In such conditions, a volley represents synchronous activity in many single fibers. Any asynchronous activity leads to cancellation of potentials and does not show up in the population discharge. The presence of an initial D wave volley is therefore not surprising because all the fibers were stimulated at the same time, but the continuing cycles of I waves are unusual and indicate the presence of a highly effective synchronizing mechanism that continues to operate for several milliseconds after the stimulus in the intrinsic circuits of the cortex.

Quite unexpectedly, the I waves appear at regular, almost clocklike, intervals of about 1.5 ms, starting about 1.5 ms after the earliest D wave. The mechanism of the I-wave periodicity is unknown. It may represent periodic synaptic inputs into pyramidal tract neurons from a reverberating neural circuit or from a cascade of inputs from multiple sources.[3] Alternatively, it may be the consequence of an intrinsic property of the neuronal membrane in response to an initial large synchronous

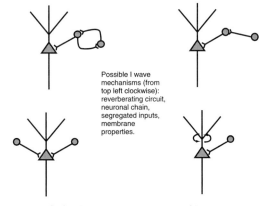

Figure 4–2 Possible mechanisms of I-wave formation. The triangular cell represents the large corticospinal neurons of the motor cortex, receiving synaptic input from other neurons in the cortex.

excitatory input[4] (Fig. 4–2). All three mechanisms may contribute. Microelectrode recording of the discharge of individual corticospinal neurons shows that some may discharge one action potential for each I wave, whereas others may discharge on only a proportion of

I waves.[5] Corticospinal neurons never appear to discharge naturally at such high frequencies, but there is a population of interneurons in the cortex known as "chattering" cells that can follow such frequencies. Whether they play any role in I-wave production is not known.[6]

The proportion of D and I waves depends on the configuration of the stimulating electrodes and on the intensity of stimulation. Stimulation with a small anode and large distant cathode (i.e., monopolar anodal stimulation) evokes D waves preferentially at lowest threshold and recruits I waves only as the intensity is raised.[1] Monopolar cathodal or bipolar stimulation (i.e., stimulation through two closely spaced small electrodes) gives rise to a mix of D and I waves even at threshold.[2] Intracortical microstimulation through a small electrode inserted into the gray matter tends to activate corticospinal neurons transynaptically.[7] Precisely what elements are stimulated to produce transsynaptic activation is still unclear but they may include cortical interneurons or afferent fibers from subcortical or cortical structures.

Experiments on Humans: Transcranial Electrical Stimulation of the Motor Cortex Hand Area

As Merton and Morton[8] demonstrated (Fig. 4–3), transcranial electrical stimulation (TES) of the brain is possible in humans although somewhat uncomfortable. The reason for this is that only a small fraction of the applied current flows into the brain; most of it travels between the electrodes along the lower resistance pathway offered by the surface of the skin and scalp where it causes local muscle contraction and pain. Merton and Morton[8] placed a small anode over the hand area of the motor cortex and an identical cathode 6 cm anterior or at the vertex of the head. Effectively, this widely spaced bipolar arrangement approximates the monopolar anodal montage in animal experiments. Other investigators used larger cathodes in a variety of arrangements.[9,10]

The results are very similar to direct anodal electrical stimulation of the exposed cortex of monkeys.[11] They are best illustrated in recordings of descending corticospinal activity that

Figure 4–3 Photograph of the first public demonstration of transcranial electrical stimulation of the motor cortex in March 1981. Dr. P. A. Merton (seated) is being stimulated by Dr. R. H. Adrian in front of a class of second-year medical students in the physiological laboratory in Cambridge, England. The circuit diagram of the stimulator is on the blackboard; Dr. Adrian discharges the electrical capacitance through the stimulating electrodes by pressing a Morse key. Dr. Merton is pointing to the left hand, which twitches when the stimulus is given.

have been taken from patients with electrodes inserted into the epidural space of the cervical spinal cord. In rare cases, where the electrodes have been inserted for treatment of chronic pain, patients are awake when the stimulation is given.[12] D and I wave recruitment is clearly visible, with the D waves being elicited first followed by I waves at higher intensities. The periodicity of the I waves is similar to that in monkeys, at about 1.5 ms or 600 Hz. Both D and I waves travel down the spinal cord at the same velocity of 60 to 80 m/s. Recordings can also be taken from the corticospinal tract during the course of surgery on the spinal column (see Fig. 4–1). These give similar results, except that the recruitment of I waves can be affected by the level of anesthesia; D waves, as might be expected, are much more resistant.

At very high intensities, Burke and colleagues[13] showed that the part of the pyramidal tract axon activated by TES shifted deeper into the brain to the approximate level of the cerebral peduncles and brainstem (see middle column of data in Fig. 4–1). They called the initial volley setup at these sites the D2 and D3 waves, respectively. Stimulation at the level of the brainstem can be achieved at more moderate stimulus intensities if the electrodes are placed 2 to 3 cm on each side of the inion.[14] Spinal epidural recordings show that this gives rise to a rapidly conducting single volley[15] that Ugawa and colleagues[14] found could be collided with the D wave from cortical stimulation. This suggests that both forms of stimulation can activate the same population of rapidly conducting fibers in the corticospinal tract.

■ Magnetic Brain Stimulation

Transcranial Magnetic Stimulation of the Motor Cortex Hand Area

Transcranial magnetic stimulation (TMS) of the human brain has probably been more intensively studied than any form of electrical stimulation in animals. The most easily interpreted results come from stimulation of the hand area of the motor cortex with a monophasic figure-of-eight stimulating coil. Two factors distinguish the results from those obtained with TES. First, recordings of descending corticospinal activity show that TMS tends to evoke I waves at a lower intensity than D waves (Fig. 4–4). Second, there is a preferred direction for recruiting descending activation (usually best if the magnetic stimulus induces electric current in the brain that flows in a posterior to anterior direction perpendicular to the central sulcus).[11,16] There is uncertainty about the reasons for both effects. The former is likely to arise because of the differences in the electric fields induced in the cortex by TES and TMS. Unlike TES, the physics of magnetic stimulation demands that induced electric current always flows parallel to the surface of the brain with no radial component of flow.[17] In the simplest case, if we imagine that all pyramidal tract neurons are aligned perpendicular to the surface of the brain, they will be activated best by radial current (see Chapter 2), whereas horizontally oriented interneurons or afferent axons would be activated best by horizontal current. Electric stimulation would tend to recruit pyramidal tract neurons

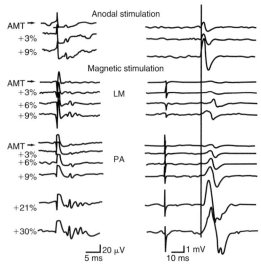

Figure 4–4 Descending corticospinal volleys (left column) and electromyographic (EMG) responses in the first dorsal interosseous muscle (right column) were recorded simultaneously in a conscious subject with an electrode implanted into the cervical epidural space for control of chronic pain. The motor cortex was stimulated with anodal transcranial electrical stimulation (TES) (top three rows) or transcranial magnetic stimulation (TMS) oriented to induce a lateromedial (LM) or posteroanterior (PA) current in the brain. The intensity of stimulation was at active motor threshold (AMT) or at different percentages of the maximal stimulator output above threshold. The vertical lines indicate the peak of the D wave from anodal stimulation and the onset of the anodal EMG response. Notice that TMS with the LM orientation evokes an I1 wave at threshold and that an earlier D wave occurs at AMT + 9%. For TMS with a PA orientation, the D wave is not seen until the intensity of stimulation is AMT + 21%. When D waves occur, the onset latency of the TMS-evoked EMG responses decreases to equal that seen after anodal TES. (Modified from Di Lazzaro V, Oliviero A, Profice P, et al. Comparison of descending volleys evoked by transcranial magnetic and electric stimulation in conscious humans. Electroencephalogr Clin Neurophysiol 1998;109:397–401.)

directly, whereas TMS would tend to recruit them indirectly by synaptic activation from horizontal connections (Fig. 4–5). Although this type of explanation is attractive, it is unclear whether these are the only factors involved in the difference between TES and

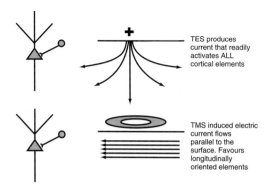

Figure 4–5 A simplified explanation of how electrical stimulation of the hand area may produce both D- and I-wave activation of the corticospinal neurons (*triangular cells on left*), whereas TMS may favor synaptic activation.

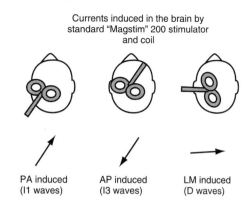

Figure 4–6 Stimulation of the hand area of the motor cortex with different directions of monophasic TMS pulses.

TMS. It might be expected that this is an oversimplification given the extensive folding of the cortical surface in the motor strip and that the hand representation is located in the anterior bank of the sulcus rather than on the surface of the gyrus.

The orientation sensitivity of the responses to TMS had not been anticipated on the basis of any of the previous work with electrical stimulation. Lowest-threshold activation at the hand area occurs with a posterior to anterior directed current flow and this tends to recruit initially the I1-wave.[18] If the coil current is reversed so that it induces an anterior-to-posterior current, the threshold for evoking a response is higher and there is a tendency to recruit the I3 wave before all other descending volleys.[19] If the coil is oriented to induce a lateral-to-medial (LM) current flow, in many subjects it is possible to record a D wave at lowest threshold, which is only followed by I waves as the intensity is increased. It is thought that such selectivity arises because any neuron is best excited by a potential difference along the length of its axon. The LM sensitivity of the D wave versus the anteroposterior (AP) or posteroanterior (PA) sensitivity of I waves suggests that at the point of excitation, the corticospinal axons of the hand area are traveling in a lateromedial direction, whereas the axons of the I wave inputs are oriented in a PA (I1 inputs) or AP (I3 inputs) direction (Figs. 4–4 and 4–6).

The problems in explaining the orientation selectivity of TMS are a typical example of how results from TMS can outstrip our basic knowledge. These particular data allow us to make one new and interesting conclusion—that different I waves can be produced by different synaptic inputs that have different spatial orientations. This means that not all I waves can be the result of reverberating activity in a single circuit. However, it also exposes serious gaps in our basic knowledge. Some of these gaps can only be filled by further studies on animals. For example, what classes of cortical neurons could possibly meet these requirements for spatial orientation? Will the same be true for other areas of cortex, or are the results peculiar to the hand area of cortex? Nevertheless, some gaps can and have been filled by further experiments in humans. For example, we know the answer to one question: Are the D waves produced by TES the same as those produced by TMS? This can be addressed by examining the interaction between subthreshold TMS and TES pulses.[20] There is remarkable facilitation if a subthreshold pulse of TES is followed within about 100 μs by a TMS pulse. Such a short time constant (much shorter than the millisecond time constants of a synaptic potential) suggests that the pulses are interacting at the cell membrane of an axon. If so, it means that both stimuli can excite a shared population of corticospinal axons at the same site. The same sort of approach could be used to tackle the question of whether the I3 waves from PA stimulation are caused by the same mechanism as the I3 waves from AP stimulation.

Different orientations and intensities of TMS can recruit different proportions of D and I waves. Because D waves are the result of stimulation at the axon of pyramidal tract neurons, they are not greatly influenced by changes in the level of excitability within the gray matter of the cerebral cortex.[21] However, the situation is quite different for I waves, which are highly sensitive to the level of cortical excitability at the time the stimulus is given because they are synaptically induced. If we wish to use TMS to probe the excitability of motor cortex, we should try to employ it so that it induces the greatest proportion of I-wave volleys possible.

Effects of Multiple Descending Volleys on the Form of Motor Evoked Potentials in Hand Muscles

The fact that TMS and TES evoke a complex series of repetitive discharges in the pyramidal tract of neuronal population has several implications for the electromyographic (EMG) responses that are evoked. These differ from the conventional compound muscle action potentials (CMAPs) produced by peripheral nerve stimulation in two important ways. First, the motor evoked potentials (MEPs) are always smaller, longer than in duration, and more polyphasic than a CMAP. Second, the onset latency, amplitude, and threshold of an MEP change according to whether subjects are relaxed or active[22,23] (Fig. 4–7).

The descending corticospinal volleys produced by TMS release excitatory postsynaptic potentials (EPSPs) at the motor neuron, and these develop over a period of several milliseconds, depending on the number of I waves that has been evoked. Motor neurons are discharged asynchronously, and this leads to the longer duration MEP, especially at high intensities of stimulation. There can be cancellation between phases of motor units discharging at slightly different latencies.

The different latencies of MEPs evoked in active and relaxed muscles result from the time taken for EPSPs to depolarize a motor neuron to its firing threshold (Fig. 4–8). At rest, a single EPSP (e.g., that produced by the arrival of a D-wave volley) may fail to raise a motor neuron to threshold and discharge will have to await the arrival of the I1 volley some 1.5 ms later. However, if the same stimulus is given during voluntary contraction, there will always be some motor neurons that are near enough to their firing threshold to be discharged on receipt of the first EPSP that arrives at the motor neuron pool. MEPs evoked in actively contracting muscles always have the shortest possible latency, whereas those evoked in subjects at rest usually have a longer latency. This feature is important when comparing the onset latency of MEPs evoked in healthy subjects and neurological patients.

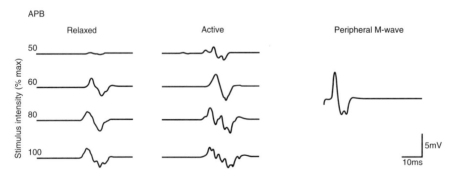

Figure 4–7 EMG records from the abductor pollicis brevis muscle of a healthy subject after different intensities of single-pulse TMS applied over the motor cortex with the subject at rest (*left column*) or preactivating the muscle (*middle column*). The responses can be contrasted with the maximum M wave evoked in the same muscle by supramaximal stimulation of the median nerve (*right column*). Notice how the motor evokes potentials are smaller in peak-to-peak amplitude and more polyphasic than the M wave. The responses are also smaller and have a later onset in relaxed compared with contracting muscle.

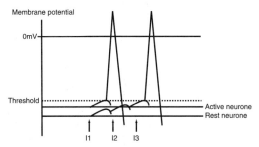

Figure 4–8 Explanation of the reason for the latency difference between responses evoked in relaxed and active muscle. The graph illustrates the notional potential difference across the membrane of a spinal motor neuron that receives synaptic input from corticospinal axons. The cell has a threshold potential that when exceeded causes the neuron to fire an action potential. In a cell that is at rest, the membrane potential is likely to be further away from threshold than in a cell that is active. At rest, a single excitatory postsynaptic potential (EPSP) from an I wave input may be insufficient to discharge the cell; it may have to await the arrival of a second or even a third EPSP from additional I waves. In the active state, it is likely that the cell may be discharged on receipt of the first EPSP; hence, the discharge latency will be shorter.

The difference in amplitude of MEPs at rest and during activation mainly results from increased excitability of spinal motor neurons during contraction. This means that a given amount of descending excitation will cause more spinal motor neurons to discharge during contraction, leading to a larger MEP. However, there is also some evidence that the response of pyramidal tract neurons to a TMS pulse is increased during voluntary contraction.[21] This may occur because the pyramidal tract neurons are more excitable and more readily discharged by synaptic input from I-wave generators during the course of voluntary contraction compared with rest.

The same factors contribute to the difference in threshold for evoking an MEP at rest and during contraction. The resting motor threshold (RMT) is always be higher than the active motor threshold (AMT) because resting spinal (and cortical) motor neurons require more excitation to reach discharge threshold than during contraction. The result of this is that RMT is a rather ill-defined measure because the basal level of "resting" excitability is not specified. Gandevia and Rothwell[24] found that subjects could voluntarily control the relative resting threshold of individual motor units in different hand muscles. From trial to trial, subjects could control whether a given transcranial stimulus recruited units in the abductor digiti minimi (ADM) or first dorsal interosseus (FDI) muscles. This illustrates the problem in defining RMT and can be used to argue that there is a well-defined sense of effort in the absence of any overt motor activity.

Distinguishing Effects on Motor Evoked Potential Amplitude Caused by Changes in Spinal and Cortical Excitability

The EMG response to TMS depends on the excitability of the cortex, which determines the amplitude and number of I waves that are recruited, and the excitability of the spinal motor neuron pool, which determines how many motor neurons are recruited by a given descending input. In many single-pulse TMS studies of motor cortex, the size and threshold of MEPs is measured before and after an intervention (e.g., learning, changes in afferent input, drug administration). Any changes that are seen could therefore be caused by alterations in excitability at the cortical or spinal level.

Four main techniques are available to distinguish between effects at these two levels, although none of them is sufficient on their own to provide an unequivocal answer. The first and most direct method is to record the descending volleys evoked by TMS before and after the intervention. This gives a direct measure of changes in the cortical response to stimulation. If the effect being studied is limited to a particular set of muscles, the disadvantage of the technique is that a spinal epidural electrode records all the descending activity set up by the stimulus and not just that destined for the target muscle group. Changes in volleys can only be said to correlate with changes in MEPs and not necessarily to cause them.

A second method is to compare the responses to TMS with the responses to TES. If low-intensity anodal stimulation is used, TES evokes a relatively pure D wave that is unaffected by changes in cortical excitability. If the response to TES is unchanged by the intervention while the response to TMS is, for example, facilitated, there might have been an increase in cortical excitability. The disadvantage is that in some individuals, a single D wave is insufficient to evoke an EMG response in relaxed muscle, and the comparison must be made during voluntary contraction. However, voluntary contraction often conceals changes in excitability that are evident at rest, and this limits the application of the method. One way around this problem is to combine TES with spinal H reflexes. Appropriately timed H-reflex input is facilitated by a subthreshold D wave if the two volleys arrive at the spinal motor neurons at the same time. The amount of H facilitation produced by subthreshold TES in resting subjects is a measure of the amplitude of the D wave input.

A third method is to use direct stimulation of the descending tracts at the level of the brainstem with the intention of evoking a single descending volley that is unaffected by changes in cortical excitability. As with TES, it may be difficult to evoke a large EMG response in relaxed muscles with this method. It is unknown to what extent activity in other pathways contributes to the size of EMG responses that are evoked. However, this may not be a serious objection if there is no change in the response to transmastoidal stimulation over the time of the study.

A fourth method is to use H reflexes or F waves to measure the excitability of spinal reflex pathways and motor neurons. If these remain unchanged by an intervention while responses to TMS differ, it is likely that the intervention has caused changes in cortical rather than spinal excitability. The disadvantage of this method is that the motor neurons within a spinal pool may be recruited in a different order by H reflexes or F waves than with TMS. Because changes in spinal excitability may be distributed differentially to motor neurons of different size or type this opens the possibility that H and F responses would fail to pick up changes in excitability of the population of spinal motor neurons recruited by TMS.

Circular versus Figure-of-Eight Coils and Monophasic versus Biphasic Stimulators

The first magnetic stimulators were supplied with circular coils with diameters of about 10 cm. Centered over the vertex of the scalp, the lateral windings lie over the hand area of each hemisphere. An anticlockwise current in the coil as viewed from above induces a PA current across the left motor area and an AP current over the right motor area. Consistent with the pattern of activation with figure-of-eight coils, this preferentially recruits an I1 wave from the left and an I3 from the right hemisphere.[25] At higher intensities, a D wave is recruited. However, this D wave has a slightly longer (0.2 ms) latency than that evoked by a figure-of-eight coil, and its amplitude is affected by the level of cortical excitation at the time the stimulus is given. These differences are thought to indicate that activation is occurring nearer the cell body than with a figure-of-eight coil, at a point where the excitability would be influenced by the membrane potential in the neuron.

Biphasic stimulus pulses are more efficient in stimulating the brain than monophasic pulses even when the initial phase of the stimulus is the same size.[26] The reason is that charge transfer is maximal in the swing between the first and second phases of the biphasic pulse.[27] One consequence of this is that the lowest threshold for activating the hand area with a biphasic stimulator occurs when the first phase of the induced stimulus is in the AP direction rather than the PA direction. With many commercial stimulators, this means that for optimal stimulation, a biphasic pulse requires a figure-of-eight coil to be held 180 degrees rotated to that when a monophasic pulse is given.

Transcranial Magnetic Stimulation of the Leg Area

Early results in which MEP latencies to TES and TMS were compared suggested that activation of the corticospinal output from the leg differed from that of the hand area. In hand and forearm muscles, TES produced earlier MEPs than TMS; in the tibialis anterior muscle, the latencies were the same.[28] The simplest explanation for this is that both forms of stimulation activate at the

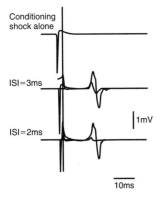
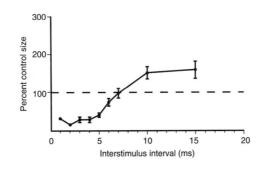

Figure 4–9 Paired-pulse intracortical inhibition in the Kujirai paradigm.[32] The *left panel* shows electromyographic responses from the first dorsal interosseus muscle after the conditioning stimulus given alone (top trace) and superimposed traces of responses to the test shock alone (*larger trace of the pair*) and when conditioned by a preceding subthreshold TMS pulse given 3 ms (*middle pair of traces*) or 2 ms (*bottom pair of traces*) earlier. The graph (*right*) shows the time course of the effect in 10 subjects. Notice the early phase of inhibition followed by facilitation at interstimulus intervals of more than 6 ms. (From Kujirai T, Caramia MD, Rothwell JC, et al. Corticocortical inhibition in human motor cortex. J Physiol Lond 1993;471:501–519.)

same site. However, which site this may be has been debated. It may be the proximal axon (e.g., a conventional D wave in the hand area), the I1 wave input (e.g., PA TMS of the hand area), or even a third site, the initial segment of the corticospinal neuron. The question is still debated, but the latest results using direct recordings of corticospinal volleys from the spinal epidural space suggest that the most likely explanation is that both TES and TMS tend to activate I1 wave inputs.[29–31] This is consistent with the fact that the initial volley from both forms of stimulation is sensitive to the level of cortical excitability when the stimulus is applied and therefore is not a conventional D wave. The latency of the response is compatible with transsynaptic rather than initial segment activation.

■ Paired-Pulse Transcranial Magnetic Stimulation Interactions

Paired-pulse experiments are designed to give insight into the nature of the cortical circuitry activated by TMS. A variety of methods exists to examine the connections within the motor cortex itself or connections to the motor cortex from other parts of the central nervous system. In all cases, the design is similar. A conditioning stimulus, S1, is given to one area of brain to test its effect on the amplitude of an MEP evoked by a test stimulus, S2, applied over the motor cortex. Direct recordings of the effects on descending volleys are the best way of documenting the nature of these effects, which in some cases involve unexpected selectivity for different components (e.g., D, I1, I2) of the response. However, after the mechanism is understood, it is possible to use MEP amplitude changes as a substitute measure of descending activity.

Short-Interval Intracortical Inhibition

Short-interval intracortical inhibition (SICI) was first reported by Kujirai and colleagues.[32] They applied both stimuli through the same coil to the motor cortex hand area and found that a small subthreshold conditioning stimulus could suppress the response to a later suprathreshold test stimulus if the interval between stimuli was 5 ms less (Fig. 4–9). Because the conditioning stimulus was below AMT, the investigators suggested that the interaction was occurring at a cortical level and that the conditioning stimulus was suppressing the recruitment of descending volleys by the test stimulus. Direct recordings of descending volleys have confirmed this.[12] A small conditioning stimulus that itself evokes no descending activity suppresses late I waves if the interval between the stimuli is between

1 to 5 ms. The I1 wave is virtually unaffected, with the most sensitive wave being the I3 and later volleys (Fig. 4–10). A similar conclusion was reached by Hanajima and coworkers[33] on the basis of studies of single motor units. Because the I1-wave is unaffected, the results suggest that inhibition does not modify directly the excitability of pyramidal neurons. It seems more likely that the inhibition in this paradigm results from reduced excitability of inputs responsible for I waves.

The threshold for producing SICI (i.e., minimum intensity of S1 needed to reduce the response to S2) is remarkably low and averages about 70% of the active motor threshold in any individual.[34] Increasing the intensity of S1 initially leads to increased inhibition, but this peaks at about 90% to 100% AMT. Higher intensities give a smaller effect and eventually lead to facilitation. The latter effect is thought to be caused by recruitment of descending corticospinal volleys at intensities above AMT that produce subthreshold excitation of spinal motor neurons. Although the threshold for inhibition is remarkably similar in different individuals (at least when expressed as a percent of that individual's AMT), the increase of inhibition at higher intensities is highly variable. This leads to a wide spread in normal values for SICI when measured with the usual S1 intensities of 80% to 100% AMT (Fig. 4–11).

SICI at 1.5 ms or less differs from that at interstimulus intervals of 2 to 5 ms.[35] The former has an even lower threshold than the more conventional longer intervals, but its mechanism is unclear. It may involve refractoriness in cortical circuits stimulated by both S1 and S2. There is one other piece of evidence that very small S1 stimuli that are below SICI threshold (at 2 to 5 ms) can nevertheless activate some cortical circuits. Pairs of S1 stimuli separated by intervals of 4 ms or less can show temporal summation and lead to clear inhibition.[36] The time course of this interaction has been suggested to reflect convergence of synaptic inputs onto a common inhibitory interneuron that leads to SICI.

Studies with subdural arrays of implanted electrodes in patients with epilepsy suggest that suppression of MEPs evoked by stimulation

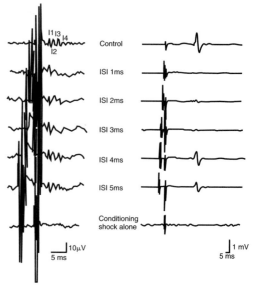

Figure 4–10 Epidural volleys (*left*) and electromyographic (EMG) responses (*right*) evoked by test stimulus alone (*upper traces*), conditioning stimulus alone (*lower traces*), and both stimuli at different interstimulus intervals (ISIs) (*middle traces*) in one subject. Recordings were performed at rest during test and paired magnetic stimulation, while conditioning stimulus alone was delivered during voluntary contraction at about 20% of maximum. Each trace is the average of five sweeps. Test stimulus evokes multiple descending waves (four waves) and an EMG response of about 1 mV. Conditioning shock alone evoked neither an EMG response nor descending volleys. When both stimuli were delivered, the EMG responses were dramatically suppressed at 1- to 3-ms ISIs. At 1-ms ISI, all the descending waves except the I1 were suppressed; at longer ISIs, there was a clear reduction of amplitude of later waves, and the last wave partially recovered only at longer ISIs. (From Di Lazzaro V, Oliviero A, Profice P, et al. Comparison of descending volleys evoked by transcranial magnetic and electric stimulation in conscious humans. Electroencephalogr Clin Neurophysiol 1998;109:397–401.)

of one cortical point comes from a relatively focal area of surrounding cortex. Conditioning stimuli applied to the face area do not suppress responses evoked from the hand area and vice versa.[37]

Kujirai and colleagues[32] originally suggested that SICI was GABAergic in origin. Administration of lorazepam, an agonist at the $GABA_A$ receptor increases the amount of

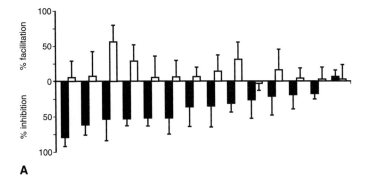

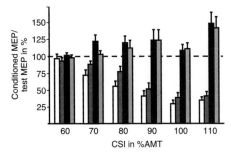

Figure 4–11 **A:** The variability of short-interval intracortical inhibition (SICI) (*black bars*, measured at an ISI of 2 ms) and short-interval intracortical facilitation (ICF) (*open bars*, measured at an ISI of 15 ms) in 14 subjects, all tested with a conditioning intensity of 80% active motor threshold. **B:** The mean amount of SICI (*open*, 2 ms; *diagonal hatching*, 3 ms) and ICF (*black*, 12 ms; *stippled*, 15 ms) is affected by the intensity of the conditioning stimulus. Notice how SICI becomes evident at about 70% active motor threshold. (From Orth M, Snijders AH, Rothwell JC. The variability of intracortical inhibition and facilitation. Clin Neurophysiol 2003;114:2362–2369.)

intracortical inhibition and increases the inhibition of later descending I waves.[38]

Short-Interval Intracortical Facilitation

Paired-pulse short-interval intracortical facilitation (SICF) was first described between EMG responses by Tokimura and coworkers[39] and Ziemann and associates.[40] If two stimuli are given at an intensity at or above active motor threshold, facilitation can be observed between them if the intervals between the shocks is approximately 1.3, 2.5, and 4.3 ms. This interaction between the stimuli is thought to be caused by interaction of I-wave inputs in the periodic bombardment of pyramidal neurons (Fig. 4–12).

Recordings of descending volleys demonstrate the action on individual I waves very clearly.[12] Because the stimuli are so closely spaced in time, the response to the second stimulus of the pair is best seen by subtracting the response to the first stimulus given alone from the response to the pair. This can then be compared with the response to the second stimulus given alone. At an interval of 1.2 ms, the I2 and I3 waves are clearly facilitated, whereas the I1 wave is unchanged. However, the apparent lack of effect on I1 deserves comment. As Amassian and colleagues[41] pointed out, at an ISI of 1.2 ms, most of the neurons activated by the first stimulus will be refractory to the second stimulus, particularly if the first stimulus is larger than the second stimulus. The fact that we obtain any response at all to the second stimulus is somewhat surprising. One possibility is that the second stimulus excites the cell bodies or initial segments of neurons that were excited but not discharged by the first stimulus. Synaptic depolarization of a neuron declines with a time constant of several milliseconds, and although cell bodies or initial segments are

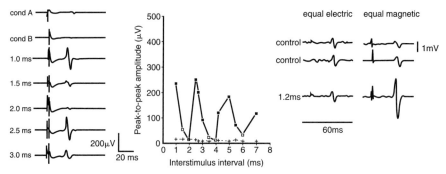

Figure 4–12 Short-interval intracortical facilitation (I-wave facilitation). The electromyographic traces (*left column*) show the responses to each transcranial magnetic stimulation (TMS) pulse given alone (i.e., stimulus pulses A and B), and the lower traces show the effect of giving both pulses together at the interstimulus intervals noted at the start of each trace. Notice the facilitation if the intervals are 1.0 and 2.5 ms. The graph in the center shows the time course of this interaction in eight subjects. The *dotted line* is the predicted sum of the response to both stimuli given together, and the *solid line* is the actual size of the response. There is no facilitation if two anodal transcranial electrical stimulation (TES) pulses are given at an interval of 1.2 ms (*right panel*), whereas there is clear facilitation between two TMS pulses at the same interval. This experiment was performed in actively contracting muscles so that the anodal TES pulse would produce a clear motor evoked potential even at an intensity of stimulation likely to evoke only a single D-wave volley. (From Tokimura H, Ridding MC, Tokimura Y, et al. Short latency facilitation between pairs of threshold magnetic stimuli applied to human motor cortex. Electroencephalogr Clin Neurophysiol 1996;101:263–272.)

not thought to be the usual sites at which TMS acts, it might be possible for the second stimulus to activate them directly if they have been facilitated.[42]

Intracortical facilitation

Kujirai and coworkers[32] originally noticed that if the interval between their pulses was greater than 6 ms or so, the MEPs evoked by the test stimulus (S2) were weakly facilitated for at least 10 ms (see Fig. 4–9). This is now known as *intracortical facilitation* (ICF). The threshold for ICF is higher than that for SICI, and unlike SICI, it is sensitive to the direction of the current induced by S1, being maximal with a PA orientation.[43] These features suggest it has a different mechanism to SICI, but more work needs to be done to clarify its characteristics fully. There are no studies of how descending corticospinal tract volleys are affected by ICF.

Long-Interval Intracortical Inhibition

Long-interval intracortical inhibition (LICI) was first described by Valls-Sole and associates.[44] They found that a conditioning stimulus strong enough to produce an MEP in the target muscle could suppress a response to a later stimulus of the same intensity if the interval was between 50 and 200 ms. Because the excitability of spinal H reflexes had recovered at this interval, the effect was thought to be cortical, probably the result of activation of $GABA_B$ receptors.[45] The cortical origin has been confirmed in recordings of corticospinal volleys (reviewed by Di Lazzaro and Rothwell[12]). These recordings show that the S1 suppresses the size and number of I waves recruited by an S2 given 100 or 150 ms later. The I waves are facilitated if the interval between S1 and S2 is 50 ms even though the MEPs are suppressed. Presumably, the effect on the MEP results from the refractoriness of spinal motor neurons that discharged in response to S1 and other potential spinal mechanisms of inhibition.

One study of the behavior of corticospinal volleys in this paradigm reported that the D wave recruited by S2 was reduced in size,[46] but this was not seen in two other studies.[47,48] There are two possible reasons for the difference. First, the single patient that showed the D-wave effect had had an avulsion of the brachial plexus, and although the investigators studied the hemisphere innervating the intact side, it is still possible that cortical changes could occur to both hemispheres after a unilateral deafferentation.[49]

Second, the patient was investigated using a round coil to stimulate the brain. This produces a D wave by activating the pyramidal axon near the cell body where it is sensitive to the level of cortical excitability. This type of D activation may be suppressed during LICI.

Interhemispheric Inhibition

Ferbert and coworkers[50] showed that TMS of the motor cortex of one hemisphere can inhibit motor responses evoked in distal hand muscles by a magnetic stimulus given 6 to 30 ms later over the opposite hemisphere. On the basis of indirect arguments, they originally suggested that the inhibition is produced at cortical level by a transcallosal route. Recordings of descending corticospinal volleys have confirmed that the conditioning stimulus suppresses I2 and later waves if given 6 to 20 ms before S2. The I1 and D waves are unaffected.[12]

There is sometimes a small early period of facilitation at ISI = 3 to 4 ms.[51] This mixture of early facilitation and predominant later inhibition may reflect the basic neurophysiology of the connection which is thought to involve a focal point-to-point facilitation with a larger surround inhibition. The amount of inhibition is task dependent, being largest when the hand contralateral to S1 is active and the opposite hand relaxed.[50] The effect of interhemispheric inhibition (IHI) can be seen in the ongoing voluntary EMG as a silent period with onset latency of about 35 ms and a duration of 30 ms. This is sometimes known as the ipsilateral silent period. Although the optimal orientation of the coil is the same, the threshold for IHI is often different from the threshold for evoking a contralateral MEP, suggesting that stimulation of a different set of neurons is required for each effect.

Although IHI is thought to involve transmission across the corpus callosum and is absent in patients with lesions that affect the transcallosal motor fibers[52] and present in some patients with corticospinal lesions in the internal capsule,[53] there is evidence that S1 may produce a small effect on spinal mechanisms,[54] indicating the importance of testing for both cortical and spinal effects.

IHI may consist of two separate phases. The original experiments of Ferbert and colleagues[50] showed that the duration of inhibition increased with the intensity of the conditioning shock, which would be compatible with recruitment of additional pathways at higher intensities. Chen and colleagues[55] investigated the effect of stimulus intensity, orientation of the stimulating coil, and amount of background muscle activation on IHI at 8 ms compared with inhibition at 40 ms or inhibition in the ipsilateral silent period and concluded that short-interval IHI had a different mechanism to the later inhibitory effects. A similar phenomenon was observed in a different set of experiments by Gilio and associates.[56] They found that a period of repetitive TMS to the conditioning hemisphere only affected IHI at short but not long intervals.

Short-Interval Afferent Inhibition of the Motor Cortex

Afferent input can modify the excitability of the motor cortex with a complex time course. There are effects on both the amplitude of MEPs as well as on SICI and ICF. Only the former are dealt with here. The most thoroughly investigated effect is the very short interval inhibition of MEPs in hand muscles that is produced by electrical stimulation of peripheral nerves innervating the hand.[57,58] This inhibition begins about 1 ms after the latency of the N20 component of the sensory evoked potential, and lasts for about 7 to 8 ms. It is most powerful after stimulation of mixed nerve but also can be observed after stimulation of cutaneous afferents in the digital nerves. EMG studies comparing responses to TMS and TES suggested that the inhibition was cortical origin, and this has been confirmed in the recordings of descending volleys (Fig. 4–13). As with other forms of inhibition, the I2 and later waves are suppressed strongly, whereas the I1 and D waves are unaffected. Administration of the muscarinic receptor blocker scopolamine reduces the amount of inhibition in both EMG and descending volley, suggesting that the pathway was under cholinergic control.[25] Di Lazzaro and colleagues[59] have shown that patients with Alzheimer's disease have reduced afferent inhibition that is increased by rivastigmine.

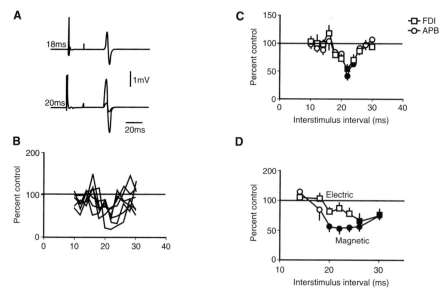

Figure 4–13 Short-interval afferent inhibition. **A:** Superimposed electromyographic traces from the first dorsal interosseus (FDI) muscle to a test transcranial magnetic stimulation (TMS) pulse given alone and when preceded by a single electrical stimulus to the median nerve at the wrist given 18 or 20 ms earlier. Notice the inhibition of the test response when the interval was 20 ms. **B:** The time course of this effect in eight subjects. Each *line* represents the result of a single subject. **C:** The grand average of data for responses evoked in the FDI and abductor pollicis brevis (APB) muscles in the same subjects. **D:** Responses to anodal transcranial electrical stimulation (TES) are not inhibited at early intervals by the median nerve stimulus, suggesting that this part of the inhibitory time course is caused by inhibition of spinal motor neurons. (From Tokimura H, Di LV, Tokimura Y, et al. Short latency inhibition of human hand motor cortex by somatosensory input from the hand. J Physiol 2000;523[Pt 2]:503–513.)

Inhibitory effects on the MEP continue for up to 200 ms and are sometimes known as long-interval afferent inhibition. Tamburin and coworkers[60] have shown that the inhibition after stimulation of digital nerves at an interval of about 30 to 50 ms has a somatotopic organization in small hand muscles: stimulation of the index has the largest effect on MEPs in FDI, whereas stimulation of the fifth digit has the largest effects on responses in ADM.

Premotor-Primary Motor Cortex Inhibition

Premotor-primary motor cortex inhibition has been recently described and is maximal at ISI = 6 ms, starting at about 2 to 3 ms and followed by a short period of facilitation with ISI > 8 to 10 ms.[61] To study the effect, two small coils must be used so that they can be placed close enough together on the scalp to activate the two areas separately. The best orientation of the conditioning coil is to induce AP currents over the premotor area, which indicates that the elements involved are arranged differently with respect to the scalp than those needed to evoke MEPs from the primary motor area (Fig. 4–14).

Cerebellocortical Inhibition

Cerebellocortical inhibition is provoked by stimulation over the cerebellum and leads to inhibition of MEPs with a latency of about 5 ms. The effect was first described using TES with electrodes placed at the same position as used to activate the corticospinal tract at the level of the brainstem.[62] However, the intensity used is below threshold for the corticospinal effect, presumably reflecting the shorter distance to the cerebellum compared with the brainstem. Cerebellar effects can also be obtained with TMS if a large stimulating coil is used.[63,64]

The inhibition of MEPs may be caused by excitation of inhibitory output from the Purkinje cells, which secondarily withdraws facilitation from cerebellothalamocortical

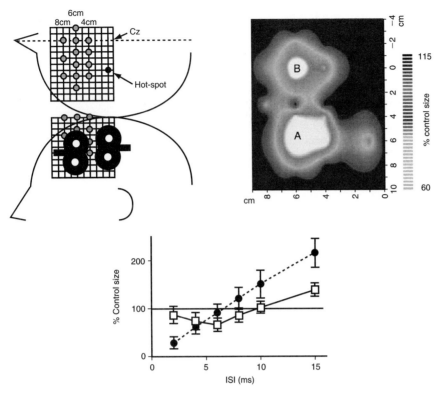

Figure 4–14 Premotor to motor cortex interactions probed with a paired-pulse design. The diagram (*left*) indicates the scalp locations at which the cortical stimuli were delivered. Views are from above and from the side of the head. The hot spot (primary motor cortex) for activating the right first dorsal interosseus (FDI) is shown by the *solid black circle* (*arrow*). The 13 points at which conditioning stimuli were applied are shown in gray. The 1 × 1 cm grid extended anterior to the interaural line. The color map on the right summarizes the effects of conditioning stimuli over the same set of frontal sites on the amplitude of the test motor evoked potential (MEP) at an ISI = 6 ms. Maximum inhibition of the test MEP is coded in bright blue. Maximum facilitation is coded in red. Letters A and B indicate the locations at which the inhibitory effects were investigated in detail. The graph (*lower panel*) shows the effects of conditioning stimuli on the size of a test MEP at ISI = 2 to 15 ms. Two locations of the conditioning coil are compared: (i) position A (*solid line*); (ii) the hot spot for the right FDI muscle (*dotted line*). Mean ± SD data are from six subjects. (Modified from Civardi C, Cantello R, Asselman P, et al. Transcranial magnetic stimulation can be used to test connections to primary motor areas from frontal and medial cortex in humans. Neuroimage 2001;14:1444–1453.)

pathways. The duration of the effect is difficult to define because the initial cerebellar inhibition is followed by a second period of inhibition thought to be caused by stimulation of afferent fibers in cervical nerve roots or even in the brainstem.[65–68]

Interactions between Pathways

Given that the predominant effect of all the interactions discussed previously (apart from SICF and ICF) is inhibitory, it is reasonable to ask whether they all converge onto the same set of inhibitory interneurons in the motor cortex or whether there are separate inhibitory circuits private to each pathway. Do the transcallosal inhibitory effects share any parts of the pathway with premotor-motor or cerebellomotor inhibition? The fact that the time courses of some of the effects are different (e.g., long- and short-latency ICI) makes it likely that separate pathways can exist. However, there is a large amount of evidence that interactions between these systems can also occur. For example, SICI appears to be inhibited by both LICI and IHI, both of which may share a common inhibitory interneuron. In contrast, LICI is inhibited by long-interval afferent inhibition (reviewed by Chen and colleagues[69]).

REFERENCES

1. Kernell D, Chien-Ping WU. Responses of the pyramidal tract to stimulation of the baboon's motor cortex. J Physiol 1967;191:653–672.
2. Patton HD, Amassian VE. Single and multiple unit analysis of the cortical stage of pyramidal tract activation. J Neurophysiol 1954;17:345–363.
3. Amassian VE, Stewart M, Quirk GJ, et al. Physiological basis of motor effects of a transient stimulus to cerebral cortex. Neurosurgery 1987;20:74–93.
4. Rothwell JC. Techniques and mechanisms of action of transcranial stimulation of the human motor cortex. J Neurosci Methods 1997;74:113–122.
5. Edgley SA, Eyre JA, Lemon RN, et al. Direct and indirect activation of corticospinal neurons by electrical and magnetic stimulation in the anesthetized macaque monkey. J Physiol (Lond) 1992;446:224.
6. Gray CM, McCormick DA. Chattering cells: superficial pyramidal neurons contributing to the generation of synchronous oscillations in the visual cortex. Science 1996;274:109–113.
7. Jankowska E, Padel Y, Tanaka R. The mode of activation of pyramidal tract cells by intracortical stimuli. J Physiol 1975;249:617–636.
8. Merton PA, Morton HB. Stimulation of the cerebral cortex in the intact human subject. Nature 1980;285:227.
9. Amassian VE, Cracco RQ. Human cerebral cortical responses to contralateral transcranial stimulation. Neurosurgery 1987;20:148–155.
10. Starr A, Caramia MD, Zarola F, et al. Enhancement of motor cortical excitability in humans by non-invasive electrical stimulation appears before voluntary movement. Electroencephalogr Clin Neurophysiol 1988;70:26–32.
11. Day BL, Dressler D, de Maertens NA, et al. Electric and magnetic stimulation of human motor cortex: surface EMG and single motor unit responses [published erratum appears in J Physiol (Lond) 1990;430:617]. J Physiol (Lond) 1989;412:449–473.
12. Di Lazzaro V, et al. The physiological basis of transcranial motor cortex stimulation in conscious humans. Clin Neurophysiol 2004;115:255.
13. Burke D, Hicks R, Gandevia SC, et al. Direct comparison of corticospinal volleys in human subjects to transcranial magnetic and electrical stimulation [published erratum appears in J Physiol (Lond) 1994;476:553]. J Physiol (Lond) 1993;470:383–393.
14. Ugawa Y, Rothwell JC, Day BL, et al. Percutaneous electrical stimulation of corticospinal pathways at the level of the pyramidal decussation in humans. Ann Neurol 1991;29:418–427.
15. Rothwell J, Burke D, Hicks R, et al. Transcranial electrical stimulation of the motor cortex in man: further evidence for the site of activation. J Physiol 1994;481(Pt 1):243–250.
16. Mills KR, Boniface SJ, Schubert M. Magnetic brain stimulation with a double coil: the importance of coil orientation. Electroencephalogr Clin Neurophysiol 1992;85:17–21.
17. Tofts PS. The distribution of induced currents in magnetic stimulation of the nervous system. Phys Med Biol 1990;35:1119–1128.
18. Di Lazzaro V, Oliviero A, Saturno E, et al. The effect on corticospinal volleys of reversing the direction of current induced in the motor cortex by transcranial magnetic stimulation. Exp Brain Res 2001;138: 268–273.
19. Sakai K, Ugawa Y, Terao Y, et al. Preferential activation of different I waves by transcranial magnetic stimulation with a figure-of-eight-shaped coil. Exp Brain Res 1997;113:24–32.
20. Rothwell JC, Day BL, Amassian VE. Near threshold electrical and magnetic transcranial stimuli activate overlapping sets of cortical-neurons in humans. J Physiol (Lond) 1992;452:109.
21. Di Lazzaro V, Oliviero A, Profice P, et al. Effects of voluntary contraction on descending volleys evoked by transcranial electrical stimulation over the motor cortex hand area in conscious humans. Exp Brain Res 1999;124:525–528.
22. Day BL, Rothwell JC, Thompson PD, et al. Motor cortex stimulation in intact man. 2. Multiple descending volleys. Brain 1987;110:1191–1209.
23. Thompson PD, Day BL, Rothwell JC, et al. The interpretation of electromyographic responses to electrical stimulation of the motor cortex in diseases of the upper motor neurone. J Neurol Sci 1987;80:91–110.
24. Gandevia SC, Rothwell JC. Knowledge of motor commands and the recruitment of human motoneurons. Brain 1987;110:1117–1130.
25. Di Lazzaro V, Oliviero A, Pilato F, et al. Descending volleys evoked by transcranial magnetic stimulation of the brain in conscious humans: effects of coil shape. Clin Neurophysiol 2002;113:114–119.
26. Kammer T, Beck S, Thielscher A, et al. Motor thresholds in humans: a transcranial magnetic stimulation study comparing different pulse waveforms, current directions and stimulator types. Clin Neurophysiol 2001;112:250–258.
27. Maccabee PJ, Nagarajan SS, Amassian VE, et al. Influence of pulse sequence, polarity

and amplitude on magnetic stimulation of human and porcine peripheral nerve. J Physiol 1998;513(Pt 2):571–585.
28. Priori A, Bertolasi L, Dressler D, et al. Transcranial electric and magnetic stimulation of the leg area of the human motor cortex: single motor unit and surface EMG responses in the tibialis anterior muscle. Electroencephalogr Clin Neurophysiol 1993;89:131–137.
29. Di Lazzaro V, Oliviero A, Profice P, et al. Descending spinal cord volleys evoked by transcranial magnetic and electrical stimulation of the motor cortex leg area in conscious humans. J Physiol 2001;537:1047–1058.
30. Nielsen J, Petersen N, Ballegaard M. Latency of effects evoked by electrical and magnetic brain stimulation in lower limb motoneurones in man. J Physiol (Lond) 1995;484:791–802.
31. Terao Y, Ugawa Y, Sakai K, et al. Transcranial stimulation of the leg area of the motor cortex in humans. Acta Neurol Scand 1994;89:378–383.
32. Kujirai T, Caramia MD, Rothwell JC, et al. Corticocortical inhibition in human motor cortex. J Physiol (Lond) 1993;471:501–519.
33. Hanajima R, Ugawa Y, Terao Y, et al. Paired-pulse magnetic stimulation of the human motor cortex: differences among I waves. J Physiol (Lond) 1998;509:607–618.
34. Orth M, Snijders AH, Rothwell JC. The variability of intracortical inhibition and facilitation. Clin Neurophysiol 2003;114:2362–2369.
35. Fisher RJ, Nakamura Y, Bestmann S, et al. Two phases of intracortical inhibition revealed by transcranial magnetic threshold tracking. Exp Brain Res 2002;143:240–248.
36. Bestmann S, et al. Inhibitory interactions between pairs of subthreshold conditioning stimuli in the human motor cortex. Clin Neurophysiol 2004;115:755.
37. Ashby P, Reynolds C, Wennberg R, et al. On the focal nature of inhibition and facilitation in the human motor cortex. Clin Neurophysiol 1999;110:550–555.
38. Di Lazzaro V, Oliviero A, Meglio M, et al. Direct demonstration of the effect of lorazepam on the excitability of the human motor cortex. Clin Neurophysiol 2000;111:794–799.
39. Tokimura H, Ridding MC, Tokimura Y, et al. Short latency facilitation between pairs of threshold magnetic stimuli applied to human motor cortex. Electroencephalogr Clin Neurophysiol 1996;101: 263–272.
40. Ziemann U, Tergau F, Wassermann EM, et al. Demonstration of facilitatory I wave interaction in the human motor cortex by paired transcranial magnetic stimulation. J Physiol (Lond) 1998;511:181–190.
41. Amassian VE, Rothwell JC, Cracco RQ, et al. What is excited by near-threshold twin magnetic stimuli over human cerebral cortex? J Physiol (Lond) 1998;506P:122.
42. Ilic TV, Meintzschel F, Cleff U, et al. Short-interval paired-pulse inhibition and facilitation of human motor cortex: the dimension of stimulus intensity. J Physiol 2002;545:153–167.
43. Ziemann U, Rothwell JC, Ridding MC. Interaction between intracortical inhibition and facilitation in human motor cortex. J Physiol 1996;496(Pt 3):873–881.
44. Valls-Sole J, Pascual-Leone A, Wassermann EM, et al. Human motor evoked responses to paired transcranial magnetic stimuli. Electroencephalogr Clin Neurophysiol 1992;85:355–364.
45. Werhahn KJ, Kunesch E, Noachtar S, et al. Differential effects on motor cortical inhibition induced by blockade of GABA uptake in humans. J Physiol (Lond) 1999;517:591–597.
46. Chen R, Lozano AM, Ashby P. Mechanism of the silent period following transcranial magnetic stimulation. Evidence from epidural recordings. Exp Brain Res 1999;128:539–542.
47. Di Lazzaro V, Oliviero A, Mazzone P, et al. Direct demonstration of long latency cortico-cortical inhibition in normal subjects and in a patient with vascular parkinsonism. Clin Neurophysiol 2002;113:1673–1679.
48. Nakamura H., Kitagawa H, Kawaguchi Y, et al. Intracortical facilitation and inhibition after transcranial magnetic stimulation in conscious humans. J Physiol (Lond) 1997;498:817–823.
49. Werhahn KJ, Mortensen J, Kaelin-Lang A, et al. Cortical excitability changes induced by deafferentation of the contralateral hemisphere. Brain 2002;125:1402–1413.
50. Ferbert A, Priori A, Rothwell JC, et al. Interhemispheric inhibition of the human motor cortex. J Physiol (Lond) 1992;453:525–546.
51. Hanajima R, Ugawa Y, Machii K, et al. Interhemispheric facilitation of the hand motor area in humans. J Physiol 2001;531:849–859.
52. Meyer BU, Roricht S, von Grafin EH, et al. Inhibitory and excitatory interhemispheric transfers between motor cortical areas in normal humans and patients with abnormalities of the corpus callosum. Brain 1995;118:429–440.
53. Boroojerdi B, Diefenbach K, Ferbert A. Transcallosal inhibition in cortical and subcortical cerebral vascular lesions. J Neurol Sci 1996;144:160–170.
54. Gerloff C, Cohen LG, Floeter MK, et al. Inhibitory influence of the ipsilateral motor cortex on responses to stimulation of the human cortex and pyramidal tract. J Physiol (Lond) 1998;510:249–259.
55. Daskalakis ZJ, Christensen BK, Fitzgerald PB, et al. The mechanisms of interhemispheric inhibition in the human motor cortex. J Physiol 2002;543:317–326.

56. Gilio F, Rizzo V, Siebner HR, et al. Effects on the right motor hand-area excitability produced by low-frequency rTMS over human contralateral homologous cortex. J Physiol 2003;551:563–573.
57. Delwaide PJ, Olivier E. Conditioning transcranial cortical stimulation (TCCS) by exteroceptive stimulation in parkinsonian patients. Adv Neurol 1990;53:175–181.
58. Tokimura H, Di Lazzaro V, Tokimura Y, et al. Short latency inhibition of human hand motor cortex by somatosensory input from the hand. J Physiol 1990;523(Pt 2):503–513.
59. Di Lazzaro V, Oliviero A, Tonali PA, et al. Noninvasive in vivo assessment of cholinergic cortical circuits in AD using transcranial magnetic stimulation. Neurology 2002;59:392–397.
60. Tamburin S, Manganotti P, Zanette G, et al. Cutaneomotor integration in human hand motor areas: somatotopic effect and interaction of afferents. Exp Brain Res 2001;141:232–241.
61. Civardi C, Cantello R, Asselman P, et al. Transcranial magnetic stimulation can be used to test connections to primary motor areas from frontal and medial cortex in humans. Neuroimage 2001;14:1444–1453.
62. Ugawa Y, Day BL, Rothwell JC, et al. Modulation of motor cortical excitability by electrical stimulation over the cerebellum in man. J Physiol (Lond) 1991;441:57–72.
63. Ugawa Y, Uesaka Y, Terao Y, et al. Magnetic stimulation over the cerebellum in humans. Ann Neurol 1995;37:703–713.
64. Pinto AD, Chen R. Suppression of the motor cortex by magnetic stimulation of the cerebellum. Exp Brain Res 2001;140:505–510.
65. Meyer BU, Roricht S, Machetanz J. Reduction of corticospinal excitability by magnetic stimulation over the cerebellum in patients with large defects of one cerebellar hemisphere. Electroencephalogr Clin Neurophysiol 1994;93:372–379.
66. Meyer BU, Roricht S. Scalp potentials recorded over the sensorimotor region following magnetic stimulation over the cerebellum in man: considerations about the activated structures and their potential diagnostic use [letter] [see comments]. J Neurol 1995;242:109–112.
67. Rothwell JC, Werhahn KJ, Amassian VE. Additional source of potentials recorded from the scalp following magnetic stimulation over the lower occiput and adjoining neck. J Neurol 1995;242:713–714.
68. Werhahn KJ, Taylor J, Ridding M, et al. Effect of transcranial magnetic stimulation over the cerebellum on the excitability of human motor cortex. Electroencephalogr Clin Neurophysiol 1996;101:58–66.
69. Chen R. Interactions between inhibitory and excitatory circuits in the human motor cortex. Exp Brain Res 2004;154(1):1–10.
70. Rothwell JC, Thompson PD, Day BL, et al. Stimulation of the human motor cortex through the scalp. Exp Physiol 1991;76:159–200.
71. Di Lazzaro V, Oliviero A, Profice P, et al. Comparison of descending volleys evoked by transcranial magnetic and electric stimulation in conscious humans. Electroencephalogr Clin Neurophysiol 1998;109:397–401.

5 The Physiology and Safety of Repetitive Transcranial Magnetic Stimulation

Zafiris J. Daskalakis and Robert Chen

Repetitive transcranial magnetic stimulation (rTMS) was first introduced in the late 1980s, making noninvasive, repetitive cortical stimulation in humans possible. Since then, numerous studies have involved rTMS as an investigational tool to elucidate cortical physiology and to probe cognitive processes. rTMS is also being investigated as a potential treatment for a variety of neurological and psychiatric disorders, including Parkinson's disease, depression, and schizophrenia. rTMS has been approved in Canada as a treatment for patients with medication-resistant depression. However, with the development of new technologies come questions of safety and tolerability. In this chapter, we explain the principle of rTMS technology and its applications to neuroscience research. We also explore the neurophysiological mechanisms through which rTMS mediates these effects and discuss concerns about safety, tolerability, and potential contraindications.

■ Principles of Repetitive Transcranial Magnetic Stimulation Technology

In 1831, Michael Faraday demonstrated that a current was induced in a secondary circuit when it was brought in close to the primary circuit in which a time-varying current was flowing. Here, a changing electrical field **E** produces a changing magnetic field **B** that, consistent with Faraday's law, induces an electrical field in a nearby conducting material. With TMS, electrical charge is stored in capacitors. Periodic discharge of this stored energy from the capacitors through the conducting coil produces a time-varying electric field. This electric field produces a transient magnetic field that causes current to flow in a secondary conducting material, such as neurons. TMS discharge over the scalp induces a depolarization of the conducting neural tissue located just under the coil. Because intervening tissue between the coil and the cortex (i.e. scalp and skull) is largely nonconducting, the magnetic field that is produced penetrates these tissues virtually unattenuated. Both electrical and magnetic fields decrease as the distance from the stimulating coil increases. The electric current generated in conduction material is better focussed when the stimulating coil is small.

rTMS involves regularly repeated stimulation of the cerebral cortex by a train of magnetic pulses.[1] By convention, stimulating at frequencies greater than 1 Hz is referred to as high-frequency rTMS, whereas stimulating at frequencies at lower than 1 Hz is referred to as low-frequency rTMS. Most commercially available stimulators and coils produce a magnetic field of approximately 1.5 to 2 T at the site of greatest stimulation. Such energy may produce a depth of excitation of 1.5 to 2 cm beneath the scalp,[1,2] although greater penetration to subcortical sites has been reported occurring, in part as a function of the orientation of the coil to the

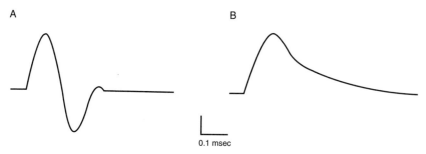

Figure 5–1 The time course of a biphasic magnetic field (**A**) and a monophasic magnetic field (**B**).

skull surface.[3,4] Commercially available stimulators produce two pulse types: biphasic pulse or monophasic pulse (Fig. 5–1). Biphasic pulses are sinusoidal and are generally of shorter duration than monophasic pulses, with the latter involving a rapid rise from zero, followed by a slow decay back to zero. Most commercially available rTMS stimulators use biphasic pulses, and single-pulse stimulators can use either monophasic or biphasic pulses. Higher frequencies are achieved with a bipolar stimulus rather than monopolar stimulus, because the bipolar stimulus is shorter and requires less energy to excite neuronal elements. Moreover, in bipolar stimulators, approximately 40% of the original energy stored is returned to it,[5] thereby requiring less time to recharge compared with the monopolar counterpart. Capacitors can charge and discharge rapidly, achieving high stimulation rates. The ability to achieve such high stimulation rates has made rTMS a valuable tool in the investigation and treatment of many neuropsychiatric disorders.

The effects of coil orientation are different for monophasic and biphasic pulses and may depend on the direction of the peak induced charge accumulation.[6,7] With a monophasic pulse and a figure-of-eight coil, the lowest motor threshold is obtained with coil held tangentially on the head with the handle pointing backward and about 45 degrees laterally from midline. The induced current flows forward and perpendicular to the central sulcus and is optimal for producing transsynaptic activation of corticospinal neurons.[8] Similarly, circular coils produce the lowest motor threshold when the induced current flows in a posteroanterior direction. In contrast, with biphasic pulses used in most rTMS studies, the motor threshold was lower with the first upstroke inducing current in the anteroposterior direction.[7] Circular coils are infrequently used in rTMS studies, in part because the area of stimulation produced by round coils is large, limiting the ability to closely link a cortical stimulation site with the physiological response. Figure-of-eight coils produce more focal stimulation as the greatest intensity of stimulation occurs at the midpoint between the two loops and not across the entire surface as with circular coils. Transcranial magnetic stimulation results in transsynaptic activation of pyramidal neurons at the cell body, in contrast to transcranial electrical stimulation (TES), in which the majority excitation occurs along the axon, largely located within the white matter.[9]

Studies involving co-registration of MRI and scalp positions showed that the optimal position for eliciting motor evoked potentials (MEPs) corresponds well with the location of the motor cortex.[10,11] Finding the motor cortex for rTMS studies is relatively simple. However, many rTMS studies involve stimulation of nonmotor cortical areas. Locating the target area is more difficult because surface landmarks and underlying cortical structures can vary from subject to subject. For example, attempting to activate the dorsolateral prefrontal cortex based on the surface landmark may be unreliable, and increasingly, these types of TMS studies are relying on co-registration of the cortical surface with underlying cortical structures using magnetic resonance imaging (MRI). This is particularly important in rTMS studies for several reasons. First, the correct identification of underlying cortical structures should be identified before proceeding with rTMS treatment, otherwise conclusions regarding efficacy may be difficult to interpret.

Second, the activation of unintended areas with rTMS may predispose patients to unwanted side effects. Third, activation of accessory areas other than those intended may produce effects that confound both investigational and treatment studies. Consequently, in some rTMS studies MRI co-registration may be desirable to locate or confirm the intended cortical surface is being stimulated.

Motor Physiology of Repetitive Transcranial Magnetic Stimulation

The neurophysiological effects of rTMS on the cortex have recently garnered much attention, in part, because the utility of rTMS as a treatment tool in several neuropsychiatric disorders is gaining wider acceptance and understanding its mechanisms of action should help guide treatment efforts. Moreover, rTMS can be used to create a "virtual lesion" and is as a potential physiological probe to elucidate complex neural processes. In this regard, rTMS has been shown to result in changes in several physiological parameters in the motor cortex, including changes in the motor threshold (MT), MEP, silent period (SP), paired-pulse inhibition and facilitation (ppTMS), and cortical plasticity. These changes appear to be frequency and intensity dependent.

Effects on Motor Evoked Potentials

Low-frequency rTMS (1 Hz or less) has been shown to decrease corticospinal excitability. Table 5–1 provides a review of the parameters used in these studies and their results. For example, Chen and colleagues[12] demonstrated that a 15-minute train of suprathreshold 0.9-Hz rTMS applied to the motor cortex reduced MEP size. This reduction lasted at least 15 minutes after the end of the stimulation train. Siebner and coworkers[13] found a similar reduction in MEP size in healthy controls but not in writer's cramp patients when 1-Hz rTMS was applied over the left primary motor hand area. These findings have been confirmed in later studies. Subthreshold 1-Hz stimulation for 4[14] or 10[15] minutes resulted in significant reduction in MEP size. Muellbacher and associates[16] found that 1-Hz suprathreshold TMS for 15 minutes increased the motor threshold and reduced MEP amplitude. Fitzgerald and colleagues[17] demonstrated an increase in RMT when stimulating the motor cortex at either 85% RMT or 115% RMT at 1 Hz for 15 minutes but only stimulation at 115% RMT resulted in a reduction in MEP amplitude. In a study by Gangitano and coworkers,[18] the motor cortex was stimulated 1 Hz at 90% RMT for a total 1,600 pulses or 20-Hz rTMS at 90% RMT for a total 1,600 pulses. The 1-Hz stimulation produced the expected increase in RMT and decrease in MEP amplitude, whereas 20-Hz stimulation had opposite effects. However, a distinct subgroup of healthy controls in this study demonstrated the opposite effects. In this subgroup of subjects, 1-Hz stimulation was associated with increased cortical excitability, whereas 20-Hz stimulation was associated with reduced cortical excitability.

Gerschlager and associates[19] used considerably lower stimulation intensity than previous studies (90% active motor threshold [AMT]), approximately corresponding to 60% to 70% resting motor threshold (RMT) (see Table 5–1) and demonstrated that prolonged 1-Hz stimulation of the premotor cortex, but not the primary motor, parietal, or prefrontal cortex, resulted in MEP suppression for at least 15 minutes after the stimulus train. These investigators suggest the effects of premotor cortex stimulation is due to its rich connection to the primary motor cortex and premotor cortex stimulation can suppress primary motor cortex excitability even more so than stimulation of the motor cortex itself. Wassermann and colleagues[20] found that 1-Hz rTMS reduced the excitability of the contralateral, unstimulated motor cortex as demonstrated by a reduction in the slope of the MEP recruitment curve. Low-frequency stimulation of a cortical area may evoke cortical inhibition in interconnected areas.[21] Iyer and coworkers[22] used a priming stimulation paradigm to enhance inhibition induced by 1-Hz rTMS. In this study, 6 Hz of rTMS at 90% RMT was applied for 10 minutes to the motor cortex, followed by 10 minutes of 1-Hz rTMS at 115% of RMT. The MEP size was significantly reduced in the priming condition compared with 1-Hz rTMS for

Table 5–1 EFFECTS OF LOW- AND HIGH-FREQUENCY REPETITIVE TRANSCRANIAL MAGNETIC STIMULATION ON MOTOR CORTICAL EXCITABILITY

Studies	rTMS Intensity	Frequency	No. of rTMS Pulses	No. of Subjects	Finding
Chen et al., 1997[12]	115% RMT	0.9 Hz	900	14	↓ MEP amplitude
Wassermann et al., 1998[20]	Produce 50–500 µV MEPs in active muscle	1 Hz	900	11	↓ slope of recruitment curve of opposite motor cortex
Siebner et al., 1999[25]	105% RMT	1 Hz	60	14	↓ MEP amplitude in healthy controls but not in dystonic patients
Romero et al., 2002[15]	90% RMT	1 Hz	600	20	↓ MEP amplitude
Maeda et al., 2000[14]	90% RMT	1 Hz	240	36	↓ MEP amplitude
Muellbacher et al., 2000[16]	115% RMT	1 Hz	900	7	↑ RMT, reduced MEP recruitment
Gerschlager et al., 2001[19]	90% AMT	1 Hz	1,500	8	↓ MEP amplitude with premotor but not motor cortex stimulation
Fitzgerald et al., 2002[17]	85% RMT 115% RMT	1 Hz	900	9	↑ RMT at both intensities, ↓ MEP amplitude with higher intensity
Gangitano et al., 2002[18]	90% RMT	1, 20 Hz	1,600	16	1 Hz: ↑ RMT and ↓ MEP amplitude 20 Hz: ↓ RMT and ↑ MEP amplitude; but subgroup of subjects showed opposite effects
Pascual-Leone et al., 1994[27]	150% RMT 110% RMT 150% RMT	5 Hz 10, 20 Hz 10, 20 Hz	10	14	↑ MEP amplitude ↑ MEP amplitude ↑↓ MEP amplitude
Maeda et al., 2000[14]	90% RMT	10 Hz 20 Hz	1,600 240	36	↑ MEP amplitude ↑ MEP amplitude
Berardelli et al., 1998[29]	120% RMT	5 Hz	100	13	↑ MEP amplitude
Jahanshahi et al., 1997[28]	105–110% AMT	20 Hz	400	6	↑ MEP amplitude
Modugno et al., 2001[31]	100% RMT 130% RMT 150% RMT	5, 10, 20 Hz	13	11	↓ MEP amplitude; facilitation with higher intensity

AMT, active motor threshold; MEP, motor evoked potential; RMT, resting motor threshold; rTMS, repetitive transcranial magnetic stimulation.

10 minutes without priming. Reduction of cortical excitability with low-frequency rTMS may be relevant in the treatment of a number of neurological and psychiatric disorders with increased cortical excitability such as schizophrenia,[23] depression,[24] dystonia,[25] and epilepsy.[26]

High-frequency rTMS may have effects opposite to that of low-frequency rTMS and may result in increased motor cortical excitability. Table 5–1 provides a review of the parameters used in these studies and their results. For example, Pascual-Leone and associates[27] demonstrated a pattern of facilitation of MEPs produced during a train of rTMS that varied with stimulus intensity and frequency. With 5-Hz rTMS at 150% RMT, there was clear facilitation of MEPs. At 10- and 20-Hz rTMS with lower intensities (110% RMT), there was also a consistent pattern of facilitation. With 150% RMT at 10- and 20-Hz rTMS, an alternating pattern of MEP inhibition and facilitation was demonstrated. Several other studies have also demonstrated that MEP size increases using high-frequency rTMS (>1 Hz).[28-30] Modugno and colleagues[31] reported that 20 stimuli of 5, 10, and 20 Hz applied to the motor cortex at 100% RMT resulted in brief MEP suppression that lasted for about 1 second after rTMS. This suppression in the post-train interval was prolonged with longer trains or higher frequencies. Increasing the intensity of the rTMS to 130% and 150% of RMT resulted in facilitation rather than suppression of the MEP, consistent with previous findings.[29] Therefore, high-frequency rTMS at low intensity may cause inhibition for 1 to 2 seconds after the rTMS train, whereas at higher intensity, high-frequency rTMS consistently produced facilitation. The mechanisms by which altered excitability occurs in the cortex are unclear although some have suggested that decreased excitability is related to long-term depression,[12] whereas increased excitability has been related to long-term potentiation.[32] Increased excitability may be caused by a transient increase in the efficacy of excitatory synapses.[27] Although these studies posit that decreased excitability is cortically mediated, spinal mechanisms cannot be ruled out. Further studies that explore the possible contribution of spinal mechanisms to changes in excitability through measurement of spinal reflexes or TES studies would be helpful.

Cortical Inhibition and Facilitation

rTMS can induce changes in cortical inhibition and facilitation. These measures can be demonstrated experimentally using single and paired-pulse TMS paradigms, including short-interval intracortical inhibition (SICI) (i.e., paired pulse TMS (ppTMS) at inhibitory interstimulus intervals [ISIs]< 5 ms)[33]; long-interval intracortical inhibition (LICI)[34,35]; and silent period (SP).[36] These inhibitory paradigms may measure different subtypes of inhibitory GABAergic neurotransmission.[37] Cortical facilitation can be measured using paired-pulse TMS with interstimulus intervals between 10 and 20 ms and referred to as intracortical facilitation (ICF).[33] ICF may be mediated by glutamatergic neurotransmission.[38]

Silent Period

The effects of rTMS on the silent period (SP) have been investigated in several studies that examined the effects of number of pulses, frequency, and intensity. Table 5–2 reviews the parameters used and the results. Berardelli and coworkers[39] demonstrated that lengthening of the SP with 3- and 5-Hz rTMS at 110% and 120% of the resting motor threshold. Di Lazzaro and associates[40] reported no effects of 5 Hz for 10 seconds in 10 healthy control subjects. Romeo and colleagues[41] tested rTMS frequencies of 1, 2, 3, 5, 10, and 15 Hz at intensities just above the resting motor threshold. They found that trains that delivered at 2 Hz resulted in a prolongation of the SP, whereas trains that delivered at 1 Hz had no effect. In both studies, the investigators suggested that the SP becomes prolonged because rTMS activated cortical inhibitory interneurons. By contrast, Cincotta and coworkers[42] demonstrated that 0.3-Hz stimulation delivered at 115% of RMT for 30 minutes resulted in a prolongation of the silent period for up to 90 minutes. Fierro and associates[43] explored the effects of 1- and 7-Hz rTMS at intensities of 100%, 115%, and 130% of the resting motor threshold. They found that 1-Hz rTMS applied to the motor cortex near or above the motor threshold reduced the SP that was interpreted as a decrease in cortical inhibition. In contrast, 7-Hz rTMS resulted in an inconsistent pattern of smaller and larger values for the SP. The investigators concluded that rTMS applied at low 1-Hz frequency decreases the excitability of inhibitory interneurons. It therefore appears the effects of rTMS on the SP are dependent on frequency and intensity of stimulation. High-frequency rTMS (>1 Hz) results in lengthening of the SP, whereas low-frequency rTMS (about 1 Hz) results in shortening of the SP. However, very low frequency (0.3 Hz) may result in SP prolongation. Moreover, at high frequencies, increasing the stimulus intensity results in further lengthening of the SP.

Table 5-2 REVIEW OF STUDIES EXAMINING THE EFFECTS OF REPETITIVE TRANSCRANIAL MAGNETIC STIMULATION ON THE SILENT PERIOD

Studies	rTMS Parameter				Effect of SP
	Frequency (Hz)	Intensity (% RMT)	No. of Pulses	No. of Subjects	
Berardelli et al., 1999[39]	3	110	11, 20	8	↑
	5	120	11, 20		↑
Romeo et al., 2000[41]	1, 2, 3, 5, 10, 15	100	20		↑
	3, 5	110, 140	10	8	↑ 140 > 110
	5, 10	90	10		↑
Fierro et al., 2001[43]	1	100, 115, 130	50	8	↓
	7	100, 115, 130	50		↑↓
Fitzgerald et al., 2002[17]	1	85% RMT	900	9	↔
		115% RMT			
Di Lazzaro et al., 2002[40]	5	100% AMT	50	10	↔
Cincotta et al., 2003[42]	0.3	115% RMT	540	9	↑

AMT, active motor threshold; RMT, resting motor threshold; rTMS, repetitive transcranial magnetic stimulation; SP, silent period.

Paired-Pulse Inhibition and Facilitation

The effects of rTMS on cortical inhibition and facilitation as measured through ppTMS have also been evaluated. Table 5–3 provides a review of the parameters used in these studies and their results. Pascual-Leone and colleagues[44] demonstrated SICI was significantly reduced with 1,600 subthreshold rTMS stimuli at 10 Hz applied to the motor cortex. In contrast, the same number of subthreshold stimuli applied at 1Hz had no effect on SICI. Similarly, Di Lazzaro and coworkers[40] showed a reduction in SICI with no change in ICF after 10 seconds of stimulation at 5 Hz. In this study, descending corticospinal volleys were recorded in two subjects with spinal epidural electrodes. In these subjects, inhibition of the I4 wave by SICI was significantly reduced post rTMS compared with baseline recordings suggesting that this suppression of SICI was mediated at a cortical level. Similarly, Peinemann and associates[45] demonstrated that 5-Hz TMS applied to the motor cortex at 90% RMT caused a significant reduction in SICI but no change in ICF. SICI was significantly reduced after only 30 stimuli of 120% RMT at 5 Hz.[46] When the stimulus frequency was increased to 15 Hz, SICI was again reduced, whereas ICF was increased. Modugno and colleagues[47] reported that 900 stimuli delivered at 1 Hz and at 115% RMT over the primary motor cortex resulted in a reduction in SICI 16 to 30 minutes after the end of the stimulus train. Romero and coworkers[15] found that subthreshold stimulation with 1 Hz for 10 minutes significantly decreased ICF, without a concomitant change in SICI. These investigators suggest that these effects occur through cortical disfacilitation. As ICF may be associated with activity of excitatory glutamatergic circuits,[38] disfacilitation may result in decreased cortical excitation. The investigators suggest that these effects could hold potential promise in the treatment of neuropsychiatric disorders (e.g., schizophrenia) in which excessive cortical excitability has been posited.[48] There studies suggest that high-frequency rTMS decreased SICI and perhaps increased ICF, whereas low-frequency rTMS may decrease ICF, although the effects on SICI at low frequencies are inconsistent.

Repetitive Transcranial Magnetic Stimulation on the Contralateral Cortex

rTMS on the contralateral motor cortex can result in changes in the excitability of the ipsilateral motor cortex. For example, Gilio and associates[49] reported that 1-Hz rTMS (i.e., 900 stimuli at 115% to 120% of RMT)

Table 5–3 STUDIES EXAMINING THE EFFECTS OF REPETITIVE TRANSCRANIAL MAGNETIC STIMULATION ON SHORT-INTERVAL INTRACORTICAL INHIBITION AND INTRACORTICAL FACILITATION

Studies	rTMS Parameter				Effect on SICI/ICF
	Frequency (Hz)	Intensity (% RMT)	No. of Pulses	No. of Subjects	
Pascual-Leone et al., 1998[44]	10	Subthreshold	1,600	5	↓ / ↔
Peinemann et al., 2000[45]	5	90	1,250	10	↓ / ↔
Wu et al., 2000[46]	5	120	30	8	↓ / ↔
	15	120	30		↓ / ↑
Romero et al., 2002[15]	1	90	900	20	↔ / ↓
Modugno et al., 2003[47]	1	90	900	14	↓ / ↔
Fitzgerald et al., 2002[17]	1	85% RMT, 115% RMT	900	9	↔
Di Lazzaro et al., 2002[40]	5	100% AMT	50	10	↓ / ↔

AMT, active motor threshold; ICF, intracortical facilitation; RMT, resting motor threshold; rTMS, repetitive transcranial magnetic stimulation; SICI, short-interval intracortical inhibition.

applied to the left motor hand area in 10 healthy controls resulted in an increase in the MEP amplitude from stimulation of the right hemisphere and a reduction of interhemispheric inhibition (IHI)[50] with a left motor cortex conditioning stimulus and a right motor cortex test stimulus at intervals of 7- and 10-ms ISI. The investigators suggest that this reduction of IHI was mediated by an increase in the excitability of the right motor cortex. Similarly, Schambra and colleagues[51] demonstrated in 10 healthy volunteers that active stimulation of the left motor cortex for 30 minutes at 1 Hz and at intensity 110% of RMT resulted in a significant increase in the MEP amplitude in the contralateral motor cortex compared with sham stimulation, suggesting an increase in excitability of the contralateral cortex by contralateral repetitive stimulation.

Cortical Plasticity

Repetitive rTMS has also been shown to induce cortical plasticity. In this context, plasticity refers to the reorganization of the central nervous system (CNS) through changes in internal connections, representational patterns, or neuronal properties.[52] Ziemann and coworkers[52] demonstrated that the deafferented motor cortex becomes modifiable by inputs that are normally subthreshold for inducing plastic changes. In these experiments, cortical excitability was measured after ischemic nerve block (INB) with and without very-low-frequency rTMS (0.1 Hz). Ischemic nerve block results in increased cortical output to muscles that are proximal to those that have been blocked.[53–55] It was demonstrated that INB plus rTMS to the motor cortex contralateral to INB increased MEP size and ICF and reduced SICI compared with INB alone. In contrast, INB plus rTMS to the motor cortex ipsilateral to INB had the opposite effect with increased SICI and decreased ICF compared with INB alone. These findings suggest that rTMS can modulate plasticity, through changes to cortical inhibition and excitation, and may be used to enhance cortical plasticity when it is beneficial and suppress it when it is detrimental. In a subsequent study,[56] lorazepam (enhances $GABA_A$ergic neurotransmission) and lamotrigine (voltage-gated Na^+ and Ca^{2+} channel blocker) were found to abolish the increase in MEP size and decrease in SICI associated with INB and rTMS to the contralateral motor cortex, whereas dextromethorphan (NMDA receptor antagonist) reduced the SICI suppression but had no effect on MEP. These results provide evidence that increase in MEP size induced by rTMS and deafferentation involves GABA-related inhibitory circuits and voltage-gated Na^+ or Ca^{2+} channels mediated mechanisms. rTMS-induced reduction in SICI also appears to involve NMDA receptor activation.

Effects of Repetitive Transcranial Magnetic Stimulation on Cognition

rTMS can alter cognitive processes. Although a broad review of the cognitive literature pertaining to rTMS is beyond the scope of this chapter, we will review several selected studies.

The ability of high-frequency rTMS to selectively disrupt cortical areas during the period of stimulation makes its effects similar to a transient and reversible lesion and is a useful method to investigate the functions of different cortical areas. In the motor system, high-frequency rTMS has been used to investigate the role of different motor areas in sequential finger movement. Inactivation of the contralateral motor cortex impaired the performance of complex sequences more than simple sequences, suggesting the motor cortex is not only an executive motor area but can also contribute to movement sequence organization.[57] Movement errors induced by rTMS of the supplementary motor area occurred about 1 second later than those induced by rTMS of the motor cortex, supporting a role for the supplementary motor area in the organization of forthcoming movement.[58] Inactivation of the ipsilateral motor cortex also affects the timing of complex sequences with greater effect for stimulation of the left than right hemisphere in right-handed subjects, suggesting that the ipsilateral motor cortex is also involved in complex motor sequences with greater involvement of the left hemisphere.[59]

Effects of rTMS on speech areas have also attracted much attention. Pascual-Leone and associates[60] produced speech arrest with 8- to 25-Hz rTMS to the left frontotemporal cortex. Claus and colleagues[61] demonstrated interference in language comprehension using 50-Hz rTMS to the left hemisphere but not to the right hemisphere. Similarly, Flitman and coworkers[62] and Wassermann and associates[63] found that rTMS at a point 2 cm above the top of the left ear, which corresponded to the left middle temporal gyrus, produced naming errors and impaired memory. Epstein and colleagues[64] stimulated the lateral frontal region of each hemisphere in 17 presurgical epilepsy patients and demonstrated that 4-Hz rTMS to the left side consistently produced speech arrest. However, rTMS to the right hemisphere also produced significant speech arrest in five subjects, and rTMS to the speech area did not replicate the results of the Wada test. Other rTMS studies have provided information about cortical specialization. Pascual-Leone and coworkers[65] investigated the role of the dorsolateral prefrontal cortex (DLPFC) in implicit procedural learning. In this study, low-intensity rTMS was applied to the DLPFC, the supplementary motor area, or directly to the ipsilateral hand used in testing. They demonstrated that DLPFC stimulation markedly impaired implicit procedural learning, whereas stimulation of the other areas did not. Similarly, Terao and associates[66] used rTMS for mapping the topography of cortical regions active during saccadic eye movement as well as for constructing a physiological map to visualize the temporal evolution of functional activities in the relevant cortical regions. Cohen and colleagues[67] demonstrated that 10-Hz rTMS for 3 seconds applied to the occipital cortex in blind subjects resulted in increased errors and distorted tactile perceptions in blind subjects on a Braille reading task. In contrast, normal sighted subjects did not experience any difficulties with tactile performance. These findings suggest that blindness at an early age can result in cross-modal plasticity that alters the function of the visual cortex to a somatosensory role.

The depression of cortical activity by low-frequency rTMS is also a useful method to investigate cortical functions. For example, impairment of visual perception and imagery after 1-Hz rTMS to the medial occipital cortex suggests that the medial occipital cortex is necessary for imagery task.[68] Similarly, 1-Hz rTMS of the motor cortex was able to disrupt the improvement of motor performance that followed a brief period of motor training, suggesting that the motor cortex is involved in the consolidation of the motor learning.[69]

There is evidence to suggest that rTMS may enhance cognitive function in some circumstances. For example, Mottaghy and coworkers[70] demonstrated that 5-Hz rTMS at applied to Wernicke's area in right-handed healthy subjects led to a brief facilitation of picture naming immediately after rTMS by shortening linguistic processing time. rTMS applied to the right hemisphere homolog of Wernicke's area, Broca's area, or the visual cortex had no effect. Similarly, Boroojerdi and associates[71] demonstrated enhanced analogic reasoning (the ability to determine the similarity between different stimuli, scenes or events) during rTMS of the left prefrontal cortex in 16 right-handed healthy volunteers. rTMS can potentially be used to enhance cognition. However, these studies are limited by a small sample size and need to be replicated before firm conclusions are made.

Repetitive Transcranial Magnetic Stimulation and Imaging Studies

Part of the allure for studying the motor cortex is that outcome variables can be easily assessed with surface EMG. Exploring nonmotor regions, however, requires the combining TMS with other methods of measurement (e.g., positron-emission tomography (PET), electroencephalography (EEG), and magnetic resonance imaging (MRI)). The ability to measure complex neural process in nonmotor cortical regions has important implications, particularly in the investigation of neuropsychiatric illnesses such as schizophrenia and depression. rTMS combined with such methods has the advantage of capitalizing on the ability to directly activate the cortex to assess neural connectivity rather than indirect coactivations that are recorded from subjects who are asked to perform various cognitive tasks (e.g., PET studies). For example, Niehaus and colleagues[72] applied rTMS to the left-hand association area of the motor cortex and measured cerebral blood flow (CBF) using transcranial Doppler sonography. They found that short trains of 10-Hz suprathreshold stimuli increased CBF in the middle and posterior cerebral arteries. Paus and coworkers[73] stimulated the frontal eye fields with 10-Hz rTMS at an intensity of 70% of the maximum stimulator output while PET scans were acquired. They found significant positive correlation with regional cerebral blood flow (rCBF), as measured with ^{15}O-labeled H_2O PET over the same cortical regions, as well as concomitant excitation in the visual cortex of the superior parietal and medial parieto-occipital regions. To ensure that the same cortical regions were being stimulated in all subjects, frameless stereotaxy was used. Paus and associates[74] demonstrated that sensorimotor cortex stimulation with brief, subthreshold 10-Hz rTMS trains decreased rCBF in the target area that covaried with the number of stimulus trains delivered. It was posited that this inhibition was mediated through activation of local inhibitory neurons and therefore related to the rTMS-induced prolongation of the SP. In contrast, Fox and colleagues[75] stimulated the primary motor cortex with 1-Hz rTMS at 120% RMT for 30 minutes (1,800 pulses) while PET scans were acquired. These stimulation parameters resulted in activation of the hand area of the motor cortex being stimulated. Moreover, Fox and coworkers[75] demonstrated concomitant activation in the ipsilateral primary and secondary somatosensory areas as well as ipsilateral ventral and lateral premotor cortex, with inhibition of the contralateral motor cortex. The latter finding suggests a contralateral inhibitory connectivity between these homologous cortical regions. This is consistent with previous TMS studies demonstrating contralateral inhibition after TMS of the ipsilateral motor cortex.[50,76,77]

Cortical activity and connectivity have also been assessed with functional MRI (fMRI). Preliminary studies demonstrated that combining rTMS with blood oxygen level–dependent (BOLD) fMRI was useful in tracing neural circuits and in measuring cortical excitability.[78,79] For example, Bohning and associates[80] demonstrated that 1-Hz rTMS applied for 18 seconds at 110% RMT produced an increase in BOLD fMRI activation at local and remote sites compared with lower intensities (80% RMT). Nahas and colleagues[81] used 1-Hz rTMS to stimulate the prefrontal cortex. This was done in an effort to measure connectivity and clarify the intensities that are required to produce activation in this cortical region. Previous studies measuring activation in nonmotor cortical areas pegged intensities to those required to produce activation to the motor cortex.[79] Given the cytoarchitectural differences between these regions,[82] however, such parameters may not be accurate. As such, the prefrontal cortex was stimulated with a range of intensities (e.g., 80%, 100%, and 120% of RMT). Their results were summarized into four main findings: Greater intensities produced greater local and contralateral activation; stimulation of the left prefrontal cortex resulted in bilateral effects; rTMS of the left prefrontal cortex produced greater activation on right side than the left side; and stimulating at 80% of the RMT for 20 seconds failed to produce significant prefrontal activation. rTMS therefore can be successfully combined with fMRI to obtain measures of excitability and to aid in deciphering the complex connectivity of the cortex.

Strafella and coworkers[83] demonstrated that rTMS applied to mid-dorsolateral prefrontal cortex resulted in dopamine release in the ipsilateral caudate nucleus, whereas rTMS of the motor cortex released dopamine in the ipsilateral putamen[84] of the human brain. This finding has several promising implications. First, rTMS applied to the cortex can induce neurochemical change in subcortical brain regions. Second, rTMS induced neurochemical changes may be used to target treatment in a variety of neuropsychiatric conditions that may be associated with subcortical neurochemical dysfunction (e.g., schizophrenia, Parkinson's disease, Tourette's syndrome). In this regard, attenuation of a given neurochemical pathway, as measured by combining rTMS with PET-ligand imaging, may one day be used to rectify aberrant brain neurochemistry that is currently available only with the use of neuropharmacologic medications (e.g., antidopaminergics in schizophrenia). Third, rTMS represents a neurophysiologic probe to assess cortical connectivity and a potential neurochemical probe by which to test the integrity of various neurochemical pathways.

■ Effects of Repetitive Transcranial Magnetic Stimulation on the Sensory and Visual Systems

Enomoto and associates[85] explored effects of rTMS of different cortical areas on the excitability of the sensory cortex. It was demonstrated that 1-Hz rTMS at 110% of the active motor threshold applied to the motor cortex suppressed the N20-P25 and P25-N33 somatosensory evoked potentials (SEPs) amplitudes from median nerve stimulation, but rTMS to the sensory or premotor cortex had no effect. There were no changes to N20 onset latency, N20 onset to peak, or sensory nerve action potential (SNAP) amplitudes. Suppression of the N20-P25 amplitude without changes to N20 onset latency and N20 onset to peak suggests that suppression occurred in the sensory cortex because the N20 component is regarded as activation of the sensory cortex by thalamocortical fibers.[85] The absence of any change in SNAP amplitudes suggests that peripheral input remained constant. The subject who showed the greatest suppression of SEP experienced numbness and temperature sensation of temperature in the right hand after rTMS to the left motor cortex. They suggest that rTMS over the motor cortex results in long lasting suppression of the ipsilateral sensory cortex through corticocortical inhibitory pathways between motor and sensory cortex.

Boroojerdi and colleagues[86] explored the effects of low-frequency rTMS on visual cortex excitability by measuring the perception of phosphenes. The minimum TMS intensity to elicit phosphenes was considered the phosphene threshold. It was demonstrated that 1-Hz rTMS at the phosphene threshold intensity over the visual cortex increased the phosphene threshold without any effect on the MT. These results suggest that low-frequency rTMS to the visual cortex may decrease cortical excitability and, as such, holds promise as a potential treatment in conditions associated with visual hallucinations, including Charles-Bonnet syndrome and schizophrenia.

■ Electroencephalography

The combination of rTMS and EEG affords a promising new approach to study cortical connectivity and reactivity. EEG has much higher temporal resolution than functional imaging methods such as PET and MRI. Moreover, stimulation can be applied to any cortical area, and detailed corticocortical connections can be studied. Such recordings were previously limited because the combination of rTMS with EEG was hampered with technical barriers that included large artifacts due to saturation of the recording amplifier and heating of recording electrodes. Ilmoniemi and coworkers[87] have overcome such technical restrictions by developing an amplifier that avoids saturation by using a sample-and-hold circuit that pins the amplifier output to a constant level during the pulse. Such an amplifier was shown to recover in about 100 μs after the end of the magnetic pulse. Moreover, electrode heating was avoided by using specially designed low-conductivity, small, Ag-AgCl–pellet electrodes. It was demonstrated rTMS stimulation of the motor cortex at a frequency of 0.8 Hz resulted in spread of neuronal activation to premotor areas then to contralateral parietal areas in about 24 ms. Moreover, stimulation of visual areas caused activation of ipsilateral occipital areas with the spread of excitation traveling to contralateral occipital areas. Such measure may be useful in the exploration of connectivity and reactivity in many neuropsychiatric conditions where connectivity abnormalities are thought to underlie their pathophysiological processes.[88]

Mood Effects of Repetitive Transcranial Magnetic Stimulation

Although a review of the effects of rTMS on neuropsychiatric illness is beyond the scope of this chapter, we review this information as it relates to the effects of rTMS on the neurophysiological deficits in these disorders. For example, several investigators have suggested that depression is associated with altered activity of various cortical areas[89,90] and with altered cortical excitability.[91] rTMS may be used to excite or inhibit cortical activity, presumably rectifying underlying abnormalities. Moreover, rTMS research has demonstrated that mood regulation may be lateralized to the right or left cerebral hemispheres. For example, high-frequency rTMS applied to the left prefrontal cortex induced transient decreases in mood, whereas rTMS to the right prefrontal cortex induced transient increases in mood.[92–94] This suggests that rTMS applied to the right or left prefrontal cortex can improve mood and alleviate depressive symptoms.

There is a robust literature demonstrating the efficacy of rTMS in depression. For example, George and associates[95] initially reported modest improvement (i.e., mean Hamilton Depression Scale Score decreased from 23.8 to 17.5) in six treatment-refractory, depressed patients in an open study using rTMS. Pascual-Leone and colleagues[96] showed that rTMS was effective for treating depressive symptoms in 17 patients with medication-resistant depression with psychotic features when applied daily for 1 week. Similarly, George and coworkers[97] and Figiel and associates[98] reported significant improvement in depressive symptoms in a group of patients with major depression in 2-week, placebo-controlled, crossover trials of real and sham high-frequency rTMS. Grunhaus and colleagues[99] compared rTMS with electroconvulsive therapy (ECT) in 40 patients with major depressive disorder in an unblinded study, which did not include a placebo control group. They found that rTMS was less effective than ECT in patients with major depressive disorder and psychosis, but it was equal to ECT in patients without psychosis. Fitzgerald and coworkers[100] demonstrated that high frequency (i.e., 10-Hz stimulation) to the left prefrontal cortex and low frequency (i.e., 1-Hz stimulation) to the right prefrontal cortex were superior to sham stimulation after 4 weeks of treatment to a total of 60 patients with refractory major depressive disorder.

Several lines of investigation have explored the mechanisms through which rTMS exerts it antidepressive effects. For example, Hausmann and associates[101] demonstrated that magnetic stimulation of rat cortical slices induced FOS expression. FOS is a transcription factor that is used as a rapid marker for neuronal activation. In this study, the parietal cortex of postnatal day 10 rats was stimulated with rTMS at frequencies of 1, 20, and 50 Hz. They demonstrated a significant increase in FOS expression after 1-Hz stimulation for 10 minutes at 75% of the maximum stimulator output. Moreover, 12 hours after 14 days of rTMS treatment at 20 Hz, it was demonstrated that FOS was markedly increased in neurons in layers I through IV and VI of the parietal cortex and in few scattered neurons in the hippocampus in rats.[102] The protooncogene *FOS* affects the expression of transmitters, receptors, and neuropeptides that play a central role in depression.[103] Such studies demonstrate that acute and chronic rTMS induces changes that may represent suitable markers for antidepressant effects in humans.[104]

rTMS was shown to result in changes to several other neurobiological markers of antidepressant action in humans. For example, Muller and colleagues[105] demonstrated that long-term rTMS increased the expression of brain-derived neurotrophic factor and cholecystokinin mRNA, a response that is similar to responses after antidepressant medications and ECT.[106] Moreover, Keck and coworkers[106] demonstrated that in selectively bred rats with high-anxiety–related behavior, rTMS allowed these animals to reach the performance of low-anxiety–related behavior rats in the forced-swim test, a test thought to reliably predict the efficacy of antidepressant drugs.[107] In a similar series of experiments, Keck and associates[108] demonstrated that rTMS blunted the response of the hypothalamic-pituitary-adrenocortical (HPA) system to the forced swim test stressor, also shown to be a marker for antidepressant drug effects. This blunted HPA response included decreased stress-induced elevation of plasma adrenocorticotropic hormone (ACTH). rTMS has demonstrated efficacy in the treatment of depression in humans and has been shown to reproduce the neurochemical changes that predict response to antidepressant medications in animal studies.

Safety of Repetitive Transcranial Magnetic Stimulation

Seizures

rTMS has been reported to induce seizures dating back to the first cortical stimulation studies. Dhuna and colleagues[109] initially reported a seizure after high-frequency rTMS in a patient with temporal lobe epilepsy. rTMS induced seizures have been shown to spread in a somatotopic pattern akin with the body representation in the sensorimotor cortex.[110] The risk is probably highest with motor cortex stimulation, because this area is one of the most epileptogenic regions of the brain.[111] A proposed mechanism for seizure induction with rTMS is that excessive activation of cortical pyramidal neurons result in a spread of excitation to neighboring pyramidal cells through excitatory axon collaterals,[112] as well as to inhibitory interneurons that terminate on these same excitatory neurons to attenuate their firing.[113] Provided that stimulus trains are of sufficient intensity and duration, repetitive stimulation of cortical pyramidal cells may eventually overwhelm compensatory inhibitory inputs and result in seizure activity even in subjects without known risk factors.[110]

Reports of seizures in healthy control subjects after rTMS[1] have prompted researchers to establish rTMS treatment parameters that minimize seizure risk.[110] In general, high-frequency rTMS at high intensity for a prolonged duration results in increased seizure risk. Chen and coworkers[114] and Wassermann and associates[1] introduced the most recent safety guidelines that take into consideration stimulus frequency, intensity, duration, inter-train intervals and number of pulses delivered. These guidelines are presented in Tables 5–4 and 5–5. These safety guidelines were derived from an analysis of multiple rTMS parameters that were shown to elicit seizures, to result in a spread of motor excitation to muscles not targeted by TMS, or to have post-TMS EMG activity. These signs demonstrate or suggest a high risk of inducing seizures and are considered unsafe.[27,110] To our knowledge, induction of seizures within these safety guidelines has not been reported.

Most subjects experiencing seizures from rTMS stimulation make a full recovery. Wassermann and colleagues[115] evaluated measures sensitive to subclinical seizures, including immediate and delayed memory, verbal fluency, prolactin levels, and EEG at the beginning of the study and after each day of rTMS. They found these measures were unaffected within their recommended stimulus parameters. Despite these guidelines, some investigators advocated additional monitoring to prevent or abort seizures. For example, the use of simultaneous EEG during rTMS to

Table 5–4 SAFE TRAIN DURATIONS/NUMBER OF PULSES FOR SINGLE TRAINS OF REPETITIVE TRANSCRANIAL MAGNETIC STIMULATION IN NORMAL VOLUNTEERS

Frequency (Hz)	rTMS Intensity (% of Resting Motor Threshold)*												
	100	110	120	130	140	150	160	170	180	190	200	210	220
1	>270/270[†]	>270/270[†]	180/180[‡]	50/50[§]	50/50[§]	50/50[§]	50/50[§]	20/20	8/8	8/8	6/6	5/5	4/4
5	10/50[§]	10/50[§]	10/50[§]	10/50[§]	5.7/28	3.9/19	2.7/13	1.95/9	1.8/9	1.2/6	1.1/5	1.2/6	0.9/4
10	5/50[§]	5/50[§]	3.2/32	2.2/22	1.0/10	0.6/6	0.7/7	0.6/6	0.4/4	0.5/5	0.3/3	0.2/2	0.2/2
20	1.5/30	1.2/24	0.8/16	0.4/8	0.3/6	0.2/4	0.2/4	0.1/2	0.2/4	0.2/4	0.2/4	0.1/2	0.1/2
25	1.0/25	0.7/17	0.3/7	0.2/5	0.2/5	0.2/5	0.2/5	0.1/2	0.1/2	0.1/2	0.1/2	0.1/2	0.1/2

*The maximum safe train duration (s) is shown followed by the number of pulses (see also Wasserman et al.[1]).
[†]Based on Chen R, Classen J, Gerloff C, et al. Depression of motor cortex excitability by low-frequency transcranial magnetic stimulation. Neurology 1997;48:1398–1403.
[‡]Based on Wassermann EM, Grafman J, Berry C, et al. Use and safety of a new repetitive transcranial magnetic stimulator. Electroencephalogr Clin Neurophysiol 1996;101:412–417.
[§]No spread of excitation or post-TMS EMG activity was observed at these train durations (based on Pascual-Leone et al.[110]).
rTMS, repetitive transcranial magnetic stimulation.
Adapted from Chen R, Gerloff C, Classen J, et al. Safety of different inter-train intervals for repetitive transcranial magnetic stimulation and recommendations for safe ranges of stimulation parameters. Electroencephalogr Clin Neurophysiol 1997;105:415–421.

Table 5–5 SAFETY RECOMMENDATION FOR INTER-TRAIN INTERVALS FOR 10 TRAINS OF REPETITIVE TRANSCRANIAL MAGNETIC STIMULATION AT LESS THAN 20 HZ

Inter-Train Intervals*	Stimulus Intensity (% of Motor Threshold)			
	100%	105%	110%	120%
5	Safe	Safe	Safe	Insufficient data
1	Unsafe (3)	Unsafe†	Unsafe (2)	Unsafe (2)
0.25	Unsafe†	Unsafe†	Unsafe (2)	Unsafe (3)

*The minimum number of trains that caused spread of excitation of post-TMS EMG activity are indicated in the parentheses. The maximum duration/number of pulses for individual rTMS trains at each stimulus intensity could not exceed that listed in Table 5–4. Stimulus parameters produced by reducing a set of parameters that is considered safe (reduction in stimulus intensity, train duration or increases in inter-train interval) is also considered safe. rTMS at 25 Hz, 120% of MT (0.4 s duration) is unsafe at inter-train intervals of 1 s or less. The safety of longer inter-train intervals at 25 Hz has not been determined.
†These stimulus parameters are considered unsafe because adverse events occurred with stimulation of lower intensity or longer inter-train interval, but no adverse event was observed with these parameters.
MT, motor threshold; rTMS, repetitive transcranial magnetic stimulation.

monitor for early seizure activity has been suggested.[116] Very few laboratories, however, are equipped with the necessary EEG amplifiers because magnetic pulses tend to saturate amplifiers for seconds to minutes. Such amplifiers are currently commercially available.[87] EMG activity after TMS stimulation and EMG activity in nontargeted muscles are considered good predictors of early seizure activity. Wassermann and coworkers[1] suggest that EMG activity should be monitored continuously from abductor pollicis brevis or the first dorsal interosseus muscles on the side contralateral to rTMS in nonmotor cortex stimulation. For motor cortex stimulation, it is recommended that muscles adjacent to those being stimulated be monitored for spreading of motor activity (e.g., monitoring of right biceps brachii for rTMS stimulation of the left-hand area). The appearance of MEPs in these muscles may suggest spread of seizures from the site of stimulation site to adjacent areas of the primary motor cortex.

Using rTMS to induce seizure as an alternative to ECT with the hope for attenuated side effects (e.g., memory impairment) was investigated. Lisanby and associates[117,118] demonstrated that seizures could be induced in nonhuman primates and human subjects by modifying stimulus parameters. Magnetic seizure therapy was obtained with a custom-modified magnetic stimulator producing a broader pulse width and increased charging units that permit stimulation frequencies of 40 Hz at maximum intensity. In a patient with depression, a generalized tonic-clonic seizure was induced under general anesthesia in all four treatment sessions, and the seizure duration ranged from 30 to 270 seconds. The seizures were reliably obtained at maximum stimulator output at a frequency of 40 Hz for 4 seconds.

Cognition

Concerns about the potential deleterious cognitive effects that may potentially lead to permanent cognitive impairment have also been explored. For example, Little and colleagues[119] showed that there were no gross deleterious cognitive effects after 2 weeks of 1- or 20-Hz rTMS at 80% of motor threshold over the left prefrontal cortex in a 10 subjects with depression who received 2 weeks of low-frequency (1-Hz) or high-frequency (20-Hz) treatment and then were crossed over to the other treatment condition. Moreover, in a study involving 46 healthy volunteers, Koren and coworkers[120] reported that low-frequency rTMS applied to the left or right prefrontal cortex improved processing speed and efficiency but had no deleterious cognitive changes compared with subjects receiving sham rTMS.

Neurotoxicity and Structural Brain Changes

Because rTMS results in the stimulation of cortical neurons, the possibility of causing neuronal damage has been investigated. In a neuroanatomical investigation, Zyss and associates[121] examined the influence of prolonged rTMS (e.g., 1.5 T at 30 Hz for 5.5 minutes) and standard ECT (e.g., 150 mA at 50-Hz stimulation for 0.5 seconds) on rat brain tissue. On electron microscopic and light microscopic examinations, rTMS showed no evidence of neurotoxicity. Moreover, Post and colleagues[122] failed to find structural alterations in rat brains after 11 weeks of long-term treatment with rTMS. In contrast, in vitro studies showed that magnetic stimulation analogous to rTMS increased the overall viability of mouse monoclonal hippocampal cells and had a neuroprotective effect against oxidative stressors (e.g., beta-amyloid and glutamate).[122]

Nahas and coworkers[123] found no structural change with T1- and T2-weighted volumetric magnetic resonance (MR) imaging scans before and after 10 days of daily left prefrontal rTMS in 22 depressed patients. Patients with epilepsy treated with rTMS demonstrated no evidence of structural damage in the temporal lobes.[124] The peak magnetic field strength induced by magnetic stimulation is about 2 T, similar to the static magnetic field strength of clinical MRI scanners. The total time of exposure to the magnetic field for TMS procedures is extremely brief compared with clinical MRI. The maximum magnetically induced charge density (expected to be in the range of 2 to 3 $\mu C/cm^2$) is below the threshold for neuronal injury of about 40 to 100 $\mu C/cm^2$.[125] At the macroscopic and microscopic level rTMS applied to the cortex has shown no evidence of permanent brain changes.

Neuroendocrine Axis

Initial studies suggested that rTMS did not result in any permanent changes in prolactin, ACTH, thyroid-stimulating hormone (TSH), luteinizing hormone, and follicle-stimulating hormone (FSH).[110] As expected, one subject who experienced a seizure had increased serum prolactin level. Wassermann and associates[115] demonstrated that rTMS at 1 Hz and 20 Hz had no effect on serum prolactin levels. George and colleagues[92] reported that increases in TSH occurred after 5-Hz rTMS to the right prefrontal cortex. This changed was significantly correlated with mood improvement in these depressed patients. Evers and coworkers[126] showed that in patients receiving rTMS for depression, rTMS had negligible effects on serum cortisol, prolactin, FSH, and TSH with stimulation at 10 and 20 Hz to the right and left dorsolateral prefrontal cortex at suprathreshold and subthreshold intensities.

Pain and Headache

The most commonly reported side effect of TMS is headache (about 5%). Magnetic stimulation results in activation of conducting neural tissue including nerves and muscles overlying the skull. Stimulation may result in tension-type headaches, particularly after repeated stimulation as occurs with rTMS. This type of pain varies with the intensity and frequency of stimulation.[1] Fortunately, this type of discomfort usually responds well to mild analgesics (i.e., acetaminophen).

Hearing

A discharging coil produces audible clicks that some subjects may find uncomfortable particularly at high stimulus frequencies and intensities. The question of whether these clicks cause any short- or long-term changes to hearing has been explored. Pascual-Leone and associates[127] found no evidence for temporary or permanent auditory threshold shifts in 20 subjects receiving rTMS. These studies used audiograms, tympanograms, acoustic reflexes, and auditory evoked potentials to evaluate hearing. Moreover, they included some subjects who had received rTMS for several years. They concluded that the risk from for hearing impairment from the TMS-induced acoustic artifact was small. Loo and colleagues[128] demonstrated no significant change to hearing after 2 weeks of rTMS in 18 patients with depression. Five subjects went on to receive rTMS for 6 weeks. Of these five, two patients had small bilateral increases in hearing threshold after 6 weeks of rTMS. When one of these subjects was retested after 1 month, these hearing threshold changes had disappeared. No subjects reported a subjective change in their hearing. It is generally accepted, however, that in some situations subjects would benefit from earplugs as some subjects find loud auditory clicks uncomfortable.

Heating of Electroencephalographic Electrodes and Stimulator Coils

Several investigators have advocated for careful EEG monitoring during rTMS stimulation to avoid seizure induction.[116] Unfortunately, rTMS stimulation causes heating of EEG electrodes that may result in discomfort and, potentially, scalp burns.[110] Methods have been devised to avoid such untoward effects. Ilmoniemi and coworkers[87] have managed to overcome electrode heating by using small, notched Ag/AgCl pellet electrodes that avoid the generation of eddy currents within the electrodes that contribute to their heating. Coil heating can also be a risk within the stimulation parameters to be used in some protocols. However, most stimulators are equipped with temperature sensors that disable the device if the coil temperature exceeds a threshold level.

Contraindications to Repetitive Transcranial Magnetic Stimulation

Although considered safe in most cases, there are circumstances in which rTMS is contraindicated. The fact that magnetic stimuli exert their effects on metallic objects makes it obvious that people with metal in their head (e.g., plates, screws, aneurysm clips) should be excluded from rTMS. The exception here is the mouth where rTMS studies have been conducted safely without any untoward effects. Patients with serious medical illness, who are either at heightened risk of having a seizure or at risk from the consequences of a seizure, should generally be excluded from rTMS studies unless supported by anesthesia, in a model akin with ECT in which there are no absolute contraindications.[129] Patients with implanted devices or cardiac pacemakers should also be excluded because of the potentially serious adverse effects of magnetic stimulation on these devices. In patients with implanted deep brain stimulators for treatment of Parkinson's disease, single- and paired-pulse TMS was found to be safe,[130] but the safety of rTMS has not been established. In cases of intractable or severe depression for which rTMS may be beneficial, ECT should remain the preferred therapeutic modality in patients with implanted devices. When such patients have a strong preference for rTMS or refuse ECT, it is highly recommended that the manufacturer of such devices be consulted about the effects of magnetic stimulation on their product.[1]

Additional monitoring or ancillary support staff (i.e., anesthesia) should be available in the treatment of patients who are on medications that lower seizure threshold (e.g., antidepressants, neuroleptics). If possible, such medications should be withdrawn before treatment with rTMS. It is generally recommended that children and pregnant women be excluded from rTMS except in extraordinary circumstances where rTMS has proven efficacy, in part because the long-term side effects in this population have not been fully investigated. It is advisable that laboratories administer screening questionnaires, with a particular emphasis on the contraindications to rTMS, for all subjects enrolled in rTMS studies.

Conclusion

rTMS represents a formidable investigative and treatment tool to study healthy and diseased brain tissue. Research is exploring the mechanisms through which rTMS exerts its potential therapeutic effects. A greater understanding of cortical processes is being elucidated, broadening our knowledge of the connectivity within the cortex and between cortical and subcortical structures. Future efforts aimed at combining rTMS with other investigative modalities may help us to better understand cortical processes and to avoid the potential hazards that may be associated with this neuroscience tool.

Acknowledgments

Dr. Daskalakis is supported through a research training fellowship from the Ontario Mental Health Foundation the Canadian Psychiatric Research Foundation and is a Canadian Institutes of Health Research INMHA Clinician Scientist. Dr. Chen is a Canadian Institutes of Health Research Scholar.

REFERENCES

1. Wassermann EM. Risk and safety of repetitive transcranial magnetic stimulation: report and suggested guidelines from the International Workshop on the Safety of Repetitive Transcranial Magnetic Stimulation, June 5–7, 1996. Electroencephalogr Clin Neurophysiol 1998;108:1–16.
2. Epstein CM, Schwartzberg DG, Davey KR, Sudderth DB. Localizing the site of magnetic brain stimulation in humans. Neurology 1990;40:666–670.
3. Amassian VE, Maccabee PJ, Cracco RQ. Focal stimulation of human peripheral nerve with the magnetic coil: a comparison with electrical stimulation. Exp Neurol 1989;103:282–289.
4. Priori A, Bertolasi L, Dressler D, et al. Transcranial electric and magnetic stimulation of the leg area of the human motor cortex: single motor unit and surface EMG responses in the tibialis anterior muscle. Electroencephalogr Clin Neurophysiol 1993;89:131–137.
5. Barker AT. The history and basic principles of magnetic nerve stimulation. Electroencephalogr Clin Neurophysiol Suppl 1999;51:3–21.
6. Corthout E, Barker AT, Cowey A. Transcranial magnetic stimulation. Which part of the current waveform causes the stimulation? Exp Brain Res 2001;141:128–132.
7. Kammer T, Beck S, Thielscher A, et al. Motor thresholds in humans: a transcranial magnetic stimulation study comparing different pulse waveforms, current directions and stimulator types. Clin Neurophysiol 2001;112:250–258.
8. Kaneko K, Kawai S, Fuchigami Y, et al. The effect of current direction induced by transcranial magnetic stimulation on the corticospinal excitability in human brain. Electroencephalogr Clin Neurophysiol 1996;101:478–482.
9. Di Lazzaro V, Oliviero A, Profice P, et al. Comparison of descending volleys evoked by transcranial magnetic and electric stimulation in conscious humans. Electroencephalogr Clin Neurophysiol 1998;109:397–401.
10. Wassermann EM, Wang B, Zeffiro TA, et al. Locating the motor cortex on the MRI with transcranial magnetic stimulation and PET. Neuroimage 1996;3:1–9.
11. Classen J, Knorr U, Werhahn KJ, et al. Multimodal output mapping of human central motor representation on different spatial scales. J Physiol 1998;512:163–179.
12. Chen R, Classen J, Gerloff C, et al. Depression of motor cortex excitability by low-frequency transcranial magnetic stimulation. Neurology 1997;48:1398–1403.
13. Siebner HR, Auer C, Conrad B. Abnormal increase in the corticomotor output to the affected hand during repetitive transcranial magnetic stimulation of the primary motor cortex in patients with writer's cramp. Neurosci Lett 1999;262:133–136.
14. Maeda F, Keenan JP, Tormos JM, et al. Modulation of corticospinal excitability by repetitive transcranial magnetic stimulation. Clin Neurophysiol 2000;111:800–805.
15. Romero JR, Anschel D, Sparing R, et al. Subthreshold low frequency repetitive transcranial magnetic stimulation selectively decreases facilitation in the motor cortex. Clin Neurophysiol 2002;113:101–107.
16. Muellbacher W, Ziemann U, Boroojerdi B, et al. Effects of low-frequency transcranial magnetic stimulation on motor excitability and basic motor behavior. Clin Neurophysiol 2000;111:1002–1007.
17. Fitzgerald PB, Brown TL, Daskalakis ZJ, et al. Intensity-dependent effects of 1 Hz rTMS on human corticospinal excitability. Clin Neurophysiol 2002;113:1136–1141.
18. Gangitano M, Valero-Cabre A, Tormos JM, et al. Modulation of input-output curves by low and high frequency repetitive transcranial magnetic stimulation of the motor cortex. Clin Neurophysiol 2002;113:1249–1257.
19. Gerschlager W, Siebner HR, Rothwell JC. Decreased corticospinal excitability after subthreshold 1 Hz rTMS over lateral premotor cortex. Neurology 2001;57:449–455.
20. Wassermann EM, Wedegaertner FR, Ziemann U, et al. Crossed reduction of human motor cortex excitability by 1-Hz transcranial magnetic stimulation. Neurosci Lett 1998;250:141–144.
21. Chen R, Seitz RJ. Changing cortical excitability with low-frequency magnetic stimulation. Neurology 2001;57:379–380.
22. Iyer MB, Schleper N, Wassermann EM. Priming stimulation enhances the depressant effect of low-frequency repetitive transcranial magnetic stimulation. J Neurosci 2003;23:10867–10872.
23. Hoffman RE, Boutros NN, Hu S, et al. Transcranial magnetic stimulation and auditory hallucinations in schizophrenia. Lancet 2000;355:1073–1075.
24. Klein E, Kreinin I, Chistyakov A, et al. Therapeutic efficacy of right prefrontal slow repetitive transcranial magnetic stimulation in major depression: a double-blind controlled study. Arch Gen Psychiatry 1999;56:315–320.
25. Siebner HR, Tormos JM, Ceballos-Baumann AO, et al. Low-frequency repetitive transcranial magnetic stimulation of the motor cortex in writer's cramp. Neurology 1999;52:529–537.
26. Tergau F, Naumann U, Paulus W, et al. Low-frequency repetitive transcranial magnetic

stimulation improves intractable epilepsy. Lancet 1999;353:2209.
27. Pascual-Leone A, Valls-Sole J, Wassermann EM, et al. Responses to rapid-rate transcranial magnetic stimulation of the human motor cortex. Brain 1994;117:847–858.
28. Jahanshahi M, Ridding MC, Limousin P, et al. Rapid rate transcranial magnetic stimulation-a safety study. Electroencephalogr Clin Neurophysiol 1997;105:422–429.
29. Berardelli A, Inghilleri M, Rothwell JC, et al. Facilitation of muscle evoked responses after repetitive cortical stimulation in man. Exp Brain Res 1998;122:79–84.
30. Maeda F, Keenan JP, Pascual-Leone A. Interhemispheric asymmetry of motor cortical excitability in major depression as measured by transcranial magnetic stimulation. Br J Psychiatry 2000;177:169–173.
31. Modugno N, Nakamura Y, MacKinnon CD, et al. Motor cortex excitability following short trains of repetitive magnetic stimuli. Exp Brain Res 2001;140:453–459.
32. Wang H, Wang X, Scheich H. LTD and LTP induced by transcranial magnetic stimulation in auditory cortex. Neuroreport 1996;7:521–525.
33. Kujirai T, Caramia MD, Rothwell JC, et al. Corticocortical inhibition in human motor cortex. J Physiol (Lond) 1993;471:501–519.
34. Valls-Sole J, Pascual-Leone A, Wassermann EM, et al. Human motor evoked responses to paired transcranial magnetic stimuli. Electroencephalogr Clin Neurophysiol 1992;85:355–364.
35. Wassermann EM, Samii A, Mercuri B, et al. Responses to paired transcranial magnetic stimuli in resting, active, and recently activated muscles. Exp Brain Res 1996;109:158–163.
36. Cantello R, Gianelli M, Civardi C, et al. Magnetic brain stimulation: the silent period after the motor evoked potential. Neurology 1992;42:1951–1959.
37. Sanger TD, Garg RR, Chen R. Interactions between two different inhibitory systems in the human motor cortex. J Physiol 2001;530:307–317.
38. Ziemann U, Chen R, Cohen LG, et al. Dextromethorphan decreases the excitability of the human motor cortex. Neurology 1998;51:1320–1324.
39. Berardelli A, Inghilleri M, Gilio F, et al. Effects of repetitive cortical stimulation on the silent period evoked by magnetic stimulation. Exp Brain Res 1999;125:82–86.
40. Di Lazzaro V, Oliviero A, Mazzone P, et al. Short-term reduction of intracortical inhibition in the human motor cortex induced by repetitive transcranial magnetic stimulation. Exp Brain Res 2002;147:108–113.
41. Romeo S, Gilio F, Pedace F, et al. Changes in the cortical silent period after repetitive magnetic stimulation of cortical motor areas. Exp Brain Res 2000;135:504–510.
42. Cincotta M, Borgheresi A, Gambetti C, et al. Suprathreshold 0.3 Hz repetitive TMS prolongs the cortical silent period: potential implications for therapeutic trials in epilepsy. Clin Neurophysiol 2003;114:1827–1833.
43. Fierro B, Piazza A, Brighina F, et al. Modulation of intracortical inhibition induced by low- and high- frequency repetitive transcranial magnetic stimulation. Exp Brain Res 2001;138:452–457.
44. Pascual-Leone A, Tormos JM, Keenan J, et al. Study and modulation of human cortical excitability with transcranial magnetic stimulation. J Clin Neurophysiol 1998;15:333–343.
45. Peinemann A, Lehner C, Mentschel C, et al. Subthreshold 5-Hz repetitive transcranial magnetic stimulation of the human primary motor cortex reduces intracortical paired-pulse inhibition. Neurosci Lett 2000;296:21–24.
46. Wu T, Sommer M, Tergau F, et al. Lasting influence of repetitive transcranial magnetic stimulation on intracortical excitability in human subjects. Neurosci Lett 2000;287:37–40.
47. Modugno N, Curra A, Conte A, et al. Depressed intracortical inhibition after long trains of subthreshold repetitive magnetic stimuli at low frequency. Clin Neurophysiol 2003;114:2416–2422.
48. Olney JW, Farber NB. Glutamate receptor dysfunction and schizophrenia [see comments]. Arch Gen Psychiatry 1995;52:998–1007.
49. Gilio F, Rizzo V, Siebner HR, et al. Effects on the right motor hand-area excitability produced by low-frequency rTMS over human contralateral homologous cortex. J Physiol 2003;551:563–573.
50. Ferbert A, Priori A, Rothwell JC, et al. Interhemispheric inhibition of the human motor cortex. J Physiol 1992;453:525–546.
51. Schambra HM, Sawaki L, Cohen LG. Modulation of excitability of human motor cortex (M1) by 1 Hz transcranial magnetic stimulation of the contralateral M1. Clin Neurophysiol 2003;114:130–133.
52. Ziemann U, Corwell B, Cohen LG. Modulation of plasticity in human motor cortex after forearm ischemic nerve block. J Neurosci 1998;18:1115–1123.
53. Brasil-Neto JP, Cohen LG, Pascual-Leone A, et al. Rapid reversible modulation of human motor outputs after transient deafferentation of the forearm: a study with transcranial magnetic stimulation. Neurology 1992;42:1302–1306.

54. Ridding MC, Rothwell JC. Reorganisation in human motor cortex. Can J Physiol Pharmacol 1995;73:218–222.
55. Sadato N, Zeffiro TA, Campbell G, et al. Regional cerebral blood flow changes in motor cortical areas after transient anesthesia of the forearm. Ann Neurol 1995;37:74–81.
56. Ziemann U, Hallett M, Cohen LG. Mechanisms of deafferentation-induced plasticity in human motor cortex. J Neurosci 1998;18:7000–7007.
57. Gerloff C, Corwell B, Chen R, et al. The role of the human motor cortex in the control of complex and simple finger movement sequences. Brain 1998;121:1695–1709.
58. Gerloff C, Corwell B, Chen R, et al. Stimulation over the human supplementary motor area interferes with the organization of future elements in complex motor sequences. Brain 1997;120:1587–1602.
59. Chen R, Gerloff C, Hallett M, et al. Involvement of the ipsilateral motor cortex in finger movements of different complexities. Ann Neurol 1997;41:247–254.
60. Pascual-Leone A, Gates JR, Dhuna A. Induction of speech arrest and counting errors with rapid-rate transcranial magnetic stimulation. Neurology 1991;41:697–702.
61. Claus D, Weis M, Treig T, et al. Influence of repetitive magnetic stimuli on verbal comprehension. J Neurol 1993;240:149–150.
62. Flitman SS, Grafman J, Wassermann EM, ct al. Linguistic processing during repetitive transcranial magnetic stimulation. Neurology 1998;50:175–181.
63. Wassermann EM, Blaxton TA, Hoffman EA, et al. Repetitive transcranial magnetic stimulation of the dominant hemisphere can disrupt visual naming in temporal lobe epilepsy patients. Neuropsychologia 1999;37:537–544.
64. Epstein CM, Woodard JL, Stringer AY, et al. Repetitive transcranial magnetic stimulation does not replicate the Wada test. Neurology 2000;55:1025–1027.
65. Pascual-Leone A, Wassermann EM, Grafman J, et al. The role of the dorsolateral prefrontal cortex in implicit procedural learning. Exp Brain Res 1996;107:479–485.
66. Terao Y, Fukuda H, Ugawa Y, et al. Visualization of the information flow through human oculomotor cortical regions by transcranial magnetic stimulation. J Neurophysiol 1998;80:936–946.
67. Cohen LG, Celnik P, Pascual-Leone A, et al. Functional relevance of cross-modal plasticity in blind humans. Nature 1997;389:180–183.
68. Kosslyn SM, Pascual-Leone A, Felician O, et al. The role of area 17 in visual imagery: convergent evidence from PET and rTMS. Science 1999;284:167–170.
69. Muellbacher W, Ziemann U, Wissel J, et al. Early consolidation in human primary motor cortex. Nature 2002;415:640–644.
70. Mottaghy FM, Hungs M, Brugmann M, et al. Facilitation of picture naming after repetitive transcranial magnetic stimulation. Neurology 1999;53:1806–1812.
71. Boroojerdi B, Phipps M, Kopylev L, et al. Enhancing analogic reasoning with rTMS over the left prefrontal cortex. Neurology 2001;56:526–528.
72. Niehaus L, Roricht S, Scholz U, et al. Hemodynamic response to repetitive magnetic stimulation of the motor and visual cortex. Electroencephalogr Clin Neurophysiol Suppl 1999;51:41–47.
73. Paus T, Jech R, Thompson CJ, et al. Transcranial magnetic stimulation during positron emission tomography: a new method for studying connectivity of the human cerebral cortex. J Neurosci 1997;17:3178–3184.
74. Paus T, Jech R, Thompson CJ, et al. Dose-dependent reduction of cerebral blood flow during rapid-rate transcranial magnetic stimulation of the human sensorimotor cortex. J Neurophysiol 1998;79:1102–1107.
75. Fox P, Ingham R, George MS, et al. Imaging human intra-cerebral connectivity by PET during TMS. Neuroreport 1997;8:2787–2791.
76. Ziemann U, Lonnecker S, Steinhoff BJ, et al. The effect of lorazepam on the motor cortical excitability in man. Exp Brain Res 1996;109:127–135.
77. Daskalakis ZJ, Christensen BK, Fitzgerald PB, et al. The mechanisms of interhemispheric inhibition in the human motor cortex. J Physiol 2002;543:317–326.
78. Bohning DE, Pecheny AP, Epstein CM, et al. Mapping transcranial magnetic stimulation (TMS) fields in vivo with MRI. Neuroreport 1997;8:2535–2538.
79. Bohning DE, Shastri A, Nahas Z, et al. Echoplanar BOLD fMRI of brain activation induced by concurrent transcranial magnetic stimulation. Invest Radiol 1998;33:336–340.
80. Bohning DE, Shastri A, McConnell KA, et al. A combined TMS/fMRI study of intensity-dependent TMS over motor cortex. Biol Psychiatry 1999;45:385–394.
81. Nahas Z, Lomarev M, Roberts DR, et al. Unilateral left prefrontal transcranial magnetic stimulation (TMS) produces intensity-dependent bilateral effects as measured by interleaved BOLD fMRI. Biol Psychiatry 2001;50:712–720.
82. Christensen BK, Bilder RM. Dual cytoarchitectonic trends: an evolutionary model of frontal lobe functioning and its

application to psychopathology. Can J Psychiatry 2000;45:247–256.
83. Strafella AP, Paus T, Barrett J, et al. Repetitive transcranial magnetic stimulation of the human prefrontal cortex induces dopamine release in the caudate nucleus. J Neurosci 2001;21:RC157.
84. Strafella AP, Paus T, Fraraccio M, et al. Striatal dopamine release induced by repetitive transcranial magnetic stimulation of the human motor cortex. Brain 2003;126:2609–2615.
85. Enomoto H, Ugawa Y, Hanajima R, et al. Decreased sensory cortical excitability after 1 Hz rTMS over the ipsilateral primary motor cortex. Clin Neurophysiol 2001;112:2154–2158.
86. Boroojerdi B, Prager A, Muellbacher W, et al. Reduction of human visual cortex excitability using 1-Hz transcranial magnetic stimulation. Neurology 2000;54:1529–1531.
87. Ilmoniemi RJ, Virtanen J, Ruohonen J, et al. Neuronal responses to magnetic stimulation reveal cortical reactivity and connectivity. Neuroreport 1997;8:3537–3540.
88. Daskalakis ZJ, Christensen BK, Chen R, et al. Evidence for impaired cortical inhibition in schizophrenia using transcranial magnetic stimulation. Arch Gen Psychiatry 2002;59:347–354.
89. Baxter LR, Jr., Schwartz JM, Phelps ME, et al. Reduction of prefrontal cortex glucose metabolism common to three types of depression. Arch Gen Psychiatry 1989;46:243–250.
90. Bench CJ, Frackowiak RS, Dolan RJ. Changes in regional cerebral blood flow on recovery from depression. Psychol Med 1995;25:247–261.
91. Shajahan PM, Glabus MF, Gooding PA, et al. Reduced cortical excitability in depression. Impaired post-exercise motor facilitation with transcranial magnetic stimulation. Br J Psychiatry 1999;174:449–454.
92. George MS, Wassermann EM, Williams WA, et al. Changes in mood and hormone levels after rapid-rate transcranial magnetic stimulation (rTMS) of the prefrontal cortex. J Neuropsychiatry Clin Neurosci 1996;8:172–180.
93. Pascual-Leone A, Catala MD, Pascual-Leone Pascual A. Lateralized effect of rapid-rate transcranial magnetic stimulation of the prefrontal cortex on mood. Neurology 1996;46:499–502.
94. Martin JD, George MS, Greenberg BD, et al. Mood effects of prefrontal repetitive high-frequency TMS in healthy volunteers. CNS Spectrums Int J Neuropsychiatric Med 1997;2:53–68.
95. George MS, Wassermann EM, Williams WA, et al. Daily repetitive transcranial magnetic stimulation (rTMS) improves mood in depression. Neuroreport 1995;6:1853–1856.
96. Pascual-Leone A, Rubio B, Pallardo F, et al. Rapid-rate transcranial magnetic stimulation of left dorsolateral prefrontal cortex in drug-resistant depression. Lancet 1996;348:233–237.
97. George MS, Wassermann EM, Kimbrell TA, et al. Mood improvement following daily left prefrontal repetitive transcranial magnetic stimulation in patients with depression: a placebo-controlled crossover trial. Am J Psychiatry 1997;154:1752–1756.
98. Figiel GS, Epstein C, McDonald WM, et al. The use of rapid-rate transcranial magnetic stimulation (rTMS) in refractory depressed patients. J Neuropsychiatry Clin Neurosci 1998;10:20–25.
99. Grunhaus L, Dannon PN, Schreiber S, et al. Repetitive transcranial magnetic stimulation is as effective as electroconvulsive therapy in the treatment of nondelusional major depressive disorder: an open study. Biol Psychiatry 2000;47:314–324.
100. Fitzgerald PB, Brown TL, Marston NA, et al. Transcranial magnetic stimulation in the treatment of depression: a double-blind, placebo-controlled trial. Arch Gen Psychiatry 2003;60:1002–1008.
101. Hausmann A, Marksteiner J, Hinterhuber H, et al. Magnetic stimulation induces neuronal c-fos via tetrodotoxin-sensitive sodium channels in organotypic cortex brain slices of the rat. Neurosci Lett 2001;310:105–108.
102. Hausmann A, Weis C, Marksteiner J, et al. Chronic repetitive transcranial magnetic stimulation enhances c-fos in the parietal cortex and hippocampus. Brain Res Mol Brain Res 2000;76:355–362.
103. Post RM. Transduction of psychosocial stress into the neurobiology of recurrent affective disorder. Am J Psychiatry 1992;149:999–1010.
104. Ji RR, Schlaepfer TE, Aizenman CD, et al. Repetitive transcranial magnetic stimulation activates specific regions in rat brain. Proc Natl Acad Sci USA 1998;95:15635–15640.
105. Muller MB, Toschi N, Kresse AE, et al. Long-term repetitive transcranial magnetic stimulation increases the expression of brain-derived neurotrophic factor and cholecystokinin mRNA, but not neuropeptide tyrosine mRNA in specific areas of rat brain. Neuropsychopharmacology 2000;23:205–215.
106. Keck ME, Welt T, Post A, et al. Neuroendocrine and behavioral effects of repetitive transcranial magnetic stimulation in a psychopathological animal model are suggestive of antidepressant-like effects. Neuropsychopharmacology 2001;24:337–349.

107. Porsolt RD, Le Pichon M, Jalfre M. Depression: a new animal model sensitive to antidepressant treatments. Nature 1977;266:730–732.
108. Keck ME, Engelmann M, Muller MB, et al. Repetitive transcranial magnetic stimulation induces active coping strategies and attenuates the neuroendocrine stress response in rats. J Psychiatr Res 2000;34:265–276.
109. Dhuna A, Gates J, Pascual-Leone A. Transcranial magnetic stimulation in patients with epilepsy. Neurology 1991;41:1067–1071.
110. Pascual-Leone A, Houser CM, Reese K, et al. Safety of rapid-rate transcranial magnetic stimulation in normal volunteers. Electroencephalogr Clin Neurophysiol 1993;89:120–130.
111. Gottesfeld BH, Lesse SM, Herskovitz H. Studies in electroconvulsive shock therapy of varied electrode applications. J Nerv Mental Dis 1944;99:56–64.
112. DeFelipe J, Conley M, Jones EG. Long-range focal collateralization of axons arising from corticocortical cells in monkey sensory-motor cortex. J Neurosci 1986;6:3749–3766.
113. Stefanis C, Jasper H. Recurrent collateral inhibition in pyramidal tract neurons. Journal of Neurophysiology 1964;27:855–877.
114. Chen R, Gerloff C, Classen J, et al. Safety of different inter-train intervals for repetitive transcranial magnetic stimulation and recommendations for safe ranges of stimulation parameters. Electroencephalogr Clin Neurophysiol 1997;105:415–421.
115. Wassermann EM, Grafman J, Berry C, et al. Use and safety of a new repetitive transcranial magnetic stimulator. Electroencephalogr Clin Neurophysiol 1996;101:412–417.
116. Boutros NN, Berman RM, Hoffman R, et al. Electroencephalogram and repetitive transcranial magnetic stimulation. Depress Anxiety 2000;12:166–169.
117. Lisanby SH, Schlaepfer TE, Fisch HU, et al. Magnetic seizure therapy of major depression. Arch Gen Psychiatry 2001;58:303–305.
118. Lisanby SH, Luber B, Sackeim HA, et al. Deliberate seizure induction with repetitive transcranial magnetic stimulation in nonhuman primates. Arch Gen Psychiatry 2001;58:199–200.
119. Little JT, Kimbrell TA, Wassermann EM, et al. Cognitive effects of 1- and 20-Hertz repetitive transcranial magnetic stimulation in depression: preliminary report. Neuropsychiatry Neuropsychol Behav Neurol 2000;13:119–124.
120. Koren D, Shefer O, Chistyakov A, et al. Neuropsychological effects of prefrontal slow rTMS in normal volunteers: a double-blind sham-controlled study. J Clin Exp Neuropsychol 2001;23:424–430.
121. Zyss T, Adamek D, Zieba A, et al. [Transcranial magnetic stimulation versus electroconvulsive shocks: neuroanatomical investigations in rats]. Psychiatr Pol 2000;34:655–675.
122. Post A, Muller MB, Engelmann M, et al. Repetitive transcranial magnetic stimulation in rats: evidence for a neuroprotective effect in vitro and in vivo. Eur J Neurosci 1999;11:3247–3254.
123. Nahas Z, DeBrux C, Chandler V, et al. Lack of significant changes on magnetic resonance scans before and after 2 weeks of daily left prefrontal repetitive transcranial magnetic stimulation for depression. J ECT 2000;16:380–390.
124. Gates JR, Dhuna A, Pascual-Leone A. Lack of pathologic changes in human temporal lobes after transcranial magnetic stimulation. Epilepsia 1992;33:504–508.
125. Barker AT. An introduction to the basic principles of magnetic nerve stimulation. J Clin Neurophysiol 1991;8:26–37.
126. Evers S, Hengst K, Pecuch PW. The impact of repetitive transcranial magnetic stimulation on pituitary hormone levels and cortisol in healthy subjects. J Affect Disord 2001;66:83–88.
127. Pascual-Leone A, Cohen LG, Shotland LI, et al. No evidence of hearing loss in humans due to transcranial magnetic stimulation. Neurology 1992;42:647–651.
128. Loo C, Sachdev P, Elsayed H, et al. Effects of a 2- to 4-week course of repetitive transcranial magnetic stimulation (rTMS) on neuropsychologic functioning, electroencephalogram, and auditory threshold in depressed patients. Biol Psychiatry 2001;49:615–623.
129. Kaplan HI, Sadock BJ, Grebb JA. Kaplan and Sadock's Synopsis of Psychiatry: Behavioral Sciences, Clinical Psychiatry. Baltimore, Williams & Wilkins, 1994:xvi, 1257.
130. Kumar R, Chen R, Ashby P. Safety of transcranial magnetic stimulation in patients with implanted deep brain stimulators. Mov Disord 1999;14:157–158.

6
Central Motor Conduction and Its Clinical Application

Christian W. Hess

■ Methodological Comments Concerning Central Motor Conduction

In clinical medicine, single-pulse TMS is primarily used to evoke motor responses in slightly activated target muscles (Fig. 6–1A). The patient is asked to exert a steady, small, voluntary contraction. If the patient is not capable of contracting the target muscle, reflex activation induced by appropriately manipulating the limb or strong contraction of the homologous contralateral muscle will usually suffice. The procedure records single stimulus-induced muscle twitches called *motor evoked potentials* (MEP). The straightforward performance and easy interpretation of the MEPs compare favorably with the afferent evoked potentials or reflex studies and are suitable for assessing the pyramidal motor system.

For diagnostic purposes, the use of a nonfocal, *large circular coil* with a diameter of 10 to 12 cm is usually preferable. It is placed with its center near the vertex to make the edge of the coil to lie over the hand-arm area and to cross the precentral gyrus perpendicularly (Fig. 6–2A). When using a *double coil*, the point of contact must lie over the target area (see Fig. 6–2A, *right panel*). For exciting the facial muscles' area, the coil has to be shifted laterally (i.e., on the scalp side contralateral to the target muscle) by few centimeters, leaving the coil orientation unchanged.[1] There is, however, an important exception to this rule. For exciting the masseter muscle, the relevant segment of the stimulating coil must be placed parallel rather than perpendicularly to the central sulcus or precentral gyrus[2] (see Fig. 6–2B). The reason for this probably lies in the weak (presynaptic) excitability of this cortical target area so as to require direct corticobulbar tract stimulation.[2] The leg area is best stimulated with the big circular coil shifted a bit rostrally making the posterior segment lie over the precentral gyri so as to cross the mid-sagittal line perpendicularly (see Fig. 6–2A, *lower panels*).[3]

When using a monophasic stimulator, the direction of coil current determines which hemisphere is preferentially excited, and the coil has to be turned over for exciting the other hemisphere. When using a bipolar stimulus, the current direction does not matter, and both hemispheres may be excited simultaneously. Sometimes, it is difficult to excite a lower limb muscle in an elderly person even when using high stimulus intensities and carefully searching for the optimal coil position on the scalp. In such a situation, the use of a very large double coil is advantageous, because it penetrates deeper into the brain.

A conspicuous feature of the MEP is its facilitation by voluntary background contraction, and this goes along with a shorter-onset latency by about 3 ms compared with responses from relaxed muscle (see Fig. 6–1B). In a clinical setting, it is preferable to obtain MEPs with an active target muscle for two reasons. First, lower stimulus intensities are

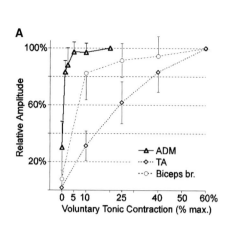
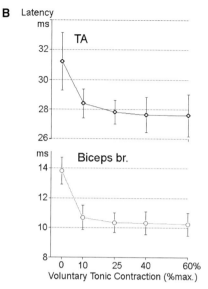

Figure 6–1 Influence of voluntary preactivation on motor evoked potentials (MEPs). **A:** MEP amplitudes related to the degree of voluntary tonic background contraction from three different target muscles in 12 and 34 normal subjects. The stimulus intensity was 1.2 times resting threshold. **B:** MEP onset latencies related to the degree of voluntary tonic background contraction in TA and biceps brachii muscle. Notice the similar latency jump from the relaxed to the contracted state, with virtually stable values when contraction increases up to 60% of maximum force. ADM, abductor digiti minimi; TA, tibials anterior. (Data from Kischka U, Fajfr R, Fellenberg T, et al. Facilitation of motor evoked potentials from magnetic brain stimulation in man: A comparative study of different target muscles. J Clin Neurophysiol 1993;10:505–512 and from Hess CW, Mills KR, Murray NMF. Responses in small hand muscles from magnetic stimulation of the human brain. J Physiol 1987;388:397–412.)

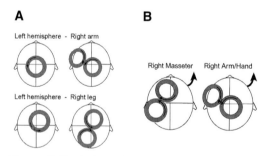

Figure 6–2 A: Approximate coil placement on scalp when using a large, round coil and a double coil (i.e., figure-of-eight coil) to excite upper and lower limb muscles. For exciting facial muscles, the coil must be moved more laterally to the target area, keeping the coil orientation (i.e., direction of the inducing current) the same. Arrows indicate the inducing current in the coil. **B:** Coil placement on the scalp for exciting the right masseter muscle with a double coil (left panel). A short latency peripheral response is evoked in the left masseter muscle. For comparison, the appropriate placement for upper limb muscles is shown on the right side.

required, and second, patients often have difficulty completely relaxing their muscles, producing facilitated responses. Because facilitated responses have shorter onset latencies, different sets of normal values must be used for active and relaxed muscles. In intrinsic hand muscles, the degree of facilitation is usually complete with a background force of 10% to 15% of maximum force (see Fig. 6–1A). In more proximal arm (e.g., biceps brachii) and leg muscles (e.g., tibialis anterior [TA]), the degree of facilitation rises more gradually with increasing force, but the latency remains constant above a force of about 10%.[4] Exerting a tonic contraction of about 20% of maximal force is sufficient for clinical purposes and is easily achieved when patients are asked to make the appropriate isotonic movement without resistance. However, because in proximal muscles the amplitude depends on the degree of preactivation, amplitudes of proximal muscles cannot be used to define abnormality without monitoring the voluntary contraction. In patients who cannot activate the target muscle, reflex activation helps. Alternatively, a strong contraction of the same muscle on the opposite side results in the same degree of latency reduction and useful amplitude increase.[5] Otherwise, the procedure is done without background contraction requiring higher stimulus intensities and the corresponding normal values, because the latencies are then longer by about 2 to 3 ms.

The small hand muscles are particularly convenient for studies of central motor conduction (CMC), because large responses are readily obtained and the peripheral nerve component of the motor pathway to the hand muscles is easily accessible. Other than MEPs from proximal muscles, MEPs from hand muscles do not much depend in size on the level of voluntary pre-innervation, making the amplitude a usable parameter.[4] Responses from lower limb muscles are at higher stimulus intensity and are proportionately smaller and of longer duration than those obtained in the hand muscles. However, in neurologic diseases, abnormality tends to be more common in lower limb muscles.[6]

In contrast to peripherally elicited supramaximal motor responses, MEPs show a considerable *inherent variability* under constant stimulation and recording conditions. The variability in onset latency is best dealt with by taking the shortest latency from a series of four stimuli, because the shortest of four is similar to the shortest of 15.[7] This procedure has the practical advantage that a single response may suffice, provided its latency is comfortably within normal limits. Analogously, the greatest amplitude encountered in a given number of trials can be taken.[8,9] Given the greater variability of conventionally assessed MEP amplitudes, a single large response could, however, be an unrepresentative estimate of the amplitude. Measurement of several individual trials and then taking the mean or median of these values may be preferred.[10] However, McDonnell and coworkers[9] did not find a significant difference between various methods of assessing conventional MEP amplitudes of first dorsal interosseous muscle, when comparing mean values, the greatest potential, and averaging of 20 responses considering peak-to-peak measure and area. Averaging is rarely needed to make visible a possible MEP, when a patient is not capable of activating only slightly and steadily, producing a large, irregular electromyographic (EMG) pattern that may obscure a small response. Otherwise, averaging does not seem to offer an advantage over other methods to asses MEP amplitudes.[9] In any case, it is important to assess patients and normal values using the same procedure.

To reduce confounding factors of peripheral pathology when measuring MEP amplitudes, the examiner preferably uses the ratio of the compound motor action potential (CMAP) from cortical stimulation to that from maximal peripheral (distal) nerve stimulation. In principle, these relative MEP amplitudes should reflect the number of conducting central motoneurons. However, MEP amplitudes are usually much smaller than those of motor responses to maximal peripheral nerve stimulation, although virtually all motoneurons supplying a target muscle have been shown to be excited by TMS in normal subjects.[11] The MEPs show marked amplitude variation between normal subjects and from one stimulus to another. The MEP amplitudes are primarily degraded by the great dispersion of the descending volley over the long motor route from brain to muscle with intercalated synapses, ensuing phase cancellation phenomena of the biphasic CMAP. The maximum relative MEPs from active small hand muscle were found to be as small as 18% of the distally evoked muscle response in some normal subjects.[8] For the reasons mentioned previously, amplitudes of more proximal muscle cannot be reliably quantified.

Some recording devices automatically provide the *area under the curve*, which can reflect the number of excited motoneurons more faithfully than the amplitude, compensating partially for the temporal dispersion. However, cancellation phenomena are not made up for by taking the area. Measuring the area instead of amplitude may be inaccurate for other reasons. With strong cortical stimuli and high voluntary background contraction, some motoneurons tend to fire more than once in response to a stimulus induced

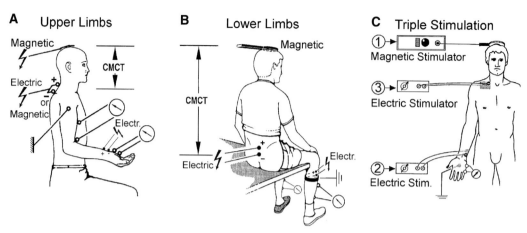

Figure 6–3 Motor evoked potential (MEP) recording setting for upper limb muscles (**A**), lower limb muscles (**B**), and the triple stimulation technique (**C**). Motor root stimulation is used to assess the peripheral conduction time which is subtracted from the total corticomuscular conduction time (**A, B**). A distal peripheral stimulation site is used to calculate relative MEP amplitudes (**A, B**) or to antidromically collide the descending stimulus induced motor volley in the peripheral nerve (**C**).

corticospinal volley, and this enlarges the measured MEP area. Multiple firing of motoneurons particularly occurs in some lower limb MEPs, where it has been shown to significantly affect the measured MEP area.[12]

The limited utility of MEP amplitudes is a major drawback of the method. A much more accurate estimate of the proportion of excited motoneurons is possible using the *triple-stimulation technique*.[13] This technique uses a double-collision paradigm to resynchronize the corticomuscular volley, making it possible to directly quantify the upper motoneuron loss (Fig. 6–3C). The TMS on the scalp is followed by two successive maximal electrical stimuli, one to the ulnar nerve at the wrist and one to the brachial plexus at Erb's point, with appropriate delays to make the induced volleys collide in the ulnar nerve (TSTtest: cortex-wrist-Erb). A first collision takes place when the distal stimulus at the wrist sets up an antidromic volley to meet the descending impulses from cortical stimulation, letting through only the ascending impulses that are traveling in a motor axon the motoneuron of which has not fired in response to the cortical shock. Subsequently, the brachial plexus stimulus sends a volley down to meet the remaining uncollided antidromic impulses, of which there are virtually none in a normal subject. The resulting test potential is compared with the control response obtained from a similar triple-stimulation procedure in which the initial stimulus is applied at the brachial plexus instead of the scalp (TSTcontrol: Erb-wrist-Erb). The ratio of TSTtest/TSTcontrol reflects the percentage of cortically activated spinal motoneurons. This method has been shown to enhance diagnostic sensitivity greatly in multiple sclerosis (MS), spinal cord disorders, and amyotrophic lateral sclerosis (ALS).[13–14] However, the method can be used only in distal limb muscles and complicates the procedure.

The *latency of MEPs* has a considerable peripheral component that is strongly influenced by body stature or limb length and possible abnormality of peripheral nerve conduction. To minimize influence from arm or leg length, body stature, and possible peripheral conduction slowing as a confounding factor, the *central motor conduction time* (CMCT), defined as time from the motor cortex to the spinal motoneurons, is usually assessed. There are two methods of estimating the peripheral conduction time. The first depends on eliciting F waves; the second on stimulating the motor roots at the vertebral column (see Fig. 6–3A,B). The F-wave method assumes that conduction is normal in the proximal segments of the motor roots and is obviously only applicable in nerves where F waves are elicitable. The second method is done with electrical or magnetic stimulation over the vertebral column, a procedure that excites motor roots at their exit foramina.[15] This method involves a small proximal root segment between cord and exit foramen and therefore overestimates CMCT somewhat. In patients with greatly reduced peripheral nerve conduction velocity or proximal conduction block (e.g., Guillain-Barré), the small root segment included in CMCT can introduce significant inaccuracy. When using a high-voltage electrical device, supramaximal root stimulation should not be aimed at because this makes the stimulus reach out into the periphery producing artifactual latencies. With the magnetic stimulator, supramaximal root stimulation can usually not be achieved. For lower limb studies, magnetic stimulation of the lumbosacral roots is not always satisfactorily possible, necessitating high-voltage electrical stimulation (see Fig. 6–3B). For the mere purpose of estimating a peripheral conduction time, supramaximal stimulation of the motor roots is not needed.[15] If the peripheral conduction time cannot be measured for technical or medical reasons (e.g., vertebral column instability), the total cortical-muscle latency must be related to arm lengths or body stature for upper or lower limb muscles respectively.

The *facial nerve* is the only cranial nerve that can be reliably stimulated by TMS at a constant site within the skull. Given the relatively proximal stimulation site, this is a useful technique to calculate the central motor conduction time of facial muscle MEPs. The actual site of intracranial facial nerve stimulation can be localized to the inner part of the facial petrosal canal,[16] hence the term *canalicular stimulation*. For this canalicular stimulation, the optimal coil position and orientation is distinct from that of cortical facial area stimulation: The stimulating coil is placed ipsilateral to the target nerve, preferably posterior to the ear with the inducing current in the relevant coil segment flowing anterolaterally (i.e., with a vector of about 120 degrees when the sagittal backward direction on the scalp is defined as 0 degrees).[1] By electrically stimulating the nerve at the stylomastoid foramen and comparing with the response from transcranial canalicular stimulation, a transosseal conduction time of 1.2 ms (SD 0.18) could be calculated.[16] When using TMS to the (contralateral) facial area to obtain facial MEPs and using (ipsilateral) transcranial magnetic canalicular stimulation, a central conduction time of 5.1 ms (SD 0.60) was assessed for the nasalis muscle. However, the central motor conduction time assessed in such a way still comprises a short peripheral segment of the facial nerve, the portion between exit from brain stem and entrance into the petrosal (fallopian) canal.

Cortical Silent Period

In a contracting muscle, the TMS-induced MEP is immediately followed by an electrical silence lasting 40 to 300 ms that interrupts the ongoing EMG pattern. The early part of this cortical silent period (CSP) is thought to reflect spinal inhibition, whereas the later part originates from cortical inhibition which hence determines its duration.[17–20] The inhibitory phenomenon is independent of the preceding excitatory MEP, because it can occur in isolation in some situations, such as in ALS patients.[21] The duration of the SP depends on the specific muscle tested,[22] the stimulus intensity,[19] the instruction set,[23] and the possible intake of certain central nervous system active substances, particularly dopaminergic, GABAergic drugs and ethanol.[20,24] The CSP duration probably reflects $GABA_B$ function,[24] which makes it an attractive phenomenon to investigate certain neurological disorders. The duration is not much influenced by the degree of tonic voluntary activation.[19] The CSP duration has shown statistically significant group differences between normal subjects and various neurological disorders, which along with cortical threshold and short-interval paired stimulation testing, have enabled interesting insights into disease mechanisms. Apart from cerebral disorders, the SP also seems to be altered in spinal cord pathology.[25,26]

As diagnostic tool for the individual patient, the CSP duration is hampered by the very large interindividual variability[27] and its dependence on the examiner,[27] the stimulation intensity,[19] and instruction set.[23] The latter two factors make very strict standardization of the test procedure and the use of high stimulation strength mandatory. Even then, the range of normal limits remains quite broad. Deciding at what time the silent period starts and ends may be a problem.[10] An evaluation of the CSP for diagnostic purposes in neurological disease is still awaited. However, the CSP duration shows a high degree of re-test reliability within the same subject,[27] making it potentially suitable for longitudinal monitoring of the course of a disease and of treatment effects.

An ipsilateral silent period (ISP) can be recorded, with an onset latency of about 35 ms after the TMS and lasting about 25 ms.[28–30] It is thought to be conveyed by trans-callosal connections, although this may not be the only route.[31] Several studies showed an abnormal ISP in patients with corpus callosum pathology.[29,30]

Corticomotor Threshold

The corticomotor threshold (CMT) is the lowest stimulation intensity able to evoke a MEP of minimal size and is usually assessed in a small hand muscle. The CMT depends on the excitability of spinal motoneurons and motor cortex neurons.[10] In theory, to define CMT, a minimal MEP should be elicited in 50% of the trials. In practice, the stimulus intensity is increased at 5% steps until reaching a level that induces approximately 100-μV responses in about 50% of 10 consecutive trials. There is a CMT at rest (i.e., resting threshold) and one during contraction (i.e., active threshold) that is lower. Distal muscles have a lower CMT than proximal ones.

The CMT augments with increasing age[32] and is enhanced by sodium or calcium channel blocking anticonvulsants[24] but not by drugs acting on GABA or glutamate transmission. The CMT is based on mechanisms different from those of the CSP, which it ideally complements when investigating disease mechanisms. The CMT has, for example, is raised in advanced cases of ALS,[33] MS,[34] in spinal injury above the lesion,[35] and it is reduced in early cases of ALS,[36] idiopathic generalized epilepsy,[37] and progressive myoclonic epilepsy.[38] Normal values for resting and active threshold for various hand muscles have been compiled in specialized monographs (see page 180 of reference 10). Based on a large cohort of 89 normal subjects (age range, 12 to 49 years) using a circular coil and abductor digiti minimi (ADM) recordings, Reutens and coworkers[38] found a mean CMT of 55.8 ± 12.9 (SD). Because of the large normal range, CMT is more useful for statistically comparing groups of patients and normal subjects rather than for determining abnormalities in individual cases.

Normal Values of Central Motor Conduction

To judge whether a CMCT is within or beyond normal limit, published normal values of the specific target muscle assessed with the same method may be used. Although the exact stimulator and coil type is not so critical, it is important to use the same method of assessing peripheral conduction time (i.e., root stimulation or F-wave technique) and consider presence or absence of facilitation by voluntary contraction, because MEPs from relaxed target muscles have longer CMCTs. Magnetic and high-voltage electrical root stimulation provide similar latencies at cervical level as long as not very high electrical currents are used in an attempt to stimulate supramaximally. Supramaximal stimulation is however essential when amplitude measurements are used (e.g., in a search of conduction block). At the lumbosacral level, magnetic stimulation often does not suffice and an electrical high-voltage device is preferred. If for some reason the peripheral conduction time cannot be assessed and a CMCT cannot be calculated, normal values for the total corticomuscular latency must be taken. Normal values of the total corticomuscular latency are preferably related to the arm length (upper limbs) or height (lower limbs). The CMCT to lower limbs is slightly height dependent, and when recording from the TA, for example, the upper limit of a normal CMCT value has been described by CMCT–0.076 × height (cm) + 3.4 ms.[39] There is also a weak correlation with age, which usually can be neglected in routine diagnostic work. Some normal values of the most

Table 6–1 MEP NORMAL VALUES OF CONDUCTION TIMES (ms) OF THE MOST IMPORTANT TARGET MUSCLES

Target Muscle	Normal Subjects' Age Range	Cx-Muscle Latency Mean ± SD	Range	CMCT Using F Wave* Mean ± SD	Range	CMCT Proximal Nerve/Root Stim. Mean ± SD	Range	Reference
ADM	19–59		6.2 ± 0.9	5.8 ± 0.8	4.2–7.4	6.3 ± 0.8[†]	4.6–8.0	39
ADM	17–35	19.3 ± 1.2				7.0 ± 1.0[†]		131
APB	20–83	20.4 ± 1.5	16.8–23.8			6.7 ± 1.2[†]	4.9–8.8	132
FDI	17–74			6.5 ± 0.9	4.3–7.6			41
Biceps br.	20–83	11.8 ± 1.2	9.1–14.7			6.1 ± 1.3[†]	4.3–8.4	132
Trapezius	23–72	9.6	7.4–12.0			7.5 ± 1.1[‡]	5.3–9.9	40
Nasalis (fac)	24–42	10.0 ± 0.96				5.1 ± 0.60[§]		16
Tongue, right	20–53	8.8 ± 0.9	7.4–10.8			6.2 ± 0.9	4.7–7.9	133
Tongue, left	"	8.6 ± 0.9	7.3–10.2			6.4 ± 1.0	4.4–8.6	134
Tongue, right		8.9 ± 0.9	7.6–11.2					134
Tongue, left		9.1 ± 1.1	7.5–11.6					135
Masseter	19–25	6.9 ± 0.71	6.2–7.3					136
Masticatory[¶]	23–50	5.5 ± 0.7						39
TA	19–59			13.1 ± 3.8	10.1–16.3	12.5 ± 1.7[†]	9.0–16.7	132
TA	20–76	27.7 ± 2.4	20.2–32.5					137
AH	19–74	39.3 ± 2.4				17.3 ± 1.8[†]		

*Peripheral CT assessed by the F wave technique according to the formula PCT = (F+M−1)/2.
[†]Peripheral CT assessed by motor root stimulation using a magnetic or high-voltage electrical device.
[‡]Stimulation of the accessory nerve at the neck.
[§]Magnetic "canalicular" stimulation of the facial nerve.
[¶]M. pterygoideus (enoral).
ADM, abductor digiti minimi; AH, abductor hallucis; APB, abductor pollicis brevis; FDI, first dosal interosseus; TA, tibialis anterior.

important target muscles are summarized in Table 6–1.

As long as the degree of voluntary background pre-innervation is not monitored, an *MEP amplitude reduction* can be taken as abnormal only when it is very pronounced and when recording from a distal muscle. As a rule of thumb, an MEP amplitude equal to 20% of the distally evoked CMAP must be considered normal, and an amplitude of ≤15% is always abnormal (valid also for TA recordings). For the trapezius muscle, the limit of normal is more than 44%.[40] Absent responses are abnormal also in recordings from proximal muscles. The relatively high corticomotor threshold of lower limb muscles in elderly subjects must be taken into account, necessitating maximum stimulator output and thorough search for the optimal coil placement on the scalp before an absent response can be taken as abnormal.

Interpretation of Motor Evoked Potential Abnormalities

The MEP abnormalities usually encountered include delay in onset latency, amplitude reduction, or absence of response to brain stimulation. Dispersion of the response (i.e., a prolonged MEP potential[41]) and an increased variability of the responses have also been considered and found in central nervous system disorders by some investigators.[42-44] Various combinations of these can also occur. When interpreting MEP results, it should be recalled that prolonged CMCT and reduced MEP amplitude are nonspecific findings and can be caused by a variety of mechanisms, including hypoexcitability of spinal or cortical motoneurons. In particular, CMCT prolongations do not necessarily imply a demyelinating pathology, because marked delays are also encountered in disorders of pure axonal degenerations. Prolongation of CMCT can also occur in motoneuron disorders or hereditary disorders of neuronal degeneration. However, there is a limited specificity in that absent responses is a rare finding in MS and spondylotic myelopathy but readily found in disorders of axonal degeneration such as ALS, hereditary spastic paraplegia (in lower limbs), and cerebrovascular disorder. Conversely, great CMCT prolongations are a rare finding in cerebrovascular disorders.[13]

Although *slowed conduction along demyelinated fibers* is an obvious mechanism of CMCT prolongation, the mechanism of *prolongation in axonal loss* or *conduction block* is not precisely known, and various possibilities exist. First, it must be remembered that MEP onset latency reflects conduction of the largest myelinated pyramidal fibers, which constitute a minority of all corticospinal tract fibers. In case of conduction failure of these large fibers, transmission may occur by small, slowly conducting fibers or by some alternative, oligosynaptic pathway, such as the corticorubrospinal tracts. This mechanism may account for the great delays sometimes encountered in axonal and demyelinating (conduction block) disorders.

Another mechanism probably accounts for small prolongations of up to 5 ms. The rapidly conducting corticospinal neurons use high frequency activity (about 600 to 800 Hz) to excite bulbar and spinal neurons, and they are known to send short *descending volleys* in response to a transcranial cortical stimulus. The spinal motoneurons, particularly the larger ones, require much excitatory input to reach firing threshold. Drop-out of corticospinal fibers reduces the spatial summation of excitatory input on spinal motoneurons, necessitating more temporal summation to get the spinal motoneurons to discharge. It is likely that fewer functioning corticospinal neurons take longer to get the spinal motoneurons to fire (if at all) in response to the impinging excitatory burst. A burst of four to five impulses, each about 1.3 ms apart, makes a difference of about 3 to 5 ms when the motoneuron fires in response to the last rather than first or second impulse. A similar delay can theoretically also result from pathological hypoexcitability of spinal motoneurons.

A mechanism of *apparent CMCT prolongation* without true slowing operates when the cortical stimulus is unable to activate the large, fast-conducting spinal motoneurons, but these fibers are still excitable by cervical root stimulation. Whether this mechanism comes into play very frequently (which may be the case) and whether it also operates when using the F-wave technique to assess peripheral conduction is unknown.

The *MEP amplitudes* are much degraded from the beginning because of the great temporal dispersion of the descending activity over a long pathway with intercalated synapses. In a much dispersed biphasic response, cancellation phenomena of the negative and positive deflections further abate the amplitude. This makes appreciation of MEP amplitudes difficult. Normal values for conventional MEP amplitudes are only valid for distal target muscles as long as the voluntary pre-activation is not monitored. Because pure slowing of conduction due to demyelination increases temporal dispersion, further attenuation of amplitude may result without drop-out of active fibers. However, in MS, this mechanism does not seem reduce MEP amplitude additionally to a relevant extent.[45] In practical terms, an *absent MEP response* strongly indicates conduction failure due to axonal degeneration, because it is a rare finding in demyelinating disorders.

Demyelinating lesions can cause conduction block, which also results in drop-out of active fibers. However, demyelinating diseases probably always comprise a certain degree of axonal loss as well. This becomes obvious when using the triple-stimulation MEP to quantify amplitude, which eliminates the temporal dispersion of the descending volleys and still demonstrates considerable degree of "genuine" amplitude reduction also in MS.[13] The high frequency descending activity of the thickly myelinated corticospinal neurons is particularly vulnerable to slight myelin damage because a little increased refractoriness may suffice to block the descending volley leading again to a drop-out of active fibers.

Using the triple-stimulation technique, it was also shown that, irrespective of the precise mechanism (i.e., conduction block or axonal damage), conduction failure is the clinically relevant abnormality in the motor system leading to weakness, whereas pure latency prolongation does not translate to much in the way of clinical symptoms or signs, if at all.[13] For this reason, latency prolongation is more likely to represent a *subclinical abnormality* than amplitude reduction. Because slowed conduction is, however, unlikely to exist in isolation very often, a positive correlation between clinical signs and prolonged CMCT can nevertheless be expected, although it may not be a very close one. The MEP findings in various neurological disorders are summarized in a qualitative way in Table 6–2.

■ Multiple Sclerosis

MS was the first and probably still is the best-studied neurological condition using magnetic brain stimulation. Clear-cut CMC abnormalities of hand muscles were readily disclosed in MS patients, demonstrating the usefulness of the method in early days of TMS,[46,47] and confirmed earlier studies with electrical stimulation of the motor cortex.[48,49] A larger follow-up study in 83 MS patients[8] established the typical MEP findings of MS, which have since been reproduced. Moderate to marked prolongation of CMCT with additionally reduced amplitude in about one half of the cases was shown in 79% patients with definitive MS and in 55% with probable MS categorized according to the Poser criteria[50] and was demonstrated in 50% patients with possible MS by the McAlpine criteria.[51] The prolonged CMCT and abnormally small MEP correlated with increased finger flexor reflexes, and marked CMCT prolongation correlated with impaired dexterity.[8] Because weakness of the target muscle was infrequent, no clear correlation between paresis and CMC findings emerged. However, if there was weakness of the target muscle, CMC was mostly abnormal. In particular, the rare finding of a completely absent MEP was associated with a weak target muscle along with impaired fine finger movements.

Ingram and coworkers[52] for the first time measured CMC to a lower limb muscle (i.e., TA muscle) in MS, which proved even more sensitive, and CMC abnormality correlated more closely with clinical signs of upper motoneuron disturbance such as hyperreflexia, spasticity, and the Babinski sign. Weak muscles were almost invariably associated with abnormal central conduction, but increased CMCT was also found for one half of the muscles with normal strength. Increased CMCT for lower limb muscles was directly

Table 6-2 MEP IN CLINICAL DIAGNOSIS: CMC ABNORMALITY IN VARIOUS NEUROLOGICAL DISORDERS

Disease	% Abnormal CMC Findings[‡]	Prolonged Latencies	Reduced Amplitudes	Duration Cortical Silent Period	Cortico-motor Threshold	Subclinical CMC Findings[‡‡]
Spondylotic Cervical Myelopathy	80–100%	+++	++	shortened		+
Multiple Sclerosis (MS)	80–90%	+++	++	prolonged	enhanced	+
Amyotrophic Lateral Sclerosis (ALS)	50–85%*	++	+++	shortened	enhanced[Φ]	++
Friedreich's Ataxia	~90%	++	+++		enhanced	–
Early Onset A. with retained reflexes	60–70%	++	+			–
Late Onset Cerebellar Ataxia (e.g., OPCA)	20–80%*	+	(+)		SCA1: enhanced	–
Hereditary Spastic Paraplegia (HSP)	80–100%**	(+)	+++[†]			(+)
Spinal Muscular Atrophy (SMA)	none	–	–			–
Parkinson's Disease	none	–	–		±	–
Multiple System Atrophy (MSA)	10–45%°	+	+	shortened		–
Progressive Supranuclear Palsy	40% (el Stim)					
Wilson's Disease	30–65%	+	++[†]			–
Huntington	0–10%	(+)	–	prolonged[§]		–
Dystonia	none	–	–	shortened	normal	–
Stroke	70%	+	++[†]	±	(enhanced)	–
Mitochondrial Myopathy	~25%	++	?			++
Functional Weakness	none[Ψ]	–	–			–

°Only to lower limbs abnormal; [§]Westphal variant: shortened; [†]Often absent responses;
*More frequent or more pronounced abnormality to lower limb muscles; **CMC to upper limbs usually normal;
[Ψ]Only in plegic limb reliable evidence; [‡]Only conduction times/latencies and amplitudes considered/approximate values;
[Φ]Lowered threshold at very early stages of the disease.

related to functional motor disability, which has been amply confirmed since.[53–56] Using recordings from three muscles in the upper limbs and two in the lower limbs in 68 patients, Ravnborg and colleagues,[57] found 83% abnormal CMC in the 40 patients who were ultimately definitely diagnosed as having MS. In a larger study of 101 MS patients, combining CMC to two upper limb (i.e., ADM and biceps brachii) and one lower limb muscle (i.e., TA), the additional use of TA significantly increased the proportion of abnormal CMC findings in the probable and possible MS cases (as defined earlier) to 64% and in definite MS to 82%.[5]

In view of the broad range of normal values of MEP amplitudes (normal > 15% of peripherally evoked CMAP), the frequency of abnormal amplitudes is remarkable,[8] and we have good reason to assume that this is caused by axonal loss and conduction block of demyelinated pyramidal fibers,[58] some of it probably as frequency-dependent conduction block.[59] Using the novel triple-stimulation method to better quantify MEP amplitude reduction,[13] conduction failure without much CMCT prolongation could be disclosed in early relapsing-remitting MS patients. Because of the methodical shortcoming of the conventional MEP technique in appropriately assessing MEP size, most clinical neurophysiologists feel more comfortable relying on the unequivocal latency measurements. In theory, CMCT prolongation is more likely to disclose subclinical involvement than amplitude reduction, of which only the latter should translate into weakness. In practice, prominent CMCT prolongation should nevertheless frequently occur along with impaired muscle performance, because severe central involvement is likely to be associated with some conduction block or axonal degeneration, or both. It must, however, be remembered that conventional CMC assesses only the relatively small proportion of rapidly conducting, thickly myelinated pyramidal fibers. Because tonic muscle contraction probably uses slower conducting corticospinal pathways than phasic contractions,[60] the weak correlation between CMC and tonic force is not surprising. Van der Kamp and coworkers[61] did find a strong inverse correlation between prolonged CMCT to the adductor pollicis muscle and voluntary phasic force of that muscle.

Although earlier studies included a mixture of MS types, Kidd and coworkers[56] specifically looked at progressive MS by measuring MEPs to several upper limb muscles and TA and found a weak correlation of CMCT to TA with disability, whereas CMCT to upper limb muscles did not correlate with any clinical measurement. The CMCT to upper limb muscles correlated with the lesion load in the cervical cord as assessed by MRI, but this was not the case for CMCT to the TA. The modest CMCT changes during a 1-year follow-up period correlated with new MRI cervical lesions rather than clinical deterioration. There also was no difference in MEP parameters between primary and secondary MS. These investigators concluded that progressive clinical impairment in such patients might be caused by fiber tract degeneration in the spinal cord that is not reflected by CMCT prolongation or MRI plaques.

When comparing secondary progressive with relapsing-remitting MS patients, Facchetti and coworkers[55] found a significantly longer spinal motor conduction time in progressive MS. This finding may at first glimpse contradict the important role of axonal degeneration recently attributed to the progressive MS. However, it was confirmed by a large study comparing 90 relapsing-remitting MS patients with 51 progressive MS patients,[58] showing significantly longer CMCT to upper and lower limb muscles in progressive MS, and this difference was also true when patients with similar clinical motor deficit or similar disease duration were compared. Although CMCT did not correlate with clinical signs and deficits in either group, the quantified MEP size as assessed by triple simulation was significantly related to clinical signs in the relapsing-remitting and the progressive MS groups. In this study, many relapsing-remitting MS patients showed a considerable degree of conduction failure (i.e., quantified amplitude reduction) in the presence of only slight or moderate CMCT prolongation, and this was explained as caused by conduction block. However, in the progressive MS patients, the prominent conduction failure often found was assumed to be mainly caused by loss of axons. Axonal degeneration is thought to be the cause of ongoing progression of disability without remittance in progressive MS. The conspicuously long CMCT primarily found in progressive MS cases is presumably caused by persistently demyelinated fibers when the capacity for remyelination has exhausted. However, unmasking of alternative, slowly conducting motor pathways (e.g., corticoreticulospinal tract) also may contribute to greatly prolonged CMCT in progressive MS patients.[58]

When comparing with afferent evoked potentials studies in MS, conventional MEPs clearly show the highest yield of abnormality and are surpassed only by MRI, with which MEPs correlate rather closely.[57,62] Despite the high sensitivity, conventional MEPs do not play a major role in ascertaining the diagnosis of MS. MEPs tend to reflect pyramidal disability in MS and therefore do not very often reveal silent lesions. Depending on the precise method used, subclinical involvement was detected in 4% to 13.5% of patients,[46,63,64] which is clearly inferior to visual evoked potentials (VEPs). In a study of 189 consecutive patients referred for suspected MS,[62] conventional MEPs ranked after VEPs in their capacity to enhance the diagnostic certainty in MS, while MRI and cerebrospinal fluid oligoclonal bands were the most powerful ones with this respect. In selected cases, however, when MRI and oligoclonal bands are not diagnostic, MEP may nevertheless be very helpful.[62] The high sensitivity makes MEPs a suitable tool to follow the course of MS with or without treatment.[62]

By measuring additional MEP parameters, the sensitivity can be somewhat enhanced, as shown for measuring the MEP latency variability[42] and the MEP potential duration,[41] but no further elaboration on these refinements was done. Other techniques to increase MEP sensitivity in MS include standardized muscle preactivation to improve MEP amplitude assessment[66]; long-interval paired stimulation when recording from relaxed target muscle to improve CMCT sensitivity[67]; TMS-induced silent period, which was shown to be prolonged in MS[68]; and the use of transcallosal (ipsilateral) inhibition in tonically activated muscle.[69] However, the diagnostic power of these procedures has not been evaluated in a larger patient cohort.

A different picture emerges when the more sophisticated TMS technique of triple-stimulation MEPs is added. It has been shown to be 2.75 times more sensitive than conventional MEPs in disclosing corticospinal conduction failure,[13] but it complicates the MEP procedure somewhat.

The strength of conventional MEPs as a simple procedure is not that of establishing or rejecting the definite diagnosis of MS, but rather that of confirming doubtful and "soft" neurological signs such as equivocal plantar responses or brisk reflexes without increased tone in an early stage of suspected disease.

■ Motoneuron Diseases

The diagnosis of amyotrophic laterals sclerosis (ALS) usually poses no problem when the course of the disease is followed and the typical clinical picture eventually emerges. With the advent of medical therapy and when more effective treatment becomes available, early diagnosis of ALS will become more important. MEPs should have the capacity to detect conduction failure in the fast corticomotoneuronal system, enabling the differentiation of ALS from motor syndromes that can clinically mimic ALS, such as multifocal motor neuropathy, monomelic amyotrophy, postpolio muscular atrophy, inclusion body myositis, and spinal muscular atrophy (e.g., bulbospinal Kennedy disease). The first report using TMS in ALS measured CMC to the ADM muscle and found a CMC abnormality in 64% of patients with ALS.[70] Eisen and coworkers who measured CMC to three upper limb muscles in 40 definite ALS patients found the rate of CMC abnormality approaching 100%.[33] As might be expected from consideration of the pathology with axonal degeneration, abnormally small and absent responses to brain stimulation occurred at higher incidence than found in MS or compressive myelopathy. However, CMCT prolongation is also found in about one half the ALS patients, sometimes to a considerable degree. It is clear from these studies that CMC findings are nonspecific and of little discriminant value in the individual case.[71] A markedly prolonged CMCT has been found in the slowly progressive familial ALS with the autosomal recessively inherited D90A CuZn–superoxide dismutase mutation.[72]

The reported yield of abnormal MEP findings in ALS has been conspicuously variable, and this is primarily caused by the different patient samples (early versus late stages), the number and type of assessed target muscles, and the considered MEP parameters. Sensitivity tends to be greater with recordings from lower limb muscles. Miscio and colleagues,[73] who looked at various stages of ALS found CMC to be abnormal in 95.4% of definite ALS patients, in 72.2% of suspected ALS cases with probable upper motoneuron signs (most of them later developed ALS), in 50% of pure lower motoneuron syndromes, and in 20% of progressive bulbar palsy. These investigators recorded from the ADM and flexor hallucis muscles and only considered CMC prolongation and absent MEP as abnormal (no reduced amplitudes considered). Di Lazzaro and coworkers[64] measured CMC to the biceps brachii, ADM, rectus femoris, TA, and abductor hallucis and found a sensitivity of 74% and a rate of subclinical involvement of 26% in ALS, which was the highest yield in their large cohort of 1,023 patients of various neurological disorders. Pohl and coworkers[74] looked at 49 ALS patients of various stages according to the 1994 El Escorial Criteria and recorded from the ADM and TA. They found an abnormal measurement (i.e., absent response or prolonged CMCT) in definite ALS in 50% (ADM) and 35% (TA), in probable ALS in 43% (ADM) and 64% (TA), and in possible and suspected ALS taken together in 25% (ADM) and 25.5% (TA).

Several investigators found assessing CMC to cranial muscles particularly rewarding. Trompetto and coworkers[75] found delayed or absent MEP to the masseter muscle in 63% of their patients also when there were no clinical bulbar signs. Urban and coworkers[76] assessed CMC to the tongue and orofacial muscles in addition to ADM and TA muscles in 51 ALS patients and ended up with 82% of patients having CMC abnormality. Truffert and colleagues[40] used the technically simple recording from trapezius muscle in 10 ALS patients and found a CMC abnormality to this muscle in all of them.

Because the pathology of ALS is axonal loss and conventional MEPs are not very sensitive in assessing reduction in MEP size, the amplitude quantification by triple-stimulation MEP was expected to be more sensitive. In a triple-stimulation study of 48 ALS patients of various diagnostic categories (19 with definitive ALS), conduction failure to the ADM was found in 24 patients (38 sides), 12 (20 sides) of which were normal in conventional MEPs.[14] The increased sensitivity of the triple-stimulation MEPs in ALS was confirmed in 19 ALS patients by Komissarow and coworkers,[77] who found more MEP abnormality by triple-stimulation MEPs than with conventional MEPs, particularly in suspected and possible ALS cases, in which a triple-stimulation MEP abnormality was found in all patients.

Apart from central motor conduction time and amplitude, assessment of increased cortical motor threshold, of shortened silent period, and of altered peristimulus time histograms have been demonstrated in ALS, sometimes also detecting subclinical abnormality in patients with doubtful or missing upper motoneuron signs.[36,78-81] In some patients with enhanced threshold and absent MEP, the cortical silent period could nevertheless be evoked.[21]

From a pathophysiological point of view, it is interesting that the corticomotor threshold has a tendency to be abnormally low in the early stages when there are only few clinical signs and rises to abnormally high levels later during the advanced stages.[36,82] This dynamic feature makes assessment of corticomotor threshold useless for clinical diagnosis in the early stage. When comparing with groups of MS and treated Parkinson's disease, the discriminative value of the CMT was found to be low,[78] because these disorders also tend to have an elevated CMT. Because of the relatively large scatter and large range and standard deviations in normal subjects, CMT is not very helpful as a diagnostic indicator in the individual patient.

The CSP has been found to be reduced in ALS to various degrees,[78,83-87] which often makes group comparisons with normal subjects significantly different. Differences become more distinct with greater stimulus intensity, because this reduces the variability of CSP duration. In ALS, the CSP duration depends less on the stimulus strength (i.e., it does not prolong to the same degree as in normal subjects when stimuli are enhanced).[78,86] In normal subjects, the CSP duration is positively related to the stimulus intensity.

Mills,[87] who followed 76 ALS patients (49 until death) with serial measurements of corticomotor threshold, central motor conduction time, silent period duration, and the amplitude of compound muscle action potentials from both first dorsal interosseous muscles, concluded that none of the measures of central motor function in ALS is likely to be useful for monitoring patients in a clinical trial setting.

Studies of only few patients with *primary lateral sclerosis* have been published, and CMC to upper and lower limb muscles was abnormal in virtually all of them.[88,89] Responses were frequently absent, and very marked prolongations of CMCT were found.

In *hereditary spastic paraplegia* (HSP), CMC is usually abnormal to the lower limbs only, with the typical pattern being an abnormally small and moderately prolonged CMCT with absent responses in one third of the patients.[90-92] With a sensitivity of approaching 100%, the MEPs were closely related to the physical signs but without revealing subclinical clear-cut CMC abnormality, which obviously limits their use in diagnostic workup.[92] Schnider and coworkers[90] nevertheless found an abnormally small MEP amplitude in the unaffected juvenile member of a family with HSP, for whom using MEP amplitude quantification might have revealed subclinical involvement.

In the *spinal muscular atrophies*, the MEPs have invariably been found to be normal.[10]

■ Myelopathy

Compressive myelopathy due to *spondylosis* or *disc herniation* is the one condition for which MEPs are most sensitive. When recording from the TA muscle, the sensitivity approaches 100% in some studies, with often considerable CMCT prolongations,[93-95] making it a suitable test for monitoring and helping to decide for surgery. Additional amplitude reduction is frequently found in about 50% of MEPs to small hand muscles. After surgical decompression, MEPs do not normalize, except for the very mild cases without much clinical disability.[96] Subclinical lesions are found in 10% to 15% of patients.[64,97,98] When combined with somatosensory evoked potentials (SEPs), neurophysiological abnormalities were reported in up to 50% of patients with "silent" compression.[99] CMC prolongations in spondylotic cervical myelopathy are often considerable, whereas absent MEP responses are rarely encountered.

It was hoped that *recording from multiple muscles* supplied by motor roots of varied segments would allow precise localization of the crucial compression level. Because the cervical column is the most frequently affected segment, recording from the trapezius, biceps brachii, intrinsic hand muscle, and TA should allow narrowing down the lesion to, for example, the mid-cervical level and determine whether a radiological narrowing is functionally relevant. This is a practically important question, because many elderly subjects have benign spondylotic alterations and imaging by MRI tends to overestimate these narrowings. Several investigators did show that the pattern of prolonged CMCT below and normal CMC above the segment in question was frequently encountered providing valuable confirmatory information.[93,100,101] The encountered patterns are, however, not completely reliable. Mathis and coworkers[93] compared 72 patients with compressive myelopathy and 101 patients with MS and found a "mid-cervical compressive pattern" (i.e., CMCT to biceps normal, to ADM abnormal) in 10% of the MS patients and 19% of patients who had their proved compression above C4 level. Truffert and colleagues,[40] who compared cervical myelopathy with ALS, found an abnormal CMC to the trapezius in one of nine patients with myelopathy. For suspected thoracic myelopathy, multiple-level paraspinal recordings were suggested.[102,103] Taniguchi and coworkers[102] found false-negative results in paraspinal recordings when the lesion was below T10.

MEPs are less sensitive in intramedullary lesions such as syringomyelia or spinal cord tumors.[98] They seem to be sensitive and perhaps of prognostic value in patients with human immunodeficiency virus (HIV), because subclinical involvement was evidenced in HIV-positive patients who tended to progress to acquired immunodeficiency syndrome (AIDS) more rapidly.[104,105] In spinal injuries, MEP measurements have given mixed results, but they appear to be of some prognostic value.[35,106]

Degenerative Ataxic Disorders

Classification of the hereditary ataxias is usually based on clinical features with the age of onset as the crucial factor, but genetic determination is becoming an essential supplement. The most important of the early-onset hereditary ataxias (i.e., those with symptoms before the age of 20 years) is *Friedreich's ataxia,* an autosomal recessive, triplet-repeat disorder. Clinically, it is characterized by progressive gait difficulty, loss of tendon reflexes, a Babinski sign, and cardiomyopathy. MEPs in Friedreich's ataxia usually have been abnormal in lower and upper limbs, with moderate to marked CMCT prolongations and often with diminished amplitudes.[107-109] The CMAP duration of the MEPs were significantly longer in Friedreich's ataxia than in the other disorders.[107]

Friedreich's ataxia is distinguished from *early-onset cerebellar ataxia* with retained tendon reflexes with better prognosis, in which tendon reflexes may be normal or increased and in which there is no cardiomyopathy. MEPs were found to be abnormal in about 60% (upper limbs) to 70% (lower limbs), and abnormalities were a bit less pronounced than in Friedreich's ataxia.[107-109]

The *late-onset cerebellar degenerations* are a complex and heterogenous group of autosomal dominant disorders, many of them genetic with the triplet-repeat expansion mechanism. Pyramidal tract involvement is variable, as are such diverse features as ophthalmoplegia, dementia, and myoclonus. In genetically unclassified patients, a MEP abnormality was found in about one half of the cases, and CMCT was only moderately prolonged, with amplitudes often normal. In genetically classified spinocerebellar ataxia type 1 patients (SCA1), MEPs to lower limbs were mostly abnormal, whereas in SCA2 and SCA3, this was only the case in 18% and 28%.[110,111]

Cerebrovascular Disease

CMC abnormalities are often found in the acute stage of stroke, but the sensitivity of MEPs in cerebrovascular disease is generally low. For instance, MEPs were found normal in one third of middle cerebral artery strokes,[64] and no subclinical CMC abnormality has been reported. MEPs are usually absent only in total middle cerebral artery strokes with profound hemiplegia. The correlation between MEP findings and site of pathology does not seem to disclose more than what can be expected from the clinical signs.[10] Using electrical brain stimulation, Abbruzzese and coworkers[112] found an abnormal CMC in 18 of 32 patients with focal deficits due to minor cerebral ischemia of the lacunar type.

Most investigators found that early MEP testing is of prognostic value in strokes in that absent MEPs predicted a persistent motor deficit rather more precisely than the clinical examination at the time of MEP testing.[113-118] Evocable MEPs in a *locked in syndrome* probably is a good prognostic sign.[119] CMC prolongations are found in about 20% of strokes, but a great prolongation is the exception. Intracerebral hemorrhage tends to prolong CMC more than ischemic stroke.[120] Ipsilateral MEPs from stimulating the unaffected hemisphere are occasionally evoked with remarkably ease in stroke patients, a phenomenon that seems to bear an unfavorable prognosis.[115]

Measuring the CSP in strokes has produced conflicting results, which may be because of the divers locations of the infarct in the brain. Several investigators found an abnormally and persistently prolonged CSP on the affected side,[68,121,122] and Uozumi[25] found a shortened CSP in patients with spastic hyperreflexia due to cerebral infarction. It appears that the CSP is only shortened when the ischemic zone lies within the motor cortex.[123] Catano and colleagues[124] have shown that in stroke patients other than in normal subjects, the duration of the CSP depends on the exerted tonic background force, making it abnormally short with great tonic contraction, and this might explain some of the discrepancies in literature. In normal subjects, the CSP duration does not depend much on the degree of tonic contraction. Measuring CSP is impeded by a great interindividual variability and requires a highly standardized procedure.

■ Extrapyramidal Disorders

CMC as conventionally assessed has been found normal in *Parkinson's disease, dystonias*, and largely in *Huntington's* disease.[125] In the latter disorder, a modest CMCT prolongation and MEP amplitude reduction was found in some patients, and an increased latency variability was found in many patients.[126]

In *multiple system atrophy* and sporadic olivopontocerebellar atrophy, CMC to the lower limbs often was abnormal.[127] Abbruzese and associates[128] found abnormal CMC in 40% of *progressive supranuclear palsy* (i.e., Steele-Richardson-Olszewski syndrome) when using electrical brain stimulation. MEPs in *Wilson's disease* were found abnormal in about one half of the patients[129] showing prolonged CMCT, reduced amplitude, or absent responses. CMCT was more often abnormal to the hand (first dorsal interosseus) than leg (TA) muscle.

■ Psychogenic Weakness

In assessing psychogenic weakness, MEPs can be helpful in three ways. First and most importantly, an abnormal CMC in a clinically frank psychogenic paresis may give the crucial hint about an underlying organic disorder that would otherwise be easily missed. Second, an normal CMC in a completely plegic limb confirms the psychogenic nature of the weakness.[130] Strictly speaking, it only confirms the weakness being largely functional and cannot rule out a minor concomitant organic component. The normal MEP in a mildly paretic muscles does not rule out an organic cause. Third, the overt jerking of the paralyzed limb during the MEP procedure in a psychogenic plegia can be used therapeutically if the examiner acts skillfully. It is important in such a situation to avoid any trace of triumphant outwitting, but rather to show relief and optimism in the face of the still functioning pathways and muscles. The latter must be learned to be properly controlled again. Cooperation of the patient with psychogenic weakness is sometimes a problem, because we need some pre-innervation of the target muscle. This is usually achieved when manipulating the limb while having the EMG loudspeakers switched off.

REFERENCES

1. Dubach P, Guggisberg AG, Rösler KM, et al. Significance of coil orientation for motor evoked potentials from nasalis muscle elicited by transcranial magnetic stimulation. Clinical Neurophysiology 2004;115: 862–870.
2. Guggisberg AG, Dubach P, Hess CW, et al. Motor evoked potentials from masseter muscle induced by transcranial magnetic stimulation of the pyramidal tract: the importance of coil orientation. Clinical Neurophysiology 2001;112:2312–2319.
3. Rösler KM, Hess CW, Heckmann R, et al. Significance of shape and size of the stimulating coil in magnetic stimulation of the human motor cortex. Neurosci Lett 1989;100:347–352.
4. Kischka U, Fajfr R, Fellenberg T, et al. Facilitation of motor evoked potentials from magnetic brain stimulation in man: A comparative study of different target muscles. J Clin Neurophysiol 1993;10:505–512.
5. Hess CW, Mills KR, Murray NMF. Responses in small hand muscles from magnetic stimulation of the human brain. J Physiol 1987; 388:397–412.
6. Mathis J, Hess CW. Motor-evoked potentials from multiple target muscles in multiple sclerosis and cervical myelopathy. Eur J Neurol 1996;3:567–573.
7. Hess CW, Mills KR, Murray NMF. Methodological considerations for magnetic brain stimulation. In Barber C, Blum T (eds): Evoked Potentials, III. London, Butterworth, 1988:456–461.
8. Hess CW, Mills KR, Murray NMF, et al. Magnetic brain stimulation: central motor conduction studies in multiple sclerosis. Ann Neurol 1987;22:744–752.

9. McDonnell MN, Ridding MC, Miles TS. Do alternate methods of analysing motor evoked potentials give comparable results? J Neurosci Methods 2004;136:63–67.
10. Mills KR. Magnetic Stimulation of the Human Nervous System. New York, Oxford University Press, 1999.
11. Magistris MR, Rösler KM, Truffert A, et al. Transcranial stimulation excites virtually all motor neurons supplying the target muscle. A demonstration and a method improving the study of motor evoked potentials. Brain 1998;121:437–450.
12. Bühler R, Magistris MR, Truffert A, et al. The triple stimulation technique to study central motor conduction to the lower limbs. Clin Neurophysiol 2001;112:938–949.
13. Magistris MR, Rösler KM, Truffert A, et al. A clinical study of motor evoked potentials using a triple stimulation technique. Brain 1999;122:265–279.
14. Rösler KM, Truffert A, Hess CW, et al. Quantification of upper motor neuron loss in amyotrophic lateral sclerosis. Clin Neurophysiol 2000;111:2208–2218.
15. Schmid UD, Walker G, Hess CW, et al. Magnetic and electrical stimulation of cervical motor roots: technique, site and mechanisms of excitation. J Neurol Neurosurg Psychiatry 1990;53:770–777.
16. Rösler KM, Hess CW, Schmid UD. Investigation of facial motor pathways by electrical and magnetic stimulation: sites and mechanisms of excitation. J Neurol Neurosurg Psychiatry 1989;52:1149–1156.
17. Fuhr P, Agostino R, Hallett M. Spinal motor neuron excitability during the silent period after cortical stimulation. Electroencephalogr Clin Neurophysiol 1991;81:257–262.
18. Inghilleri M, Berardelli A, Cruccu G, et al. Silent period evoked by transcranial stimulation of the human cortex and cervicomedullary junction. J Physiol 1993;466:521–534.
19. Wilson SA, Lockwood RJ, Thickbroom GW, et al. The muscle silent period following transcranial magnetic cortical stimulation. J Neurol Sci 1993;114:216–222.
20. Ziemann U, Netz J, Szelenyi A, et al. Spinal and supraspinal mechanisms contribute to the silent period in the contracting soleus muscle after transcranial magnetic stimulation of human motor cortex. Neurosci Lett 1993;156:167–171.
21. Triggs WJ, Macdonell RAL, Cros D, et al. Motor inhibition and excitation are independent effects of magnetic cortical stimulation. Ann Neurol 1992;32:345–351.
22. Roick H, von Giesen HJ, Benecke R. On the origin of the postexcitatory inhibition seen after transcranial magnetic brain stimulation in awake human subjects. Exp Brain Res 1993;94:489–498.
23. Mathis J, de Quervain D, Hess CW. Dependence of the transcranially induced silent period on the 'instruction set' and the individual reaction time. Electroencephalogr Clin Neurophysiol 1998;109:426–435.
24. Ziemann U, Lonnecker S, Steinhoff BJ, et al. Effects of antiepileptic drugs on motor cortex excitability in humans: a transcranial magnetic stimulation study. Ann Neurol 1996;40:367–378.
25. Uozumi T, Tsuji S, Murai Y. Motor potentials evoked by magnetic stimulation of the motor cortex in normal subjects and patients with motor disorders. Electroencephalogr Clin Neurophysiol 1991;81:251–256.
26. Kaneyama O, Shibano K, Kawakita H, et al. Transcranial magnetic stimulation of the motor cortex in cervical spondylosis and spinal canal stenosis. Spine 1995;20:1004–1010.
27. Fritz C, Braune HJ, Pylatiuk C, et al. Silent period following transcranial magnetic sitmulation: a study of intra- and inter-examiner reliability. Electroencephalogr Clin Neurophysiol 1997;105:235–240.
28. Taylor JL, Fogel W, Day BL, et al. Ipsilateral cortical stimulation inhibited the long-latency response to stretch in the long finger flexors in humans. J Physiol 1995;488:821–831.
29. Hoppner J, Kunesch E, Buchmann J, et al. Demyelination and axonal degeneration in corpus callosum assessed by analysis of transcallosally mediated inhibition in multiple sclerosis. Clin Neurophysiol 1999;110:748–756.
30. Meyer BU, Röricht S, Gräfin von Einsiedel H, et al. Inhibitory and excitatory interhemispheric transfers between motor cortical areas in normal humans and patients with abnormalities of the corpus callosum. Brain 1995;118:429–440.
31. Gerloff C, Cohen LG, Floeter MK, et al. Inhibitory influence of the ipsilateral motor cortex on responses to stimulation of the human cortex and pyramidal tract. J Physiol 1998;510:249–259.
32. Rossini PM, Desiato MT, Caramia MD. Age related changes of motor evoked potentials in healthy humans: Noninvasive evaluation of central and peripheral motor tracts excitability and conductivity. Brain Res 1992;593:14–19.
33. Eisen A, Shtybel W, Murphy K, et al. Cortical magnetic stimulation in amyotrophic lateral sclerosis. Muscle Nerve 1990;13:146–151.
34. Ravnborg M, Blinkenberg M, Dahl KAD. Standardization of facilitation of compound muscle action potentials evoked by magnetic stimulation of the cortex. Results in healthy volunteers and in patients with multiple

sclerosis. Electroencephalogr Clin Neurophysiol 1991;81:195–201.
35. Macdonell RA, Donnan GA. Magnetic cortical stimulation in acute spinal cord injury. Neurology 1995;45:303–306.
36. Mills KR, Nithi KA. Corticomotor threshold is reduced in early sporadic amyotrophic lateral sclerosis. Muscle Nerve 1997;20:1137–1141.
37. Reutens DC, Berkovic SF. Increased cortical excitability in generalised epilepsy demonstrated with transcranial magnetic stimulation. Lancet 1992;339:362–363.
38. Reutens DC, Puce A, Berkovic SF. Cortical hyperexcitability in progressive myoclonus epilepsy: A study with transcranial magnetic stimulation. Neurology 1993;43:186–192.
39. Claus D. Central motor conduction: method and normal results. Muscle Nerve 1990;13:1125–1132.
40. Truffert A, Rösler KM, Magistris MR. Amyotrophic lateral sclerosis versus cervical spondylotic myelopathy: a study using transcranial magnetic stimulation with recordings from the trapezius and limb muscles. Clin Neurophysiol 2000;111:1031–1038.
41. Kukowski B. Duration, configuration and amplitude of the motor response evoked by magnetic brain stimulation in patients with multiple sclerosis. Electromyogr Clin Neurophysiol 1993;33:295–299.
42. Britton TC, Meyer BU, Benecke R. Variability of cortically evoked motor responses in multiple sclerosis. Electroencephalogr Clin Neurophysiol Electromyogr Motor Control 1991;81:186–194.
43. Kiers L, Cros D, Chiappa KH, et al. Variability of motor potentials evoked by transcranial magnetic stimulation. Electroencephalogr Clin Neurophysiol 1993;89:415–423.
44. Brouwer B, Qiao J. Characteristics and variability of lower limb motoneuron responses to transcranial magnetic stimulation. Electroencephalogr Clin Neurophysiol 1995;97:49–54.
45. Rösler KM, Petrow E, Mathis J, et al. Effect of discharge desynchronization on the size of motor evoked potentials: an analysis. Clin Neurophysiol 2002;113:1680–1687.
46. Barker AT, Freeston IL, Jalinous R, et al. Clinical evaluation of conduction time measurements in central motor pathways using magnetic stimulation of human brain. Lancet 1986;1:1325–1326.
47. Hess CW, Mills KR, Murray NMF. Measurement of central motor conduction in multiple sclerosis by magnetic brain stimulation. Lancet 1986;2:355–358.
48. Cowan JMA, Dick JPR, Day BL, et al. Abnormalities in central motor pathway conduction in multiple sclerosis. Lancet 1984;2:304–307.
49. Mills KR, Murray NMF. Corticospinal tract conduction time in multiple sclerosis. Ann Neurol 1985;18:601–605.
50. Poser CM, Paty DW, Scheinberg L, et al. New diagnostic criteria for multiple sclerosis: Guidelines for research protocols. Annals of Neurology 1983;13:227–231.
51. Matthews WB, Acheson ED, Batchelor JR, et al. McAlpine's multiple sclerosis. Edinburgh, Churchill Livingstone, 1985:3–46.
52. Ingram DA, Thompson AJ, Swash M. Central motor conduction in multiple sclerosis: evaluation of abnormalities revealed by transcutaneous magnetic stimulation of the brain. J Neurol Neurosurg Psychiatry 1988;51:487–494.
53. Jones SM, Streletz LJ, Raab VE, et al. Lower extremity motor evoked potentials in multiple sclerosis. Arch Neurol 1991;48:944–948.
54. Kandler RH, Jarratt JA, Davies-Jones GA, et al. The role of magnetic stimulation as a quantifier of motor disability in patients with multiple sclerosis. J Neurol Sci 1991;106:31–34.
55. Facchetti D, Mai R, Micheli A, et al. Motor evoked potentials and disability in secondary progressive multiple sclerosis. Can J Neurol Sci 1997;24:332–337.
56. Kidd D, Thompson PD, Day BL, et al. Central motor conduction time in progressive multiple sclerosis. Correlations with MRI and disease activity. Brain 1998;121:1109–1116.
57. Ravnborg M, Liguori R, Christiansen P, et al. The diagnostic reliability of magnetically evoked motor potentials in multiple sclerosis. Neurology 1992;42:1296–1301.
58. Humm AM, Magistris MR, Truffert A, et al. Central motor conduction differs between acute relapsing remitting and chronic progressive MS. Electroencephalogr Clin Neurophysiol 2003;114:2196–2203.
59. Boniface SJ, Mills K R, Schubert M. Responses of single spinal motoneurons to magnetic brain stimulation in healthy subjects and patients with multiple sclerosis. Brain 1991;114:643–662.
60. Lemon RN, Mantel GWH, Muir RB. Corticospinal facilitation of hand muscles during voluntary movement in the conscious monkey. J Physiol 1986;381:497–527.
61. van der Kamp W, Maertens de Noordhout A, Thompson PD, et al. Correlation of phasic muscle strength and corticomotoneuron conduction time in multiple sclerosis. Ann Neurol 1991;29:6–12.
62. Beer S, Rösler KM, Hess CW. Diagnostic value of paraclinical tests in multiple sclerosis Relative sensitivities and specificities for reclassification according to Poser committee

criteria. J Neurol Neursurg Psychiatry 1995;59:152–159.
63. Kandler RH, Jarratt JA, Gumpert EJ, et al. The role of magnetic stimulation in the diagnosis of multiple sclerosis. J Neurol Sci 1991;106:25–30.
64. Di Lazzaro V, Oliviero A, Profice P, et al. The diagnostic value of motor evoked potentials. Clin Neurophysiol 1999;110:1297–1307.
65. Fuhr P, Borggrefe-Chappuis A, Schindler C, et al. Visual and motor evoked potentials in the course of multiple sclerosis. Brain 2001;124:2162–2168.
66. Nielsen JF. Improvement of amplitude variability of motor evoked potentials in multiple sclerosis patients and in healthy subjects. Electroencephalogr Clin Neurophysiol 1996;101:404–411.
67. Nielsen JF. Frequency-dependent conduction delay of motor-evoked potentials in multiple sclerosis. Muscle Nerve 1997;20:1264–1274.
68. Haug BA, Kukowski B. Latency and duration of the muscle silent period following transcranial magnetic stimulation in multiple sclerosis, cerebral ischemia, and other upper motoneuron lesions. Neurology 1994;44:936–940.
69. Schmierer K, Irlbacher K, Grosse P, et al. Correlates of disability in multiple sclerosis detected by transcranial magnetic stimulation. Neurology 2002;59:1218–1224.
70. Schriefer TN, Hess CW, Mills KR, et al. Central motor conduction studies in motor neurone disease using magnetic brain stimulation. Electroencephalogr Clin Neurophysiol 1989;74:431–437.
71. Claus D, Brunhölzl C, Kerling FP, et al. Transcranial magnetic brain stimulation as a diagsnotic and prognostic test in amyotrophic lateral sclerosis. J Neurol Sci 1995;129(Suppl):30–34.
72. Weber M, Eisen A, Stewart HG, et al. The physiological basis of conduction slowing in ALS patients homozygous for the D90A CuZn-SOD mutation. Muscle Nerve 2001;24:89–97.
73. Miscio G, Pisano F, Mora G, et al. Motor neuron disease: usefulness of transcranial magnetic stimulation in improving the diagnosis. Clin Neurophysiol 1999;110:975–981.
74. Pohl C, Block W, Traber F, et al. Proton magnetic resonance spectroscopy and transcranial magnetic stimulation for the detection of upper motor neuron degeneration in ALS patients. J Neurol Sci 2001;190:21–27.
75. Trompetto C, Caponnetto C, Buccolieri A, et al. Responses of masseter muscles to transcranial magnetic stimulation in patients with amyotrophic lateral sclerosis. Electroencephalogr Clin Neurophysiol 1998;109:309–314.
76. Urban PP, Wicht S, Hopf HC. Sensitivity of transcranial magnetic stimulation of cortico-bulbar vs. cortico-spinal tract involvement in amyotrophic lateral sclerosis (ALS). J Neurol 2001;248:850–855.
77. Komissarow L, Rollnik JD, Bogdanova D, et al. Triple stimulation technique (TST) in amyotrophic lateral sclerosis; Clin Neurophysiol 2004;115:356–360.
78. Desiato MT, Caramia MD. Towards a neurophysiological marker of amyotrophic lateral sclerosis as revealed by changes in cortical excitability. Electroencephalogr Clin Neurophysiol 1997;105:1–7.
79. Enterzari-Taher M, Eisen A, Stewart H, et al. Abnormalities of cortical inhibitory neurons in amyotrophic lateral sclerosis. Muscle Nerve 1997;20:65–71.
80. Triggs WJ, Menkes D, Onorato J, et al. Transcranial magnetic stimulation identifies upper motor neuron involvement in motor neuron disease. Neurology 1999;53:605–611.
81. Pouget J, Trefouret S, Attarian S. Transcranial magnetic stimulation (TMS): compared sensitivity of different motor response parameters in ALS. Amyotroph Lateral Scler Other Motor Neuron Disord 2000;1(Suppl 2):S45–S49.
82. Eisen A, Pant B, Stewart H. Cortical excitability in amyotrophic lateral sclerosis: A clue to pathogenesis. Can J Neurol Sci 1993;20:11–16.
83. Prout AJ, Eisen AA. The cortical silent period and amyotrophic lateral sclerosis. Muscle Nerve 1994;17:217–223.
84. Salerno A, Georgesco M. Modifications des divers paramètres du potentiel évoque moteur dans la sclérose latérale amyotrophique. Neurophysiol Clin 1996;26:227–235.
85. Siciliano G, Manca ML, Sagliocco L, et al. Cortical silent period in patients with amyotrophic lateral sclerosis. J Neurol Sci 1999;169:93–97.
86. Desiato MT, Bernardi G, Hagi HA, et al. Transcranial magnetic stimulation of motor pathways directed to muscles supplied by cranial nerves in amyotrophic lateral sclerosis. Clin Neurophysiol 2002;113:132–140.
87. Mills KR. The natural history of central motor abnormalities in amyotrophic lateral sclerosis. Brain 2003; 126:2558–2566.
88. Brown WF, Ebers GC, Hudson AJ, et al. Motor evoked responses in primary lateral sclerosis. Muscle Nerve 1992;15:626–629.
89. Kuipers-Upmeijer J, de Jager AE, Hew JM, et al. Primary lateral sclerosis: clinical, neurophysiological, and magnetic resonance findings. J Neurol Neurosurg Psychiatry 2001;71:615–620.

90. Schnider A, Hess CW, Koppi S. Central motor conduction in a family with hereditary motor and sensory neuropathy with pyramidal signs (HMSN V). J Neurol Neurosurg Psychiatry 1991;54:511–515.
91. Pelosi L, Lanzillo B, Perretti A, et al. Motor and somatosensory evoked potentials in hereditary spastic paraplegia. J Neurol Neurosurg Psychiatry 1991;54:1099–1102.
92. Schady W, Dick JP, Sheard A, et al. Central motor conduction studies in hereditary spastic paraplegia. J Neurol Neurosurg Psychiatry 1991;54:775–779.
93. Mathis J, Hess CW. Motor-evoked potentials from multiple target muscles in multiple sclerosis and cervical myelopathy. Eur J Neurol 1996;3:567–573.
94. Maertens de Noordhout A, Myressiotis S, Delvaux V, et al Motor and somatosensory evoked potentials in cervical spondylotic myelopathy. Electroencephalogr Clin Neurophysiol 1998;108:24–31.
95. Lyu RK, Tang LM, Chen CJ, et al. The use of evoked potentials for clinical correlation and surgical outcome in cervical spondylotic myelopathy with intramedullary high signal intensity on MRI. J Neurol Neurosurg Psychiatry 2004;75:256–261.
96. Chang CW, Lien IN. Predictability of surgical results of herniated disc-induced cervical myelopathy based on spinal cord motor conduction study. Neurosurg Rev 1999;22:107–111.
97. Maertens de Noordhout A, Remacle JM, Pepin JL, et al. Magnetic stimulation of the motor cortex in cervical spondylosis. Neurology 1991;41:75–80.
98. Brunhölzl C, Claus D. Central motor conduction time to upper and lower limbs in cervical cord lesions. Arch Neurol 1994;51:245–249.
99. Bednarik J, Kadanka Z, Vohanka S, et al. The value of somatosensory and motor evoked potentials in pre-clinical spondylotic cervical cord compression. Eur Spine J 1998;7:493–500.
100. Di Lazzaro V, Restuccia D, Colosimo C, et al. The contribution of magnetic stimulation of the motor cortex to the diagnosis of cervical spondylotic myelopathy. Correlation of central motor conduction to distal and proximal upper limb muscles with clinical and MRI findings. Electroencephalogr Clin Neurophysiol 1992;85:311–320.
101. Tavy DLJ, Wagner GL, Keunen RWM, et al. Transcranial magnetic stimulation in patients with cervical spondylotic myelopathy: Clinical and radiological correlations. Muscle Nerve 1994;17:235–241.
102. Taniguchi S, Tani T, Ushida T, et al. Motor evoked potentials elicited from erector spinae muscles in patients with thoracic myelopathy. Spinal Cord 2002;40:567–573.
103. Cariga P, Catley M, Nowicky AV, et al. Segmental recording of cortical motor evoked potentials from thoracic paravertebral myotomes in complete spinal cord injury. Spine 2002;27:1438–1443.
104. Zandrini C, Ciano C, Alfonsi E, et al. Abnormalities of central motor conduction in asymptomatic HIV-positive patients. Significance and prognostic value. Acta Neurol 1990;12:296–300.
105. Moglia A, Zandrini C, Alfonsi E, et al. Neurophysiological markers of central and peripheral involvement of the nervous system in HIV infection. Clin Electroencephalogr 1991;22:193–198.
106. Curt A, Keck ME, Dietz V. Functional outcome following spinal cord injury: significance of motor-evoked potentials and ASIA scores. Arch Phys Med Rehabil 1998;79:81–86.
107. Claus D, Harding AE, Hess CW, et al. Central motor conduction in degenerative ataxic disorders: a magnetic stimulation study. J Neurol Neurosurg Psychiatry 1988;51:790–795.
108. Cruz Martinez A, Anciones B. Central motor conduction to upper and lower limbs after magnetic stimulation of the brain and peripheral nerve abnormalities in 20 patients with Friedreich's ataxia. Acta Neurol Scand 1992;85:323–326.
109. Mondelli M, Rossi A, Scarpini C, et al. Motor evoked potentials by magnetic stimulation in hereditary and sporadic ataxia. Electromyogr Clin Neurophysiol 1995;35:415–424.
110. Perretti A, Santoro L, Lanzillo B, et al. Autosomal dominant cerebellar ataxia type I: multimodal electrophysiological study and comparison between SCA1 and SCA2 patients. J Neurol Sci 1996;142:45–53.
111. Abele M, Burk K, Andres F, et al. Autosomal dominant cerebellar ataxia type I. Nerve conduction and evoked potential studies in families with SCA1, SCA2 and SCA3. Brain 1997;120:2141–2148.
112. Abbruzzese G, Morena M, Dall'Agata D, et al. Motor evoked potentials (MEPs) in lacunar syndromes. Electroencephalogr Clin Neurophysiol 1991;81:202–208.
113. Heald A, Bates D, Cartlidge NE, et al. Longitudinal study of central motor conduction time following stroke. 2. Central motor conduction measured within 72 h after stroke as a predictor of functional outcome at 12 months. Brain 1993;116:1371–1385.
114. Catano A, Houa M, Caroyer JM, et al. Magnetic transcranial stimulation in non-haemorrhagic sylvian strokes: interest of facilitation for early functional prognosis.

Electroencephalogr Clin Neurophysiol 1995;97:349–354.
115. Turton A, Wroe S, Trepte N, et al. Contralateral and ipsilateral EMG responses to transcranial magnetic stimulation during recovery of arm and hand function after stroke. Electroencephalogr Clin Neurophysiol 1996;101:316–328.
116. Rapisarda G, Bastings E, Maertens de Noordhout A, et al. Can motor recovery in stroke patients be predicted by early transcranial magnetic stimulation? Stroke 1996;27:2191–2196.
117. Cruz Martinez A, Tejada J, Diez Tejedor E. Motor hand recovery after stroke. Prognostic yield of early transcranial magnetic stimulation. Electromyogr Clin Neurophysiol 1999;39:405–410.
118. Schwarz S, Hacke W, Schwab S. Magnetic evoked potentials in neurocritical care patients with acute brainstem lesions. J Neurol Sci 2000;172:30–37.
119. Bassetti C, Mathis J, Hess CW. Multimodal electrophysiological studies including motor evoked potentials in patients with locked-in syndrome: report of six patients. J Neurol Neurosurg Psychiatry 1994;57:403–406.
120. Tsai SY, Tchen PH, Chen JD. The relation between motor evoked potential and clinical motor status in stroke patients. Electromyogr Clin Neurophysiol 1992;32:615–620.
121. Braune HJ, Fritz C. Transcranial magnetic stimulation-evoked inhibition of voluntary muscle activity (silent period) is impaired in patients with ischemic hemispheric lesion. Stroke 1995;26:550–553.
122. Ahonen JP, Jehkonen M, Dastidar P, et al. Cortical silent period evoked by transcranial magnetic stimulation in ischemic stroke. Electroencephalogr Clin Neurophysiol 1998;109:224–229.
123. Werhahn KJ, Classen J, Benecke R. The silent period induced by transcranial magnetic stimulation in muscles supplied by cranial nerves: normal data and changes in patients. J Neurol Neurosurg Psychiatry 1995;59:586–596.
124. Catano A, Houa M, Noël P. Magnetic transcranial stimulation: dissociation of excitatory and inhibitory mechanisms in acute strokes. Electroencephalogr Clin Neurophysiol 1997;105:29–36.
125. Hömberg V, Lange HW. Central motor conduction to hand and leg muscles in Huntington's disease. Mov Disord 1990;5:214–218.
126. Meyer BU, Noth J, Lange HW, et al. Motor responses evoked by magnetic brain stimulation in Huntington's disease. Electroencephalogr Clin Neurophysiol 1992;85:197–208.
127. Abbruzzese G, Marchese R, Trompetto C. Sensory and motor evoked potentials in multiple system atrophy: a comparative study with Parkinson's disease. Mov Disord 1997;12:315–321.
128. Abbruzzese G, Tabaton M, Morena M, et al. Motor and sensory evoked potentials in progressive supranuclear palsy. Mov Disord 1991;6:49–54.
129. Hefter H, Roick H, von Giesen HJ, et al. Motor impairment in Wilson's disease. 3. The clinical impact of pyramidal tract involvement. Acta Neurol Scand 1994;89:421–428.
130. Meyer BU, Britton TC, Benecke R, et al. Motor responses evoked by magnetic brain stimulation in psychogenic limb weakness: diagnostic value and limitations. J Neurol 1992;239:251–255.
131. Chu NS. Motor evoked potentials with magnetic stimulation: correlations with height. Electroencephalogr Clin Neurophysiol 1989;74:481–485.
132. Eisen AA, Shtybel W. Clinical experience with transcranial magnetic stimulation. Muscle Nerve 1990;13:995–1011.
133. Muellbacher W, Mathis J, Hess CW. Electrophysiological assessment of central and peripheral motor routes to the lingual muscles. J Neurol Neurosurg Psychiatry 1994;57:309–315.
134. Urban PP, Heimgärtner I, Hopf HC. Stimulation der Zungenmuskulatur bei Gesunden und Patienten mit Encephalomyelitis disseminata. Z EEG EMG 1994;25:254–258.
135. Pavesi G, Macaluso GM, Tinchelli S, et al. Magnetic motor evoked potentials (MEPs) in masseter muscles. Electromyogr Clin Neurophysiol 1991;31:303–309.
136. Türk U, Rösler KM, Mathis J, et al. Assessment of motor pathways to masticatory muscles: an examination technique using electrical and magnetic stimulation. Muscle Nerve 1994;17:1271–1277.
137. Tobimatsu S, Sun SJ, Fukui R, et al. Effects of sex, height and age on motor evoked potentials with magnetic stimulation. J Neurol 1998;245:256–261.

Evaluation of Myelopathy, Radiculopathy, and Thoracic Nerve

Vincenzo Di Lazzaro and Antonio Oliviero

Myelopathies and radiculopathies are common disorders. About 14% of 1,688 patients referred to our laboratory for motor evoked potential (MEPs) studies over a period of 3 years had a disorder of one of these types.[1]

The diagnosis of myeloradiculopathies often requires the execution of a neuroradiological study, and the enormous progress in the field of neuroradiology has been associated with an expanding role of neuroimaging techniques at the expense of neurophysiological techniques. However, if it is true that neuroradiologic examination represents the procedure of choice for the recognition of the causes of most of these disorders, it is also true that a correct interpretation of the data provided by neuroradiological studies can be achieved only if morphological findings are correlated with functional data. The first step in the diagnosis of spinal and radicular disorders should be, as for any other disorder, history taking and the clinical examination, and neuroradiological study should be considered as the final step that may demonstrate or exclude extrinsic or intrinsic lesions involving spinal cord or roots. The bridge between these two approaches is the neurophysiological evaluation. The main role of neurophysiology is to define the locus of the disease within the nervous system, particularly when this is not possible on a clinical basis alone, to perform a targeted neuroradiological study when this is indicated.

Consider the case of a patient with a pure spastic paraparesis without any other neurological symptom or sign. Many different disorders, such as spinal cord compression at any level, multiple sclerosis, motoneuron disease with predominant upper motoneuron involvement, or conditions such as hereditary spastic paraparesis, may manifest in this way. In this case, MEP studies can help in localizing the structures of the nervous system involved, showing, for instance, a subclinical involvement of upper limb central motor pathways and thereby pointing to brain and cervical spinal cord as the site of main interest for a neuroradiological study. MEP studies also can reveal characteristic patterns of nonspecific abnormalities that are commonly associated with certain neurological disorders. Examples are given for disorders such as cervical spondylotic myelopathy, motoneuron diseases, and hereditary spastic paraparesis.

Neurophysiological evaluation also helps the clinician ascertain if there is a functional correlate for any minor change demonstrated by neuroradiological studies and whether this change may be responsible for the patient's illness. This is particularly relevant for the early phases of many myeloradicular disorders that are characterized by very mild symptoms or when only minor radiological changes are evident.

For example, the common complaint of stiff and heavy legs and lower limb fatigability in an old patient with no clear neurological signs on examination may have many different causes involving different parts of the nervous system. Performing an extensive neuroradiological study to seek the cause of this symptom may yield information that is difficult to interpret. If the patient undergoes magnetic resonance imaging (MRI) of the brain and spine, it may demonstrate cervical and lumbar spondylosis, which are common conditions in elderly people, or small cerebrovascular lesions that may be present even when there is no clear history of cerebral ischemia. On the basis of clinical findings alone, it would be very difficult to establish whether one or a combination of two minor changes shown by MRI could be responsible for the symptoms of the patient. It is well known that there is no direct correlation between the degree of spinal canal stenosis and the degree of spinal cord and root dysfunction, and severe stenosis of the spinal canal does not necessarily entail a significant myeloradicular dysfunction. Moreover, despite the presence of myeloradiculopathy and cerebrovascular disease, other diseases such as motoneuron disease or subacute combined degeneration may still be responsible for the symptoms.

The neurophysiological study, providing a functional exploration of central and peripheral nervous system, may identify the site of the dysfunction (e.g., brain, cervical, thoracic or lumbar spinal cord, peripheral nervous system) and confirm the etiologic role of any of the abnormalities eventually demonstrated by neuroradiological study.

In an integrated diagnostic process based on clinical, neurophysiological, and neuroimaging data, MEPs represent a fundamental test (for brain and spinal disorders), a useful ancillary test (for polyradiculopathies), or a study of doubtful value (for monoradiculopathies). In the latter case, other well-established neurophysiological methods such as electromyography should be employed.

■ Myelopathy

Myelopathy is the neurological disorder most extensively investigated by MEPs (Table 7–1).[1–33] All MEP studies have demonstrated a high sensitivity for revealing corticospinal tract involvement in different spinal cord disorders. A study including a large number of patients with myelopathies[1] (>170 patients) confirmed the high sensitivity (about 0.85) of MEPs in assessing myelopathies. This series also demonstrated that MEP studies might document a subclinical involvement of central motor pathways in 12% to 13% of patients. However, the main contribution of MEPs in the diagnosis of myelopathies is to reveal the site of spinal cord lesion, allowing a multilevel assessment of the spinal cord through the determination of central motor conduction to different myotomes from the cervical to sacral level.

To improve the capability of MEPs in demonstrating the site of the functional involvement in the nervous system, it is important to record from multiple muscles corresponding to different spinal levels and to choose the correct technique for central motor conduction time determination. Central motor conduction time is usually calculated by subtracting the latency of muscle responses evoked by magnetic paravertebral stimulation from the latency of responses evoked by brain stimulation. The calculated central motor conduction time includes the time of conduction along the most proximal part of the motor root because magnetic paravertebral stimulation activates the motor axons of peripheral nerves near their exit from the intervertebral neuroforamina.[34–36] The component of central motor conduction time due to conduction along motor roots is represented by the conduction time between the anterior horn cells and the actual site of activation of peripheral axons at the neuroforaminal level. A delay of conduction along the motor roots therefore may determine a prolongation of central motor conduction time that is peripheral in origin rather than pyramidal. For this reason, when the central motor conduction time calculated using magnetic paravertebral stimulation is prolonged in patients with suspected motoneuron disease, radiculopathies, and in general in all patients with clinical or neurophysiologic (i.e., neurographic or electromyographic) evidence of lower motoneuron involvement, it is important to evaluate if a delay of conduction along the motor roots is responsible for or contributes to the abnormality. For this reason, it is useful to calculate the central motor conduction time in a different way—as the difference between the latency of responses

Table 7-1 MOTOR EVOKED POTENTIAL STUDIES IN MYELOPATHY

Studies	Year	Diagnosis	Patient Number	MEP Sensitivity	MEP Subclinical Abnormalities (%)
Thompson et al.[2]	1987	Cervical compressive myelopathy	6	0.83	0
Abbruzzese et al.[3]	1988	Cervical spondylosis	30	0.77	NR
Maertens de Noordhout et al.[4]	1991	Cervical spondylosis	67	0.51	11
Di Lazzaro et al.[5]	1992	Cervical spondylotic myelopathy	24	1	0
Herdmann[6]	1992	Cervical spondylotic myelopathy	15	1	0
Tavy et al.[7]	1994	Cervical spondylotic myelopathy	28	0.96	25
Chistyakov et al.[8]	1995	Cervical spondylotic myelopathy	22	0.86	0
Maertens de Noordhout et al.[9]	1998	Cervical spondylotic myelopathy	55	0.93	0
Bednarik et al.[10]	1998	Cervical spondylotic myelopathy, preclinical	30	0.37	37
Travlos et al.[11]	1992	Cervical spondylotic myelopathy, preclinical	23	0.65	52
Tavy et al.[12]	1999	Cervical cord compression, asymptomatic	25	0.08	8
De Mattei et al.[13]	1993	Compressive myelopathies	42	0.74	10
Linden et al.[14]	1994	Myelopathies	40	0.68	20
Di Lazzaro et al.[1]	1999	Myelopathies	176	0.85	12.5
Brunholzl et al.[15]	1994	Cervical cord lesions	47	0.63	17
Dvorak et al.[16]	1990	Cervical spine disorders	268	0.68	NR
Misra et al.[17]	1996	Pott's paraplegia	7	1	0
Kalita et al.[18]	2000	Myelitis	39	0.9	NR
Linden et al.[19]	1996	Spinal arteriovenous malformations	18	0.78	0
Di Lazzaro et al.[20]	1997	Ischemic myelopathy	3	1	0
Clarke et al.[21]	1994	Spinal cord injury	10	1	NR
Bondurant et al.[22]	1997	Spinal cord injury	9	1	NR
Curt et al.[23]	1998	Spinal cord injury	70	0.97	NR
Di Lazzaro et al.[24]	1992	Subacute combined degeneration	2	1	0
Nogues et al.[25]	1992	Syringomyelia	13	0.77	0
Masur et al.[26]	1992	Syringomyelia	22	0.68	23
Botzel et al.[27]	1993	Syringomyelia	17	0.44	NR
Stetkarova et al.[28]	2001	Syringomyelia	4	0.25	NR
Claus et al.[29]	1990	Hereditary spastic paraplegia	10	0.2	0
Caramia et al.[30]	1991	Hereditary spastic paraplegia	3	1	0
Pelosi et al.[31]	1991	Hereditary spastic paraplegia	11	1	0
Schady et al.[32]	1991	Hereditary spastic paraplegia	25	0.83	0
Restuccia et al.[33]	1997	Adrenomyeloneuropathy	6	1	0

MEP, motor evoked potential; NR, not reported.

evoked by brain magnetic stimulation and the total peripheral conduction time from the anterior horn cells to muscles. The total peripheral conduction time can be estimated from the latency of the F wave and of the distal motor response (M response) using the formula of Kimura,[37] whereby the F-wave and M-response latencies are added; 1 ms, corresponding to the turn-around time at the anterior horn cells, is subtracted from the total; and the remainder is divided by two. The central motor conduction calculated using the F-wave method enables a selective exploration of fast-conducting corticospinal projections to all muscles in which an F wave is elicitable. The difference between the two differently calculated central conduction times represents the *root motor conduction time*. When the central motor conduction time calculated with respect to the latency of motor responses evoked by magnetic paravertebral stimulation is prolonged while the central motor conduction time calculated with respect to the latency of the F wave is normal, the delay may be attributed to proximal motor root damage with a normal central motor pathway function. This is particularly relevant for lower limb muscles because the roots supplying these muscles have a long intrathecal segment, and the radicular component of the central conduction evaluated using magnetic paravertebral stimulation therefore represents a significant percentage of the total central conduction time.

Compressive Myelopathy

The most frequently observed myelopathy is cervical spondylotic myelopathy, which represents in our experience about 50% of all forms of spinal cord disorders. For assessment of the cervical spinal cord, it is useful to record MEPs from multiple upper limb muscles corresponding to different segmental levels.

Almost all patients with cervical spondylotic myelopathy present an abnormality of the central motor conduction time for distal upper limb muscles and for lower limb muscles. When the abnormality of central motor conduction for distal upper limb muscles and lower limb muscles is associated with a normal central motor conduction time for a muscle supplied by higher cervical segments such as biceps brachii, a lower cervical spinal cord lesion is highly probable. An example in one patient with cervical cord spondylotic compression at the C5-6 level is shown in Figure 7–1.

However, MEP studies demonstrate only the segmental level of cord dysfunction, and this does not necessarily correspond exactly to the level of cord compression. A strict correspondence between the site of cervical cord compression documented by the neuroradiological study and the level of cord lesion documented by neurophysiological study is not clear-cut in all patients. In some patients, the abnormalities may be confined to distal

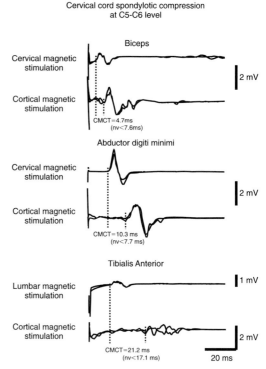

Figure 7–1 Motor evoked potentials were recorded from the biceps brachii, abductor digiti minimi, and tibialis anterior muscles after cortical and paravertebral magnetic stimulation in a patient with cervical cord spondylotic compression at the C5 to C6 level. The latency of responses evoked by magnetic paravertebral stimulation is normal for all muscles. Central motor conduction time is normal for the biceps brachii, but it is prolonged for the abductor digiti minimi and tibialis anterior muscles. nv, normal value.

upper limb muscles even when the narrowing of the spinal canal also involves the upper cervical segments.[5] In some patients with a high cervical cord compression, the central motor conduction time is prolonged for biceps brachii and distal hand muscles, whereas in others, the abnormalities may be confined to distal upper limb muscles. The former usually have a single compression level such as a central disc herniation, and the latter usually have a multilevel compression of the cervical cord such as a cervical canal stenosis documented by MRI.[4] The discrepancies sometimes observed between the levels of cord compression and the levels of cord dysfunction documented by MEP study suggest that there may be more than one mechanism of cord damage in cases of cervical spondylotic myelopathy. Segmental demyelination of central motor pathways due to mechanical cord compression is probably the most important factor in the patients for whom there is a strict correlation between neuroradiological and neurophysiological findings. The more caudal electrophysiological involvement of the cervical spinal cord with respect to the site of cord compression documented by MRI as observed in some patients is difficult to explain on the basis of direct compression only, and other mechanisms should be considered, such as decreased blood supply to the cord. The more frequent involvement of lower cervical segments probably reflects their higher vulnerability to ischemic damage. The anterolateral regions of lower cervical segments receive their blood supply almost exclusively from the anterior spinal artery, whereas the higher cervical segments are located between the cervical and intracranial arterial territories and therefore have more sources of blood supply. When cervical spondylotic compression involves the anterior spinal artery, the major damage is vascular in nature and localized to lower cervical segments, irrespective of the levels of spondylotic change.

Because MEPs can reveal only the site and not the nature of cord dysfunction, besides cervical cord spondylotic compression, several different causes such as neoplastic compression, segmental myelitis, a spinal form of multiple sclerosis, and intraspinal tumors should also be considered when there is electrophysiological evidence of lower cervical cord dysfunction. However, these disorders have much lower incidences. It should be also considered that other disorders such as motoneuron disease rarely may result in a similar pattern of central conduction abnormalities. Of the last 100 consecutive patients with motoneuron disease observed in our laboratory, only 6 presented with abnormal central motor conduction for distal upper limb and lower limb muscles with normal conduction for proximal upper limb muscles. Conversely, this pattern was found in 80 of the last 100 consecutive patients with unequivocal cervical spondylotic myelopathy documented by MRI (unpublished observations).

MEPs can provide evidence of cervical spinal cord damage even in patients with cervical spondylotic myelopathy presenting with pure spastic paraparesis. In these patients, who represent about 15% of patients with cervical spondylotic myelopathy,[5] MEP study may reveal a subclinical involvement of corticospinal projections to cervical cord myotomes. The abnormality of central motor conduction for distal upper limb muscles together with all muscles supplied by more caudal myelomeres in the presence of a normal conduction for biceps muscle is more frequently observed in cervical spondylotic myelopathy, and in any case, this pattern of abnormality makes mandatory an MR examination of the cervical cord.

When central motor conduction is also abnormal for the biceps muscle, the test yields limited information about the site of the lesion because there are several conditions compatible with this pattern, such as a high cervical cord compression, an intracranial disorder involving central motor pathways, motoneuron disease, or a neurodegenerative disorder involving spinal cord. To differentiate a high cervical cord lesion from an intracranial disorder, it is useful to study central conduction for more cranial myotomes through, for instance, the study of MEPs of the trapezius muscle. In the case of high cervical cord compression, conduction to the trapezius is normal. An example in one patient with C3-C4 spondylotic cord compression is shown in Figure 7–2.

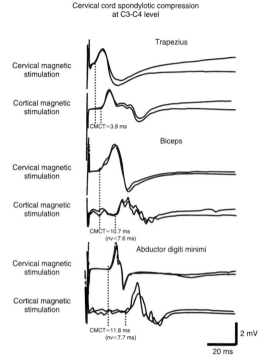

Figure 7–2 Motor evoked potentials were recorded from the trapezius, biceps brachii, and abductor digiti minimi muscles after cortical and paravertebral magnetic stimulation in a patient with cervical cord spondylotic compression at the C3 to C4 level. The latency of responses evoked by magnetic paravertebral stimulation is normal for all muscles. Central motor conduction time is normal for trapezius, but it is prolonged for biceps and abductor digiti minimi muscles.

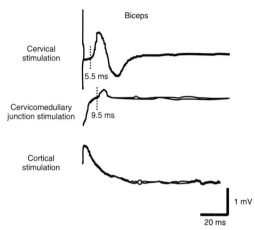

Figure 7–3 Motor evoked potentials were recorded from the biceps brachii after cortical and paravertebral magnetic stimulation and after cervicomedullary junction electrical stimulation in a patient with multiple cortical subcortical vascular lesion in the brain and cervical canal stenosis. The latency of the response evoked by magnetic paravertebral stimulation is normal. Cervicomedullary junction stimulation evokes a response with normal latency, but there is no response after cortical stimulation.

Another method for differentiating a high cervical cord lesion from an intracranial disorder is provided by the technique of cervicomedullary junction stimulation. Using this technique introduced by Ugawa and coworkers in 1991,[38,39] it is possible to evaluate selectively the intracranial and the spinal segments of central motor pathways of a muscle such as biceps brachii. After cervicomedullary junction stimulation, it is possible to calculate the conduction time from the cervicomedullary junction to the spinal segments supplying the muscle under study by subtracting the latency of the response evoked in the same muscle by paravertebral magnetic stimulation from the latency of responses evoked by cervicomedullary junction stimulation. An example is shown in Figure 7–3. This patient presented with a tetraparesis, and an MR study demonstrated abnormalities at the brain level, with multiple cortical and subcortical vascular lesions, and at cervical level, with a severe cervical canal stenosis. The relevance of the cervical canal stenosis was difficult to evaluate from the neuroradiological data. In this patient, cortical brain stimulation demonstrated the absence of motor response in the biceps brachii, and cervicomedullary junction stimulation evoked a small motor response of normal latency. Taken together, these data indicate that the intracranial lesions are more relevant as the cause of tetraparesis in this patient.

Much more challenging can be the differential diagnosis between a high cord lesion and motoneuron disease. Both conditions may manifest with a variable combination of upper and lower motoneuron signs. Any progress in the diagnosis obtainable with electrophysiological studies is therefore particularly relevant.

In the differential diagnosis between cord compression and motoneuron disease, the multilevel assessment of central motor conduction can be extremely useful. In the case of spinal compressive disorders, central motor conduction abnormalities are confined to all the muscles situated below the site of the spinal lesion sparing muscles supplied by more cranial myelomeres. In motoneuron disease, abnormalities demonstrated in certain muscles may be associated with a normal conduction to muscle innervated by more caudal myelomeres, resulting in a "suspended" abnormality of the central motor conduction. The most common type of suspended abnormality is represented by an abnormal conduction for the biceps brachii with normal central motor conduction for distal upper limb muscles and lower limb muscles.[1] An example in one patient with amyotrophic lateral sclerosis is shown in Figure 7–4. The relatively high incidence of the suspended abnormality of central motor conduction for biceps brachii, which represents about one third of all subclinical abnormalities of central motor conduction demonstrated by MEPs in patients with motoneuron diseases (26% of patients[1]), is probably caused by the limited number of fast-conducting corticospinal projections to the biceps brachii, which were explored by transcranial brain stimulation. Even the loss of a restricted number of corticospinal fibers in the early stage of the disease is sufficient to give an abnormality of central motor conduction for this muscle. Theoretically, because central motor conduction time evaluated by subtracting the latency of MEPs after magnetic paravertebral stimulation from the latency after brain stimulation also includes a small part due to conduction along the most proximal part of the motor root, a suspended abnormality of central motor conduction for biceps brachii can also be caused by a delay of conduction along the C5 and C6 roots. However, a multiradicular involvement at cervical level is uncommon, and in any case, a concomitant lesion of the C5 and C6 roots can be ruled out by clinical examination (i.e., retained biceps reflex) and electromyographic study of muscles supplied by C5 and C6 roots. In the absence of clear signs of C5 and C6 root lesion, the suspended abnormality of central motor conduction for biceps brachii strongly supports the diagnosis of a degenerative disorder of corticospinal projections.

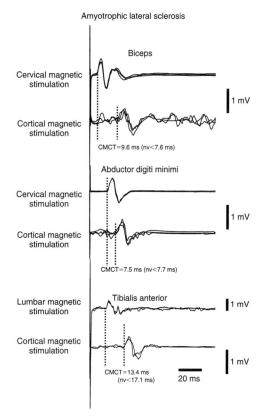

Figure 7–4 Motor evoked potentials were recorded from biceps brachii, abductor digiti minimi, and tibialis anterior muscles after cortical and paravertebral magnetic stimulation in a patient with amyotrophic lateral sclerosis. The latency of responses evoked by magnetic paravertebral stimulation is normal for all muscles. Central motor conduction time is prolonged for biceps brachii, but it is normal for abductor digiti minimi and tibialis anterior muscles.

The abnormality of central motor conduction for proximal upper limb muscles with normal conduction to distal upper limb muscles is exactly the opposite of the pattern of abnormalities more commonly observed in cervical spondylotic myelopathy. A suspended abnormality of central motor conduction also can be observed in distal upper limb muscles with a normal conduction for lower limb muscles (Fig. 7–5).

In all cases of suspended abnormality of central motor conduction for distal upper limb muscles, the evaluation of the central motor conduction with the F-wave method is mandatory for a correct localization of

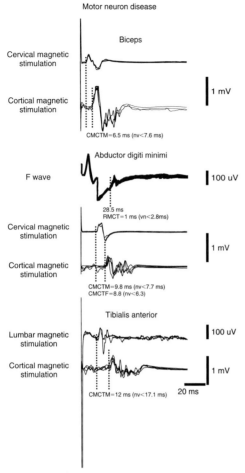
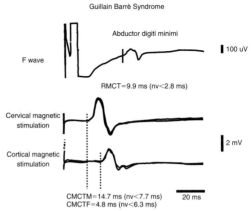

Figure 7–5 Motor evoked potentials were recorded from the biceps brachii, abductor digiti minimi, and tibialis anterior muscles after cortical and paravertebral magnetic stimulation in a patient who presented clinically with a lower motoneuron syndrome. The latency of responses evoked by magnetic paravertebral stimulation is normal for all muscles. The central motor conduction time of abductor digiti minimi muscle, calculated as the difference between the latency of responses evoked by cortical and paravertebral magnetic stimulation ($CMCT^M$) and calculated using the F-wave method ($CMCT^F$), is prolonged. The difference between these two parameters, corresponding to root motor conduction time (RMCT), is normal. Central motor conduction time is normal for the biceps and tibialis anterior muscles.

Figure 7–6 Motor evoked potentials were recorded from the abductor digiti minimi muscle after cortical and paravertebral magnetic stimulation in a patient with a Guillain-Barré syndrome. The central motor conduction time of abductor digiti minimi muscle, calculated as the difference between the latency of responses evoked by cortical and paravertebral magnetic stimulation ($CMCT^M$), is prolonged, but the central motor conduction time calculated using the F-wave method ($CMCT^F$) is normal. The difference between these two parameters, corresponding to root motor conduction time (RMCT), is extremely prolonged.

motor pathway dysfunction. If the suspended abnormality of central motor conduction for distal upper limb muscles is confirmed with the F-wave method, a diagnosis of amyotrophic lateral sclerosis is highly probable.

Figures 7–5 and 7–6 provide two examples of suspended abnormality of central motor conduction in distal hand muscles. In the first patient, the abnormality of central conduction is confirmed using the F-wave method, whereas in the second, the latter method results a finding of a normal central conduction, suggesting a peripheral delay. The first patient presented clinically with a pure lower motoneuron disease localized in cervical myotomes, a clinical presentation compatible with cervical spondylotic compression, but had a disease progression consistent with the diagnosis of amyotrophic lateral sclerosis, whereas the second patient presented with right hand paresthesias at the time of the MEP study and subsequently developed a clear Guillain-Barré syndrome.

MEPs may also be particularly useful in the differential diagnosis between spinal cord compression and primary lateral sclerosis. This condition, clinically characterized by variable degrees of tetraparesis, can simulate a spinal cord compressive disorder when bulbar signs are lacking. In primary lateral

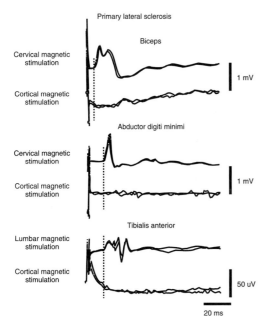

Figure 7–7 Motor evoked potentials were recorded from the biceps brachii, abductor digiti minimi, and tibialis anterior muscles after cortical and paravertebral magnetic stimulation in a patient with primary lateral sclerosis. The latency of responses evoked by magnetic paravertebral stimulation is normal for all muscles, but there are no responses after cortical stimulation.

sclerosis, the integration of MEP findings with clinical findings gives a very peculiar picture that can be of considerable value in the differential diagnosis. In almost all patients with this disorder, MEPs are extremely delayed or absent (Fig. 7–7), even in the early stage of the disease and in muscles with minimal motor deficit. The pronounced alteration of MEPs evoked by cortical stimulation in association with a disproportionate minimal degree of motor deficit represent a pattern of MEP abnormalities that can be considered quite specific for primary lateral sclerosis.

In the diagnosis of compressive myelopathies, together with the localization of the site of lesion, MEPs may also be useful in quantifying the degree of functional involvement of spinal cord. This is particularly relevant in patients with a suspected cervical spondylotic myelopathy and MR evidence of minor compression of the cord. Because the spinal cord shows a high capacity to adapt to spondylotic changes, a minor compression documented by MRI can have no significant effect on the spinal cord. Spondylotic changes are often demonstrated in patients undergoing a cervical MR study for cervical pain. When there is no clear evidence of cord damage and the patient shows only dubious clinical findings such as a tendency to hyperexcitable reflexes, MEPs may be extremely useful in revealing or ruling out cord dysfunction (Fig. 7–8). In all cases of cervical spondylotic myelopathy, functional exploration of spinal cord with MEPs demonstrates the actual dysfunction of the cord and may help in the management of patients. It has not been proved whether the outcome after surgery is better than the natural history of the disease in these patients.[40,41] MEP recording may help in the identification of patients with severe dysfunction of central motor pathways that are probably better candidates for surgery. Moreover, serial MEP recording may be useful in ascertaining the progressive forms of cervical spondylotic myelopathy and selecting patients who may benefit from surgical treatment. MEP studies may be useful in suggesting the surgical site in patients who have multilevel disc herniations (i.e., cervical and thoracic).

In the case of thoracic cord compression or lesion, MEP studies result in a normal conduction for upper limb muscles and abnormal conduction for all lower limb muscles. This pattern can be considered nonspecific because it can also be found in mainly intracranial disorders such as multiple sclerosis and in all conditions characterized by an axonal type of involvement of central motor pathways. The axonal involvement of central motor pathways that can be found in conditions such as hereditary spastic paraparesis and in a few patients with diabetes mellitus causes only a minor degree of conduction slowing along central motor pathways caused by the loss of large diameter axons. The abnormality of central conduction becomes evident only for the lower limb long tracts, whereas central conduction is normal for upper limbs. In cases of axonal central motor disorders, the central motor conduction for lower limbs exceeds the normal upper limits by few milliseconds, whereas in cord

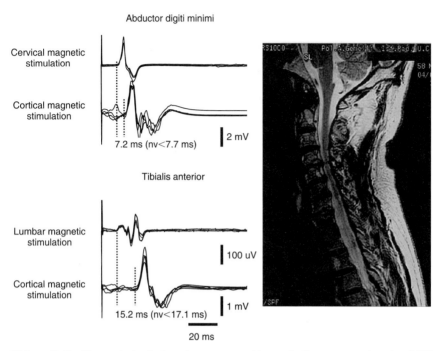

Figure 7–8 The cervical spinal cord was assessed by magnetic resonance imaging (MRI), and motor evoked potentials were recorded from the biceps brachii, abductor digiti minimi, and tibialis anterior muscles after cortical and paravertebral magnetic stimulation in a 58-year-old man with a 3-year history of cervical pain and a tendency to hyperexcitable tendon reflexes on clinical examination. Sagittal T2-weighted MRI shows degenerative changes of the cervical spine that result in moderate narrowing of the spinal canal with multiple disc protrusions. There is no signal modification of the spinal cord itself. MEPs are completely normal.

compression, prolongation of central conduction to the lower limbs is usually more pronounced because it is due to segmental demyelination of the cord. In the last 100 patients with cord compression, the central motor conduction time to the lower limbs exceeded the normal upper limit by a mean value of 6.5 ms (±4.5 SD). Much longer delays of central motor conduction to lower limbs or absences of cortical MEPs are usually observed in multiple sclerosis. The mean delay was 14 ms (±16.3 SD) of central motor conduction time in the last 100 consecutive patients with multiple sclerosis studied in our laboratory who had recordable lower limb responses (unpublished observations).

In an attempt to differentiate between a thoracic cord disorder and an intracranial demyelinating disorder presenting with an isolated abnormality of central motor conduction to lower limbs, it may be useful to record MEPs evoked by brain stimulation from paravertebral muscles at different levels along the spine.[42,43] For paraspinal recording, needle electrodes are preferable because they give clearer responses and help to avoid superficial back muscles.[43] In this way, it may be possible to verify if there is a definite level below which the latency of responses evoked by brain stimulation in paravertebral muscles is abnormal, suggesting a focal cord lesion or compression. However, because of the short distance from the spine, it is not possible to record radicular responses from paravertebral muscles, and the results obtained with this technique should be interpreted with caution because the central and peripheral delay in conduction may result in prolongation of latencies of paraspinal responses evoked by brain stimulation. Only delays of paravertebral MEPs associated with a delay of central motor conduction for lower limb muscles can be considered indicative of central motor pathway involvement.

To assess thoracic cord conduction, responses can also be recorded from intercostal muscles.[44,45] However, when these responses are recorded with surface electrodes, the onset is difficult to determine, whereas when they are recorded with needle electrodes, the onset may be clear, but there is the risk of pneumothorax. The study of different muscles innervated by the thoracic cord, such as the rectus abdominis or external oblique, may be useful because the roots of these muscles can be stimulated using magnetic stimulation,[46] and central motor conduction to the corresponding thoracic segments can be measured.

Clinical diagnosis of lumbosacral cord lesions is often difficult because upper motoneuron signs may be masked by the prevalence of lower motoneuron involvement, causing a misdiagnosis of a cauda equina syndrome. Conversely, lower motoneuron signs may be absent or very mild because of the prevalence of central motor pathway involvement, making it difficult to distinguish these lesions from those located above the lumbar enlargement. MEPs may be useful in localizing the site of the lesion if they are recorded from multiple lower limb muscles, corresponding to different cord levels, and if the technique of central motor conduction determination can differentiate a slowing along central motor pathways from a delay along lumbosacral motor roots. This usually requires the double determination of central motor conduction time, as the difference between the latencies of responses evoked by cortical and paravertebral magnetic stimulation and as the difference between the latency of cortical MEP and the total peripheral conduction time calculated from the F-wave latency. The root motor conduction time can be evaluated from the difference between the two differently calculated central conduction times. Several patterns of abnormalities can be recognized using this double determination.

When the central motor conduction time calculated from the latency of the response evoked by magnetic paravertebral stimulation and from the F wave and the root conduction time are prolonged, a myeloradicular

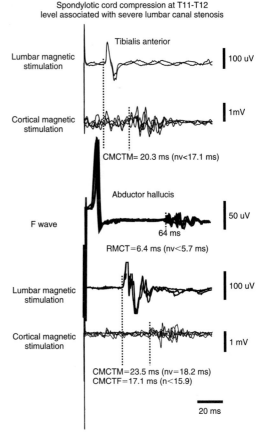

Figure 7–9 Motor evoked potentials were recorded from the tibialis anterior and abductor hallucis muscles after cortical and paravertebral magnetic stimulation in a patient with spondylotic cord compression at the T11 to T12 level associated with a severe lumbar canal stenosis. The central motor conduction time to the abductor hallucis muscle, calculated as the difference between the latency of responses evoked by cortical and paravertebral magnetic stimulation ($CMCT^M$) and calculated using the F-wave method ($CMCT^F$), is prolonged. The difference between these two parameters, corresponding to root motor conduction time (RMCT) and the latency of the F wave, is also prolonged. Central motor conduction time also is prolonged for the tibialis anterior muscle.

lumbosacral lesion is highly probable. An example in a patient with cord compression at T11 to T12 level associated with a severe lumbar canal stenosis is shown in Figure 7–9. In this patient, central motor conduction time is prolonged for the tibialis anterior and abductor hallucis. In the latter muscle, the root conduction time is also prolonged, suggesting a root dysfunction together with lumbosacral cord compression.

An abnormal central motor conduction for distal lower limb muscles confirmed by the F-wave method in association with normal root conduction time and normal central motor conduction for more proximal lower limb muscles suggests a lesion of the lowest part of the lumbosacral cord. An example of this pattern in one patient with spondylotic cord compression at the T12 to L1 level is shown in Figure 7–10.

An abnormal central motor conduction evaluated from the latency of the responses evoked by magnetic paravertebral stimulation in lower limb muscles, with a normal central motor conduction calculated from the latency of the F wave and a prolonged root conduction time, strongly suggests a dysfunction of multiple lumbosacral roots with a normal cord. Recordings from multiple muscles may help in the identification of roots involved.

An example in one patient with a severe central lumbar canal stenosis is shown in Figure 7–11. In this patient, central motor conduction is normal for the rectus femoris, but it is prolonged for the tibialis anterior and abductor hallucis when calculated from the latency of magnetic paravertebral stimulation. However, central motor conduction for these muscles is normal when calculated from the latency of the F wave, and the root conduction time is prolonged. This pattern strongly suggests a (L4)-L5-S1 radicular dysfunction. This pattern of abnormalities can be found even in patients with minor clinical manifestations as in neurogenic claudication due to lumbar canal stenosis. The contribution of MEPs can be particularly important in that this condition is difficult to diagnose clinically because it is characterized by intermittent symptoms and normal clinical examination, and it is also difficult to diagnose radiologically because abnormal lumbosacral MR is common even in asymptomatic subjects.

It is also possible to find a suspended abnormality of central motor conduction for a few myotomes with a normal conduction for more cranial and more caudal myotomes. In this situation, it is critical to evaluate central motor conduction using the F-wave method.

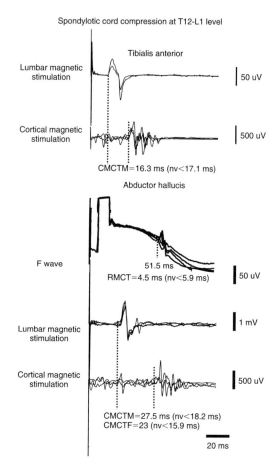

Figure 7–10 Motor evoked potentials were recorded from the tibialis anterior and abductor hallucis muscles after cortical and paravertebral magnetic stimulation in a patient with spondylotic cord compression at the T12 to L1 level. The central motor conduction time of abductor hallucis muscle, calculated as the difference between the latency of responses evoked by cortical and paravertebral magnetic stimulation (CMCT^M) and calculated using the F-wave method (CMCT^F), is prolonged. The difference between these two parameters, corresponding to root motor conduction time (RMCT), is normal. Central motor conduction time for the tibialis anterior muscle is normal.

In the case of root dysfunction, the findings are normal, but the root conduction time is prolonged. An example in one patient with a severe multilevel lumbar canal lateral recess stenosis more pronounced at the L4 and L5 levels is provided in Figure 7–12. In this patient, central motor conduction time is within normal limits for rectus femoris and

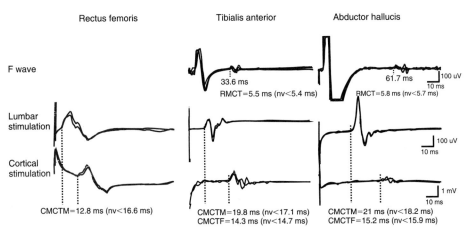

Figure 7-11 Motor evoked potentials were recorded from the rectus femoris, tibialis anterior, and abductor hallucis muscles after cortical and paravertebral magnetic stimulation in a patient with a severe central lumbar canal stenosis. The central motor conduction time of tibialis anterior and abductor hallucis muscles, calculated as the difference between the latency of responses evoked by cortical and paravertebral magnetic stimulation ($CMCT^M$), is prolonged, whereas the central motor conduction, calculated using the F-wave method ($CMCT^F$), is normal. The difference between these two parameters, corresponding to root motor conduction time (RMCT), is prolonged. The latency of the F wave recorded from the abductor hallucis also is prolonged. Central motor conduction time for the rectus femoris muscle is normal.

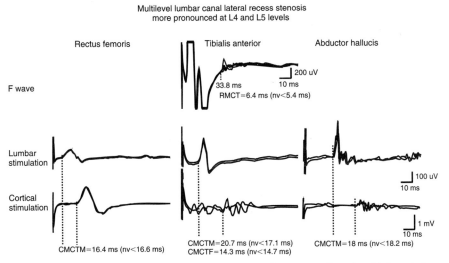

Figure 7-12 Motor evoked potentials were recorded from rectus femoris, tibialis anterior, and abductor hallucis muscles after cortical and paravertebral magnetic stimulation in a patient with a severe multilevel lumbar canal lateral recess stenosis that was more pronounced at the L4 and L5 levels. The central motor conduction time of tibialis anterior muscle, calculated as the difference between the latency of responses evoked by cortical and paravertebral magnetic stimulation ($CMCT^M$), is prolonged, whereas the central motor conduction, calculated using the F-wave method ($CMCT^F$), is normal. The difference between these two parameters, corresponding to root motor conduction time (RMCT), is prolonged. Central motor conduction time for the rectus femoris and abductor hallucis muscles is normal.

abductor hallucis, whereas central motor conduction to tibialis anterior muscle is prolonged when evaluated from the latency of the response evoked by magnetic paravertebral stimulation and within normal limits when calculated from the latency of the F wave. The root conduction time is prolonged. This pattern suggests L4 and L5 root dysfunction.

If MEP studies had been limited to the determination of central motor conduction to the tibialis anterior muscle using magnetic stimulation alone, three of the previously illustrated cases would result in an identical pattern with an abnormality of central motor conduction to the tibialis anterior muscle. It is therefore mandatory to perform a multilevel assessment of lower limb muscle central motor conduction and determine the multimodal central motor conduction time.

Spinal Cord Vascular Disorders

Because in the early phases of spinal cord ischemia MRI can be normal,[47] MEP study may be useful, along with clinical examination, in demonstrating spinal cord involvement and in localizing the site of central motor pathway dysfunction.[20] After hypoperfusion in the territory of the artery of Adamkiewicz, only lower limb MEPs are abnormal, whereas in case of an ischemic lesion in the territory of the anterior spinal artery, upper and lower limb MEPs are abnormal. MEPs may disclose spinal cord involvement even in the case of spinal transitory ischemic attacks.[20] However, spinal cord involvement may be obvious clinically in most cases, and the MEP contribution to the diagnosis therefore may be mainly confirmatory. The contribution of MEPs may be more relevant for spinal disorders resulting from dural arteriovenous malformations,[19] the clinical diagnosis of which may be very difficult. Together with a spinal cord lesion, there is often root damage caused by venous congestion, and the clinical manifestation in these cases is characterized by a preponderance of lower motoneuron signs and pain very similar to that of a lumbar canal stenosis. Because spondylotic changes of the lumbosacral column are a common condition in elderly people, an MR study limited to the lumbosacral column may lead to an incorrect attribution of the cause of the disorder to spondylotic changes. In these cases, MEPs are useful to document a lower spinal cord involvement necessitating the extension of the neuroradiological study to this region.

Syringomyelia

MEP abnormalities have been reported in 44%[27] to 77%[25] of patients with syringomyelia. MEPs may be normal, even in a patient with an extensive syrinx (Fig. 7–13), and a normal MEP therefore does not exclude this diagnosis.

Genetic Disorders Involving Spinal Cord

MEPs may be useful in the functional assessment of the spinal cord in genetic disorders that mainly involve the spinal cord, such as hereditary spastic paraparesis and spinal forms of adrenoleukodystrophy. The clinical presentation of hereditary spastic paraparesis may be similar to that of cord compression, and MEPs may be particularly useful in the differential diagnosis of sporadic cases of hereditary spastic paraparesis.

Although hyperreflexia of upper limbs is a common finding in these patients,[48] central motor conduction to hand muscles is completely normal in most cases.[32] Moreover, central motor conduction for lower limb muscles is usually only slightly prolonged, despite the often severe lower limb spasticity. An example from one patient is shown in Figure 7–14. The isolated and mild abnormality of lower limb central conduction can be explained on the basis of the major neuropathological feature of hereditary spastic paraparesis represented by axonal degeneration that is maximal in the terminal portions of the corticospinal tract. The marked discrepancy between clinical and electrophysiological findings that is peculiar to hereditary spastic paraparesis may be useful in the differential diagnosis with cervical spondylotic myelopathy. Although in cervical spondylotic myelopathy central motor conduction to hand muscles may be abnormal even in patients with mild or no upper motoneuron signs in the upper limbs, in hereditary spastic paraparesis, the opposite is generally the case. Because MEPs, like clinical examination, may be completely normal in

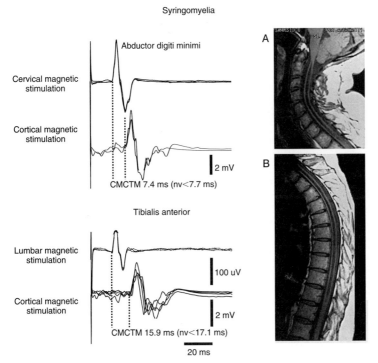

Figure 7–13 The spinal cord was assessed by magnetic resonance imaging (MRI), and motor evoked potentials were recorded from the abductor digiti minimi and tibialis anterior muscles after cortical and paravertebral magnetic stimulation in a 54-year-old woman. She had a 5-year history of sensory loss in her arms. Sagittal T1-weighted MRI of the cervical and thoracic spine showed a cervicothoracic syrinx demonstrated by minimal swelling of the cervical (**A**) and dorsal (**B**) cord from level C1 to T11 of the spine. MEPs are completely normal.

affected relatives of a patient, the test cannot be used to exclude the condition.

In spinal forms of adrenoleukodystrophy, MEPs may disclose neurological dysfunction even in the early stage of the disease.[33] In adrenomyeloneuropathy, MEPs demonstrate a higher percentage of abnormalities in lower limbs, and this correlates well with the pathologic finding of more pronounced involvement of longer axons of the spinal tracts with a "dying back" pattern.

■ Myelitis

In inflammatory disorders of the spinal cord such as acute transverse myelitis, MEPs can be used to localize the site of the lesion. An example for a segmental myelitis at the thoracic level is shown in Figure 7–15. MEPs are normal for upper limbs and prolonged for lower limb muscles; the recording from paravertebral muscles shows a normal conduction to paraspinal muscles at the T6 level, but no response is recorded at more caudal paravertebral sites. MEPs therefore may identify the site of cord lesion. In cases of myelitis, MEPs may show abnormalities even when MRI is normal.[13]

MEPs evaluating conduction along the spinal cord and roots may help to define the exact distribution of the inflammatory process, demonstrating whether there is isolated myelitis, myeloradiculitis, or isolated radiculitis. An example of myeloradiculitis with an abnormality of central and radicular conduction is given in Figure 7–16. In this patient, central motor conduction time to distal hand muscles is prolonged with both techniques of determination; the root conduction time is also prolonged.

The demonstration of spinal cord involvement may be useful in the diagnosis, because in the early phases of the disease, an acute

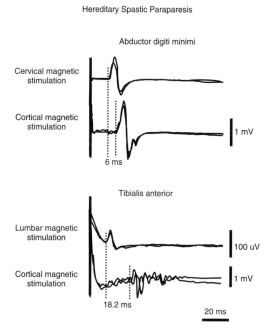

Figure 7–14 Motor evoked potentials were recorded from the abductor digiti minimi and tibialis anterior muscles after cortical and paravertebral magnetic stimulation in a patient with hereditary spastic paraparesis. Central motor conduction for abductor digiti minimi muscle is normal, whereas central motor conduction for tibialis anterior is slightly prolonged.

myelitis may exceptionally mimic the pattern of Guillain-Barré syndrome.

In certain forms of myelitis, such avascular myelopathy associated with acquired immunodeficiency syndrome (AIDS) and, less frequently, tropical spastic paraparesis due to HTLV-1 virus, symptoms and signs of spinal cord involvement may be obscured by a concomitant neuropathy. In these cases, MEP abnormalities demonstrate cord involvement that is usually localized at the thoracic level.[43]

Subacute Combined Degeneration

The neurological manifestations of vitamin B_{12} deficiency primarily result from the degeneration of spinal cord white matter. The functional exploration of the central motor pathways in subacute combined degeneration may reveal a spinal cord dysfunction even in the early stages of the disease,[24] when the clinical manifestation may be limited to brisk tendon reflexes.

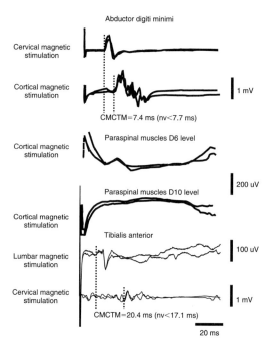

Figure 7–15 The cervical spinal cord was assessed by magnetic resonance imaging, and motor evoked potentials were recorded from the abductor digiti minimi and tibialis anterior muscles after cortical and paravertebral magnetic stimulation and from the paraspinal muscles at the T6 and T10 levels in a patient with thoracic segmental myelitis. Normal responses are recorded from abductor digiti minimi muscle and from paraspinal muscles at the T6 level, whereas no responses are recorded from paraspinal muscles at the T10 level. The central motor conduction time for tibialis anterior muscle is prolonged.

Spinal Cord Injury

MEP recording can supplement clinical examination in assessment of the level, the extent, and the severity of corticospinal tract lesions in patients with spinal cord injury. The recording from multiple muscles can be used to identify the level of spinal cord lesion, and this is particularly useful in uncooperative and unconscious patients. However, although upper limb MEPs are easily obtained in all normal subjects at rest, lower limb MEPs may be absent at rest, even in a few normal subjects.[5] This results from slight differences in the way of activation of hand and lower

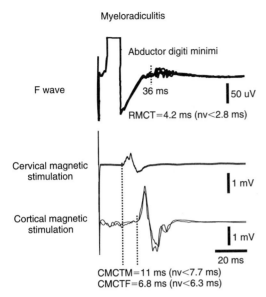

Figure 7–16 Motor evoked potentials were recorded from abductor digiti minimi muscle after cortical and paravertebral magnetic stimulation in a patient with a myeloradiculitis. The central motor conduction time, calculated as the difference between the latency of responses evoked by cortical and paravertebral magnetic stimulation (CMCT^M) and calculated using the F-wave method (CMCT^F), is prolonged. The difference between these two parameters, corresponding to root motor conduction time (RMCT) and the F wave latency, is also prolonged.

limb motor area using magnetic transcranial stimulation, with a larger effect of voluntary contraction on threshold for lower limb motor responses.[49,50] For this reason, the contribution of MEPs in the assessment of thoracolumbar spinal cord injuries is more limited than for cervical spinal cord injuries.

MEPs may be useful in the prediction of functional outcome in the acute phase of spinal cord injuries (reviewed by Curt and Dietz[51]). In a study of Curt and coworkers,[23] MEP recording from abductor digiti minimi muscle was highly correlated to the outcome of hand function. When abductor digiti minimi MEPs were absent, active hand function was not regained. For the reasons outlined previously, lower limb MEPs were less effective in predicting the recovery of ambulatory capacity, and there were some patients with an initial loss of lower limb MEPs who showed a recovery of ambulatory capacity.[23]

Radiculopathy and Thoracic Nerve

Roots are among the least accessible structures of the peripheral nervous system. Late responses can be used to assess conduction along the roots, but abnormalities of late responses are not specific for the diagnosis of radiculopathies because they are conducted over the entire length of peripheral motor pathways and lesions at any level determine similar changes of late responses. Moreover, the F-wave latency measurement is not very sensitive in revealing a slowing of conduction along the motor roots because the slowing of conduction in the short segment of compressed root is diluted in a much longer segment of normal conduction along all the remaining peripheral motor pathways. In root lesions, needle electromyographic studies may document fibrillation potentials in the corresponding paraspinal muscles or myotomes, but these electromyographic changes appear only after 2 to 3 weeks. Because the noninvasive and painless magnetic paravertebral stimulation activates the proximal part of peripheral motor pathways, it may be useful in addition to the previously mentioned traditional techniques in the diagnosis of radiculopathies and in the assessment of those peripheral nerves, such as the thoracic nerves, that cannot be directly explored using standard electrophysiological techniques.

Magnetic paravertebral stimulation activates the motor axons of peripheral nerves near their exit from the spine. This site is distal to that of root compression produced by disc disorders or spondylotic changes, and the latency of responses evoked by magnetic paravertebral stimulation is normal in the case of root compression. The conduction time along the proximal part of motor roots can be calculated by subtracting the latency of motor responses evoked by magnetic paravertebral stimulation from the total peripheral conduction time calculated using the formula of Kimura[37] $[(F + M - 1)/2]$.[52,53] The mean root motor conduction time is 1.4 ms for distal upper limb muscles and 3 ms for distal lower limb muscles. A pathological increment of the motor root conduction time suggests a demyelination in this portion of the peripheral nervous system. Its assessment may be particularly useful in pathological conditions that involve proximal motor segments, such as spondylotic compression of roots and root involvement in inflammatory peripheral nerve disorders that may cause focal demyelination.

Magnetic paravertebral stimulation cannot be used to document proximal conduction block because available magnetic stimulators evoke submaximal motor responses. Moreover, the involvement of a single motor root usually does not determine a significant delay of conduction because muscles are supplied by more than one root and a potential slowing along a single root is overcome by a normal conduction through the intact roots.

To determine a delay of conduction along the motor roots, multiple roots should be involved, and this is common in central disc herniations or lumbar canal stenosis. In these cases, determination of central motor conduction from the latency of the response evoked by magnetic paravertebral stimulation and from the latency of the F wave may demonstrate functional involvement of the lumbosacral roots. The central motor conduction time calculated from the latency of responses evoked by paravertebral magnetic stimulation is abnormal together with the motor root conduction time, whereas the central motor conduction time calculated from the latency of the F wave is within normal limits.

In lumbar canal stenosis, the abnormality of motor root conduction time is evident in multiple lower limb muscles when most lumbosacral roots are involved (see Fig. 7–11) or in a restricted number of lower limb muscles when a few lumbosacral roots are compressed (see Fig. 7–12).

The determination of conduction along lumbosacral motor roots is even more useful in cases of acute large central disc protrusions causing multiradicular compression. The clinical presentation of this syndrome is very similar to a lower cord lesion. The electromyographic study may reveal changes only after 2 to 3 weeks, but the conduction along motor roots becomes abnormal with the onset of root compression. An example in one patient with a 3-day history of acute paraplegia due to central disc herniation at the L3 level is shown in Figure 7–17. Central motor conduction time is within normal limits for the rectus femoris and abductor hallucis, whereas central motor conduction for the tibialis anterior muscle is prolonged when evaluated from the latency of the response evoked by magnetic paravertebral, but it could not be evaluated with the F-wave method because of the absence of the F wave. The abnormality of central motor conduction to the tibialis anterior muscle together with the absence of the F wave indicates an L4 and L5 root dysfunction, whereas the normal central motor conduction for a more caudal muscle such as the abductor hallucis demonstrates a normal lumbosacral cord function.

Spondylotic compression of cervical roots usually does not result in a significant delay of motor root conduction time because of the very limited segments of the roots involved by the lesion and because a multiradicular involvement at cervical level is quite uncommon. Conversely, in cervical radiculitis, the marked demyelination of the roots may result in a pronounced delay of conduction along the roots. An example in one patient with C8-T1 radiculitis associated with a spondylodiscitis is represented in Figure 7–18.

The evaluation of conduction along motor roots is especially useful in certain cases of acute and chronic inflammatory demyelinating polyneuropathies and of multifocal motor neuropathies because, particularly in the initial phases of these disorders, a root involvement may be exclusive or predominant. The most pronounced abnormalities of proximal conduction can be found in these disorders.

When demyelinating neuropathies involve multiple motor roots, the central motor conduction time calculated by subtracting the latency of the response evoked by magnetic paravertebral stimulation from the latency of MEPs evoked by cortical stimulation may be extremely prolonged, whereas the central motor conduction time evaluated from total peripheral conduction time is normal. An example of conduction slowing in motor

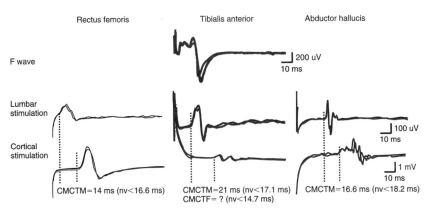

Figure 7–17 Motor evoked potentials were recorded from the rectus femoris, tibialis anterior, and abductor hallucis muscles after cortical and paravertebral magnetic stimulation in a patient with acute paraplegia due to central herniation of L3 disc. The central motor conduction time of tibialis anterior muscle calculated as the difference between the latency of responses evoked by cortical and paravertebral magnetic stimulation (CMCT^M) is prolonged. The F wave is absent. Central motor conduction time for rectus femoris and abductor hallucis muscles is normal.

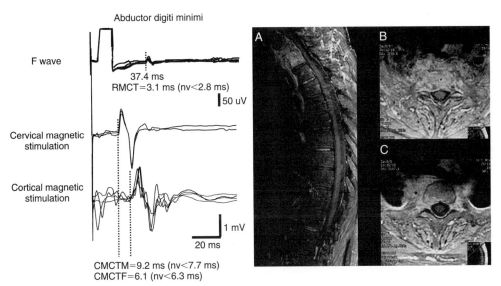

Figure 7–18 The spinal cord was assessed by magnetic resonance imaging (MRI), and motor evoked potentials were recorded from the abductor digiti minimi muscle after cortical and paravertebral magnetic stimulation in a 70-year-old man. He had had acute, severe pain in his right arm with a distribution along the C8 and T1 roots. T1-weighted axial (**B** and **C**) and sagittal (**A**) MRI show a marked contrast enhancement involving the T1 and T2 vertebral bodies and the disc space due to infectious spondylodiscitis. The infectious disease compromises the right lateral recesses and intervertebral foramen. The central motor conduction time of abductor digiti minimi muscle, calculated as the difference between the latency of responses evoked by cortical and paravertebral magnetic stimulation (CMCT^M), is prolonged, whereas the central motor conduction time, calculated using the F-wave method (CMCT^F), is normal. The difference between these two parameters, corresponding to root motor conduction time (RMCT) and the F wave latency, is prolonged.

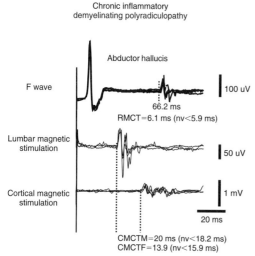

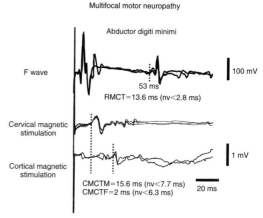

Figure 7–19 Motor evoked potentials were recorded from the abductor hallucis muscle after cortical and paravertebral magnetic stimulation in a patient with chronic inflammatory demyelinating polyradiculoneuropathy. The central motor conduction time of abductor hallucis muscle, calculated as the difference between the latency of responses evoked by cortical and paravertebral magnetic stimulation ($CMCT^M$), is prolonged, whereas the central motor conduction time, calculated using the F-wave method ($CMCT^F$), is normal. The difference between these two parameters, corresponding to root motor conduction time (RMCT) and the F wave latency, is prolonged.

Figure 7–20 Motor evoked potentials were recorded from the abductor digiti minimi muscle after cortical and paravertebral magnetic stimulation in a patient with multifocal motor neuropathy. The central motor conduction time of abductor digiti minimi muscle, calculated as the difference between the latency of responses evoked by cortical and paravertebral magnetic stimulation ($CMCT^M$), is prolonged, and the central motor conduction time, calculated using the F-wave method ($CMCT^F$), is abnormally short. The difference between these two parameters, corresponding to root motor conduction time (RMCT) and the latency of the F wave, is extremely prolonged.

roots from one patient with Guillain-Barré syndrome is shown in Figure 7–6 and from one patient with chronic inflammatory demyelinating polyneuropathy is shown in Figure 7–19.

Abnormal radicular conduction can also be demonstrated in patients with multifocal motor neuropathy. In some cases, when the conduction block is localized at proximal motor roots, the central motor conduction time calculated from the latency of the F wave may be abnormally short[54] (Fig. 7–20). A possible explanation for this finding is an impaired safety margin for repeated discharge of faster fibers of motor roots, which makes them more refractory than normal during backfiring of larger spinal motoneurons. Normally, the antidromic and orthodromic volleys generating the F-wave travel twice at a very short interval the proximal motor roots. When the excitability of the largest motor axons at proximal root segments is decreased, the conduction of the orthodromic volley may be blocked. The preserved backfiring of some slower conducting root fibers can generate an extremely prolonged F wave. Because the largest fibers can still discharge after transcranial stimulation, the population of root fibers involved in F-wave generation and the population recruited after transcranial stimulation have completely different conduction velocities and this result in an abnormally short central motor conduction time estimated from the F-wave latency together with an extremely prolonged root conduction time.

The assessment of root conduction using magnetic paravertebral stimulation together with F-wave recording is particularly useful for the peripheral nerve inflammatory disorders that involve only the root while sparing segments that are more peripheral or in diabetic multiradiculopathies. An example in one patient with inflammatory lumbosacral multiradiculopathy is shown in Figure 7–21.

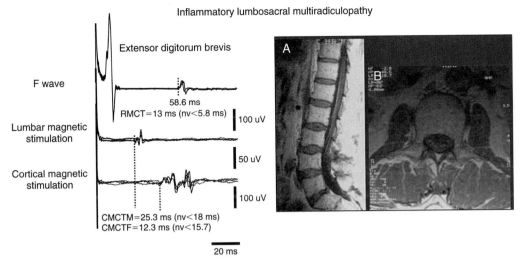

Figure 7–21 Lumbosacral assessment was done with magnetic resonance imaging (MRI), and motor evoked potentials were recorded from the extensor digitorum brevis muscle after cortical and paravertebral magnetic stimulation in a 65-year-old man with a 10-year history of muscle wasting and strength loss in his legs. Postcontrast sagittal (**A**) and axial (**B**) T1-weighted MRI shows abnormal contrast enhancement of the cauda equina roots suggestive of inflammatory lumbosacral multiradiculopathy. The central motor conduction time of abductor digiti minimi muscle, calculated as the difference between the latency of responses evoked by cortical and paravertebral magnetic stimulation (CMCT^M), is prolonged, whereas the central motor conduction time, calculated using the F-wave method (CMCT^F), is normal. The difference between these two parameters, corresponding to root motor conduction time (RMCT), is extremely prolonged.

Magnetic stimulation may be useful in the study of some peripheral nerves, such as the thoracic spinal nerves, that are not accessible to standard conduction studies. Chokroverty and colleagues[46] recorded muscle responses from the rectus abdominis, external oblique, and intercostal muscles after paravertebral root stimulation and after distal intercostal nerve stimulation. In this way, the conduction along thoracic nerves can be assessed. Using this technique, the same investigators demonstrated a thoracoabdominal radiculoneuropathy in one diabetic patient.

REFERENCES

1. Di Lazzaro V, Oliviero A, Profice P, et al. The diagnostic value of motor evoked potentials. Clin Neurophysiol 1999;110:1297–1307.
2. Thompson PD, Dick JPR, Asselman P, et al. Examination of motor function in lesions of the spinal cord by stimulation of the motor cortex. Ann Neurol 1987;21:389–396.
3. Abbruzzese G, Dall'Agata D, Morena M, et al. Electrical stimulation of the motor tracts in cervical spondylosis. J Neurol Neurosurg Psychiatry 1988;51:796–802.
4. Maertens de Noordhout MA, Remacle JM, et al. Magnetic stimulation of the motor cortex in cervical spondylosis. Neurology 1991;41:75–80.
5. Di Lazzaro V, Restuccia D, Colosimo C, et al. The contribution of magnetic stimulation of the motor cortex to the diagnosis of cervical spondylotic myelopathy. Correlation of central motor conduction to distal and proximal upper limb muscles with clinical and MRI findings. Electroencephalogr Clin Neurophysiol 1992;85:311–320.
6. Herdmann J, Dvorak J, Bock WJ. Motor evoked potentials in patients with spinal disorders: upper and lower motor neuron affection. Electromyogr Clin Neurophysiol 1992;32:323–330.
7. Tavy DLJ, Wagner GL, Keunen RWM, et al. Transcranial magnetic stimulation in patients with cervical spondylotic myelopathy: clinical and radiological correlations. Muscle Nerve 1994;17:235–241.
8. Chistyakov AV, Soustiel JF, Hafner H, et al. Motor and somatosensory conduction in cervical myelopathy and radiculopathy. Spine 1995;20:2135–2140.
9. Maertens de Noordhout AM, Myressiotis S, et al. Motor and somatosensory evoked potentials in cervical spondylotic myelopathy.

Electroencephalogr Clin Neurophysiol 1998;108:24–31.
10. Bednarik J, Kadanka Z, Vohanka S, et al. The value of somatosensory and motor evoked potentials in pre-clinical spondylotic cervical cord compression. Eur Spine J 1998;7:493–500.
11. Travlos A, Pant B, Eisen A. Transcranial magnetic stimulation for detection of preclinical cervical spondylotic myelopathy. Arch Phys Med Rehabil 1992;73:442–446.
12. Tavy DL, Franssen H, Keunen RW, et al. Motor and somatosensory evoked potentials in asymptomatic spondylotic cord compression. Muscle Nerve 1999;22:628–634.
13. De Mattei M, Paschero B, Sciarretta A, et al. Usefulness of motor evoked potentials in compressive myelopathy. Electromyogr Clin Neurophysiol 1993;33:205–216.
14. Linden D, Berlit P. Magnetic motor evoked potentials (MEP) in diseases of the spinal cord. Acta Neurol Scand 1994;90:348–353.
15. Brunholzl C, Claus D. Central motor conduction time to upper and lower limbs in cervical cord lesions. Arch Neurol 1994;51:245–249.
16. Dvorak J, Herdmann J, Janssen B, et al. Motor-evoked potentials in patients with cervical spine disorders. Spine 1990;15:1013–1016.
17. Misra UK, Kalita J. Somatosensory and motor evoked potential changes in patients with Pott's paraplegia. Spinal Cord 1996;34:272–276.
18. Kalita J, Misra UK. Neurophysiological studies in acute transverse myelitis. J Neurol 2000;247:943–948.
19. Linden D, Berlit P. Spinal arteriovenous malformations: clinical and neurophysiological findings. J Neurol 1996;243:9–12.
20. Di Lazzaro V, Restuccia D, Oliviero A, et al. Ischaemic myelopathy associated with cocaine: clinical, neurophysiological, and neuroradiological features. J Neurol Neurosurg Psychiatry 1997;63:531–533.
21. Clarke CE, Modarres-Sadeghi H, Twomey JA, et al. Prognostic value of cortical magnetic stimulation in spinal cord injury. Paraplegia 1994;32:554–560.
22. Bondurant CP, Haghighi SS. Experience with transcranial magnetic stimulation in evaluation of spinal cord injury. Neurol Res 1997;19:497–500.
23. Curt A, Keck ME, Dietz V. Functional outcome following spinal cord injury: significance of motor-evoked potentials and ASIA scores. Arch Phys Med Rehabil 1998;79:81–86.
24. Di Lazzaro V, Restuccia D, Fogli D, et al. Central sensory and motor conduction in vitamin B_{12} deficiency. Electroencephalogr Clin Neurophysiol 1992;84:433–439.
25. Nogues MA, Pardal AM, Merello M, et al. SEPs and CNS magnetic stimulation in syringomyelia. Muscle Nerve 1992;15:993–1001.
26. Masur H, Oberwittler C, Fahrendorf G, et al. The relation between functional deficits, motor and sensory conduction times and MRI findings in syringomyelia. Electroencephalogr Clin Neurophysiol 1992;85:321–330.
27. Botzel K, Witt TN. Transcranial cortical stimulation in syringomyelia: correlation with disability? Muscle Nerve 1993;16:537–541.
28. Stetkarova I, Kofler M, Leis AA. Cutaneous and mixed nerve silent periods in syringomyelia. Clin Neurophysiol 2001;112:78–85.
29. Claus D, Waddy HM, Harding AE, et al. Hereditary motor and sensory neuropathies and hereditary spastic paraplegia: a magnetic stimulation study. Ann Neurol 1990;28:43–49.
30. Caramia MD, Cicinelli P, Paradiso C, et al. "Excitability" changes of muscular responses to magnetic brain stimulation in patients with central motor disorders. Electroencephalogr Clin Neurophysiol 1991;81:243–250.
31. Pelosi L, Lanzillo B, Perretti A, et al. Motor and somatosensory evoked potentials in hereditary spastic paraplegia. J Neurol Neurosurg Psychiatry 1991;54:1099–1102.
32. Schady W, Dick JPR, Sheard A, et al. Central motor conduction studies in hereditary spastic paraplegia. J Neural Neurosurg Psychiatry 1991;54:775–779.
33. Restuccia D, Di Lazzaro V, Valeriani M, et al. Abnormalities of somatosensory and motor evoked potentials in adrenomyeloneuropathy: comparison with magnetic resonance imaging and clinical findings. Muscle Nerve 1997;20:1249–1257.
34. Ugawa Y, Rothwell JC, Day BL, et al. Magnetic stimulation over spinal enlargements. J Neurol Neurosurg Psychiatry 1989;52:1025–1032.
35. Britton TC, Meyer BU, Herdmann J, et al. Clinical use of the magnetic stimulator in the investigation of peripheral conduction time. Muscle Nerve 1990;13:396–406.
36. Schmidt UD, Walker G, Schmid-Sigron J, et al. Transcutaneous magnetic and electrical stimulation over the cervical spine: excitation of plexus rots rather than spinal roots? J Neurol Neurosurg Psychiatry 1990;53:770–777.
37. Kimura J. F wave velocity in the central segment of the median and ulnar nerves. A study in normal subjects and in patients with Charcot-Marie-Tooth disease. Neurology 1974;24:539–546.
38. Ugawa Y, Rothwell JC, Day BL, et al. Percutaneous electrical stimulation of corticospinal pathways at the level of the pyramidal decussation in humans. Ann Neurol 1991;29:418–427.
39. Ugawa Y, Uesaka Y, Terao Y, et al. Clinical utility of magnetic corticospinal tract

stimulation at the foramen magnum level. Electroencephalogr Clin Neurophysiol 1996;101:247–254.
40. Rowland LP. Surgical treatment of cervical spondylotic myelopathy: time for a controlled trial. Neurology 1992;42:5–13.
41. Fouyas IP, Statham PF, Sandercock PA, et al. Surgery for cervical radiculomyelopathy. Cochrane Database Syst Rev 2001;3:CD001466.
42. Urban PP, Vogt T. Conduction times of cortical projections to paravertebral muscles in controls and in patients with multiple sclerosis. Muscle Nerve 1994;17:1348–1349.
43. Hashimoto T, Uozumi T, Tsuji S. Paraspinal motor evoked potentials by magnetic stimulation of the motor cortex. Neurology 2000;55:885–888.
44. Gandevia SC, Plassman BL. Responses in human intercostal and truncal muscles to motor cortical and spinal stimulation. Respir Physiol 1988;73:325–337.
45. Misawa T, Ebara S, Kamimura M, et al. Evaluation of thoracic myelopathy by transcranial magnetic stimulation. J Spinal Disord 2001;14:439–444.
46. Chokroverty S, Deutsch A, Guha C, et al. Thoracic spinal nerve and root conduction: a magnetic stimulation study. Muscle Nerve 1995;18:987–991.
47. Cheshire WP, Santos CC, Massey EW, et al. Spinal cord infarction: etiology and outcome. Neurology 1996;47:321–330.
48. McDermott CJ, White K, Bushby K, et al. Hereditary spastic paraparesis: a review of new developments. J Neurol Neurosurg Psychiatry 2000;69:150–160.
49. Di Lazzaro V, Restuccia D, Oliviero A, et al. Effects of voluntary contraction on descending volleys evoked by transcranial stimulation in conscious humans. J Physiol 1998;508 (Pt 2):625–633.
50. Di Lazzaro V, Oliviero A, Profice P, et al. Descending spinal cord volleys evoked by transcranial magnetic and electrical stimulation of the motor cortex leg area in conscious humans. J Physiol 2001;537(Pt 3):1047–1058.
51. Curt A, Dietz V. Electrophysiological recordings in patients with spinal cord injury: significance for predicting outcome. Spinal Cord 1999;37:157–165.
52. Macdonell RAL, Cros D, Shahani BT. Lumbosacral nerve root stimulation comparing electrical with surface magnetic coil techniques. Muscle Nerve 1992;15:885–890.
53. Banerjee TK, Mostofi MS, Us O, et al. Magnetic stimulation in the determination of lumbosacral motor radiculopathy. Electroencephalogr Clin Neurophysiol 1993;89:221–226.
54. Molinuevo JL, Cruz-Martinez A, Graus F, et al. Central motor conduction time in patients with multifocal motor conduction block. Muscle Nerve 1999;22:926–932.

Cranial Nerves

Giorgio Cruccu and Mark Hallett

Transcranial magnetic stimulation (TMS)-evoked motor potentials in cranial nerve muscles are particularly useful because, besides yielding information on corticobulbar pathways, they provide information on extra-axial nerve pathways that cannot be easily accessed by standard nerve conduction studies. Cortical control of eye movements is a complex and well-studied area that is not considered in this chapter.[1,2]

■ Anatomical and Functional Organization

The general principle for cortical control of cranial nerve motor nuclei 5, 7, 9, 10, 11, and 12 is that there is predominant contralateral influence and variable but important ipsilateral influence.[3] Although the degree of lateralization seems to be an individual characteristic,[4] the bilaterality of innervation usually depends on muscle function. The contralateral cortex has a predominant control over cranial muscles that produce lateral movements and can act independently. Conversely, symmetrically bilateral cortical projections reach the motoneurons of muscles (e.g., laryngeal, pharyngeal, suprahyoid) that are strictly synergistic and always coactivated on the two sides. The importance of the bilaterality of innervation is clearly different from that for limb motor control and is helpful in understanding clinical phenomena such as recovery from stroke.

The best understood situation is the seventh nerve, thanks in part to an anatomical study in the rhesus monkey by Morecraft and colleagues.[5] The corticobulbar projection was defined by injecting anterograde tracers into the face representation of each motor cortex, and in the same animals, the musculotropic organization of the facial nucleus was defined by injecting fluorescent retrograde tracers into individual muscles of the upper and lower face. Perioral muscles, prototypical for the lower face, are innervated largely contralaterally from the primary motor cortex (M1), ventral and dorsal regions of the lateral premotor cortex, and caudal cingulate (M4). For upper face muscles, orbicularis oculi are innervated mostly by rostral cingulate (M3) and auricular muscles by the supplementary motor area (M2), both bilaterally. Such a pattern explains the upper face sparing in typical middle cerebral artery stroke. Recovery from stroke in any location is explained by at least a minimal projection from all cortical face areas to all parts of the face. In addition, that one component of the corticofacial projection arises from the limbic pro-isocortices (M3 and M4) and another component arises from frontal isocortices (M1, M2, ventral lateral premotor cortex, and dorsal lateral premotor cortex) suggests an anatomical substrate that may contribute to the clinical dissociation of emotional and volitional facial movement.[3]

Another characteristic of the cranial nerve motoneurons is that they lack axon collaterals and do not undergo recurrent inhibition.[6] Many muscles are devoid of muscle spindles and consequently antagonist muscles do not undergo reciprocal inhibition. Neck muscles are a noteworthy exception to this rule. Cranial nerve motoneurons seem to be mostly regulated by the superficial reflex input and by the motor cortices. The facial motor system is the extreme example of this organization, because facial motoneurons do not undergo any kind of intrinsic or reflex inhibition. The TMS-induced silent period in facial muscles is entirely produced by inhibition of the motor cortex.[7]

TMS fails to evoke motor potentials in several cranial muscles (discussed later) unless they are voluntarily contracted. In some instances, even a strong contraction is necessary. This may result from a small cortical representation, the need for a high level of temporal summation, or a high threshold of lower motoneurons, with these characteristics varying from muscle to muscle. In the case of perioral muscles, for instance, the cortical representation is by no means small, and lower motoneurons are very low threshold, at least to reflex excitatory inputs. In the case of masticatory muscles, which do have a very small cortical representation, even trains of cortical stimuli fail to activate fully relaxed muscles (unpublished data).

Masticatory Muscles: Cranial Nerve V

Responses in jaw-closing muscles are heavily contralateral but have some ipsilateral influence.[8–10] The latency for the contracting masseter muscle is approximately 6 ms. The ipsilateral response can be inhibitory as well as excitatory. Post-stimulus time histograms of single motor unit responses show the earliest responses to be monosynaptic.[9] TMS responses in the anterior digastric muscle are also bilateral, and although there is a slight contralateral predominance, there is much less asymmetry than for the masseter.[9,11] The latency is about 10 ms when the muscle is relaxed and 7 ms when it is contracting. The ipsilateral projection is mostly monosynaptic, but the stronger contralateral projection has a slightly longer latency and broader post-stimulus time response and is therefore more likely to be oligosynaptic.[9,11]

Technique and Normal Values

The masseter is the muscle most frequently and most easily studied to investigate the trigeminal motor pathways. Surface recordings from the muscle motor point on the lower third of the muscle belly provide a standard negative–positive biphasic potential (Fig. 8–1, 120 degrees). Using a large round coil the best position is with the coil centered on the midline, slightly anterior to the vertex; lateral positions entail activation of the ipsilateral trigeminal root and elicit M waves and trigeminal reflexes. The best results are obtained with an 8-shaped coil oriented as shown in Figure 8–1. The subject must preinnervate the masseter muscle by clenching the teeth from a moderate to intense level.[8,12] In the largest studies, the mean latency and peak-to-peak amplitude were 6.0 ± 0.4 ms and 1.3 ± 0.5 mV, respectively, in 75 subjects (Table 8–1).

Although the trigeminal root is far more easily excited by electrical than magnetic transcranial stimuli, when the coil is positioned laterally, it does excite the trigeminal motor root near the foramen ovale (its exit from the middle cranial fossa)[8,13] and evokes a direct response at 2 to 3 ms latency (see Table 8–1). The central conduction time is therefore about 3 to 4 ms, rather shorter than that of the other cranial nerves.

Clinical Applications

The motor evoked potentials (MEPs) of the masseter are usually dampened on the affected side of patients with hemiplegic stroke, even though the masticatory weakness is not manifested clinically.[8] In amyotrophic lateral sclerosis, masseter MEPs have been used to differentiate dysfunction of upper and lower motoneurons and assess bulbar involvement; cortical or root responses, or both, are suppressed, providing preclinical information.[14,15] In hemimasticatory spasm, a rare condition due to a trigeminal motor neuropathy most often associated to facial hemiatrophy, TMS demonstrated a focal slowing of conduction.[16]

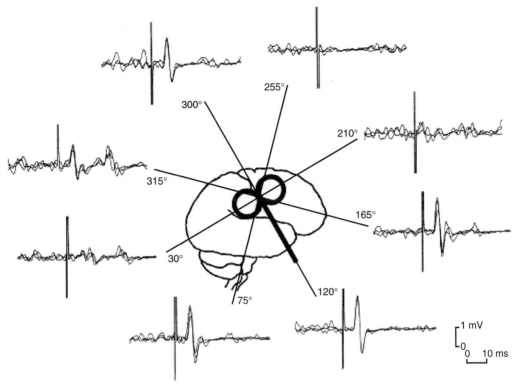

Figure 8–1 Motor potentials of the masseter muscle evoked by contralateral transcranial magnetic stimulation. For each of the eight different coil orientations, three motor evoked potentials (MEPs) are superimposed. The greatest and most consistent amplitudes are found at 120 degrees, whereas the smallest MEPs are found at 30, 210, and 255 degrees. (From Guggisberg AG, Dubach P, Hess CW, et al. Motor evoked potentials from masseter muscle induced by transcranial magnetic stimulation of the pyramidal tract: the importance of coil orientation. Clin Neurophysiol 2001;112:2312–2319.)

Masseter MEPs are also being used in experimental and clinical pain studies about temporomandibular dysfunction. However, they appear insensitive to experimental or clinical pain.[17,18]

■ Facial Muscles: Cranial Nerve VII

Transcranial magnetic and electrical stimulation studies have explored cortical connections to muscles of the lower and upper face (Fig. 8–2). Only a few studies have reported data for upper facial muscles. Latencies are reported to be about 10 ms and bilateral in the orbicularis oculi[19,20] and frontalis muscles,[21] but given that the R1 component of the blink reflex has about the same latency, this response may be contaminated by the blink reflex,[7] as is further suggested by the fact that responses in the orbicularis oculi muscle are easily elicited in the fully relaxed muscle. Stimulation was done over M1 regions for the face, and as observed in the anatomical studies,[5] there is only very sparse projection to the eyelid muscles from M1. A large round coil positioned more anteriorly, yields responses at about 7 ms in the orbicularis oculi muscle that may well result from activation of the anterior cingulate cortex.[22]

TMS studies of the orbicularis oris in the lower face are much better documented. The effects are primarily contralateral, with latencies of 9 to 10 ms. Studies of single motor unit latency histograms show that the earliest responses are very likely to be monosynaptic.[23] The relative slowness of this projection is caused by slow central conduction and a relatively longer length of cranial nerve VII compared with other cranial motor nerves.[23] It is also possible to find inhibitory responses

Table 8–1 NORMAL VALUES (MEAN ± SD)

Muscle (Cranial Nerve)	Root Latency (ms)	MEP Latency (ms)	Amplitude (mV)	Ipsi*	Pre-inn†	Coil‡	n	Studies
Masseter (V)	2.1 ± 0.3	5.9 ± 0.4	2.0 ± 0.6	+	+	O	25	Cruccu et al., 1989[8]
	3.4 ± 0.4	6.3 ± 0.5	0.6 ± 0.4	+	+	O	22	Trompetto et al., 1998[14]
	—	6.7 ± 0.3	0.7 ± 0.1	++	+	8	12	Butler et al., 2001[54]
	1.9 ± 0.3	5.2 ± 0.5	1.4 ± 0.7	++	+	8	16	Guggisberg et al., 2001[12]
		6.0 ± 0.4	**1.3 ± 0.5**				**75**	Grand average
Suprahyoid (V)	3.1**	6.9	1.0	+++	+		6	Cruccu et al., 1989[8]
	2.7 ± 0.3	7.7 ± 1.1	0.5 ± 0.3	+++	—	8	12	Gooden et al., 1999[11]
Lower facial (VII)								
Nasalis	4.8 ± 0.7	9.8 ± 1.9	1.0 ± 0.5	—	+	O	83	Glocker et al., 1994[24]
Buccinator	4.3 ± 0.5	9.9 ± 1	1.6 ± 1.1	63%	+	O	43	Urban et al., 1997[28]
Mentalis	4.9 ± 0.7	9.9 ± 2.4	1.0 ± 0.7	—	+	O	68	Rosler et al., 1995[30]
		9.8 ± 1.5	**1.2 ± 0.8**				**194**	Grand average
Orbicularis oculi (VII)	4.6 ± 0.4	—	—	—	—	O	120	Cocito et al., 2003[55]
	4.5 ± 0.6	9.5 ± 1.3	1.5 ± 0.6	+++	—	O	10	Ghezzi et al., 1992[56]
	—	10.2 ± 1.3	0.8 ± 0.4	100%	—	8	17	Roedel et al., 2003[20]
		11		86%	—	O	14	Benecke et al., 1988[19]

Muscle					Pre-inn	Coil	n	Reference
Pharyngeal (IX, X)	5.1 ± 1.5	10.7 ± 0.5	0.8 ± 0.2	++	—	O	14	Ertekin et al., 2001[37]
Laryngeal superior (X)	3.8 ± 1.2	8.4 ± 0.7	1.0 ± 0.3	++	+	O	6	Ertekin et al., 2001[37]
	2.9 ± 0.5	8.8 ± 2.5		+++	—	8	10	Khedr, 2002[40]
Sternomastoid (XI)	2.4 ± 0.4	6.9 ± 0.7	4.8 ± 2	80%	±	O	35	Berardelli et al., 1991[44]
		9.3 ± 0.6	1.2 ± 0.9	++	+	8	8	Thompson et al., 1997[45]
Trapezius (XI)	3.7 ± 0.4	9.0 ± 0.9	4.6 ± 2.1	0%	±	O	35	Berardelli et al., 1991[44]
		10.5 ± 1.3	7.8 ± 7.2	53%	+	O	15	Strenge and Jahns, 1998[46]
Tongue (XII)	4.1 ± 0.3	7.9 ± 0.8	1.4 ± 0.9	++	+	O	20	Ghezzi et al., 1998[53]
	3.4 ± 0.9	8.5 ± 1	1.7 ± 0.8	100%	—	8	21	Meyer et al., 1997[49]
	—	8.8 ± 0.9	1.8 ± 1.1	++	+	O	20	Muellbacher et al., 1994[52]
	3.6 ± 0.5	8.8 ± 1.1	—	100%	+	O	43	Urban et al., 1997[28]
		8.6 ± 1.0	1.6 ± 0.9				104	**Grand average**

*The Ipsi column reports the symmetry and frequency of ipsilateral responses in a scale from + to +++ or, when available, the exact percentage (0–100).

†In the Pre-inn column, the plus sign indicates that voluntary pre-innervation is indispensable and the minus sign that most subjects also have responses in fully relaxed muscles.

‡In the Coil column, O and 8 indicate whether the investigators used a round or a figure-of-eight coil.

**Indicates electrical stimulation.

MEP, motor evoked potential after magnetic stimulation.

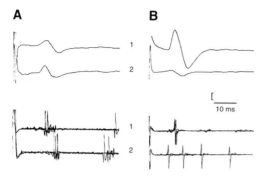

Figure 8–2 Motor potentials evoked by transcranial magnetic stimulation (TMS) in the upper and lower facial muscles. *Upper panel*: Simultaneous surface recordings from the right (1) and left (2) frontalis muscle (**A**) and from the right (1) and left (2) lower facial muscles (**B**) during voluntary contraction. TMS of the left hemisphere was conducted, and findings of eight trials were averaged. The vertical calibration was 250 μV. The motor evoked potentials are bilateral in the frontalis muscle and contralateral in the lower facial muscles. The small deflection in B2 (ipsilateral lower facial electrodes) is probably picked up by volume conduction. *Lower panel*: Bipolar concentric needle recording of single motor unit potentials in the right frontalis muscle (**A**) and the right mentalis muscle (**B**) during voluntary contraction. TMS of the left (1) and right hemisphere (2) was conducted, and four representative trials were superimposed. The vertical calibration was 100 μV. The same single motor unit of the right frontalis muscle is repeatedly activated by both left and right TMS (A1 and A2). In B1, a single motor unit of the right mentalis muscle is repeatedly activated by contralateral TMS. B2 shows occasional (voluntary) activation of the same motor unit. (From Cruccu G, Berardelli A, Inghilleri M, et al. Corticobulbar projections to upper and lower facial motoneurons. A study by magnetic transcranial stimulation in man. Neurosci Lett 1990;117:68–73.)

in orbicularis oris, and these can be seen even without a preceding motor evoked potential. Because the facial muscles remain responsive to trigeminal stimulation during this silent period, the origin of the inhibition must be intracortical.[7,23]

Technique and Normal Values

The nature of responses in upper facial muscles is still debatable. In clinical studies, the facial MEPs are most frequently recorded from perioral muscles. These muscles, however, entail some disadvantage. Because the motor point slightly varies between subjects and surface recording electrodes pick up signals from nearby muscles, even contralateral (see Fig. 8–2B2), to obtain a proper MEP starting with a negative deflection, it is necessary to adjust the position of the electrodes by controlling the M wave after stimulation of the facial nerve at the stylomastoid foramen (Fig. 8–3). The optimal muscle is probably the nasalis, with the active electrode just above the ala nasi and the reference on the nasal bone.[24] If the subject is able to contract selectively the nasalis muscle, the need for pre-innervation becomes an advantage, because all the other muscles remain silent (with the exception of the orbicularis oculi, for which blink reflexes are easily elicited by excitation of supraorbital nerve fibers or even the simple noise of the magnetic stimulus). The coil should be positioned to activate the facial area in M1, taking care not to shift too laterally, because the induced current may easily spread in the cerebrospinal fluid to excite the facial and trigeminal roots. It would be useful to monitor a possible excitation of intracranial trigeminal afferents, as revealed by the appearance of an M-like response in the masseter. The activation of the intracranial facial nerve fibers, inducing a direct motor response in the relaxed facial muscle, cannot be missed. To study this response and assess facial nerve conduction, the optimal position of the coil is temporo-occipital, behind and above the ear (see Fig. 8–3); the facial nerve is excited in a tract between the entrance into the internal acoustic meatus and the labyrinthine segment of the facial canal.[24–26]

In the largest studies, the mean latency and peak-to-peak amplitude of cortical MEPs were 9.8 ± 1.5 ms and 1.2 ± 0.8 mV, respectively, and the latency of the root MEPs was 4.8 ± 0.6 ms in 194 subjects (see Table 8–1). The central conduction time is therefore 5 ms, similar to that of most cranial nerves. Using a 12-cm round coil and the highest stimulus intensity that the individual could tolerate, the cortical silent period in lower face muscles lasted 191 ± 47 ms in 19 subjects.[7,27]

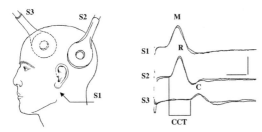

Figure 8–3 Conduction along the central and peripheral facial nerve pathways. *Left panel*: Sites of stimulation are shown. Electrical stimulation of the left facial nerve was conducted near its exit from the stylomastoid foramen (S1). The position of the magnetic coil for transcranial stimulation of the left facial root (S2) and position of the magnetic coil for transcranial stimulation of the face area of the right motor cortex (S3) are indicated. *Right panel*: Surface recordings were made from the nasalis muscle. The M wave after stimulation of the nerve (M) and motor responses after magnetic stimulation of the intracranial root (R) are indicated. Motor evoked potentials after stimulation of the contralateral motor cortex (C) were recorded during voluntary contraction. Two averages of eight trials were superimposed. The calibration is 5 ms and 2 mV, and the central conduction time (CCT) is 5.3 ms.

Clinical Applications

Because of the interest in the corticofacial projections and the frequency of facial neuropathy, the facial MEPs are those having the strongest clinical relevance and the largest number of published studies. In patients with central lesions, most studies relied on MEPs in lower facial muscles.

In patients with hemispheric or brainstem infarctions affecting the supranuclear projections, contralateral facial responses were either absent or delayed.[28,29] In patients with amyotrophic lateral sclerosis, TMS evoked abnormal responses despite the lack of a clinical involvement of the facial motor pathway, indicating the presence of early and, in most cases, subclinical involvement in the pathways to the orofacial muscles.[29]

Using magnetic stimulation of the proximal part of the facial nerve, Rosler and associates[30] studied 174 patients with facial paresis due to peripheral lesion. They found that in intracanalicular lesion, as in Bell's palsy, the electrical stylomastoid stimulation often yielded normal facial responses, whereas magnetic stimulation of the more proximal part of facial nerve evoked low amplitude responses indicating a local canalicular nerve hypoexcitability. In facial paresis due to demyelinating disease, the more common abnormality was a prolonged latency. In facial paresis caused by herpes zoster infection, which causes axonal degeneration, the amplitude of the responses to canalicular stimulation was reduced in the same range as that with stylomastoidal electrical stimulation. The TMS-induced silent period was shortened in patients with central facial paresis, amyotrophic lateral sclerosis, and orofacial focal dystonias.[15,17,31]

■ Pharyngeal and Laryngeal Muscles: Cranial Nerves IX and X

TMS studies using pharyngeal muscle responses have demonstrated that swallowing has a bilateral but asymmetric hemispheric representation.[32–34] Although the cortical representation of swallowing can also be seen with positron-emission tomography on the inferior precentral gyrus, this asymmetry is not as prominent.[35] The asymmetry is clearly critical, however, and explains the variability of appearance of dysphagia after stroke and the remarkable ability to recover when there is dysphagia. Damage to the hemisphere that has the greater swallowing output appears to predispose that individual to swallowing problems, whereas damage to the hemisphere with the smaller swallowing output does not affect swallowing.[36] However, when there is dysphagia, because there is additional substrate for swallowing in the undamaged hemisphere, the capacity for compensatory reorganization in the contralateral motor cortex can be increased, leading to a greater likelihood of recovery. This finding has been demonstrated in a study of 28 patients who had a unilateral hemispheric stroke.[33] Dysphagia was initially present in 71% of patients and in 46% and 41% of the patients at 1 and 3 months, respectively. Nondysphagic and persistently dysphagic patients showed little change in pharyngeal representation in either hemisphere at 1 and 3 months compared with presentation, but dysphagic patients who recovered had an increased pharyngeal representation in the unaffected hemisphere at 1 and 3 months without change in the affected hemisphere. Return of swallowing is associated with increased pharyngeal representation in the unaffected hemisphere, indicating a critical role for the intact hemisphere reorganization in recovery.

Needle recordings from the cricopharyngeal muscle, even relaxed, showed motor potentials elicited by magnetic stimulation with a round coil positioned at the vertex; during voluntary activation the mean latency was 10.7 ± 0.5 ms; magnetic stimulation near the mastoid, close to the nerve exit from the skull, evoked direct motor responses with a latency of about 5 ms (Fig. 8–4).[37]

Notwithstanding the obvious physiological and clinical interest in the motor control of the vocal cords, very few studies have so far been published on laryngeal MEPs, probably because of the great technical difficulties in laryngeal muscle recordings. Corticobulbar fibers directed to nucleus ambiguus in the medulla are thought to be mostly contralateral. Ludlow and associates,[38,39] however, showed bilateral activation of the intrinsic laryngeal muscles, suggesting bilateral projections from each motor cortex to the laryngeal motoneurons in humans. Khedr and associates[40] inserted concentric needle electrodes in the cricothyroid muscle (which is innervated by the superior laryngeal nerves) and the thyroarytenoid muscle (which is innervated by the recurrent laryngeal nerves). These investigators obtained the best results placing the midpoint of a figure-of-eight coil 8 cm lateral and 1 cm anterior to the vertex. TMS evoked bilateral responses. The latency was shorter in ipsilateral than contralateral muscle and, in the thyroarytenoid muscle, shorter on the right than left side (because the left recurrent laryngeal nerve has a longer course below the aortic arch): 9.6 ± 1.0 ms on the right and 11.1 ± 1.0 ms on the left side. On peripheral stimulation at the mastoid, the latency was 2.7 ± 0.4 and 3.1 ± 0.4 ms in the right and left muscles, respectively. Responses evoked by cisternal magnetic stimulation of the vagus nerve (2 cm behind and 6 cm above the external auditory meatus), however, have a

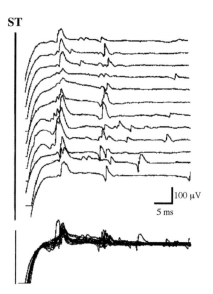

Figure 8–4 Laryngeal motor evoked potentials. Concentric needle recordings were made from the cricothyroid muscle. Motor evoked potentials evoked by transcranial magnetic stimulation of the contralateral hemisphere during voluntary activation. Twelve consecutive trials are superimposed at the bottom. (From Ertekin C, Turman B, Tarlaci S, et al. Cricopharyngeal sphincter muscle responses to transcranial magnetic stimulation in normal subjects and in patients with dysphagia. Clin Neurophysiol 2001;112:86–94.)

latency of 4 to 6 ms,[41] which leaves about 5 ms for central conduction time.

Khedr and associates[40] studied 16 patients with unilateral or bilateral disorders of vocal cord mobility due to cardiomegaly or other mediastinal disease, thyroidectomy or other neck disease, or "idiopathic paralysis." TMS helped to localize the lesion. Nevertheless, given the difficulties in obtaining selective and accurate laryngeal muscle recordings and the wide variability of latency of responses to peripheral magnetic stimulation, these techniques still need confirmation on their reliability.[39]

■ Neck Muscles: Cranial Nerve XI

There has been some ambiguity in the cortical control of neck muscles, particularly because of the somewhat unusual action of the sternocleidomastoid muscle. When rotating the head to the right side, the left sternocleidomastoid muscle is active. With hemiplegia, there is weakness of head turning to the side of the hemiplegia, and this has suggested that the sternocleidomastoid gets important ipsilateral cortical control.[42] In addition to the sternocleidomastoid muscle rotating the head, lateral bending of the neck, as opposed to rotation, uses the sternocleidomastoid muscle of the same side. TMS and transcranial electrical stimulation studies have shown bilateral, but much more contralateral, innervation of the sternocleidomastoid.[43,44] The latency in the contralateral sternocleidomastoid is approximately 7 ms when activated and 10 ms when relaxed. Latencies in the ipsilateral sternocleidomastoid are slightly longer. Similar results were found for the splenius, but innervation of the trapezius seems exclusively contralateral.[43,44]

Technique and Normal Values

The motor responses from trapezius and sternocleidomastoid muscles are usually recorded by bipolar surface electrodes.[44] However, because surface recordings from the sternocleidomastoid muscle (with the active electrode placed between the upper third and the lower two thirds of the muscle and the reference electrode 2 cm below this position) may detect activity from the platysma muscle (which partly overlays the sternocleidomastoid), the origin of sternocleidomastoid responses must be controlled by needle electrodes recording.[45]

The projection to sternocleidomastoid muscles probably arises from a cortical area high up on the cerebral convexity close to the trunk representation and at a comparable level to the sensory representation of the neck in the postcentral cortex. Consistent with this view, Thompson and associates[45] found the exact positioning of a figure-of-eight coil about 4 cm from the vertex and 0.6 cm posterior to the interaural line. A similar position also activates the contralateral trapezius muscle.[46] Other investigators used a round coil positioned on the vertex.[44]

Although sternocleidomastoid and trapezius responses can also be evoked in relaxed muscles,[44] MEPs are usually recorded during voluntary activation. The latency and amplitude of MEPs were, respectively, 9.5 ± 1.0 ms and 5.6 ± 3.6 mV in the trapezius and 7.3 ± 0.7 ms and 4.1 ± 1.8 mV in the sternocleidomastoid in 43 subjects (see Table 8–1). The latency difference between trapezius and sternocleidomastoid MEPs results from the peripheral conduction distance. With the coil positioned below the mastoid process, Berardelli and associates[44] found direct responses at 3.7 ± 0.5 ms in the trapezius and 2.4 ± 0.4 ms in the sternocleidomastoid muscle. The central conduction time was 5 to 6 ms.

The clinical applications of neck muscle MEPs are so far very few. There are, however, consistent reports that neck muscle MEPs are enhanced in patients with cervical dystonias.[47,48]

Tongue Muscles: Cranial Nerve XII

In the classic representation of the M1 homunculus, the tongue area is immediately below (i.e., lateral) to the lip area. Corticobulbar projections from these two areas run close together through the internal capsule toward the cerebral peduncle; in patients with unilateral lesions rostral to the pons, TMS responses in contralateral perioral and tongue muscles are affected in parallel[28] (Fig. 8–5). However, the projections to tongue muscles are bilateral, although contralateral responses have a lower threshold and a larger amplitude, TMS of either hemisphere invariably produces contralateral and ipsilateral MEPs.[36,49,50] Ipsilateral projections run through the dorsolateral and mediolateral medulla to reach the hypoglossal nucleus from its lateral aspect; contralateral projections branch off the main ventral pyramidal tract and cross the midline at the pontomedullary junction.[51]

Technique and Normal Values

Bipolar surface recordings from tongue muscles are taken with adhesive electrodes (or electrodes kept in place by clips or special devices), positioned laterally on the right and left sides of the tongue.[28,49,52] Given the close vicinity of right and left electrodes, care should be taken to identify signal generated

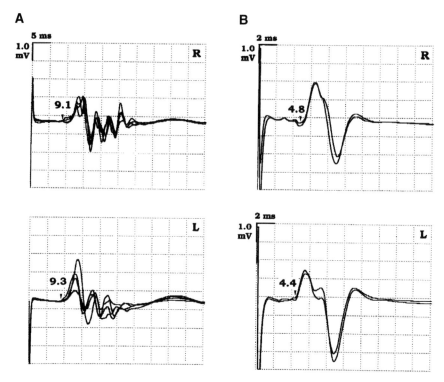

Figure 8–5 Tongue responses after cortical and peripheral transcranial magnetic stimulation (TMS). Surface recordings were made from the left (L) and right (R) half of the tongue. **A:** During voluntary contraction, TMS of the left hemisphere evokes bilateral motor potentials. **B:** Motor responses after magnetic stimulation of the right *(upper traces)* and left *(lower traces)* hypoglossal nerve roots. Notice the time base changes in **A** and **B**. The central conduction time is less than 5 ms. (From Urban PP, Hopf HC, Fleischer S, et al. Impaired cortico-bulbar tract function in dysarthria due to hemispheric stroke. Functional testing using transcranial magnetic stimulation. Brain 1997;120[Pt 6]:1077–1084.)

by ipsilateral and contralateral muscles by controlling the direct motor responses after peripheral stimulation of the hypoglossal nerve.

To stimulate the motor cortex, most investigators used a round coil placed 4 to 6 cm lateral and 2 to 4 cm anterior to the vertex.[28,36,53] Although tongue MEPs are usually recorded during voluntary activation, responses have also been reported in fully relaxed muscles.[49] In the largest studies, the latency and amplitude of tongue MEPs were 8.6 ± 1.0 ms and 1.6 ± 0.9 mV, respectively, in 104 subjects (see Table 8–1). With the coil positioned over the parieto-occipital skull (about the same positioning used for facial nerve stimulation), magnetic stimuli excite hypoglossal nerve fiber probably at their exit from the skull; the mean latency was 3.6 ± 0.5 in 84 subjects (see Table 8–1). The central conduction time was about 5 ms.

Clinical applications

Tongue MEPs are absent unilaterally in patients with hypoglossal nerve lesions. In patients with unilateral hemispheric stroke or brainstem infarctions affecting the supranuclear projections, tongue MEPs are often delayed or absent in the acute stage and may remain abnormal after clinical recovery, suggesting that the recovery of lingual movements must be attributed to the intact hemisphere.[28,36] Interruption of the corticolingual pathways to the tongue seems to be crucial in the pathophysiology of dysarthria after extracerebellar lacunar strokes.[51]

In a study in 51 patients with amyotrophic lateral sclerosis, Urban and associates[29] found that tongue MEPs demonstrated upper motoneuron dysfunction in 53% of patients, a frequency higher than that yielded by limb MEPs.

REFERENCES

1. Gaymard B, Ploner CJ, Rivaud S, et al. Cortical control of saccades. Exp Brain Res 1998;123:159–163.
2. Tehovnik EJ, Sommer MA, Chou IH, et al. Eye fields in the frontal lobes of primates. Brain Res Brain Res Rev 2000;32:413–448.
3. Hallett M. Cortical control of brainstem motor systems. Mov Disord 2002;17 (Suppl 2):S23–S26.
4. Kuypers HGJM. Corticobulbar connections to the pons and lower brainstem in man. Brain 1958;81:364–388.
5. Morecraft RJ, Louie JL, Herrick JL, et al. Cortical innervation of the facial nucleus in the non-human primate: a new interpretation of the effects of stroke and related subtotal brain trauma on the muscles of facial expression. Brain 2001;124:176–208.
6. De Nò Lorente R. Vestibulo-ocular reflex arc. Arch Neurol Psychiatry 1933;30:245–291.
7. Cruccu G, Inghilleri M, Berardelli A, et al. Cortical mechanisms mediating the inhibitory period after magnetic stimulation of the facial motor area. Muscle Nerve 1997;20:418–424.
8. Cruccu G, Berardelli A, Inghilleri M, et al. Functional organization of the trigeminal motor system in man. A neurophysiological study. Brain 1989;112:1333–1350.
9. Nordstrom MA, Miles TS, Gooden BR, et al. Motor cortical control of human masticatory muscles. Prog Brain Res 1999;123:203–214.
10. Pearce SL, Miles TS, Thompson PD, et al. Responses of single motor units in human masseter to transcranial magnetic stimulation of either hemisphere. J Physiol 2003;1;549:583–596.
11. Gooden BR, Ridding MC, Miles TS, et al. Bilateral cortical control of the human anterior digastric muscles. Exp Brain Res 1999;129:582–591.
12. Guggisberg AG, Dubach P, Hess CW, et al. Motor evoked potentials from masseter muscle induced by transcranial magnetic stimulation of the pyramidal tract: the importance of coil orientation. Clin Neurophysiol 2001;112:2312–2319.
13. Schmid UD, Moller AR, Schmid J. Transcranial magnetic stimulation of the trigeminal nerve: intraoperative study on stimulation characteristics in man. Muscle Nerve 1995;18:487–494.
14. Trompetto C, Caponnetto C, Buccolieri A, et al. Responses of masseter muscles to transcranial magnetic stimulation in patients with amyotrophic lateral sclerosis. Electroencephalogr Clin Neurophysiol 1998;109:309–314.
15. Desiato MT, Bernardi G, Hagi HA, et al. Transcranial magnetic stimulation of motor pathways directed to muscles supplied by cranial nerves in amyotrophic lateral sclerosis. Clin Neurophysiol 2002;113:132–140.
16. Cruccu G, Inghilleri M, Berardelli A, et al. Pathophysiology of hemimasticatory spasm. J Neurol Neurosurg Psychiatry 1994;57:43–50.
17. Cruccu G, Frisardi G, Pauletti G, et al. Excitability of the central masticatory pathways in patients with painful temporomandibular disorders. Pain 1997;73:447–454.
18. Romaniello A, Cruccu G, McMillan AS, et al. Effect of experimental pain from trigeminal muscle and skin on motor cortex excitability in humans. Brain Res 2000;882:120–127.
19. Benecke R, Meyer BU, Schonle P, et al. Transcranial magnetic stimulation of the human brain: responses in muscles supplied by cranial nerves. Exp Brain Res 1988;71:623–632.
20. Roedel RM, Laskawi R, Markus H. Cortical representation of the orbicularis oculi muscle as assessed by transcranial magnetic stimulation (TMS). Laryngoscope 2001;111:2005–2011.
21. Cruccu G, Berardelli A, Inghilleri M, et al. Corticobulbar projections to upper and lower facial motoneurons. A study by magnetic transcranial stimulation in man. Neurosci Lett 1990;117:68–73.
22. Sohn YH, Voller B, Dimyan M, et al. Cortical control of voluntary blinking: a transcranial magnetic stimulation study. Clin Neurophysiol 2004;115:341–347.
23. Liscic RM, Zidar J, Mihelin M. Evidence of direct connection of corticobulbar fibers to orofacial muscles in man: electromyographic study of individual motor unit responses. Muscle Nerve 1998;21:561–566.
24. Glocker FX, Magistris MR, Rosler KM, et al. Magnetic transcranial and electrical stylomastoidal stimulation of the facial motor pathways in Bell's palsy: time course and relevance of electrophysiological parameters. Electroencephalogr Clin Neurophysiol 1994;93:113–120.
25. Rimpilainen I, Pyykko I, Blomstedt G, et al. The site of impulse generation in transcranial magnetic stimulation of the facial nerve. Acta Otolaryngol 1993;113:339–344.
26. Wolf SR, Strauss C, Schneider W. On the site of transcranial magnetic stimulation of the facial nerve: electrophysiological observations in two patients after transection of the facial nerve during neuroma removal. Neurosurgery 1995;36:346–349.
27. Curra A, Romaniello A, Berardelli A, et al. Shortened cortical silent period in facial muscles of patients with cranial dystonia. Neurology 2000;54:130–135.

28. Urban PP, Hopf HC, Fleischer S, et al. Impaired cortico-bulbar tract function in dysarthria due to hemispheric stroke. Functional testing using transcranial magnetic stimulation. Brain 1997;120(Pt 6):1077–1084.
29. Urban PP, Wicht S, Hopf HC. Sensitivity of transcranial magnetic stimulation of cortico-bulbar vs. cortico-spinal tract involvement in amyotrophic lateral sclerosis (ALS). J Neurol 2001;248:850–855.
30. Rosler KM, Magistris MR, Glocker FX, et al. Electrophysiological characteristics of lesions in facial palsies of different etiologies. A study using electrical and magnetic stimulation techniques. Electroencephalogr Clin Neurophysiol 1995;97:355–368.
31. Werhahn KJ, Classen J, Benecke R. The silent period induced by transcranial magnetic stimulation in muscles supplied by cranial nerves: normal data and changes in patients. J Neurol Neurosurg Psychiatry 1995;59:586–596.
32. Hamdy S, Aziz Q, Rothwell JC, et al. Explaining oropharyngeal dysphagia after unilateral hemispheric stroke. Lancet 1997;350:686–692.
33. Hamdy S, Aziz Q, Rothwell JC, et al. Recovery of swallowing after dysphagic stroke relates to functional reorganization in the intact motor cortex. Gastroenterology 1998;115:1104–1112.
34. Hamdy S, Rothwell JC. Gut feelings about recovery after stroke: the organization and reorganization of human swallowing motor cortex. Trends Neurosci 1998;21:2781–2782.
35. Zald DH, Pardo JV. The functional neuroanatomy of voluntary swallowing. Ann Neurol 1999;46:281–286.
36. Muellbacher W, Artner C, Mamoli B. The role of the intact hemisphere in recovery of midline muscles after recent monohemispheric stroke. J Neurol 1999;246:250–256.
37. Ertekin C, Turman B, Tarlaci S, et al. Cricopharyngeal sphincter muscle responses to transcranial magnetic stimulation in normal subjects and in patients with dysphagia. Clin Neurophysiol 2001;112:86–94.
38. Ludlow C, Cohen L, Yin SG, et al. Studies of the laryngeal motor system in humans. Presented at the Conference on Neurological Disorders of Laryngeal Function, UCLA, Los Angeles, January 8, 1990.
39. Ludlow C, Yeh J, Cohen L, et al. Limitations in electromyography and magnetic stimulation for assessing laryngeal muscle control. Ann Otol Rhinol Laryngol 1994;103:16–27.
40. Khedr EM, Aref EE. Electrophysiological study of vocal-fold mobility disorders using a magnetic stimulator. Eur J Neurol 2002;9:259–267.
41. Thumfart W, Potoschnig C, Zorowka P, et al. Electrophysiological investigation of lower cranial nerve diseases by means of magnetically stimulated neuromyography of the larynx. Ann Otol Rhinol Laryngol 1992;101:629–634.
42. Mastaglia FL, Knezevic W, Thompson PD. Weakness of head turning in hemiplegia: a quantitative study. J Neurol Neurosurg Psychiatry 1986;49:195–197.
43. Gandevia SC, Applegate C. Activation of neck muscles from the human motor cortex. Brain 1988;111:801–813.
44. Berardelli A, Priori A, Inghilleri M, et al. Corticobulbar and corticospinal projections to neck muscle motoneurons in man. A functional study with magnetic and electric transcranial brain stimulation. Exp Brain Res 1991;87:402–406.
45. Thompson ML, Thickbroom GW, Mastaglia FL. Corticomotor representation of the sternocleidomastoid muscle. Brain 1997;120:245–255.
46. Strenge H, Jahns R. Activation and suppression of the trapezius muscle induced by transcranial magnetic stimulation. Electromyogr Clin Neurophysiol 1998;38:141–145.
47. Odergren T, Rimpilainen I, Borg J. Sternocleidomastoid muscle responses to transcranial magnetic stimulation in patients with cervical dystonia. Electroencephalogr Clin Neurophysiol 1997;105:44–52.
48. Amadio S, Panizza M, Pisano F, et al. Transcranial magnetic stimulation and silent period in spasmodic torticollis. Am J Phys Med Rehabil 2000;79:361–368.
49. Meyer BU, Liebsch R, Roricht S. Tongue motor responses following transcranial magnetic stimulation of the motor cortex and proximal hypoglossal nerve in man. Electroencephalogr Clin Neurophysiol 1997;105:15–23.
50. Muellbacher W, Boroojerdi B, Ziemann U, et al. Analogous corticocortical inhibition and facilitation in ipsilateral and contralateral human motor cortex representations of the tongue. J Clin Neurophysiol 2001;18:550–558.
51. Urban PP, Hopf HC, Zorowka PG, et al. Dysarthria and lacunar stroke: pathophysiologic aspects. Neurology 1996 Nov;47:1135–1141.
52. Muellbacher W, Mathis J, Hess CW. Electrophysiological assessment of central and peripheral motor routes to the lingual muscles. J Neurol Neurosurg Psychiatry 1994;57:309–315.
53. Ghezzi A, Baldini S. A simple method for recording motor evoked potentials of lingual muscles to transcranial magnetic and peripheral electrical stimulation. Electroencephalogr Clin Neurophysiol 1998;109:114–118.
54. Butler SL, Miles TS, Thompson PD, et al. Task-dependent control of human masseter muscles from ipsilateral and contralateral motor cortex. Exp Brain Res 2001;137:65–70.

55. Cocito D, Isoardo G, Migliaretti G, et al. Intracranial stimulation of the facial nerve: normative values with magnetic coil in 240 nerves. Neurol Sci 2003;23:307–311.
56. Ghezzi A, Callea L, Zaffaroni M, et al. Motor potentials of inferior orbicularis oculi muscle to transcranial magnetic stimulation. Comparison with responses to electrical peripheral stimulation of facial nerve. Electroencephalogr Clin Neurophysiol 1992;85:248–252.

9

Transcranial Magnetic Stimulation and Brain Plasticity

Adriana B. Conforto and Leonardo G. Cohen

For decades, it had been thought that the organization of the nervous system of adult mammals was not able to change. It was assumed that the number of available synapses and the way in which they were organized was fixed. Consequently, the capacity of the nervous system to adapt to environmental challenges or reorganize in response to lesions was very limited. Advances in recent years have shed new light over the mechanisms underlying the function and capabilities of the central and the peripheral nervous system. Discoveries such as the expression of nerve growth factor (NGF) in the adult brain,[1] demonstration of sprouting in the central nervous system,[2-4] and identification of long-term potentiation (LTP) in the hippocampus,[5,6] among others,[7,8] have challenged classic conceptions of the nervous system as rigidly wired structures that are unable to change. We now know that the central nervous system (CNS) is able to adapt to changes in the environment.[7,9-12] Such capacity has been described as *plasticity*,[13] which occurs during development and in the adult CNS. These studies led to the more modern conception that the nervous system is permanently changing, adapting to changes in the internal or external environments. These changes, although crucial in the intact nervous system for functions such as learning and memory, may play also a fundamental role underlying recovery of function after lesions.[14-31]

Mechanisms underlying plastic changes include modulation of neuronal excitability and synaptic efficacy, as well as disinhibition. For example, under certain conditions, synapses that already exist but are physiologically inactive may be disinhibited.[13,15,25,32,33] This phenomenon is called *unmasking*, and it may be caused by release from tonic inhibition, often of GABAergic origin in the cerebral cortex. Changes in synaptic efficacy may develop through activity-dependent modifications, including LTP and long-term depression (LTD).[11,27,34-36] Generalized increase in postsynaptic excitability,[37] which does not have the synaptic specificity of LTD or LTP, may underlie other forms of reorganization.[13,38] Generation of new dendritic connections leading to increase in the number of dendritic ramifications[2] or increase in axonal collaterals through horizontal pathways[4] (i.e., sprouting) constitute morphological changes that may have functional implications.

■ Evaluation of Plastic Changes in the Human Central Nervous System

In the past 30 years, different techniques became available for the noninvasive evaluation of functional activity in the human brain. These techniques were soon used to study the ability of the human brain to reorganize.

For the first time, it was possible to measure changes in for example electrical activity (e.g., with electroencephalography [EEG]) or blood flow (e.g., with functional magnetic resonance imaging [fMRI]) in various settings of plasticity. These techniques allowed investigators to formulate questions geared to understand the mechanisms underlying the ability of the human brain to reorganize. One of these techniques was transcranial magnetic stimulation (TMS). Consequently, it has become possible to stimulate in a relatively focal manner specific cortical regions. Applications of TMS to the motor cortex and measurement of the responses from contralateral muscles permits the evaluation of changes in motor cortex excitability and mapping of body part representations in the motor cortex.[39] When single TMS pulses are applied in rapid succession (i.e., trains, repetitive TMS [rTMS]), they disrupt in a relatively focal manner activity in cortical sites under the stimulating coil. This feature led to the use of rTMS to interfere with cortical functions by inducing "virtual lesions." For the first time, it became possible to evaluate the behavioral consequences of transient, noninvasive disruption of activity in specific cortical sites.[40] Overall, each technique allows the study of specific aspects of human brain function. In many cases, all together lead to a more integral understanding of physiological processes in health and disease.[41–49] This chapter describes some ways in which TMS can contribute to the understanding of plastic changes in the human brain.

Motor Learning

Motor practice is a potent trigger of cortical reorganization,[50–53] and disuse appears to reorganize in an opposite manner.[54] For example, practice of a motor task in nonhuman primates leads to expansions in the cortical representation of the practicing fingers in the motor cortex, studied with microstimulation techniques.[26,55] In human healthy volunteers, implicit motor learning is accompanied by characteristic increases in motor cortex excitability as evaluated with TMS. When explicit knowledge is achieved, motor cortical excitability changes return to pre-learning values.[50]

Learning of complex finger motor sequences in healthy volunteers leads to characteristic changes in motor cortex organization. Performance of a one-handed, five-finger exercise over the course of 5 days leads to enlarged cortical motor output targeting long finger flexor and extensor muscles involved in the motor training.[56] Acquisition of the motor skills needed for successful performance of a motor sequence is associated with enhanced excitability of the representation of that body part in the motor cortex. Although the mechanisms underlying this form of learning remain to be determined, previous TMS studies have shown that mental practice of a similar sequence leads to milder performance improvements and milder excitability changes. This finding is consistent with reports of enhanced motor evoked potential (MEP) to TMS of the motor cortex when human subjects imagine a movement.[57,58] Overall, these results suggest that voluntary motor drive is a crucial trigger of plasticity.

Use-dependent plasticity can also be elicited by repetitive performance of stereotyped, relatively simple finger movements.[59,60] In these studies, TMS of the motor cortex has been used to evoke isolated and directionally consistent thumb movements.[59] The investigators then evaluated the influence of a 30-minute practice period on the directions of TMS-evoked thumb movements. It was reported that practice led to a transient change in TMS-evoked movement directions toward the training direction (Fig. 9–1). Training led to encoding of a memory trace in the primary motor cortex, representing the kinematic details of the practiced movements. TMS, in combination with pharmacological interventions provided additional understanding of the mechanisms underlying this form of plasticity. Premedication with single doses of dextromethorphan, a N-methyl-D-aspartate (NMDA) receptor blocker; lorazepam, a γ-aminobutyric acid (GABA) type A receptor–positive allosteric modulator; and scopolamine, an anticholinergic agent, have had clear deleterious effects on encoding of this memory trace, supporting the hypothesis that NMDA, GABA, and muscarinic receptor

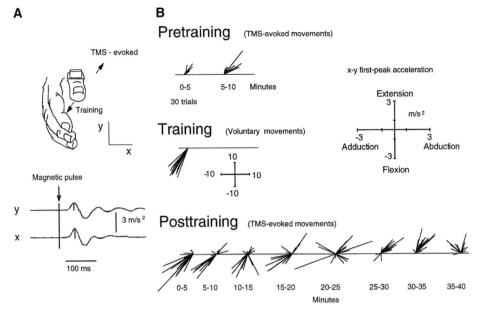

Figure 9–1 **A:** Diagram of the experimental setting. The hand was immobilized with the thumb in a neutral position. An accelerometer mounted on the thumb recorded acceleration traces after transcranial magnetic stimulation (TMS) and voluntary movements. Original acceleration signals in the horizontal (abduction and adduction) and vertical (extension and flexion) axes of thumb movements are indicated. The direction of TMS-evoked or voluntary movement was derived from the first peak acceleration in the two major axes of the movement. **B:** Directional change of the first peak acceleration vector of movements evoked by TMS after 30 minutes of training in a representative subject. Pretraining, TMS-evoked extension and abduction thumb movements are indicated. Training consisted of repetitive stereotyped brisk thumb movements in a flexion and adduction direction. TMS-derived vectors are grouped in intervals of 5 minutes. For clarity, vectors representing only the last 3 minutes of training are shown for training movements. Post-training, the direction of TMS-evoked thumb movements changed from the pretraining direction to the trained direction. The movement angle gradually changed back toward the pretraining direction after about 15 to 25 minutes. Calibration bars *(right)* refer to the pretraining and post-training vectors. (From Classen J, Liepert J, Wise SP, et al. Rapid plasticity of human cortical movement representation induced by practice. J Neurophysiol 1998;79:1117–1723.)

function are mechanisms operating in use-dependent plasticity in the intact human motor cortex.[60,61] A similar pharmacological strategy has been used to study the mechanisms underlying acquisition and recall of motor memories when healthy humans hold a robotic arm against various loads.[62] In this setting, the hand's trajectory deviates from a straight path as unpredicted forces are applied to the robotic arm. As the subject practices, an internal model of the experienced forces is created and motor commands are adjusted to compensate. GABAergic agents and NMDA receptor antagonists have blocked acquisition but not recall of the motor memory,[62] pointing to the involvement of LTP-like mechanisms in this form of plasticity.

In reports by Butefisch and colleagues,[63] it has been shown that premedication with a single oral dose of D-amphetamine can enhance the effect of motor practice on use-dependent plasticity. These results are consistent with findings in animal studies[64] and in human stroke patients[65–67] with cortical lesions.

Overall, some of these paradigms used in healthy volunteers could contribute to test strategies to promote recovery of function in patients. For example, in a recent study, Lotze and colleagues[68] evaluated the ability of different training strategies, passive and

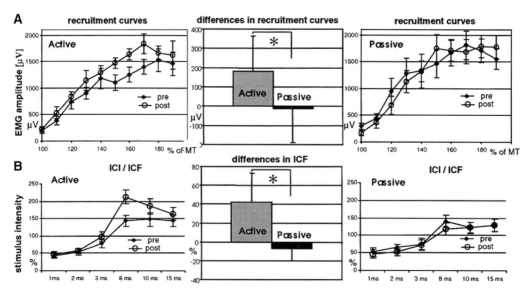

Figure 9–2 **A:** Effects of voluntary (active training) and passively elicited (passive training) movements on cortical reorganization. Recruitment curves after transcranial magnetic stimulation (TMS) increased significantly after active (left) but not after passive (right) training. Training-dependent recruitment curve changes were significantly larger for active than for passive training (center; *$P < .05$). Bars indicate standard errors. **B:** Intracortical facilitation (ICF; interstimulus interval of 8 to 15 ms) increased significantly after active training (left) but did not change after passive training (right). Differences in ICF were significantly larger for active than for passive training (center; *$P < .05$). Bars indicate standard errors. (From Lotze M, Braun C, Birbaumer N, et al. Motor learning elicited by voluntary drive. Brain 2003;126:866–872.)

active, to elicit reorganizational changes in the human motor cortex.[68] Subjects participated in two training sessions consisting of performance of voluntary and passively elicited movements. It was shown that active motor training led to greater behavioral gains than passive training, in parallel with a greater increase in contralateral sensorimotor cortex activation in fMRI. Active motor training led to an increase in recruitment curves and intracortical facilitation measured with TMS applied to the motor cortex, but passive training did not (Fig. 9–2). These results are consistent with the concept of a pivotal role of active motor training as the most effective strategy for motor recovery in neurorehabilitation.

Motor training is thought to contribute to motor learning in the intact brain and to recovery of motor function in patients with brain injury. Liepert and colleagues[69–71] evaluated the effects of a neurorehabilitative technique called *constraint-induced movement therapy*, which relies on immobilization of the intact arm of hemiparetic patients, forcing them to use the paretic arm during waking hours. This intervention results in an increase in amount of use of the paretic hand and is associated with an increase in excitability in the motor cortex of the affected hemisphere.[72] Muellbacher and colleagues[73] also reported an increase in MEP amplitude after training in a muscle involved in the practiced task (flexor pollicis brevis) but not in a muscle unrelated to the task (abductor digiti minimi) in healthy volunteers.[73] These findings have led to the intriguing hypothesis that changes in cortical excitability may contribute to behavioral gains elicited by motor training.

■ Influence of Somatosensory Input on Cortical Reorganization

Somatosensory input is required for motor learning[74] and for recovery of function after cortical lesions.[75] Deafferentation in the form of interruption and reconnection of peripheral nerves,[76] moving islands of skin to new locations across the hand,[77] operant conditioning such as discrimination of surface roughness,[78] maintenance of finger contact pressure for food reward,[79] and after training in discrimination of vibratory frequencies[17–20] leads to reorganizational changes. Nerve lesions induce a rapid expansion of motor cortical representations of intact body parts into representational zones disconnected from the periphery.[80–84]

In humans, TMS has demonstrated increases in corticomotor excitability in the body part representation nearby the deafferented cortex.[54,85–88] Transient forearm deafferentation induced by ischemic nerve block (INB) leads to an enlargement in MEP amplitudes recorded from a muscle immediately proximal to the ischemic level (i.e., biceps brachii) and disappearance of MEP from muscles distal to this level (i.e., motor block).[85] The human motor system is capable of rapid and selective facilitation of motor outputs to muscles immediately above the deafferentation level. Pharmacological interventions in combination with TMS showed that administration of lorazepam (a GABAergic agent) and lamotrigine (a voltage-gated sodium and calcium channel blocker) blocked this form of plasticity.[89] Together with the finding of decrease in GABA levels in the sensorimotor cortex during INB,[90] these results are highly suggestive of the involvement of GABA-related disinhibition as one of the mechanisms operating in this form of plasticity. Ziemann and colleagues[91] also demonstrated that application of rTMS to the arm representation of the motor cortex could upregulate these plastic changes, whereas rTMS to the face representation downregulated the INB-dependent effect. The relevance of this study is that it demonstrated the possibility of modulation of cortical plasticity by selective application of TMS to focal body part representations within the motor cortex. In a different experiment, Ziemann and colleagues[92] tested the hypothesis that INB-dependent decrease in GABA in the motor cortex contralateral to the INB could facilitate training-dependent improvements in motor function. When practice of a motor task in the upper arm was coupled with forearm and hand deafferentation by INB, MEPs recorded in the biceps brachii proximal to the INB as well as paired-pulse excitability increased compared with motor practice alone or INB alone in healthy subjects (Fig. 9–3). More importantly, the increase in biceps MEP size induced by motor practice coupled with INB was paralleled by an increase in peak acceleration of elbow flexion movements.

In addition to changes in body part representations near the deafferented one, INB conditions an increase in motor cortical excitability for muscles targeting the nondeafferented hand (contralateral to INB), an effect consistent with a decrease in interhemispheric inhibition.[93] It is intriguing that in addition to these changes in the motor domain, INB elicits an improvement in tactile spatial acuity in the nondeafferented hand that is accompanied by increased processing in the somatosensory cortex.[94]

On the other side of the spectrum of deafferentation, somatosensory stimulation timed to the application of a cortical stimulus[95] or in the form of prolonged peripheral nerve stimulation[96–99] results in an increase in motor cortical excitability of the stimulated body part. The finding that this effect can be blocked by premedication with lorazepam and dextromethorphan has suggested the involvement of GABAergic and NMDA-dependent mechanisms.[99,100]

Longer-lasting modifications in sensory input were seen, for example, in patients with complete upper limb paralysis secondary to traumatic cervical root avulsion who had undergone surgical anastomosis of the intercostal nerve to the musculocutaneous nerve to restore biceps brachii function and therefore allow arm movements.[101] In the first 4 to 6 months after surgery, motor unit discharges were recorded in biceps in association with respiration. Between 1 and 2 years later, motor unit discharges became independent of respiration and more susceptible to voluntary contraction. With TMS mapping, immediately after the anastomosis, the biceps motor map was located in the region of intercostal muscles. With time and training to control

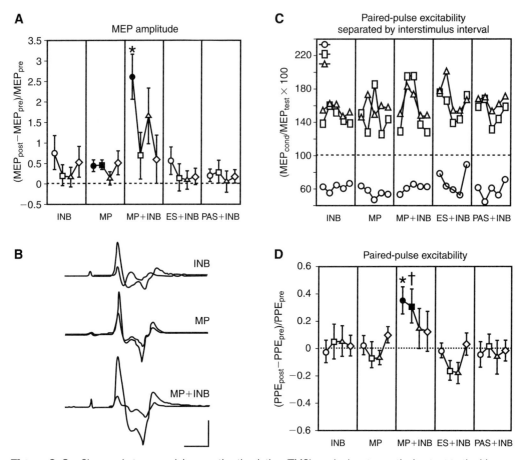

Figure 9–3 Changes in transcranial magnetic stimulation (TMS)–evoked motor cortical output to the biceps muscle as induced by different interventions. **A:** Motor evoked potential (MEP) amplitude at the end of intervention *(circles)* and 20 minutes *(squares)*, 40 minutes *(triangles)*, and 60 minutes *(diamonds)* later are given as increments of the pre-intervention measurements (mean ± standard error). Solid symbols indicate significant differences from zero ($P < .05$). *Different from all other interventions at this time point ($P < .05$). Notice that hand deafferentation in combination with practice in performance of an upper arm movement led to a substantial enlargement of biceps MEP amplitudes. **B:** Biceps MEP (average of 10 trials) in one subject before *(thin lines)* and at the end of intervention *(thick lines)*. Calibration bars are 15 ms *(horizontal)* and 0.25 mV *(vertical)*. **C:** Intervention-induced changes in paired-pulse excitability (PPE) are shown separately for the three interstimulus intervals of 4 ms *(circles)*, 10 ms *(squares)*, and 15 ms *(triangles)*. The five data points for each intervention are before intervention, late into intervention, and 20, 40, and 60 minutes after the end of intervention. **D:** Intervention-induced changes in PPE. †Different from all interventions except INB ($P < .05$). Other conventions are as in **A**. INB, ischemic nerve block at the forearm; MP, motor practice; MP + INB, ES + INB, and PAS + INB, motor practice, electrical stimulation of the biceps muscle, and passive elbow flexion movements, respectively, during INB. (From Ziemann U, Muellbacher W, Hallett M, et al. Modulation of practice-dependent plasticity in human motor cortex. Brain 2001;124:1171–1181.)

biceps voluntarily, the biceps map moved laterally toward the cortical regions normally representing the arm. The change in map representation suggests that cortical reorganization occurring months after nerve anastomosis has mediated the improved biceps motor control.

■ Cross-Modal Plasticity

Cortical reorganization takes place within a specific sensory modality and across modalities (i.e., cross-modal plasticity). For example, congenitally or early blind patients activate the visual cortex to a larger extent than sighted

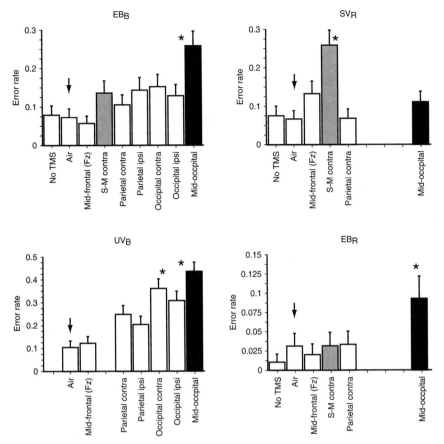

Figure 9–4 Error rates (mean ± SE) for stimulation of different positions in early-blind subjects identifying Braille letters (EB$_B$), early-blind subjects identifying embossed Roman letters (EB$_R$), sighted individuals identifying embossed Roman letters (SV$_R$), and a different group of early-blind subjects identifying Braille letters (UV$_B$). Missing bars indicate that stimulation at that position was not performed in that group. *Black bars* indicate error rates induced by stimulation of the mid-occipital position, and *gray bars* indicate the error rates induced by stimulation of the contralateral sensorimotor cortex. In both groups of early blind subjects, stimulation of the mid-occipital position induced more errors in reading Braille and Roman letters than stimulation of any other position, whereas in the sighted volunteers, stimulation of the contralateral primary sensorimotor region induced more errors than stimulation of any other position. Asterisks indicate scalp positions where significantly more errors occurred than control (air, marked with *arrows*). S-M, sensorimotor cortex; contra, contralateral; ipsi, ipsilateral. *$P < .001$. (Modified from Cohen LG, Celnik P, Pascual-Leone A, et al. Functional relevance of cross-modal plasticity in blind humans. Nature 1997;389:180–183.)

controls when performing a tactile discriminative task or when reading Braille.[102] TMS has been used in this case to determine the functional relevance of this occipital activation. Cohen and colleagues[103] evaluated the behavioral consequences of disruption of activity in the occipital cortex of blind and sighted individuals. It was shown that TMS applied to the occipital cortex, but not other cortical regions, resulted in a higher number of accuracy errors in tactile discrimination in early blind than in sighted controls[103] (Fig. 9–4) or late blind subjects.[104] These results indicate that the critical period for this form of functionally relevant cross-modal plasticity does not extend beyond the teens. Overall, these observations indicate that a cortical function that is normally present but restricted to specific modalities can change its specificity in response to how it is used and to the magnitude of sensory inputs.[104]

Hemispheric Lesions

TMS has also been used to evaluate cortical reorganization after cortical lesions. Perhaps the most dramatic example of unilateral cortical lesion is hemispherectomy. In some patients with hemispherectomy at an early age who recovered motor control of the intact arm, stimulation with TMS of the remaining hemisphere elicited MEP in the affected hand. This finding suggests that the intact hemisphere is capable of at least partially taking over control of ipsilateral motor function.[105] In the setting of lesions acquired later in life, such as stroke, it is still controversial whether uncrossed motor pathways are important for functional recovery after damage of the corticospinal pathway. Imaging studies have suggested a role for the unaffected hemisphere in recovery after stroke.[106-109] However, TMS studies have identified the presence of ipsilateral MEPs in the paretic arm, predominantly in cases with poor recovery.[110] However, the presence of contralateral responses after stimulation of the affected hemisphere is consistently associated with better chances of recovery.[111-112] It is conceivable that nonprimary motor regions such as dorsal or ventral premotor cortex could contribute to recovery of motor function.[113,114] TMS applied to the ipsilateral dorsal premotor cortex 100 ms after a cue to move has slowed the performance of the paretic hand in a simple reaction-time task in stroke patients with focal lesions of corticospinal outflow originated in M1.[113] The same parameters of stimulation have not affected task performance in healthy individuals.

Conclusion

TMS is one of several techniques that have contributed to the study of plasticity in cortical circuits in health and disease. Integration of TMS with other neurophysiological and neuroimaging techniques will likely shed light on the mechanisms underlying cortical plasticity. TMS also can be used to modulate plastic processes of possible relevance for neurorehabilitation.

REFERENCES

1. Levi-Montalcini R, Skaper SD, Dal Toso R, et al. Nerve growth factor: from neurotrophin to neurokine. Trends Neurosci 1996;19:514–520.
2. Trachtemberg JT, Chen BE, Knott GW, et al. Long-term in vivo imaging of experience-dependent synaptic plasticity in adult cortex. Nature 2002;420:788–794.
3. Fischer M, Kaech S, Knutti D, et al. Rapid actin-based plasticity in dendritic spines. Neuron 1998;20:847–854.
4. Darian-Smith C, Gilbert CD. Axonal sprouting accompanies functional reorganization in adult cat striate cortex. Nature 1994;368:737–740.
5. Bear MF, Malenka RC. Synaptic plasticity: LTP and LTD. Curr Opin Neurobiol 1994;4:389–399.
6. Kelso SR, Brown TH. Differential conditioning of associative synaptic enhancement in hippocampal brain slices. Science 1986;232:85–87.
7. Kaas JH. Plasticity of sensory and motor maps in adult mammals. Annu Rev Neurosci 1991;14:137–167.
8. Klintsova AY, Greenough WT. Synaptic plasticity in cortical systems. Curr Opin Neurobiol 1999;9:203–208.
9. Jenkins WM, Merzenich MM. Reorganization of neocortical representations after brain injury: a neurophysiological model of the bases of recovery from stroke. Prog Brain Res 1987;71:249–266.
10. Pons TP, Garraghty PE, Mishkin M. Lesion-induced plasticity in the somatosensory cortex of adult macaques. Proc Natl Acad Sci U S A 1988;85:5279–5281.
11. Buonomano DV, Merzenich MM. Cortical plasticity: from synapses to maps. Annu Rev Neurosci 1998;21:149–186.
12. Sanes JN, Donoghue JP. Plasticity and primary motor cortex. Annu Rev Neurosci 2000;23:393–415.
13. Donoghue JP, Hess G, Sanes JN. Substrates and mechanisms for learning in motor cortex. In Bloedel J, Ebner T, Wise SP (eds): Acquisition of Motor Behavior in Vertebrates. Cambridge, MA, MIT Press, 1996:363–386.
14. Garraghty PE, LaChica EA, Kaas JH. Injury-induced reorganization of somatosensory cortex is accompanied by reductions in GABA staining. Somatosens Mot Res 1992;8:347–354.
15. Sanes JN, Donoghue JP. Static and dynamic organization of motor cortex. Adv Neurol 1997;73:277–296.
16. Kaas JH, Florence SL. Mechanisms of reorganization in sensory systems of primates

after peripheral nerve injury. Adv Neurol 1997;73:147–158.
17. Recanzone GH, Merzenich MM, Schreiner CE. Changes in the distributed temporal response properties of SI cortical neurons reflect improvements in performance on a temporally based tactile discrimination task. J Neurophysiol 1992;67:1071–1091.
18. Recanzone GH, Merzenich MM, Jenkins WM. Frequency discrimination training engaging a restricted skin surface results in an emergence of a cutaneous response zone in cortical area 3a. J Neurophysiol 1992;67:1057–1070.
19. Recanzone GH, Merzenich MM, Jenkins WM, et al. Topographic reorganization of the hand representation in cortical area 3b owl monkeys trained in a frequency-discrimination task. J Neurophysiol 1992;67:1031–1056.
20. Recanzone GH, Jenkins WM, Hradek GT, et al. Progressive improvement in discriminative abilities in adult owl monkeys performing a tactile frequency discrimination task. J Neurophysiol 1992;67:1015–1030.
21. Nicolelis MAL. Dynamic and distributed somatosensory representations as the substance for cortical and subcortical plasticity. Semin Neurosci 1997;9:24–33.
22. Jones EG, Pons TP. Thalamic and brain stem contributions to large-scale plasticity of primate somatosensory cortex. Science 1998;282:1121–1125.
23. Jones TA, Schallert T. Overgrowth and pruning of dendrites in adult rats recovering from neocortical damage. Brain Res 1992;581:156–160.
24. Jones TA, Schallert T. Use-dependent growth of pyramidal neurons after neocortical damage. J Neurosci 1994;14:2140–2152.
25. Jacobs KM, Donoghue JP. Reshaping the cortical motor map by unmasking latent intracortical connections. Science 1991;251:944–947.
26. Nudo RJ, Wise BM, SiFuentes F, et al. Neural substrates for the effects of rehabilitative training on motor recovery afterischemic infarct. Science 1996;272:1791–1794.
27. Hess G, Donoghue JP. Long-term potentiation of horizontal connections provides a mechanism to reorganize cortical motor maps. J Neurophysiol 1994;71: 2543–2547.
28. Hess G, Aizenman CD, Donoghue JP. Conditions for the induction of long-term potentiation in Layer II/III horizontal connections of the rat motor cortex. J Neurophysiol 1996;75:1765–1777.
29. Stroemer RP, Kent TA, Hulsebosch CR. Neocortical neural sprouting, synaptogenesis, and behavioral recovery after neocortical infarction in rats. Stroke 1995;26:2135–2144.
30. Risedal A, Mattsson B, Dahlqvist P, et al. Environmental influences on functional outcome after a cortical infarct in the rat. Brain Res Bull 2002;58:315–321.
31. Knott GW, Quairiaux C, Genoud C, et al. Formation of dendritic spines with GABAergic synapses induced by whisker stimulation in adult mice. Neuron 2002;34:265–273.
32. Donoghue JP, Hess G, Sanes JN. Substrates and mechanisms for learning in motor cortex. In Bloedel J, Ebner T and Wise SP (eds): Acquisition of Motor Behavior in Vertebrates. Cambridge, MA, MIT Press, 1996:363–386.
33. Huntley GW. Correlation between patterns of horizontal connectivity and the extent of short-term representational plasticity in rat motor cortex. Cereb Cortex 1997;7:143–156.
34. Asanuma H, Pavlides C. Neurobiological basis of motor learning in mammals. Neuroreport 1997;8:i/vi.
35. Cruikshank SJ, Weinberger NM. Receptive-field plasticity in the adult auditory cortex induced by Hebbian covariance. J Neurosci 1996;16:861–875.
36. Kirkwood A, Bear MF. Hebbian synapses in visual cortex. J Neurosci 1994;14:1634–1645.
37. Woody CD, Gruen E, Birt D. Changes in membrane currents during Pavlovian conditioning of single cortical neurons. Brain Res 1991;539:76–84.
38. Donoghue JP. Plasticity of adult sensorimotor representations. Curr Opin Neurobiol 1995;5:749–754.
39. Cohen LG, Ziemann U, Chen R, et al. Studies of neuroplasticity with transcranial magnetic stimulation. J Clin Neurophysiol 1998;15:305–324.
40. Walsh V, Rushworth M. A primer of magnetic stimulation as a tool for neuropsychology. Neuropsychologia 1999;37:125–135.
41. Ilmonemi R, Ruohonen J, Karhu J. Transcranial magnetic stimulation—a new tool for functional imaging of the brain. Crit Rev Biomed Eng 1999;27:241–284.
42. Wassermann EM, Wang B, Zeffiro TA, et al. Locating the motor cortex on the MRI with transcranial magnetic stimulation and PET. Neuroimage 1996;3:1–9.
43. Foltys H, Kemeny S, Krings T, et al. The representation of the plegic hand in the motor cortex: a combined fMRI and TMS study. Neuroreport 2000;11:147–150.
44. Krings T, Buchbinder BR, Butler WE, et al. Functional magnetic resonance imaging and transcranial magnetic stimulation: complementary approaches in the evaluation of cortical motor function. Neurology 1997;48:1406–1416.
45. Rossini PM, Narici L, Martino G, et al. Analysis of interhemispheric asymmetries of somatosensory evoked magnetic fields to right and left median nerve stimulation. Electroencephalogr Clin Neurophysiol 1994;91:476–482.

46. Paus T, Wolforth M. Transcranial magnetic stimulation during PET: reaching and verifying the target site. Hum Brain Mapp 1998;6:399–402.
47. Strenz LH, Oliviero A, Bloem BR, et al. The effects of subthreshold 1 Hz repetitive TMS on cortico-cortical and interhemispheric coherence. Clin Neurophysiol 2002;113:1279–1285.
48. Armin C, Alkadhi H, Crelier GR, et al. Changes of non-affected upper limb cortical representation in paraplegic patients as assessed by fMRI. Brain 2002;125:2567–2578.
49. Lotze M, Braun C, Birbaumer N, et al. Motor learning elicited by voluntary drive. Brain 2003;126:866–872.
50. Pascual-Leone A, Grafman J, Hallett M. Modulation of cortical motor output maps during development of implicit and explicit knowledge. Science 1994;263:1287–1289.
51. Pascual-Leone A, Tarazona F, Catala MD. Applications of transcranial magnetic stimulation in studies on motor learning. Electroencephalogr Clin Neurophysiol 1999;51(Suppl):157–161.
52. Karni A, Meyer E, Rey-Hypolito C, et al. The acquisition of skilled motor performance: fast and slow experience-driven changes in primary motor cortex. Proc Natl Acad Sci U S A 1998;95:861–868.
53. Shadmehr R, Holcomb HH. Neural correlates of motor memory consolidation. Science 1997;277:821–825.
54. Liepert J, Tegenthoff M, Malin JP. Changes of cortical motor area size during immobilization. Electroencephalogr Clin Neurophysiol 1995;97:382–386.
55. Nudo RJ, Milliken GW, Jenkins WM, et al. Use-dependent alterations of movement representations in primary motor cortex of adult squirrel monkeys. J Neurosci 1996;16:785–807.
56. Pascual-Leone A, Dang N, Cohen LG, et al. Modulation of muscle responses evoked by transcranial magnetic stimulation during the acquisition of new fine motor skills. J Neurophysiol 1995;74:1037–1045.
57. Rossi S, Pasqualetti P, Tecchio F, et al. Corticospinal excitability modulation during mental simulation of wrist movements in human subjects. Neurosci Lett 1998;243:147–151.
58. Rossini PM, Rossi S, Pasqualetti P, et al. Corticospinal excitability modulation to hand muscles during movement imagery. Cereb Cortex 1999;9:161–167.
59. Classen J, Liepert J, Wise SP, et al. Rapid plasticity of human cortical movement representation induced by practice. J Neurophysiol 1998;79:1117–1723.
60. Butefisch CM, Davis C, Wise SP, et al. Mechanisms of use-dependent plasticity in the human motor cortex. Proc Natl Acad Sci U S A 2000;97;3661–3665.
61. Sawaki L, Boroojerdi B, Kaelin-Lang A, et al. Cholinergic influences on use-dependent plasticity. J Neurophysiol 2002;87:166–171.
62. Donchin O, Sawaki L, Madupu G, et al. Mechanisms influencing acquisition and recall of motor memories. J Neurophysiol 2002;88:2114–2123.
63. Butefisch CM, Davis BC, Sawaki L. Modulation of use-dependent plasticity by D-amphetamine. Ann Neurol 2002;51:59–68.
64. Feeney DM, Hovda DA. Amphetamine and apomorphine restore tactile placing after motor cortex injury in the cat. Psychopharmacology 1983;79:67–71.
65. Cristostomo EA, Duncan PW, Propst M, et al. Evidence that amphetamine with physical therapy promotes recovery of motor function in stroke patients. Ann Neurol 1988;23:94–97.
66. Walker-Batson D, Curtis S, Natarajan R, et al. A double-blind, placebo-controlled study of the use of amphetamine in the treatment of aphasia. Stroke 2001;32:2093–2098.
67. Walker-Batson D, Smith P, Curtis S, et al. Amphetamine paired with physical therapy accelerates motor recovery after stroke. Further evidence. Stroke 1995;26:2254–2259.
68. Lotze M, Braun C, Birbaumer N, et al. Motor learning elicited by voluntary drive. Brain 2003;126:866–872.
69. Taub E. Constraint-induced movement therapy and massed practice. Stroke 2000;31:986–988.
70. Kopp B, Kunkel A, Muhlnickel W, et al. Plasticity in the motor system related to therapy-induced improvement of movement after stroke. Neuroreport 1999;10:807–810.
71. Kunkel A, Kopp B, Muller G, et al. Constraint-induced movement therapy: a powerful new technique to induce motor recovery in chronic stroke patients. Arch Phys Med Rehabil 1999;80:624–628.
72. Liepert J, Bauder H, Miltner WHR, et al. Treatment-induced cortical reorganization after stroke in humans. Stroke 2000;31:1210–1216.
73. Muellbacher W, Ziemann U, Boroojerdi B, et al. Role of the human motor cortex in rapid motor learning. Exp Brain Res 2001;136:431-438.
74. Pavlides C, Miyashita E, Asanuma H. Projection from the sensory to the motor cortex is important in learning motor skills in the monkey. J Neurophysiol 1993;70:733–741.
75. Reding MJ, Potes E. Rehabilitation outcome following initial unilateral hemispheric stroke. Life table analysis approach. Stroke 1988;19:1354–1358.
76. Wall JT, Kaas JH, Sur M, et al. Functional reorganization in somatosensory cortical areas 3b and 1 of adult monkeys after median nerve repair: possible relationships

to sensory recovery in humans. J Neurosci 1986;6:218–233.
77. Merzenich MM, Recanzone G, Jenkins WM, et al. Cortical representation of plasticity. In Rakic P, Singer W (eds): Neurobiology of Neocortex. New York, Wiley S Bernhard, 1988:41–67.
78. Guic E, Rodriguez E, Caviedes P, et al. Use-dependent reorganization of the barrel field in adult rats. Soc Neurosci Abstr 1993;19:163.
79. Jenkins WM, Merzenich MM, Recanzone G. Neocortical representational dynamics in adult primates: implications for neuropsychology. Neuropsychologia 1990;28:573–584.
80. Sanes JN, Suner S, Lando JF, et al. Rapid reorganization of adult rat motor cortex somatic representation patterns after motor nerve injury. Proc Natl Acad Sci U S A 1988;85:2003–2007.
81. Donoghue JP, Suner S, Sanes JN. Dynamic organization of primary motor cortex output to target muscles in adult rats. II. Rapid reorganization following motor nerve lesions. Exp Brain Res 1990;79:492–503.
82. Merzenich MM, Kaas JH, Wall J, et al. Topographic reorganization of somatosensory cortical areas 3b and 1 in adult monkeys following restricted deafferentation. Neuroscience 1983;8:33–55.
83. Donoghue JP, Sanes JN. Peripheral nerve injury in developing rats reorganizes representation pattern in motor cortex. Proc Natl Acad Sci U S A 1987;84:1123–1126.
84. Sanes JN, Suner S, Donoghue JP. Dynamic organization of primary motor cortex output to target muscles in adult rats. I. Long-term patterns of reorganization following motor or mixed peripheral nerve lesions. Exp Brain Res 1990;79:479–491.
85. Brasil-Neto JP, Cohen LG, Pascual-Leone A, et al. Rapid reversible modulation of human motor outputs after transient deafferentation of the forearm: a study with transcranial magnetic stimulation. Neurology 1992;42:1302–1306.
86. Rossini PM, Rossi S, Tecchio F, et al. Focal brain stimulation in healthy humans: motor maps changes following partial hand sensory deprivation. Neurosci Lett 1996;214:191–195.
87. Zanette G, Tinazzi M, Bonato C, et al. Reversible changes of motor cortical outputs following immobilization of the upper limb. Electroencephalogr Clin Neurophysiol 1997;105:269–279.
88. Ridding MC, Rothwell JC. Stimulus/response curves as a method of measuring motor cortical excitability in man. Electroencephalogr Clin Neurophysiol 1997;105:340–344.
89. Ziemann U, Hallett M, Cohen LG. Mechanisms of deafferentation-induced plasticity in human motor cortex. J Neurosci 1998;18:7000–7007.
90. Levy LM, Ziemann U, Chen R, et al. Rapid modulation of GABA in sensorimotor cortex induced by acute deafferentation. Ann Neurol 2002;52:755–761.
91. Ziemann U, Wittenberg G, Cohen LG. Stimulation-induced within-representation and across-representation plasticity in human motor cortex. J Neurosci 2002;22:5563–5571.
92. Ziemann U, Muellbacher W, Hallett M, et al. Modulation of practice-dependent plasticity in human motor cortex. Brain 2001;124:1171–1181.
93. Werhahn K, Mortensen J, Kaelin-Lang A, et al. Cortical excitability changes induced by deafferentation of the contralateral hemisphere. Brain 2002;125:1402–1413.
94. Werhahn KJ, Mortensen J, VanBoven RW, et al. Enhanced tactile spatial perception and cortical processing during acute hand deafferentation. Nat Neurosci 2002;5:936–938.
95. Stefan K, Kunesch E, Cohen LG, et al. Induction of plasticity in the human motor cortex by paired associative stimulation. Brain 2000;123:572–584.
96. Hamdy S, Rothwell JC, Aziz Q, et al. Long-term reorganization of human motor cortex driven by short-term sensory stimulation. Nat Neurosci 1998;1;64–68.
97. Ridding MC, Brouwer B, Miles TS, et al. Changes in muscle responses to stimulation of the motor cortex induced by peripheral nerve stimulation in human subjects. Exp Brain Res 2000;131:135–143.
98. Ridding MC, McKay DR, Thompson PD, et al. Changes in corticomotor representations induced by prolonged peripheral nerve stimulation in humans. Clin Neurophysiol 2001;112:1461–1469.
99. Kaelin-Lang A, Luft AR, Sawaki L, et al. Modulation of human corticomotor excitability by somatosensory input. J Physiol 2002;540.2:623–633.
100. Wolters A, Sandbrink F, Schlottmann A, et al. A temporally asymmetric Hebbian rule governing plasticity in the human motor cortex. J Neurophysiol 2003; 89:2239–2245.
101. Mano Y, Nakamuro T, Tamura R, et al. Central motor reorganization after anastomosis of the musculocutaneous and intercostal nerves following cervical root avulsion. Ann Neurol 1995;38:15–20.
102. Sadato N, Okada T, Honda M, et al. Critical periods for cross-modal plasticity in blind humans: a functional MRI study. Neuroimage 1996;16:389–400.
103. Cohen LG, Celnik P, Pascual-Leone A, et al. Functional relevance of cross-modal plasticity in blind humans. Nature 1997;389:180–183.

104. Cohen LG, Weeks RA, Sadato N, et al. Period of susceptibility for cross-model plasticity in the blind. Ann Neurol 1999;45:451–460.
105. Shimizu T, Nariai T, Maehara T, et al. Enhanced motor cortical excitability in the unaffected hemisphere after hemispherectomy. Neuroreport 2000;11:3077–3084.
106. Seitz RJ, Azari NP, Knorr U, et al. The role of diaschisis in stroke recovery. Stroke 1999;30:1844–1850.
107. Chollet F, DiPiero V, Wise RJ, et al. The functional anatomy of motor recovery after stroke inhumans: a study with positron emission tomography. Ann Neurol 1991;29:63–71.
108. Weiller C, Chollet F, Friston KJ, et al. Functional reorganization of the brain in recovery from striatocapsular infarction in man. Ann Neurol 1992;31:463–472.
109. Marshall RS, Perera GM, Lazar RM, et al. Evolution of cortical activation during recovery from corticospinal tract infarction. Stroke 2000;31:656–661.
110. Turton A, Wroe S, Trepte N, et al. Contralateral and ipsilateral EMG responses to transcranial magnetic stimulation during recovery of arm and hand function after stroke. Electroencephalogr Clin Neurophysiol 1996;101:316–328.
111. Trompetto C, Assini A, Buccolieri A, et al. Motor recovery following stroke: a transcranial magnetic stimulation study. Clin Neurophysiol 2000;111:1860–1867.
112. Cicinelli P, Traversa R., Rossini PM. Post-stroke reorganization of brain motor output to the hand: a 2–4 month follow-up with focal magnetic transcranial stimulation. Electroencephalogr Clin Neurophysiol 1997;105:438–450.
113. Johansen-Berg H, Rushworth MF, Bogdanovic MD, et al. The role of ipsilateral premotor cortex in hand movement after stroke. Proc Natl Acad Sci U S A 2002;99:14518–14523.
114. Fridman EA, Hanakawa T, Chung M, et al. Reorganization of the human ipsilesional premotor cortex after stroke. *Brain* 2004;127(Pt 4):747–758.

10 Transcranial Magnetic Stimulation in Amyotrophic Lateral Sclerosis

Ryuji Kaji and Nobuo Kohara

Amyotrophic lateral sclerosis (ALS) is a disease of upper and lower motoneurons, with a peculiar sparing of extraocular and sphincter muscles. Its pathogenesis is unclear, but findings of recent studies indicated glutamate-induced excitatory cell deathas one of the possible mechanisms of neuronal loss.[1,2] With the discovery of a genetic mutation of Cu-Zn superoxide dismutase 1 *(SOD1)* in familial ALS, much attention has been drawn to oxidative stress as a mechanism,[3] because the gene product SOD1 copes with free radicals. Subsequent transgenic animal studies, however, showed that the gain of function by abnormal human *SOD1* gene, but not the deficiency of the function, causes the disease.[4]

Electrophysiologic methods are useful in clinical diagnosis of ALS. The most important is to rule out treatable causes of muscle weakness or atrophy. Needle electromyography (EMG) can detect denervation of clinically normal muscles. Nerve conduction studies may reveal conduction block or slowing, which are characteristic of treatable demyelinating neuropathies, such as multifocal motor neuropathy.[5] A diagnostic problem may arise when pure lower motoneuron signs are present without conduction block. This may be a form of ALS or spinal progressive muscular atrophy. Otherwise, the case may be chronic motor axonal neuropathy, which is treatable with intravenous immunoglobulins or immunosuppressants.[6] It would be helpful if transcranial magnetic stimulation (TMS) could uncover subclinical upper motoneuron involvement in this case. TMS could be a tool for differential diagnosis of ALS from other treatable neuropathies.

In this chapter, we review the conventional motor evoked potential (MEP) findings in ALS and their diagnostic usefulness and then depict a more detailed method of analyzing the excitatory postsynaptic potentials (EPSPs) at the lower motoneuron using peri-stimulus time histogram (PSTH). The latter was used to explore the pathophysiology of ALS.

■ Motor Evoked Potential Recording Using Surface Electrodes

MEPs are usually recorded from the first dorsal interosseous (FDI) muscle. FDI is favored because surface electrodes placed over this muscle allow minimal contamination from other muscles. In patients with severe lower motoneuron involvement, such as those with compound muscle action potential amplitudes of less than 1 mV (baseline to peak), the central conduction may not be precisely reflected in MEPs. In such cases, another muscle should be sought for recording. A round stimulating coil can be placed over the vertex so that induced current in the brain be passed anteriorly at the motor cortex. For testing the resting motor threshold (RMT), MEP should be monitored using an amplifier gain of 50 μV per division. Stimulus intensities of minimal increments and decrements should be tested, and the intensity just to elicit an all or none response in 5 of 10 trials is defined as the threshold. The latency and the waveform of MEP should be recorded using large or supramaximal stimulus intensities at rest and at weak muscle contraction, although supramaximal stimulation is not always possible.

Motoneuron Firing Evoked by Transcranial Magnetic Stimulation

MEPs are evoked through a sequence of events. Electric current induced by transcranial magnetic stimulation stimulates the upper motoneurons in the primary motor cortex directly or indirectly. Impulses are transmitted through the axons down to the anterior horn cells of the spinal cord, and EPSPs are generated at the postsynaptic membrane of the lower motoneuron. If temporally and spatially summated EPSPs reach the threshold, the lower motoneuron fires, and the impulse along the peripheral motor axon activates the muscle through neuromuscular transmission, evoking MEPs. Abnormalities at any step in these could produce MEP changes in ALS. MEPs evoked by TMS are mostly mediated by monosynaptic direct corticospinal projections.[7,8]

Motor Evoked Potential Abnormalities in Amyotrophic lateral sclerosis

Resting Motor Evoked Potential Threshold

In our series,[9] 35 patients with ALS who later fulfilled the clinical criteria[10,11] were divided into those with pure lower motoneuron signs (PMA, n = 5), those with normal or decreased finger flexion reflexes but with upper motoneuron signs elsewhere (ALS-ND, n = 12), those with exaggerated finger flexor reflexes (ALS-U, n = 14), and those with pure upper motoneuron signs (PLS, n = 4) at the initial stage of diagnostic work-up. The other group was those who did not fulfill the ALS criteria but turned out to have other diseases with lower motoneuron involvement (LMN, n = 12). All PLS patients showed increased

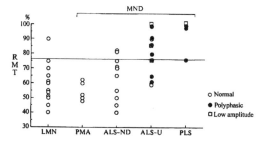

Figure 10–1 Plots of resting motor threshold (RMT) in diseases affecting lower motoneurons (LMN) excluding amyotrophic lateral sclerosis (ALS), progressive muscular atrophy type of ALS (PMA), ALS with normal or decreased finger flexion reflex of the tested hand (ALS-ND), ALS with increased finger flexion reflex of the tested hand (ALS-U), and primary lateral sclerosis type of ALS (PLS). Cases with polyphasic motor evoked potentials (MEPs) are shown with *solid circles*, those with low amplitude MEPs with *rectangles*, and normal MEP waveforms with *open circles*. (Adapted from Kohara N, Kaji R, Kojima Y, et al. An electrophysiological study of the corticospinal projections in amyotrophic lateral sclerosis. Clin Neurophysiol 1999;110:1123–1132.)

RMTs, and all except one from the PMA and LMN groups had normal thresholds (Fig. 10–1). It was increased in 8 (57%) of ALS-U patients and in 4 (33%) of ALS-ND patients. These findings suggest that RMT is a good indicator of upper motoneuron involvement.

What are the mechanisms of increased RMT in TMS? The resting membrane potential of the lower motoneuron is usually several millivolts below the threshold potential. If EPSPs generated by TMS exceeds this level, the lower motoneuron fires. Increased RMT may be caused by the increased gap between the resting membrane and threshold potentials or by the loss of upper motoneurons that demands compensatory excitation of the larger cortical area. The fact that increased RMT was seen mostly in those with upper motoneuron signs suggests the latter being the mechanism.

For understanding the generation of EPSPs at the lower motoneuron, the concept of *multiple descending volleys* is essential. The cortical motoneurons discharge several times

When stimulated with higher intensities, I waves are preceded by an earlier peak (i.e., D wave), which represents direct excitation of the cortical neuron.[7,14] The I and D waves are highly synchronized volleys generated by numerous cortical neurons, and EPSPs generated by these volleys are summated spatially and temporally.

If cortical neurons are decreased in number, the spatial summation of EPSPs becomes incomplete. When volleys reaching the lower motoneuron vary in time, the temporal summation is impaired. In both cases, the threshold should increase or MEP amplitudes decrease despite large stimulus intensities. However, ALS patients could have small MEP amplitudes because of lower motoneuron involvement. The threshold increase therefore is a better clinical indicator of upper motoneuron involvement.

Motor Evoked Potential Latency

The *central motor conduction time* (CMCT) is defined by the time interval between MEP latencies after cortical and cervical stimulation. If cervical stimulation is not possible, F-wave latencies are used to calculate the peripheral conduction time. Previous studies showed prolonged CMCT and normal peripheral conduction time in ALS.[15] In this section, we discuss MEP latencies after cortical stimulation in ALS.

In our series of 27 ALS patients whose MEPs were recorded from FDI, only 20% had prolonged latencies[9] (Fig. 10–3). In ALS, the rise time of EPSPs at the lower motoneuron is the major factor in determining the latency. It may become slower if the synchrony of the cortical motoneuron firing is impaired or if the loss of axons causes temporal dispersion of the volleys. The efficacy of the synaptic transmission could also result in the loss of temporal summation. Although comparisons between groups of normal subjects and ALS patients could produce a significant difference, the latency alone is rarely useful in the diagnosis of individual cases. This lack of sensitivity is partly because large stimulus intensities used in ALS can resynchronize the volley by stimulating larger number of cortical neurons.

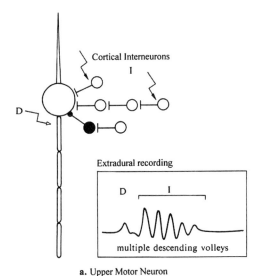

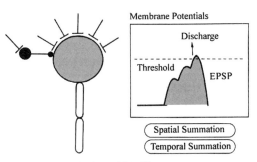

Figure 10–2 An illustration of the multiple descending volleys after transcranial magnetic stimulation (TMS). **A:** Mechanism at the upper motoneuron level. I waves are mediated by the cortical interneurons and D wave generated directly at the axon hillock of the pyramidal neuron. In epidural recordings, the D wave (if present) is followed by multiple descending volleys of I waves. **B:** Mechanism at the lower motoneuron level. Individual excitatory postsynaptic potentials (EPSPs) conveyed by multiple descending volleys summate temporally and spatially to give total EPSP amplitudes enough to reach the threshold potential for firing the lower motoneuron.

at a high frequency (>500 Hz or intervals of <2 ms) after single TMS. Using epidural electrodes at the cervical level, this repetitive firing can be recorded as multiple wavelets (i.e., I waves).[12,13] These are considered as volleys indirectly generated through activation of intracortical interneurons (Fig. 10–2).

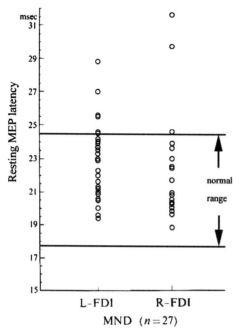

Figure 10–3 Plots of motor evoked potential (MEP) latencies in amyotrophic lateral sclerosis (ALS). Latencies measured from the left and right first dorsal interossei muscles (FDI) are shown. (Adapted from Kohara N, Kaji R, Kojima Y, et al. An electrophysiological study of the corticospinal projections in amyotrophic lateral sclerosis. Clin Neurophysiol 1999; 110:1123–1132.)

Single Motor Unit Recording

The previous consideration illustrates the importance of EPSPs at the lower motoneuron for interpreting MEP abnormalities. It is not possible to directly record EPSPs in humans, but they can be inferred by analyzing the effect of peri-threshold TMS on the pattern of lower motoneuron firing at weak voluntary activation[16] (Fig. 10–4).

The lower motoneuron discharges at a low frequency (5 to 10 Hz) under weak voluntary activation, and motor unit potentials (MUPs) represent this firing. The regular firing process includes the membrane potential trajectory of post-spike refractory periods, membrane hyperpolarization, and gradual increase to the level of the threshold potential. EPSPs induced by randomly applied TMS are superimposed on this trajectory. If the sum of these exceeds the threshold, a premature

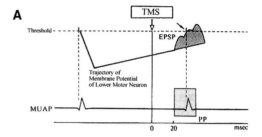

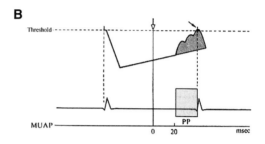

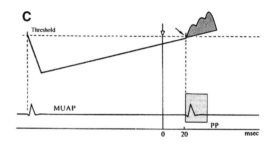

Figure 10–4 The effect of transcranial magnetic stimulation (TMS) on the membrane potential trajectory and the firing probability of the lower motoneuron under weak and steady voluntary activation. In a neuron discharging regularly by voluntary activation, the membrane potential undergoes serial changes of the post-spike refractory period, hyperpolarization, and gradual depolarization that reaches the level of the threshold at the expected time of the next firing. When TMS is given randomly as to the discharges (*open arrows*), excitatory postsynaptic potentials (EPSPs) mediated by corticospinal projections are added on the trajectory at various timings. **A:** If the sum of the trajectory and EPSPs exceeds the threshold, a premature firing (*solid arrows*) occurs before the expected time of the regular discharge. **B:** If the sum is equal to the threshold, a premature firing occurs at the peak of EPSPs. **C:** If the membrane potential is near the threshold, a premature firing occurs at the onset of EPSPs. MUAP, motor unit action potential.

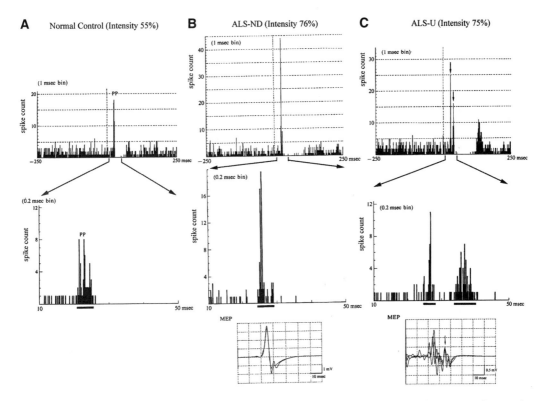

Figure 10–5 Peri-stimulus time histograms (PSTHs) constructed from single motor unit analyses. Concentric needle electrodes placed in the first dorsal interosseus (FDI) muscle were used for recording. Single-unit discharge was extracted from multiple-unit recording using a window discriminator. Off-line analysis confirmed that spike count was made from a single unit. Upper PSTHs were drawn using the 1-ms bin. Broken *vertical lines* indicate the timing of TMS application (time 0). Lower PSTHs were plotted with the 0.2-ms bin (10 to 50 ms is shown). *Solid bar* below indicate the duration of PP. **A:** PSTHs obtained from a motor unit in a normal subject. In 0.2-ms bin plots, the primary peak (PP) showed 3 subpeaks with about 2-ms intervals. **B:** PSTHs obtained from a patient with early stage ALS without exaggerated finger flexion reflex in the tested limb. Increased height of PP suggested enlarged EPSPs generated at the lower motoneuron. The PP duration was normal. The motor evoked potential (MEP) recording showed a normal waveform *(bottom)*. **C:** PSTHs obtained from a patient with upper motoneuron signs in the limb tested. Double PPs were observed in the 1-ms bin plots, the later of which showed significantly delayed peak onset as seen in the 0.2-ms plot. MEPs recorded with surface electrodes showed polyphasic waveforms *(bottom)*. (Adapted from Kohara N, Kaji R, Kojima Y, et al. Abnormal excitability of the corticospinal pathway in patients with amyotrophic lateral sclerosis: a single motor unit study using transcranial magnetic stimulation. Electroencephalogr Clin Neurophysiol 1996; 101:32–41.)

MUP occurs before the expected time (see Fig. 10–4A). If the sum is equal to the threshold, MUP is seen at the peak of EPSPs (see Fig. 10–4B). When EPSPs are generated immediately before the expected MUP, the premature one is recorded near the onset of EPSPs (see Fig. 10–4C). If the sum falls short of the threshold, no premature MUPs are observed.

It follows that the probability of motor unit firing is increased from the onset to the peak of EPSPs. If the MUP discharge pattern is plotted as to the timing of TMS (time 0) using a PSTH, the discharge probability would steeply increase at 20 to 30 ms (i.e., the delay expected for induced EPSPs to affect motor unit firing). This peak of probability is called the *primary peak* (PP) (Fig. 10–5). The duration of PP therefore represents the time between the onset and the peak (or the rise time) of EPSPs. The degree of the probability increase or the height of the peak also reflects the magnitude of EPSPs. Ashby and colleagues[17,18] tested this paradigm using group Ia afferent as the source of EPSPs in an animal model and demonstrated a close correlation between peaks in PSTH and the first differential of intracellularly recorded EPSPs. This method enables noninvasive estimation of EPSPs at the lower motoneuron in humans.

Normal Peri-stimulus Time Histogram

Figure 10–5A depicts a PSTH obtained from a normal subject. The histogram displays the counts of firing of a single motor unit in two time bins (1 or 0.2 ms), and the time 0 represents the timing of TMS. The intensity is just below the threshold of eliciting MEP during weak voluntary activation (i.e., peri-threshold stimulation). TMS was repeated about 150 times. In PSTH of the 1-ms bin (see Fig. 10–5A), a surge of probability occurs around 20 ms (i.e., PP). The PP is thought to be mediated by fast-conducting monosynaptic (direct) corticospinal tract fibers. If the PP is analyzed with the 0.2-ms bin, three subpeaks of PP are revealed (see Fig. 10–5A). These subpeaks of about 1.7 ms inter-peak intervals (or 588 Hz) most likely represent I wave in the multiple descending volleys. The mean duration of PP in normal is 4.5 ms, and our normal range does not exceed 6 ms.

Peri-stimulus Time Histogram in Early Amyotrophic Lateral Sclerosis

Using the previously described paradigm, we compared PSTHs obtained from six ALS patients with short disease duration (<1.3 years) and six with long duration (>1.8 years) and from normal control subjects.[19,20] Motor units from ALS patients with short duration of illness showed significantly increased firing probability at PP ($P < .01$) compared with those from the other subjects (see Fig. 10–5B), although stimulus intensities were similar among the groups. This is interpreted as the increased magnitude of EPSPs after TMS in the early stage of ALS.

Fasciculations Evoked by Transcranial Magnetic Stimulation

Fasciculation is a characteristic finding in ALS, although it is also seen in peripheral neuropathies such as multifocal motor neuropathy. The origin of fasciculations in ALS has been located mostly at distal motor axons but rarely at the nerve trunk or at the soma of the lower motoneuron. We showed that potentials closely reminiscent of spontaneous fasciculation potentials were sometimes evoked by TMS in the early stage of ALS[21] (Fig. 10–6). This is interpreted as EPSPs induced by TMS triggers firing of a lower motoneuron with increased excitability. Another robust explanation is that such TMS-evoked fasciculation potential consist of multiple motor units with increased excitability that share common EPSP input. Our recordings suggested that the latter might be the case on some occasions (see Fig. 10–6).

These evoked fasciculations, together with increased height of PP, indicate increased excitability of the lower motoneurons and support the view that glutamate-induced excitotoxicity may cause hyperexcitability of the corticospinal projection in early ALS.

Peri-stimulus Time Histogram in Amyotrophic Lateral Sclerosis with Upper Motoneuron Signs

Patients with upper motoneuron signs frequently showed delayed PP or bifid peaks (see Fig. 10–5C). This indicates a prolonged rise time of EPSPs at the lower motoneuron or the temporal dispersion of the monosynaptically activating fast-conducting volleys. Because the later of the bifid peaks or the second peak of PP often showed a delay of as much as 7 ms, it is also likely that slower-conducting monosynaptically or polysynaptically activating volleys contribute to these peaks.[20,22] These findings suggest that ALS tends to preferentially affect the larger and faster conducting fibers of the corticospinal projection.[23]

■ Motor Evoked Potential Waveforms in Amyotrophic Lateral Sclerosis

Most of the previous studies of MEPs have not dealt with the abnormalities of MEP waveforms. However, polyphasic and low-amplitude MEPs are frequently seen in ALS (see Fig. 10–5C, *lower box*), in contrast with

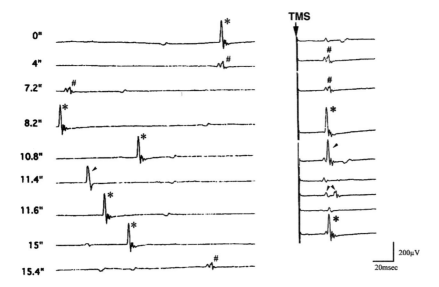

Figure 10–6 Cortically evoked fasciculation potentials. **Left:** Spontaneous fasciculation potentials recorded from the right first dorsal interosseus (FDI) muscle in a patient with early amyotrophic lateral sclerosis (ALS) (duration of 6 months). Digits on the left show the elapsed time from the beginning of recording. **Right:** Motor evoked potentials after transcranial magnetic stimulation (TMS) using peri-threshold intensities. Potentials marked with an *asterisk* and *hatch mark* were identical and were not activated voluntarily. This means that the origin of fasciculation was proximal to the axon branching. TMS at the peri-threshold intensity normally triggers units that are voluntarily activated first (i.e., lowest threshold), but those evoked fasciculations were not among them. The fasciculations must arise from the lower motoneuron with abnormally increased excitability at the soma because they were readily evoked by TMS. Another interpretation is that they were from a group of multiple motor units with extraordinarily increased excitability. A line of evidence in favor was that various parts of the fasciculation potentials were occasionally seen on careful scrutiny *(arrowheads)*.

the normal biphasic MEPs. In our series,[9] as many as 10 of 31 ALS patients with MEP amplitudes of more than 0.5 mV showed polyphasic waveforms with increased RMT (see Fig. 10–1). It is therefore concluded that the waveform is a good indicator of upper motoneuron involvement.

In most of the ALS patients with polyphasic MEPs, single motor unit studies showed bifid PPs (see Fig. 10–5C). The second peak of PP coincided with the later component of polyphasic MEP in this recording. Loss of the early peak again points to the selective vulnerability of the fastest conducting monosynaptic corticospinal projections. Because of the loss of these, the slower-conducting ones may become apparent in the recording or even compensate for the loss of the faster ones. This mechanism of latency and waveform abnormalities of MEPs may be operative in other diseases affecting upper motoneurons and should be taken into account for interpreting the findings in general.

Pathophysiology of Amyotrophic Lateral Sclerosis

ALS has a peculiar lack of involvement of extraocular and sphincter muscles, although these may also be involved in advanced patients connected to a respirator. The TMS findings of preferential involvement of fast-conducting monosynaptic corticospinal system in ALS could account for these negative findings, because lower motoneurons supplying extraocular and sphincter muscles are the only ones not having fast-conducting monosynaptic corticospinal or corticobulbar projections.[24] This direct corticospinal projection, a relatively new system in evolution, is seen only in humans and monkeys. The finding may be a consequence of a general biologic principle that the most newly acquired feature in phylogeny is likely to be lost first in aging or disease.

Hyperexcitability of the direct corticospinal tract system in the early ALS, as seen in increased height of PP, is followed by bifid or low amplitude PP, which indicated loss of this system. Cortically evoked fasciculations, which reflect increased excitability of the lower motoneuron or the corticospinal tract, were seen only in the early ALS. These findings are consonant with the glutamate-induced neuroexcitatory cell death. It may be useful to artificially reduce the neuronal excitability, such as by antiepileptic drugs, in the early phase of ALS. For this to be practical, we should seek for a method of diagnosing ALS earlier than the conventional method using the current clinical criteria,[10,11] because as many as 50% of the lower motoneurons could be lost at the clinical onset of weakness in ALS.[25]

■ Conclusion

TMS is a useful tool for diagnosing subclinical upper motoneuron involvement in possible ALS patients and for exploring the pathophysiology of the corticospinal or corticobulbar system. This technique was unique in unveiling serial events of the initial increase of excitability and the subsequent loss of the fast, direct corticospinal system, whereas pathologic studies can only detect residual findings at autopsy.

Acknowledgments

This work was supported in part by grants from Japanese Ministry of Health, Welfare and Labor and those from Japanese Ministry of Science, Culture and Sports.

REFERENCES

1. Rothstein JD, Kuncl R, Chaudhry V, et al. Excitatory amino acids in amyotrophic lateral sclerosis: an update. Ann Neurol 1991;30:224–225.
2. Rothstein JD, Martin LJ, Kuncl RW. Decreased glutamate transport by the brain and spinal cord in amyotrophic lateral sclerosis. N Engl J Med 1992;326:1464–1468.
3. Rosen DR, Siddique T, Patterson D, et al. Mutations in Cu/Zn superoxide dismutase gene are associated with familial amyotrophic lateral sclerosis. Nature 1993;362:59–62.
4. Siddique T, Nijhawan D, Hentati A. Molecular genetic basis of familial ALS. Neurology 1996;47(Suppl 2):S27–S34; discussion S34–S25.
5. Kaji R. Physiology of conduction block in multifocal motor neuropathy and other demyelinating neuropathies. Muscle Nerve 2003;27:285–296.
6. Kaji R, Kusunoki S, Mizutani K, et al. Chronic motor axonal neuropathy associated with antibodies monospecific for N-acetylgalactosaminyl GD1a. Muscle Nerve 2000;23:702–706.
7. Day BL, Dressler D, Maertens de Noordhout A, et al. Electric and magnetic stimulation of human motor cortex: surface EMG and single motor unit responses. J Physiol 1989;412:449–473.
8. Rothwell JC, Thompson PD, Day BL, et al. Motor cortex stimulation in intact man. 1. General characteristics of EMG responses in different muscles. Brain 1987;110 (Pt 5):1173–1190.
9. Kohara N, Kaji R, Kojima Y, et al. An electrophysiological study of the corticospinal projections in amyotrophic lateral sclerosis. Clin Neurophysiol 1999;110:1123–1132.
10. Brooks BR. El Escorial World Federation of Neurology criteria for the diagnosis of amyotrophic lateral sclerosis. Subcommittee on Motor Neuron Diseases/Amyotrophic Lateral Sclerosis of the World Federation of Neurology Research Group on Neuromuscular Diseases and the El Escorial Clinical Limits of Amyotrophic Lateral Sclerosis Workshop contributors. J Neurol Sci 1994;124(Suppl):96–107.
11. Brooks BR, Miller RG, Swash M, et al. El Escorial revisited: revised criteria for the diagnosis of amyotrophic lateral sclerosis. Amyotroph Lateral Scler Other Motor Neuron Disord 2000;1:293–299.
12. Thompson PD, Day BL, Crockard HA, et al. Intra-operative recording of motor tract potentials at the cervico-medullary junction following scalp electrical and magnetic stimulation of the motor cortex. J Neurol Neurosurg Psychiatry 1991;54:618–623.
13. Nakamura H, Kitagawa H, Kawaguchi Y, et al. Intracortical facilitation and inhibition after transcranial magnetic stimulation in conscious humans. J Physiol 1997;498(Pt 3):817–823.
14. Day BL, Rothwell JC, Thompson PD, et al. Motor cortex stimulation in intact man.

2. Multiple descending volleys. Brain 1987; 110(Pt 5):1191–1209.
15. Ingram DA, Swash M. Central motor conduction is abnormal in motor neuron disease. J Neurol Neurosurg Psychiatry 1987;50:159–166.
16. Mills KR. Corticomotoneuronal PSTH studies. Muscle Nerve 1999;22:297–298.
17. Ashby P, Zilm D. Relationship between EPSP shape and cross-correlation profile explored by computer simulation for studies on human motoneurons. Exp Brain Res 1982;47:33–40.
18. Ashby P, Zilm D. Characteristics of postsynaptic potentials produced in single human motoneurons by homonymous group 1 volleys. Exp Brain Res 1982;47:41–48.
19. Kohara N, Kaji R, Kojima Y, et al. Magnetic stimulation in ALS–a single motor unit study. Electroencephalogr Clin Neurophysiol Suppl 1996;46:327–336.
20. Kohara N, Kaji R, Kojima Y, et al. Abnormal excitability of the corticospinal pathway in patients with amyotrophic lateral sclerosis: a single motor unit study using transcranial magnetic stimulation. Electroencephalogr Clin Neurophysiol 1996;101:32–41.
21. Kaji R, Kohara N, Kimura J. Fasciculations evoked by magnetic cortical stimulation in patients with ALS. Neurology 1993;43:A257.
22. Mills KR. Motor neuron disease. Studies of the corticospinal excitation of single motor neurons by magnetic brain stimulation. Brain 1995;118(Pt 4):971–982.
23. Sobue G, Hashizume Y, Mitsuma T, et al. Size-dependent myelinated fiber loss in the corticospinal tract in Shy-Drager syndrome and amyotrophic lateral sclerosis. Neurology 1987;37:529–532.
24. Iwatsubo T, Kuzuhara S, Kanemitsu A, et al. Corticofugal projections to the motor nuclei of the brainstem and spinal cord in humans. Neurology 1990;40:309–312.
25. Aggarwal A, Nicholson G. Detection of preclinical motor neurone loss in SOD1 mutation carriers using motor unit number estimation. J Neurol Neurosurg Psychiatry 2002;73:199–201.

11

Motor System Physiology

Joseph Classen

The cortical motor system of primates is formed by a mosaic of anatomically and functionally distinct areas[1,2] that is extensively interconnected with other cortical and subcortical regions. Anatomically, the most caudal region of the frontal cortex is characterized by its lack of granular cells (i.e., agranular frontal cortex). Collectively, this region has been referred to as the *motor cortex*. It may be subdivided by its cytoarchitectonic features or by its extrinsic connections.[1,2] The posterior motor areas in the frontal cortex receive their main cortical input from the parietal lobe, and the anterior motor areas receive their main cortical connections from the prefrontal lobe. Many transcranial magnetic stimulation (TMS) studies in humans have focused on the intrinsic organization, output pathways, and function of a frontal motor region bordering the central sulcus. This region, called the *primary motor cortex*, is essentially equivalent to area F1 (nomenclature according to Matelli and coworkers[3] in monkey cortex or to cytoarchitectonically defined Brodmann area 4 in humans). It contains the largest output neurons of the cortex, the Betz cells, and it is probably the single most important source of crossed direct corticospinal projections controlling hand movements. It is noteworthy that the other frontal motor areas—the dorsal premotor cortex (superior part of Brodmann area 6, monkey areas F2 and F7[3]), the ventral premotor cortex (inferior part of Brodmann area 6 and Brodmann area 44, monkey areas F4 and F5[3]), and the mesial area 6 (also called the *supplementary motor area*, monkey areas F3 and F6[3])—also are important sources of corticospinal projections. A part of the dorsal premotor cortex (F7) and a part of the mesial cortical area (F6) do not contain corticospinal neurons, but send their efferent projections exclusively to the brainstem. Many of the motor areas contain complete representations of body movements. By virtue of their distinctive connections, each motor area may be involved in different aspects of sensorimotor transformations, in motor planning, and in motor execution.[2] Studies have revealed that the motor system creates internal representations of actions in addition to these classically accepted functions. The identification of parietofrontal circuits suggests that the motor system contains many functional units that act in parallel. A view of the motor system as being exclusively hierarchically and serially organized is no longer tenable based on a multitude of anatomical and physiological data.

This chapter attempts to summarize how TMS has been used to improve our understanding of the human motor system. Many of the deductions originating from TMS studies in humans have depended on knowledge gained from nonhuman primate experiments.[4] However, many observations can only be

Table 11–1 PATHWAYS ACCESSIBLE TO TESTING BY TRANSCRANIAL MAGNETIC STIMULATION

Corticofugal pathways
    Monosynaptic corticomotoneuronal projections
    Presumably oligosynaptic corticomotoneuronal projections—ipsilateral responses
    Contralateral responses possibly representing activation of the propriospinal system
    Corticospinal projections targeting segmental inhibitory spinal interneurons
Pathways afferent to the primary motor cortex
    Transcallosal pathway from homologous motor cortex
    Premotor-primary motor cortical pathway
    Cerebello-thalamo-cortical pathway
    Pathways afferent to the primary motor cortex carrying somatosensory information

made in humans because several aspects of motor behavior are unique to humans. The complexity of tasks requiring cooperation of the subjects is frequently beyond the limits of animal experiments. There may be important differences between the organization of the motor system in human and nonhuman mammals. Although TMS studies have provided information about cortical connectivity and sensorimotor transformations of the primary motor cortex in surprising detail, TMS studies have successfully been extended to study motor areas outside the primary motor cortex and to address more complex cortical functions.

Pathways Accessible to Testing by Transcranial Magnetic Stimulation

TMS has been used to probe corticofugal and corticopetal pathways. An overview of various pathways that are accessible to TMS is provided in Table 11–1.

Corticofugal Pathways

Monosynaptic Corticomotoneuronal Projections. TMS of the primary motor cortex evokes motor potentials in contralateral hand muscles. Suprathreshold stimulation excites spinal alpha-motoneurons primarily, if not exclusively, by the fastest-conducting crossed corticomotoneuronal fibers (reviewed by Rothwell[5]). The density of monosynaptic cortical projections arising from the primary motor cortex appears to be highest for distal hand muscles and decreases over a distal-proximal gradient, with the deltoid muscle receiving the least dense projection.[6] Magnetic stimulation of the motor cortex gives rise to a sequence of downgoing activity that matches closely the sequence of a D wave followed by I waves observed with electric stimulation of the exposed motor cortex of experimental animals.[7] Experiments employing post-stimulus time histograms constructed from activity of single motor units,[8] recordings from epidural electrodes in conscious patients,[9] and studies employing paired-pulse stimulation[10,11] all show that the precise temporal segmentation of subpeaks within the main motor evokes potential (MEP) fits exactly with that of the D-I–wave complex. In healthy subjects, the site where MEPs in small hand muscles can be most readily elicited by focal magnetic stimulation coincides with the anatomical location of the primary motor cortex in the anterior bank of the central sulcus.[12] This "hot spot" position has usually been taken as surrogate marker for the location of the primary motor cortex in numerous studies of healthy subjects and patients. The implicit assumption underlying this approach is that corticospinal projections arising from other brain regions, such as the premotor cortex, cannot be activated by TMS.

Two studies suggest that this idea may be incorrect.[13,14] In a small group of stroke patients, very focal subcortical lesions affected the corticospinal outflow originating in the primary motor cortex.[14] TMS was applied to the primary motor cortex, dorsal premotor cortex, and ventral premotor cortex of the affected and intact hemispheres of patients. MEPs after stimulation of the dorsal premotor cortex of the damaged hemisphere were, on average, larger and had a shorter latency than after stimulation of the primary motor cortex of the affected hemisphere. These results are consistent with the notion that the dorsal premotor cortex contains corticospinal projections that can be recruited by TMS and likely are functionally important.[14] Although Fridman and coworkers failed to elicit MEPs by stimulation of the ventral premotor cortex, another study suggested that this region (which would largely overlap with Brodmann area 44 and to some extent inferior area 6 and possibly area 45) contains corticospinal projections to hand muscles that can be activated by TMS.[13] This observation fits well with neuroimaging evidence that Brodmann area 44 (i.e., Broca's area), traditionally considered to be exclusively devoted to speech production, also is involved in the control of hand movements. However, the monosynaptic nature of the corticospinal projections arising from the dorsal and from the ventral premotor cortex has not been demonstrated unequivocally. In monkeys, responses in M1 may be facilitated by stimulation of premotor cortex at intervals of 1 to 2 ms, suggesting that signals may have traveled from the premotor cortex to the spinal cord with an intermediate relay in M1 without substantially prolonging the latency of responses. Anatomical studies have not revealed direct corticospinal projections to intrinsic hand muscles arising from premotor cortex.[15] It will be important to replicate and extend the evidence for the existence and to define the functional relevance of the corticospinal projections arising from different parts of the premotor cortex that have so long escaped detection by TMS studies.

Other, Presumably Oligosynaptic Corticomotoneuronal Projections: Ipsilateral Responses. In humans, it is possible to elicit by TMS responses in ipsilateral muscles in addition to contralateral responses. Ipsilateral responses can be most readily evoked in bulbar muscles,[16] suggesting that these muscles receive strong inputs from both hemispheres. As a rule, only in children younger than 10 years[17] and in patients having suffered an extensive brain lesion[18] can responses in muscles of the ipsilateral extremities be evoked relatively easily at short latencies. In patients with mirror movements due to a congenital abnormality of the corpus callosum, the latency of ipsilateral MEPs was identical to that of the contralateral MEPs.[19] In contrast, the latency of ipsilateral MEPs in patients with brain damage acquired later in life[20,21] and in healthy subjects[21-23] exceeded that of the contralateral MEPs by about 5 ms. These responses cannot be caused by monosynaptic excitation of spinal motoneurons by branching fibers of the crossed corticospinal tract at a segmental level, because in this case, ipsilateral MEPs would occur at the same latency as contralateral ones. In resting, healthy adults, ipsilateral responses in upper extremity muscles are generally far more difficult to elicit than in children or in brain-lesioned patients.[22] However, when the target muscle is contracted, ipsilateral responses can be elicited in all subjects.[23] In the study by Ziemann and coworkers,[23] ipsilateral MEPs could preferentially be elicited in finger abductors, finger and wrist extensors, and elbow flexors but were absent or could less readily be elicited in finger and wrist flexors and elbow extensors. The threshold TMS intensity for ipsilateral MEPs was substantially higher compared with size-matched contralateral MEPs. The optimal position for eliciting ipsilateral MEPs[23,24] and the preferred stimulating current direction[23] were different for ipsilateral and contralateral MEPs. Large ipsilateral MEPs could be evoked in a patient with complete agenesis of the corpus callosum. These observations virtually excluded that action potentials giving rise to ipsilateral MEPs traveled by the corpus callosum over a crossed monosynaptic corticospinal pathway. The size of the ipsilateral MEPs could be modulated by turning of the head toward the side of the recorded muscle or away from it.[23] This finding suggested that pathways involved in the tonic neck reflex had access to the neuronal structures subserving ipsilateral MEP responses. Ipsilateral MEPs may be mediated by a fast-conducting uncrossed corticomotoneuronal pathway targeting upper extremity muscles. Although the nature of the ipsilateral pathways is unknown, a (cortico-)reticulospinal[25] fiber tract may be involved. It seems unlikely that ipsilateral responses are exclusively generated by a propriospinal pathway because intrinsic hand muscles, in which ipsilateral responses could be obtained, do not receive projections from the cortico-propriospinal system.[26] These observations suggest that in human cortex, just as in nonhuman motor cortex, there is a distinct ipsilateral cortical representation of the upper extremity[27] that is accessible to TMS investigations.

Contralateral Responses Possibly Representing Activation of the Propriospinal System. In cats, the corticospinal command to forelimb motoneurons is exclusively transmitted through oligosynaptic pathways with intercalated spinal interneurons. Some of these interneurons are located at the level of vertebrae C3 and C4 (i.e., rostral to the alpha-motoneurons of arm muscles). These C3 to C4 pre-motoneurons, frequently referred to as propriospinal neurons, have attracted special attention. In cats, there is extensive convergence onto propriospinal neurons of descending excitation and inhibition and of peripheral inputs from the moving limb. This makes this system ideally suited to perform the necessary computations that allow descending commands for target-reaching movements to be updated at a pre-motoneuronal level.[28] Pierrot-Deseilligny and coworkers[29] found evidence for a similar system in humans. The TMS-evoked MEP in lower arm muscles was facilitated when weak TMS was conditioned by weak volleys to musculocutaneous, ulnar, and superficial radial nerves. This facilitation was qualitatively similar to that observed by conditioning the flexor carpi radialis H reflex by mixed nerve (e.g., ulnar, musculocutaneous) and cutaneous (e.g., afferents from both sides of the hand) inputs. In post-stimulus time histograms constructed from recordings of single motor units, the central delay of the peripheral facilitation of the peak of corticospinal excitation in motoneurons located at end of the cervical enlargement was longer the more caudal the motoneuron pool. This strongly suggested an interaction in premotoneurons located rostral to the tested motoneurons. This property closely resembles that of the propriospinal system as characterized in cats.[28] Small increases in the strength of the TMS pulse caused the facilitation to disappear and then to be reversed to inhibition. This observation suggested that the facilitation of cervical, presumably propriospinal pre-motoneurons is paralleled by corticospinal activation of inhibitory projections to these premotoneurons. The facilitatory and inhibitory effects had the same latencies and spared the initial 0.5 to 1 ms of the corticospinal excitatory response. The reversal of the facilitation to inhibition by stronger corticospinal volleys is consistent with a system of "feedback inhibitory interneurons" activated by corticospinal and afferent inputs inhibiting the presumed propriospinal excitatory premotoneurons. Such a system would also explain why responses attributable to a propriospinal system have only rarely been found in monkeys[30] and in human single motor unit studies.[31] In these studies, a significant disynaptic excitation of propriospinal neurons from the corticospinal system could have been masked by strong corticospinal activation of inhibitory interneurons. In flexor carpi radialis H-reflex experiments, the distribution of the increased facilitation depended on the muscles involved in the contraction; ulnar nerve-evoked facilitation was increased much more at the onset of voluntary wrist flexion than voluntary elbow flexion and vice versa for the musculocutaneous-induced facilitation.[29] This finding is consistent with the view that there are subsets of propriospinal-like neurons, specialized with regard to afferent input. Descending excitation may be directed preferentially to the subset of neurons that receive excitatory feedback from the contracting muscle. The wide convergence found between different inputs onto common neurons and the finding that descending excitation during contraction of a given muscle reaches subsets of neurons projecting to motor nuclei of muscles operating at other joints suggested that the propriospinal-like system would be operative during complex multi-joint movements. The significance of these findings for human studies in clinical neurophysiology has yet to be fully appreciated.[32] In a patient with a limited lesion of the spinal cord at the

C6-7 junction, ulnar and superficial radial-induced modulations of MEPs and of ongoing EMG activity were observed in the biceps (above the lesion) but not in the triceps (below the lesion), a finding consistent with interruption of the axons of cervical propriospinal neurons.[33]

Corticospinal Projections Targeting Segmental Inhibitory Spinal Interneurons. Inhibition of antagonist muscles during activation of agonists is essential for performing coordinated, smooth movements. One important mechanism to achieve this depression is *disynaptic reciprocal inhibition*, mediated by glycinergic Ia interneurons in the spinal cord.[34] An important input to these interneurons comes from Ia afferents.[35] A disynaptic corticospinal input to the same population of reciprocal glycinergic inhibitory interneurons can be assessed by TMS over the primary motor cortex.[35] Nielsen and coworkers[35] investigated the effect of magnetic stimulation of the human motor cortex on the excitability of soleus, tibialis anterior, and flexor carpi radialis motoneurons by H-reflex testing. When the H reflexes were conditioned by prior magnetic cortex stimulation, the first event seen at rest in some muscles and with contraction in all muscles was facilitation, lasting a few milliseconds. This was followed by inhibition of H reflexes of all three motoneuron pools, lasting another 3 to 4 ms. This inhibition was most consistently observed during antagonist contraction.[35]

Using elaborate physiological stimulation techniques *presynaptic inhibition* of Ia afferents may be explored selectively.[36,37] A differential control has been disclosed during voluntary movements among various motoneuronal pools. At the onset of a selective voluntary contraction, presynaptic inhibition of Ia afferents projecting to the "contracting" motoneurons is strongly decreased, whereas presynaptic inhibition of Ia afferents to antagonist or synergistic motoneuronal pools, not involved in the contraction, is increased. TMS studies have shown that a corticospinal volley preferentially *activates* presynaptic interneurons at the cervical spinal level,[38,39] whereas it preferentially *inhibits* presynaptic interneurons at the lumbar spinal level.[40]

Studies in experimental animals suggest that the segmental and suprasegmental spinal interneuronal circuits are powerful computational units that subserve complex movements.[41–43] This field has been heavily underrepresented in studies of normal human motor behavior.

Cortical Pathways Afferent to the Motor Cortex

TMS has been used to study long-range interconnections between different brain regions. Although even complex functional circuits, such as those encompassing the analysis of visual information[44] are accessible to neurophysiological analysis by TMS, most studies have focused on pathways projecting more or less directly to the primary motor cortex. The effects induced by TMS in remote brain regions may be visualized by neuroimaging, electroencephalographic,[45,46] and behavioral methods or by employing conditioning-test magnetic stimulation paradigms. In the latter approach, a test pulse, usually over the primary motor cortex, is conditioned by one or two TMS pulses or by extended trains of TMS pulses that are delivered over a remote brain region. Interconnections of the primary motor cortex with the homologous contralateral primary motor cortex, the ipsilateral premotor cortex, and the cerebellum (by means of the thalamus) have been investigated. Modulation of the primary motor cortex by afferent somatosensory information has been examined intensively to gain insight into sensorimotor transformations that would likely be relevant for normal motor behavior.

Transcallosal Pathway from Homologous Motor Cortex. Transcallosal connections between homologous parts of the primary motor cortex can readily be assessed by TMS. If a TMS pulse over the motor cortex of one hemisphere is conditioned by a TMS pulse over the other hemisphere, the test response evoked in a contralateral small hand muscle is reduced when the conditioning-test interval exceeds 5 ms.[47,48] Intraoperative recordings of the TMS-evoked downgoing activity[49] have shown that this inhibition is mediated by the corpus callosum although one report suggests a substantial contribution from a subcortical site.[48] Single motor unit studies suggested that the true interhemispheric conduction time is likely to be around 10 to 11 ms. If the intensity of both stimuli is just above the MEP threshold in the contracting target muscle, interhemispheric facilitation can be observed at interstimulus intervals of about 8 ms.[50] TMS over the hand motor cortex also induces inhibition in the ongoing voluntary EMG activity recorded from ipsilateral hand muscles.[51] In normal subjects, this phenomenon, called the *ipsilateral silent period*, had an onset latency of about 35 ms and a duration of about 25 ms. The ipsilateral silent period depends on the intactness of the anterior part of the trunk of the corpus callosum as demonstrated by observations in patients with various lesions to the corpus callosum.[51] The calculated mean transcallosal conduction time for the ipsilateral silent period was 13 ms. The threshold of the ipsilateral silent period exceeded that for eliciting excitatory contralateral motor responses.

Fibers from the homologous hand motor cortex may interact with inhibitory phenomena that are testable by double-shock TMS applied over the primary motor cortex.[52] Modulation by transcallosal fibers has been demonstrated both for short-latency inhibition as well as for long-latency inhibition produced by conditioning-test pulse TMS arrangements. These studies have also revealed that the ipsilateral silent period and interhemispheric inhibition (as tested by delivering the conditioning stimulation to one hemisphere and the test magnetic stimulation to the other hemisphere) are phenomena mediated by different neuronal populations.[52] In humans, each motor cortex exerts an important modulatory influence on a number of local excitatory and inhibitory circuits located in its homologous counterpart.

Premotor-Primary Motor Cortical Pathway. Test responses elicited by TMS over the hand area of the primary motor cortex were suppressed when conditioned by a subthreshold stimulus applied 2 to 15 ms beforehand over skull regions overlying the premotor and frontal cortex. The largest effect was seen when the interstimulus interval was 6 ms. Facilitatory effects on the test responses could be elicited when the conditioning stimulus intensity was increased to levels suprathreshold at the hand motor cortex.[53] A similar inhibitory effect could be elicited, when trains of subthreshold 1-Hz repetitive TMS (rTMS) were delivered over the premotor cortex.[54] This suppression outlasted the rTMS treatment for at least 15 minutes. Conditioning rTMS over the prefrontal or parietal cortex or ipsilateral (with respect to the recorded muscle) premotor cortex did not change the size of MEP.[54] Conditioning premotor rTMS also modulated lasting short-latency intracortical inhibition and facilitation in primary motor cortex.[55]

Cerebellothalamocortical Pathway. Magnetic stimulation performed with a double-cone coil placed over the back of the head reduced the size of electromyographic responses evoked by magnetic cortical stimulation in the first dorsal interosseous muscle when it preceded the cortical stimulus by 5 to 7 ms for a duration of little more than 3 ms. The most effective position for magnetic stimulation over the back of the head was slightly rostral to the foramen magnum level on the ipsilateral side of the muscle studied. This position was most consistent with activation of the cerebellum.[56] Suppression of motor cortical excitability was reduced or absent in patients with dysfunction of the cerebellum or the cerebellothalamocortical pathway.[57] In a study on essential tremor, this pathway displayed normal physiological properties suggesting a role for afferents to the cerebellum but not a disturbance of the cerebellum itself, in the pathogenesis of this disorder.[58]

Pathways Afferent to the Primary Motor Cortex Carrying Somatosensory Information. Animal studies have shown that afferent information from mechanoreceptors and muscle spindles reaches the primary motor cortex, although the exact route, by which this information is transmitted, is still a matter of debate. TMS-evoked test responses in hand muscles were suppressed by a single electrical stimulus delivered to the median nerve (at the level of the wrist) 19 to 25 ms beforehand.[59] Epidural recordings obtained from patients undergoing spinal column surgery showed that conditioning median nerve stimulation reduced the size and number of descending corticospinal volleys evoked by magnetic stimulation. This indicated that the reduction of corticomotor excitability by median nerve stimulation resulted from cortical interactions. The MEP amplitude in resting hand muscles was modulated[60,61] by conditioning digital nerve stimulation homotopic to the hand muscle whose central representation was stimulated. The duration of the silent period recorded from voluntary contracting hand muscles was shortened by conditioning digital nerve stimulation ipsilateral to the hand muscle.[60] The magnitude of silent period shortening induced by nerve stimulation depended on the spatial distance of the conditioned digit from the target muscle.[60] This finding showed that, at least with voluntary contraction, somatotopy is an important organizational principle in sensorimotor integration. Because the somatotopical gradient was lost when the subjects performed a power grip, the specifics of sensorimotor integration depended on the task employed.[60] Input from muscle spindles has also been shown to exert a high degree of topographical specificity of modulating motor cortical output circuits.[61a] Low-amplitude muscle vibration was applied in turn to each of three different intrinsic hand muscles to test its effect on the TMS-evoked MEP and on short-interval intracortical inhibition and long-interval intracortical inhibition. Vibration increased the amplitude of MEPs evoked in the vibrated muscle. Of greater interest, MEP amplitudes were suppressed in the two nonvibrated hand muscles. There was less short-interval intracortical inhibition in the vibrated muscle and more in the nonvibrated hand muscles, whereas opposite effects were observed for long-interval intracortical inhibition.[61a]

The previously mentioned studies have shown that afferent input to the motor cortex modulates excitability in homonymous muscle representations. Bertolasi and colleagues[62] demonstrated that low-intensity conditioning stimulation of the median or radial nerve can suppress the TMS-evoked EMG response in the muscle representations that act as *antagonist* with respect to the afferent nerve. Control experiments showed that this effect, which could be elicited at conditioning test intervals of 13 to 19 ms, was generated in the cortex. This observation is of special interest as it reveals principles of the cortical organization of input-output relationships that are deeply similar to those first described in the spinal cord. Cortical antagonist inhibition presumably acts in parallel with spinal reciprocal inhibition in preventing inappropriate coactivation of antagonists.[62] Muscular afferent stimulation produced effects on intracortical inhibition and facilitation (as probed by double-shock TMS) in antagonist representations that were directed oppositely to those observed in MEPs elicited by single-pulse TMS.[63] This pattern of antagonist sensory afferent effects may be of significance for control of the wrist extensor and flexor muscles when used as synergists during manipulatory finger movements and gripping tasks.[63]

Normal Motor Behavior

Simple Movements

Buildup of Excitability. Making a simple movement requires preparatory activity in a number of brain areas. Much of this preparation is concerned with selecting and then sending an appropriate pattern of excitation to the spinal cord so that muscles are activated in the correct order and by the appropriate amount at the right time.[64] When MEPs were elicited at varying intervals before the onset of a voluntary ballistic movement[65–68] facilitation of MEPs from the prime mover occurred as early as 200 ms before the end of the reaction time. Facilitation of the prime mover preceded facilitation of MEPs from a nonprime mover. Excitability changes preceded directional kinematic changes occurring in TMS-evoked thumb movements about 100 ms before the end of the reaction time.[66] With self-paced movements, corticospinal excitability increase began about 20 ms earlier compared with simple reaction time movements.[65] Corticospinal excitability was decreased from about 500 to 1,000 ms after EMG offset for both self-paced and stimulus-triggered movements. The period of decreased corticospinal excitability after movement corresponded to the onset of event-related synchronization of electroencephalographic signals in the 20-Hz band compatible with a view that this state represents an inactive, idling state of the motor cortex.[65]

A study employing a conditioning-test TMS paradigm showed that movements involve increasing the amount of excitation in parts of the motor cortex and reducing the amount of inhibition and that these two processes are distinct.[69] When subjects made a tonic wrist extension, paired-pulse inhibition of the forearm extensor MEP decreased, whereas that of the forearm flexors remained unchanged. When the opposite movement direction was tested, similar changes in the intracortical inhibition of the prime mover were observed, but no changes were detected in the antagonist. Inhibition of the movement agonist began to decline about 100 ms before the onset of the agonist EMG activity in a reaction time task employing an auditory go stimulus. These findings suggested that the balance of excitation and inhibition of corticospinal neurons associated with a voluntary movement changes before the movement. Because these changes were specifically directed at the corticospinal neurons projecting to the agonists, they may provide a mechanism to select the population of cortical neurons responsible for the movement.[69] This view was supported by findings studying intracortical inhibition by TMS in a task requiring volitional *inhibition* in response to NoGo stimuli interspersed in Go stimuli. Task-related relaxation of the target muscle was associated with enhanced short-latency inhibition in the prime mover exclusively in the NoGo condition.[70] The change in short-latency intracortical inhibition in NoGo was a mirror image of the selective reduction in intracortical short-latency inhibition in the agonists before and possibly during voluntary movements. Short-latency intracortical inhibition therefore may have a role in providing nonselective suppression of voluntary movement in addition to focusing the subsequent excitatory drive to produce the intended movement.[64] This pattern of short-latency inhibition change appears to be compatible with the concept of focused disinhibition with tonic background inhibition in voluntary movement.[71,74]

Modulation of Activity in Contralateral Primary Motor Cortex by Ipsilateral Motor Cortex. A number of local neuronal circuits in the primary motor cortices are under powerful mutual control from the other primary motor cortex. Several studies employing functional imaging methods have provided evidence for enhanced activity in ipsilateral motor cortex associated with contralateral hand movements. A number of TMS studies suggested that the ipsilateral motor cortex might be of considerable importance for motor behavior. In a simple reaction time paradigm, suprathreshold TMS delayed the execution of ipsilateral finger movement when the cortex stimulus preceded the onset of the intended movement by about 25 to 65 ms.[72] Phasic pinch grips performed with one hand with low force induced a significant decrease of MEP amplitudes elicited by TMS in the contralateral first dorsal interosseus muscle. The effect lasted for about 100 ms after reaching the force level. In contrast, tonic contractions at a higher force level enhanced MEPs in the homologous first dorsal interosseus muscle. This observation suggests that the motor cortex may exert differential effects on its homologous counterpart depending on the motor task.[73] Further experiments indicated that inhibition is organized functionally and not strictly somatotopically.[74] The inhibition produced by phasic ipsilateral muscle activation was diffuse, affecting adjacent muscles (i.e., those near the homologous muscle in the same extremity) and homologous muscles. More inhibition was observed in adjacent and distal muscles than homologous and proximal muscles. Paired-pulse TMS (at 2- and 10-ms interstimulus intervals) showed a significant increase in intracortical facilitation selectively in the homologous muscle when triggered by self-paced movement of the opposite hand, but no change was observed in intracortical inhibition. These findings demonstrate that voluntary hand movement exerts an inhibitory influence on a diffuse area of the ipsilateral motor cortex. This inhibitory influence is nonselective, while the facilitatory influence appears to act selectively on the homologous muscles.[74] Taken together, the findings cited above suggest that one role of interhemispheric inhibition is to suppress unwanted mirror movements in the contralateral hand. It appears likely that transcallosally mediated communication between the primary motor cortices is important in the organization of bimanual movements. Generally, these findings are compatible with a view that the ipsilateral motor cortex is one important component of the network of motor areas, possibly centered by the contralateral motor cortex that controls hand movements. The clinical importance of the ipsilateral motor cortex in motor control has been demonstrated in multiple sclerosis[75] and stroke patients.[76]

Complex Movements

Pointing and Translational Movements. The previously described studies have shown that corticospinal excitability changes steadily toward the onset of a simple, unidirectional, voluntary movement, usually about a single joint. A more complex spatiotemporal pattern of cortical excitability changes emerged with a reach-and-grasp movement and pointing movements.[77,78] TMS directed to the hand area of the motor cortex at different phases of the reach-and-grasp task revealed a striking phase-related modulation in the amplitude of the short-latency EMG responses elicited by TMS in six arm- and hand muscles. Muscles subserving hand transport and orienting the hand and fingertips appeared to receive their strongest excitatory corticospinal drive throughout the reach. In contrast, cortical input to the intrinsic hand muscles was strongest as the digits closed around and first touched the object. Recording of corticospinal cells in the primary motor cortex of the awake monkey suggested that the phasic modulation of corticomotoneuronal excitability as observed in humans may reflect the phasic-tonic pattern of corticomotoneuronal cell discharge during the task.[79] Devanne and colleagues[77,78] sought to obtain physiological evidence on the mode of operation of the primary motor cortex during a pointing movement of the arm. Focal TMS was directed to representations of various proximal and distal arm muscles during isolated contraction of one of these muscles or during selective coactivation with other muscles involved in pointing. The recruitment curve of all arm muscles, except for that of the interosseus dorsalis muscle, was influenced by activation of the more proximal muscles. Several control experiments suggested a cortical site for these effects. These results indicate that activation of shoulder, elbow, and wrist muscles subserving a pointing movement appear to involve, at least in part, common motor cortical circuits. In contrast, at least in the pointing task, the motor cortical circuits involved in activation of the first dorsal interosseus muscle, a muscle not participating in the pointing task per se, appear to act independently.[77]

Uozumi and coworkers[13] provided evidence that stimulation over the ventral premotor cortex (essentially equivalent to Brodmann area 44) has facilitatory and inhibitory influences on tonic and phasic finger movements of the contralateral and ipsilateral hand. Target-oriented hand movements in particular were shown to be disrupted with magnetic stimulation.[13]

Timing of Sequential Movements. The organization of sequential motor actions is a central component of human motor behavior. High-frequency rTMS over the primary motor cortex had a differential effect on performing contralateral sequences of different complexity. Stimulus intensities capable of disrupting the performance of a complex sequence did not affect simple sequences. These findings suggested that the human primary motor cortex might be the site of complex neuronal computations related to movement sequence organization,[80] independent of its role in sending commands to the spinal cord. rTMS induced timing errors in finger movement sequences of different complexity also when directed to the ipsilateral motor cortex.[81] Timing of finger movements was more easily disrupted by ipsilateral stimulation in the left hand than in the right hand. Errors of the left hand with a complex sequence occurred in the stimulation and post-stimulation periods. These results indicate that the left motor cortex plays a greater role in timing ipsilateral complex sequences than the right motor cortex.[81] Stimulation by rTMS over the supplementary motor area induced accuracy errors in a complex sequence of contralateral finger movements. Errors induced by stimulation over the supplementary motor area occurred approximately 1 second later into performance of the finger sequence than with stimulation over the primary motor cortex. These findings are consistent with a critical role of the supplementary motor area in the organization of forthcoming movements in complex motor sequences.[82] Timing errors were not observed with stimulation sites outside the primary motor cortex or the supplementary motor area.

■ Higher Cortical Motor Functions

Neurophysiological data show that motor areas play a role in behavior and are involved in functions traditionally considered proper of higher-order associative cortical areas. Areas considered nonmotor by conventional teaching are shown to be involved in certain aspects of higher motor control.

Perception of Movement, Motor Attention, Awareness of Movement Intention, and Motor Volition

That the primary motor cortex has a role in controlling voluntary action and in perceptual aspects of movements was suggested by a study by Naito and coworkers.[83] The tendon of the wrist extensor muscle on one side was vibrated while both hands had mutual skin contact. In this situation, an illusion was created. The subjects felt that both hands—the vibrated one and the one not receiving vibration—were bending. The primary motor cortex contralateral to the nonvibrated hand was activated as demonstrated by functional MRI. TMS showed enhanced excitability specifically of the wrist flexor representation that would act as an agonist to the perceived movement. These findings indicate that activation of the primary motor cortex is compulsory for the somatic perception of hand movement. Although previous studies have reported that TMS over the primary motor cortex may elicit a sense of movement,[84] later results support the concept that TMS-evoked sensations of movements result from peripheral sensory feedback rather than activation of cortical structures.[85]

Neuroimaging studies revealed activation of the parietal cortex, when subjects covertly prepared to make hand movements.[86] Activation by motor attention was restricted to the anterior parietal areas in the supramarginal gyrus and the adjacent anterior intraparietal region during hand movement preparation. TMS was applied over the anterior parietal region to interfere locally with these neuronal activations and thereby to test their functional significance.[87] Subjects were precued about which movement they would be instructed to make by a subsequent imperative cue. On a few trials, the precue was invalid, the subsequent cue instructed a different movement, and the subject had to redirect attention from one intended movement to making a quite different movement. For these invalid trials, TMS over the left anterior parietal cortex impaired the redirecting of motor attention, regardless of which hand was used to respond.[88] These findings suggest that motor attention, unlike visuospatial attention, is predominantly associated with activation in the left rather than the right parietal cortex.

Reports of patients with different brain lesions attribute the awareness of motor intention to the parietal lobe.[89] Patients with parietal lesions were able to report when they started moving, but not when they first became aware of their intention to move. In contrast, performance of cerebellar patients did not differ from normal subjects who became aware of their intention to move roughly 200 ms before to movement onset. These findings may indicate that when a movement is planned, activity in the parietal cortex, as part of a corticocortical sensorimotor processing loop, generates a predictive internal model of the upcoming movement. This internal model has been proposed to form the neural correlate of motor awareness.[89] This concept receives substantial support from a study using rTMS to temporarily deactivate the superior parietal lobule.[90] Volunteers performed a finger extension actively and passively. A data glove recorded these finger movements and used this information in real time to move a virtual hand displayed on a computer screen. The onset of the virtual movement was delayed with respect to the actual movement by various intervals. After the stimulation over the superior parietal lobule, the subjects performed poorly in assessing asynchrony for active but not for passive movements.[90] Another study suggested a role for brain structures located anterior to the primary motor cortex in motor awareness.[91]

Motor volition (i.e., the perception of being the subject of willed action) has been linked with activity in a network comprising the presupplementary motor area, the right dorsal prefrontal cortex and the left intraparietal cortex, but not the primary motor cortex in a neuroimaging study.[92] A role for the primary motor cortex in motor volition was suggested by two early TMS studies. Single magnetic stimuli over the primary motor cortex, subthreshold for movement, produced significant preference for selection of one hand in a forced-choice task in which subjects were instructed to move a finger of either hand at their own will after a Go stimulus. The hand preference depended on the cortical hemisphere being stimulated by a focal figure-of-eight coil[93] or on the direction of the induced current when a round coil was used.[94] A response bias was not induced when the coil was positioned over prefrontal or occipital cortex and was not mimicked by weak direct current stimulation. Response preference was seen with movements exhibiting response times shorter than 200 ms but not with longer response times.[93] However, these remarkable findings have not been replicated in a similar study from one of the original groups.[95] The role of the primary motor cortex in motor volition therefore remains an open issue.

Action Observation/Action Execution Matching System and Movement Imagery. According to Victor von Weizsäcker,[96] action and perception are intimately related, building a *Gestaltkreis*. Similar ideas were advanced by William James in the late 19th century.[97] The discovery of a movement observation and movement execution matching system (i.e., mirror neuron system) by Rizzolatti and coworkers[98] provided neurophysiological support for this general behavioral principle. The monkey premotor cortex (area F5 according to Matelli and colleagues[3]) contains neurons that discharge when the monkey performs specific hand actions and when it observes another individual performing the same action.[99,100] These mirror neurons may enable individuals to recognize actions made by others, a capacity that may play a crucial role in nonverbal communication and procedural learning. Neurons with similar properties were also found in other anatomically interconnected cortical areas such as the anterior inferior parietal cortex[98] and those located adjacent to the superior temporal sulcus. The mirror neuron system emerges as an extended network of brain regions. Brain imaging, encephalographic or magnetoencephalographic, and TMS studies[101,102] provided evidence that such a system also exists in humans. TMS-evoked potentials increased during the conditions in which subjects observed movements. The pattern of MEP facilitation[101] and modulation of short-latency inhibition and facilitation, but not that of long-latency inhibition,[102] reflected the pattern of muscle activity recorded when the subjects executed the observed actions. Significant differences were found between effects produced by different hand orientations[103] but not by self versus nonself hand movement observation.[104] These findings underline a fundamental similarity between action execution and action observation. Because the primary motor cortex has not been found to contain mirror neurons in invasive studies on nonhuman primates, the TMS results, obtained by stimulation of the primary motor cortex, possibly reflect the influence of projections from the ventral premotor cortex onto the primary motor cortex. However, premotor mirror neurons are preferentially active with object-related activity. Because this was not a property of modulation of excitability in primary motor cortex, the possibility of local genuine mirror neurons in the human primary motor cortex cannot be ruled out. Other components of the mirror neuron system, such as the anterior inferior parietal cortex have not so far been studied in humans by TMS.

Motor imagery produced effects similar to those of action observation or action execution.[104] Subtle hints as to a greater involvement of the left versus the right hemisphere were found. Although magnetic stimulation of the left motor cortex revealed increased corticospinal excitability when subjects imagined ipsilateral and contralateral hand movements, the stimulation of the right motor cortex revealed a facilitatory effect induced by imagery of contralateral hand movements only.[105] Imagination of suppressing movements led to suppression of the excitatory corticospinal drive in a Go/NoGo task. MEP amplitudes of the first dorsal interosseus muscle were significantly suppressed in negative motor imagery but were unchanged during positive motor imagery. During negative motor imagery, resting motor threshold was significantly increased, but short and long intracortical inhibition and intracortical facilitation remained unchanged.[95]

Abstract Cognitive Operations in Motor Cortex

Neuroimaging studies of humans and neurophysiological studies of experimental animals have shown that motor structures are activated during complex cognitive tasks that require no overt motor activity, such as mental rotation. Single-pulse TMS was delivered to the representation of the hand in left primary motor cortex while participants performed mental rotation of pictures of hands and feet.[106] Response times were slower when TMS was delivered at 650 ms but not at 400 ms after stimulus onset. The magnetic stimulation effect at 650 ms was larger for hands than for feet. This finding suggests that activation of the primary motor cortex is needed for mental rotation of pictures of hands and that this role is stimulus specific.[106] The excitability of cortical representations of muscles used in a serial reaction time task increased steadily with increasing proficiency of task performance as long as the subjects were unaware of the sequence displayed to them.[107] When explicit knowledge was reached, excitability returned to baseline. These results suggest a role of the primary motor cortex in motor sequence learning and awareness of knowledge about finger movement sequences.

■ Conclusion

Because of occasionally ingenious paradigms, the depth of insight into physiological principles underlying motor control that has been gained from TMS studies has truly been remarkable. However, we are still far from understanding fully the operation of the motor system. In view of the amazing success of applying TMS in motor control studies, we must beware of a situation illustrated by the story of a drunken man who lost his keys in a dark street. He stands under a lantern searching for them, when a passerby asks what he was looking for. "I have lost my keys," the man says. "Here, under the lantern?" "No, no, way down in the dark." "But why for God's sake are you searching them here then?" "Because here's more light."

While we have been happy (and partially successful) in searching under the lantern, we must begin to look in the dark. Areas of darkness are readily identified. For instance, anatomical areas underrepresented in human motor physiology research are the descending spinal projections outside the part of the crossed monosynaptic corticomotoneuronal pathway that arises from Brodmann area 4, the intrinsic neuronal circuits of the spinal cord, the cerebellum and the basal ganglia, all of which are difficult but not impossible to assess neurophysiologically. Although we have studied the physiology of the motor system at rest (an ill-defined motor state), we are only beginning to explore how these findings translate into the physiology of natural movements. The dynamic operation in large-scale and small-scale networks that likely underlies motor control of even simple movements remains a challenge that will require the integration of many neuroscience methods.

REFERENCES

1. Rizzolatti G, Luppino G. The cortical motor system. Neuron 2001;31:889–901.
2. Rizzolatti G, Luppino G, Matelli M. The organisation of the cortical motor system: new concepts. Electroencephalogr Clin Neurophysiol 1998;106:283–296.
3. Matelli M, Luppino G, Rizzolatti G. Patterns of cytochrome oxidase activity in the frontal

agranular cortex of the macaque monkey. Behav Brain Res 1985;18:125–136.
4. Lemon RN. Basic physiology of transcranial magnetic stimulation. In Pascual-Leone A, Davey NJ, Rothwell J, et al (eds): Handbook of Transcranial Magnetic Stimulation. London, Arnold, 2002.
5. Rothwell JC. Techniques and mechanisms of action of transcranial stimulation of the human motor cortex. J Neurosci Methods 1997;74:113–122.
6. Chen R, Tam A, Butefisch C, et al. Intracortical inhibition and facilitation in different representations of the human motor cortex. J Neurophysiol 1998;80:2870–2881.
7. Patton HD, Amassian VE. Single- and multiple-unit analysis of cortical stage of pyramidal tract activation. J Neurophysiol 1954;17:345–363.
8. Day BL, Dressler D, Maertens de Noordhout A, et al. Electric and magnetic stimulation of human motor cortex: surface EMG and single motor unit responses. J Physiol (Lond) 1989;412:449–473.
9. Di Lazzaro V, Oliviero A, Pilato F, et al. The physiological basis of transcranial motor cortex stimulation in conscious humans. Clin Neurophysiol 2004;115:255–266.
10. Tokimura H, Ridding MC, Tokimura Y, et al. Short latency facilitation between pairs of threshold magnetic stimuli applied to human motor cortex. Electroencephalogr Clin Neurophysiol 1996;101:263–272.
11. Ziemann U, Tergau F, Wassermann EM, et al. Demonstration of facilitatory I wave interaction in the human motor cortex by paired transcranial magnetic stimulation. J Physiol 1998;511(Pt 1):181–190.
12. Classen J, Knorr U, Werhahn KJ, et al. Multimodal output mapping of human central motor representation on different spatial scales. J Physiol (Lond) 1998;512:163–179.
13. Uozumi T, Tamagawa A, Hashimoto T, et al. Motor hand representation in cortical area 44. Neurology 2004;62:757–761.
14. Fridman EA, Hanakawa T, Chung M, et al. Reorganization of the human ipsilesional premotor cortex after stroke. Brain, Accessed January 28, 2004, 10.1093/brain/awh082.
15. He SQ, Dum RP, Strick PL. Topographic organization of corticospinal projections from the frontal lobe: motor areas on the lateral surface of the hemisphere. J Neurosci 1993;13:952–980.
16. Benecke R, Meyer BU, Schönle P, et al. Transcranial magnetic stimulation of the human brain: responses in muscles supplied by cranial nerves. Exp Brain Res 1988;71:623–632.
17. Muller K, Kass-Iliyya F, Reitz M. Ontogeny of ipsilateral corticospinal projections: a developmental study with transcranial magnetic stimulation. Ann Neurol 1997;42:705–711.
18. Turton A, Wroe S, Trepte N, et al. Contralateral and ipsilateral EMG responses to transcranial magnetic stimulation during recovery of arm and hand function after stroke. Electroencephalogr Clin Neurophysiol 1996;101:316–328.
19. Mayston MJ, Harrison LM, Quinton R, et al. Mirror movements in X-linked Kallmann's syndrome. I. A neurophysiological study. Brain 1997;120(Pt 7):1199–1216.
20. Benecke R, Meyer BU, Freund HJ. Reorganisation of descending motor pathways in patients after hemispherectomy and severe hemispheric lesions demonstrated by magnetic brain stimulation. Exp Brain Res 1991;83:419–426.
21. Netz J, Lammers T, Homberg V. Reorganization of motor output in the non-affected hemisphere after stroke. Brain 1997;120(Pt 9):1579–1586.
22. Wassermann EM, Fuhr P, Cohen LG, et al. Effects of transcranial magnetic stimulation on ipsilateral muscles. Neurology 1991;41:1795–1799.
23. Ziemann U, Ishii K, Borgheresi A, et al. Dissociation of the pathways mediating ipsilateral and contralateral motor-evoked potentials in human hand and arm muscles. J Physiol 1999;518(Pt 3):895–906.
24. Wassermann EM, Pascual-Leone A, Hallett M. Cortical motor representation of the ipsilateral hand and arm. Exp Brain Res 1994;100:121–132.
25. Nathan PW, Smith M, Deacon P. Vestibulospinal, reticulospinal and descending propriospinal nerve fibres in man. Brain 1996;119(Pt 6):1809–1833.
26. Pierrot-Deseilligny E. Transmission of the cortical command for human voluntary movement through cervical propriospinal premotoneurons. Prog Neurobiol 1996;48:489–517.
27. Aizawa H, Mushiake H, Inase M, et al. An output zone of the monkey primary motor cortex specialized for bilateral hand movement. Exp Brain Res 1990;82:219–221.
28. Lundberg A. Descending control of forelimb movements in the cat. Brain Res Bull 1999;50:323–324.
29. Pierrot-Deseilligny E. Propriospinal transmission of part of the corticospinal excitation in humans. Muscle Nerve 2002;26:155–172.
30. Lemon RN. Neural control of dexterity: what has been achieved? Exp Brain Res 1999;128:6–12.
31. Maertens de Noordhout A, Rapisarda G, et al. Corticomotoneuronal synaptic connections in normal man: an electrophysiological study. Brain 1999;122(Pt 7):1327–1340.

32. Burke D. Clinical relevance of the putative C-3-4 propriospinal system in humans. Muscle Nerve 2001;24:1437–1439.
33. Marchand-Pauvert V, Mazevet D, Pradat-Diehl P, et al. Interruption of a relay of corticospinal excitation by a spinal lesion at C6-C7. Muscle Nerve 2001;24:1554–1561.
34. Crone C, Nielsen J. Central control of disynaptic reciprocal inhibition in humans. Acta Physiol Scand 1994;152:351–363.
35. Nielsen J, Petersen N, Deuschl G, et al. Task-related changes in the effect of magnetic brain stimulation on spinal neurones in man. J Physiol 1993;471:223–243.
36. Katz R. Presynaptic inhibition in humans: a comparison between normal and spastic patients. J Physiol Paris 1999;93:379–385.
37. Stein RB. Presynaptic inhibition in humans. Prog Neurobiol 1995;47:533–544.
38. Meunier S. Modulation by corticospinal volleys of presynaptic inhibition to Ia afferents in man. J Physiol Paris 1999;93:387–394.
39. Meunier S, Pierrot-Deseilligny E. Cortical control of presynaptic inhibition of Ia afferents in humans. Exp Brain Res 1998;119:415–426.
40. Iles JF. Evidence for cutaneous and corticospinal modulation of presynaptic inhibition of Ia afferents from the human lower limb. J Physiol 1996;491(Pt 1):197–207.
41. Bizzi E, Tresch MC, Saltiel P, et al. New perspectives on spinal motor systems. Nat Rev Neurosci 2000;1:101–108.
42. Fetz EE, Perlmutter SI, Prut Y, et al. Functional properties of primate spinal interneurones during voluntary hand movements. Adv Exp Med Biol 2002;508:265–271.
43. Poppele R, Bosco G. Sophisticated spinal contributions to motor control. Trends Neurosci 2003;26:269–276.
44. Amassian VE, Cracco RQ, Maccabee PJ, et al. Transcranial magnetic stimulation in study of the visual pathway. J Clin Neurophysiol 1998;15:288–304.
45. Paus T. Imaging the brain before, during, and after transcranial magnetic stimulation. Neuropsychologia 1998;37:219–224.
46. Siebner HR, Rothwell J. Transcranial magnetic stimulation: new insights into representational cortical plasticity. Exp Brain Res 2003;148:1–16.
47. Ferbert A, Priori A, Rothwell JC, et al. Interhemispheric inhibition of the human motor cortex. J Physiol 1992;453:525–546.
48. Gerloff C, Cohen LG, Floeter MK, et al. Inhibitory influence of the ipsilateral motor cortex on responses to stimulation of the human cortex and pyramidal tract. J Physiol 1998;510(Pt 1):249–259.
49. Di Lazzaro V, Oliviero A, Profice P, et al. Direct demonstration of interhemispheric inhibition of the human motor cortex produced by transcranial magnetic stimulation. Exp Brain Res 1999;124:520–524.
50. Ugawa Y, Hanajima R, Kanazawa I. Interhemispheric facilitation of the hand area of the human motor cortex. Neurosci Lett 1993;160:153–155.
51. Meyer BU, Roricht S, Grafin von Einsiedel H, et al. Inhibitory and excitatory interhemispheric transfers between motor cortical areas in normal humans and patients with abnormalities of the corpus callosum. Brain 1995;118(Pt 2):429–440.
52. Chen R. Interactions between inhibitory and excitatory circuits in the human motor cortex. Exp Brain Res 2004;154:1–10.
53. Civardi C, Cantello R, Asselman P, et al. Transcranial magnetic stimulation can be used to test connections to primary motor areas from frontal and medial cortex in humans. Neuroimage 2001;14:1444–1453.
54. Gerschlager W, Siebner HR, Rothwell JC. Decreased corticospinal excitability after subthreshold 1 Hz rTMS over lateral premotor cortex. Neurology 2001;57:449–455.
55. Münchau A, Bloem BR, Irlbacher K, et al. Functional connectivity of human premotor and motor cortex explored with repetitive transcranial magnetic stimulation. J Neurosci 2002;22:554–561.
56. Ugawa Y, Uesaka Y, Terao Y, et al. Magnetic stimulation over the cerebellum in humans. Ann Neurol 1995;37:703–713.
57. Ugawa Y, Terao Y, Hanajima R, et al. Magnetic stimulation over the cerebellum in patients with ataxia. Electroencephalogr Clin Neurophysiol 1997;104:453–458.
58. Pinto AD, Lang AE, Chen R. The cerebellothalamocortical pathway in essential tremor. Neurology 2003;60:1985–1987.
59. Tokimura H, Di Lazzaro V, Tokimura Y, et al. Short latency inhibition of human hand motor cortex by somatosensory input from the hand. J Physiol 2000;523(Pt 2):503–513.
60. Classen J, Steinfelder B, Liepert J, et al. Cutaneomotor integration in humans is somatotopically organized at various levels of the nervous system and is task dependent. Exp Brain Res 2000;130:48–59.
61. Kobayashi M, Ng J, Theoret H, et al. Modulation of intracortical neuronal circuits in human hand motor area by digit stimulation. Exp Brain Res 2003;149:1–8.
61a. Rosenkranz K, Rothwell JC. Differential effect of muscle vibration on intracortical inhibitory circuits in humans. J Physiol 2003;551:649–660.
62. Bertolasi L, Priori A, Tinazzi M, et al. Inhibitory action of forearm flexor muscle afferents on corticospinal outputs to

antagonist muscles in humans. J Physiol 1998; 511(Pt 3):947–956.
63. Aimonetti JM, Nielsen JB. Changes in intracortical excitability induced by stimulation of wrist afferents in man. J Physiol (Lond) 2001;534:891–902.
64. Floeter MK, Rothwell JC. Releasing the brakes before pressing the gas pedal. Neurology 1999;53:664–665.
65. Chen R, Yaseen Z, Cohen LG, et al. Time course of corticospinal excitability in reaction time and self-paced movements. Ann Neurol 1998;44:317–325.
66. Sommer M, Classen J, Cohen LG, et al. Time course of determination of movement direction in the reaction time task in humans. J Neurophysiol 2001;86:1195–1201.
67. Starr A, Caramia M, Zarola F, et al. Enhancement of motor cortical excitability in humans by non-invasive electrical stimulation appears before voluntary movement. Electroencephalogr Clin Neurophysiol 1988;70:26–32.
68. Tomberg C, Caramia MD. Prime mover muscle in finger lift or finger flexion reaction times: identification with transcranial magnetic stimulation. Electroencephalogr Clin Neurophysiol 1991;81:319–322.
69. Reynolds C, Ashby P. Inhibition in the human motor cortex is reduced just before a voluntary contraction. Neurology 1999;53:730–735.
70. Sohn YH, Wiltz K, Hallett M. Effect of volitional inhibition on cortical inhibitory mechanisms. J Neurophysiol 2002;88:333–338.
71. Mink JW. The basal ganglia: focused selection and inhibition of competing motor programs. Prog Neurobiol 1996;50:381–425.
72. Meyer BU, Voss M. Delay of the execution of rapid finger movement by magnetic stimulation of the ipsilateral hand-associated motor cortex. Exp Brain Res 2000;134:477–482.
73. Liepert J, Dettmers C, Terborg C, et al. Inhibition of ipsilateral motor cortex during phasic generation of low force. Clin Neurophysiol 2001;112:114–121.
74. Sohn YH, Jung HY, Kaelin-Lang A, et al. Excitability of the ipsilateral motor cortex during phasic voluntary hand movement. Exp Brain Res 2003;148:176–185.
75. Schmierer K, Irlbacher K, Grosse P, et al. Correlates of disability in multiple sclerosis detected by transcranial magnetic stimulation. Neurology 2002;59:1218–1224.
76. Murase N, Duque J, Mazzocchio R, et al. Influence of interhemispheric interactions on motor function in chronic stroke. Ann Neurol 2004;55:400–409.
77. Devanne H, Cohen LG, Kouchtir-Devanne N, et al. Integrated motor cortical control of task-related muscles during pointing in humans. J Neurophysiol 2002;87:3006–3017.
78. Lemon RN, Johansson RS, Westling G. Corticospinal control during reach, grasp, and precision lift in man. J Neurosci 1995;15:6145–6156.
79. Lemon RN, Johansson RS, Westling G. Modulation of corticospinal influence over hand muscles during gripping tasks in man and monkey. Can J Physiol Pharmacol 1996;74:547–558.
80. Gerloff C, Corwell B, Chen R, et al. The role of the human motor cortex in the control of complex and simple finger movement sequences. Brain 1998;121:1695–1709.
81. Chen R, Gerloff C, Hallett M, et al. Involvement of the ipsilateral motor cortex in finger movements of different complexities. Ann Neurol 1997;41:247–254.
82. Gerloff C, Corwell B, Chen R, et al. Stimulation over the human supplementary motor area interferes with the organization of future elements in complex motor sequences. Brain 1997;120:1587–1602.
83. Naito E, Roland PE, Ehrsson HH. I feel my hand moving: a new role of the primary motor cortex in somatic perception of limb movement. Neuron 2002;36:979–988.
84. Pascual-Leone A, Cohen LG, Brasil-Neto JP, et al. Differentiation of sensorimotor neuronal structures responsible for induction of motor evoked potentials, attenuation in detection of somatosensory stimuli, and induction of sensation of movement by mapping of optimal current directions. Electroencephalogr Clin Neurophysiol 1994;93:230–236.
85. Ellaway PH, Prochazka A, Chan M, et al. The sense of movement elicited by transcranial magnetic stimulation in humans is due to sensory feedback. J Physiol 2004;556:651–660.
86. Rushworth MF, Krams M, Passingham RE. The attentional role of the left parietal cortex: the distinct lateralization and localization of motor attention in the human brain. J Cogn Neurosci 2001;13:698–710.
87. Rushworth MF, Ellison A, Walsh V. Complementary localization and lateralization of orienting and motor attention. Nat Neurosci 2001;4:656–661.
88. Rushworth MF, Johansen-Berg H, Gobel SM, et al. The left parietal and premotor cortices: motor attention and selection. Neuroimage 2003;20(Suppl 1):S89–S100.
89. Sirigu A, Daprati E, Ciancia S, et al. Altered awareness of voluntary action after damage to the parietal cortex. Nat Neurosci 2004;7:80–84.
90. MacDonald PA, Paus T. The role of parietal cortex in awareness of self-generated movements: a transcranial magnetic stimulation study. Cereb Cortex 2003;13:962–967.

91. Haggard P, Magno E. Localising awareness of action with transcranial magnetic stimulation. Exp Brain Res 1999;127:102–107.
92. Lau HC, Rogers RD, Haggard P, et al. Attention to intention. Science 2004;303:1208–1210.
93. Brasil-Neto JP, Pascual-Leone A, Valls-Sole J, et al. Focal transcranial magnetic stimulation and response bias in a forced-choice task. J Neurol Neurosurg Psychiatry 1992;55:964–966.
94. Ammon K, Gandevia SC. Transcranial magnetic stimulation can influence the selection of motor programmes. J Neurol Neurosurg Psychiatry 1990;53:705–707.
95. Sohn YH, Kaelin-Lang A, Hallett M. The effect of transcranial magnetic stimulation on movement selection. J Neurol Neurosurg Psychiatry 2003;74:985–987.
96. Von Weizsäcker V. Der Gestaltkreis. Stuttgart, Thieme, 1940.
97. James W. Principles of Psychology. New York, Holt, 1890.
98. Rizzolatti G, Fogassi L, Gallese V. Neurophysiological mechanisms underlying the understanding and imitation of action. Nat Rev Neurosci 2001;2:661–670.
99. Gallese V, Fadiga L, Fogassi L, et al. Action recognition in the premotor cortex. Brain 1996;119(Pt 2):593–609.
100. Rizzolatti G, Fadiga L, Gallese V, et al. Premotor cortex and the recognition of motor actions. Brain Res Cogn Brain Res 1996;3:131–141.
101. Fadiga L, Fogassi L, Pavesi G, et al. Motor facilitation during action observation: a magnetic stimulation study. J Neurophysiol 1995;73:2608–2611.
102. Strafella AP, Paus T. Modulation of cortical excitability during action observation: a transcranial magnetic stimulation study. Neuroreport 2000;11:2289–2292.
103. Maeda F, Kleiner-Fisman G, Pascual-Leone A. Motor facilitation while observing hand actions: specificity of the effect and role of observer's orientation. J Neurophysiol 2002;87:1329–1335.
104. Patuzzo S, Fiaschi A, Manganotti P. Modulation of motor cortex excitability in the left hemisphere during action observation: a single- and paired-pulse transcranial magnetic stimulation study of self- and non-self-action observation. Neuropsychologia 2003;41:1272–1278.
105. Fadiga L, Buccino G, Craighero L, et al. Corticospinal excitability is specifically modulated by motor imagery: a magnetic stimulation study. Neuropsychologia 1999;37:147–158.
106. Ganis G, Keenan JP, Kosslyn SM, et al. Transcranial magnetic stimulation of primary motor cortex affects mental rotation. Cereb Cortex 2000;10:175–180.
107. Pascual-Leone A, Grafman J, Hallett M. Modulation of cortical motor output maps during development of implicit and explicit knowledge. Science 1994;263:1287–1289.

12

Transcranial Magnetic Stimulation in Movement Disorders

Nicola Modugno, Antonio Currà, Francesca Gilio, Cinzia Lorenzano, Sergio Bagnato, and Alfredo Berardelli

Transcranial magnetic stimulation (TMS) of the human brain is a safe, noninvasive technique widely used to study patients with clinical or subclinical abnormalities of central motor pathways. TMS provides useful information on the cortical motor function of patients with movement disorders and on the pathophysiology of their underlying motor disturbance.

This chapter reviews the findings from brain stimulation studies using TMS in patients with Parkinson's disease or parkinsonism, Wilson's disease, dystonia, Huntington's disease, tremor, tics, myoclonus and ataxias.

Parkinson's Disease

Motor Threshold

In patients with Parkinson's disease (PD), the motor threshold (MT) is often difficult to investigate because of incomplete muscle relaxation and resting tremor in the target muscle. These features may be associated with increased spinal excitability that lowers the threshold for eliciting motor evoked potentials (MEPs).[1,2] Rigid parkinsonian muscles may also exhibit periods of electromyographic (EMG) silence,[3–5] during which threshold may diminish on the more affected side.[3] The MEP threshold diminishes because subliminally excited motoneurons fire more readily in response to the cortical stimulus.[6]

Despite these possible confounding factors, most studies on the relaxed threshold have reported no difference between patients with PD and controls.[7–15] Active motor threshold was also normal in patients with PD.[6] In parkinsonian patients, neither active nor relaxed thresholds vary as a function of treatment with levodopa, anticholinergic agents and pergolide.[9,15,16] In healthy subjects, neither dopaminergic (e.g., levodopa, selegiline, bromocriptine, pergolide) nor anti-dopaminergic drugs (e.g., sulpiride, haloperidol) influence the motor threshold.[17,18]

Motor Evoked Potentials and Central Motor Conduction Time

A number of studies initially with electrical[19,20] and later with magnetic stimulation[6,10–12,16,21,22] have demonstrated that the central motor conduction time (CMCT) is normal in patients with PD. Some investigators reported a shortened CMCT together with increased amplitude MEPs[3,23–26] and larger MEPs in active muscles,[27] interpreted as indicating increased excitability of cortical neurons. Others concluded that shorter MEP latency and increased size might simply reflect the increased spinal excitability due to the difficulty parkinsonian patients have in relaxing the target muscles.[24]

Some evidence favors enhanced MEP amplitude at rest and reduced facilitation during voluntary activation. Valls-Solé and colleagues[7] found that some patients with PD had larger MEPs at rest than controls and that voluntary muscle activation elicited a smaller increase in MEP area, amplitude, and duration in patients than in controls. Enhanced drive from descending fibers at "rest" would coexist with diminished activation when higher force amounts are requested. An increase of MEP size at rest was also found by Chen and coworkers[28] in severely dyskinetic PD patients. An increase of MEP size at rest may express an imbalance toward disinhibition of the corticomotoneuronal system, including the spinal cord as supported by studies showing F wave and H reflex abnormalities in PD.[29,30]

In parkinsonian patients, rigidity generated by impaired muscle relaxation and tested at rest is associated with an abnormal firing of motor units.[31] Cantello and associates[4] observed that motor units that fire at rest have a high coincidence with those first recruited by TMS. TMS may influence cortical modulation of motor unit firing. Investigating the peri-stimulus time histograms (PSTHs), Kleine and colleagues[32] found that TMS significantly increased the duration of PSTH in PD patients. Because the MEP size reflects the number and synchronization of alpha-motoneurons discharging in response to a magnetic shock, PSTH prolongation may reflect desynchronization of the postsynaptic potential evoked by TMS in alpha-motoneurons.

MEP size can be inhibited in normal subjects by cutaneous stimuli.[33] After a conditioning cutaneous stimulus MEP inhibition was significantly reduced, absent or even reversed in akinetic-rigid PD patients[33] and in PD patients off treatment and increased after apomorphine administration.[34] The lack of MEP modulation probably reflects an abnormal function of long-loop reflexes at motor cortical level and the apomorphine-induced change suggests that dopaminergic treatment potentiates the various inhibitory inputs at cortical level.[34] A paper investigating the inhibitory effect on MEP size exerted by electrical stimulation of finger contiguous and noncontiguous to the target muscles showed that PD patients have a normal somatotopic inhibitory effect. Patients showed a more powerful MEP inhibition by stimulation of the contiguous than the noncontiguous finger, as normal subjects did. Patients showed a significant increase of the MEP size when stimulation of the contiguous finger preceded the test stimulus at ISIs of 20 to 40 ms.[35]

The Silent Period after Transcranial Magnetic Stimulation

When a TMS stimulus is delivered during voluntary contraction, a period of EMG silence (i.e., silent period [SP]) follows the MEP. The duration of the SP is an index of the function of inhibitory mechanisms in cortical motor areas.[36]

SP duration has been reported to be shorter on the more affected side of hemiparkinsonian, rigid patients.[3,7,9,10,12,13,37] The SP duration can be prolonged by L-Dopa, anticholinergics,[16] and pergolide therapy.[17] The normalization of the SP in PD after antiparkinsonian treatment has been attributed to changes in cortical inhibitory mechanisms that are impaired in PD during the off-therapy state.

The shortening of the ipsilateral SP—due in part to the activation of interhemispheric inhibitory pathways and not preceded by a MEP—and its normalization after antiparkinsonian treatment suggest that spinal mechanisms are not involved in the shortening of the cortical SP[16] and that the basal ganglia influence not only excitatory but also inhibitory mechanisms bilaterally in the motor cortex.

Paired-Pulse Studies

When the paired-pulse technique is used, a suprathreshold stimulus is preceded by a conditioning subthreshold TMS stimulus at various interstimulus intervals (ISIs). At short ISIs (2 to 5 ms) the MEP evoked by the suprathreshold stimulus (i.e., test MEP) is reduced in amplitude, a phenomenon called *intracortical inhibition* (ICI). At 8 to 20 ms, ISIs the test MEP has increased amplitude (i.e., intracortical facilitation [ICF]).[38] Intracortical inhibition can also be tested at long ISIs (100 to 200 ms).[10] ICI appears to be under the influence of the GABA$_A$ receptor population, whereas the long-interval ICI is thought to be mediated by GABA$_B$ receptors.[39]

In patients with PD, ICI tested at rest is significantly reduced,[9,17,40] but it is normal during contraction.[10] Ridding and coworkers[9] demonstrated that ICI was reduced in the Off periods, whereas in the On periods, it increased although not significantly. These investigators postulated that the abnormal activity of inhibitory connections causes an excess firing probability of some populations of descending area 4 neurons. Defective ICI would result in a lack of "focusing" activity onto appropriate groups of neurons for each movement. Cortical inhibition has been found abnormal also with paired-pulse at long ISIs under various experimental conditions: 40 to 75 ms ISI and low stimulus intensity, at 75 to 150 ms ISI and high stimulus intensity,[12] and at ISI of 100 of 250 ms and during contraction.[10] All these studies found increased inhibition in patients with PD, suggesting that overall cortical excitability is reduced. Although greater inhibition at long ISIs appears to be in contrast with the shortening of the SP, these two TMS variables are likely subserved by different interneuron populations.[10,39] One possible reason why enhanced inhibition at long ISIs and reduced inhibition at short ISIs coexist is that circuits activated by long ISIs depress those activated by short ISIs by activating presynaptic GABA$_B$ receptors.[39]

ICI at short (1 to 6 ms) and long (20 to 200 ms) ISIs is influenced by apomorphine[41] and levodopa but not by pergolide administration.[15] Changes in ICI after dopaminergic stimulation can be seen even in normal subjects: dopamine receptor agonists such as bromocriptine, pergolide, and levodopa enhance inhibition, whereas dopamine antagonists such as haloperidol and sulpiride reduce it.[18] Dopamine agonists therefore enhance ICI, whereas dopamine antagonists reduce it. These changes may be caused by a direct action on striatal dopaminergic receptors or to a direct influence on the dopaminergic innervation of motor cortical neurons.[17,18,42]

Interference of Transcranial Magnetic Stimulation with Movement Preparation and Execution

In normal subjects the excitability of primary motor cortex increases for about 100 ms before and 160 ms after the movement.[43-45] Chen and associates[46] observed that excitability began to increase earlier in patients with PD than in controls (350 versus 150 ms before EMG onset), although the rate of the increase itself was lower. After the end of the movement, the MEP size remained increased for much longer in patients than in controls. Abnormal cortical activation and deactivation could underlie bradykinesia or be a compensatory mechanism for bradykinesia.

In normal subjects a subthreshold TMS pulse reduced simple reaction times (RTs) to an acoustic, visual, or somatosensory input.[47] In patients with PD, the RT was prolonged, especially in the Off state,[48] but a TMS stimulus shortened the reaction time,[48] and normalized the premovement cortical excitability with changes in the agonist-antagonist EMG pattern of voluntary movement.[48]

In normal subjects, a high-intensity TMS stimulus applied over the supplementary motor areas (SMAs) before the onset of a self-paced sequential finger movement task leaves the movement unchanged. However, given during the early stages of movement in patients with PD, it disrupts the movement by increasing the execution time.[49] These findings confirm the role of the SMA in motor planning processes. They also show that in PD these processes are unstable and can be easily impaired by perturbations of cortical processing.

Repetitive Transcranial Magnetic Stimulation

The effect of repetitive TMS (rTMS) on MEPs has been investigated in parkinsonian patients by Gilio and colleagues.[50] Although in normal subjects at rest, the MEP to each stimulus in a train of magnetic stimuli progressively increased in size during the train, in parkinsonian patients in the Off state, they did not (Fig. 12-1). In normal subjects and patients, 5-Hz rTMS trains delivered during a voluntary contraction of the target muscle left the MEP unchanged in size, but the SP that

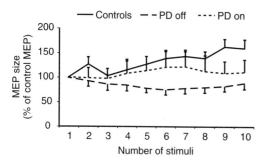

Figure 12–1 Effect of repetitive magnetic stimulation (rTMS) delivered at rest at a 5-Hz frequency and 120% stimulation intensity on the muscle evoked potential (MEPs) size in controls and in parkinsonian patients in Off and On conditions. Each point corresponds to the mean amplitude of the MEP. Bars represent the standard error (SE). PD, Parkinson's disease. (From Gilio F, Currà A, Inghilleri M, et al. Repetitive transcranial magnetic stimulation of cortical motor areas in Parkinson's disease. Mov Disord 2002;147:86–92.)

followed the MEPs increased in duration during the course of the train.[50,51] The investigators proposed that the reduced rTMS-induced facilitation of MEPs together with normal SP lengthened in patients with PD reflects a decreased facilitation of the excitatory cells in the cortical motor areas. In a similar experiment conducted in 10 patients, Siebner and coworkers[14] reported that 5-Hz rTMS induced a short (10-minute) prolongation of the cortical SP.

Numerous studies have investigated the effects of rTMS on the clinical conditions of patients with movement disorders. For a group of six patients, Pascual-Leone and associates[52] reported that rTMS delivered over the hand motor area significantly reduced RTs and movement times and improved performance in the grooved pegboard test, especially when patients were in the Off state. Conversely, studying a group of 11 parkinsonian patients, Ghabra and colleagues[53] failed to reproduce the rTMS results originally obtained by the same group with nearly identical methods.[52] Negative results also came from the work of Tergau and coworkers,[54] who studied seven patients before and after long trains (500 stimuli) at four different frequencies (1 to 20 Hz) in separate experimental sessions.

When patients received stimulation with short rTMS trains at low intensity, repeated twice each day for 10 days, the Unified Parkinson's Disease Rating Scale (UPDRS) scores significantly improved, and the improvement persisted for nearly 3 months.[55] In a second study in a larger group of medicated PD patients, the same investigators described similar results with similar trains of rTMS but with a different range of intensity.[56] All groups of patients except for those receiving low-intensity rTMS showed significant clinical improvement. Other investigators reported a similar rTMS-induced improvement lasting for about 9 months.[57]

Siebner and associates[58] studied 12 drug-free PD patients and analyzed the kinematics of pointing movements with the index finger. Real, but not sham, 5-Hz rTMS was associated with a significant reduction of movement time assessed before and 20 minutes after the stimulation session. The same stimulus protocol, but not the sham stimulation, induced a clinical improvement in 10 PD patients off medication as assessed with the items of the UPDRS.[59] In 10 PD patients off medication, Boylan and colleagues[60] investigated the effects of rTMS over the SMA and found a clinical worsening of a number of motor task.

Deep Brain Stimulation and Lesions

An early report by Pascual-Leone and coworkers[61] described clinical improvement of the cardinal parkinsonian signs in nine patients together with a significant reduction of the MEP threshold after high-frequency stimulation of the internal globus pallidus. Conversely, lower stimulation frequencies (<25 Hz) were associated with a worsening of clinical symptoms and an increase in MEP threshold. After pallidotomy, MT remained unchanged, whereas MEP size diminished.[62]

In a group of dyskinetic patients, Chen and associates[28] showed that MT was normal and the MEP size was larger in patients than in controls at rest. The difference remained irrespective of whether patients had their internal globus pallidus stimulator set at the optimal parameters (On) or switched off (Off). The cortical SP was longer in the Off and in one half of the optimal amplitude (half-Amp) conditions than in normal control subjects. The SP was significantly shorter in the stimulator On condition than in the Off condition and approached that in normal control subjects. Measurement of intracortical excitability with paired stimuli at short and long ISIs showed no changes in intracortical excitability.[28] Thalamotomy[63] and pallidotomy[62] increase the duration of the SP.

Summary

The most consistent finding from TMS studies in PD is a shortened SP. Other abnormalities are changes in the input-output curve, reduced ICI at rest, increased inhibition at long ISIs in active and resting muscles, and during rTMS, a reduction of the usual MEP facilitation. Some of these changes can be reversed by levodopa or dopamine agonists, proving that a defect in dopaminergic modulation includes complex changes in the interneuronal pathways that modulate the corticospinal output. The beneficial effects produced by rTMS is controversial and still need to be confirmed in further research.

■ Parkinsonian Syndromes

In patients with multiple system atrophy (MSA), some investigators have reported a prolongation of the mean CMCT in lower limbs,[11] whereas CMCT in the upper limbs was normal.[11,64] These findings were not confirmed in a subsequent paper.[65]

Although apomorphine injection lengthens the duration of the cortical SP in patients with PD, it does not improve the SP in patients with MSA,[13] suggesting that the relation between clinical parkinsonian symptoms (i.e., akinesia and rigidity) and the SP duration differs in the two diseases.

ICI (at an ISI of 3 ms) was significantly reduced in 10 patients with predominantly parkinsonian MSA (MSA-P) and in seven patients with vascular parkinsonism (VP) but not in four with predominantly cerebellar MSA, whereas ICF (at ISI of 12 ms) remained unchanged.[66]

In a patient with corticobasal degeneration (CBD), MT was higher on the affected side and further increased as the disease progressed,[67] whereas a different study in a larger group of patients reported normal MT.[68] The SP duration was also shorter on the affected side and further decreased as the disease progressed.[67] In patients with CBD, there was a high incidence of bilateral responses in hand muscles after stimulation of the hemisphere ipsilateral to the alien hand. Ipsilateral responses may arise from transcallosal activation of the contralateral hand area. Moreover, in these patients, cortical maps had a larger extension to stimulation of the hemisphere contralateral to the alien hand than to stimulation of the ipsilateral hemisphere. This difference suggested an enhanced overall excitability of the motor area contralateral to the alien hand.[68] Stimulation of peripheral afferents induced a marked MEP facilitation that could be attributed to hyperexcitability of the sensorimotor cortex.[69]

In patients with progressive supranuclear palsy (PSP), TMS also showed a prolonged CMCT,[70] whereas in a later study, CMCT was normal.[71] In patients with PSP, the TMS-mediated RT shortening typical of PD patients was missing.[71] The absence of this phenomenon was interpreted as a disruption of facilitating systems descending to the spinal cord, possibly corticoreticulospinal fibers. In patients with drug-induced parkinsonism, the cortical SP also was abnormally short, and its duration was lengthened by levodopa.[16]

■ Wilson's Disease

Studies using electrical and magnetic transcranial stimulation have shown that patients with Wilson's disease sometimes have prolonged CMCTs,[72–74] and MEPs may be abnormal or absent, whereas MEPs evoked by electrical stimulation are of normal threshold and amplitude.[74,75] These abnormalities can be reversed by chelating therapy.[76,77] In patients with Wilson's disease, MEPs after TMS are often followed by long-latency responses.[78] To explain these long-latency responses, Chu[78] proposed that corticospinal fibers were damaged in the context of the subcortical white matter degeneration typical of Wilson's disease.

In another study, most patients had an increased SP threshold, and in some cases, the SP was absent.[75] Experiments using the paired-pulse technique disclosed reduced ICI but normal ICF in Wilson's disease.[40]

■ Dystonia

Motor Threshold, Motor Evoked Potentials, and Central Motor Conduction Time

Several researchers have reported normal MTs at rest and during contraction.[79–93] MT was also normal in the affected and the unaffected hemisphere in patients with writer's cramp[81,83,87,94] or before and after botulinum toxin injection.[94]

Studies with electrical stimulation[20] and TMS[79] showed that patients with primary dystonia have normal CMCT. In focal dystonia involving the arm or cranial dystonia, CMCT is normal in the hand[79,87] and cranial muscles.[85,90] In cervical dystonia, MEPs of shorter latencies have been observed in the sternomastoid muscle, but this result has been attributed to incomplete relaxation.[84] MEP latencies remained unchanged after muscular injection of botulinum toxin[94] or muscle vibration.[92] In patients with cervical dystonia, MEPs are normal at rest,[84,85,90] but they may be facilitated excessively during contraction.[90] Abnormalities of input-output curve have been also demonstrated in patients with arm dystonia[79,81] and interpreted as evidence of motor cortical hyperexcitability.

Mapping the motor cortex projections with TMS stimuli in patients with writer's cramp has provided evidence of an altered corticomotor representation, more evident in patients with long-standing cramps. Injection of botulinum toxin into the affected muscles transiently reversed the topographic changes during the period of greatest clinical benefit, but they reappeared when the effects of the injection wore off.[94] This effect probably results from the action of botulinum toxin on the spindle afferents, which would produce secondary changes at cortical level.

A study investigating the effects of peripheral stimulation on the MEP size in patients with cranial dystonia provided evidence that the inhibitory effect induced by conditioning stimulation of the median nerve is preserved, whereas in patients with focal hand dystonia, it is lost and substituted by a paradoxical facilitation.[93] Afferent peripheral feedback seems to be a physiological mechanism for activating inhibitory mechanisms at cortical or subcortical levels (i.e., sensorimotor integration). One study aimed to detect abnormalities of sensorimotor interactions and their topographic distribution in the hand muscles of patients with hand and cervical dystonia.[35] The effect of conditioning electrical stimulation of contiguous and noncontiguous fingers on the MEP size evoked by TMS or transcranial electrical stimulation in contiguous and noncontiguous muscles was investigated. In contrast to normal subjects, dystonic patients had their MEPs consistently inhibited by contiguous and noncontiguous finger stimulation, but the somatotopically distributed input-output organization of the sensorimotor interactions was lost. The investigators conclude that abnormalities of sensorimotor integration in dystonia also take place at the spinal and cortical level.[35] Using the technique of paired associative stimulation (PAS), Quartarone and colleagues[95] found that stimulation-induced facilitation of the MEPs amplitude was stronger in patients with writer's cramp than in normal subjects. This result was taken as evidence of an abnormal sensorimotor plasticity in patients with dystonia.

Silent Period

Investigations of the SP in patients with focal dystonias have disclosed various abnormalities. At low stimulus intensities, patients with focal dystonia have normal SP duration but with increasing stimulus intensity the cortical inhibitory mechanisms saturate more rapidly than those of normal subjects.[79,81] High stimulus intensities induce a slightly shorter SP in patients with upper limb dystonia than in controls[79] and an even shorter SP in patients with focal task-specific dystonia.[82,86] In patients with writer's cramp, the SP during dystonic contraction is significantly shorter than the SP obtained during comparable voluntary contraction in the same muscle.[83]

In patients with cranial dystonia, Odergren and colleagues[84] reported a SP of normal duration in the sternomastoid muscle, but with a delayed onset. Amadio and coworkers[90] found a bilaterally shortened SP when recording from the sternomastoid and trapezius muscles. In patients with blepharospasm or oromandibular dystonia as an isolated manifestation, the SP recorded in the orbicularis oculi and perioral muscles has an abnormally short duration.[96] The SP is even shorter in patients with dystonia affecting upper and lower facial muscles. This finding is of particular interest because the SP recorded in the facial muscle originates solely from intracortical inhibitory mechanisms.[97] The shortening of the SP indicates reduced excitability of cortical inhibitory interneurons in the motor cortex, in agreement with the reduced activation of the primary motor area observed with positron-emission tomography (PET) scans during movement.[98]

Paired Shocks

The paired-pulse technique at short ISIs in muscles at rest showed reduced ICI in dystonic patients, suggesting decreased cortical inhibition.[80,91] Studies at long ISIs also yielded abnormal, although controversial, results. In patients with writer's cramp, Chen and coworkers[82] reported that the amount of suppression is reduced at ISIs of 60 to 80 ms during muscle activation, whereas Rona and colleagues,[86] who studied active muscles in patients with different dystonic syndromes, found a normal short-interval inhibition but abnormally high inhibition at long ISIs. Hanajima and associates[85] described a reduced inhibition in the sternomastoid muscle in patients with cervical dystonia.

In patients with focal task-specific dystonia defective ICI can be normalized by injections of botulinum toxin type A. When the clinical effect vanishes 3 months after the injection, inhibition again increases. A plausible explanation is that the toxin modifies afferent input from muscles spindles, transiently altering motor cortex excitability.[91] This finding parallels the results of Byrnes and coworkers[94] on mapping motor cortex before and after botulinum toxin injection.

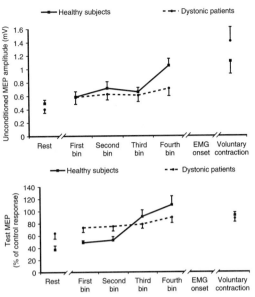

Figure 12–2 Motor evoked potential (MEP) amplitude *(upper panel)* and intracortical inhibition *(lower panel)* at rest, in the bins preceding rapid wrist extension, and during contraction in healthy subjects *(solid line)* and dystonic patients *(dashed line)*. The first bin for a stimulus-to-EMG interval corresponds to 139 to 120 ms; the second bin corresponds to 119 to 100 ms; the third bin corresponds to 99 to 80 m; and the fourth bin corresponds to 79 to 60 ms before the movement. Data are presented as the mean ± SE mV. (From Gilio F, Currà A, Inghilleri M, et al. Abnormalities of motor cortex excitability preceding movement in patients with dystonia. Brain 2003;126:1745–1754.)

The excitability of cortical motor areas can be studied with TMS by testing the MEP size and the degree of ICI during the RT preceding movement execution.[45] In patients with dystonia involving the upper limb, the MEP size and ICI tested before the onset of wrist extension lacks the changes observed in normal subjects. In particular, MEP size increase is lower and delayed, whereas the decreased ICI normally seen up to 60 ms before movement onset is absent[99] (Fig. 12–2). These findings suggest that patients with dystonia lack the normal cortical activation needed to produce movement.

Repetitive Transcranial Magnetic Stimulation

Few studies have used rTMS in patients with dystonia. Different from normal subjects, in patients with writer's cramp, suprathreshold

1-Hz rTMS applied over the primary motor hand area increases the mean MEP area of the train. This enhanced output of the primary motor cortex controlling the affected side could reflect a disease-related state of increased excitability of motor cortical structures in patients with writer's cramp.[89] In patients with writer's cramp, long trains of subthreshold 1-Hz rTMS normalized ICI and prolonged the SP without affecting the stimulus-response curve. Writing transiently improved, with reduced writing pressure.[89]

In a patient who had paroxysmal dystonia induced by exertion, with normal corticomotoneuronal system excitability between the attacks, rTMS induced dystonic attacks, as did a series of other stimuli, except for thermal and tactile cutaneous ones. These findings implied that the attacks were triggered by activation of proprioceptive afferents rather than cutaneous stimuli or the motor command itself.[100]

Summary

The most consistent finding in dystonia is reduced ICI. Other abnormalities include shortening of the SP, abnormal input-output curve, and abnormal mapping of cortical motor areas. Experiments designed to study the sensorimotor integration have demonstrated clear abnormalities. Like the effect seen in patients with PD, the improvement in motor performance seen in dystonic patients after rTMS needs to be confirmed in further studies.

■ Huntington's Disease

Motor Threshold, Motor Evoked Potentials, and Central Motor Conduction Time

In patients with Huntington's disease (HD), MEPs had normal threshold, latency, amplitude, and corticospinal conduction.[20,101–105] In a study of 34 patients with HD, Meyer and colleagues[106] found abnormal MEPs in about 72% of patients. The MT was higher in one third of the patients, and MEP latency was prolonged or the amplitude decreased in another third. The investigators attributed this result to an abnormal modulation of motor cortex secondary to disrupted input from the basal ganglia. However, some of these results may reflect the difficulty these patients have in relaxing their muscles appropriately during the tests performed at rest and in maintaining a constant level of background contraction.

Silent Period and Paired-Pulse Studies

The results of studies investigating the SP and the responses to paired-pulse TMS in HD remain controversial. In contrast to the slight shortening of the SP originally described in a small number of patients with HD,[101] others later reported that the cortical SP in HD can be abnormally long and of variable duration.[103,105,107] Whereas some investigators found reduced ICI and increased ICF,[108] others reported normal corticocortical inhibition and facilitation at short and long intervals in patients with HD.[109,110] Possible explanations for these controversial results in patients with HD include the numerous clinical variables known to influence motor cortical excitability, including age of patients, age at onset of the disease, severity of disease, and medical treatment. Some of the discordant findings may arise from differences in the experimental approach, such as the methods used for the collection and selection of the raw data during the experimental session and the method used for assessing the duration of the cortical SP. In patients with HD, the modality of data collection is especially important for documenting abnormalities in the cortical SP. Abnormal duration of the SP is best sought by collecting unselected consecutive traces.[105] The prolonged SPs in patients with HD may reflect changes in the activity of cortical inhibitory interneurons caused by altered output from the basal ganglia thalamocortical circuits.[103] An alternative explanation is that HD patients have an excessive delay in restarting a motor task when it has been interrupted, and they therefore need a longer time to resynchronize the activity of cortical motor and sensory areas.[105]

In a case report, Ziemann and associates[111] described a single diabetic patient with hemiballism-hemichorea who had markedly reduced intracortical excitability and an unusually long SP. A possible explanation for the prolonged SP in patients with HD is increased GABA content in the neocortex.

Summary

In patients with HD, the most common finding seems to be a prolonged cortical SP. However, this finding is limited by the variability due to the phenotypic heterogeneity, the various stages of the disease, and the different methods for data collecting.

Tremor

Patients with essential tremor have a normal CMCT, a normal duration of the cortical SP, and normal ICI at short and long ISIs.[112] SP duration has higher variability in patients with essential tremor than in control subjects, a finding interpreted as a sign of periodic fluctuations in motor cortical excitability related to tremor. Patients with primary writing tremor also have a normal MT, MEP size, SP duration, ICI, ICF, and cortical inhibition at long ISIs.[113]

Single TMS pulses reset essential tremor,[114,115] parkinsonian tremor,[114] and symptomatic palatal tremor[116] but not primary orthostatic tremor.[117,118] These results lend support to the involvement of the primary motor cortex in the pathophysiology of tremor. In contrast, Manto and coworkers[119] reproduced a phase-resetting phenomenon in three subjects with orthostatic tremor secondary to cerebellar cortical atrophy, and Wu and colleagues[118] found that rather weak electrical stimuli over the posterior fossa were capable of resetting the tremor.

In a double-blind, crossover, placebo-controlled design, low-frequency suprathreshold rTMS of the cerebellum was tested in patients with essential tremor.[120] Tremor improvement (clinical and accelerometric) was seen at 5 minutes but vanished at 1 hour after rTMS. However, rTMS may induce a cerebellar-like and postural tremor in normal subjects by interfering with the adaptive cerebellar afferent inflow to the motor cortex.[121] The frequency of rTMS-induced tremor is independent of stimulus variables.

Myoclonus

Although some reports have described normal or increased MT in patients with cortical myoclonus, the patients studied were treated with various antiepileptic drugs that can enhance MT, and many of them had cerebellar atrophy, which may have a similar effect.[122,123]

In a single case of cortical myoclonus epilepsy, Inghilleri and associates[124] observed a shorter SP on the side contralateral to the myoclonus. Patients with progressive myoclonic epilepsy have decreased ICI[40,123,124] and decreased inhibition at long ISIs.[125]

Patients with benign myoclonic epilepsy had increased interhemispheric facilitation and reduced interhemispheric inhibition (tested with conditioning stimulation of the motor cortex followed by test stimulation of the contralateral motor cortex). TMS studies have also provided evidence of cortical involvement underlying myoclonus seen in patients with corticobasal degeneration.[126] These findings are consistent with hyperexcitability of the sensorimotor cortex in cortical myoclonus.

Tics

In Tourette's syndrome, the MT is normal, but the SP is shortened, and ICI at short ISIs is reduced. These abnormalities are found in patients in whom the tics involve the distal target muscle and in patients without neuroleptic treatment.[127]

Patients with obsessive-compulsive disorder (OCD) have decreased active and resting motor thresholds, normal duration of the SP, and reduced ICI. In these patients, the recruitment curves show slightly larger MEP amplitude at most stimulus intensities. Decreased ICI and motor threshold were greater in patients with OCD than in those with comorbid tics.[128]

These findings support the hypothesis that patients with Tourette's syndrome and

OCD have overlapping dysfunctions and that abnormalities of motor cortex are particularly evident when the two conditions coexist.[128]

Ataxias

Threshold and Central Motor Conduction Time

MT is normal in ataxias,[129] except in SCA1 and SCA2 types,[130] and in patients with unilateral cerebellar lesions,[131] in whom MT is increased. CMCT is prolonged in patients with spinocerebellar ataxia type 1 (SCA1) and Friedreich's ataxia[129,132,133] but not in those with SCA2 or SCA3. Restivo and coworkers[134] nevertheless observed that CMCT was abnormal in SCA2 if MEPs were recorded from lower limbs. These results suggest that pyramidal conduction is impaired in ataxias.

Silent Period

Although most studies of patients with ataxia (of heterogeneous classification) have reported a prolonged cortical SP,[135] in a study separating the various subtypes of genetic ataxias, Schwenkreis and colleagues[129] failed to replicate this finding.

Paired-Pulse Studies

ICI is normal in various forms of ataxias.[129,136,137] Conversely, intracortical facilitation was described as deficient,[136] possibly linked to reduced excitatory drive from deep cerebellar nuclei. Analyzing distinct genotypic groups, Schwenkreis and associates[129] concluded that facilitation was reduced in patients with SCA2 and SCA3 but not in those with Friedreich's ataxia, SCA1, or SCA6. The investigators attributed the reduced facilitation to the widespread expression of the pathologic SCA2- and SCA3-related proteins in the cortex. This abnormality disrupts the function of motor cortical interneurons.

Repetitive Transcranial Magnetic Stimulation

In a sham-controlled study, Shiga and coworkers[138] observed that after low-frequency suprathreshold rTMS delivered once each day for 3 weeks, several indexes of motor impairment significantly improved. The improvement lasted on average a week and was more pronounced in the subgroup of patients with the cerebellar type than in those with the olivopontocerebellar type of ataxia.

Summary

Abnormal findings on TMS probably differ in the various forms of ataxia. In some, such as Friedreich's ataxia, SCA1, and possibly SCA2, a slowing of pyramidal conduction corresponds to the anatomic damage. The MT can be higher and the cortical SP prolonged. Deficient intracortical facilitation may be seen in SCA2 and SCA3.

Conclusion

From a clinical point of view, TMS has a potential role in demonstrating abnormalities of central motor conduction in patients with secondary forms of movement disorders in whom pyramidal signs may be clinically equivocal. Testing investigational TMS measures (i.e., paired-pulse studies, input-output curves, MT measurements, and SP) frequently disclose abnormal findings, but these abnormalities are not specific for any disease. The most important role of TMS methods is to provide information on the physiological role of cortical motor areas in the pathophysiology of movement disorders.[139]

REFERENCES

1. Day BL, Rothwell JC, Thompson PD, et al. Motor cortex stimulation in intact man. II. Multiple descending volleys. Brain 1987;110:1191–1209.
2. Hess W, Mills KR, Murray NMF. Responses in small hand muscles from magnetic stimulation of the human brain. J Physiol 1987;388:397–419.
3. Cantello R, Gianelli M, Civardi C, et al. Magnetic brain stimulation: the silent period after the motor evoked potential. Neurology 1991;42:1951–1959.
4. Cantello R, Gianelli M, Civardi C, et al. Parkinson's disease rigidity: EMG in a small hand muscle at "rest." Electroencephalogr Clin Neurophysiol 1995;97:215–222.
5. Landau WM, Struppler A, Mehls O. A comparative electromyographic study of the reactions to passive movement in parkinsonism. Neurology 1966;16:34–48.

6. Ellaway PH, Davey NJ, Maskill DW, et al. The relation between bradykinesia and excitability of the motor cortex assessed using transcranial magnetic stimulation in normal and parkinsonian subjects. Electroencephalogr Clin Neurophysiol 1995;97:169–178.
7. Valls-Solé J, Pascual-Leone A, Brasil-Neto JP, et al. Abnormal facilitation of the response to transcranial magnetic stimulation in patients with Parkinson's disease. Neurology 1994;44:735–741.
8. Nakashima K, Wang Y, Shimoda M, et al. Shortened silent period produced by magnetic cortical stimulation in patients with Parkinson's disease. J Neurol Sci 1995;130:209–214.
9. Ridding MC, Inzelberg R, Rothwell JC. Changes in excitability of motor cortical circuitry in patients with Parkinson's disease. Ann Neurol 1995;37:181–188.
10. Berardelli A, Rona S, Inghilleri M, et al. Cortical inhibition in Parkinson's disease. A study with paired magnetic stimulation. Brain 1996;119:71–77.
11. Abbruzzese G, Marchese R, Trompetto C. Sensory and motor evoked potentials in multiple system atrophy: a comparative study with Parkinson's disease. Mov Disord 1997;12:315–321.
12. Valzania F, Strafella AP, Quatrale R, et al. Motor evoked responses to paired cortical magnetic stimulation in Parkinson's disease. Electroencephalogr Clin Neurophysiol 1997;105:37–43.
13. Manfredi L, Garavaglia P, Beretta S, et al. Increased cortical inhibition induced by apomorphine in patients with Parkinson's disease. Clin Neurophysiol 1998;28:31–38.
14. Siebner HR, Mentschel C, Auer C, et al. Repetitive transcranial magnetic stimulation causes a short-term increase in the duration of the cortical silent period in patients with Parkinson's disease. Neurosci Lett 2000;284:147–150.
15. Strafella AP, Valzania F, Nassetti SA, et al. Effects of chronic levodopa and pergolide treatment on cortical excitability in Parkinson's disease: a transcranial magnetic stimulation study. Clin Neurophysiol 2000;111:1198–1202.
16. Priori A, Berardelli A, Inghilleri M, et al. Motor cortical inhibition and the dopaminergic system. Pharmacological changes in the silent period after transcranial brain stimulation in normal subjects, patients with Parkinson's disease and drug-induced parkinsonism. Brain 1994;117:317–323.
17. Ziemann U, Lonnecker S, Steinhoff BJ, et al. Effects of antiepileptic drugs on motor cortex excitability in humans: a transcranial magnetic stimulation study. Ann Neurol 1996;40:367–378.
18. Ziemann U, Tergau F, Burns D, et al. Changes in human motor cortex excitability induced by dopaminergic and antidopaminergic drugs. Electroencephalogr Clin Neurophysiol 1997;105:430–437.
19. Dick JPR, Cowan JMA, Day BL, et al. The corticomotoneurone connection is normal in Parkinson disease. Nature 1984;310:407–409.
20. Thompson PD, Dick JPR, Day BL, et al. Electrophysiology of the corticomotoneurone pathways in patients with movement disorders. Mov Disord 1986;1:113–118.
21. Eisen AA, Shtybel W. AAEM minimonograph #35: clinical experience with transcranial magnetic stimulation. Muscle Nerve 1990;13:995–1011.
22. Valls-Solé J, Pascual-Leone A, Wassermann EM, et al. Human motor evoked responses to paired transcranial magnetic stimulation. Electroencephalogr Clin Neurophysiol 1992;85:355–364.
23. Kandler RH, Jarrat JA, Sagar HJ, et al. Abnormalities of central motor conduction in Parkinson's disease. J Neurol Sci 1990;100:94–97.
24. Shimamoto H, Morimitsu H, Sugita S, et al. Motor evoked potentials of transcranial magnetic stimulation for Parkinson's disease. No To Shinkei 1996;48:825–829.
25. Diószeghy P, Hidasi E, Mechler F. Study of central motor functions using magnetic stimulation in Parkinson's disease. Electromyogr Clin Neurophysiol 1999;39:101–105.
26. Hu MTM, Bland J, Clough C, et al. Limb contractures in levodopa-responsive parkinsonism: a clinical and investigational study of seven new cases. J Neurol 1999;246:671–676.
27. Eisen A, Siejka S, Schulzer M, et al. Age-dependent decline in motor evoked potential MEP. amplitude: with a comment on changes in Parkinson's disease. Electroencephalogr Clin Neurophysiol 1991;81:209–215.
28. Chen R, Garg RR, Lozano AM, et al. Effects of internal globus pallidus stimulation on motor cortex excitability. Neurol 2001;56:716–723.
29. Sax DS, Johnson TL, Feldman RG. L-Dopa effects on H-reflex recovery in Parkinson's disease. Ann Neurol 1977;2:120–124.
30. Abbruzzese G, Vische M, Ratto S, et al. Assessment of motor neuron excitability in parkinsonian rigidity by the F wave. J Neurol 1985;232:246–249.
31. Marsden CD. The mysterious motor function of the basal ganglia: the Robert Wartenberg lecture. Neurology 1982;32:514–539.
32. Kleine BU, Praamstra P, Stegeman DF. Impaired motor cortical inhibition in

Parkinson's disease: motor unit responses to transcranial magnetic stimulation. Exp Brain Res 2001;138:477–483.
33. Delwaide PJ, Olivier E. Conditioning transcranial cortical stimulation TCCS by exteroceptive stimulation in parkinsonian patients. In Striefler MB, Korczyn AD, Melamed E (eds): Parkinson's Disease: Anatomy, Pathology and Therapy. Advances in Neurology. New York, Raven Press, 1994:175–181.
34. Clouston PD, Lim CL, Sue C, et al. Apomorphine can increase cutaneous inhibition of motor activity in Parkinson's disease. Electroencephalogr Clin Neurophysiol 1996;101:8–15.
35. Tamburin S, Manganotti P, Marzi CA, et al. Abnormal somatotopic arrangement of sensorimotor interactions in dystonic patients. Brain 2002;125:2719–2730.
36. Inghilleri M, Berardelli A, Cruccu G, et al. Silent period evoked by transcranial stimulation of the human cortex and cervicomedullary junction. J Physiol 1993;466:521–534
37. Haug BA, Schonle PW, Knobloch C, et al. Silent period revives as a valuable diagnostic tool with transcranial magnetic stimulation. Electroencephalogr Clin Neurophysiol 1992;85:158–160.
38. Kujirai T, Caramia MD, Rothwell JC, et al. Corticocortical inhibition in human motor cortex. J Physiol 1993;471:501–519.
39. Sanger TD, Garg RR, Chen R. Interactions of two different inhibitory systems in the human motor cortex. J Physiol 2001;530:307–317.
40. Hanajima R, Ugawa Y, Terao Y, et al. Ipsilateral cortico-cortical inhibition of the motor cortex in various neurological disorders. J Neurol Sci 1996;140:109–116.
41. Pierantozzi M, Calmieri MG, Marciani MG, et al. Effect of apomorphine on cortical inhibition in Parkinson's disease patients: a transcranial magnetic stimulation study. Exp Brain Res 2001;141:52–62.
42. Goldman-Rakic PS, Lidow MS, Smiley JF, et al. The anatomy of dopamine in monkey and human prefrontal cortex. J Neural Transm Suppl 1992;36:163–177.
43. Rossini PM, Zarola F, Stalberg E, et al. Pre-movement facilitation of motor-evoked potentials in man during transcranial stimulation of the central motor pathways. Brain Res 1998;458:20–30.
44. Chen R, Yaseen Z, Cohen LG, et al. Time course of corticospinal excitability in reaction time and self-paced movements. Ann Neurol 1998;44:317–325.
45. Reynolds C, Ashby P. Inhibition in the human motor cortex is reduced just before a voluntary contraction. Neurology 1999;53:730–735.
46. Chen R, Kumar S, Garg RG, et al. Impairment of motor cortex activation and deactivation in Parkinson's disease. Clin Neurophysiol 2001;112:600–607.
47. Pascual-Leone A, Valls-Solé J, Wassermann EM, et al. Effect of focal transcranial magnetic stimulation on simple reaction time to visual, acoustic and somatosensory stimuli. Brain 1992;115:1045–1059.
48. Pascual-Leone A, Valls-Solé J, Brasil-Neto JP, et al. Akinesia in Parkinson's disease. I. Shortening of simple reaction time with focal, single-pulse transcranial magnetic stimulation. Neurology 1994;44:884–891.
49. Cunnington R, Iansek R, Thickbroom GW, et al. Effects of magnetic stimulation over supplementary motor area on movement in Parkinson's disease. Brain 1996;199:815–822.
50. Gilio F, Currà A, Inghilleri M, et al. Repetitive transcranial magnetic stimulation of cortical motor areas in Parkinson's disease. Mov Disord 2002;147:86–92.
51. Berardelli A, Inghilleri M, Rothwell JC, et al. Facilitation of muscle evoked responses after repetitive cortical stimulation in man. Exp Brain Res 1998;122:79–84.
52. Pascual-Leone A, Valls-Solé J, Brasil-Neto JP, et al. Akinesia in Parkinson's disease. II. Effects of subthreshold repetitive transcranial motor cortex stimulation. Neurology 1994;44:892–898.
53. Ghabra MB, Hallett M, Wassermann EM. Simultaneous repetitive transcranial magnetic stimulation does not speed fine movement in PD. Neurology 1999;52:768–770.
54. Tergau F, Wassermann EM, Paulus W, et al. Lack of clinical improvement in patients with Parkinson's disease after low and high-frequency repetitive transcranial magnetic stimulation. Clin Neurophysiol Suppl 1999; 51:281–288.
55. Mally J, Stone TW. Improvement in parkinsonian symptoms after repetitive transcranial magnetic stimulation. J Neurol Sci 1999;162:179–184.
56. Mally J, Stone TW. Therapeutic, "dose dependent" effect of repetitive microelectroshock induced by transcranial magnetic stimulation in Parkinson's disease. J Neurosci Res 1999;57:935–940.
57. Shimamoto H, Morimatsu H, Sugita S, et al. Therapeutic effect of repetitive transcranial magnetic stimulation in Parkinson's disease. Rinsho Shinkeigaku 1999;39:264–267.
58. Siebner HR, Mentschel C, Auer C, et al. Repetitive transcranial magnetic stimulation has a beneficial effect on bradykinesia in Parkinson's disease. Neuroreport 1999;10:589–594.
59. Siebner HR, Rossmeier C, Mentschel C, et al. Short-term motor improvement after

59. subthreshold 5-Hz repetitive transcranial magnetic stimulation of the primary motor hand area in Parkinson's disease. J Neurol Sci 2000;178:91–94.
60. Boylan LS, Pullman SL, Lisanby SH, et al. Repetitive transcranial magnetic stimulation to SMA worsens complex movements in Parkinson's disease. Clin Neurophysiol 2001;112:259–264
61. Pascual-Leone A, Barcia-Salorio JL, Arcusa MJ, et al. Effects of stimulation of the globus pallidus internus on motor excitability in Parkinson's disease [abstract]. Neurology 1996;46:A372.
62. Young MS, Triggs WJ, Bowers D, et al. Stereotactic pallidotomy lengthens the transcranial magnetic cortical stimulation silent period in Parkinson's disease. Neurology 1997;49:1278–1283.
63. van der Linden C, Bruggeman R, Goldman WH. Alterations of motor evoked potentials by thalamotomy. Electromyogr Clin Neurophysiol 1993;33:329–334.
64. Cruz Martinez A, Arpa J, Alonso M, et al. Transcranial magnetic stimulation in multiple system and late onset cerebellar atrophies. Acta Neurol Scand 1995;92:218–224.
65. Abele M, Schulz JB, Burk K, et al. Evoked potentials in multiple system atrophy. Acta Neurol Scand 2000;101:111–115.
66. Marchese R, Trompetto C, Buccolieri A, et al. Abnormalities of motor cortical excitability are not correlated with clinical features in atypical parkinsonism. Mov Disord 2000;15:1210–1214.
67. Fujimoto K, Salama S, Shisuma N, et al. Intracranial inhibitory mechanisms in clinically diagnosed corticobasal degeneration: a study of a silent period followed by transcranial magnetic stimulation. Rinsho Shinkeigaku 2000;40:701–706.
68. Valls-Solé J, Tolosa E, Marti MJ, et al. Examination of motor output pathways in patients with corticobasal degeneration using transcranial magnetic stimulation. Brain 2001;124:1131–1137.
69. Yokota T, Saito Y, Shimizu Y. Increased corticomotoneuronal excitability after peripheral nerve stimulation in dopa-nonresponsive hemiparkinsonism. J Neurol Sci 1995;129:34–39.
70. Abbruzzese G, Tabaton M, Dall'Agata D, et al. Motor and sensory evoked potentials in progressive supranuclear palsy. Mov Disord 1991;6:49–54.
71. Molinuevo JL, Valls-Solé J, Valldeoriola F. The effect of transcranial magnetic stimulation on reaction time in progressive supranuclear palsy. Clin Neurophysiol 2000;111:2008–2013.
72. Caramia MD, Bernardi G, Zarola F, et al. Neurophysiological evaluation of the central nervous impulse propagation in patients with sensorimotor disturbances. Electroencephalogr Clin Neurophysiol 1988;70:16–25.
73. Berardelli A, Inghilleri M, Priori A, et al. Involvement of corticospinal tracts in Wilson's disease. Mov Disord 1990;5:334–337.
74. Meyer BU, Britton TC, Bischoff C, et al. Abnormal conduction in corticospinal pathways in Wilson's disease: investigation of nine cases with magnetic brain stimulation. Mov Disord 1991;6:320–323.
75. Perretti A, Pellecchia MT, Lanzillo B, et al. Excitatory and inhibitory mechanisms in Wilson's disease: investigation with magnetic motor cortex stimulation. J Neurol Sci 2001;192:35–40.
76. Meyer BU, Britton TC, Benecke R. Wilson's disease: normalisation of cortically evoked motor responses with treatment. J Neurol 1991;238:327–330.
77. Hefter H, Roick H, van Giesen HJ, et al. Motor impairment in Wilson's disease. 3. The clinical impact of pyramidal tract involvement. Acta Neurol Scand 1994;89:421–428.
78. Chu N. Motor evoked potentials in Wilson's disease: early and late motor responses. J Neurol Sci 1990;99:259–269.
79. Mavroudakis N, Caroyer JM, Brunko E, et al. Abnormal motor evoked responses to transcranial magnetic stimulation in focal dystonia. Neurology 1995;45:1671–1677.
80. Ridding MC, Sheean G, Rothwell JC, et al. Changes in the balance between motor cortical excitation and inhibition in focal, task specific dystonia. J Neurol Neurosurg Psychiatry 1995;59:493–498.
81. Ikoma K, Samii A, Mercuri B, et al. Abnormal cortical motor excitability in dystonia. Neurology 1996;46:1371–1376.
82. Chen R, Wassermann EM, Canos M, et al. Impaired inhibition in writer's cramp during voluntary muscle activation. Neurology 1997;49:1054–1059.
83. Filipovic SR, Ljubisavljevic M, Svetel M, et al. Impairment of cortical inhibition in writer's cramp as revealed by changes in electromyographic silent period after transcranial magnetic stimulation. Neurosci Lett 1997;222:167–170.
84. Odergren T, Rimpilainen I, Borg J. Sternocleidomastoid muscle responses to transcranial magnetic stimulation in patients with cervical dystonia. Electroencephalogr Clin Neurophysiol 1997;105:44–52.
85. Hanajima R, Ugawa Y, Terao Y, et al. Cortico-cortical inhibition of motor cortical area projecting to sternocleidomastoid muscle in normal subjects and patients with spasmodic torticollis or essential tremor. Electroencephalogr Clin Neurophysiol 1998;109:391–396.
86. Rona S, Berardelli A, Vacca L, et al. Alterations of motor cortical inhibition in

patients with dystonia. Mov Disord 1998;13:118–124.
87. Schwenkreis P, Vorgerd M, Malin JP, et al. Assessment of postexcitatory inhibition in patients with focal dystonia. Acta Neurol Scand 1999;100:260–264.
88. Siebner HR, Auer C, Conrad B. Abnormal increase in the corticomotor output to the affected hand during repetitive transcranial magnetic stimulation of the primary motor cortex in patients with writer's cramp. Neurosci Lett 1999;262:133–136.
89. Siebner HR, Tormos JM, Ceballos-Baumann AO, et al. Low-frequency repetitive transcranial magnetic stimulation of the motor cortex in writer's cramp. Neurology 1999;52:529–537.
90. Amadio S, Panizza M, Pisano F, et al. Transcranial magnetic stimulation and silent period in spasmodic torticollis. Am J Phys Med Rehabil 2000;79:361–368.
91. Gilio F, Currà A, Lorenzano C, et al. Effects of botulinum toxin type A on intracortical inhibition in patients with dystonia. Ann Neurol 2000;48:20–26.
92. Rosenkranz K, Altenmuller E, Siggelkow S, et al. Alteration of sensorimotor integration in musician's cramp: impaired focusing of proprioception. Clin Neurophysiol 2000;111:2040–2045.
93. Abbruzzese G, Marchese R, Buccolieri A, et al. Abnormalities of sensorimotor integration in focal dystonia: a transcranial magnetic stimulation study. Brain 2001;124:537–545.
94. Byrnes ML, Thickbroom GW, Wilson SA, et al. The corticomotor representation of upper limb muscles in writer's cramp and changes following botulinum toxin injection. Brain 1998;121:977–988.
95. Quartarone A, Bagnato S, Rizzo V, et al. Abnormal associative plasticity of the human motor cortex in writer's cramp. Brain 2003;126:2586–2596.
96. Currà A, Romaniello A, Berardelli A, et al. Shortened cortical silent period in facial muscle of patients with cranial dystonia. Neurology 2000;54:130–135.
97. Cruccu G, Inghilleri M, Berardelli A, et al. Cortical mechanisms mediating the inhibitory period after magnetic stimulation of the facial motor area. Muscle Nerve 1997;20:418–424.
98. Ceballos-Baumann AO, Passingham RE, Warner T, et al. Overactive prefrontal and underactive motor cortical areas in idiopathic dystonia. Ann Neurol 1995;37:363–372.
99. Gilio F, Currà A, Inghilleri M, et al. Abnormalities of motor cortex excitability preceding movement in patients with dystonia. Brain 2003;126:1745–1754.
100. Meyer BU, Irlbacher K, Meierkord H. Analysis of stimuli triggering attacks of paroxysmal dystonia induced by exertion. J Neurol Neurosurg Psychiatry 2001;70:247–251.
101. Eisen A, Bohlega S, Bloch M, et al. Silent periods, long-latency reflexes and cortical MEPs in Huntington's disease and at risk relatives. Electroencephalogr Clin Neurophysiol 1989;74:444–449.
102. Homberg V, Lange HW. Central motor conduction to hand and leg muscles in Huntington's disease. Mov Disord 1990;5:214–218.
103. Priori A, Berardelli A, Inghilleri M, et al. Electromyographic silent period after transcranial brain stimulation in Huntington's disease. Mov Disord 1994;9:178–182.
104. Abbruzzese G, Marchese R, Trompetto C. Motor cortical excitability in Huntington's disease. J Neurol Neurosurg Psychiatry 2000;68:120–121.
105. Modugno N, Currà A, Giovannelli M, et al. The prolonged cortical silent period in patients with Huntington's disease. Clin Neurophysiol 2001;112:1470–1474.
106. Meyer BU, Noth J, Lange HW, et al. Motor responses evoked by magnetic brain stimulation in Huntington's disease. Electroencephalogr Clin Neurophysiol 1992;85:197–208.
107. Tegenthoff M, Vorgerd M, Juskowiak F, et al. Postexcitatory inhibition after transcranial magnetic single and double brain stimulation in Huntington's disease. Electroencephalogr Clin Neurophysiol 1996;101:298–303.
108. Abbruzzese G, Buccolieri A, Marchese R, et al. Intracortical inhibition and facilitation are abnormal in Huntington's disease: a paired magnetic stimulation study. Neurosci Lett 1998;228:87–90.
109. Hanajima R, Ugawa J, Terao Y, et al. Intracortical inhibition of the motor cortex is normal in chorea. J Neurol Neurosurg Psychiatry 1999;66:783–786.
110. Priori A, Polidori L, Rona S, et al. Spinal and cortical inhibition in Huntington's chorea. Mov Disord 2000;15:938–946.
111. Ziemann U, Koc J, Reimers CD, et al. Exploration of motor cortex excitability in a diabetic patient with hemiballism-hemichorea. Mov Disord 2000;15:1000–1005.
112. Romeo S, Berardelli A, Pedace F, et al. Cortical excitability in patients with essential tremor. Muscle Nerve 1998;21:1304–1308.
113. Modugno N, Nakamura Y, Bestmann S, et al. Neurophysiological investigations in patients with primary writing tremor. Mov Disord 2002;17:1336–1340
114. Britton TC, Thompson PD, Day BL, et al. Modulation of postural wrist tremors by magnetic stimulation of the motor cortex in

114. patients with Parkinson's disease or essential tremor and in normal subjects mimicking tremor. Ann Neurol 1993;33:473–479.
115. Pascual-Leone A, Valls-Solé J, Toro C, et al. Resetting of essential tremor and postural tremor in Parkinson's disease with transcranial magnetic stimulation. Muscle Nerve 1994;17:800–807.
116. Chen J, Yu H, Wu Z, et al. Modulation of symptomatic palatal tremor by magnetic stimulation of the motor cortex. Clin Neurophysiol 2000;111:1191–1197.
117. Mills KR, Nithi KA. Motor cortex stimulation does not reset primary orthostatic tremor. J Neurol Neurosurg Psychiatry 1997;63:553.
118. Wu YR, Ashby P, Lang AE. Orthostatic tremor arises from an oscillator in the posterior fossa. Mov Disord 2001;16:272–279.
119. Manto M, Setta F, Legros B, et al. Resetting of orthostatic tremor associated with cerebellar cortical atrophy by transcranial magnetic stimulation. Arch Neurol 1999;56:1497–1500.
120. Gironell A, Kulisevsky J, Lorenzo J, et al. Transcranial magnetic stimulation of the cerebellum in essential tremor: a controlled study. Arch Neurol 2002; 59:413–417.
121. Topka H, Mescheriakov S, Boose A, et al. A cerebellar-like terminal and postural tremor induced in normal man by transcranial magnetic stimulation. Brain 1999;122:1551–1562.
122. Cantello R, Gianelli M, Civardi C, et al. Focal subcortical reflex myoclonus. A clinical and neurophysiological study. Arch Neurol 1997;54:187–196.
123. Manganotti P, Tamburin S, Zanette G, et al. Hyperexcitable cortical responses in progressive myoclonic epilepsy. Neurology 2001;57:1793–1799.
124. Inghilleri M, Mattia D, Berardelli A, et al. Asymmetry of cortical excitability revealed by transcranial stimulation in a patient with focal motor epilepsy and cortical myoclonus. Electroencephalogr Clin Neurophysiol 1998;109:70–72.
125. Valzania F, Strafella AP, Tropeani A, et al. Facilitation of rhythmic events in progressive myoclonus epilepsy: a transcranial magnetic stimulation study. Clin Neurophysiol 1999;110:152–157.
126. Thompson PD, Day BL, Rothwell JC, et al. The myoclonus in corticobasal degeneration. Evidence for two forms of cortical reflex myoclonus. Brain 1994;117:1197–1207.
127. Ziemann U, Paulus W, Rothenberger A. Decreased motor inhibition in Tourette disorder: evidence from transcranial magnetic stimulation. Am J Psychiatry 1997;154:1277–1284.
128. Greenberg BD, Ziemann U, Corà-Locatelli G, et al. Altered cortical excitability in obsessive-compulsive disorder. Neurology 2000;54:142–147.
129. Schwenkreis P, Tegenthoff M, Witscher K, et al. Motor cortex activation by transcranial magnetic stimulation in ataxia patients depends on the genetic defect. Brain 2002;125:301–309.
130. Cruz Martinez A, Palau F. Central motor conduction time by magnetic stimulation of the cortex and peripheral nerve conduction follow-up studies in Friedreich's ataxia. Electroencephalogr Clin Neurophysiol 1997;105:458–461.
131. Di Lazzaro V, Restuccia D, Molinari M, et al. Excitability of the motor cortex to magnetic stimulation in patients with cerebellar lesions. J Neurol Neurosurg Psychiatry 1994;57:108–110.
132. Schöls L, Amoiridis G, Langkafel M, et al. Motor evoked potentials in the spinocerebellar ataxias type 1 and type 3. Muscle Nerve 1997;20:226–228.
133. Yokota T, Sasaki H, Iwabuchi K, et al. Electrophysiological features of central motor conduction in spinocerebellar atrophy type 1, type 2 and Machado-Joseph disease. J Neurol Neurosurg Psychiatry 1998;65:530–534.
134. Restivo DA, Giuffrida S, Rapisarda G, et al. Central motor conduction to lower limb after transcranial magnetic stimulation in spinocerebellar ataxia (type 2 SCA2). Clin Neurophysiol 2000;111:630–635.
135. Wessel K, Tegenthoff M, Vorgerd M, et al. Enhancement of inhibitory mechanisms in the motor cortex of patients with cerebellar degeneration: a study with transcranial magnetic brain stimulation. Electroencephalogr Clin Neurophysiol 1996;101:273–280.
136. Liepert J, Wessel F, Schwenkreis P, et al. Reduced intracortical facilitation in patients with cerebellar degeneration. Acta Neurol Scand 1998;98:318–323.
137. Ugawa Y, Hanajima R, Kanazawa I. Motor cortex inhibition in patients with ataxia. Electroencephalogr Clin Neurophysiol 1994;93:225–229.
138. Shiga Y, Tsuda T, Itoyama Y, et al. Transcranial magnetic stimulation alleviates truncal ataxia in spinocerebellar degeneration. J Neurol Neurosurg Psychiatry 2002;72:124–126.
139. Currà A, Modugno N, Inghilleri M, et al. Transcranial magnetic stimulation techniques in clinical investigation. Neurology 2002;59:1851–1859.

13

Cerebellar Stimulation in Normal Subjects and Ataxic Patients

Yoshikazu Ugawa and Nobue K. Iwata

It is well known that the cerebellum plays an important role in motor execution. Its function and functional connections with other areas have been studied in animals.[1] However, because we have had only limited physiological ways to study central nervous system function directly in humans until the invention of high-voltage electrical[2] or magnetic stimulation[3] methods, details of human cerebellar function remained to be studied. After transcranial electrical stimulation (TES)[2] and transcranial magnetic stimulation (TMS)[3] were introduced, they were first used to investigate physiology of the corticospinal tract (CST). Several modulatory effects on CST neurons were also physiologically studied with these methods in humans. One of them is a cerebellar modulatory effect on the motor cortex. In this chapter, we first briefly summarize effects of cerebellar stimulation on the contralateral motor cortex in normal subjects and later describe changes of this effect in patients with ataxia and other disorders.

Cerebellar Stimulation in Normal Subjects

The dentatothalamocortical pathway has a facilitatory influence on the motor cortex, and low-frequency tonic facilitatory inputs are continuously going to the motor cortex from the contralateral cerebellum. We therefore anticipated seeing a facilitation of the motor cortex by cerebellar stimulation. Unexpectedly, however, we elicited a suppressive effect on the motor cortex by cerebellar stimulation, probably because suppression lasts longer than facilitation and it is more easily seen. A facilitatory effect is also evoked under a very restricted condition in which the intensity of the cerebellar stimulus is adjusted to be just below the threshold for activation of the CST and the test stimulus elicits I3 waves. We first describe inhibition and then facilitation.

Inhibitory Effect

Activation of the human cerebellum was first performed with high-voltage electrical stimulation[4] and later with TMS.[5] Cerebellar stimuli reduced the excitability of the contralateral motor cortex at several milliseconds after the stimulus. A randomized conditioning-test design was used in the experiment of cerebellar stimulation. The conditioning stimulus is TES or TMS over the cerebellum, and the test stimulus is TMS over the contralateral motor cortex. Intervals between these two stimuli were 4 to 10 ms. Motor cortical excitability changes were evaluated by size changes of electromyographic (EMG) responses to motor cortical stimulation.

Figure 13–1A shows an example of responses in the cerebellar stimulation experiment. A control response elicited by the test stimulus alone is in the first row, and conditioned responses at several interstimulus intervals (ISIs) are in the lower rows. Conditioned responses at ISIs of 5, 6, and 7 ms were smaller than the control response,

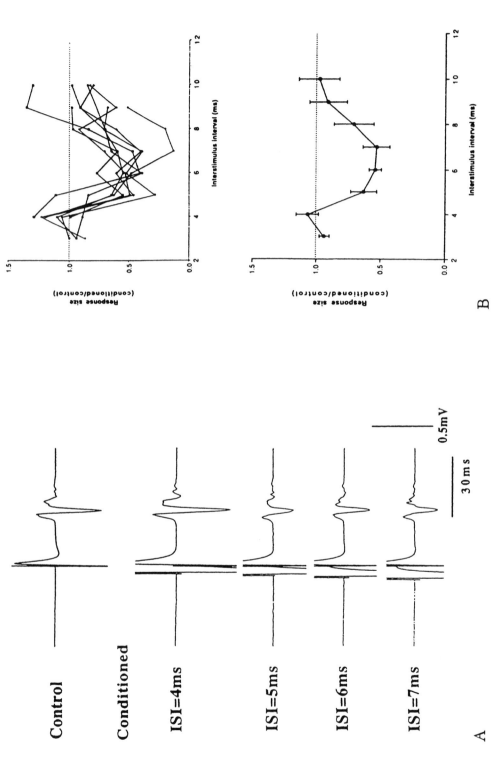

Figure 13-1 Average electromyographical responses from the first dorsal interosseous muscle in a normal subject during cerebellar stimulation (**A**) and time courses of the cerebellar suppression (**B**). The *top trace* shows a control response to transcranial magnetic stimulation (TMS) over the motor cortex, and the *lower rows* are conditioned responses at interstimulus intervals (ISIs) of 4, 5, 6, and 7 ms. As a conditioning stimulus, TMS was given over the cerebellum. Conditioned responses at ISIs of 5, 6, and 7 ms were smaller than the control response, whereas it is almost the same in size as the control response at an ISI of 4 ms. In time courses for single subjects or the mean time course from normal subjects, suppression began at an ISI of 5 ms and lasted for a few milliseconds.

whereas that at an ISI of 4 ms was almost the same size as the control response. Time courses of this effect are shown in Figure 13–1B. The ordinate indicates size ratios of the conditioned response to the control response, and the abscissa shows ISIs. Suppression occurs at an ISI of 5 ms and lasts a few milliseconds. To show that this suppressive effect is produced by activation of the cerebellum, we studied the best position and polarity of the conditioning stimulus for eliciting the effect in normal subjects. The best position of the conditioning stimulus was at the level of the inion in vertical direction and 3 to 6 cm lateral to ipsilateral side to the studied muscle in horizontal direction. In electrical stimulation, the effective polarity was that the anode was over the target cerebellum.[4] In magnetic stimulation, a double-cone coil should be centered over the cerebellum.[5] These physiological characteristics of the best conditioning stimulus are consistent with the idea that a conditioning stimulus activates cerebellum. The cerebellum may affect the cortex through a cerebellothalamocortical pathway, the brainstem structures through direct connection from the cerebellum, or the spinal cord through some descending pathways from the cerebellum. Suppression should occur at the cortex if the first candidate is the mechanism of action. However, if either of the latter two candidates produces suppression, inhibition should occur at a subcortical level.

To study at which level the suppression occurs, we compared suppressive effects on responses to motor cortical magnetic stimulation with those on responses to motor cortical electrical stimulation.[4,5] In electrical and magnetic cerebellar stimulation experiments, conditioning stimuli suppressed responses to TMS even though they did not affect responses to TES. This indicates that the motor cortex contralateral to the activated cerebellum is suppressed by the conditioning stimulus. Based on these arguments, we conclude that cerebellar stimulation with TMS or TES transiently suppresses the contralateral motor cortex in humans, probably through cerebellothalamocortical pathways.

There were controversies about our conclusion that the suppression is produced by cerebellar activation. Meyer and associates[6] reported that CST excitability was reduced by TMS over the cerebellum in patients with large defects of a cerebellar hemisphere. However, in their patients, suppression was evoked at ISIs longer than 9 ms. Because our suppression is evoked at ISIs of 5 to 8 ms, their suppression may be a different effect from that evoked by cerebellar activation. Werhahn and colleagues[7] observed two types of motor cortex suppression elicited by stimulation over the back of the head. One is a cerebellar effect, and the other is produced by cervical root stimulation and suppression occurs at the spinal cord. The former begins at an ISI of 5 ms and lasts a few milliseconds, and the latter begins at an ISI of 8 ms. Di Lazzaro and coworkers[8] supported our conclusion that cerebellar suppression is seen at ISIs of 5 to 7 ms by showing no inhibition of the motor cortex at ISIs of 5 to 7 ms in patients with focal cerebellar lesions. The observation that activation of cerebellar thalamus elicited a transient motor cortical suppression in humans[9] also supports an inhibitory effect on the motor cortex by cerebellar stimulation. Based on these arguments, we conclude that cerebellar stimulation inhibits the contralateral motor cortex at ISIs of 5 to 7 ms.

We studied which descending volleys are suppressed by cerebellar stimulation using a preferential activation[10] of a certain I or D wave as the test motor cortical stimulus. Figure 13–2 shows cerebellar effects on I1 and I3 waves. Top rows are control responses and the other rows are conditioned responses. The test stimulus was given over the left motor cortex. In the left column, the test stimulus was inducing posteriorly directed currents in the brain that preferentially evoke I3 waves, and in the right column, it was inducing anteriorly directed currents that preferentially produce I1 waves. The conditioning stimulus was given over the right cerebellum. Two control responses from the right first dorsal interosseous muscle (FDI) are similar in size but their onset latencies are different. The latency of the left control response was 23.6 ms, which corresponded to an I3 wave, and that of the right control response was 20.8 ms, which was consistent with an I1 wave.

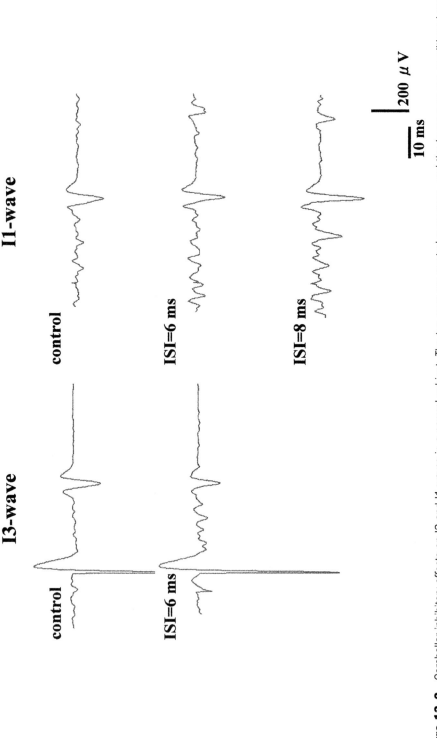

Figure 13–2 Cerebellar inhibitory effects on I3 and I1 waves in a normal subject. The *top rows* are control responses, and the *lower rows* are conditioned responses. Transcranial magnetic stimulation over the cerebellum reduced electromyographical responses to I3 waves at an interstimulus interval (ISI) of 6 ms, whereas it did not affect responses to I1 waves. It also had no effect on responses at an ISI of 8 ms.

The same conditioning stimulus over the right cerebellum, given at an ISI of 6 ms, reduced responses elicited by I3 waves, whereas it did not affect responses to I1 waves. These results suggest that reduction of EMG response sizes is caused by suppression of I3 waves by cerebellar stimulation at this intensity. When using higher-intensity cerebellar stimuli, even I1 waves were suppressed by cerebellar stimulation.[4]

Facilitatory Effect

Because there was a small facilitation at short intervals just before the onset of inhibition in the previous experiment, we studied if the early facilitation is constantly evoked by cerebellar stimulation in normal subjects. We used a high-voltage electrical stimulus over the cerebellum that was strictly adjusted just below the threshold for activation of the CST at the level of foramen magnum.[11] In this experiment, we again used preferential activation of I1- or I3 waves as a test stimulus. Figure 13–3 shows control and conditioned responses in this experiment. Responses to I3 waves are on the left column and those to I1 waves on the right. Latencies of control responses corresponded to I3 and I1 waves, respectively. At an ISI of 3 ms, a conditioned response to I3 waves was larger than the control response, whereas that to I1 waves was not affected by the same conditioning stimulus. Similar modulation of responses was seen in all normal subjects. In the mean time course of this effect on I3 waves, facilitation began at an ISI of 3 ms and lasted less than 2 ms. The conditioning stimulus did not affect electrical cortical responses, and this indicates that facilitation is not a spinal event but occurs at the motor cortex. The best position and polarity of the conditioning stimulus were the same as those for the inhibitory cerebellar effect. This suggests that the facilitation is produced by activation of the cerebellum. Based on all these results in normal subjects, we conclude that stimuli over the cerebellum have a transient, facilitatory influence on the contralateral motor cortex when the conditioning stimulus intensity is strictly adjusted and the coil is positioned over the cerebellum.[12]

Intervals between Cerebellar-Cerebral Interactions in Humans

We have demonstrated that facilitation occurs in the motor cortex at an ISI of 3 ms and inhibition at an ISI of 5 ms in cerebellar stimulation in human subjects. These intervals seem to be too short for a cerebellar impulse to reach the motor cortex through the thalamus. However, these intervals were the intervals between when the conditioning stimulus was given and when the test stimulus was given. The time when the stimulus is given is not always the time when some part of the central nervous system responsible for an observed effect is activated. In such cases, the interval between two stimuli does not indicate the time during which responsible impulses are mediated from one part of the brain that is activated by the conditioning stimulus to another part where the observed effect occurs. In motor cortical stimulation, activation of the corticospinal tract neurons (CTNs) usually does not occur at the same time when cortical stimuli are given.[13] Activation of CTNs occurs a few milliseconds later than the time when the stimulus is given through activation of axons of the motor cortical interneurons. This pattern of activation of CTN is called *indirect activation*, and descending volleys produced by this pattern of activation are called *indirect waves* (I waves).[14] The waves are numbered in order of appearance as I1, I2, and I3 and so on. These I waves are almost differentially elicited with an 8-shaped coil placed in a certain direction.[10] Using this method, we have shown that I3 waves are facilitated by cerebellar stimulation. The cerebellar facilitation effect must act on the target neuron for I3 interneuron (probably an I2 interneuron), and this must occur at 5 ms (3-ms interval + 2 ms for one synapse) after the cerebellar stimulation. In cerebellar stimulation, does such indirect activation occur in the cerebellum similarly to motor cortical stimulation? Although TMS and TES over the motor cortex elicited EMG responses at different onset latencies, TMS and TES over the cerebellum evoke the suppressive effect on the motor cortex at the same intervals from stimulation. This suggests that direct activation of a responsible structure occurs at the

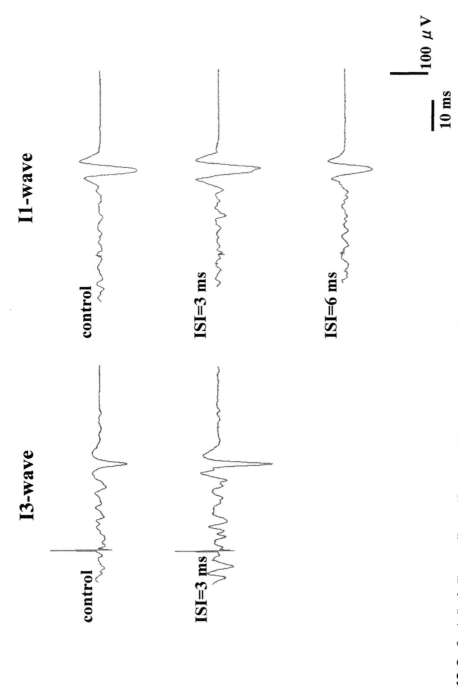

Figure 13-3 Cerebellar facilitatory effect on I3 waves and I1 waves in a normal subject. Responses are shown similar to those in Figure 13-2. The conditioning stimulus was placed over the cerebellum, and its intensity was precisely adjusted at just less than the threshold for corticospinal activation at the foramen magnum level. Cerebellar stimuli facilitate motor cortical responses to I3 waves at an interstimulus interval (ISI) of 3 ms, but it had no effect on I1 waves. It did not facilitate responses to I1 waves at an ISI of 6 ms, which is the interval compensated for the latency difference between I1 and I3 waves.

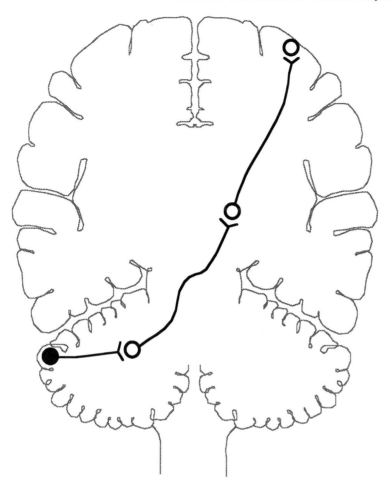

Figure 13-4 Schematic diagram of the cerebellomotor-cortical interaction. Dentate nuclei have a facilitatory connection with contralateral thalamic nuclei, which puts facilitatory inputs to the motor cortex. Through this system, the cerebellum has a tonic facilitatory influence on the motor cortex. Cerebellar stimulation may activate the dentate nucleus or superior cerebellar peduncle. This should produce a phasic facilitation to the motor cortex that lasts a few milliseconds or less. Purkinje cells inhibit this facilitatory system that finally suppresses the motor cortex. Inhibition of the tonic facilitatory influence on the motor cortex through the described system by activation of Purkinje cells lasts longer than the monosynaptic facilitation of the motor cortex. The onset latency of the inhibitory effect should be a few milliseconds longer than the facilitatory effect that is consistent with one synaptic delay. Our results of inhibitory and facilitatory effects are compatible with these connections.

same interval from the stimulus in TMS and TES of the cerebellum, although we cannot completely exclude other possibilities. The value of 5 ms fits well the time interval between cerebellum and motor cortex in humans. Our proposed pathway responsible for the facilitation is shown in Figure 13-4.

On the basis of the same arguments presented earlier, the finding that inhibition occurs at an ISI of 5 ms in cerebellar stimulation must mean that the motor cortex is inhibited 7 ms $(5+2=7)$ after cerebellar stimuli. This latency suggests that the pathway for the inhibition needs one more synapse than that for the facilitation, which is consistent with the idea that cerebellar stimulus activates Purkinje cells. Activation of Purkinje cells should inhibit the dentate nucleus which itself tonically facilitates the motor cortex through the thalamus. All results in patients support this idea of activation of Purkinje cells by cerebellar stimulation. Our proposed pathway for cerebellar inhibition of the motor cortex is also shown in Figure 13–4.

Intracortical Inhibition of the Motor Cortex in Ataxic Patients

In addition to our cerebellar inhibitory effect on the motor cortex, another well-known inhibitory effect is the intracortical inhibition of the motor cortex elicited by paired stimulation over the motor cortex.[15] In this experiment, a subthreshold first stimulus decreases the size of responses to a second suprathreshold stimulus when both of them are given at short intervals. That inhibition is considered to reflect functions of GABAergic inhibitory system of the motor cortex.[16] To investigate whether cerebellar inhibitory effect is independent of this intracortical inhibitory system, we performed paired-pulse stimulation experiments and cerebellar stimulation experiments in the same patients with ataxia and compared results.[17] The intracortical inhibition was completely normal in those patients whereas cerebellar inhibition was abnormally reduced or absent in all of them. Moreover, in patients with extrapyramidal signs, such as Parkinson's disease (PD) or dystonia, normal cerebellar inhibition of the motor cortex was seen even though they had abnormally reduced intracortical inhibition.[18,19] If the inhibitory interneurons of the motor cortex which are responsible for the intracortical inhibition are also involved in the cerebellar inhibitory effect, cerebellar inhibition should be reduced in PD or dystonia. However, in the present investigation, normal cerebellar inhibition was evoked in such patients with extrapyramidal signs, and this suggests an independence of these two effects. These comparisons between the cerebellar inhibition and intracortical inhibition in patients suggest that the two effects are independent of each other and cerebellar inhibitory effect must not be mediated by GABAergic interneurons involved in short interval intracortical inhibition of the motor cortex. It is also compatible with our previous conclusion that the inhibition of the motor cortex by cerebellar stimulation is caused by inhibition of the dentate nucleus by the activation of Purkinje cells produced by cerebellar stimuli.

Cerebellar Stimulation in Patients

There have been no studies about cerebellar facilitation of the motor cortex in patients. We here summarize cerebellar inhibition in patients.[20,21] Our results of normal subjects just suggest that the observed inhibitory effect must be produced through an activation of the cerebellum because of the best position and polarity of the conditioning stimulus. One way to prove this proposition correct is to examine this suppression effect in patients with ataxia of various causes and those without ataxia. If suppression is not evoked only in patients with cerebellar ataxia and is evoked in patients with lesions other than cerebellar structures or its related pathways, the suppression effect should be elicited by an activation of the cerebellum. We describe such results confirming our idea.

Patients without Ataxia

We studied patients who had no ataxia (as a disease control group) to show that systems other than the cerebellar system do not much contribute to the suppressive effect. This group consisted of basal ganglia disorders, including PD, motoneuron diseases, and focal cerebrovascular lesions sparing cerebellar systems, such as those with lesions in the cerebral cortex, thalamus sparing cerebellar thalamus, and brainstem lesions sparing cerebellar pathways. In all of them, normal degree of suppression was elicited by cerebellar stimulation.

A patient with a large cortical infarction was not able to make any voluntary contraction in hand muscles. Infarction of this patient involved the supplementary motor area (SMA), primary leg motor area, and frontal cortex but did not involve the primary hand motor area, cerebellum, and motor thalamus. In this patient, no voluntary movements occurred. This lack of movement was not caused by CST dysfunction because TMS evoked normal motor evoked potentials (MEPs) in muscles in which no voluntary movements occurred. No commands must go into the intact primary motor cortex from higher motor cortical areas, because such cortical areas were affected or fibers connecting between such areas and the primary motor cortex were involved. We considered that the isolation of the intact primary motor cortex from other motor areas should cause no voluntary movement in this patient and called it *primary motor cortex isolation*.[22] In this patient with no movements, normal cerebellar suppression was elicited. This finding indicates two important points. One is that the present suppression must be evoked by cerebellar activation. The other is that we can evaluate cerebellar function, at least partially, using this method even in patients in whom we cannot evaluate cerebellar function clinically because of lack of voluntary movements.

Noncerebellar Ataxic Patients

Clinical ataxia is seen in patients with cerebellar dysfunction and in those with sensory disturbance (i.e., sensory ataxia), vestibular dysfunction, or frontal ataxia. If our method can clearly differentiate this noncerebellar ataxia from cerebellar ataxia, it will have clinical usefulness. We therefore studied such patients with noncerebellar ataxia using this method. Patients with vestibular dysfunction usually have truncal ataxia but do not have limb ataxia. However, our method has been applied to only hand muscles and has not been applied to lower limb or truncal muscles. Because of this limitation of muscles studied with our method, vestibular ataxia was excluded from our study. A group of sensory ataxia included patients with tabes dorsalis, pure sensory neuropathy due to Sjögren syndrome or paraneoplastic syndrome, and a lesion of the sensory thalamus. In all of them, normal suppression was evoked by cerebellar stimulation. This is consistent with the idea that the suppression is evoked by inputs to the motor cortex from the cerebellum.

Cerebellar Ataxia Caused by a Lesion in Cerebellar Systems

To know whether the suppression is abnormal in any patients with cerebellar ataxia due to a lesion anywhere in the cerebellum, efferent pathways from or afferent pathways to the cerebellum, and also to see whether certain structures of the cerebellum or certain pathways through cerebellum are responsible for the suppression, we studied patients with a focal cerebellar lesion which was suspected to cause ataxia. This group consisted of patients with a lesion responsible for ataxia in the cerebellar hemisphere, Purkinje cells, dentate nucleus, superior cerebellar peduncle, middle cerebellar peduncle, pontine nucleus, and motor thalamus. All of them had limb ataxia and clinical cerebellar signs. The cerebellar hemisphere or Purkinje cells were affected in patients with CVD, cerebellitis,[23] cerebellar cortical atrophy (CCA) (one kind of degenerative ataxia), spinocerebellar ataxia type 6 (SCA6), paraneoplastic cerebellar cortical atrophy with anti-Yo antibody, and ataxia due to intoxication of antiepileptic drugs. Damage of the dentate nucleus or superior cerebellar peduncle should be mainly responsible for ataxia in patients with a CVD lesion at those sites, dentatorubropallidoluysian atrophy, and Wilson's disease (Fig. 13–5). In those patients, dysfunction of cerebellar efferent outputs produces ataxia. In patients with a CVD lesion at the motor thalamus, abnormal outputs from the cerebellum should cause ataxia. In patients with a plaque of multiple sclerosis in the middle cerebellar peduncle[24] or a CVD lesion at the same peduncle, ataxia was caused by abnormal afferent inputs to the cerebellum through the middle cerebellar peduncle. Reduced afferent inputs to the cerebellum must be also responsible for ataxia in patients with a CVD lesion at the pontine nucleus.

Cerebellar suppression of the motor cortex was reduced or absent in patients in whom cerebellar hemisphere, Purkinje cells, dentate nucleus, superior cerebellar peduncle, or motor thalamus is involved. In contrast, a normal pattern of suppression was seen in patients with a lesion in the middle cerebellar peduncle or pontine nucleus even though they all had definite clinical ataxia. These results suggest that pathologic mechanisms for ataxia are partly differentiated with our cerebellar stimulation method in true cerebellar ataxia (i.e., ataxia caused by a lesion of the cerebellar hemisphere or cerebellar efferent pathways) from that caused by a lesion of the cerebellar afferent pathways.

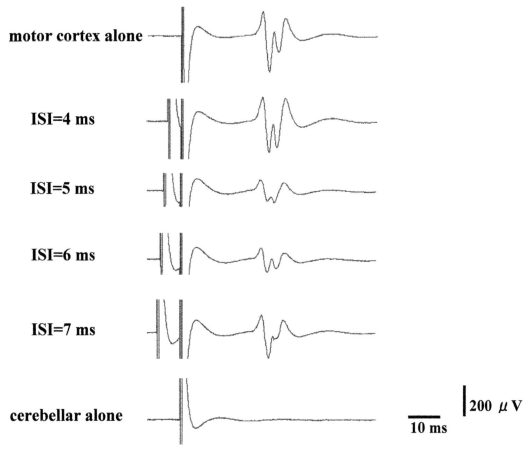

Figure 13–5 Responses in a patient with ataxia due to hypothyroidism. The *top trace* shows a control response elicited by motor cortical stimulation alone, and the *bottom trace* shows when the cerebellar stimulus was given alone. The others are conditioned responses at interstimulus intervals (ISIs) of 4, 5, 6, and 7 ms. They were smaller than the control response at ISIs of 5, 6, and 7 ms, whereas it was almost the same as the control response at an ISI of 4 ms. This indicates that the incoordination often seen in hypothyroidism is an ataxia due to dysfunction of cerebellar afferent systems or noncerebellar ataxia.

Ataxia of Unknown Pathophysiology

Several disorders have ataxia whose mechanisms remain to be determined. We applied the stimulation method to two kinds of such disorders: Fisher's syndrome[25] and ataxia due to hypothyroidism. Seven patients with Fisher's syndrome had an acute-onset ataxia, ocular movement disturbances, and hyporeflexia, all of which are typical of Fisher's syndrome. They all had an anti-GQ1b IgG antibody, which confirmed the diagnosis. In all of them, cerebellar stimulation provoked normal suppression of the motor cortex even when they had ataxic signs. This suggests that dysfunction of the cerebellar afferent systems or noncerebellar mechanisms should explain ataxia in Fisher's syndrome.

Two patients with hypothyroidism showed limb ataxia when we studied the cerebellar inhibitory effect on the motor cortex. Figure 13–5 shows EMG responses from FDI in one of the patients. Conditioned responses at an ISIs of 5, 6, and 7 ms were smaller than the control response. In this patient, normal motor cortical suppression was evoked by cerebellar stimulation. This indicates that ataxia in hypothyroidism must also be caused by dysfunction of the cerebellar afferent systems or noncerebellar systems.

Unmasking of Cerebellar Signs Masked by Hypertonia of the Studied Muscle

Clinical cerebellar signs, such as dysmetria, hypermetria, decomposition, and hypotonia, usually appear in patients in whom neither basal ganglia nor CST is involved. In patients who have also an involvement of basal ganglia and CST in addition to cerebellar involvement, rigidity seen in basal ganglia disorders or spasticity seen in pyramidal tract involvement often mask cerebellar signs. In such patients, we cannot detect the clinical cerebellar signs due to hypertonia, such as rigidity or spasticity. If we can judge cerebellar involvement with our stimulation method in these patients, this is another merit of the cerebellar stimulation. A few patients whose cerebellar dysfunction was clinically masked by hypertonus were detected by cerebellar stimulation.

The dentate nucleus is sometimes affected in Wilson's disease, and the basal ganglia are another main target of this disease. One patient with Wilson's disease showed rigidity, tremor, and other signs of basal ganglia involvement but had no cerebellar signs. Magnetic resonance images imaging (MRI) of his brain revealed involvement of the dentate nucleus (Fig. 13–6). This finding suggested that clinical cerebellar signs were masked by extrapyramidal signs in this patient. The cerebellar stimulation provoked no motor cortical suppression, which unmasked this clinically masked cerebellar involvement. Similar unmasking of cerebellar signs often occurs in multiple system atrophy (MSA), which often affects the cerebellum and basal ganglia. In some of them, parkinsonian symptoms appeared earlier than cerebellar symptoms. In such patients, we cannot evaluate cerebellar involvement clinically. We studied three patients with MSA who had parkinsonian symptoms but no cerebellar signs clinically. In all of them, there was no cerebellar inhibition of the motor cortex. Cerebellar stimulation experiments again unmasked an involvement of the cerebellum in these patients. This kind of unmasking is one of clinical utilities of the present cerebellar stimulation.

Follow-up of Cerebellar Function

Another clinical utility of this method is to monitor the degree of cerebellar involvement

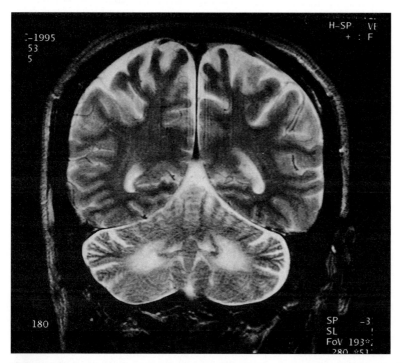

Figure 13–6 Magnetic resonance image of the brain of a patient with Wilson's disease. Bilateral dentate nuclei were affected in this patient. Motor cortical suppression was not evoked in this patient, which is consistent with suppression elicited by activation of the cerebellothalamocortical pathways.

objectively. One report followed two patients with acute cerebellar ataxia.[23] In these patients, cerebellar stimulation did not evoke any suppression of the motor cortex when he or she had ataxia. After recovery from the acute ataxia, when they had no ataxia, normal suppression of the motor cortex was elicited by cerebellar stimulation. This supports our proposition that the suppression is produced by activation of the cerebellum. Degenerative ataxia usually worsens slowly. If we study the same patient twice at a long interval, we should objectively detect progression of ataxia by cerebellar stimulation. For this purpose, we studied five patients with degenerative ataxia twice at an interval of 2 years. In all of them, cerebellar suppression was abnormally reduced at both times of examination, and the amount of suppression at the second examination was significantly less than that at the first examination. These results indicate that progression of disease can be evaluated by magnetic cerebellar stimulation method.

Other Transcranial Magnetic Stimulation Studies on Cerebellum

There are several other TMS studies of ataxic patients or about cerebellar function in normal subjects. Effects of cerebellar stimulation on eye movements were studied by one group of investigators.[26,27] Focal TMS was given over the cerebellum with an 8-shaped coil. In the first experiment,[26] TMS was given during visually guided saccades. TMS over the posterior cerebellar vermis accelerated saccades directed ipsilateral to the stimulated side and decelerated those directed contralateral to the stimulated side. This is also compatible with an activation of Purkinje cells by TMS. Purkinje cells in the posterior vermis inhibit ipsilateral fastigial ocular motoneurons that activate brainstem structures producing saccades directed contralateral to the fastigial neurons. Activation of the Purkinje cells therefore should reduce the saccades directed contralateral to the stimulated side and enhance those to the ipsilateral side. The investigators also stimulated the cerebellum during smooth pursuit eye movement.[27] TMS produced abrupt acceleration of the smooth pursuit directed ipsilateral to the stimulated side and abrupt deceleration of contralaterally directed smooth pursuit. These results are compatible with an activation of Purkinje cells in the posterior vermis by TMS.

Di Lazzaro and colleagues[28] evaluated the motor cortex excitability in patients with unilateral cerebellar lesions with TMS techniques. They demonstrated higher thresholds of the motor cortex contralateral to the impaired hemicerebellum. This suggests the existence of a facilitating tonic action of the cerebellum on central motor circuits that is consistent with our knowledge of the tonic cerebellar facilitatory connection with the motor cortex in animals.

MEP size changes before reaction time tasks were studied in normal subjects and patients with degenerative ataxia.[29] In normal subjects, MEPs were significantly facilitated before voluntary movement compared with the relaxed condition (i.e., premovement facilitation). Premovement facilitation was significantly reduced in patients with degenerative ataxia compared with normal subjects. This indicates incomplete cerebellar regulation of voluntary movement in patients with ataxia.

A few long-term effects on the cerebellum were studied by rTMS over the cerebellum. One effect is the rhythm generation of finger tapping before and after rTMS over the cerebellum.[30] Variability of paced finger tapping was increased after rTMS over the medial cerebellum, whereas rTMS over the lateral cerebellum, motor cortex, or sham stimulation had any effect on the performance of finger tapping. This is compatible with the finding obtained by neuroimaging methods that the medial cerebellum is involved in the self-generation of timed movement. A few investigators have used rTMS over the cerebellum as a tool for the treatment of degenerative ataxia.[31] However, the efficacy remains to be questionable, and additional studies should be done to make a conclusion.

■ Limitation of Magnetic Cerebellar Stimulation

Magnetic cerebellar stimulation is a new method that gives us a chance to study cerebellar function objectively. The most important limitation of this stimulation is that we cannot study cerebellar regulation of the trunk and lower limb muscles. Gait disturbance due to truncal or lower limb ataxia often creates serious problems in ataxic patients. However, we have not been able to get a constant effect on truncal or lower limb muscles by cerebellar stimulation. This point should be solved in the future.

REFERENCES

1. Ito M. The cerebellum and neural control. New York, Raven Press, 1984.
2. Merton PA, Morton HB. Stimulation of the cerebral cortex in the intact human subject. Nature 1980;285:227–228.
3. Barker AT, Jallinous R, Freeston IL. Non-invasive magnetic stimulation of the human motor cortex. Lancet 198;2:1106–1107.
4. Ugawa Y, Day BL, Rothwell JC, et al. Modulation of motor cortical excitability by electrical stimulation over the cerebellum in man. J Physiol 1991;441:57–72.
5. Ugawa Y, Uesaka Y, Terao Y, et al. Magnetic stimulation over the cerebellum in humans. Ann Neurol 1995;37:703–713.
6. Meyer BU, Roricht S, Machetanz J. Reduction of corticospinal excitability by magnetic stimulation over the cerebellum in patients with large defects of one cerebellar hemisphere. Electroencephalogr Clin Neuropysiol 1994;93:372–379.
7. Werhahn KJ, Taylor J, Ridding M, et al. Effect of transcranial magnetic stimulation over the cerebellum on the excitability of human motor cortex. Electroencephalogr Clin Neuropysiol 1996;101:58–66.
8. Di Lazzaro V, Restuccia D, Nardone R, et al. Motor cortex changes in a patient with hemicerebellectomy. Electroencephalogr Clin Neuropysiol 1995;97:259–263.
9. Ashby P, Lang AE, Lozano AM, et al. Motor effects of stimulating the human cerebellar thalamus. J Physiol 1995;489:287–298.
10. Sakai K, Ugawa Y, Terao Y, et al. Preferential activation of different I waves by transcranial magnetic stimulation with a figure-of-eight shaped coil. Exp Brain Res 1997;113:24–32.
11. Ugawa Y, Rothwell JC, Day BL, et al. Percutaneous electrical stimulation of corticospinal pathways at the level of the pyramidal decussation in humans. Ann Neurol 1991;441:57–72.
12. Iwata NK, et al. Facilitatory effect on the motor cortek by electrical stimulation over the cerebellum in humans. Exp Brain Res 2004 (in press).
13. Day BL, Dressler D, Maertens de Noordhout A, et al. Electric and magnetic stimulation of human motor cortex: surface EMG and single motor unit responses. J Physiol 1989;412:449–473.
14. Amassian VE, Stewart M, Quirk GJ, et al. Physiological basis of motor effects of a transient stimulus to cerebral cortex. Neurosurgery 1987;20:74–93.
15. Kujirai T, Caramia MD, Rothwell JC, et al. Cortico-cortical inhibition in human motor cortex. J Physiol 1993;471:501–519.
16. Ziemann U, Lonnecker S, Steinhoff BJ, et al. Effects of antiepileptic drugs in motor cortex excitability in humans: a transcranial magnetic stimulation study. Ann Neurol 1996;40:367–378.
17. Ugawa Y, Hanajima R, Kanazawa I. Motor cortex inhibition in patients with ataxia. Electroencephalogr Clin Neuropysiol 1994;93:225–229.
18. Hanajima R, Ugawa Y, Terao Y, et al. Ipsilateral cortico-cortical inhibition of the motor cortex in various neurological disorders. J Neurol Sci 1996;140:109–116.
19. Ridding MC, Inzelberg R, Rothwell JC. Changes in excitability of motor cortical circuitry in patients with Parkinson's disease. Ann Neurol 1995;37:181–188.
20. Ugawa Y, Genba-Shimizu K, Rothwell JC, et al. Suppression of motor cortical excitability by electrical stimulation over the cerebellum in ataxia. Ann Neurol 1994;36:90–96.
21. Ugawa Y, Terao Y, Hanajima R, et al. Magnetic stimulation over the cerebellum in patients with ataxia. Electroencephalogr Clin Neuropysiol 1997;104:453–458.
22. Sakai K, Kojima E, Suzuki M, et al. Primary motor cortex isolation: complete paralysis with preserved primary motor cortex. J Neurol Sci 1998;155:115–119.
23. Matsunaga K, Uozumi T, Hashimoto T, et al. Cerebellar stimulation in acute cerebellar ataxia. Clin Neurophysiol 2001;112:619–622.
24. Ugawa Y, Terao Y, Nagai C, et al. Electrical stimulation over the cerebellum normally suppresses motor cortical excitability in a patient with ataxia due to a lesion of the middle cerebellar peduncle. Eur Neurol 1995;35:243–244.
25. Ugawa Y, Genba-Shimizu K, Kanazawa I. Suppression of motor cortical excitability by electrical stimulation over the cerebellum in Fisher's syndrome. J Neurol Neurosurg Psychiatry 1994;57:1275–1276.
26. Hashimoto M, Ohtsuka K. Transcranial magnetic stimulation over the posterior cerebellum during visually guided saccades in man. Brain 1995;118:1185–1193.
27. Ohtsuka K, Enoki T. Transcranial magnetic stimulation over the posterior cerebellum

28. Di Lazzaro VD, Restuccia D, Molinari M, et al. Excitability of the motor cortex to magnetic stimulation in patients with cerebellar lesions. J Neurol Neurosurg Psychiatry 1994;57:108–110.
29. Nomura T, Takeshima T, Nakashima K. Reduced pre-movement facilitation of motor evoked potentials in spinocerebellar degeneration. J Neurol Sci 2001;187:41–47.
30. Theoret H, Haque J, Pacsual-Leone A. Increased variability of paced finger tapping accuracy following repetitive magnetic stimulation of the cerebellum in humans. Neurosci Lett 2001;306:29–32.
31. Shiga Y, Tsuda T, Itoyama Y, et al. Transcranial magnetic stimulation alleviates truncal ataxia in spinocerebellar degeneration. J Neurol Neurosurg Psychiatry 2002;72:124–126.

during smooth pursuit eye movements in man. Brain 1998;121:429–435.

The Role of Transcranial Magnetic Stimulation in the Study of Fatigue

Paul Sacco, Gary W. Thickbroom, and Frank L. Mastaglia

Fatigue is a frequently reported complaint in a wide variety of conditions, particularly in disorders of the nervous system. Two issues complicate the study of fatigue in humans. The first is that fatigue may be defined in a number of ways, depending on the approach taken. From a functional perspective, fatigue can be described as "a failure in the capacity of the neuromuscular system to generate force or power."[1] Alternatively, fatigue may be expressed in purely subjective terms ("a sustained sense of exhaustion and decreased capacity for physical and mental work"[2]). This is not purely a semantic issue because it has important implications for measuring fatigue. For the purpose of clarity, the term *fatigue* is used here primarily in its functional context. Both of these issues are discussed later in relation to findings using transcranial magnetic stimulation (TMS).

■ Historical Perspective of Fatigue

Although from a physiological point of view fatigue is generally measured as a failure to maintain a given level of physical performance, the causes of the failure of humans to perform has intrigued investigators for some time. Mosso[3] is generally considered to be the first person to have systematically studied muscle fatigue in humans by recording movements of the middle finger using an ergograph. He found that the fatigue curves obtained when subjects performed repeated contractions could vary considerably. He attributed changes in performance to differences in the "nervous arousal" of subjects, who performed better at times when they had greater "mental energy" (Fig. 14–1). His conclusion was that muscle fatigue was primarily of nervous origin. Numerous studies using isolated muscle preparations have shown that considerable fatigue occurs in the absence of nervous input, and it is now generally agreed that under specific conditions, all levels

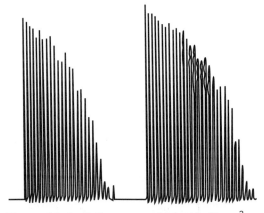

Figure 14–1 Fatigue curves obtained by Mosso[3] using a finger ergograph of a subject before (*left*) and after (*right*) performing a lecture. The improvement in work capacity was ascribed to an increase in "mental energy" associated with a mentally stimulating task.

of the motor pathway can contribute to this phenomenon.[1] For convenience, these multifactorial processes have been designated *central* and *peripheral fatigue mechanisms*, with the former referring to changes proximal to the neuromuscular junction. Central fatigue can be defined as any loss of voluntary muscle activation associated with exercise.

Merton[4] was the first to evaluate the role of central fatigue. In his classic study, the ability of subjects to achieve maximum force output of the thumb adductors during a sustained maximum voluntary contraction (MVC) was evaluated using supramaximal tetanic electrical stimulation of the ulnar nerve. He found that no extra force was produced with addition of the stimulus, even when voluntary force had declined by more than 50%. He reached the conclusion that fatigue occurred only in the muscle, opposite to that of Mosso. Many investigators have subsequently used motor nerve stimulation to evaluate voluntary muscle activation, with variations in measurement techniques, contraction types, and protocols and muscle groups used. It is now generally accepted that the inability to achieve full volitional activation of a muscle may contribute to fatigue in a variety of circumstances. Although this idea of a role of central fatigue has been generally accepted, its importance for exercise under normal conditions has been downplayed, and little information has been obtained for its contribution to abnormal fatigue in pathological conditions. Gandevia[5] has provided a detailed review of the role of central mechanisms in muscle fatigue.

The advent of TMS enabled study of the characteristics of the motor pathway upstream from the motor axon. Alterations in the excitability and function of the corticomotor system from interneurons targeting corticospinal neurons of the pyramidal tract to muscle during activity may be investigated in relatively free-living conditions. This chapter is concerned with studies that have used TMS to examine changes in the motor pathway associated with fatiguing exercise in normal subjects, and TMS findings in individuals suffering from conditions associated with activity- and nonactivity-related fatigue are discussed.

Transcranial Magnetic Stimulation in the Study of Fatigue

Brasil-Neto and colleagues[6] were the first to describe the use TMS of the motor cortex to study responses to fatigue. They reported a transient decrease in resting motor evoked potential (MEP) amplitude of the flexor carpi radialis muscle after a bout of weighted wrist flexions repeated to exhaustion. They found a serial decrease in MEP size over the course of four stimuli at 5-second intervals. At least 20 studies subsequently used TMS to study responses in the corticomotor system during and after fatigue, and studies that have dealt with changes occurring in parallel to fatigue are described initially. Subsequently, investigations that have used TMS to study responses during recovery from fatiguing exercise are reviewed.

Transcranial Magnetic Stimulation during Maximal Sustained Contractions

Sustained MVCs are extremely tiring and require a high level of motivation and compliance by the subject. For this reason, the duration of contractions tend to be short (usually 1 to 2 minutes). Mills and Thomson[7] compared changes in MEP amplitude and M waves during a 90-second, sustained MVC of the first dorsal interosseus (FDI) in which the average force declined to 25% of initial values. They found increases in MEP amplitude (not significant) and silent period (SP) duration after the first 5 seconds of contraction in contrast to a large decline in the size of the M wave. They concluded that TMS elicited responses in a group of motor units whose excitability was not affected during fatigue or that changes in cortical and spinal excitability occurred in equal and opposite directions. The next year, McKay and coworkers[8] reported TMS changes associated with a sustained ankle dorsiflexion MVC (120 seconds). Their findings were consistent with those from the FDI in that the MEP amplitude showed no change and SP duration increased over the course of the contraction. Moreover, the fact that MEP amplitude increased in the contralateral (nonexercised) tibialis anterior supported the concept of an increased cortical excitatory drive during the contraction compensating for a decrease in peripheral excitability.

Taylor and associates[9] followed changes in MEP responses of the biceps brachii and brachioradialis muscles during a 3-minute sustained MVC of the elbow flexors. They found simultaneous increases in MEP amplitude and SP duration during the contraction (and in subsequent studies, an increase in the ratio of MEP/M-wave amplitude in this model).[10] The increased MEP excitability was said to be of cortical origin, because responses to electrical stimulation of the corticospinal tract at the cervicomedullary junction were unchanged.[9] This has been confirmed in a small number of subjects by Di Lazzaro and colleagues who recorded descending volleys from electrodes implanted in the cervical epidural space during fatiguing efforts.[10a] The evidence suggests that sustained MVCs have a potentiating effect on the responses of the motor cortex to TMS, but that the net effect (in terms of resulting output from spinal motoneurons) also is influenced by factors operating downstream from the motor cortex. Increases in SP duration during a sustained MVC most likely result from potentiation in the responses of inhibitory interneurons synapsing onto the corticospinal neurons.[9,11]

An important finding from studies of TMS responses during MVCs is that cortical stimulation often results in muscle force output increasing above that which can be voluntarily sustained. Under nonfatigued conditions, most subjects show no extra force output when TMS is applied during an MVC[12,12a] (Fig. 14–2). Just as central fatigue, quantified by an increase in force output associated with electrical stimulation of a motor nerve during a maximal effort, can be explained by an inability to maintain complete activation of the muscle, TMS responses can be interpreted in a similar way. Despite apparent increases in cortical motor excitability during an MVC, corticomotor output is still insufficient to achieve full muscle activation, meaning that input to the primary motor cortex is suboptimal. Gandevia and colleagues[12] expressed this concept in terms of the site of fatigue being "upstream" from the motor cortex. The functional significance of such changes needs to be determined, although some evidence indicates that this phenomenon may be influenced by the type of muscle contraction preformed. Loscher and Nordlund[13] found a greater reduction in motor output (as judged by TMS-induced twitch torque) during repeated eccentric versus concentric MVCs of the elbow flexors, although both types of contraction showed suboptimal motor drive.

Transcranial Magnetic Stimulation during Submaximal Contractions

The use of sustained maximal voluntary efforts to study fatigue has a number of disadvantages. They are difficult and stressful for subjects to complete, and they are usually of relatively short duration under unstable physiological conditions. More importantly, their utility for describing exercise-related changes associated with everyday activities is questionable, because most activities of daily living involve repeated muscle contractions of modest intensity.

An alternative approach to the study of fatigue involves the study of submaximal efforts. During a sustained submaximal contraction, there is a gradual increase in gross electromyographic amplitude as additional motor units are recruited to compensate for active units whose force output is declining.[14] At the same time, there is good evidence that the firing frequency of active units declines during a sustained submaximal contraction.[15] The time course and extent of these responses will vary with the muscle being studied and the strength (i.e., percent of MVC) of the sustained contraction.

Ljubisavljevic and coworkers,[16] using sustained contractions of the adductor pollicis (60% MVC for 2 minutes), found a decrease in the MEP amplitude and SP duration over the course of activity. In contrast, Sacco and associates[17] showed that during a sustained 20% MVC of the elbow flexors to exhaustion (average time, 19 minutes), increases in excitability (i.e., potentiated MEP amplitude) and inhibition (i.e., extended SP duration) of the corticomotor system occurred. No increase in SP duration occurred in responses to TES, indicating that the inhibitory effects were located cortically. Prolongation of the SP duration took place only during the second

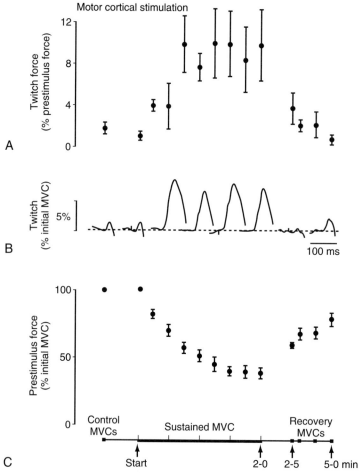

Figure 14–2 Effects of transcranial magnetic stimulation (TMS) on force output during fatigue and recovery for elbow flexion. **A:** An increase in force output in response to TMS occurs during a sustained maximum voluntary contraction in four subjects. **B:** Twitch responses for a single subject. **C:** Changes in voluntary force over the 3-minute contraction measured immediately before TMS. (From Gandevia SC, Allen GM, Butler JE, et al. Supraspinal factors in human muscle fatigue: evidence for suboptimal output from the motor cortex. J Physiol [Lond] 1996;490(Pt 2):529–536.)

half of the task, when the MEP potentiation had plateaued (Fig. 14–3). Similar increases in MEP amplitude were observed by Taylor and colleagues[9] during a sustained 30% MVC of the biceps brachii. In both studies, TMS applied during the end of the contraction period (when volitional intensity was maximal) resulted in extra force output, indicating suboptimal motor output. The discrepancy in the MEP findings between studies of the hand versus upper arm muscles may be a consequence of the differences in task intensities or a reflection of the fact that the contraction intensity or MEP response curves are different. In the biceps brachii, MEP amplitude increases for the same TMS intensity up to contraction levels of at least 50% MVC.[18] In contrast, intrinsic muscles of the hand show less additional facilitation above 10% to 15% MVC.[19] This is to be expected because control of force output of proximal and distal muscles involves different strategies of activation (recruitment versus frequency modulation, respectively).[15]

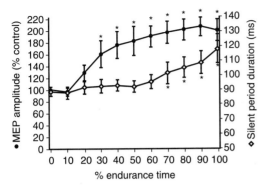

Figure 14–3 Changes in transcranial magnetic stimulation–induced MEP amplitude (percent of pre-exercise control) and post–motor evoked potential silent period duration (in milliseconds) during a sustained 20% maximum voluntary contraction of the elbow flexors. Results are given as the mean ± SEM for eight subjects (*$P < .05$) for comparison with pre-exercise values. The endurance time was normalized to 100% for all subjects. (From Sacco P, Thickbroom GW, Thompson ML, et al. Changes in corticomotor excitation and inhibition during prolonged submaximal muscle contractions. Muscle Nerve 1997;20:1158–1166.)

Transcranial Magnetic Stimulation under Resting Conditions after Fatiguing Contractions

The first report of TMS changes associated with fatigue[6] documented reduced MEP amplitudes during the minutes after cessation of a bout of fatiguing exercise. The lack of any consistent change in TES responses, H responses, or M waves during the postcontraction period was evidence that the depression occurred at the cortical level. The investigators reported a serial reduction in MEP amplitude over groups of four MEPs 5 seconds apart. Subsequent studies by the same group, using series of four stimuli at differing interstimulus intervals (0.1 to 6 Hz) interspersed with repeated bouts of exercise (10 × 30-second wrist flexion at 50% MVC), found that the serial decrement was greatest at 0.3 Hz and leveled off after the fourth exercise bout.[20] The results were attributed to a decline in the cortical synaptic safety factor as a consequence of the fatiguing exercise. The serial decrement has not been consistently reproduced in other studies, and Samii and coworkers[21] suggest that this phenomenon probably reflects the transient facilitation of resting MEPs that takes place *immediately* after a voluntary contraction. Augmentation of the MEP postcontraction has been found to be independent of contraction intensity, to peak within the first few seconds of relaxation, and to last up to several minutes.[21,22] It may persist for 30 minutes or more.[20,23,24] In addition to the short-term facilitation of the MEP amplitude after exercise, there exists a more pronounced and long-lasting depression of the MEP. Samii and associates[24] found that the degree of MEP depression was related to the duration of exercise, and Sacco and colleagues[25] showed that MEP diminution occurred to a similar extent (30% to 40% of the pre-exercise level) and duration (at least 20 minutes) after the same subjects performed a long-duration, low-intensity task or (on a separate occasion) a 60-second MVC. Zanette and associates[26] reported a 40% reduction in the spatial area of the cortex over which responses could be elicited with TMS.

Reduced resting MEP amplitudes after fatiguing exercise have been demonstrated in hand muscles,[26] wrist flexors,[20] elbow flexors,[25] and ankle dorsiflexors.[8] Most findings point to the cortex as the primary site of postexercise depression of resting MEP amplitude, because H responses or F waves and TES or transmastoid stimulation show minimal changes over the course of the recovery period. Peripheral factors may also influence MEP attributes. A number of studies have characterized changes in the compound muscle action potential (M wave) after exercise,[27] although these tend to be more pronounced after fatigue induced by electrical stimulation. Lentz and Nielsen[28] found that changes in peripheral excitability under certain circumstances could partially account for postexercise facilitation and depression of the MEP. This contrasts with earlier studies demonstrating minimal changes in M-wave characteristics after exercise.[6,17]

A number of corticospinal changes may also account for postexercise MEP depression, including an increase in the threshold of corticomotor units to TMS (through a reduction in cortical motoneuron excitability and an increase in corticocortical inhibition).

Alternately, a decrease in the output of spinal motoneurons to a given descending volley may occur. Zanette and associates[26] reported no change in motor threshold during postexercise MEP depression of the thenar muscles (although this finding remains to be confirmed for other muscle groups). Intracortical inhibition, as measured by prolongation of the latter portion of the post-MEP SP, is not obtainable under resting conditions. Paired-pulse TMS, which measures the effect of a subthreshold preconditioning stimulus on the response to a suprathreshold stimulus delivered a short time later, has been used to quantify levels of cortical excitability and inhibition at various interstimulus intervals (ISI).[29] Tergau and colleagues[30] used the paired-pulse stimulation paradigm of Ziemann and coworkers[31] to study the effects of repeated sets of chin pull-up exercises performed to exhaustion in 23 subjects on MEP responses of the biceps brachii during recovery. They found a significant reduction in the facilitatory effects of a preconditioning stimulus at ISIs of 8 to 15 ms (indicating a reduced cortical excitability) lasting for 8 to 10 minutes after exercise. In contrast, responses at shorter ISIs were unchanged, indicating that the exercise task had no effect on short-latency intracortical inhibition. The investigators reported a significant correlation between the amount of work subjects performed during the task and the decrease in potentiating effect of the conditioning stimulus at long ISI. The investigators posit that the exercise task resulted in a decrease in the excitatory input to cortical motoneurons upstream from the motor cortex. More recently, Verin and colleagues[31a] found that, following exhaustive treadmill running, both diaphragm and quadriceps MEP amplitude were reduced. Responses to long-latency paired pulse TMS were reduced, indicating a reduced intracortical facilitation. The use of "real-life" type activities to study corticomotor changes associated with muscle fatigue presents a series of additional problems regarding control of experimental variables. Hollge and associates[31b] studied MEP responses in a number of muscle groups in response to prolonged aerobic (i.e., submaximal) versus high-intensity anaerobic activities. They found that only high-intensity actions resulted in post-exercise MEP depression.

Transcranial Magnetic Stimulation under Active Conditions after Fatiguing Exercise

Although marked reductions in resting MEP amplitude can be demonstrated after a fatiguing task, it appears that this phenomenon is readily reversed if the subject elicits a voluntary effort.[10] For example, Figure 14–4A illustrates the extent and durability of post-exercise depression of the resting biceps brachii MEP (reduction of about 70% of the baseline value) after exercise. This effect was not influenced by the duration of the fatiguing task, because a 60-second MVC and a sustained 20% contraction to exhaustion produced similar responses. However, during a weak contraction (see Fig. 14–4B), the MEP depression was entirely reversed after the 60-second MVC and only partially reversed in the case of the sustained contraction.[25] Given the evidence that much of the (nonfatigued) facilitating effect of a voluntary contraction on the size of the evoked MEP can be explained by an increase in excitability at the segmental level,[32,33] a volitional increase in motoneuron excitability could account for at least part of this reversal of postexercise depression in MEP amplitude. Whether the reduced resting MEP amplitude after a sustained MVC can be explained by a postcontraction increase in the threshold of spinal motoneurons, which can then be reversed by a weak voluntary effort serving to raise the level of excitability sufficiently to overcome the depression, is unproved. The absence of any reliable test for segmental excitability in the elbow flexors limits the information that can be gained from this fatigue model. Evidence from the intrinsic hand muscles (F waves) and wrist flexors (H responses) suggest that this may not be the case. In this context, the level of voluntary facilitation used is likely to have an influence, because there was still appreciable loss in MEP amplitude of the biceps brachii for several minutes when the level of activation was 10% MVC. In contrast, Gandevia[12] found that the MEP amplitude fully recovered within 30 seconds of exercise during maximal activation of the same muscle group.

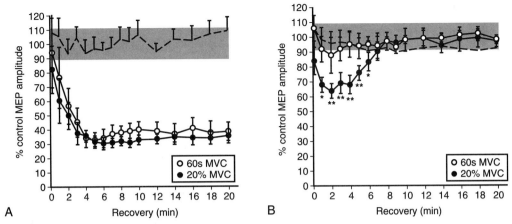

Figure 14–4 Mean changes in evoked responses (± SEM) for biceps brachii muscles after a 60-second maximum voluntary contraction *(open circles)* and a sustained 20% maximum voluntary contraction (MVC) *(solid circles)*. The pre-exercise control range is *shaded*, and the *dashed line* represents the mean (± SEM) responses of the nonexercised contralateral muscle after a 20% MVC. **A:** Resting motor evoked potential (MEP) amplitude (n = 10 subjects) as a percentage of the pre-exercise control value. **B:** Facilitated MEP amplitude (n = 10 subjects) as a percentage of the pre-exercise control. *$P < .05$; **$P < .01$ compared with to pre-exercise values. (From Sacco P, Thickbroom GW, Byrnes ML, et al. Changes in corticomotor excitability after fatiguing muscle contractions. Muscle Nerve 2000;23:1840–1846.)

The fact that the post-MEP SP after exercise returns rapidly to prefatigue levels[12,25] argues against an increase in intracortical inhibition after fatiguing activity. However, all muscle groups and exercise paradigms may not have similar effects. For example, McKay and colleagues[8] showed that prolongation of the post-MEP SP in the foot dorsiflexors occurred for several minutes after exercise, although only if the voluntary contraction was at a low level. The apparent reversal of post-exercise MEP depression with a weak contraction may also reflect tonic afferent input from the active muscle. Supporting this idea, Rollnik and associates[33a] showed that post-exercise MEP depression of the ankle dorsiflexors at rest can be reversed using a single electrical stimulus to the ipsilateral sural nerve 50 to 80 ms prior to TMS.

Although the physiological changes identified by the study of TMS changes associated with muscle fatigue are of interest, their sensitivity as a diagnostic tool for quantifying central fatigue and their functional significance in relation to normal or pathological fatigue states remain to be determined. Thus, Lazarski and colleagues[33b] showed that, despite large post-exercise decrements in resting MEPs of FDI muscles, finger dexterity was unaffected. Also, Humphry and associates[33c] showed that post-exercise MEP depression of the biceps brachii was associated with a reduction of the MEP in the (non-exercise) contralateral muscle group, but no change in contralateral muscle strength. A more complete understanding of neurophysiological changes associated with muscle fatigue will require the use of complimentary techniques to study neuronal activation (such as functional magnetic resonance imaging, positron emission tomography, and electroencephalography). The application of TMS to fatigue in dynamic muscle contractions and those of increased complexity (e.g., functional activity) are likely to provide useful information, as suggested by the findings reviewed above.

■ Fatigue in Neuromuscular Disorders

The complaint of fatigue is common in many patients with neurological disorders.

Approximately 80% to 95% of patients with multiple sclerosis (MS) report fatigue that is frequently disabling.[34] It is also common in Parkinson's disease,[35] post-polio syndrome,[36] depression,[24] traumatic brain injury,[37] and by definition, the chronic fatigue syndromes (CFSs). Commonly, mental and physical fatigue are closely associated, and although it is important to differentiate between them, there is little understanding of whether the causal mechanisms are similar or how they may relate to one another. Although TMS is widely used as a clinical and research tool in neurology, relatively few studies have examined whether changes in the properties of the corticomotor system are related to the symptoms of fatigue or can help explain functional deficits in affected individuals.

Transcranial Magnetic Stimulation–Fatigue Studies of Multiple Sclerosis

Given the pathophysiology of MS involving impaired conduction in myelinated nerve fibers, and the prevalence of fatigue in this condition, it would seem likely that any activity-related conduction failure would be readily identifiable using TMS. Nonactivity-related alterations have been described in the latency, amplitude, and waveform of the MEP in this condition.[11] Sandroni and associates[38] studied MEP characteristics in ten MS patients suffering from symptomatic fatigue in rested and fatigue conditions. They found no alterations in MEP amplitude when subjects were in the fatigued state. Studies of TMS changes associated with activity-related fatigue in MS have yielded equivocal findings. Liepert and colleagues[23] studied the time for MEP amplitude to return to baseline after exhaustive exercise of the fifth finger abductor. They found that recovery time was significantly prolonged in MS patients (n = 8) compared with healthy controls. Sheean and coworkers[39] compared corticomotor responses during a 45-second sustained MVC of the adductor pollicis in 21 MS patients and 19 controls. They found no difference in muscle strength between groups, but MS patients had a greater force loss during the fatigue task and normal corticomotor responses. This occurred despite the fact that peripheral nerve stimulation showed MS patients had a greatly reduced capacity for muscle activation during the exercise task. Petajan and White[40] compared responses to a 3-minute handgrip MVC in MS patients with and without muscle weakness and controls. Time to 50% strength loss was shorter in the MS groups, and potentiation of MEP amplitude was significantly smaller for the MS subjects who were clinically weak. All groups showed similar levels of postexercise MEP depression (although this was only followed for 6.5 minutes after exercise).

To study the fatigue responses to an exercise paradigm more functionally relevant than a sustained MVC, Archer and colleagues[41] used repeated submaximal contractions of the FDI in 22 MS subjects without hand weakness. They found that compared to controls, MS subjects showed greater MVC decline after the task, but this was not associated with a reduction in MEP amplitude during or after activity. Schubert and coworkers[42] studied MEP responses of the tibialis anterior after a walking task in 11 MS subjects. MEP depression was greater than normal after exercise, but responses were highly variable between individuals with MS. In contrast, Perretti and coworkers[42a] found that TMS responses in the thenar muscles of 41 MS patients showed no post-exercise depression following muscle fatigue. They attribute this finding to intracortical motor dysfunction in MS, although the MEP findings were similar in patients with and without symptomatic fatigue. Attempts to relate fatigue symptoms with activity-related abnormalities should consider using muscle groups and exercise paradigms that reflect naturally occurring patterns of functional activity.

Transcranial Magnetic Stimulation–Fatigue Studies of Chronic Fatigue Syndromes

Several researchers have made use of TMS to investigate whether abnormalities in corticomotor excitability might play a role in the pathophysiology of this heterogeneous and contentious group of disorders. Brouwer and Packer[43] identified increased variability in the MEP responses of CFS subjects to sustained activity compared with controls. The responses of subjects with CFS to the repeated 30-second wrist flexion fatigue protocol of

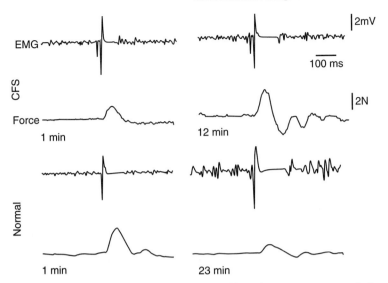

Figure 14–5 Comparison of electromyographic and torque responses of elbow flexors to transcranial magnetic stimulation in a subject with chronic fatigue syndrome (CFS) and a healthy control at the beginning and end of a sustained 20% maximum voluntary contraction to exhaustion. Notice the greater interpolated twitch torque of the CFS subject near the point of exhaustion compared with the normal subject.

Samii[21] were similar to those of patients suffering from depression; they were normal for exercise capacity and postexercise MEP depression but had reduced MEP facilitation during the task. Abnormalities in the pattern of corticomotor changes associated exercise have also been supported by Starr and associates,[44] who found that the postexercise facilitation of MEP amplitudes immediately after exercise of the FDI was absent in CFS subjects. Sacco and colleagues[45] compared the responses of 10 CFS subjects with controls during a sustained 20% MVC of the elbow flexor muscles to exhaustion (inability to achieve 20% MVC). CFS subjects had a markedly reduced time to exhaustion and an increased perception of effort during the task. The MEP increase during exercise tended to be smaller for CFS subjects, despite increased effort, than the control group. This is in accord with earlier findings, suggesting a reduced facilitatory response of the motor cortex to fatiguing activity. CFS subjects had a greater increase in TMS-induced twitch force interpolated at the end of the sustained contraction (i.e., maximal effort) (Fig. 14–5). This is indicative of impairment in the ability to achieve full muscle activation volitionally (and commonly occurs to a lesser extent in healthy control subjects with this type of exercise paradigm) (see Fig. 14–5). Regarding CFS, the evidence from multiple studies clearly suggests that the symptoms of fatigue and reduced exercise capacity are not the result of intrinsic dysfunction of the motor cortex or corticospinal pathway but that they probably involve executive centers "upstream" from the primary motor cortex.

Transcranial Magnetic Stimulation–Fatigue Studies of Parkinson's Disease

A number of studies using TMS have shown abnormally elevated corticomotor excitability in Parkinson's disease[45a,b] and this has been associated with a reduced short-latency intracortical inhibition.[45c] In the first systematic study of the effects of fatiguing muscle contraction on TMS responses in Parkinson's disease, Lou and colleagues[45d] found raised MEP excitability and a lack of significant post-exercise depression compared to controls (although patients did not show more objective fatigue during the exercise task). The fact that the MEP abnormalities were normalized following levodopa administration indicates that this phenomenon is disease related.

Transcranial Magnetic Stimulation–Fatigue Studies of Post-polio Syndrome

The only study to investigate fatigue-related TMS changes in post-polio syndrome[36] found that the 13 patients performed normally in a fatigue task consisting of repeated 30-second wrist extension contractions. There were no differences in MEP amplitude changes during the exercise (facilitation) or postexercise (depression) between patients and controls, indicating no exercise-related abnormality in corticomotor responses in this condition.

Transcranial Magnetic Stimulation–Fatigue Studies of Depression

Although repetitive TMS has been widely applied in the treatment of depression,[46] its utility in the study of exercise responses in the corticomotor system in this disorder has received less attention. Samii and colleagues[21] studied MEP responses to wrist flexion contractions in 10 patients with medication-free clinically depressed subjects. They found that facilitatory responses to exercise were reduced in depressed subjects despite normal exercise performance and that postexercise MEP depression was similar to that of controls. These findings were supported by Shajahan and coworkers,[47,48] who found similar abnormalities in MEP facilitation during exercise in the adductor pollicis of medicated depressed patients. The same group also found that this difference was not apparent in subjects who had recovered from their depressive illness.[47,48] Chroni and associates[48a] compared TMS responses to non-fatiguing exercise in a group of patients with depression, schizophrenia and mania, with those of controls. They showed a reduced post-exercise facilitation of MEP responses in patients, but this was not associated with any particular disorder or medication regimen. The findings suggest that abnormalities of central neurotransmission associated with mood and motivation may be reflected in the inability to enhance excitatory responses of the primary motor cortex.

■ Conclusion

Transcranial magnetic stimulation provides a unique tool for the noninvasive investigation of changes in the characteristics of the corticomotor pathway in response to prolonged activity in normal and pathological conditions. Findings indicate that during voluntary fatigue and recovery in healthy subjects, complex excitatory and inhibitory adaptations take place at multiple levels of the corticomotor system. These responses probably are influenced by the muscle group in question, as well as the intensity, duration, and nature of the fatiguing task. Many questions remain regarding the role of altered corticomotor function in the pathophysiology of fatigue in conditions where fatigue is a significant problem.

REFERENCES

1. Bigland-Ritchie B, Rice CL, Garland SJ, et al. Task-dependent factors in fatigue of human voluntary contractions. Adv Exp Med Biol 1995;384:361–380.
2. Piper BF. Fatigue: current bases for practice. In Funk SG et al (eds): Key Aspects of Comfort: Management of Pain, Fatigue, and Nausea. New York, Springer, 1989.
3. Mosso A. Fatigue. London, Swan Sonnenschein, 1904.
4. Merton PA. Voluntary strength and fatigue. J Physiol 1954;128:553–564.
5. Gandevia SC. Spinal and supraspinal factors in human muscle fatigue [review]. Physiol Rev 2001;81:1725–1789.
6. Brasil-Neto JP, Pascual-Leone A, Valls-Sole J, et al. Postexercise depression of motor evoked potentials: a measure of central nervous system fatigue. Exp Brain Res 1993;93:181–184.
7. Mills KR, Thomson CC. Human muscle fatigue investigated by transcranial magnetic stimulation. Neuroreport 1995;6:1966–1968.
8. McKay WB, Stokic DS, Sherwood AM, et al. Effect of fatiguing maximal voluntary contraction on excitatory and inhibitory responses elicited by transcranial magnetic motor cortex stimulation. Muscle Nerve 1996;19:1017–1024.
9. Taylor JL, Butler JE, Allen GM, et al. Changes in motor cortical excitability during human muscle fatigue. J Physiol (Lond) 1996;490(Pt 2):519–528.
10. Taylor JL, Butler JE, Gandevia SC. Altered responses of human elbow flexors to peripheral-nerve and cortical stimulation

during a sustained maximal voluntary contraction. Exp Brain Res 1999;127:108–115.
10a. Di Lazzaro V, Oliviero A, et al. Direct demonstration of reduction of the output of the human motor cortex induced by a fatiguing muscle contraction. Exp Brain Res 2003;149:535–538.
11. Ravnborg M. The role of transcranial magnetic stimulation and motor evoked potentials in the investigation of central motor pathways in multiple sclerosis. Dan Med Bull 1996;43:448–462.
12. Gandevia SC, Allen GM, Butler JE, et al. Supraspinal factors in human muscle fatigue: evidence for suboptimal output from the motor cortex. J Physiol (Lond) 1996;490(Pt 2):529–536.
12a. Todd G, Taylor JL, Gandevia SC. Measurement of voluntary activation of fresh and fatigued human muscle using transcranial magnetic stimulation. J Physiol 2003;551:661–671.
13. Loscher WN, Nordlund MM. Central fatigue and motor cortical excitability during repeated shortening and lengthening actions. Muscle and Nerve 2002;25:864–872.
14. Garland SJ, Enoka RM, Serrano LP, et al. Behavior of motor units in human biceps brachii during a submaximal fatiguing contraction. J Appl Physiol 1994;76:2411–2419.
15. DeLuca CJ, Lefever RS, et al. Behaviour of human motor units in different muscles during linearly varying contractions. J Physiol 1982;329:113–28.
16. Ljubisavljevic M, Milanovic S, Radovanovic S, et al. Central changes in muscle fatigue during sustained submaximal isometric voluntary contraction as revealed by transcranial magnetic stimulation. Electroencephalogr Clin Neurophysiol 1996;101:281–288.
17. Sacco P, Thickbroom GW, Thompson ML, et al. Changes in corticomotor excitation and inhibition during prolonged submaximal muscle contractions. Muscle Nerve 1997;20:1158–1166.
18. Kischka U, Fajfr R, Fellenberg T, et al. Facilitation of motor evoked potentials from magnetic brain stimulation in man: a comparative study of different target muscles. J Clin Neurophysiol 1993;10:505–512.
19. Taylor JL, Allen GM, Butler JE, et al. Effect of contraction strength on responses in biceps brachii and adductor pollicis to transcranial magnetic stimulation. Exp Brain Res 1997;117:472–478.
20. Brasil-Neto JP, Cohen LF, Hallett M. Central fatigue as revealed by postexercise decrement of motor evoked potentials. Muscle Nerve 1994;17:713–719.
21. Samii A, Wassermann EM, Ikoma K, et al. Characterization of postexercise facilitation and depression of motor evoked potentials to transcranial magnetic stimulation. Neurology 1996;46:1376–1382.
22. Norgaard P, Nielsen JF, Andersen H, et al. Post-exercise facilitation of compound muscle action potentials evoked by transcranial magnetic stimulation in healthy subjects. Exp Brain Res 2000;132:517–522.
23. Liepert J, Kotterba S, Tegenthoff M, et al. Central fatigue assessed by transcranial magnetic stimulation. Muscle Nerve 1996;19:1429–1434.
24. Samii A, Wassermann EM, Hallett M. Post-exercise depression of motor evoked potentials as a function of exercise duration. Electroencephalogr Clin Neurophysiol 1997;105:352–356.
25. Sacco P, Thickbroom GW, Byrnes ML, et al. Changes in corticomotor excitability after fatiguing muscle contractions. Muscle Nerve 2000;23:1840–1846.
26. Zanette G, Bonato C, Polo A, et al. Long-lasting depression of motor-evoked potentials to transcranial magnetic stimulation following exercise. Exp Brain Res 1995;107:80–86.
27. Cupido CM, Galea V, McComas AJ. Potentiation and depression of the M wave in human biceps brachii. J Physiol 1996;491:541–550.
28. Lentz M, Nielsen JF. Post-exercise facilitation and depression of M wave and motor evoked potentials in healthy subjects. Clin Neurophysiol 2002;113:1092–1098.
29. Kujirai T, Sato M, Rothwell JC, et al. The effect of transcranial magnetic stimulation on median nerve somatosensory evoked potentials. Electroencephalogr Clin Neurophysiol 1993;89:227–234.
30. Tergau F, Geese R, Bauer A, et al. Motor cortex fatigue in sports measured by transcranial magnetic double stimulation. Med Sci Sports Exerc 2000;32:1942–1948.
31. Ziemann U, Rothwell JC, Ridsding MC. Interaction between intracortical inhibition and facilitation in human motor cortex. J Physiol 1996;496:873–881.
31a. Verin E, Ross E, et al. Effects of exhaustive incremental treadmill exercise on diaphragm and quadriceps motor potentials evoked by transcranial magnetic stimulation. J Applied Physiol 2004;96:253–259.
31b. Hollge J, Kunkel M, et al. Central fatigue in sports and daily exercises. A magnetic stimulation study. Int J Sports Med 1997;18:614–617.
32. Maertens de Noordhout A, Pepin JL, et al. Facilitation of responses to motor cortex stimulation: effects of isometric voluntary contraction. Annals of Neurology 1992;32:365–370.

33. Di Lazzaro V, Restuccia D, Oliviero A, et al. Effects of voluntary contraction on descending volleys evoked by transcranial magnetic stimulation in conscious humans. J Physiol 1998;508:625–633.
33a. Rollnik JD, Schubert M, et al. Effects of somatosensory input on central fatigue: a pilot study. Clin Neurophysiol 2000;111:1843–1846.
33b. Lazarski JP, Ridding MC, Miles TS. Dexterity is not affected by fatigue-induced depression of human motor cortex excitability. Neuroscience Lett 2002;321:69–72.
33c. Humphry AT, Lloyd-Davies EJ, et al. Specificity and functional impact of post-exercise depression of cortically evoked motor potentials in man. Eur J Applied Physiol 2004;92:211–218.
34. Freal JE, Kraft GH, Coryell JK. Symptomatic fatigue in multiple sclerosis. Arch Phys Med Rehabil 1984;65:135–138.
35. Friedman JH, Friedman H. Fatigue in Parkinson's disease. Neurology 1993;43:2016–2018.
36. Samii A, Lopez-Devine J, Wasserman EM, et al. Normal postexercise facilitation and depression of motor evoked potentials in postpolio patients. Muscle Nerve 1998;21:948–950.
37. Chistyakov AV, Soustiel JF, Hafner H, et al. Altered excitability of the motor cortex after minor head injury revealed by transcranial magnetic stimulation. Acta Neurochir (Wien) 1998;140:467–472.
38. Sandroni P, Walker C, Starr A. 'Fatigue' in patients with multiple sclerosis. Motor pathway conduction and event-related potentials. Arch Neurol 1992;49:517–524.
39. Sheean GL, Murray NM, Rothwell JC, et al. An electrophysiological study of the mechanism of fatigue in multiple sclerosis. Brain 1997;120(Pt 2):299–315.
40. Petajan JH, White AT. Motor-evoked potentials in response to fatiguing grip exercise in multiple sclerosis patients. Clin Neurophysiol 2000;111:2188–2195.
41. Archer SA, Sacco P, Thickbroom GW, et al. Physiological correlates of fatigue in multiple sclerosis. J Clin Neurosci 2001;8:493.
42. Schubert M, Wohlfarth K, Rollnik JD, et al. Walking and fatigue in multiple sclerosis: the role of the corticospinal system. Muscle Nerve 1998;21:1068–1070.
42a. Perretti A, Balbi P, et al. Post-exercise facilitation and depression of motor evoked potentials to transcranial magnetic stimulation: a study in multiple sclerosis. Clin Neurophysiol 2004;115:2128–2133.
43. Brouwer B, Packer T. Corticospinal excitability in patients diagnosed with chronic fatigue syndrome. Muscle Nerve 1994;17:1210–1212.
44. Starr A, Scalise A, Gordon R, et al. Motor cortex excitability in chronic fatigue syndrome. Clin Neurophysiol 2000;111:2025–2031.
45. Sacco P, Hope PA, Thickbroom GW, et al. Corticomotor excitability and perception of effort during sustained exercise in the chronic fatigue syndrome. Clin Neurophysiol 1999;110:1883–1891.
45a. Cantello R, Gianelli M, et al. Parkinson's disease rigidity: magnetic motor evoked potentials in a small hand muscle. Neurology 1991;41:1449–1456.
45b. Valls-Sole J, Pascual-Leone A, et al. Abnormal facilitation of the response to transcranial magnetic stimulation in patients with Parkinson's disease. Neurology 1994;44:735–741.
45c. Ridding MC, et al. Changes in excitability of motor cortical circuitry in patients with Parkinson's disease. Ann Neurol 1995;37:181–188.
45d. Lou JS, Benice T, et al. Levodopa normalizes exercise related cortico-motoneuron excitability abnormalities in Parkinson's disease. Clin Neurophysiol 2003;114:930–937.
46. Kirkcaldie MT, Pridmore SA, Pascual-Leone A. Transcranial magnetic stimulation as therapy for depression and other disorders. Aust N Z J Psychiatry 1997;31:264–272.
47. Shajahan PM, Glabus MF, Gooding PA, et al. Reduced cortical excitability in depression. Impaired post-exercise motor facilitation with transcranial magnetic stimulation. Br J Psychiatry 1999;174:449–454.
48. Shajahan PM, Glabus MF, Jenkins JA, et al. Postexercise motor evoked potentials in depressed patients, recovered depressed patients, and controls. Neurology 1999;53:644–646.
48a. Chroni E, Lekka NP, et al. Effect of exercise on motor evoked potentials elicited by transcranial magnetic stimulation in psychiatric patients. J Clin Neurophysiol 2002;19:240–244.

15 Treatment of Movement Disorders

Hartwig R. Siebner

■ Background

Soon after its introduction in 1985 by Barker and colleagues,[1] single-pulse and paired-pulse transcranial magnetic stimulation (TMS) of the primary motor cortex (M1) emerged as a feasible tool to assess the integrity of the corticospinal motor tract and to probe intracortical excitability. TMS also was successfully employed to investigate functional interactions between frontal motor cortical areas and to evaluate sensorimotor integration within the central nervous system. By applying these methods to patients with motor dysfunction, TMS has provided revealing new insights in the pathophysiology of movement disorders (see Chapter 12).

The introduction of stimulating devices for repetitive transcranial magnetic stimulation (rTMS) added a new dimension to the conventional use of TMS in patients with movement disorders. TMS emerged from a diagnostic probe to become an interventional tool for noninvasive modulation of brain function. Early studies on healthy volunteers demonstrated that the regularly repeated application of magnetic stimuli to the M1 can induce a lasting change in corticomotor excitability.[2,3] For instance, a 10-pulse rTMS train at 150% resting motor threshold and 20 Hz caused an increase in the size of the motor evoked potential (MEP) that lasted for about 3 minutes after the administration of rTMS.[2] Conversely, a 15-minute train of 0.9 Hz rTMS at 115% of motor resting threshold resulted in inhibition of the motor evoked responses.[3] Prompted by these studies on healthy subjects (see Chapter 5), the conditioning effects of rTMS have been subsequently explored in various neuropsychiatric disorders with the ultimate goal of using rTMS as a therapeutic intervention.[4,5] Compared with numerous studies focusing on conditioning effects of prefrontal rTMS in psychiatric disorders, relatively few studies have looked at lasting effects of rTMS in movement disorders.[5] This chapter provides a critical overview of current applications for rTMS in movement disorders, with an emphasis on Parkinson's disease (PD) and writer's cramp. The goal is to exemplify the potential and the limitations of rTMS as an interventional procedure to improve motor dysfunction in patients.

Movement disorders are a particularly suitable target for research focusing on lasting conditioning effects of rTMS on cerebral dysfunction. The rTMS-induced changes in motor performance can be readily assessed by kinematic recordings. Another advantage derives from the fact that most motor cortical areas are easily accessible to TMS. This especially applies for the M1 and the lateral premotor cortex. In addition to cortical targets, some subcortical components of the motor network (i.e., the cerebellum) can be stimulated with rTMS. However, some motor regions cannot be targeted directly by rTMS, such as the basal ganglia, thalamus, and the anterior cingulate cortex, which are located deep in the brain. This does not mean that no after-effects can be induced in these regions.

At least for the M1, conditioning effects of rTMS on cortical excitability can be assessed with a wide range of noninvasive electrophysiological methods. The detailed study of these after-effects will help to shed some light on the basic mechanisms that mediate the conditioning effects of rTMS. Because the size of the MEP can be used to reliably define the optimal site for M1 stimulation and to individually adjust stimulus intensity, rTMS over the M1 is easy to titrate in functional terms. Patients with movement disorders demonstrate distinct patterns of electrophysiological abnormalities in the corticospinal motor system, which provide a framework to interpret after-effects of rTMS (see Chapter 12).

■ Basic Concepts

When studying after-effects of rTMS in patients, several issues are important. Apart from the direct conditioning effects induced by the rapidly changing magnetic field, focal rTMS can cause to a variable extent an orthodromic and antidromic activation of cortico-cortical and cortico-subcortical connections. For instance, 2,250 stimuli of subthreshold 5-Hz rTMS applied to the left M1 induced a lasting increase in regional glucose metabolism in M1 and the caudal supplementary motor area.[6] Conversely, rTMS over a remote brain region can induce a powerful modulation of cortical excitability in the left M1. A total of 1,500 pulses of 1-Hz rTMS applied at 90% active motor threshold over the left lateral premotor cortex can produce a marked decrease in corticospinal excitability in ipsilateral M1.[7,8] This after-effect is not caused by a spread of stimulation to adjacent M1, because direct 1-Hz rTMS of the M1 at 90% active motor threshold failed to induce an inhibition of corticospinal excitability. rTMS of a superficially located motor cortical area, such as the M1 or the lateral premotor cortex, can shape the level of excitability in

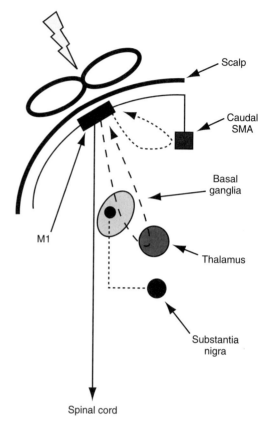

Figure 15–1 Schematic diagram of the spatial distribution of possible after-effects that are associated with focal repetitive transcranial magnetic stimulation (rTMS) over the primary motor cortex (M1). Focal rTMS conditions the directly stimulated M1. Focal rTMS may exert remote conditioning effects through the activation of cortico-subcortical connections, causing after-effects in the basal ganglia, spinal cord, red nucleus, or cerebellum. A transsynaptic spread of neuronal excitation by corticocortical connections may induce lasting conditioning effects in remote cortical areas that are functionally interconnected with M1 (i.e., the caudal supplementary motor area, the dorsal and ventral lateral premotor cortex, and the primary and secondary somatosensory areas). It is reasonable to assume that these remote effects in interconnected brain regions reflect back to the stimulated M1 by means of corticocortical and cortico–sub-corticocortical reentry loops.

connected brain regions by rTMS-induced repetitive activation of corticocortical and cortico-subcortical pathways (Fig. 15–1). These distant conditioning effects may become crucial in determining the lasting functional effects of rTMS.

In addition to the rTMS protocol, the functional state at the time of stimulation has a profound impact on the rTMS-induced response in the stimulated cortex.[9,10] The underlying pathophysiology and current medication are likely to cause substantial functional cortical changes, which alter the susceptibility of the stimulated cortex to the conditioning effects of rTMS. The physician needs to exercise caution when predicting the after-effects in patients based on previous studies in healthy controls. This notion is corroborated by studies on the immediate effects of suprathreshold 1-Hz rTMS of the M1 on corticomotor excitability in patients with writer's cramp.[11] In accordance with the well-known inhibitory net effect of suprathreshold 1-Hz rTMS, healthy controls demonstrated a decrease in MEP size during a 10-minute period of rTMS. In contrast, patients with writer's cramp showed an abnormal increase in corticospinal excitability during the course of 1-Hz rTMS. A discrepant response between patients and healthy controls was also reported for high-frequency rTMS. During a 10-pulse train of suprathreshold 5-Hz rTMS, the amplitude of the MEP remained unchanged in unmedicated patients with PD, whereas healthy subjects showed a progressive increase in MEP size during rTMS.[12]

As a rule of thumb, high-frequency rTMS over the M1 (i.e., rTMS applied at frequencies of 5 to 20 Hz) induces a lasting increase in corticospinal excitability as indexed by a significant facilitation of motor evoked responses, whereas low-frequency rTMS (i.e., at stimulation rates of about 1 Hz) results in a lasting reduction in corticospinal excitability. There is evidence that this rule applies also for other cortical areas.[13] However, this rule is too simplistic to fully characterize the conditioning effects of rTMS.[10] A more complex picture has emerged, demonstrating that, in addition to the rate of stimulation, the stimulus intensity and the number of magnetic stimuli per session influence the direction and the magnitude of the after-effects.[14,15]

Subsets of neurons within a given cortical area may be modulated differentially by a given rTMS protocol. The same holds true for different processing properties in the stimulated cortex. Some subsets of cortical neurons may show an increase in excitability, whereas other sets of neurons may demonstrate reduced excitability or no change at all. Moreover, a distinct function (i.e., processing capacity) may be impaired after rTMS, but another function may actually improve. This is of relevance in relation to movement disorders as patients may show "negative" symptoms (i.e., reduction in motor activity such as bradykinesia) or "positive" symptoms (i.e., abnormal increase in motor activity such as tremor or dyskinesia). It is conceivable that a rTMS protocol that is capable of improving positive motor symptoms may give rise to a deterioration of negative motor symptoms and vice versa.

It may be intuitively assumed that rTMS-induced net inhibition in the stimulated cortex equals a lasting rTMS-induced impairment in function. However, the relationship between lasting changes in cortex excitability and processing abilities are likely to be more complex.

Conditioning Effects of Repetitive Transcranial Magnetic Stimulation in Parkinson's Disease

Physiological Rationale for Repetitive Transcranial Magnetic Stimulation in Parkinson's Disease

At first glance, the idea to stimulate the motor cortex to modify motor dysfunction in PD may not appear very plausible because symptoms in PD are primarily related to a dysfunction in the basal ganglia caused by a deficient dopaminergic input from the substantia nigra.[16] However, the basal ganglia do not receive any direct sensory input, nor do they give rise to any direct motor output to the spinal cord. The motor deficit in PD patients therefore must be caused by secondary influences on other structures.[16] There is converging evidence that an abnormal output from the basal ganglia through the thalamus to frontal premotor and primary motor areas plays a crucial role in the pathophysiology of motor symptoms in PD.[17,18]

Several TMS studies have demonstrated a link between corticospinal motor dysfunction and motor impairment in PD. Tonic voluntary contraction caused a smaller increase in MEP size in PD patients than in healthy controls, indicating a reduced "gain function" (i.e., a reduced responsiveness) of the corticospinal motor system in PD.[19] Ellaway and colleagues[20] demonstrated a positive correlation between the threshold to TMS during weak voluntary contraction and the degree of bradykinesia. Moreover, a TMS study by Chen and coworkers[21] suggested a close relationship between prolonged movement time (e.g., bradykinesia) and impaired activation and deactivation of the M1 before and after a brisk voluntary finger movement in PD. The premovement increase in corticospinal excitability started earlier in PD patients, but the slope of increase was flattened. Moreover, the postmovement increase in excitability was also prolonged. In accordance with the notion of a deficient movement-related activation of the M1, PD patients fail to scale appropriately the initial agonist burst of a ballistic movement to the parameters of movement.[22] Inappropriate scaling leads to corrective adjustments at the end of a ballistic movement, resulting in multiple cycles of alternating agonist-antagonist electromyographic bursts and a prolonged movement time.[22]

The excitability of multiple inhibitory circuits in the M1 is significantly reduced in PD patients, resulting in an attenuation of paired-pulse intracortical inhibition,[23] a shortened duration of the cortically evoked contralateral and ipsilateral silent period,[24] and a reduced suppression of the TMS-evoked motor response by peripheral nerve stimulation.[16] Deficient intracortical inhibition in the M1 may hamper neuronal processes that are crucial to an appropriate channeling of the motor command. In accordance with this notion, the duration of TMS-induced modulation in motor unit firing is increased whereas the synchrony is decreased in PD patients.[25] This may also explain why patients with PD show a steeper slope of the stimulus-response curve than healthy controls.[19,26]

These multiple abnormalities at the motor cortical level form the physiological basis for attempts to modulate motor symptoms with rTMS over the M1. Most studies on the after-effects of rTMS in parkinsonian patients gave high-frequency rTMS to the M1 to increase the voluntary gain function in the corticospinal system and to facilitate the release of corticospinal motor commands through the M1. Alternatively, because intracortical inhibition is deficient in PD, there is also a rationale to apply "inhibitory" 1-Hz rTMS over M1 to enhance deficient intracortical inhibition to improve the "channeling function" of the M1. However, the physiological and clinical effects of 1-Hz rTMS have not been studied in detail.

In addition to a direct modulatory effect on the M1 and the corticospinal motor output, rTMS of the M1 may induce conditioning effects in remote brain regions caused by rhythmic activation of corticocortical connections and cortical–subcortico-cortical reentry loops (Fig. 15–1). Functional imaging studies provided strong evidence that high-frequency rTMS is capable of inducing remote effects in the basal ganglia and cortical areas.[6,27–31] These remote effects occurred only in brain areas that were functionally interconnected with the stimulated cortex.

There is no evidence to suggest that the M1 is the most suitable cortical target for rTMS in PD. On the contrary, the supplementary motor area (SMA) and the lateral premotor cortex (LPC) are equally plausible candidate structures for rTMS because a deficient movement-related activation of the SMA and a compensatory overactivation of the LPC are consistent features of motor pathophysiology in PD.[17,18,32] In contrast to rTMS of M1 or LPC, effective rTMS of the SMA poses several methodological problems. Because rTMS is located in the depth of the interhemispheric fissure, a high intensity is required for an effective rTMS. Moreover, it should be assumed that the medial aspect of the dorsal LPC is more effectively activated by fronto-central midline rTMS, because it is located closer to the magnetic coil. Effective magnetic stimulation of the SMA always results in a combined SMA-LPC stimulation, which renders the interpretation of conditioning effects in PD patients rather difficult.[33]

After-effects of Repetitive Transcranial Magnetic Stimulation on Motor Symptoms in Parkinson's Disease

Table 15–1 summarizes previously published studies of rTMS-induced after-effects on

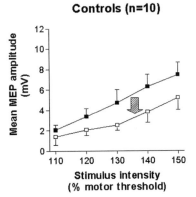 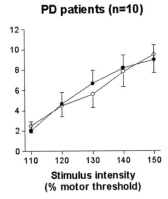

Figure 15–2 Stimulus-response curve of the relaxed contralateral first dorsal interosseus (FDI) muscle recorded in 10 patients with Parkinson's disease (*circles*) and 10 age-matched healthy controls (*squares*). The mean (±SEM) stimulus-response curve after subthreshold 5-Hz repetitive transcranial magnetic stimulation (rTMS) of the contralateral primary motor hand area (M1) (*open symbols*) is superimposed on the stimulus-response curve at baseline (*solid symbols*). In accordance with previous studies,[16,23] Parkinson's disease patients (*black circles*) show larger MEP amplitudes at baseline compared with healthy controls (*black squares*), although this difference did not reach statistical significance. After 2,250 stimuli of subthreshold 5-Hz rTMS over the M1, the stimulus-response curve is suppressed in healthy controls (*left panel*), whereas there is no change in the patient group (*right panel*).

motor function in PD. Most studies focused on short-term effects (lasting for minutes or hours) induced by a single session of high-frequency subthreshold rTMS to M1. Nonetheless, a direct comparison of these studies is problematic because there are considerable differences across studies in relation to the clinical features of the patients, the study design, and the rTMS protocol. Although most of the studies used a crossover design, including a control condition, none of these studies fulfilled the criteria of a placebo-controlled, double-blind study. Moreover, it is still an unresolved issue how to define an appropriate control condition that matches the nonspecific sensory stimulation associated with rTMS. One study used a "realistic" sham stimulation to overcome this problem.[34] The sham rTMS was combined with electric stimulation of the scalp to mimic the sensation induced by real rTMS. However, even this approach does not result in a perfect match of real and sham rTMS, and patients easily notice that the real and sham interventions are different.

Two studies that were primarily designed to assess changes in motor performance during the administration of rTMS looked also at possible after-effects.[35,36] Pascual-Leone and associates[35] reported no lasting beneficial effect of subthreshold 5-Hz rTMS on the serial-choice reaction-time task in six parkinsonian subjects, although there was an improvement in task performance during 5-Hz rTMS. PD patients were in a clinically defined On state during the experiment (i.e., at the time of peak L-Dopa effect). Ghabra and colleagues[36] applied 5-Hz rTMS to the M1 while PD patients performed a grooved peg-board task. The intensity of rTMS was set at 80% to 85% of resting motor threshold because higher intensities produced muscle jerking. Moreover, PD patients were off medication and the total number of rTMS stimuli delivered to M1 was varied across patients. The investigators found an unspecific practice effect on motor performance but no evidence for a specific improvement after real rTMS to the M1.[36]

A series of studies that was specifically designed to study lasting effects of rTMS on motor dysfunction in PD were performed by Siebner and coworkers[5,37,38] (see Table 15–1). In these studies, 2,250 stimuli of focal subthreshold 5-Hz rTMS were applied to the hand area of M1 with the intention of inducing a short-term increase in intrinsic corticospinal excitability and facilitating movement-related

Table 15–1 SYNOPSIS OF THE LITERATURE ON THE CONDITIONING EFFECTS OF REPETITIVE TRANSCRANIAL MAGNETIC STIMULATION IN PARKINSON'S DISEASE

Studies, No. of Patients	Study Design	Medication	Motor Assessment	rTMS Protocol Intensity (% rMT)	Rate (Hz)	No. of Stimuli	Coil	Area	Effects
Studies on Short-Term Effects of rTMS on Motor Function in Parkinson's Disease (Single Session of rTMS)									
Pascual-Leone et al.[35] n=6	Crossover	On/off	Choice RT task, procedural learning, pegboard test	90%	5	?	E	M1	No after-effect
Ghabra et al.[36] n=11	Crossover	Off	Pegboard test	80–85%	5	>350	E	M1	No specific after-effect
Pascual-Leone et al.[41] n=7	Crossover	Off	Walking, simple RT task, pegboard test	90%	10	1,000	E	M1	Improved performance
Siebner et al.[37] n=12	Crossover	Off	Ballistic aiming task	90%	5	2,250	E	M1	Improved performance
Tergau et al.[44] n=7	Crossover	On	Motor section of UPDRS, walking, simple RT task	90%	1,5,10,20	1,000	R	M1	No improvement of clinical signs and reaction time
Sommer et al.[43] n=11	Crossover	Off	Finger tapping	90%	1	900	E	M1	Improved tapping rate
Siebner et al.[39] n=10	Crossover	Off	Motor section of UPDRS	90%	5	2,250	E	M1	Improved motor signs in the contralateral upper limb
Von Raison et al.[42] n=5	Crossover	Off	Motor section of UPDRS	90%	0.5, 10	600, 2,000	E	M1	Improved motor signs in the contralateral upper limb after 10 Hz but not after 0.5 Hz rTMS

Study	Design	State	Task	Intensity	Frequency (Hz)	Pulses per session	Coil	Site	Outcome
Siebner et al.[5] n=14	Crossover	Off	Circle drawing	90%	5	2,250	E	M1	Improved circle drawing
Boylan et al.[33] n=8	Crossover	Off	Cued RT task, spiral drawing, pronation/supination, motor section of UPDRS	68–110%	10	2,000	E	SMA	No effect on clinical signs, worsening of spiral drawing and prolongation of RT
De Groot et al.[40] n=8	Crossover	Off	Motors section of UPDRS, pointing, finger tapping	90%	5	2,250	E	M1	Improved motor signs but no improvement of kinematics

Studies on Long-Term Effects of rTMS on Motor Function in Parkinson's Disease (Multiple Session of rTMS)

Study	Design	State	Task	Intensity	Frequency (Hz)	Pulses per session	Coil	Site	Outcome
Mally et al.[46] n=49	Parallel group (4 groups)	On	Motor section of UPDRS, short-term memory test	15%, 25%, 35% of MSO	1	30 or 60 per day for 1 week	R	M1	Long-lasting improvement of motor signs; no effect on short-term memory test
Shimamoto et al.[47] n=18	Parallel group (2 groups)	On	Total UPDRS, Schwab and England ADL score	770 V	0.2	60 per week for 2 months	R	M1	Improvement of motor signs and activities of daily living
Okabe et al.[34] n=85	Parallel group (3 groups)	On	Total UPDRS, Hamilton Rating Scale for Depression	110% of aMT	0.2	60 per week for 2 months	R	M1	No add-on effects relative to realistic sham stimulation

aMT, active motor threshold; E, figure-of-eight shaped coil; M1, primary motor hand area, primary motor cortex; MSO, maximal stimulator output; R, round coil; rTMS, repetitive transcranial magnetic stimulation; RT, reaction time; SMA, supplementary motor area; UPDRS, Unified Parkinson's Disease Rating Scale; % rMT, percentage of resting motor threshold.

activation of M1. A stimulus intensity of 90% resting motor threshold was selected to avoid any confounding effect related to rTMS-induced movements. In 12 PD patients with predominant bradykinesia, Siebner and associates[37] examined ballistic pointing movements before and 20 minutes after 5-Hz rTMS. Patients were tested in a clinically defined Off state after overnight withdrawal of dopaminergic medications. On 2 separate days, fifteen 30-second trains of focal 5-Hz rTMS were delivered to the hand area of M1 contralaterally to the more severely affected hand (i.e. real-rTMS) or to the medial prefrontal cortex (control-rTMS). Compared with prefrontal rTMS, real-rTMS over the M1 resulted in a decrease in movement time of pointing movements without affecting end-point accuracy. There was a significant reduction in inversions of sagittal velocity peaks per movement, indicating an improved fluency of pointing movements.

A subsequent study by Siebner and colleagues[5] investigated the conditioning effects of 5-Hz rTMS applied to the left M1 on open-loop cycle drawing in 14 right-handed PD patients. As in the first study, all patients had prominent bradykinesia of the right hand without significant tremor. Before and 30 minutes after 5-Hz rTMS, all subjects were instructed to draw continuously superimposed circles "as fluent as possible at their own speed and size." Real-rTMS of the left M1 led to a significant increase in mean drawing velocity, which was associated with an increase in fluency. The mean circle diameter and the mean vertical writing pressure were not changed after real-rTMS. There were no lasting effects of prefrontal control-rTMS on kinematic measures. Although, on average, drawing movements were faster and smoother after real-rTMS, there was considerable interindividual variability of rTMS-related changes in kinematics, and movement kinematics were still impaired after real-rTMS of the contralateral M1. A total of 60 stimuli of subthreshold 5-Hz rTMS applied to the left M1 produced no enduring improvement of circle drawing in PD patients.[38] This implies that a certain number of magnetic stimuli needs to be given to induce a beneficial after-effect on drawing movements in PD.

Siebner and coworkers[39] also investigated the effect of subthreshold 5-Hz rTMS on motor disability in 10 unmedicated PD. Clinical outcome was assessed by two unblinded examiners using the motor section of the Unified Parkinson's Disease Rating Scale (UPDRS). Compared with control-rTMS, patients showed a significant short-term improvement in motor function one hour after real-rTMS over M1. The beneficial effect was most prominent in the upper limb contralaterally to the stimulated M1. Adopting the same stimulation protocol, De Groot and associates[40] confirmed a short-term benefit of contralateral rTMS on parkinsonian symptoms of the upper limb in nine unmedicated PD patients as assessed by the motor subscore of the UPDRS. In contrast to the study by Siebner and colleagues,[39] the examiner was blinded to the treatment condition, and evaluation was carried out 24 hours after rTMS. Despite clinical improvement, movement analysis revealed no significant changes in kinematics during a repetitive pointing task or finger tapping, which were assessed 20 minutes after each rTMS session.[40] Two other groups have published data in abstract form that showed a short-term beneficial effect of subthreshold high-frequency rTMS of the M1 on bradykinesia in PD patients while off medication.[41,42] Only one study examined conditioning effects of rTMS on tremor. In 11 tremor-dominant PD patients, Sommer and coworkers[43] observed an increase in tapping rate after a 900-pulse train of suprathreshold 1-Hz rTMS applied over the contralateral M1. Taken together, the published rTMS data suggest that there is a threshold for inducing motor after-effects in unmedicated PD patients, because a short-term beneficial effect of rTMS on parkinsonian symptoms was only found in the studies that applied a high number of subthreshold stimuli or used a high (suprathreshold) stimulus intensity.

No beneficial after-effect has been reported in medicated patients with PD. Applying 1,000 pulses of nonfocal rTMS, Tergau and associates[44] observed no short-term effect on global motor score, walking, and simple reaction time in seven PD patients who continued their dopaminergic medication.

In contrast to studies on short-term effects of rTMS in PD, studies on long-lasting effects of rTMS gave short-trains of nonfocal low-frequency rTMS on separate days while patients continued their medication.[34,45–47] The studies by Mally and Jones[45,46] used a parallel-group design, and investigators were not blind to the rTMS condition. Mally and Jones included 49 PD patients who continued to take their medication during the course of the study.[45] Each patient was assigned to one of four rTMS protocols. Using different stimulation intensities, one or two 30-pulse trains of nonfocal 1-Hz rTMS were applied on 7 consecutive days over the vertex. Mally and Stone observed an intensity-dependent improvement in motor disability after a 1-week course of 1-Hz rTMS (see Table 15–1).

Two Japanese studies examined the conditioning effects of weekly sessions of very-low-frequency rTMS. The investigators gave 0.2-Hz rTMS through a round coil centered over the vertex. Stimulus intensity was 10% above active motor threshold.[34,47] The rationale to use a frequency of 0.2 Hz was motivated only by the fact that this protocol could be applied with a standard magnetic stimulator. However, there is no evidence that short trains of very-low-frequency rTMS (well below 1 Hz) have any conditioning effects on the corticospinal system in humans. Applying short trains of weekly rTMS at 0.2 Hz for 2 months, Shimamoto and colleagues[47] reported a long-term benefit only in patients who had received effective stimulation. Using the same protocol, Okabe and coworkers[34] reported only a modest placebo effect but no add-on effects of real rTMS compared with a realistic sham stimulation in PD patients. Given these conflicting data, it is still not clear whether repeated sessions of rTMS can specifically induce a lasting improvement in motor symptoms in PD patients.

The M1 may not be the most suitable target for rTMS in PD, and other areas, particularly premotor areas, should be considered. However, only one study has chosen a cortical target outside the M1. Boylan and associates[33] studied 10 PD patients after at least 12 hours without dopaminergic medication. Five patients received 2,000 stimuli of 10-Hz rTMS over the frontocentral cortex at a stimulus intensity of 110% of resting motor threshold of an intrinsic hand muscle. Two patients were excluded from the study because they did not tolerate the protocol. In another three patients, the stimulation intensity was reduced to 69% to 78% of resting motor threshold because of considerable unpleasantness of the stimulation protocol. The investigators found worsening of spiral drawing and a prolongation of reaction time 45 minutes after real-rTMS over the frontocentral cortex, whereas no changes in motor performance were observed after sham-rTMS. Neither real-rTMS nor sham-rTMS altered motor symptoms. This subclinical worsening after centrofrontal rTMS serves as a reminder that lasting effects of rTMS may not always be beneficial. However, the study provides an instructive example for the methodological challenges that are associated when studying conditioning effects of rTMS. For placebo-rTMS, the coil was angled at 90 degrees, with one wing of the coil having contact with the scalp. This approach caused less rTMS-associated peripheral sensory stimulation and resulted in an ineffective blinding of the patients. Moreover, given the low stimulus intensity, it remains unclear whether rTMS effectively stimulated the SMA in the interhemispheric fissure. It appears more likely that the reported worsening of motor performance after frontocentral 10-Hz rTMS was caused by after-effects in medial aspects of the "overactive" dorsal lateral premotor cortex.[5]

After-effects of Repetitive Transcranial Magnetic Stimulation on Corticomotor Excitability in Parkinson's Disease

Just as for psychiatric disorders,[4] the basic mechanisms underlying the conditioning effects of rTMS in PD remain yet to be clarified. There are only limited data available about the after-effects of rTMS over the M1 on corticomotor excitability in PD. In 10 patients with PD, Siebner and colleagues[48] studied the duration of the cortical silent period before and after the administration of fifteen 30-second trains of subthreshold 5-Hz rTMS over the left M1. Stimulus intensity was set at 90% of resting motor threshold. Patients were tested in a clinically defined Off state

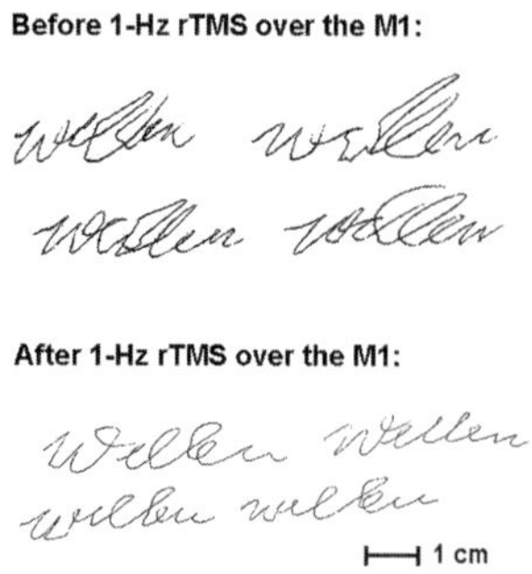

Before 1-Hz rTMS over the M1:

[handwriting sample: "Willen willen willen willen"]

After 1-Hz rTMS over the M1:

[handwriting sample: "Willen Willen willen willen"]

⊢——⊣ 1 cm

Figure 15–3 Representative examples of handwriting in a patient with writer's cramp before and 20 minutes after 1-Hz repetitive transcranial magnetic stimulation (rTMS) over the contralateral primary motor hand area (M1) showing a marked improvement in handwriting after 1-Hz rTMS. The rTMS protocol consisted of a 10-minute train of subthreshold 1-Hz rTMS applied at 90% active motor threshold. The rTMS train was delivered while the patient performed meaningless scribbling movements with the affected hand.

after overnight withdrawal of dopaminergic medication. Ten minutes after rTMS, PD patients showed a significant prolongation of the transcranially evoked silent period in the contralateral first dorsal interosseus muscle, whereas the silent period remained unchanged in healthy subjects. Because the duration of the transcranially evoked silent period is a well-established measure of intracortical (presumably $GABA_B$-ergic) inhibition,[49,50] this finding demonstrates that rTMS is capable of inducing a short-term increase in one form of intracortical inhibition in PD.

In the same experiment, Siebner and coworkers[48] measured the so-called stimulus-response curve at rest, which relates stimulus intensity to the mean amplitude of the MEP (Fig. 15–2). The stimulus-response curve provides a measure of the *gain function* of corticospinal excitability.[51] To this end, the stimulus intensity was increased in 10% steps, ranging from 110% to 150% of resting motor threshold, and five consecutive MEPs for each stimulus intensity were recorded from the relaxed contralateral first dorsal interosseus (FDI) muscle before and 5 minutes after rTMS. At variance with a facilitatory after-effect of a short-train of 5-Hz rTMS in young healthy subjects,[2] elderly controls showed an overall depression of the stimulus-response curve after prolonged 5-Hz rTMS, indicating a decreased gain in corticospinal excitability (see Fig. 15–2A). In contrast, PD patients showed no rTMS-related change in the stimulus-response curve (see Fig. 15–2B). Repeated measures analysis of variance revealed a significant interaction between the within-subjects factor "time" and the between-subjects factor "group" ($F_{[1,18]} = 5,12$; $p = 0,036$; Greenhouse-Geissler corrected). The electrophysiological data strongly suggest that the beneficial after-effect of subthreshold 5-Hz rTMS on motor dysfunction in PD cannot be attributed to a lasting increase in the gain of corticospinal excitability as indexed by the stimulus-response curve at rest. Future studies need to address whether an "inhibitory" 1-Hz rTMS is even more effective in enhancing deficient intracortical inhibition and alleviating motor dysfunction in patients suffering from PD.

These results demonstrate substantial differences in the susceptibility to 5-Hz rTMS between PD patients and healthy age-matched controls. PD patients appear to be relatively resistant to the inhibitory 5-Hz rTMS effect on corticomotor excitability seen in healthy age-matched controls (i.e., the "gain function" of the M1). However, PD patients appear to be more susceptible to the facilitatory effects of rTMS on intracortical inhibition (i.e., channeling function of the M1). These discrepancies between healthy controls and PD patients lend further support to the notion that after-effects observed in healthy volunteers cannot reliably predict after-effects of rTMS in patients.

Conditioning Effects of Repetitive Transcranial Magnetic Stimulation in Focal Dystonia of the Hand

This chapter considers the after-effects of rTMS only in writer's cramp, because the conditioning effects of rTMS on dystonia have been studied only in patients with writer's cramp.[52,53] In most patients with dystonia, involuntary co-contractions of the muscles can be found at rest. Therefore, it is difficult to ensure constant stimulation conditions during rTMS and to assess the neurophysiological consequences of rTMS. In contrast, in task-specific dystonias (e.g., writer's cramp) dystonia is present during the voluntary use of the affected body part but not at rest. In simple writer's cramp, dystonia is limited to the very act of handwriting, whereas other manual skills are affected by dystonia in complex writer's cramp.[54] The absence of dystonic symptoms at rest allows standardization of rTMS and neurophysiological measurements in patients with writer's cramp.

Physiological Rationale for Repetitive Transcranial Magnetic Stimulation in Hand Dystonia

The pathophysiology of writer's cramp is still poorly understood. Like other forms of idiopathic dystonias, writer's cramp is considered to arise from alterations in basal ganglia motor loops.[55] Although the primary dysfunction is thought to be located within the basal ganglia, there is converging evidence that writer's cramp is also associated with significant dysfunction at the cortical level, especially in the premotor cortex and the M1.[55] A functional activation study has shown decreased motor activity of M1 and caudal SMA and overactivity of lateral premotor areas during handwriting.[56] In line with abnormal motor activity of frontal executive motor areas, patients with writer's cramp show an abnormal negative slope component of the Bereitschaftpotential during voluntary contraction and relaxation[57,58] and deficient premovement desynchronization of high-frequency electroencephalographic activity over contralateral and midline central regions.[59] Moreover, the cortical representation of the MEPs of tonically active hand muscles is displaced in patients with writer's cramp.[60] Regarding cortical excitability of the M1, the excitability of inhibitory circuitry is reduced as indexed by a decrease in paired-pulse inhibition and a shorter duration of the TMS-evoked silent period.[61,62] The gain of corticospinal excitability is increased in focal dystonia, because the size of TMS-evoked electromyographic responses increases more steeply with increasing stimulus intensity or with increasing level of background contraction than in healthy subjects.[63,64]

These excitability changes are similar to the abnormalities that have been reported in PD patients. Considering the increased gain in corticospinal excitability and the reduced excitability of intracortical inhibitory circuits, Siebner and associates[52] tested the hypothesis that 1-Hz rTMS of the M1 results in a beneficial after-effect on handwriting in writer's cramp by temporarily reducing the gain in corticospinal excitability and reinforcing intracortical inhibition.

After-effects of Repetitive Transcranial Magnetic Stimulation on Motor Symptoms in Hand Dystonia

In 16 right-handed patients with writer's cramp and 11 age-matched controls, Siebner and colleagues[52] investigated handwriting before and 20 minutes after 1-Hz rTMS with the use of a digitizing tablet and a stroke-based approach to analyze the kinematics of handwriting movements. A total of 1800 stimuli of 1-Hz rTMS was applied by an 8-shaped coil over the left M1 at 90% of resting motor threshold while the participants kept the right arm relaxed. Patients with writer's cramp showed a significant reduction of mean writing pressure after subthreshold 1-Hz rTMS, which was associated in six patients with clear symptomatic improvement.[52] In a follow-up study, Siebner and coworkers[53] explored whether 1-Hz rTMS of the left M1 applied during unstructured writing movements results in a more consistent clinical effect on writer's cramp. The rationale behind this approach was that task-specific voluntary activation of the M1 would render the cortical circuits that are involved in the production of dystonia particularly susceptible to the inhibitory effects of 1-Hz rTMS and, hence, result in a more consistent clinical benefit. In twelve patients with writer's cramp, a 10-minute train of 1-Hz rTMS was applied to the left M1. Stimulus intensity was set just below active motor threshold, while the patients performed meaningless scribbling with their affected right hand. In 10 of these patients, a sham rTMS was performed on a separate day in a counterbalanced order. Kinematic analysis revealed an increase in regularity and fluency of circle drawing one hour after 1-Hz rTMS.[53] Moreover, 10 patients noticed a reduction in dystonia during handwriting that lasted for about 1 day (see Figure 15–3). Compared with the sham condition, the after-effects of real-rTMS over the M1 on motor performance were statistically significant. These results suggest that subthreshold 1-Hz rTMS of M1 can produce a short-term beneficial effect on writer's cramp, especially when being applied during meaningless scribbling. The beneficial effects of 1-Hz rTMS applied during scribbling [53] were more consistent than with 1-Hz rTMS applied at rest,[52] although rTMS during scribbling was only applied for 10 minutes, as opposed to 30 minutes in the first study.

After-effects of Repetitive Transcranial Magnetic Stimulation on Corticomotor Excitability in Hand Dystonia

A single session of 1,800 pulses of subthreshold 1-Hz rTMS over the left M1 applied during relaxation induced a prolongation of the transcranially evoked silent period (like 5-Hz rTMS in PD patients) and an increase in paired-pulse intracortical inhibition.[52] The finding that 1-Hz rTMS over the M1 reinforced deficient intracortical paired-pulse inhibition was confirmed in a second study in which a 10-minutes train of 1-Hz rTMS was applied during scribbling with the affected hand.[53] This study also showed that 1-Hz rTMS to the left M1 flattened the slope of the stimulus-response curve without affecting resting motor threshold.[53]

These data lend additional support to the concept of deficient intracortical inhibition as an important factor in the pathophysiology of writer's cramp. However, rTMS over the primary motor hand area implies concurrent stimulation of the primary sensory cortex, and the modulatory effects of rTMS on processing in the primary sensory cortex may have contributed to short-term clinical benefit. Moreover, a spread of excitation to the adjacent lateral premotor cortex needs to be considered as a contributing factor, because it has been shown that premotor areas are overactive in primary dystonia.[56]

Considering the strong inhibitory effect of 1-Hz rTMS over the lateral premotor cortex on corticospinal excitability,[7,8] it is conceivable that premotor 1-Hz rTMS will be more efficient for normalizing M1 excitability and improving dystonia. This assumption was not supported by a study that combined rTMS and positron-emission tomography (PET) of regional cerebral blood flow (rCBF) to assess the after-effects of premotor 1-Hz rTMS.[31] Seven patients with focal arm dystonia and seven healthy controls received 1,800 stimuli of subthreshold 1-Hz rTMS or sham stimulation to the left lateral dorsal premotor cortex. Stimulus intensity was 90% of individual resting motor threshold. The 1-Hz rTMS was given at rest. In both groups, real rTMS caused widespread bilateral decreases in neuronal activity (as indexed by rCBF) in prefrontal, premotor, primary motor cortex, and left putamen that lasted for at least 1 hour after the end of rTMS.[31] After-effects on neuronal activity were larger in patients than in healthy subjects; there was a greater decrease of rCBF in lateral and medial premotor areas, putamen, and thalamus, including the stimulated premotor cortex, and a larger increase in cerebellar rCBF. Despite these strong effects on rCBF, blinded assessment of handwriting revealed no after-effect of premotor rTMS on hand dystonia and handwriting.[31] However, Murase and associates[65] reported in abstract form that subthreshold 0.2-Hz rTMS given to the lateral premotor cortex improved performance in a writing task and prolonged the duration of the cortical silent period in nine patients with writer's cramp.

Quartarone and colleagues[66] used paired associative stimulation of the right median nerve and left M1 to assess associative (Hebbian) plasticity of the M1 in 10 patients with writer's cramp and 10 age-matched controls. In healthy subjects, paired associative stimulation led to a lasting increase in corticospinal excitability (as indexed by an increase in MEP amplitudes at rest) and intracortical inhibition (as indexed by a prolongation of the cortical silent period). In patients with writer's cramp, paired associative stimulation provoked an abnormal increase in corticospinal excitability and an attenuated reinforcement of intracortical inhibition relative to healthy controls.[66] Quartarone and coworkers[66] argued that this altered pattern of sensorimotor plasticity may favor maladaptive plasticity during repetitive skilled hand movements and may contribute to the pathophysiology of writer's cramp and other task-specific dystonias. These results are of relevance for future attempts to improve task-specific arm dystonia with TMS, indicating that the normal response pattern to TMS conditioning is markedly changed in focal hand dystonia. This abnormal plasticity underscores the necessity to closely monitor the physiological after-effects of any conditioning protocol in patients with movement disorders.

Outlook for the Use of Repetitive Transcranial Magnetic Stimulation in Movement Disorders

The therapeutic use of rTMS in movement disorders is still in the early stages, and no predictions can yet be made whether rTMS will acquire a well-established role in the neurorehabilitation of movement disorders. Effective blinding of patients and examiners remains a problem. Keeping in mind that the beneficial effects of rTMS on parkinsonian symptoms are short-lived and that there are a number of highly effective treatment options available in PD, it may be some time before rTMS plays a significant role in the treatment of PD. Things may differ in dystonia, because treatment options are more limited, especially in the subgroup showing no major clinical improvement after botulinum toxin injections.

The use of rTMS may be extended to other movement disorders such as ataxia,[67] tics in Tourette's syndrome,[68] chorea in Huntington's disease, cortical myoclonus,[69] or tremor.[70] Moreover, it is predicted that future studies will target additional brain regions such as the premotor cortex, prefrontal cortex, or the cerebellum. Transcranial direct current stimulation (TDCS) has been reintroduced by Nitsche and Paulus[71] as an additional technique to induce a lasting change in cortical function, and TDCS can be readily interfaced with rTMS.[10] It would also be interesting to explore the functional after-effects of TDCS alone or the combination of TDCS and rTMS in patients with movement disorders.

The major obstacle to the therapeutic use of rTMS is a lack of knowledge regarding its principles of action. By means of modern functional imaging techniques, it is possible to study the lasting impact of rTMS on various aspects of brain function in great detail.[27-31] This will provide a closer understanding of the neuronal mechanisms that mediate rTMS-induced after-effects in the brain. This issue is of fundamental importance for future applications of rTMS in movement disorders. A deeper insight into the basic principles of rTMS-mediated neuromodulation will allow more reliable predictions of the magnitude, direction, and duration of a conditioning effect in an individual patient. Moreover, it will help to refine the currently used rTMS protocols by adjusting stimulation parameters to the individual requirements of a given patient. Regardless of its therapeutic potential, the potential of rTMS to temporarily modulate the function of the stimulated cortex and functionally interconnected areas has already opened up unprecedented possibilities for in vivo studies on the pathophysiology of movement disorders.[31,66]

REFERENCES

1. Barker AT, Jalinous R, Freeston IL. Non-invasive magnetic stimulation of human motor cortex. Lancet 1985;1:1106–1107.
2. Pascual-Leone A, Valls-Sole J, Wassermann EM, et al. Responses to rapid-rate transcranial magnetic stimulation of the human motor cortex. Brain 1994;117:847–858.
3. Chen R, Classen J, Gerloff C, et al. Depression of motor cortex excitability by low-frequency transcranial magnetic stimulation. Neurology 1997;48:1398–1403.
4. George MS, Lisanby SH, Sackeim HA. Transcranial magnetic stimulation: applications in neuropsychiatry. Arch Gen Psychiatry 1999;56:300–311.
5. Siebner HR, Loeer C, Mentschel C, et al. Repetitive transcranial magnetic stimulation in Parkinson's disease and focal dystonia. Clin Neurophysiol Suppl 2002;54:399–409.
6. Siebner HR, Peller M, Willoch F, et al. Lasting cortical activation after repetitive TMS of the motor cortex: a glucose metabolic study. Neurology 2000;54:956–963.
7. Gerschlager W, Siebner HR, Rothwell JC. Decreased corticospinal excitability after subthreshold 1 Hz rTMS over lateral premotor cortex. Neurology 2001;57:449–455.

8. Rizzo V, Siebner HR, Modugno N, et al. Shaping the excitability of human motor cortex with premotor rTMS. J Physiol 2004;554:483–495.
9. Ziemann U, Hallett M, Cohen LG. Mechanisms of deafferentation-induced plasticity in human motor cortex. J Neurosci 1998;18:7000–7007.
10. Siebner HR, Lang N, Rizzo V, et al. Preconditioning of low-frequency repetitive transcranial magnetic stimulation with transcranial direct current stimulation: evidence for homeostatic plasticity in the human motor cortex. J Neurosci 2004;24:3379–3385.
11. Siebner HR, Auer C, Conrad B. Abnormal increase in the corticomotor output to the affected hand during repetitive transcranial magnetic stimulation of the primary motor cortex in patients with writer's cramp. Neurosci Lett 1999;262:133–136.
12. Gilio F, Curra A, Inghilleri M, et al. Repetitive transcranial magnetic stimulation of cortical motor areas in Parkinson's disease: implications for the pathophysiology of cortical function. Mov Disord 2002:467–473.
13. Boroojerdi B, Prager A, Muellbacher W, et al. Reduction of human visual cortex excitability using 1-Hz transcranial magnetic stimulation. Neurology 2000;54:1529–1531.
14. Modugno N, Nakamura Y, MacKinnon CD, et al. Motor cortex excitability following short trains of repetitive magnetic stimuli. Exp Brain Res 2001;140:453–459.
15. Siebner HR, Rothwell J. Transcranial magnetic stimulation: new insights into representational cortical plasticity. Exp Brain Res 2003;148:1–16.
16. Delwaide PJ, Olivier E. Conditioning transcranial cortical stimulation (TCCS) by exteroceptive stimulation in Parkinson patients. Adv Neurol 1990;53:175–181.
17. Playford ED, Jenkins IH, Passingham RE, et al. Impaired mesial frontal and putamen activation in Parkinson's disease: a positron emission tomography study. Ann Neurol 1992;32:151–161.
18. Samuel M, Ceballos-Baumann AO, et al. Evidence for lateral premotor and parietal overactivity in Parkinson's disease during sequential and bimanual movements: a PET study. Brain 1997;120:963–976.
19. Valls-Sole J, Pascual-Leone A, Brasil-Neto JP, et al. Abnormal facilitation of the response to transcranial magnetic stimulation in patients with Parkinson's disease. Neurology 1994;44:735–741.
20. Ellaway PH, Davey NJ, Maskill DW, et al. The relation between bradykinesia and excitability of the motor cortex assessed using transcranial magnetic stimulation in normal and parkinsonian subjects. Electroencephalogr Clin Neurophysiol 1995;97:169–178.
21. Chen R, Kumar S, Garg RR, et al. Impaired motor cortex activation and deactivation in Parkinson's disease. Clin Neurophysiol 2001,112:600–607.
22. Hallett M, Khoshbin S. A physiological mechanism of bradykinesia. Brain 1980;103:301–314.
23. Ridding MC, Inzelberg R, Rothwell JC. Changes in excitability of motor cortical circuitry in patients with Parkinson's disease. Ann Neurol 1995;37:181–188.
24. Priori A, Berardelli A, Inghilleri M. Motor cortical inhibition and the dopaminergic system. Pharmacological changes in the silent period after transcranial brain stimulation in normal subjects, patients with Parkinson's disease and drug-induced parkinsonism. Brain 1994;117:317–323.
25. Kleine BU, Praamstra P, Stegeman DF, et al. Impaired motor cortical inhibition in Parkinson's disease: motor unit responses to transcranial magnetic stimulation. Exp Brain Res 2001;138:477–483.
26. Chen R, Garg RR, Lozano AM, et al. Effects of internal globus pallidus stimulation on motor cortex excitability. Neurology 2001;56:716–723.
27. Paus T, Castro-Alamancos MA, Petrides M. Cortico-cortical connectivity of the human mid-dorsolateral frontal cortex and its modulation by repetitive transcranial magnetic stimulation. Eur J Neurosci 2001;14:1405–1411.
28. Strafella AP, Paus T, Barrett J, et al. Repetitive transcranial magnetic stimulation of the human prefrontal cortex induces dopamine release in the caudate nucleus. J Neurosci 2001;21:RC157.
29. Lee L, Siebner HR, Rowe JB, et al. Acute remapping within the motor system induced by low-frequency repetitive transcranial magnetic stimulation. J Neurosci 2003;23:5308–5318.
30. Strafella AP, Paus T, Fraraccio M, et al. A. Striatal dopamine release induced by repetitive transcranial magnetic stimulation of the human motor cortex. Brain 2003;126:2609–2615.
31. Siebner HR, Filipovic SR, Rowe JB, et al. Patients with focal arm dystonia have increased sensitivity to slow-frequency repetitive TMS of the dorsal premotor cortex. Brain 2003;126:2710–2725.
32. Cunnington R, Lalouschek W, Dirnberger G, et al. A medial to lateral shift in pre-movement cortical activity in hemi-Parkinson's disease. Clin Neurophysiol 2001;112:608–618.
33. Boylan LS, Pullman SL, Lisanby SH, et al. Repetitive transcranial magnetic stimulation to SMA worsens complex movements in

Parkinson's disease. Clin Neurophysiol 2001;112:259–264.
34. Okabe S, Ugawa Y, Kanazawa I. 0.2-Hz repetitive transcranial magnetic stimulation has no add-on effects as compared to a realistic sham stimulation in Parkinson's disease. Mov Disord 2003;18:382–388.
35. Pascual-Leone A, Valls-Sole J, Brasil-Neto JP, et al. Akinesia in Parkinson's disease. II. Effects of subthreshold repetitive transcranial motor cortex stimulation. Neurology 1994;44:892–898.
36. Ghabra MB, Hallett M, Wassermann EM. Simultaneous repetitive transcranial magnetic stimulation does not speed fine movement in PD. Neurology 1999;52:768–770.
37. Siebner HR, Mentschel C, Auer C, et al. Repetitive transcranial magnetic stimulation has a beneficial effect on bradykinesia in Parkinson's disease. Neuroreport 1999;10:589–594.
38. Siebner HR, Auer C, Weindl D, et al. Low-intensity repetitive transcranial magnetic stimulation of the motor cortex has no beneficial effect on skilled drawing movements in Parkinson's disease [abstract]. J Neurol 1997;244(Suppl 3):S51–S52.
39. Siebner HR, Rossmeier C, Mentschel C, et al. Short-term motor improvement after subthreshold 5-Hz repetitive transcranial magnetic stimulation of the primary motor hand area in Parkinson's disease. J Neurol Sci 2000;15:91–94.
40. De Groot M, Hermann W, Steffen J, et al. Contralateral and ipsilateral repetitive transcranial magnetic stimulation in Parkinson patients. Nervenarzt 2001;72:932–938.
41. Pascual-Leone A, Catala D. Lasting beneficial effect of rapid-rate transcranial magnetic stimulation on slowness in Parkinson's disease [abstract]. Neurology 1995;45(Suppl 4):A315.
42. Von Raison F, Drouot X, Nguyen JP, et al. The clinical effects of repetitive transcranial magnetic stimulation on PD depend on stimulation frequency [abstract]. Neurology 2000;54(Suppl 3):A281.
43. Sommer M, Tergau F, Paulus W. TMS in hypokinetic movement disorders. In George M, Belmaker RH (eds): Transcranial Magnetic Stimulation in Neuropsychiatry. Washington, DC, American Psychiatric Press, 2000:163–172.
44. Tergau F, Wassermann EM, Paulus W, et al. Lack of clinical improvement in patients with PD after low and high frequency repetitive transcranial magnetic stimulation. Electroencephalogr Clin Neurophysiol Suppl 1999;51:281–288.
45. Mally J, Stone TW. Improvement in Parkinsonian symptoms after repetitive transcranial magnetic stimulation. J Neurol Sci 1999;162:179–184.
46. Mally J, Stone TW. Therapeutic and "dose-dependent" effect of repetitive microelectroshock induced by transcranial magnetic stimulation in Parkinson's disease. J Neurosci Res 1999;57:935–940.
47. Shimamoto H, Takasaki K, Shigemori M, et al. Therapeutic effect and mechanism of repetitive transcranial magnetic stimulation in Parkinson's disease. J Neurol 2001;248(Suppl 3):48–52.
48. Siebner HR, Mentschel C, Auer C, et al. Repetitive transcranial magnetic stimulation causes a lasting increase in the duration of the cortical silent period in patients with PD. Neurosci Lett 2000;284:147–150.
49. Werhahn KJ, Kunesch E, Noachtar S, et al. Differential effects on motorcortical inhibition induced by blockade of GABA uptake in humans. J Physiol 1999;517:591–597.
50. Siebner HR, Dressnandt J, Auer C, et al. Continuous intrathecal baclofen infusions induced a marked increase of the transcranially evoked silent period in a patient with generalized dystonia. Muscle Nerve 1998;21:1209–1212.
51. Ridding MC, Rothwell JC. Stimulus/response curves as a method of measuring motor cortical excitability in man. Electroencephalogr Clin Neurophysiol 1997;105:340–344.
52. Siebner HR, Tormos JM, Ceballos-Baumann AO, et al. Low-frequency repetitive transcranial magnetic stimulation of the motor cortex in writer's cramp. Neurology 1999;52:529–537.
53. Siebner HR, Auer C, Ceballos-Baumann A, et al. Has repetitive transcranial magnetic stimulation of the primary motor hand area a therapeutic application in writer's cramp? Electroencephalogr Clin Neurophysiol Suppl 1999;51:265–275.
54. Marsden CD, Sheehy MP. Writer's cramp. Trends Neurosci 1990;13:148–153.
55. Hallett M. Physiology of dystonia. Adv Neurol 1998;78:11–18.
56. Ceballos-Baumann AO, Sheean G, Passingham RE, et al. Botulinum toxin does not reverse the cortical dysfunction associated with writer's cramp. A PET study. Brain 1997;120:571–582.
57. Deuschl G, Toro C, Matsumoto J, et al. Movement-related cortical potentials in writer's cramp. Ann Neurol 1995;38:862–868.
58. Yazawa S, Ikeda A, Kaji R, et al. Abnormal cortical processing of voluntary muscle relaxation in patients with focal hand dystonia studied by movement-related potentials. Brain 1999;122:1357–1366.
59. Toro C, Deuschl G, Hallett M. Movement-related electroencephalographic desynchronization in patients with hand cramps: evidence for motor cortical involvement in focal dystonia. Ann Neurol 2000;47:456–461.

60. Byrnes ML, Thickbroom GW, Wilson SA, et al. The corticomotor representation of upper limb muscles in writer's cramp and changes following botulinum toxin injection. Brain 1998;121:977–988.
61. Ridding MC, Sheean G, Rothwell JC, et al. Changes in the balance between motor cortical excitation and inhibition in focal, task specific dystonia. J Neurol Neurosurg Psychiatry 1995;59:493–498.
62. Filipovic SR, Ljubisavljevic M, Svetel M, et al. Impairment of cortical inhibition in writer's cramp as revealed by changes in electromyographic silent period after transcranial magnetic stimulation. Neurosci Lett 1997;222:167–170.
63. Ikoma K, Samii A, Mercuri B, et al. Abnormal cortical motor excitability in dystonia. Neurology 1996;46:1371–1376.
64. Mavroudakis N, Caroyer JM, Brunko E, et al. Abnormal motor evoked responses to transcranial magnetic stimulation in focal dystonia. Neurology 1995;45:1671–1677.
65. Murase N, Kaji R, Murayama N, et al. Subthreshold low-frequency repetitive transcranial magnetic stimulation over the premotor cortex in writer's cramp [abstract]. Mov Disord 2002;17:S304.
66. Quartarone A, Bagnato S, Rizzo V, et al. Abnormal associative plasticity of the human motor cortex in writer's cramp. Brain 2003;126:2586–2596.
67. Shiga Y, Tsuda T, Itoyama Y, et al. Transcranial magnetic stimulation alleviates truncal ataxia in spinocerebellar degeneration. J Neurol Neurosurg Psychiatry 2003;72:124–126.
68. Munchau A, Bloem BR, Thilo KV, et al. Repetitive transcranial magnetic stimulation for Tourette syndrome. Neurology 2002;59:1789–1791.
69. Rossi S, Ulivelli M, Bartalini S, et al. Reduction of cortical myoclonus-related epileptic activity following slow-frequency rTMS. A case study. Neuroreport 2004;15:293–296.
70. Gironell A, Kulisevsky J, Lorenzo J, et al. Transcranial magnetic stimulation of the cerebellum in essential tremor: a controlled study. Arch Neurol 2002;59:413–417.
71. Nitsche MA, Paulus W. Sustained excitability elevations induced by transcranial DC motor cortex stimulation in humans. Neurology 2002;57:1899–1901.

16 Transcranial Magnetic Stimulation in Stroke

Paolo M. Rossini and Flavia Pauri

It has been traditionally thought that the adult brain has no significant ability for self-repair or reorganization after injuries that cause neuronal death. However, it is common in clinical practice to see slow but consistent recovery over a period of weeks and months after stroke.[1] Such spontaneous recovery ranges widely at 1 year, even in light of very similar clinical pictures acutely after stroke.

Compensatory plasticity of residual tissues in language areas or extension of language function to areas previously not involved in language processing must exist to explain the nearly universal improvement seen in postlesional aphasias despite persistence of the focal neuronal damage.[2,3] Evidence from studies shows that the areas most likely to take over in aphasia recovery are the undamaged portions of the dominant hemisphere or the homologous areas in the nondominant one.[4] A better understanding of mechanisms possibly improving on naturally occurring restorative phenomena are needed to help rehabilitative efforts during postlesional function retraining. Several interpretations have been offered to explain spontaneous recovery. They include reabsorption of perilesional edema, interindividual variability of perfusion patterns and arterial collaterals, and prestroke organization of the neuronal network subserving the initially lost function and the amount of its corticocortical and transcallosal connections.[5] In this respect, it is worth remembering that multiple representations of the same function in the cortex (e.g., primary motor cortex),

presence and amount of alternative neural routes (e.g., ipsilateral corticospinal fibers), unmasking of functionally silent connections, and changes in synaptic efficacy and remodeling are all important contributors to the *plasticity* of the human central nervous system (CNS).[5,6]

Additional and more general mechanisms for human CNS functional recovery are progressively and more faithfully reconstructed in animal models,[7-9] and they can now be directly addressed by modern techniques for the study of brain functional organization in vivo.[10-15] A better understanding of the mechanisms underlying recovery or deterioration of function after a CNS lesion, as well as those leading to maladaptive or unfavorable outcomes, is essential for directing specific and effective rehabilitative strategies and for avoiding potentially harmful interventions.

In this chapter, we discuss the anatomical and physiological aspects of the nervous system that likely play a major role in postlesional functional improvement, with special emphasis on the motor system. Methods that allow in vivo evaluation of plastic phenomena also are reviewed.

The multiple representations of each muscle and contiguous joints, the overlap of outputs to different motoneurons, and the large cortical representation of hand and fingers are probably the anatomical substrate for the extraordinary repertoire of possible movement strategies and the coordination of action initiated by many muscles at different joints.[16] Functionally, strict coordination of different muscle groups is essential. For instance, the stability of the proximal arm and girdle is necessary for the successful execution of finger movements. Movements occur in a three-dimensional space with several degrees of freedom, and the body surface may be envisioned as a three-dimensional volume in the cortex. Multiple representations underlying different motor functions of separate body parts along with distributed and cooperating neural networks may therefore overlap in space and time,[17,18] offering flexibility in motor learning and progressively substituting a related dysfunctional area more easily than highly specialized and unique groups of cells would, thereby being fundamental for functional (or maladaptive) plasticity.

Transcranial brain stimulation (TMS),[19] a safe, painless, and noninvasive technique, is increasingly used in investigations of brain function, including plastic phenomena, and permits mapping of underlying cortical representation areas.[10,20-25] Changes in cortical maps that are use dependent or result from a lesion usually demonstrate two main characteristics. The first is the enlargement or restriction of the excitable area without changes in the amplitude-weighted center of the motor output maps (i.e., center of gravity), possibly from the recruitment or inhibition of a fringe of adjacent neurons. The second is the migration of the excitable area outside the usual boundaries, possibly due to a lesion affecting the brain region where the "hot spot" was originally located. Such migration may be apparent and caused by the activation of a secondary hot spot previously hidden by the predominant one, or it may be real and caused by the progressive activation of new synaptic connections.[26,27] TMS is a powerful tool for studying mechanisms of intracortical inhibition and excitation.[28,29]

Numerous examples of neural plasticity can be found in healthy subjects, such as skilled or repeated training, and in instances of permanent or transient deafferentation of cerebral cortical areas. The role for TMS in studying plastic changes of the human motor system in the acquisition of new fine motor skills is further evidenced by a study in which a brief training session of synchronous movements of the ipsilateral thumb and foot induced a transient shift of the center of gravity of the hand muscle examined (i.e., abductor pollicis brevis) of about 7 mm medially toward the foot area. The changes observed were ascribed to corticocortical modulation between hand and foot representation areas in motor cortex.[30] Cortical modulation phenomena may be in part the basis for commonplace performance improvements in speed and accuracy after practice of complex motor tasks. Similarly, the cortical motor areas targeting the contralateral long finger flexor and extensor muscles in subjects learning a one-handed, five-finger piano exercise were shown to increase in size and the activation threshold to decrease after only 5 days of training.[31]

Mental imagery of movements (not evoking a contraction) may increase the capability of acquiring new motor skills. Such a technique has been successfully applied as a training procedure to improve the actual performance without receiving any feedback about the results, and TMS represents an excellent tool to demonstrate motor output changes taking place during the ideation of movement. Izumi and colleagues[32] have described facilitatory effects of motor imagery, as evidenced by a decrease in the excitability threshold, induced by "thinking about" a movement. Nonmotor mental activity and alerting stimuli per se do potentiate the amplitude of MEPs in the target muscles, in strict time relationship with a desynchronization of the background electroencephalographic activity of the stimulated brain areas.[33] Meanwhile, movement imagery can focus a specific facilitation on the prime-mover muscle involved in a mentally simulated movement.[34-36] This process is specifically and asymmetrically affected when the motor cortex contralateral to the affected limbs is stimulated with focal TMS in early, unilateral Parkinson's disease patients,[37] suggesting a crucial role of motor imagery in normal movement initiation and execution; however, such ability is usually preserved in stroke patients with good recovery.[38] Better understanding of motor imagery mechanisms in physiologic and pathologic conditions is therefore likely to be of help in designing more effective rehabilitative approaches.

Use-dependent aberrant plastic reorganization of brain may represent the substrate for some post-stroke neurological deficits or dysfunctions. Post-stroke, spastic focal dystonia, for example, tends to occur with repetitive synchronous movements of the fingers.

Monohemispheric Lesions and Stroke

Numerous factors can influence degree and mode of reorganization after a stroke, such as site and extent of the lesion, diaschisis, and prestroke organization of the motor areas output, particularly the amount of ipsilateral uncrossed corticospinal fibers.[39]

When damage to a functional system is partial, a within-system recovery is possible, whereas after complete destruction, substitution by functionally related systems is the main mechanism for functional recovery.[40] Experimental and clinical studies have shown that approximately one fifth of the pyramidal fibers are sufficient to ensure restitution of fractionated hand-finger movement. In this case, within–pyramidal system reorganization is probably justifying functional recovery of the motor control of hand and upper limb.[27,34,41–47]

Early studies using positron-emission tomography (PET) and functional magnetic resonance imaging (fMRI) documented abnormal activation patterns during movement of a paretic hand even after full clinical recovery.[48–52] Enhanced bilateral activation of motor pathways has been reported after striatocapsular infarctions, together with recruitment of sensory and secondary motor structures normally not involved in the task and displacement of primary motor peak activation toward the face area.[12,50,53] Because of their considerable logistic complexity, only few longitudinal studies have been published.[12,13,54–57] However, in most such studies, a finger-tapping paradigm has been employed, without taking into consideration that the attentional load for programming and execution is quite asymmetric for the affected and unaffected hand, and this is the possible reason for most of the described differences.[58]

In patients with cortical infarcts, a pattern of overactivation of bilateral noninfarcted motor-related and nonmotor areas similar to that seen in striatocapsular strokes has been reported.[51,52,59] Other interesting findings have been a strong peri-infarct activation[51] and ipsilesional premotor cortex activation.[59] The posterior shift and inferior extension of M1 activation observed in cortical and subcortical strokes may represent unmasking or disinhibition of synaptic connections within the corticospinal tract system.

Functional reorganization of the motor output after a hemispheric stroke has been consistently observed by means of TMS.[27,35,47,60,61] In stroke patients, MEPs recorded in a relaxed muscle state are often absent with TMS of the affected hemisphere.[62,63] Other abnormalities include increased excitability threshold, delayed latency of the responses, and an excessive asymmetry of the hand muscle motor maps between the affected and the unaffected hemispheres. Such asymmetries result from "map migration," usually on the mediolateral axis. An anteroposterior shift of several millimeters of the center of gravity often has been reported. This is particularly frequent and stable over time in patients in the chronic stages, whereas it can be seen as a transient dynamic phenomenon in first few months after a stroke.[26,27,47,60] Intracortical inhibitory or excitatory curves to paired-pulse TMS[64] are abnormal even in the chronic state, with the affected hemisphere having partly or totally lost the inhibitory part of the curve[65] (Figs. 16–1 and 16–2).

The remote effects of the lesioned area, or diaschisis, suggest that acute neuronal failure in the ischemic area induce modulatory effects on cortical excitability of unaffected regions of the same hemisphere by corticocortical connections and of the contralateral unaffected hemisphere by transcallosal fibers.[66,67] Transient hyperexcitability of the unaffected hemisphere contralateral to a neocortical infarction has been documented in animal models.[68] The relationship between clinical recovery and early cortical excitability to TMS after a stroke has been studied as a possible prognostic tool.[63,64,69–71] MEPs from upper limb muscles have always been obtained in

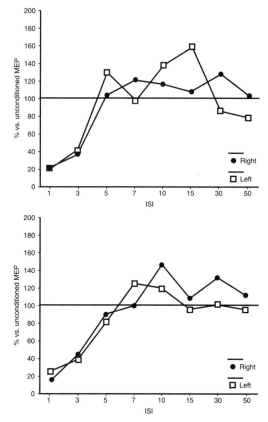

Figure 16-1 Symmetry of inhibitory and excitatory curves to paired-pulse transcranial magnetic stimulation in the two hemispheres of healthy persons. (From Cicinelli P, Pasqualetti P, Zaccagnini M, et al. Interhemispheric asymmetries of motor cortex excitability in the postacute stroke stage: a paired-pulse transcranial magnetic stimulation study. Stroke 2003;34:2653-2658.)

patients who fully recover finger control.[63,64,72] Less is known about excitability in subacute and chronic stages (weeks to months) and its correlation with clinical outcome.[73-75]

The role of other nonpyramidal motor cortex efferents, such as the bilaterally organized premotor or reticulospinal projections, is still unclear. Large-scale reorganization of M1 beyond the boundaries of the physiologic corticocortical connections probably requires long-term repetitive engagement.[76]

TMS studies performed in the acute (4 to 72 hours), subacute (1 week), and chronic stages (6 months) after a stroke showed increased excitability thresholds and prolonged MEPs latencies (when present) and CCT acutely in the affected hemisphere, whereas responses from the unaffected hemisphere remained normal in each recording session.[77-79] Interhemispheric differences between the affected and unaffected hemispheres were significant for all the parameters studied. The responses from the affected hemisphere changed significantly during follow-up, with most parameters showing a partial recovery. Latencies of MEPs decreased, and amplitudes increased, although they were still altered at the sixth month, but the clinical improvement was ongoing. Interhemispheric differences between the affected and unaffected hemispheres were the most altered neurophysiological parameters, particularly for excitability thresholds. MEP amplitudes, as measured in the acute stage had a positive correlation with the clinical picture and with long-term outcome; the higher the MEP amplitude, the better the outcome. No correlation was found between side (right versus left) or type of the lesion (cortical versus subcortical) and neurophysiological parameters. The steepest part of the recovery slopes was concentrated between 40 and 80 days after the stroke[80] (Fig. 16-3).

Hemispheric motor output was studied at two times after a stroke (T1 = 8 weeks; T2 = 16 to 18 weeks). Excitability thresholds were significantly higher and the MEP amplitudes smaller in the affected hemisphere despite a stronger TMS employed. The area of cortical output to the target muscle was asymmetrically restricted compared with controls[81] and with the patients' unaffected hemispheres. The percentage of altered parameters was significantly higher in T1 and in those suffering from subcortical lesions, possibly because of the large number of densely packed fibers affected by a subcortical lesion and to a less efficient short-term "plastic" reorganization. Anomalous hot spot sites were observed more frequently in the cortical group. In T2, the parameters of the subcortical patients were improved to the same level of the cortical group. A significant enlargement of the affected hemisphere's hand motor cortical area was found in 67% of cases in T2 compared with T1, together with clinical improvement as measured by the Canadian Neurological Scale, its Hand Motor Subscore,

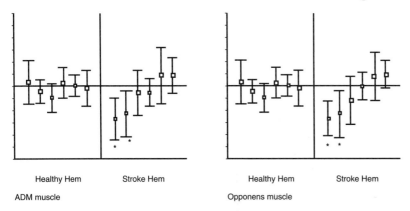

Figure 16–2 Interhemispheric differences of inhibitory and excitatory curves to paired-pulse transcranial magnetic stimulation in chronic stroke. Notice the significant reduction of the inhibitory curve *(asterisk)* on the affected hemisphere. (From Cicinelli P, Pasqualetti P, Zaccagnini M, et al. Interhemispheric asymmetries of motor cortex excitability in the postacute stroke stage: a paired-pulse transcranial magnetic stimulation study. Stroke 2003;34:2653–2658.)

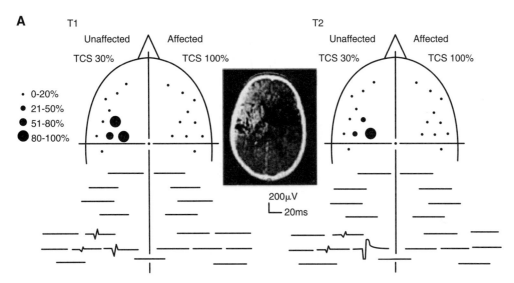

Figure 16–3 Transcranial magnetic stimulation (TMS) shows TMS mapping of the motor output to the hand in stroke patients. **A:** P.A. is 60 years old and has a mainly subcortical lesion. Motor evoked potentials (MEPs) from the affected hemisphere (AH) were missing at two times after a stroke: T1 = 2 months after the event and T2 = 8 weeks after T1. The topographical output from the unaffected hemisphere (UH) was normal in shape and unchanged at T2. **B:** C.F. is 62 years old. The AH was inexcitable at T1, but several responsive sites were evident in the second recording session. **C:** Anomalous hot spot sites were present at T2, suggesting a rearrangement of motor cortical output and possible recruitments of areas outside the usual boundaries of primary motor cortex. **D:** Abnormal MEP parameters are expressed as absolute values *(open squares)* and as interhemispheric asymmetries. It emerges that the interhemispheric differences of the examined items are more frequently abnormal than their absolute values. (Modified from Cicinelli P, Traversa R, Rossini PM. Post-stroke reorganization of brain motor output to the hand: a 2–4 month follow-up with focal magnetic transcranial stimulation. Electroencephalogr Clin Neurophysiol 1997;105:438–450.)

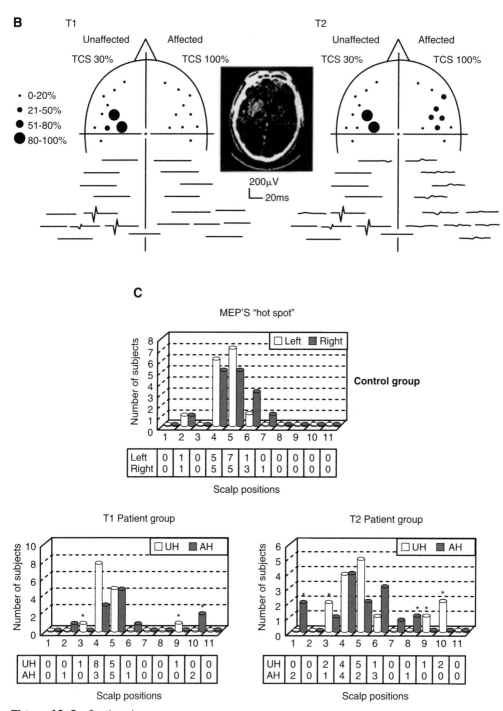

Figure 16–3 Continued.

and the Barthel Index. Patients demonstrating enlargement of the hand area also showed a correlated improvement of the hand score. In this group, the type of lesion, whether cortical or subcortical, did not influence the clinical outcome.[26,34] The role of ipsilateral corticospinal connections is somewhat controversial. Palmer and coworkers[82] were unable to obtain ipsilateral MEPs (iMEPs) in stroke patients, concluding that ipsilateral connections did not contribute to motor recovery. In other studies, iMEPs could be elicited by TMS

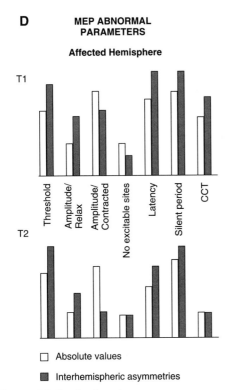

Figure 16–3 Continued.

over the unaffected hemisphere[72,83–87] or over the affected hemisphere.[87–89] The iMEPs were usually obtained by stimulating anteriorly and medially to the primary motor cortex, suggesting activation of corticospinal pathways from premotor areas. Outcome was variable, and some investigators found a correlation between iMEPs and motor recovery,[85–87] but others did not.[72,83,84] In any case, the iMEPs described by Caramia and colleagues[85] and Trompetto and associates[87] seem to differ. In Caramia's and Trompetto's patient groups, the iMEPS had rather low excitability thresholds and large amplitudes, whereas in Turton's patients, only small iMEPs could be elicited with high stimulation intensities. Alagona and coworkers[89] found an association between iMEPs produced by stimulation of the affected hemisphere and bimanual dexterity 6 months after stroke, suggesting that their existence reflects hyperexcitability of affected hemisphere premotor areas.

The study of intracortical inhibition (ICI) and facilitation[64] early after a stroke sheds light on some plasticity mechanisms. A TMS study of large MCA infarctions, with single and paired pulse TMS of the unaffected hemisphere, showed a decrease of ICI 2 weeks after the stroke.[90] This finding corresponds to results obtained in animal studies.[68,91,92] In rats, the ICI was associated with downregulation of γ-aminobutyric acid type A (GABA$_A$) receptors and enhancement of glutamatergic activity.[93,94] Because ICI changes are supposed to be modulated by GABAergic activity,[95,96] the decreased ICI in the unaffected hemisphere of patients with large infarctions may reflect GABA activity downregulation. This effect may result from damage of transcallosal fibers and loss of the physiological interhemispheric inhibitory modulation[97–100] or to enhanced use of the unaffected arm in daily activities, because ICI is modified in a task- and use-dependent manner.[101] Disinhibition in the unaffected hemisphere is also supported by the finding of enlarged MEP amplitudes when stimulating this hemisphere after a stroke.[26,47,81,87]

A loss of ICI was observed in the affected hemisphere in stroke patients with small motor deficits at onset or rapid spontaneous motor recovery, whereas ICI in the unaffected hemisphere was not different from that of age-matched controls.[28,29,65,102,103] This result corresponds to animal studies describing loss of GABAergic inhibition immediately surrounding a cortical lesion.[104] On this basis, it has been suggested that the rapid clinical improvement of some patients might be induced by motor cortical disinhibition resulting from the remote effects of structurally intact areas (i.e., diaschisis) or to the preexisting organization of the motor areas (i.e., amount of ipsilateral uncrossed corticospinal fibers).[102] In humans, as in experimental animals, there is growing evidence supporting a positive influence of brain monoamine concentrations toward rate and degree of recovery from cortical lesions.[105] Particularly if coupled with physical therapy, amphetamine, L-Dopa, and fluoxetine seem to enhance recovery,[106–108] at least in the case of fluoxetine, possibly by enhanced serotoninergic transmission.

One way to examine intervention-induced plastic reorganization is to study patients in the very chronic stages of their illness, a phase when the probability of spontaneous recovery is negligible.[90,101,103] Another possibility is to study the short-term effects of interventions (e.g., within a single day) for which the changes observed after a single therapeutic session probably are caused by the intervention itself rather than spontaneous improvements. We can consider motor reorganization of constraint-induced movement therapy (CIMT).[90,101,109–111] With TMS, repeated baseline measurements have shown highly reproducible results in strokes occurred more than 6 months earlier. Before CIMT, the patients had higher motor thresholds and smaller motor output maps in the affected hemisphere. After therapy, motor output maps in the affected hemisphere had increased by approximately 40%, whereas those in the unaffected hemisphere had decreased, although not significantly. Changes presumably reflected use-dependent mechanisms (i.e., increased use of the paretic hand during training and decreased use of the nonaffected hand) and were paralleled by a large clinical improvement of motor function. Motor thresholds had remained identical after CIMT, and because they are determined in the *center* of the cortical representation area, it was concluded that enlargements of the motor output maps were caused by excitability increase at the *borders* of the representation area, possibly through GABA-dependent modulation of horizontal intracortical inhibitory circuits.[112] After CIMT, the center of gravity of the motor output maps had shifted significantly in the affected hemisphere, mainly in the mediolateral axis, presumably indicating recruitment of additional adjacent brain areas. In the 6 months after CIMT, the motor output map sizes equalized (i.e., less area in the affected hemisphere, more area in the unaffected hemisphere), and motor performance remained unchanged. These changes were interpreted as indicators of improved connectivity between neuronal populations, allowing a reduction of cortical excitability while maintaining performance.[100] CIMT and manipulation of monoamine neuromodulators have been shown in randomized, controlled trials to enhance motor recovery through stimulation of use-dependent connections.[100,101,109,111] The facilitatory effects of different interventions studied with TMS showed that contraction of the target muscle exceeded all other facilitatory techniques such as preinnervation of more proximal or contralateral homologous muscles, or passive cutaneous stimulation.[113] These results confirm that increased or repetitive use has facilitatory effects and improves motor performance.[114]

All these findings from TMS studies suggest that reorganization of the motor output from a lesioned hemisphere is still taking place several months after a stroke. The time course and degree of motor recovery in humans could largely depend on the degree of damage to the distributed motor network, because different motor areas operate in a parallel rather than in a hierarchical fashion and parallel descending pathways might be able to compensate functionally for each other.[115] The poststroke interval of some studies was long enough to suggest that the observed modifications were caused by corticospinal tract reorganization rather than recovery from perilesional edema and "early" cortical hypoexcitability. Recovery of sensory deficits can also play a significant role, because the modulation of the tonic sensory flow from the skin enveloping the target muscle significantly affects the amount of its cortical representation.[116,117]

Conclusion

The study of neural plasticity has expanded rapidly in the past decades and has demonstrated the remarkable ability of the developing brain and of the adult and aging brain to be shaped by environmental inputs in health and after a lesion.[117] Robust experimental evidence supports the hypothesis that neuronal aggregates adjacent to a lesion in the sensorimotor brain areas can progressively assume the function previously played by the damaged neurons. It is accepted that such reorganization modifies sensibly the interhemispheric differences in somatotopic organization of the sensorimotor cortices. This reorganization largely subtends clinical recovery of motor performances and sensorimotor integration after a stroke.

Functional imaging studies of the brain show that recovery from hemiplegic strokes is associated with a marked reorganization of the activation patterns of specific brain structures.[48] To regain hand motor control, the recovery process tends over time to bring the bilateral motor network activation toward a more normal intensity or extent while overrecruiting simultaneously new areas, perhaps to sustain this process. Considerable intersubject variability exists in the activation or hyperactivation pattern changes over time. Some patients, for instance, display late-appearing dorsolateral prefrontal cortex activation, suggesting the development of executive strategies to compensate for the lost function. The affected hemisphere in stroke often undergoes a significant remodeling of sensory and motor hand somatotopy outside the normal areas and enlargement of the hand representation. The unaffected hemisphere also undergoes reorganization to a lesser degree. Although absolute values of the investigated parameters fluctuate across subjects because of individual anatomical and functional variability, variation is instead minimal with regard to interhemispheric differences because individual morphometric and functional characters—at least for the arm and hand control—are mirrored in the two hemispheres. Excessive interhemispheric asymmetry of the sensorimotor hand areas seems to be the parameter with highest sensitivity in describing brain reorganization after a monohemispheric lesion and mapping motor and somatosensory cortical areas through focal TMS, fMRI, PET, electroencephalography, and magnetoencephalography (MEG) is quite useful in studying hand representation and interhemispheric asymmetries in normal and pathologic conditions.[12,34,119–121]

Neurophysiological techniques allow detection of sensorimotor areas' reshaping due to neuronal reorganization or to recovery of the previously damaged neural network. They have the advantage of high temporal resolution but suffer from limitations. For example, TMS provides only bidimensional maps, and MEG, even if giving three-dimensional mapping of generator sources, does so by procedures that rely on the choice of a mathematical model of the head and the sources. Moreover, these techniques do not test movement execution and sensorimotor integration as used in everyday life. A multi-technological approach is the best way to test the presence and amount of plasticity phenomena underlying partial or total recovery of several functions, particularly sensorimotor.

Dynamic patterns of recovery are progressively emerging from the relevant literature. Enhanced recruitment of the affected cortex, whether it is spared perilesional tissue as in the case of cortical stroke or intact but deafferented cortex as in subcortical strokes, appears to be the rule, a mechanism especially important in the early stages after an insult. However, the transfer over time of preferential activation toward contralesional cortices, as observed in some cases, appears to reflect a less efficient type of plastic reorganization, with some aspects of maladaptive plasticity. Reinforcing the use of the affected side can cause activation to increase again in the affected side and balancing back to normality the transcallosal, interhemispheric excitatory and inhibitory modulation with a corresponding enhancement of clinical function. Meanwhile, activation of the unaffected hemisphere motor cortex may represent recruitment of direct (uncrossed) corticospinal tracts and relate more to mirror movements, but it more likely reflects activity redistribution within preexisting bilateral, large-scale motor networks. Activation of areas not normally engaged in the dysfunctional tasks, such as the dorsolateral prefrontal cortex or the superior parietal cortex in motor paralysis, may implicate compensatory cognitive strategies.

REFERENCES

1. Twitchell TE. The restoration of motor function following hemiplegia in man. Brain 1951;74:443–480.
2. Holland AL, Fromm DS, DeRuyter F, et al. Treatment efficacy: aphasia. J Speech Hear Res 1996;39:S27–36.
3. Wertz RT. Aphasia in acute stroke: incidence, determinants, and recovery. Ann Neurol 1996;40:120–130.
4. Heiss WD, Kessler J, Thiel A, et al. Differential capacity of left and right hemispheric areas for compensation of poststroke aphasia. Ann Neurol 1999;45:430–438.

5. Rossini PM, Calautti C, Pauri F, et al. Post-stroke plastic reorganisation in the adult brain. Lancet Neurol 2003;2:493–502.
6. Rossini PM. Brain redundancy: responsivity or plasticity? Ann Neurol 2001;48:128–129.
7. Nudo RJ, Milliken GW. Reorganization of movement representations in primary motor cortex following focal ischemic infarcts in adult squirrel monkeys. J Neurophysiol 1996;75:2144–2149.
8. Nudo RJ. Remodeling of cortical motor representations after stroke: implications for recovery from brain damage. Mol Psychiatry 1997;2:188–191.
9. Calabresi P, Centonze D, Pisani A, et al. Synaptic plasticity in the ischaemic brain. Lancet Neurol 2003;2:622–629.
10. Rossini PM, Rossi S. Clinical application of motor evoked potentials. Electroencephalogr Clin Neurophysiol 1998;106:180–194.
11. Baron JC. Positron tomography in cerebral ischemia. A review. Neuroradiology 1985;27:509–516.
12. Calautti C, Leroy F, Guincestre JY, et al. Dynamics of motor network overactivation after striatocapsular stroke: a longitudinal PET study using a fixed-performance paradigm. Stroke 2001;32:2534–2542.
13. Calautti C, Leroy F, Guincestre JY, et al. Sequential activation brain mapping after subcortical stroke: changes in hemispheric balance and recovery. Neuroreport 2001;12:3883–3886.
14. Calautti C, Baron JC. Functional neuroimaging studies of motor recovery after stroke in adults: a review. Stroke 2003;34:1553–1566.
15. Taub E, Uswatte G, Elbert T. New treatments in neurorehabilitation founded on basic research. Nat Rev Neurosci 2002;3:228–36.
16. Kwan HC, MacKay WA, Murphy JT, et al. Spatial organization of precentral cortex in awake primates. II. Motor outputs. J Neurophysiol 1978;41:1120–1131.
17. Kalaska JF, Crammond DJ. Cerebral cortical mechanisms of reaching movements. Science 1992; 255:1517–23.
18. Sanes JN, Donoghue JP, Thangaraj V, et al. Shared neural substrates controlling hand movements in human motor cortex. Science 1995;268:1775–1777.
19. Barker AT, Jalinous R, Freeston IL. Non-invasive magnetic stimulation of human motor cortex. Lancet 1985;1:1106–1107.
20. Rossini PM, Caramia M, Zarola F. Central motor tract propagation in man: studies with non-invasive, unifocal, scalp stimulation. Brain Res 1987;415:211–225.
21. Rothwell JC, Thompson PD, Day BL, et al. Motor cortex stimulation in intact man. 1. General characteristics of EMG responses in different muscles. Brain 1987;110:1173–1190.
22. Cohen LG, Roth BJ, Nilsson J, et al. Effects of coil design on delivery of focal magnetic stimulation. Technical considerations. Electroencephalogr Clin Neurophysiol 1990;75:350–357.
23. Fuhr P, Cohen LG, Roth BJ, et al. Latency of motor evoked potentials to focal transcranial stimulation varies as a function of scalp positions stimulated. Electroencephalogr Clin Neurophysiol 1991;81:81–89.
24. Wassermann EM, McShane LM, Hallett M, et al. Noninvasive mapping of muscle representations in human motor cortex. Electroencephalogr Clin Neurophysiol 1992;85:1–8.
25. Pascual-Leone A, Cohen LG, Brasil-Neto JP, et al. Differentiation of sensorimotor neuronal structures responsible for induction of motor evoked potentials, attenuation in detection of somatosensory stimuli, and induction of sensation of movement by mapping of optimal current directions. Electroencephalogr Clin Neurophysiol 1994;93:230–236.
26. Cicinelli P, Traversa R, Rossini PM. Post-stroke reorganization of brain motor output to the hand: a 2–4 month follow-up with focal magnetic transcranial stimulation. Electroencephalogr Clin Neurophysiol 1997;105:438–450.
27. Traversa R, Cicinelli P, Bassi A, et al. Mapping of motor cortical reorganization after stroke. A brain stimulation study with focal magnetic pulses. Stroke 1997;28:110–117.
28. Shimizu T, Filippi MM, Palmieri MG, et al. Modulation of intracortical excitability for different muscles in the upper extremity: paired magnetic stimulation study with focal versus non-focal coils. Clin Neurophysiol 1999;110:575–581.
29. Shimizu T, Oliveri M, Filippi MM, et al. Effect of paired transcranial magnetic stimulation on the cortical silent period. Brain Res 1999;834:74–82.
30. Liepert J, Terborg C, Weiller C. Motor plasticity induced by synchronized thumb and foot movements. Exp Brain Res 1999;125:435–443.
31. Pascual-Leone A, Nguyet D, Cohen LG, et al. Modulation of muscle responses evoked by transcranial magnetic stimulation during the acquisition of new fine motor skills. J Neurophysiol 1995;74:1037–1045.
32. Izumi S, Findley TW, Ikai T, et al. Facilitatory effect of thinking about movement on motor-evoked potentials to transcranial magnetic stimulation of the brain. Am J Phys Med Rehabil 1995;74:207–213.
33. Rossini PM, Desiato MT, Lavaroni F, et al. Brain excitability and electroencephalographic activation: non-invasive evaluation in healthy humans via transcranial magnetic stimulation. Brain Res 1991;567:111–119.

34. Rossini PM, Tecchio F, Pizzella V, et al. On the reorganization of sensory hand areas after mono-hemispheric lesion: a functional (MEG)/ anatomical (MRI) integrative study. Brain Res 1998;782:153–166.
35. Rossini PM, Caltagirone C, Castriota-Scandberg A, et al. Hand motor cortical area reorganization in stroke: a study with fMRI, MEG amt TMS maps. Neuroreport 1998;9:2141–2146.
36. Rossini PM, Rossi S, Pasqualetti P, et al. Corticospinal excitability modulation to hand muscles during movement imagery. Cereb Cortex 1999;9:161–167.
37. Filippi MM, Olivieri M, Pasqualetti P, et al. Effects of motor imagery on motor cortical output topography in Parkinson's disease. Neurology 2001;57:55–61.
38. Cicinelli B, et al. Imagery-induced cortical excitability changes in stroke: a transcranial magnetic stimulation study. 2004 (submitted for publication).
39. Weiller C, Rijntjes M. Learning, plasticity, and recovery in the central nervous system. Exp Brain Res 1999;128:134–138.
40. Seitz RJ, Freund HJ. Plasticity of the human motor cortex. Adv Neurol 1997;73:321–333.
41. Bucy PC. Stereotactic surgery: philosophical considerations. Clin Neurosurg 1964;11:138–149.
42. Jane JA, Yashon D, Becker DP, et al. The effect of destruction of the corticospinal tract in the human cerebral peduncle upon motor function and involuntary movements. Report of 11 cases. J Neurosurg 1968; 29:581–585.
43. Lawrence DG, Kuypers HG. The functional organization of the motor system in the monkey. I. The effects of bilateral pyramidal lesions. Brain 1968;91:1–14.
44. Lawrence DG, Kuypers HG. The functional organization of the motor system in the monkey. II. The effects of lesions of the descending brain-stem pathways. Brain 1968;91:15–36.
45. Warabi T, Inoue K, Noda H, et al. Recovery of voluntary movement in hemiplegic patients. Correlation with degenerative shrinkage of the cerebral peduncles in CT images. Brain 1990;113:177–189.
46. Fukui K, Iguchi I, Kito A. Extent of pontine pyramidal tract wallerian degeneration and outcome after supratentorial hemorrhagic stroke. Stroke 1994;25:1207–1210.
47. Traversa R, Cicinelli P, Pasqualetti P, et al. Follow-up of interhemispheric differences of motor evoked potentials from the "affected" and "unaffected" hemisphere in human stroke. Brain Res 1998;803:1–8.
48. Chollet F, Di Piero V, Wise RJS, et al. The functional anatomy of motor recovery after stroke in humans: a study with positron emission tomography. Ann Neurol 1991;29:63–71.
49. Weiller C, Chollet F, Friston KJ, et al. Functional reorganization of the brain in recovery from striatocapsular infarction in man. Ann Neurol 1992;31:463–472.
50. Weiller C, Ramsey SC, Wise RJ, et al. Individual patterns of functional reorganization in the human cerebral cortex after capsular infarction. Ann Neurol 1993;33:181–189.
51. Cramer SC, Nelles G, Benson RR, et al. A functional MRI study of subjects recovered from hemiparetic stroke. Stroke 1997;28:2518–2527.
52. CaoY, D'Olhaberriague L, Vikingstad EM, et al. Pilot study of functional MRI to assess cerebral activation of motor function after poststroke hemiparesis. Stroke 1998;29:112–122.
53. Pineiro R, Pendelbury S, Johansen-Berg H, et al. Functional MRI detects posterior shifts in primary sensorimotor cortex activation after stroke: evidence of local adaptive reorganization? Stroke 2001;32:1134–1139.
54. Marshall RS, Perera GM, Lazar RM, et al. Evolution of cortical activation during recovery from corticospinal tract infarction. Stroke 2000;31:656–661.
55. Nelles G, Spiekermann G, Jueptner M, et al. Evolution of functional reorganization in hemiplegic stroke: a serial positron emission tomographic activation study. Ann Neurol 1999;46:901–909.
56. Feydy A, Carlier R, Roby-Brami A, et al. Longitudinal study of motor recovery after stroke.Recruitment and focusing of brain activation. Stroke 2002;33:1610–1617.
57. Small SL, Hlustik P, Noll DC, et al. Cerebellar hemispheric activation ipsilateral to the paretic hand correlates with functional recovery after stroke. Brain 2002;125:1544–1557.
58. Rossini PM, Altamura C, Ferretti A, et al. Does cerebrovascular disease affect the coupling between neuronal activity and local haemodynamics? Brain 2003;21:1–12.
59. Seitz RJ, Hoflich P, Binkofski F, et al. Role of the premotor cortex in recovery from middle cerebral artery infarction. Arch Neurol 1998;55:1081–1088.
60. Byrnes ML, Thickbroom GW, Phillips BA, et al. Physiological studies of the corticomotor projection to the hand after subcortical stroke. Clin Neurophysiol 1999;110:487–498.
61. Rapisarda G, Bastings E, de Noordhout AM, et al. Can motor recovery in stroke patients be predicted by early transcranial magnetic stimulation? Stroke 1996; 27:2191–2196.
62. Heald A, Bates D, Cartlidge NEF, et al. Longitudinal study of central motor conduction time following stroke. 1. Natural

history of central motor conduction. Brain 1993;116:1355–1370.
63. Heald A, Bates D, Cartlidge NEF, et al. Longitudinal study of central motor conduction time following stroke. 2. Central motor conduction time measured within 72 h after a stroke as a predictor of functional outcome at 12 months. Brain 1993;116:1371–1385.
64. Kujirai T, Caramia MD, Rothwell JC, et al. Corticocortical inhibition in human motor cortex. J Physiol 1993;471:501–519.
65. Cicinelli P, Pasqualetti P, Zaccagnini M, et al. Interhemispheric asymmetries of motor cortex excitability in the postacute stroke stage: a paired-pulse transcranial magnetic stimulation study. Stroke 2003;34:2653–2658.
66. Andrews RJ. Transhemispheric diaschisis. Stroke 1991;22:943–949.
67. Conti F, Manzoni T. The neurotransmitters and postsynaptic actions of callosally projecting neurons. Behav Brain Res 1994;64:37–53.
68. Buchkremer-Ratzmann I, August M, Hagemann G, et al. Electrophysiological transcortical diaschisis after cortical photothrombosis in rat brain. Stroke 1996;27:1105–1111.
69. Catano A, Houa M, Caroyer JM, et al. Magnetic transcranial stimulation in non-hemorrhagic sylvian strokes: interest of facilitation for early functional prognosis. Electroencephalogr Clin Neurophysiol 1995;97:349–354.
70. Pennisi G, Rapisarda G, Bella R, et al. Absence of response to early transcranial magnetic stimulation in ischemic stroke patients. Prognostic value for hand motor recovery. Stroke 1999;30:2666–2670.
71. Benecke R, Meyer BU, Freund HJ. Reorganization of descending motor pathways in patients after hemispherectomy and severe hemispheric lesions demonstrated by magnetic brain stimulation. Exp Brain Res 1991;83:419–426.
72. Turton A, Wroe S, Trepte H, et al. Contralateral and ipsilateral EMG responses to transcranial magnetic stimulation during recovery of arm and hand function after stroke. Electroencephalogr Clin Neurophysiol 1996;41:316–328.
73. Shimizu T, Hosaki A, Hino T, et al. Motor cortical disinhibition in the unaffected hemisphere after unilateral cortical stroke. Brain 2002;125:1896–1907.
74. Muellbacher W, Richards C, Ziemann U, et al. Improving hand function in chronic stroke. Arch Neurol 2002;59:1278–1282.
75. Werhahn KJ, Conforto AB, Kadom N, et al. Contribution of the ipsilateral motor cortex to recovery after chronic stroke. Ann Neurol 2003;54:464–472.
76. Stepniewska I, Preuss TM, Kaas JH. Architectonics, somatotopic organization, and ipsilateral cortical connections of the primary motor area (M1) of owl monkeys. J Comp Neurol 1993;330:238–271.
77. Rossini PM, Barker AT, Berardelli A, et al. Non-invasive electrical and magnetic stimulation of the brain, spinal cord and roots: basic principles and procedures for routine clinical application. Report of an IFCN committee. Electroencephalogr Clin Neurophysiol 1994;91:79–92.
78. Rossini PM, Berardelli A, Deuschl G, et al. Applications of magnetic cortical stimulation.Clin Neurophysiol Suppl 1999;52:171–185.
79. Rossini PM, Pauri F. Neuromagnetic integrated methods tracking human brain mechanisms of sensorimotor areas 'plastic' reorganisation. Brain Res Rev 2000;33:131–154.
80. Traversa R, Cicinelli P, Oliveri M, et al. Neurophysiological follow-up of motor output in stroke patients. Clin Neurophysiol 2000;111:1695–1703.
81. Cicinelli P, Traversa R, Bassi A, et al. Interhemispheric differences of hand muscle representation in human motor cortex. Muscle Nerve 1997;20:535–542.
82. Palmer E, Ashby P, Hajek VE Ipsilateral fast corticospinal pathways do not account for recovery in stroke. Ann Neurol 1992;32:519–525.
83. Hendricks HT, Hageman G, van Limbeek J. Prediction of recovery from upper extremity paralysis after stroke by measuring evoked potentials. Scand J Rehabil Med 1997;29:155–159.
84. Netz J, Lammers T, Hömberg V. Reorganization of motor output in the non-affected hemisphere after stroke. Brain 1997;120:1579–1586.
85. Caramia MD, Iani C, Bernardi G. Cerebral plasticity after stroke as revealed by ipsilateral responses to magnetic stimulation. Neuroreport 1996;7:1756–1760.
86. Caramia MD, Palmieri MG, Giacomini P, et al. Ipsilateral activation of the unaffected motor cortex in patients with hemiparetic stroke. Clin Neurophysiol 2000;111:1990–1996.
87. Trompetto C, Assini A, Buccolieri A, et al. Motor recovery following stroke: a transcranial magnetic stimulation study. Clin Neurophysiol 2000;111:1860–1867.
88. Fries W, Daneck A, Witt TN. Motor responses after transcranial electrical stimulation of cerebral hemispheres with degenerated pyramidal tract. Ann Neurol 1991; 29:646–649.
89. Alagona G, Delvaux V, Gérard P, et al. Ipsilateral motor responses to focal transcranial magnetic stimulation in healthy subjects and

acute-stroke patients. Stroke 2001;32:1304–1309.
90. Liepert J, Storch P, Fritsch A, et al. Motor cortex disinhibition in acute stroke. Clin Neurophysiol 2000;111:671–676.
91. Buchkremer-Ratzmann I, Witte OW. Extended brain disinhibition following small photothrombotic lesions in rat frontal cortex. Neuroreport 1997;8:519–522.
92. Reinecke S, Lutzenburg M, Hageman G, et al. Electrophysiological transcortical diaschisis after middle cerebral artey occlusion (MCAO) in rats. Neurosci Lett 1999;261:85–88.
93. Que M, Schiene K, Witte OW, et al. Widespread up-regulation of N-methyl-D-aspartate receptors after focal photothrombotic lesion in rat brain. Neurosci Lett 1999;273:77–80.
94. Que M, Witte OW, Neumann-Haefelin T, et al. Changes in GABA(A) and GABA(B) receptor binding following cortical photothrombosis: a quantitative receptor autoradiographic study. Neuroscience 1999;93:1233–1240.
95. Chen R, Corwell B, Yaseen Z, et al. Mechanisms of cortical reorganization in lower-limb amputees. J Neurosci 1998;18:3443–3450.
96. Zieman U, Rothwell JC, Ridding MC. Interaction between intracortical inhibition and facilitation in human motor cortex. J Physiol 1996;496:873–881.
97. Boroojerdi B, Diefenbach K, Ferbert A. Transcallosal inhibition in cortical and subcortical cerebral vascular lesions. J Neurol Sci.1996;144:160–170.
98. Ferbert A, Vielhaber S, Meincke U, et al. Transcranial magnetic stimulation in pontine infarction: correlation to degree of paresis. J Neurol Neurosurg Psychiatry 1992;55:294–299.
99. Leocani L, Cohen LG, Wassermann EM, et al. Human corticospinal excitability evaluated with transcranial magnetic stimulation during different reaction time paradigms. Brain 2000;123:1161–1173.
100. Liepert J, Uhde I, Gräf S, et al. Motor cortex plasticity during forced use therapy in stroke patients. J Neurol 2001;248:315–321.
101. Liepert J, Miltner WH, Bauder H, et al. Motor cortex plasticity during constraint-induced movement therapy in stroke patients. Neurosci Lett 1998;250:5–8.
102. Weiller C. Imaging recovery from stroke. Exp Brain Res 1998;123:13–17.
103. Rossini PM, Liepert J. Lesions of cortex and post-stroke reorganisation. In Boniface S, Ziemann U (eds): Plasticity in the Human Nervous System. New York, Cambridge University Press, 2003.
104. Schiene K, Bruehl C, Zilles K, et al. Neuronal hyperexcitability and reduction of $GABA_A$-receptor expression in the surround of cerebral photothrombosis. J Cereb Blood Flow Metab 1996;16:906–914.
105. Goldstein LB. Effects of amphetamines and small related molecules on recovery after stroke in animals and man. Neuropharmacology 2000;39:852–859.
106. Scheidtmann K, Fries W, Muller F, et al. Effect of levodopa in combination with physiotherapy on functional motor recovery after stroke: a prospective, randomised, double-blind study. Lancet 2001;358:787–90.
107. Dam M, Tonin P, De Boni A, et al. Effects of fluoxetine and maprotiline on functional recovery in poststroke hemiplegic patients undergoing rehabilitation therapy. Stroke 1996;27:1211–1214.
108. Pariente J, Loubinoux I, Carel C, et al. Fluoxetine modulates motor performance and cerebral activation of patients recovering from stroke. Ann Neurol 2001;50:718–729.
109. Taub E, Miller NE, Novack TA, et al. Technique to improve chronic motor deficit after stroke. Arch Phys Med Rehabil 1993;347–354.
110. Kopp B, Kunkel A, Muhnickel W, et al. Plasticity in the motor system related to therapy-induced improvement of movement after stroke. Neuroreport 1999;10:807–810.
111. Levy CE, Nichols DS, Schmalbrock PM, et al. Functional MRI evidence of cortical reorganization in upper-limb stroke hemiplegia treated with constraint-induced movement therapy. Am J Phys Med Rehabil 2001;80:4–12.
112. Jacobs KM, Donoghue JP. Reshaping the cortical motor map by unmasking latent intracortical connections. Science 1991;251:944–947.
113. Hummelsheim H, Hauptmann B. Transcranial magnetic stimulation and motor rehabilitation. Electroencephalogr Clin Neurophysiol Suppl 1999;51:221–232.
114. Butefisch C, Hummelsheim H, Denzler P, et al. Repetitive training of isolated movements improves the outcome of motor rehabilitation of the centrally paretic hand. J Neurol Sci 1995;130:59–68.
115. Fries W, Danek A, Schedtmann K, et al. Motor recovery following capsular stroke: role of descending pathways from multiple motor areas. Brain 1993;116:369–382.
116. Rossini PM, Rossi S, Tecchio F, et al. Focal brain stimulation in healthy humans: motor maps changes following partial hand sensory deprivation. Neurosci Lett 1996;214:191–195.
117. Rossini PM, Tecchio F, Sabato A, et al. The role of cutaneous inputs during magnetic

transcranial stimulation. Muscle Nerve 1996;19:1302–1309.
118. Bavelier D, Neville HJ. Cross-modal plasticity: where and how? Nat Rev Neurosci 2002;3:443–452.
119. Rossini PM, Martino G, Narici L, et al. Short-term brain 'plasticity' in humans: transient finger representation changes in sensory cortex somatotopy following ischemic anesthesia. Brain Res 1994;642:169–177.
120. Rossini PM, Narici L, Martino G, et al. Analysis of interhemispheric asymmetries of somatosensory evoked magnetic fields to right and left median nerve stimulation. Electroencephalogr Clin Neurophysiol 1994;91:476–482.
121. Forss N, Hietanen M, Salonen O, et al. Modified activation of somatosensory cortical network in patients with right-hemisphere stroke. Brain 1999;122:1889–1899.

17 Evaluation of Epilepsy and Anticonvulsants

Ulf Ziemann

The role of transcranial magnetic stimulation (TMS) in epileptology is discussed in this chapter, and several important questions are answered. How useful is TMS as a diagnostic tool in testing altered cortical excitability in patients with epilepsy, and how useful is this to improve our understanding of pathophysiological mechanisms in epilepsy? How useful is TMS to test the modes of action of antiepileptic drugs (AEDs)? What is the current standing of TMS as a potential therapeutic tool in epilepsy? Throughout this paper, a clear distinction is made between single- or paired-pulse TMS and repetitive TMS (rTMS), because these two techniques have principally different scopes and applications.

Measures of Motor Excitability

Measures of motor excitability can be obtained with single-pulse and paired-pulse TMS. A basic understanding of these measures is important to appreciate the findings of altered motor excitability in patients with epilepsy and of the effects of AEDs on motor excitability as evaluated with TMS.

Motor threshold (MT) refers to the minimum stimulus intensity needed to elicit a small motor response in a target muscle.[1] MT is lower in a contracting muscle than in a resting muscle.[2] These conditions are referred to as the active (AMT) and resting motor threshold (RMT), respectively. The physiology of RMT is determined by several processes, including the excitability of corticocortical axons and the excitability of synaptic contacts between these axons and corticospinal neurons and between corticospinal neurons and their target motoneurons in the spinal cord. In contrast, AMT is mainly determined by the excitability of corticocortical axons and therefore mainly reflects membrane-related excitability.[3] RMT and AMT show a large intersubject variability but low within-subject variability and low within-subject interhemispheric difference.[4]

Motor evoked potential (MEP) size reflects the excitability of the whole corticospinal system and is related to the portion of the spinal motoneuron pool that is activated by TMS.[1] MEP size increases with stimulus intensity in a sigmoid manner.[5] The part of the MEP intensity curve close to MT is determined by the excitability of low-threshold corticospinal neurons, and the high-intensity part of the MEP intensity curve reflects the excitability of high-threshold neurons.[6] MEP size increases with voluntary activation of the target muscle.[2,7] MEP size may be modulated by inputs to motor cortex from the periphery or other parts of the brain.

Electrical stimulation of an upper limb mixed nerve or muscle stretch produces MEP facilitation in the stimulated or stretched muscle at inter-stimulus intervals of about 20 to 30 ms, usually followed by MEP inhibition at longer intervals.[8–11] The short-latency MEP facilitation after mixed nerve stimulation may be preceded by a short-lasting MEP inhibition at inter-stimulus intervals of 19 to 21 ms.[12]

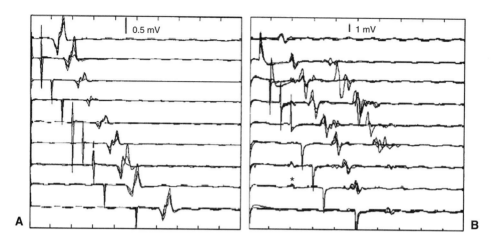

Figure 17-1 Motor evoked potential (MEP) recordings from the right thenar muscles in a healthy subject (**A**) and in a patient with progressive myoclonus epilepsy (**B**). Top traces show the unconditioned MEP; all other traces show MEP conditioned by electrical digital nerve stimulation of the right index finger. The inter-stimulus interval increases from top to bottom (10, 20, 30, 40, 50, 60, 70, and 100 ms), and the conditioning digital nerve stimulus is given at the beginning of each trace (time base = 20 ms). MEPs are inhibited at intervals of 20 to 40 ms in the healthy subject, whereas MEP facilitation was obtained in the progressive myoclonus epilepsy (PME) patient. Notice the occurrence of a C response *(asterisk)* and a second MEP in the PME patient. The C response is thought to reflect an exaggerated long-loop transcortical reflex. (From Manganotti P, Tamburin S, Zanette G, et al. Hyperexcitable cortical responses in progressive myoclonic epilepsy: A TMS study. Neurology 2001;57:1793.)

Nonpainful conditioning stimulation of digital cutaneous nerves produces short-latency MEP inhibition in muscles adjacent to the stimulated finger[10,13-16] (Fig. 17-1A). Conditioning stimulation over the cerebellum tests cerebellodentatothalamomotor cortical pathways and results in MEP inhibition over the contralateral motor cortex at inter-stimulus intervals of 5 to 10 ms.[17-19] Conditioning TMS over the opposite motor cortex tests interhemispheric connections between the motor cortexes and results in MEP inhibition over the test motor cortex at inter-stimulus intervals of 8 to 20 ms.[20-22]

The cortical silent period (CSP) refers to a period of silence in the electromyographic pattern of a voluntarily contracted target muscle (Fig. 17-2). Usually, the CSP is preceded by an MEP.[23,24] However, the CSP may have a slightly lower threshold than the MEP, and CSPs without a preceding MEP have been reported.[25] The CSP duration increases with stimulus intensity.[23,24] In hand muscles and at high stimulus intensity, the CSP may easily exceed 200 ms before voluntary muscle activity returns.[23,24] The early part of the CSP is contaminated by spinal inhibitory mechanisms, and the late part most likely reflects inhibition specifically at the level of the motor cortex.[26-28] It is thought that this late part of the CSP is determined by long-lasting cortical inhibition mediated through the γ-aminobutyric acid (GABA) type B receptor.[29,30]

Paired-pulse TMS measures refer to techniques that use two stimuli delivered through the same coil. If the first (S1) and second stimulus (S2) are above MT and of the same intensity, long-interval intracortical facilitation (LICF) occurs at inter-stimulus intervals of approximately 10 to 40 ms,[31] and long-interval intracortical inhibition (LICI) is obtained at longer intervals of approximately 50 to 200 ms.[31-33] Sometimes, LICI is followed by weak LICF at even longer inter-stimulus intervals of more than 200 ms.[32] If S1 is clearly below MT and S2 above MT, short-interval intracortical inhibition (SICI) is obtained at inter-stimulus intervals of 1 to 5 ms and intracortical facilitation (ICF) at intervals of

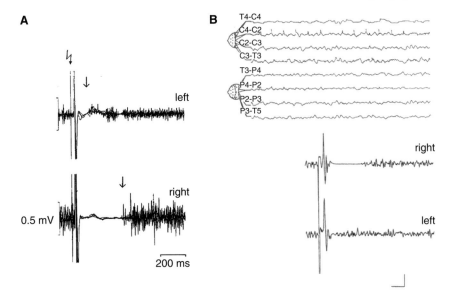

Figure 17–2 **A:** Motor evoked potential (MEP) and cortical silent period (CSP) in the first dorsal interosseus muscle of a patient with focal epilepsy due to a lesion of the left supplementary motor area. Arrows indicate the time of transcranial magnetic stimulation (TMS) and the end of the CSP. The CSP from the epileptic hemisphere is markedly longer *(lower tracing)* than that from the normal hemisphere. **B:** Electroencephalography *(upper panel)* and CSP recordings from the first dorsal interosseus muscle *(lower panel)* of a patient with focal epilepsy with left-sided myoclonus. Notice the continuous spiking over the C4 lead in the EEG tracing and a markedly shortened CSP *(bottom tracing)* elicited from the epileptic hemisphere. (**A** from Classen J, Witte OW, Schlaug G, et al. Epileptic seizures triggered directly by focal transcranial magnetic stimulation. Electroencephalogr Clin Neurophysiol 1995;94:19; **B** from Inghilleri M, Mattia D, Berardelli A, et al. Asymmetry of cortical excitability revealed by transcranial stimulation in a patient with focal motor epilepsy and cortical myoclonus. Electroencephalogr Clin Neurophysiol 1998;109:70.)

7 to 20 ms[34–36] (Fig. 17–3). If S1 and S2 are approximately equal to MT, or if S1 is above and S2 below MT, MEP facilitation occurs at discrete, very short inter-stimulus intervals of 1.1 to 1.5 ms, 2.3 to 2.9 ms, and 4.1 to 4.4 ms.[37–40] This facilitation was called short-interval intracortical facilitation, or I-wave facilitation, because it follows the periodicity of I waves. It is thought that all paired-pulse measures reflect mainly synaptic excitability of various inhibitory and excitatory neuronal circuits at the level of the motor cortex.[41,42] This synaptic excitability is controlled mainly by neurotransmission through the GABA and *N*-methyl-D-aspartate (NMDA) receptors but is also modulated by many other substances, such as the monoamines. SICI and LICI underlie separate mechanisms and may reflect inhibition mediated through the $GABA_A$ and $GABA_B$ receptors, respectively.[43]

Single-Pulse and Paired-Pulse Transcranial Magnetic Stimulation Studies of Motor Excitability in Patients with Epilepsy

Resting and Active Motor Thresholds

RMT was reported to be elevated in patients with epilepsy in a number of studies[44–57] (Table 17–1). However, many patients were under chronic AED treatment at the time of TMS testing. This suggests that the threshold increasing effect of many AEDs contributed to the finding of increased RMT in epileptic patients. This view was directly supported by the demonstration that untreated groups of patients with idiopathic generalized epilepsy (IGE)[46] or benign epilepsy with centrotemporal spikes (BECT)[56] had reduced or normal RMT values compared with healthy controls. However, RMT in the patient groups

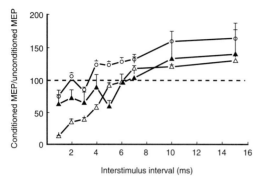

Figure 17–3 The pattern of short-interval intracortical inhibition (SICI) and intracortical facilitation (ICF) in 12 healthy controls (*open triangles*), six patients with generalized cortical myoclonus (i.e., spreaders, *open circles*), and six patients with focal cortical myoclonus (i.e., nonspreaders, *solid triangles*). The effect of the conditioning stimulus is given as ratio of conditioned motor evoked potential (MEP) over unconditioned MEP size on the y axis. MEPs were recorded from the relaxed first dorsal interosseus muscle. Error bars show the SEM. Notice that the SICI was significantly reduced in spreaders compared with controls and nonspreaders. (From Brown P, Ridding MC, Werhahn KJ, et al. Abnormalities of the balance between inhibition and excitation in the motor cortex of patients with cortical myoclonus. Brain 1996;119:309.)

increased significantly above normal level when remeasured after the commencement of treatment with valproic acid.[46,56] RMT correlated with the plasma level of valproic acid.[46,56] In a study on temporal lobe epilepsy (TLE) patients, RMT significantly increased with the number of AEDs taken by the patients.[44] In one subgroup of this study, RMT dropped significantly after tapering AED treatment, even though normalization was not reached.[44]

At variance with one previous study,[46] RMT in untreated IGE patients was found to be increased compared with healthy controls.[48] The reason for this disparity is unclear, because methods and patients were similar. Patients with autosomal dominant cortical myoclonus epilepsy (ADCME) displayed a significantly reduced RMT.[58] Patients with cortical myoclonus, treated or untreated, had an increased RMT.[50] This was particularly true for the subgroup of nonspreaders with no or relatively few seizures, as opposed to the subgroup of patients with generalized cortical myoclonus (i.e., spreaders). The investigators interpreted this association of elevated RMT with relatively fewer seizures and less spread

Table 17–1 TRANSCRANIAL MAGNETIC STIMULATION OF MOTOR EXCITABILITY IN EPILEPSY

Epilepsy/ Number of Patients	Medication	RMT	MEP	cMEP	CSP	SICI	ICF	INI	LICI	LICF	Reference
IGE/11	Untreated	↓									45
IGE/34	VPA	↑[1]									
IGE/20	Untreated	↓									46
IGE/36	VPA	↑[1]									
IGE/8	Untreated	↑									48
IGE/12	VPA/PB	↑									
IGE/14	?		↓[2]								
JME/7	VPA	↑[1]				↓					49
IGE/7	Untreated	↔							↓	↑[3]	72
IGEvc/10	VPA/PB	↔[4]									57
IGE/13	VPA/PB	↔									
JME/15	VPA/PB/ Untreated	↔[5]	↔		↔	↓	↔		↔		61
IGE/21	Untreated	↔[6]			↑						68
PME/3	VPA/CLZ	↑[1]	↑								47
IGE/8	VPA	↑			↔						
PME/3	VPA	↑[1]	↔	↑							51
PME/12	VPA/CLZ	↔			↔				↓	↑[7]	69
PME/4	VPA/PHT/CBZ/ CLZ/Untreated	↑[1]	↔	↑	↓						60

(*continued*)

Table 17-1 CONTINUED

Epilepsy/Number of Patients	Medication	RMT	MEP	cMEP	CSP	SICI	ICF	INI	LICI	LICF	Reference
ADCME/8	Various AEDs	↓			↓						58
Generalized CM/8	CLZ, others	↔				↓	↔	↓			50
Focal CM/10	CLZ, others	↑			↑	↔	↔				
CME/5	?						↓[8]				63
BECT/8	VPA	↑[1]									56
BECT/5	Untreated	↔									
BECT/7	Untreated				↑						62
First-ever GS/18	Untreated	↑[9]	↔[9]		↔[9]	↔[9]	↓[9]				54
TLE/53	CBZ/others	↑[1]	↔								44
FE(TLE)/7	Untreated	↔[10]			↔[10]	↓			↓[11]		59
FE/6	Untreated	↔[10]			↔[10]	↓	↓				
FE (motor cortex)/8	CBZ/others	↑[1,10]			↑[10]						124
CFE/10	CBZ/others	↑[1,10]			↔						
FE (motor cortex)/1	PB	↔[10]			↑[13]						67
	CBZ/PHT	↔[10]			↔[10]						
CFE/1											
FE (SMA)[12]/1	CBZ				↑[13]						65
FE (motor cortex)/1	?	↑[13]			↓[13]	↓[13]	↓[13]				55
FE (motor cortex)/7	CBZ/PHT/VPA					↓[13]	↑[13]				71
FE (motor cortex)/1	?	↔		↑[13,14]	↓[13]	↓[13,14]	↔				125
EPC/1	PHT					↓[13]		↔[10]			126
CFE/16	VPA/CBZ/others	↑[1,10]	↔[10]		↓[13]						53
CFE/18	VPA/CBZ/others				↔[10]	↓[10,15]	↑[10,15]				52

[1]Probably caused by AED treatment.
[2]MEP decrease when TMS is delivered during the slow wave of the EEG spike-slow wave complex.
[3]LICF increased at inter-stimulus intervals of 200 to 300 ms.
[4]Normal MT but increased inter-hemispheric MT difference. No lateralizing value.
[5]Mean MT not different from controls. However, untreated patients had a lower MT than treated patients.
[6]Measure includes RMT and AMT.
[7]LICF increased at inter-stimulus intervals of 200 to 300 ms.
[8]In addition to absent INI, inter-hemispheric facilitation occurred at inter-stimulus intervals of 4 to 6 ms.
[9]Measured within 48 hours after first-ever GS.
[10]In addition, no difference between epileptic and normal hemisphere.
[11]Trend for a stronger decrease of ICF in the epileptic hemisphere.
[12]Post-ictal Todd's paresis.
[13]Epileptic hemisphere only.
[14]Normalization after multiple subpial transaction.
[15]Associated with high seizure frequency and high incidence of interictal generalized epileptic EEG discharge.

AEDs, antiepileptic drugs; AMT, active motor threshold; CLZ, clonazepam; cMEP, MEP conditioned by peripheral nerve stimulation; CSP, cortical silent period; EEG, electroencephalographic; GS, generalized seizure; ICF, intracortical facilitation; INI, inter-hemispheric inhibition; LICF, long-interval intracortical facilitation; LICI, long-interval intracortical inhibition; MEP, motor evoked potential; MT, motor threshold; RMT, resting motor threshold; SICI, short-interval intracortical inhibition; SMA, supplementary motor area; TMS, transcranial magnetic stimulation; ↑ increased; ↓, decreased; ↔, normal.

of myoclonic activity as an adaptive process to compensate for existing deficiencies in motor cortex inhibition. Similarly, patients showed an elevated RMT and AMT within 48 hours after a first-ever generalized seizure.[54] When tested again 2 to 4 weeks later, these changes were no longer present. This short-lasting elevation of MT may reflect cortical dysfunction after the seizure but more likely is a protective mechanism against spread or recurrence of seizures.[54]

Whether the interhemispheric difference in MT may be used for identification of the epileptic hemisphere in patients with focal epilepsy (FE) is uncertain. A single patient with focal motor seizures showed clearly higher AMT and RMT of the epileptic motor cortex.[55] However, a later study of seven patients with TLE and six patients with extratemporal FE did not reveal significant differences of MT between the epileptic and normal hemispheres.[59] An increased interhemispheric difference in RMT was observed in IGE patients who also displayed versive or circling seizures.[57] However, the direction of this difference was not indicative for the ipsiversive or contraversive hemisphere and therefore of no lateralizing value.

In summary, the usefulness of MT measurements in epilepsy is not clear. Many variables, such as the presence or absence of AED treatment, type of epilepsy, and time from the last seizure, seem to affect MT.

Motor Evoked Potential Size

MEP size or the MEP/maximum M wave ratio were reported to be normal in patients with TLE,[44] progressive myoclonus epilepsy (PME),[51,60] juvenile myoclonic epilepsy (JME)[61] and in patients after a first-ever generalized seizure[54] (see Table 17–1). In patients with IGE, a significant decrease in MEP size was revealed, but only if TMS was triggered during the slow wave of the spike-and-wave electroencephalographic (EEG) complex.[48] This is consistent with the classic electrophysiological interpretation of the slow wave as a state of hyperpolarization of cortical pyramidal cells. The lateralizing value of MEP size in patients with FE appears to be low because no hemispheric difference was found in most of these patients[53] (see Table 17–1).

Motor Evoked Potential Size Conditioned by Peripheral Nerve Stimulation

Conditioning electrical stimulation of a peripheral mixed nerve resulted in normal MEP inhibition at inter-stimulus intervals of 20 to 60 ms in patients with IGE.[47] In contrast, patients with PME showed MEP facilitation at these intervals.[47,51] A similar MEP facilitation was observed in PME patients if the MEP was conditioned by electrical digital nerve stimulation.[60] Although healthy controls showed a significant MEP inhibition at inter-stimulus intervals of 25 to 50 ms, PME patients displayed MEP facilitation at these intervals (see Fig. 17–1). A subpopulation of patients with BECT who present with evoked rolandic paroxysmal EEG activity on conditioning stimulation of a digital nerve of the contralateral hand also show MEP facilitation if the TMS pulse is given during the peak or the ascending phase of the evoked cortical spike.[62]

These studies show that conditioned MEP size is a very interesting technique that may reveal abnormal sensorimotor integration and hyperexcitability time-locked to the afferent input in some rare forms of epilepsy.

Motor Evoked Potential Size Conditioned by Stimulation of the Contralateral Motor Cortex

Interhemispheric inhibition was normal in a subgroup with focal cortical myoclonus (i.e., nonspreaders), but it was abolished in a subgroup with generalized myoclonus (i.e., spreaders).[50,63] It is likely that this deficiency in interhemispheric inhibition explains the interhemispheric spread of myoclonic activity.[64] However, an association of this decrease in interhemispheric inhibition with the presence of epilepsy was not found.[50] The contribution of interhemispheric inhibition to the epileptic process was regarded as not important in these patients. One study even showed an interhemispheric facilitation at inter-stimulus intervals of 4 to 6 ms in the patients with benign cortical myoclonus.[63] This was interpreted as indicating a breakdown of cortical surround inhibition in the conditioned motor cortex. Under normal conditions, this surround inhibition prevents the expression of interhemispheric facilitation.

Cortical Silent Period Duration

In a single patient with focal motor seizures resulting from a lesion in the left supplementary motor area, the CSP was markedly longer (about 200 ms) when TMS was applied to the motor cortex of the epileptic hemisphere[65] (see Fig. 17–2A). The seizures were followed consistently by a Todd's paresis in the arm contralateral to the epileptic hemisphere. It is therefore possible that the dramatic lengthening of the CSP indicated a post-seizure increase in cortical inhibition that might have contributed to the paresis. In contrast, another patient with focal motor epilepsy showed a dramatic shortening of the CSP in a hand muscle contralateral to the epileptic motor cortex at the time when this motor cortex showed high EEG spiking activity[55] (see Fig. 17–2B). This was interpreted as indicating a breakdown of cortical inhibitory mechanisms in that motor cortex, leading to epileptic EEG activity. In another study of eight patients with focal motor epilepsy, the CSP was lengthened bilaterally, but more prominently in the normal hemisphere.[66] The investigators proposed that the CSP lengthening indicated increased interictal cortical inhibition to ward off epileptic activity. This view was supported by findings in one patient with a rolandic meningioma whose prolonged CSP duration normalized after surgery.[67] This normalization occurred no earlier than 3 months after surgery, at a time when the risk for seizure relapse usually subsides. The investigators suggested that the maintenance of a long CSP for 3 months after surgery helped to prevent the reoccurrence of seizures. A lengthened CSP also was found in untreated patients with IGE.[68] A normal CSP was reported in patients after a first-ever generalized seizure,[54] patients with PME,[69] JME,[61] and cryptogenic FE[52] (see Table 17–1). A shortened CSP was observed in ADCME[58] and in a different group of patients with FE.[53] In the latter study, CSP was determined at several different stimulus intensities. Whereas the CSP increased approximately linearly with intensity in the normal hemisphere, the CSP intensity curve reached an early plateau in the epileptic hemisphere.[53] The CSP intensity curve may have a lateralizing value in FE with shorter CSP in the epileptic hemisphere.

In summary, the CSP findings in patients with epilepsy are variable, but it appears that the time from the last seizure, the site of a cortical lesion, the interictal EEG spiking activity, and the type of epilepsy are determinants of CSP duration.

Short-Interval Intracortical Inhibition

A reduced SICI was found in most studies (see Table 17–1), including the patients with cortical myoclonus[50] (see Fig. 17–3), JME,[49,61] PME,[60,69] TLE and extratemporal FE,[59] focal motor epilepsy with cortical myoclonus,[55] epilepsia partialis continua,[70] and "not further classified" or cryptogenic FE.[52,71] In cortical myoclonus, the deficiency in SICI was most pronounced in the subgroup with generalized myoclonus (i.e., spreaders) compared with the subgroup with focal myoclonus (i.e., non-spreaders)[50] (see Fig. 17–3). However, in that study, a difference in SICI was not found between cortical myoclonus patients with epilepsy compared with those without epilepsy. The investigators concluded that the amount of reduction of SICI did not play a major part in the epileptic process of this patient group.

In patients with FE, the reduction in SICI was restricted to the motor cortex of the epileptic hemisphere.[55,71] This was thought to be important for the epileptogenesis in those patients. However, in a different study of patients with cryptogenic FE, the reduction in SICI was not associated with the epileptic hemisphere, but there was a clear relationship with seizure frequency and the frequency of interictal spikes in the EEG pattern.[52] In patients with TLE, the reduction in SICI was even more pronounced in the normal hemisphere, whereas the alteration of SICI in the epileptic hemisphere was not significantly different from that of healthy controls.[59]

In summary, it appears that SICI is usually reduced in even very different epileptic syndromes, and SICI therefore can be considered as a sensitive but unspecific measure of altered cortical inhibition (see Table 17–1). Some studies suggest a relation with the epileptic activity, but others do not. The reasons for these inconsistencies are unknown. The lateralizing value of SICI is limited and may vary with the type of FE. In particular, the distinction between motor and nonmotor FE may be of relevance.

Intracortical Facilitation

ICF was normal in patients with cortical myoclonus[50] (see Fig. 17–3) and patients with JME[61] (see Table 17–1). An increased ICF was found in patients with focal epilepsy in the motor cortex of the epileptic hemisphere[71] and in patients with cryptogenic FE.[52] In the latter group, there was no association of increased ICF with the epileptic hemisphere, but there was an association with seizure frequency and the number of interictal EEG spikes. These findings may support the notion that an increased ICF is related to the epileptic process in localization-related epilepsies. However, in patients with TLE, the ICF was significantly reduced in the epileptic and normal hemispheres.[59] This indicates that an epileptic focus in the temporal lobe produces remote effects on excitability in the motor cortex and that this occurs in a different way compared with the effects of a focus located within the motor cortex. ICF was significantly reduced within 48 hours after a first-ever generalized seizure.[54] These changes persisted for 2 to 4 weeks, although to a lesser extent. The investigators speculated that this finding might indicate a protective mechanism to prevent from further seizures.

Long-Interval Intracortical Inhibition and Long-Interval Intracortical Facilitation

LICI was shorter than in healthy controls and followed by a period of increased LICF at inter-stimulus intervals of 200 to 300 ms in untreated patients with IGE.[72] These findings suggest a deficiency of cortical inhibition and increased facilitatory mechanisms. The investigators observed that the LICF at intervals of 200 to 300 ms correlated with the individual mean inter-discharge interval and therefore may explain the generalized 3- to 5-Hz spike-and-wave discharge characteristic of IGE.[72] LICI (inter-stimulus intervals of 100 to 150 ms) was abolished in patients with PME.[69] These patients displayed marked LICF at an inter-stimulus interval of 50 ms. Some of these patients showed prominent 20-Hz rhythmic EEG activity over the central leads that was coupled with rhythmic muscle jerks in the upper limbs. The investigators concluded that the strong LICF at an interval of 50 ms reflected the exaggerated 20-Hz oscillatory activity in the PME motor cortex.[69] In patients with JME, LICI was found to be normal.[61] The investigators proposed that this negative finding in the presence of a clear reduction in SICI indicates a differential involvement of $GABA_A$ (affected) and $GABA_B$ (nonaffected) neurotransmission in JME.[61]

In summary, LICF may offer the interesting possibility to detect exaggerated oscillatory activity in motor cortex. The frequency of these rhythms differs between epilepsy syndromes and this may be used for diagnostic purposes. Detection of a differential involvement of $GABA_A$ and $GABA_B$ receptors may be possible, if SICI and LICI are studied in conjunction.

■ Safety of Single-Pulse and Paired-Pulse Transcranial Magnetic Stimulation

Accidental Induction of Seizures

When the generally accepted exclusion criteria (e.g., cardiac pacemakers, intracranial electronic devices, intracranial metal objects, skull defects) are respected, single-pulse TMS is regarded as a safe technique in neurologically normal subjects.[73-77] On few occasions, the accidental induction of seizures has been reported in patients with disorders of the brain,[78-80] although a causal relationship between single-pulse TMS and the occurrence of seizures remained uncertain because, with the exception of one stroke case,[78] the seizures occurred with some delay after the application of TMS.

Safety of Single-Pulse and Paired-Pulse Transcranial Magnetic Stimulation in Patients with Epilepsy: Activation of the Epileptic Focus

Two of 53 patients with an average of three to five spontaneous focal seizures per day had a typical seizure during single-pulse TMS testing.[44] In a different study, the same group reported selective activation of the previously identified EEG focus in 12 of 13 patients with medically intractable FE.[81] One of these patients developed a typical seizure immediately after TMS, and in another patient, a transition of a nonactive theta focus to a self-sustained epileptic focus occurred. In another study, six clinical seizures and two auras were reported in 140 tested patients.[82] Five ictal events occurred immediately after TMS, and the others occurred within 5 minutes after TMS testing. In a study on verbal working memory in 20 patients with medically intractable FE, three auras and two clinical seizures occurred during TMS.[83] However, in only one of these complex partial seizures, a focal build up of epileptic after-discharges preceded the seizure. In that case, it was assumed that the seizure was triggered by TMS.[83] Many of the ictal events of the previously described studies might have been coincident with the TMS testing but not caused by TMS.

Only one case has been reported of a patient in whom seizures were triggered reproducibly and promptly by focal single-pulse TMS.[65] This patient had motor FE that originated from the left supplementary motor and was frequently generalized. The seizures triggered by TMS were identical to the spontaneous ones. TMS was capable of triggering seizures only when an angular figure-of-eight coil was used over the interhemispheric sulcus, a technique considered particularly effective in activating the structures of the mesial wall.

Other investigators largely failed to provoke seizures or EEG changes during single-pulse TMS in patients with epilepsy. In one study of 58 patients with FE or generalized epilepsy, no ictal events or EEG changes were found during TMS.[84] A long-term follow-up study disclosed no deterioration of seizure frequency. Another group tried to replicate the findings by Hufnagel and colleagues[81] under identical experimental conditions in 10 patients with drug-resistant FE.[85] An activation of the epileptic focus during TMS was seen in only three patients, and hyperventilation was more effective (six patients). In two patients, single-pulse TMS even reduced the interictal EEG spike frequency.[85] Single, paired, or quadruple focal TMS was applied to the motor cortex of the epileptic hemisphere and as close as possible to the suspected epileptogenic zone in 21 patients with intractable FE during invasive presurgical monitoring.[86] A seizure was triggered in no case, and the EEG interictal discharge remained unaffected.

In conclusion, single-pulse TMS appears to be relatively safe in patients with epilepsy. A conclusive demonstration of TMS triggering seizures has been made on only a very few occasions,[65] but application of TMS in patients with epilepsy should be performed only if justified by diagnostic purposes, and the patients must be informed about the potential risk of seizure induction.

Single-Pulse and Paired-Pulse Transcranial Magnetic Stimulation Studies of Acute Antiepileptic Drugs Effects on Motor Excitability

All TMS studies discussed in this section were performed on healthy subjects. Therefore, none of the observed AED effects can be ascribed to brain pathology. However, the observed AED effects might have affected the TMS measures in patients with epilepsy when treated with AED (see Table 17–1). In all studies, a similar protocol was used. Motor excitability was measured immediately before the administration of a single AED dose (baseline). Thereafter, repeat measurements were performed at delays adjusted to the pharmacokinetics of the individual AED. A comprehensive summary of the AED effects is given in Table 17–2. The basic idea of these experiments is that AED with different modes of action may produce different patterns of effects on the various TMS measures of motor excitability. It may be possible to define patterns of effects for AED whose modes of action are unknown.

Resting and Active Motor Thresholds

AEDs (e.g., carbamazepine [CBZ], phenytoin [PHT], lamotrigine [LTG], losigamone [LSG]) with a main mode of action on voltage-gated sodium channels increased MT in most of the studies (see Table 17–2). In one study, it was shown that this increase correlates with the AED plasma level.[87] In contrast, AEDs, which increase neurotransmission at the $GABA_A$ receptor (e.g., lorazepam [LZP], diazepam [DZP], phenobarbital [PB], thioperamide [TP]) or which increase the availability of GABA in the synaptic cleft (e.g., vigabatrin [VGB], tiagabine [TGB]), did not affect the MT (see Table 17–2). These data suggest that MT (at least in the active target muscle) reflects membrane-related axon excitability, which mainly depends on ion channel conductivity.[88–90]

Table 17–2 ACUTE EFFECTS OF ANTIEPILEPTIC DRUGS ON MOTOR EXCITABILITY IN HEALTHY SUBJECTS

Drug	Mode of Action	Dose/Application	Time (hrs)	RMT	MEP	CSP	SICI	ICF	LICI	I-wave facilitation	Reference
CBZ	Na^+	600 mg/PO	**2, 5, 24, 48,** 96	↑		↑	↔	↔			3
		600 mg/PO	6	↔	↔	(↑)	↔	(↓)			127
		600 mg/PO	**2, 5, 24**	↑						↔	128
PHT	Na^+	16 mg/kg/IV	**1.5,** > 15 d	↑	↔	↔					129
		18 mg/kg/PO	**8–12**	↑¹	↔	↔	↔	↔			87
LTG	Na^+	300 mg/PO	**2, 5,** 24	↑		↔	↔	↔			3
		300 mg/PO	**2, 5,** 24	↑						↔	128
		200 mg/PO	2.5	↑	↓		↔	↔			91
VPA	Na^+/$GABA_A$	1,200 mg/PO	4, 8, 24, 48	↔		↔	↔	↔			130
LZP	$GABA_A$	2.5 mg/PO	**2, 5,** 24	↔		↑	↑	↓			131
		2.5 mg/PO	**2, 6,** 24	↔	(↓)				↓		128
		2.5 mg/PO	2.5	↔			↑			↔	132
		0.038 mg/kg/PO	2.5	↔	↓		↔	↔			91
DZP	$GABA_A$	0.17 mg/kg/IV	**5, 30,** 60 min	↔	↔	↓	↔	(↓)			92
		3.75 mg/PO	20, 30, 40, 60 min	↔		↔					133
PB	$GABA_A$	200 mg/PO	**3, 6, 24,** 48	↔	↔				↓		128
TP	$GABA_A$	2 mg/kg/IV	5, **10,** 30 min	↔	(↓)	↔					92
VGB	GABA	2,000 mg/PO	6, **24,** 48, 96	↔		↔	↔	↓			3
		50 mg/kg/PO	24, > 30 d	↔	↔	↔					134
		2,000 mg/PO	**6,** 24, 48	↔	↔				↓		128
TGB	GABA	5–15 mg/PO	90–120 min	↔	↔	↑²	↓²	↑²	↑		30
BCF	$GABA_B$	50 mg/PO	**2, 5,** 24	↔		↔	↑	↓			3
		0.6 mg/kg/PO	15, 30, 60 min	↔	↔	↔					92
		50 mg/PO	2, 5, 24	↔	↔					↔	128
TPM	Na^+/$GABA_A$/Anti-GLU	50/200 mg/PO	2	↔		↔	↑²	↓			94
LSG	? Na^+/Ca^{2+}	1,000 mg/PO	**2, 5,** 24	↑		↔	↔	↔			3
GBP	? GABA/anti-GLU	1,200 mg/PO	**2, 5,** 24	↔		↑	↑	↓			3
		1,200 mg/PO	3, 6, 24	↔	↔					↔	128
		800 mg/PO	**3,** 24	↔		(↑)	↑	↓			135
PTC	?	4,000 mg/PO	1, 3, **6,** 24						↓		136
LTC	?	3,000 mg/PO	**1,** 2	↔	↓	↔	↔	↔			93

¹Effect correlated with AED plasma level.
²Effect was dose-dependent.
anti-GLU, anti-glutamatergic; BCF, baclofen; CBZ, carbamazepine; CSP, cortical silent period; DZP, diazepam; GABA, increase of γ-aminobutyric acid concentration; $GABA_A$, increased neurotramission through the $GABA_A$ receptor; $GABA_B$, $GABA_B$ receptor agonist; GBP, gabapentin; ICF, intracortical facilitation; LICI, long-interval intracortical inhibition; LSG, losigamone; LTC, levetiracetam; LTG, lamotrigine; LZP, lorazepam; MEP, motor evoked potential; Na^+, blockade of voltage-gated sodium channels; PB, phenobarbital; PHT, phenytoin; PO, oral; IV, intravenous; PTC, piracetam; RMT, resting motor threshold; SICI, short-interval intracortical inhibition; TGB, tiagabine; Time, time when a measurement was performed relative to drug administration (boldface indicates times when significant drug effects were obtained); TP, thiopental; TPM, topiramate; VGB, vigabatrin; VPA, valproic acid; ?, unknown mode of action; ↑, increased; ↓, decreased; ↔, no effect.

Motor Evoked Potential Size

The effects of AEDs on MEP size appear to be not so clearly related to different modes of action (see Table 17–2). Sodium channel blocking drugs (e.g., LTG)[91] and $GABA_A$ receptor agonists (TP, LZP)[91,92] depressed the MEP intensity curve. This may be explained by the idea that high-threshold corticospinal neurons, which are being recruited at the plateau of the MEP intensity curve, depend on membrane-related axon excitability and synaptic excitability, whereas low-threshold corticospinal neurons, which are recruited at about MT, depend more strongly on axon excitability. In one study, MEP size was the only affected and therefore most sensitive measure for revealing AED action (e.g., levetiracetam [LTC]).[93]

Cortical Silent Period Duration

The CSP was lengthened by various AEDs with different modes of action (e.g., CBZ, LZP, TGB, gabapentin [GBP]) and shortened by DZP in one study (see Table 17–2). All other AEDs had no significant effect (see Table 17–2). This suggests that AEDs with the same mode of action (e.g., LZP, DZP) may have opposite effects on CSP duration, whereas AEDs with different modes of action (e.g., CBZ, LZP) may produce a similar lengthening effect. The reasons for those inconsistencies are unknown.

Short-Interval Intracortical Inhibition

None of the tested sodium-channel blocking AED produced a significant effect on SICI (see Table 17–2). In contrast, most of the $GABA_A$ receptor agonists (e.g., LZP, topiramate [TPM], GBP) increased SICI (see Table 17–2). One study demonstrated that these effects were dose dependent.[94] These data are consistent with the early notion that SICI reflects inhibition mediated through the $GABA_A$ receptor.[34] One notable finding is the lack of effect of VGB on SICI[3] and the decrease of SICI by TGB.[30] This is most likely explained by the idea that an increase of GABA in the synaptic cleft leads to increased neurotransmission through presynaptic $GABA_B$ receptors located on inhibitory interneurons (i.e., auto-inhibition) and an increase in neurotransmission through postsynaptic $GABA_A$ receptors.

Intracortical Facilitation

Similar to the findings for SICI, none of the sodium-channel blocking AED exerted a significant effect on ICF (see Table 17–2). Most of the $GABA_A$ receptor agonists (e.g., LZP, DZP, TPM, GBP) induced a suppression of ICF (see Table 17–2). This finding indicates that the duration of SICI merges into the inter-stimulus intervals used for ICF testing, as may be expected from the duration of the $GABA_A$ receptor–mediated inhibitory postsynaptic potential (IPSP),[95] or that the excitatory interneurons that mediate ICF are under powerful control by inhibitory interneurons. One notable exception is TGB, which increased ICF.[30] Similar to the discussion of the TGB effects on SICI, this may be explained by auto-inhibition of inhibitory interneurons.

Long-Interval Intracortical Inhibition

LICI has been tested in only one study. In accordance with the idea that LICI reflects long-lasting inhibition mediated through the $GABA_B$ receptor,[43] TGB resulted in a significant increase in LICI, probably through an enhancement of postsynaptic $GABA_B$-mediated neurotransmission (see Table 17–2).[30]

I-Wave Facilitation

The available data can be summarized by a lack of effect by ion channel blocking AEDs and a significant decrease of I-wave facilitation by $GABA_A$ receptor agonists and piracetam (PTC) (see Table 17–2). This supports the notion that the neuronal elements that are responsible for the generation of I waves are under the powerful control of GABA-related inhibitory mechanisms.

Repetitive Transcranial Magnetic Stimulation

rTMS is defined as any regularly repeated stimulation of a single scalp site with the intention of altering the excitability or function of the stimulated site.[96] High-frequency rTMS is delivered at stimulus rates higher than 1 Hz, whereas low-frequency rTMS refers to stimulus rates equal to or lower than 1 Hz.[96] This division is based on the different physiological effects and different risks of adverse effects. Low-frequency rTMS results in a long-lasting depression of the excitability of the stimulated or connected cortex[97–105] and has a low risk of adverse effects.[96] High-frequency rTMS leads to increased excitability of the stimulated cortex[100,106–108] and is associated with a higher risk for adverse effects.[96]

Therapeutic Application of Low-Frequency Repetitive Transcranial Magnetic Stimulation in Epilepsy

The ability of low-frequency rTMS to depress excitability of the stimulated or connected cortex may be of therapeutic use in disorders with increased cortical excitability. One report described a single patient with focal cortical dysplasia in the left parasagittal region and medically refractory focal motor seizures of the right leg.[109] At the time of rTMS treatment, the patient had a stable seizure frequency of 38, 35, and 37 seizures in the 3 months before rTMS treatment. For rTMS, a circular nonfocal stimulating coil was placed over the area of the cortical dysplasia, and 100 stimuli were applied with 0.5 Hz at 95% RMT biweekly for 4 consecutive weeks. Eleven seizures occurred during the month of rTMS treatment (i.e., a 70% reduction compared with pretreatment), and 20 seizures occurred in the next month. No seizures occurred during the rTMS sessions. The frequency of interictal spikes in the EEG fell by 77% in one representative recording when the 30 minutes of tracing before the rTMS session were compared with the 30 minutes of tracing immediately thereafter. The investigators concluded that low-frequency rTMS was safe and might exert a therapeutic effect in otherwise intractable epilepsy, most likely through induction of long-lasting, long-term depression-like inhibition.[109]

Nine patients with medically refractory FE were included in an open, noncontrolled trial.[110] All patients had, on average, more than seven focal or secondarily generalized seizures per week in the 6 months before rTMS treatment. For rTMS, a circular nonfocal stimulating coil was placed over the vertex, and two trains of 500 pulses were given at a frequency of 0.33 Hz on 5 consecutive days. The direction of the current in the coil was counterclockwise for one train and clockwise for the other train. Weekly seizure frequency dropped significantly from an average of 10.3 in the 4 weeks before rTMS to 5.8 in the 4 weeks after rTMS. Seizures did not occur during rTMS. After 6 to 8 weeks, seizure frequency returned to baseline level. The investigators concluded that low-frequency rTMS might temporarily improve otherwise intractable epilepsy.[110] The therapeutic effect might have been achieved by induction of long-term depression-like inhibition.

In one early study enrolling seven patients with intractable TLE, a decrease in interictal EEG spike frequency was noticed, most prominently when low-frequency rTMS (0.1 to 0.3 Hz, 21 to 100 pulses, with an intensity above MT) was applied to the normal hemisphere.[111] Effects on seizure frequency were not reported in the study.

These data are encouraging. What is needed to appreciate the true therapeutic value of low-frequency rTMS in epilepsy are placebo-controlled (sham stimulation), double-blinded trials on larger populations of patients.

Activation of the Epileptic Focus with High-Frequency Repetitive Transcranial Magnetic Stimulation

It may be expected that high-frequency rTMS is more effective in the activation the epileptic focus than single-pulse TMS (inconsistent results with single-pulse TMS were discussed earlier). Even in normal subjects, rTMS is capable of inducing spread of MEPs to muscles that were not activated at rTMS train onset[106] and of inducing seizures. Most likely, these effects are caused by a breakdown of cortical inhibitory mechanisms.[106] In one study, high-frequency rTMS was delivered to different scalp sites to test the possibility of selective activation of the epileptic focus.[112] Eight patients with medically intractable FE were investigated with high-frequency rTMS at stimulus rates of 8 to 25 Hz, stimulus intensities of 40% to 100% of maximum stimulator output, and 490 to 1,060 stimuli per coil position. Seven patients showed no short-term or long-term differences when comparing the EEG before and after rTMS. In one patient, rTMS at 100% stimulator output and 16 Hz over the right parietal cortex first induced after-discharges, and in a second rTMS train, it induced a simple motor seizure with a Jackson march, which eventually generalized. The spontaneous seizures in that patient originated exclusively from the opposite left hemisphere. In a different study,

10 patients with drug-resistant temporal lobe epilepsy were subjected to rTMS trains of 1-second duration at 30 or 50 Hz and at a stimulus intensity of 1.2 times MT.[113,114] Regardless of whether rTMS was applied to the epileptic or normal hemisphere, a significant decrease in spike frequency was observed in all patients.

In summary, the available data do not support the idea that high-frequency rTMS effectively and selectively activates the epileptic focus. Some data encourage exploration of a wider range of stimulus frequencies, even those above 1 Hz, in therapeutic trials of patients with epilepsy.

Safety of Repetitive Transcranial Magnetic Stimulation

A detailed account on the safety aspects of rTMS was given as a report and recommendations from the 1996 International Workshop on the Safety of Repetitive Transcranial Magnetic Stimulation.[96] The safety of rTMS was also addressed by several other studies.[115–122] Generally, rTMS can be considered safe if the recommendations with respect to rTMS frequency, intensity, train duration, and inter-train interval are obeyed.[96,119] Since the publication of these recommendations, no accidentally rTMS-induced seizures were reported in the literature. Before publication, seizures had been induced in five healthy subjects, two patients with epilepsy and one patient with depression, usually with rTMS protocols that are now considered unsafe.[96,119] None of the subjects developed lasting physical problems.[116,118] There is no evidence that a single provoked seizure or even a series of provoked seizures, as in electroconvulsive therapy for depression, makes spontaneous seizures more likely in otherwise healthy individuals.[123]

It is unresolved whether long-term, repeated rTMS in the same subject bears a cumulative risk for adverse side effects. It is important to address this issue if rTMS is to become a therapeutic tool in epilepsy.

■ Conclusion

Single-pulse and paired-pulse TMS can detect abnormalities of motor excitability in patients with epilepsy. In particular, LICF may disclose exaggerated rhythmic motor cortical activity in IGE and PME. CSP, SICI, and ICF reveal abnormalities that are often but not always restricted to the motor cortex of the epileptic hemisphere in patients with various forms of FE (Table 17–1). Single-pulse and paired-pulse TMS is useful in delineating specific acute AED effects in healthy subjects. Sodium channel blocking AEDs consistently increase MT but do not affect SICI and ICF. Conversely, $GABA_A$ receptor agonists leave the MT unaffected but increase SICI and decrease ICF (Table 17–2). Single-pulse TMS and rTMS are not useful for selective activation of the epileptic focus, but encouraging first therapeutic trials indicate that low-frequency rTMS may be useful for treatment of epilepsy. If the internationally accepted safety recommendations are obeyed, single-pulse TMS and rTMS can be considered safe for testing and treating patients with epilepsy.

REFERENCES

1. Rossini PM, Barker AT, Berardelli A, et al. Non-invasive electrical and magnetic stimulation of the brain, spinal cord and roots: basic principles and procedures for routine clinical application. Report of an IFCN committee. Electroencephalogr Clin Neurophysiol 1994;91:79.
2. Hess CW, Mills KR, Murray NM. Magnetic stimulation of the human brain: facilitation of motor responses by voluntary contraction of ipsilateral and contralateral muscles with additional observations on an amputee. Neurosci Lett 1986;71:235.
3. Ziemann U, Lönnecker S, Steinhoff BJ, et al. Effects of antiepileptic drugs on motor cortex excitability in humans: a transcranial magnetic stimulation study. Ann Neurol 1996;40:367.
4. Cicinelli P, Traversa R, Bassi A, et al. Interhemispheric differences of hand muscle representation in human motor cortex. Muscle Nerve 1997;20:535.
5. Devanne H, Lavoie BA, Capaday C. Input-output properties and gain changes in the human corticospinal pathway. Exp Brain Res 1997;114:329.
6. Hallett M. Transcranial magnetic stimulation and the human brain. Nature 2000;406:147.
7. Hess CW, Mills KR, Murray NM. Responses in small hand muscles from magnetic stimulation of the human brain. J Physiol (Lond) 1987;388:397.

8. Deuschl G, Michels R, Berardelli A, et al. Effects of electric and magnetic transcranial stimulation on long latency reflexes. Exp Brain Res 1991;83:403.
9. Day BL, Riescher H, Struppler A, et al. Changes in the response to magnetic and electrical stimulation of the motor cortex following muscle stretch in man. J Physiol (Lond) 1991;433:41.
10. Mariorenzi R, Zarola F, Caramia MD, et al. Non-invasive evaluation of central motor tract excitability changes following peripheral nerve stimulation in healthy humans. Electroencephalogr Clin Neurophysiol 1991;81:90.
11. Baldissera F, Leocani L. Afferent excitation of human motor cortex as revealed by enhancement of direct cortico-spinal actions on motoneurones. Electroencephalogr Clin Neurophysiol 1995;97:394.
12. Tokimura H, Di Lazzaro V, Tokimura Y, et al. Short latency inhibition of human hand motor cortex by somatosensory input from the hand. J Physiol 2000;523:503.
13. Maertens de Noordhout A, Rothwell JC, Day BL, et al. Effect of digital nerve stimuli on responses to electrical or magnetic stimulation of the human brain. J Physiol (Lond) 1992;447:535.
14. Clouston PD, Kiers L, Menkes D, et al. Modulation of motor activity by cutaneous input: inhibition of the magnetic motor evoked potential by digital electrical stimulation. Electroencephalogr Clin Neurophysiol 1995;97:114.
15. Classen J, Steinfelder B, Liepert J, et al. Cutaneomotor integration in humans is somatotopically organized at various levels of the nervous system and is task dependent. Exp Brain Res 2000;130:48.
16. Kofler M, Fuhr P, Leis AA, et al. Modulation of upper extremity motor evoked potentials by cutaneous afferents in humans. Clin Neurophysiol 2001;112:1053.
17. Ugawa Y, Day BL, Rothwell JC, et al. Modulation of motor cortical excitability by electrical stimulation over the cerebellum in man. J Physiol (Lond) 1991;441:57.
18. Ugawa Y, Uesaka Y, Terao Y, et al. Magnetic stimulation over the cerebellum in humans. Ann Neurol 1995;37:703.
19. Werhahn KJ, Taylor J, Ridding M, et al. Effect of transcranial magnetic stimulation over the cerebellum on the excitability of human motor cortex. Electroencephalogr Clin Neurophysiol 1996;101:58.
20. Ferbert A, Priori A, Rothwell JC, et al. Interhemispheric inhibition of the human motor cortex. J Physiol (Lond) 1992;453:525.
21. Netz J, Ziemann U, Hömberg V. Hemispheric asymmetry of transcallosal inhibition in man. Exp Brain Res 1995;104:527.
22. Di Lazzaro V, Oliviero A, Profice P, et al. Direct demonstration of interhemispheric inhibition of the human motor cortex produced by transcranial magnetic stimulation. Exp Brain Res 1999;124:520.
23. Cantello R, Gianelli M, Civardi C, et al. Magnetic brain stimulation: the silent period after the motor evoked potential. Neurology 1992;42:1951.
24. Wilson SA, Lockwood RJ, Thickbroom GW, et al. The muscle silent period following transcranial magnetic cortical stimulation. J Neurol Sci 1993;114:216.
25. Davey NJ, Romaiguere P, Maskill DW, et al. Suppression of voluntary motor activity revealed using transcranial magnetic stimulation of the motor cortex in man. J Physiol (Lond) 1994;477:223.
26. Fuhr P, Agostino R, Hallett M. Spinal motor neuron excitability during the silent period after cortical stimulation. Electroencephalogr Clin Neurophysiol 1991;81:257.
27. Inghilleri M, Berardelli A, Cruccu G, et al. Silent period evoked by transcranial stimulation of the human cortex and cervicomedullary junction. J Physiol (Lond) 1993;466:521.
28. Ziemann U, Netz J, Szelenyi A, et al. Spinal and supraspinal mechanisms contribute to the silent period in the contracting soleus muscle after transcranial magnetic stimulation of human motor cortex. Neurosci Lett 1993;156:167.
29. Hallett M. Transcranial magnetic stimulation. Negative effects. Adv Neurol 1995;67:107.
30. Werhahn KJ, Kunesch E, Noachtar S, et al. Differential effects on motorcortical inhibition induced by blockade of GABA uptake in humans. J Physiol (Lond) 1999;517:591.
31. Claus D, Weis M, Jahnke U, et al. Corticospinal conduction studied with magnetic double stimulation in the intact human. J Neurol Sci 1992;111:180.
32. Valls-Sole J, Pascual-Leone A, Wassermann EM, et al. Human motor evoked responses to paired transcranial magnetic stimuli. Electroencephalogr Clin Neurophysiol 1992;85:355.
33. Nakamura H, Kitagawa H, Kawaguchi Y, et al. Intracortical facilitation and inhibition after transcranial magnetic stimulation in conscious humans. J Physiol (Lond) 1997;498:817.
34. Kujirai T, Caramia MD, Rothwell JC, et al. Corticocortical inhibition in human motor cortex. J Physiol (Lond) 1993;471:501.
35. Ziemann U, Rothwell JC, Ridding MC. Interaction between intracortical inhibition and facilitation in human motor cortex. J Physiol (Lond) 1996;496:873.

36. Di Lazzaro V, Restuccia D, Oliviero A, et al. Magnetic transcranial stimulation at intensities below active motor threshold activates intracortical inhibitory circuits. Exp Brain Res 1998;119:265.
37. Tokimura H, Ridding MC, Tokimura Y, et al. Short latency facilitation between pairs of threshold magnetic stimuli applied to human motor cortex. Electroencephalogr Clin Neurophysiol 1996;101:263.
38. Ziemann U, Tergau F, Wassermann EM, et al. Demonstration of facilitatory I-wave interaction in the human motor cortex by paired transcranial magnetic stimulation. J Physiol (Lond) 1998;511:181.
39. Di Lazzaro V, Rothwell JC, Oliviero A, et al. Intracortical origin of the short latency facilitation produced by pairs of threshold magnetic stimuli applied to human motor cortex. Exp Brain Res 1999;129:494.
40. Hanajima R, Ugawa Y, Terao Y, et al. Mechanisms of intracortical I-wave facilitation elicited with paired-pulse magnetic stimulation in humans. J Physiol 2002;538:253.
41. Ziemann U. Intracortical inhibition and facilitation in the conventional paired TMS paradigm. Electroencephalogr Clin Neurophysiol Suppl 1999;51:127.
42. Ziemann U, Rothwell JC. I-waves in motor cortex. J Clin Neurophysiol 2000;17:397.
43. Sanger TD, Garg RR, Chen R. Interactions between two different inhibitory systems in the human motor cortex. J Physiol 2001;530.2:307.
44. Hufnagel A, Elger CE, Marx W, et al. Magnetic motor-evoked potentials in epilepsy: effects of the disease and of anticonvulsant medication. Ann Neurol 1990;28:680.
45. Reutens DC, Berkovic SF. Increased cortical excitability in generalised epilepsy demonstrated with transcranial magnetic stimulation [letter]. Lancet 1992;339:362.
46. Reutens DC, Berkovic SF, Macdonell RA, et al. Magnetic stimulation of the brain in generalized epilepsy: reversal of cortical hyperexcitability by anticonvulsants. Ann Neurol 1993;34:351.
47. Reutens DC, Puce A, Berkovic SF. Cortical hyperexcitability in progressive myoclonus epilepsy: a study with transcranial magnetic stimulation. Neurology 1993;43:186.
48. Gianelli M, Cantello R, Civardi C, et al. Idiopathic generalized epilepsy: magnetic stimulation of motor cortex time-locked and unlocked to 3-Hz spike-and-wave discharges. Epilepsia 1994;35:53.
49. Caramia MD, Gigli G, Iani C, et al. Distinguishing forms of generalized epilepsy using magnetic brain stimulation. Electroencephalogr Clin Neurophysiol 1996;98:14.
50. Brown P, Ridding MC, Werhahn KJ, et al. Abnormalities of the balance between inhibition and excitation in the motor cortex of patients with cortical myoclonus. Brain 1996;119:309.
51. Cantello R, Gianelli M, Civardi C, et al. Focal subcortical reflex myoclonus. A clinical and neurophysiological study. Arch Neurol 1997;54:187.
52. Cantello R, Civardi C, Cavalli A, et al. Cortical excitability in cryptogenic localization-related epilepsy: interictal transcranial magnetic stimulation studies. Epilepsia 2000;41:694.
53. Cicinelli P, Mattia D, Spanedda F, et al. Transcranial magnetic stimulation reveals an interhemispheric asymmetry of cortical inhibition in focal epilepsy. Neuroreport 2000;11:701.
54. Delvaux V, Alagona G, Gerard P, et al. Reduced excitability of the motor cortex in untreated patients with de novo idiopathic "grand mal" seizures. J Neurol Neurosurg Psychiatry 2001;71:772.
55. Inghilleri M, Mattia D, Berardelli A, et al. Asymmetry of cortical excitability revealed by transcranial stimulation in a patient with focal motor epilepsy and cortical myoclonus. Electroencephalogr Clin Neurophysiol 1998;109:70.
56. Nezu A, Kimura S, Ohtsuki N, et al. Transcranial magnetic stimulation in benign childhood epilepsy with centro-temporal spikes. Brain Dev 1997;19:134.
57. Aguglia U, Gambardella A, Quartarone A, et al. Interhemispheric threshold differences in idiopathic generalized epilepsies with versive or circling seizures determined with focal magnetic transcranial stimulation. Epilepsy Res 2000;40:1.
58. Guerrini R, Bonanni P, Patrignani A, et al. Autosomal dominant cortical myoclonus and epilepsy (ADCME) with complex partial and generalized seizures: A newly recognized epilepsy syndrome with linkage to chromosome 2p11.1-q12.2. Brain 2001;124:2459.
59. Werhahn KJ, Lieber J, Classen J, et al. Motor cortex excitability in patients with focal epilepsy. Epilepsy Res 2000;41:179.
60. Manganotti P, Tamburin S, Zanette G, et al. Hyperexcitable cortical responses in progressive myoclonic epilepsy: A TMS study. Neurology 2001;57:1793.
61. Manganotti P, Bongiovanni LG, Zanette G, et al. Early and late intracortical inhibition in juvenile myoclonic epilepsy. Epilepsia 2000;41:1129.
62. Manganotti P, Zanette G. Contribution of motor cortex in generation of evoked spikes in patients with benign rolandic epilepsy. Clin Neurophysiol 2000;111:964.
63. Hanajima R, Ugawa Y, Okabe S, et al. Interhemispheric interaction between the

64. Thompson PD, Rothwell JC, Brown P, et al. Transcallosal and intracortical spread of activity following cortical stimulation in a patient with generalized cortical myoclonus. J Physiol (Lond) [abstract] 1993;459:64P.
65. Classen J, Witte OW, Schlaug G, et al. Epileptic seizures triggered directly by focal transcranial magnetic stimulation. Electroencephalogr Clin Neurophysiol 1995;94:19.
66. Cincotta M, Borgheresi A, Lori S, et al. Interictal inhibitory mechanisms in patients with cryptogenic motor cortex epilepsy: A study with transcranial magnetic stimulation [abstract]. Electroencephalogr Clin Neurophysiol 1997;103:75.
67. Cincotta M, Borgheresi A, Benvenuti F, et al. Cortical silent period in two patients with meningioma and preoperative seizures: a pre- and postsurgical follow-up study. Clin Neurophysiol 2002;113:597.
68. Macdonell RA, King MA, Newton MR, et al. Prolonged cortical silent period after transcranial magnetic stimulation in generalized epilepsy. Neurology 2001;57:706.
69. Valzania F, Strafella AP, Tropeani A, et al. Facilitation of rhythmic events in progressive myoclonus epilepsy: a transcranial magnetic stimulation study. Clin Neurophysiol 1999;110:152.
70. Herrendorf G, Ziemann U, Kurth C, et al. Ictal cerebrovascular near infra-red spectroscopy (NIRS) and motor cortex excitability in a patient with epilepsia partialis continua [abstract]. Epilepsia 1997;38 (Suppl 3):226.
71. Fong JKY, Werhahn KJ, Rothwell JC, et al. Motor cortex excitability in focal and generalized epilepsy [abstract]. J Physiol (Lond) 1993;459:468P.
72. Brodtmann A, Macdonell RAL, Gilligan AK, et al. Cortical excitability and recovery curve analysis in generalized epilepsy. Neurology 1999;53:1347.
73. Bridgers SL, Delaney RC. Transcranial magnetic stimulation: an assessment of cognitive and other cerebral effects. Neurology 1989;39:417.
74. Dressler D, Voth E, Feldmann M, et al. Safety aspects of transcranial brain stimulation in man tested by single photon emission-computed tomography. Neurosci Lett 1990;119:153.
75. Bridgers SL. The safety of transcranial magnetic stimulation reconsidered: evidence regarding cognitive and other cerebral effects. Electroencephalogr Clin Neurophysiol Suppl 1991;43:170.
76. Hamano T, Kaji R, Fukuyama H, et al. Lack of prolonged cerebral blood flow change after transcranial magnetic stimulation. Electroencephalogr Clin Neurophysiol 1993;89:207.
77. Chokroverty S, Hening W, Wright D, et al. Magnetic brain stimulation: safety studies. Electroencephalogr Clin Neurophysiol 1995;97:36.
78. Hömberg V, Netz J. Generalised seizures induced by transcranial magnetic stimulation of motor cortex [letter]. Lancet 1989;2:1223.
79. Kandler R. Safety of transcranial magnetic stimulation [letter; comment]. Lancet 1990;335:469.
80. Fauth C, Meyer BU, Prosiegel M, et al. Seizure induction and magnetic brain stimulation after stroke [letter]. Lancet 1992;339:362.
81. Hufnagel A, Elger CE, Durwen HF, et al. Activation of the epileptic focus by transcranial magnetic stimulation of the human brain. Ann Neurol 1990;27:49.
82. Hufnagel A, Elger CE. Induction of seizures by transcranial magnetic stimulation in epileptic patients [letter]. J Neurol 1991;238:109.
83. Düzel E, Hufnagel A, Helmstaedter C, et al. Verbal working memory components can be selectively influenced by transcranial magnetic stimulation in patients with left temporal lobe epilepsy. Neuropsychologia 1996;34:775.
84. Tassinari CA, Michelucci R, Forti A, et al. Transcranial magnetic stimulation in epileptic patients: usefulness and safety. Neurology 1990;40:1132.
85. Schüler P, Claus D, Stefan H. Hyperventilation and transcranial magnetic stimulation: two methods of activation of epileptiform EEG activity in comparison. J Clin Neurophysiol 1993;10:111.
86. Schulze-Bonhage A, Scheufler K, Zentner J, et al. Safety of single and repetitive focal transcranial magnetic stimuli as assessed by intracranial EEG recordings in patients with partial epilepsy. J Neurol 1999;246:914.
87. Chen R, Samii A, Canos M, et al. Effects of phenytoin on cortical excitability in humans. Neurology 1997;49:881.
88. Hodgkin AL, Huxley A. Currents carried by sodium and potassium ions through the membrane of the giant axon of Loligo. J Physiol (Lond) 1952;116:449.
89. Hodgkin AL, Huxley AF. The components of membrane conductance in the giant axon of Loligo. J Physiol (Lond) 1952;116:473.
90. Hodgkin AL, Huxley AF. A quantitative description of membrane current and its application to conduction and excitation in nerve. J Physiol (Lond) 1952;116:500.
91. Boroojerdi B, Battaglia F, Muellbacher W, et al. Mechanisms influencing stimulus-response properties of the human corticospinal system. Clin Neurophysiol 2001;112:931.
92. Inghilleri M, Berardelli A, Marchetti P, et al. Effects of diazepam, baclofen and thiopental

on the silent period evoked by transcranial magnetic stimulation in humans. Exp Brain Res 1996;109:467.
93. Sohn YH, Kaelin-Lang A, Jung HY, et al. Effect of levetiracetam on human corticospinal excitability. Neurology 2001;57:858.
94. Reis J, Tergau F, Hamer HM, et al. Effekte des Antiepileptikums Topiramat auf die Erregbarkeit des menschlichen Motorkortex: eine Studie mit transkranieller Magnetstimulation [abstract]. Klin Neurophysiol 2001;32:193.
95. McCormick DA. GABA as an inhibitory neurotransmitter in human cerebral cortex. J Neurophysiol 1989;62:1018.
96. Wassermann EM. Risk and safety of repetitive transcranial magnetic stimulation: report and recommendations from the International Workshop on the Safety of Repetitive Transcranial Magnetic Stimulation June 5-7, 1996. Electroencephalogr Clin Neurophysiol 1998;108:1.
97. Chen R, Classen J, Gerloff C, et al. Depression of motor cortex excitability by low-frequency transcranial magnetic stimulation. Neurology 1997;48:1398.
98. Kosslyn SM, Pascual-Leone A, Felician O, et al. The role of area 17 in visual imagery: convergent evidence from PET and rTMS. Science 1999;284:167.
99. Boroojerdi B, Prager A, Muellbacher W, et al. Reduction of human visual cortex excitability using 1-Hz transcranial magnetic stimulation. Neurology 2000;54:1529.
100. Maeda F, Keenan JP, Tormos JM, et al. Modulation of corticospinal excitability by repetitive transcranial magnetic stimulation. Clin Neurophysiol 2000;111:800.
101. Muellbacher W, Ziemann U, Boroojerdi B, et al. Effects of low-frequency transcranial magnetic stimulation on motor excitability and basic motor behavior. Clin Neurophysiol 2000;111:1002.
102. Gerschlager W, Siebner HR, Rothwell JC. Decreased corticospinal excitability after subthreshold 1 Hz rTMS over lateral premotor cortex. Neurology 2001;57:449.
103. Enomoto H, Ugawa Y, Hanajima R, et al. Decreased sensory cortical excitability after 1 Hz rTMS over the ipsilateral primary motor cortex. Clin Neurophysiol 2001;112:2154.
104. Touge T, Gerschlager W, Brown P, et al. Are the after-effects of low-frequency rTMS on motor cortex excitability due to changes in the efficacy of cortical synapses? Clin Neurophysiol 2001;112:2138.
105. Tsuji T, Rothwell JC. Long lasting effects of rTMS and associated peripheral sensory input on MEPs, SEPs and transcortical reflex excitability in humans. J Physiol 2002;540:367.
106. Pascual-Leone A, Valls-Sole J, Wassermann EM, et al. Responses to rapid-rate transcranial magnetic stimulation of the human motor cortex. Brain 1994;117:847.
107. Peinemann A, Lehner C, Mentschel C, et al. Subthreshold 5-Hz repetitive transcranial magnetic stimulation of the human primary motor cortex reduces intracortical paired-pulse inhibition. Neurosci Lett 2000;296:21.
108. Wu T, Sommer M, Tergau F, et al. Lasting influence of repetitive transcranial magnetic stimulation on intracortical excitability in human subjects. Neurosci Lett 2000;287:37.
109. Menkes DL, Gruenthal M. Slow-frequency repetitive transcranial magnetic stimulation in a patient with focal cortical dysplasia. Epilepsia 2000;41:240.
110. Tergau F, Naumann U, Paulus W, et al. Low-frequency repetitive transcranial magnetic stimulation improves intractable epilepsy [letter]. Lancet 1999;353:2209.
111. Steinhoff BJ, Stodieck SR, Zivcec Z, et al. Transcranial magnetic stimulation (TMS) of the brain in patients with mesiotemporal epileptic foci. Clin Electroencephalogr 1993;24:1.
112. Dhuna A, Gates J, Pascual-Leone A. Transcranial magnetic stimulation in patients with epilepsy [see comments]. Neurology 1991;41:1067.
113. Jennum P, Winkel H, Fuglsang-Frederiksen A, et al. EEG changes following repetitive transcranial magnetic stimulation in patients with temporal lobe epilepsy. Epilepsy Res 1994;18:167.
114. Jennum P, Winkel H. Transcranial magnetic stimulation. Its role in the evaluation of patients with partial epilepsy. Acta Neurol Scand Suppl 1994;152:93.
115. Pascual-Leone A, Valls-Sole J, Brasil-Neto JP, et al. Seizure induction and transcranial magnetic stimulation [letter]. Lancet 1992;339:997.
116. Pascual-Leone A, Houser CM, Reese K, et al. Safety of rapid-rate transcranial magnetic stimulation in normal volunteers. Electroencephalogr Clin Neurophysiol 1993;89:120.
117. Wassermann EM, Grafman J, Berry C, et al. Use and safety of a new repetitive transcranial magnetic stimulator. Electroencephalogr Clin Neurophysiol 1996;101:412.
118. Wassermann EM, Cohen LG, Flitman SS, et al. Seizures in healthy people with repeated "safe" trains of transcranial magnetic stimuli [letter]. Lancet 1996;347:825.
119. Chen RC, Gerloff C, Classen J, et al. Safety of different inter-train intervals for repetitive transcranial magnetic stimulation and recommendations for safe ranges of

stimulation parameters. Electroencephalogr Clin Neurophysiol 1997;105:415.
120. Foerster A, Schmitz JM, Nouri S, et al. Safety of rapid-rate transcranial magnetic stimulation: heart rate and blood pressure changes. Electroencephalogr Clin Neurophysiol 1997;104:207.
121. Jahanshahi M, Ridding MC, Limousin P, et al. Rapid rate transcranial magnetic stimulation–a safety study. Electroencephalogr Clin Neurophysiol 1997;105:422.
122. Niehaus L, Hoffmann KT, Grosse P, et al. MRI study of human brain exposed to high-dose repetitive magnetic stimulation of visual cortex. Neurology 2000;54:256.
123. Devinsky O, Duchowny MS. Seizures after convulsive therapy: a retrospective case survey. Neurology 1983;33:921.
124. Cincotta M, Borgheresi A, Lori S, et al. Interictal inhibitory mechanisms in patients with cryptogenic motor cortex epilepsy: a study of the silent period following transcranial magnetic stimulation. Electroencephalogr Clin Neurophysiol 1998;107:1.
125. Shimizu T, Maehara T, Hino T, et al. Effect of multiple subpial transection on motor cortical excitability in cortical dysgenesis. Brain 2001;124:1336.
126. Herrendorf G, Ziemann U, Kurth C, et al. Messung der zerebralen Oxygenierung mittels Nah-Infrarot-Spektroskopie (NIRS) und der Exzitabilitat des motorischen Kortex mittels transkranieller Magnetstimulation (TMS) bei einem Patienten mit Epilepsia partialis continua. Klin Neurophysiol 1998;29:334.
127. Schulze-Bonhage A, Knott H, Ferbert A. Effects of carbamazepine on cortical excitatory and inhibitory phenomena: a study with paired transcranial magnetic stimulation. Electroencephalogr Clin Neurophysiol 1996;99:267.

128. Ziemann U, Tergau F, Wischer S, et al. Pharmacological control of facilitatory I-wave interaction in the human motor cortex. A paired transcranial magnetic stimulation study. Electroencephalogr Clin Neurophysiol 1998;109:321.
129. Mavroudakis N, Caroyer JM, Brunko E, et al. Effects of diphenylhydantoin on motor potentials evoked with magnetic stimulation. Electroencephalogr Clin Neurophysiol 1994;93:428.
130. Ziemann U, Lönnecker S, Steinhoff BJ, et al. Motor excitability changes under antiepileptic drugs. In Stefan H, Andermann F, Chauvel P, et al. (eds): Plasticity in Epilepsy: Dynamic Aspects of Brain Function, vol 81. Philadelphia, Lippincott Williams & Wilkins, 1999:291.
131. Ziemann U, Lönnecker S, Steinhoff BJ, et al. The effect of lorazepam on the motor cortical excitability in man. Exp Brain Res 1996;109:127.
132. Di Lazzaro V, Oliviero A, Meglio M, et al. Direct demonstration of the effect of lorazepam on the excitability of the human motor cortex. Clin Neurophysiol 2000;111:794.
133. Palmieri MG, Iani C, Scalise A, et al. The effect of benzodiazepines and flumazenil on motor cortical excitability in the human brain. Brain Res 1999;815:192.
134. Mavroudakis N, Caroyer JM, Brunko E, et al. Effects of vigabatrin on motor potentials with magnetic stimulation. Electroencephalogr Clin Neurophysiol 1997;105:124.
135. Rizzo V, Quartarone A, Bagnato S, et al. Modification of cortical excitability induced by gabapentin: a study by transcranial magnetic stimulation. Neurol Sci 2001;22:229.
136. Wischer S, Paulus W, Sommer M, et al. Piracetam affects facilitatory I-wave interaction in the human motor cortex. Clin Neurophysiol 2001;112:275.

18 Components of Language as Revealed by Magnetic Stimulation

Agnes P. Funk and Charles M. Epstein

Language is a highly complex function composed fundamentally of speech and gestures, reading and comprehension, writing and interpretation. Language has given humans the ability to communicate and cooperate in ways unmatched by other animal species. Unlike other animals, we communicate for reasons other than to warn of an approaching predator or the location of a food source. The complexity of language allows humans to express their thoughts and feelings at the most abstract levels, such as in poetry. The underlying neural mechanisms of language are equally complex and difficult to establish. Several neuroanatomical models of language have been proposed; however, it is generally accepted that Wernicke's area is involved in language construction and comprehension and that Broca's area controls the motor production of language (e.g., speech, writing, sign language).[1–3] This chapter reviews the contributions made by transcranial magnetic stimulation (TMS) to the investigation of neural mechanisms underlying language.

Initially, TMS studies focused on replicating the intracarotid amobarbital test (IAT), or Wada test, which is used to evaluate the lateralization of language areas before neurosurgery.[4] However, debate soon arose over whether TMS was disrupting actual language function or simply motor speech output. There was also difficulty in establishing the TMS parameters that would affect language processing with the least discomfort to subjects. In early studies, TMS was delivered at frequencies and durations that often resulted in pain[5] and even seizure.[6,7] Later studies confirmed that TMS could interrupt speech output without excess discomfort.[8] It has also become increasingly evident that TMS can disrupt selected components of language, such as verb processing, object naming, and verbal working memory.

■ How Does Transcranial Magnetic Stimulation Affect Language Processing?

The use of rapid-rate TMS (rTMS) to investigate various components of language has yielded variable results. Whether TMS exerts an inhibitory or facilitative effect on language processing does not seem to correlate with the choice of task or area of stimulation. For example, picture-naming tasks have been shown to be disrupted[9] or facilitated[10] after rTMS of temporal areas. However, it appears overall that studies that use less intense TMS parameters tend to show more facilitatory effects on language processing.

Evidence for Disruption of Language Processing

Many TMS experiments examining language processing have disrupted or facilitated some type of naming task, typically a picture naming task. Wassermann and colleagues[9] were the first to demonstrate interference with picture naming. They used 10-Hz TMS at 100% motor threshold (MT) over the left temporal lobe and found an increase in the number of naming errors but very few reading errors. A few years later, they added a comparison of picture naming and reading using rTMS in patients with temporal lobe epilepsy.[11] They again found an increase in naming errors after rTMS with no effect on the reading task. However, this time the affected cortical sites were frontal (motor speech) and the middle superior and posterior temporal areas in the left hemisphere; the difference between these sites did not reach significance. They also tested mirror right hemisphere sites, but found no effect on picture naming. TMS of the left hemisphere sites and right frontal areas did reduce the reaction time (RT), which is apparently inconsistent with the error result. The investigators explain this anomaly as being analogous to the facilitatory effect observed after low-frequency TMS of the motor cortex. They argue that the RT reduction observed in their study is most likely caused by an increase in excitation of the motor output system, rather than language processing mechanisms per se.

Subsequent studies do not report any increases in error frequency, but instead rely on increases in RT as evidence of disruption to language processing. Stewart and colleagues[12] used 10-Hz TMS at an intensity of 75% of maximum stimulator output for 600 ms; MT was not tested. They targeted posterior BA37 (inferior temporal cortex) of left and right hemispheres, and used the vertex as a control site. They found an increased RT for naming pictures but not colors after TMS of left posterior BA37; there was no effect on word or nonword reading. The investigators conclude that their finding confirms the role of left posterior BA37 in object recognition, which parallels numerous imaging studies and neuropsychological lesion reports.

Shapiro and colleagues[13] used a word generation task to show a left prefrontal bias for processing verbs in a region anterior and superior to Broca's area. Subjects were required to generate plural and singular forms of real and pseudo nouns and verbs while they received 1-Hz TMS at 110% of MT. They found that TMS increased RT only for the production of verb forms, which they argued demonstrates the role of left prefrontal cortex in the grammatical processing and retrieval of verbs. Matthews and colleagues[14] used a combination of fMRI and TMS to examine whether the anterior part of Broca's area is critical for semantic processing. An initial fMRI study revealed that there was some overlap of semantic and phonological processing in Broca's area. However, it was unclear whether semantic activation was independent or due to phonological processing implicitly engaged during word reading. They targeted the left inferior prefrontal cortex using single-pulse TMS while subjects made semantic decisions ("Is it man-made?") to single written words or judged the length of words (i.e., visual perceptual task). They found that delivery at 250 ms after stimulus onset caused a significant increase in RT for the semantic processing task but not the perceptual (control) task, suggesting that the left inferior prefrontal cortex does contribute to the semantic decision-making process.

The three latter studies used lower TMS frequencies or intensities compared with previous investigations that showed increases in error rates.[7,11] Perhaps the lack of effect of TMS on error frequency was caused by the more conservative parameters used in the later studies. Although RT is a less specific measure of cognitive performance, this limitation is partially offset by the convenience of parametric analysis and the reduced discomfort of lower TMS frequencies and intensities.

Evidence for Facilitation of Language Processing

Facilitation of various naming tasks has been found after TMS of Wernicke's area. Topper and colleagues provided the earliest evidence for facilitation; however, it appears that this effect depends on the choice of TMS parameters.[15] In their initial study, Topper and colleagues[16] found that single-pulse TMS at 35% to 55% of maximum stimulator output (Novametrix 200, Magstim, UK), delivered at 500- or 1,000-ms prestimulus onset, reduced RT for a picture naming task. After the use of higher TMS intensities, the facilitatory effect vanished. A year later, this group used the same task[10] but instead applied 20-Hz TMS at 55% of maximum stimulator output for 40 seconds over Wernicke's area. A facilitation of RT was observed when TMS was delivered immediately after the task. In their latest endeavor using the same task, they compared 20- and 1-Hz TMS and confirmed their previous result that 20-Hz TMS immediately after task performance facilitates RT; there was no effect at 1 Hz.[16] This group did not use MT to determine TMS intensity, but they did estimate that 55% of maximum stimulator output was suprathreshold in most subjects, and 35% was subthreshold in all subjects. The investigators observed that the presence or absence of effect for 1-Hz TMS might be caused by the timing of TMS delivery, with facilitation by single pulses before but not after picture presentation.

Cappa and colleagues[17] also used a picture-naming task, but their pictures comprised objects and actions, with the latter being mostly tool use. They found that 20-Hz TMS at 90% MT over the left dorsolateral prefrontal cortex (DLPFC) was facilitatory and reduced RT only for naming action pictures. The results appear complementary to those of Shapiro and coworkers,[13] who found *disruption* (increased RT) of verb (or action word) processing after rTMS of left prefrontal areas, but at a higher intensity of 110% MT (Fig. 18–1). Recall that Topper[16] also found higher TMS intensities eliminated facilitatory effects on language processing.

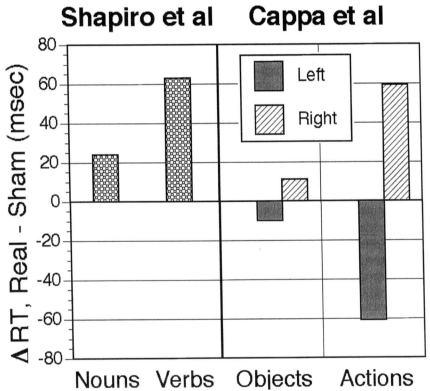

Figure 18–1 Comparison of low- and high-intensity transcranial magnetic stimulation (TMS) effects on verb or action word processing. Data are measured as a change in the reaction time (RT), for which positive (+) values indicate increased RT and negative (−) values decreased RT. The *checkered bars* show that after high-intensity TMS of the left prefrontal cortex, the RT to retrieve verbs was increased (disrupted) compared with noun retrieval.[13] The *gray, hashed bars* show hemispheric differences between the object and action word processing after dorsolateral prefrontal cortex (DLPFC) stimulation. A decrease in RT (facilitation) for action word processing occurred after low-intensity TMS of left DLPFC.[17] There were no differences for object word processing. (Adapted from Shapiro KA, Pascual-Leone A, Mottaghy FM, et al. Grammatical distinctions in the left frontal cortex. J Cogn Neurosci 2001;13:713–720 and from Cappa SF, Sandrini M, Rossini PM, et al. The role of the left frontal lobe in action naming: rTMS evidence. Neurology 2002;59:720–723.)

Sakai and colleagues[18] did not use a picture-naming task; instead, subjects were required to judge whether sentences were syntactically or semantically normal. They used paired-pulse TMS over the left frontal lobe, with an inter-stimulus interval of 2 ms. The intensity was 33% to 55% of maximum stimulator output (Magstim 200, Magstim, UK), which was equivalent to 55% to 98% of MT. They found a reduction in RT when TMS was delivered at 150 ms after stimulus onset to left BA44 (inferior frontal gyrus) but not to left BA8/9 (middle frontal gyrus); no right hemisphere sites were tested. The investigators argued that their results indicated the essential role of Broca's area in syntactic but not semantic processing.

Interference with Speech Production

Initial attempts to produce language deficits with rTMS were influenced by the belief that stimulation parameters should resemble those used in operative electrocorticography, and preceded the existence of safety guidelines. Frequencies of up to 50 Hz were employed, and intensity was commonly set between 80% and 100% of maximum stimulator output, with stimulus trains lasting up to 10 seconds. It comes as no surprise that pain, crying, and seizures were among the outcomes.[5,6,19–21] Nonetheless, speech disruption seemed difficult to induce with TMS and round stimulation coils. For example, Jennum and associates[5] achieved it in only 7 of 14 subjects, but Michelucci and colleagues[20] did not fair much better with 14 of 21. Epstein and colleagues[8] used a more powerful figure-of eight coil and consistently induced speech arrest with frequencies as low as 2 Hz, intensities of 150% MT, and durations of only 5 seconds.

Although the problem of extreme TMS parameters had been essentially resolved by the use of slower stimulation rates, there remained the issue of whether speech arrest was a consequence of motor or purely language mechanisms. An early study by Pascual-Leone and coworkers[21] applied 25-Hz TMS for 10 seconds at 15 different sites and found that laryngeal muscles were activated 1 to 5.3 cm posterior to the site for speech arrest, which they argued was induced over Broca's area. Epstein and colleagues[22] performed a mapping study with figure-of-eight coils and found that the speech arrest site was more consistent with the orbicularis oris muscle site of activation (Fig. 18–2). A possible explanation for this discrepancy between the two studies lies in the observation that the different coils (round versus figure-of-eight) induce electric fields differing in orientation by 90 degrees. They also found that speech arrest does not occur while subjects are singing, an observation later corroborated by Stewart and coworkers,[23] suggesting that different mechanisms may underlie the production of song, and consistent with many years of observations on brain-injured patients.

Stewart and colleagues[24] later addressed the question of a motoric or nonmotoric source of speech arrest, and concluded that both are possible. They recorded electromyographic activity from the mentalis muscle while they induced speech arrest with rTMS (10 Hz, 120% to 140% MT, 1-second duration). They then moved the coil approximately 5 cm anteriorly and 2 cm laterally and again induced speech arrest without activation of the mentalis muscle. They demonstrated that magnetic stimulation of two nonoverlapping cortical sites could result in speech arrest, one with and the other without mentalis muscle activity, which were motoric and nonmotoric, respectively. They found that the more posterior (motor) speech arrest site was bilateral, whereas the anterior nonmotor speech arrest site was lateralized to the left but was also more difficult to activate. It was not possible to determine if the anterior site represented interference with language as opposed to activation of a negative speech area anterior to motor cortex.

A Question of Lateralization

Many of the early speech studies were intended to replicate the IAT or Wada test. Consequently, investigators reported the concordance of language lateralization as

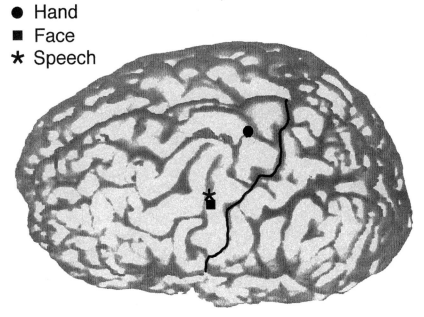

Figure 18-2 Map of behavioral responses after transcranial magnetic stimulation at different sites over the left lateral frontal region. Behavioral responses were speech interruption during counting, electromyographic (EMG) response from the first dorsal interosseous in the right hand, and EMG response from orbicularis oris in the right face. Letter placement shows center of gravity location for each type of response. The speech arrest site is congruous with F, the site for the orbicularis oris muscle.

determined by rTMS and IAT. Some of these studies suggested a high concordance between the rTMS and IAT results.[5,6,20] However, a later study by Epstein and colleagues[25] reported that rTMS does not fully replicate the IAT. They examined 17 presurgical epilepsy patients and compared the results of the IAT and rTMS-induced speech arrest. The IAT showed that all patients were left hemisphere dominant for language; however, the rTMS results suggested 70.6% were left dominant, 11.8% right dominant and 17.6% bilateral. Most patients underwent resection of the presumed epileptic focus, and it was found that the IAT results corresponded more closely with postoperative language deficits. This result suggested that language lateralization as determined by rTMS speech arrest yields results biased towards the right hemisphere.

Other studies of rTMS have used normal subjects, not allowing comparison with the IAT. Khedr and colleagues[26] examined the effect of handedness on language lateralization in normal subjects, using rTMS induced speech arrest as their language measure. They stimulated left and right fronto-temporal regions at 3 Hz using an intensity of 100% to 150% MT for 5 to 10 seconds. They found that regardless of handedness, speech arrest was typically lateralized to the left hemisphere (80.1%), less likely to be right dominant (7.4%) and sometimes bilateral (12.1%). Left-handed subjects had a much higher incidence of right (10.5%) and bilateral (15.8%) speech arrest compared with right-handers (4.2% and 8.4% respectively). They also found that ambidextrous subjects were left dominant (43%) or had bilateral representation (57%). Although they come closer to classical expectations, these results also suggest that establishing language dominance by rTMS shows some bias towards the right hemisphere.

Knecht and colleagues[27] examined language lateralization using a picture-word verification task while magnetically stimulating Wernicke's area. The investigators observed that many functional imaging studies on language had shown weak lateralization or bihemispheric activation in normal subjects. They focused on the extent to which language was represented bilaterally in normal subjects using TMS. They first established language dominance using functional transcranial Doppler sonography and then disrupted language processing using 1-Hz TMS at 110% MT for 600 seconds. They found an interesting relationship between the degree of language lateralization and the asymmetry of language disruption after unilateral TMS. They reported that subjects who were left dominant for language had increased RT for the task after rTMS of the left hemisphere but decreased RT when the right hemisphere was stimulated; the opposite pattern was observed in subjects right dominant for language. When they disrupted language processing in subjects with weak lateralization, they found that stimulation of either hemisphere had little effect on task performance.

The Words We Remember

Verbal working memory is the mechanism by which we are able to store verbal information in an active state, and manipulate it for a short period. Early TMS studies of verbal working memory yielded mixed results, demonstrating facilitation of verbal working memory[28,29] or no effect at all.[30,31]

There are two dominant models of working memory; one claims that the separation of frontal neuronal networks is based on function (i.e., maintenance or comparison versus monitoring or manipulation),[32] whereas the other suggests that different types of stimuli (or modalities) are processed by separate networks (i.e., verbal versus spatial).[33] To address this discrepancy, Mottaghy and colleagues[34,35] combined rTMS with PET and examined the cortical sites of activation, measured by regional cerebral blood flow (rCBF) before and after rTMS while subjects performed a verbal working memory task. They used the two-back letter task, in which subjects had to recall whether a letter was the same as two letters before its display. In their first study, they used very prolonged 4-Hz TMS at 110% MT for 30 seconds over left and right DLPFC.[34] They found that stimulation of either site caused an increase in the number of recall errors; however, there were differences between the sites in the amount that PET metabolic activity decreased. After left DLPFC rTMS, activity decreased only in the left prefrontal cortex; however, right DLPFC stimulation resulted in decreased activation of the right prefrontal cortex plus left and right parietal areas. They suggest that the decreased activation observed in these areas might reflect a decrease in number of outgoing signals for the verbal working memory task.

The investigators then conducted a study using the same TMS parameters to examine the role of the middle frontal gyrus (MFG), supramarginal gyrus, and inferior parietal cortex in verbal working memory.[35] They found a decrease in performance, but only after rTMS of the MFG or inferior parietal cortex, with the effect being bilateral for both sites. In their latest study, they focused their analysis on left and right MFG and on activation changes that occurred in connected sites, which presumably make up a working memory network.[36] Although their 2002 study showed bilateral involvement of MFG in verbal working memory, the subsequent results indicated an asymmetry, with greater effects on the left. However, the investigators argue that as rTMS of right MFG removes the correlation between rCBF and accuracy for left MFG, this result shows the functional interconnectivity between left and right MFG.

Motor Cortex Excitability and Language

A growing body of evidence suggests that the evolution of language in humans may have involved a "premotor mirror neuron system."[37] Much of the evidence for mirror neurons comes from monkey studies of specific neuronal populations contained in area F5. Typically, the same neurons would discharge if a monkey were to grasp an object and if it saw another monkey performing the identical action. Area F5 is the monkey homologue of Broca's area.[38]

Several TMS studies in humans have demonstrated an increase in motor cortex excitability when subjects hear language sounds or read aloud. Tokimura and colleagues[39] first demonstrated this effect by showing enhanced motor responses from hand muscles while subjects read aloud or during spontaneous speech. Fadiga and colleagues[40] reported an increase in the amplitude of motor-evoked potentials (MEPs) when subjects heard words containing a double "r" consonant. Lin and associates[41] showed a striking difference between musicians and non–musically trained (control) subjects. They recorded from the first dorsal interosseus muscle (FDI) and found an enhanced MEP for both left and right FDI while musicians read aloud, but only right FDI facilitation for the control subjects.

After the Tokimura study, Meister and colleagues[42] investigated the topographic specificity of the effect and its duration in the motor cortex. They used TMS at 120% MT and delivered it before, during, and after reading aloud while recording TMS-induced MEPs from the hand and leg areas of the motor cortex. They found that the MEP enhancement was topographically specific (i.e., it was present only after TMS of the hand area). This effect lasts only while speech is in progress, and does not occur before or after articulation. The investigators argue that their finding demonstrates the existence of "phylogenetically old connections" between the motor cortex and language areas. Seyal and colleagues[43] demonstrated that enhancement of MEPs from the first dorsal interosseous muscle (FDI) is specific for reading aloud. They used five different conditions: reading aloud or silently, repeating the word "cat," pursing the lips, and not responding at all. They delivered TMS at 110% MT in the period between a warning light at 1,500 ms before stimulus onset and the Go tone and in the period after the tone. They found enhanced MEPs predominantly during anticipation of the reading aloud condition, especially after TMS of the left hemisphere (Fig. 18–3). They argue that the MEP

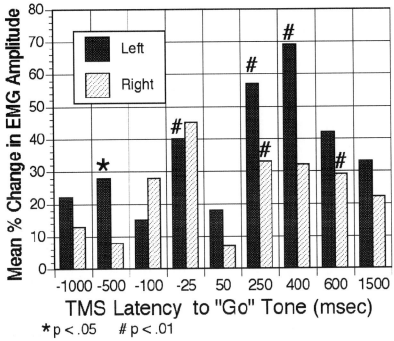

Figure 18–3 Hemispheric differences after transcranial magnetic stimulation (TMS) of the first dorsal interosseous (FDI) motor site while reading words aloud. The negative (−) values on abscissa indicate TMS delivered before the Go tone. There is a left hemisphere bias for electromyographic (EMG) amplitude enhancement, but right hemisphere TMS can also enhance EMG responses (see −100 and −25 ms). (Adapted from Seyal M, Mull B, Bhullar N, et al. Anticipation and execution of a simple reading task enhance corticospinal excitability. Clin Neurophysiol 1999;110:424–429.)

enhancement shares temporal characteristics with the contingent negative variation (CNV), readiness potential, and N400. The enhanced MEP was left hemisphere biased, sharing the same asymmetry with the CNV.

Enhancement 400 ms after stimulus onset appears to follow the time course of the N400. The N400 is modulated by the semantics of a word in relation to its sentence. The investigators suggest that if enhanced activity and the N400 share the same neural mechanism, the degree of activation should be related to semantic processing.

■ Conclusion

The studies reviewed in this chapter demonstrate that TMS has become a useful and accepted tool in the study of language. One of the original goals of TMS studies, the production of classical aphasias, has not come to pass. In the earliest of studies, the induction of speech arrest required TMS of extremely high frequency and intensity. However, as it became apparent that less extreme TMS parameters could produce behavioral effects, studies focused on investigating various components of language processing, including noun and verb production. Many, though not all, studies demonstrated a lateralization of function consistent with classical concepts of hemispheric specialization. Dissociations between different aspects of language were identified, as was the correlation between motor cortical activity and speech. Continued investigations using TMS combined with neuroimaging and neurobehavioral techniques will allow us to tease apart the complex, intricate relationships among the components of language. This can only lead to a deeper understanding of one of our most evolutionarily useful though least understood cognitive behaviors.

Acknowledgments

Dr. Epstein shares in patent rights to a magnetic stimulation coil. He is a consultant for Neuronetics, Inc., which plans to manufacture magnetic stimulators.

REFERENCES

1. Binder JR, Frost JA, Hammeke TA, et al. Human brain language areas identified by functional magnetic resonance imaging. J Neurosci 1997;17:353–362.
2. Papathanassiou D, Etard O, Mellet E, et al. A common language network for comprehension and production: a contribution to the definition of language epicenters with PET. Neuroimage 2000;11:347–357.
3. Price CJ. The anatomy of language: contributions from functional neuroimaging. J Anat 2000;197:335–359.
4. Wada J, Rasmussen T. Intracarotid injection of sodium Amytal for the lateralization of cerebral speech dominance. J Neurosurg 1960;17:266–282.
5. Jennum P, Friberg L, Fuglsang-Frederiksen A, et al. Speech localization using repetitive transcranial magnetic stimulation. Neurology 1994;44:269–273.
6. Pascual-Leone A, Gates JR, Dhuna A. Induction of speech arrest and counting errors with rapid-rate transcranial magnetic stimulation. Neurology 1991;41:697–702.
7. Flitman SS, Grafman J, Wassermann EM, et al. Linguistic processing during repetitive transcranial magnetic stimulation. Neurology 1998;50:175–181.
8. Epstein CM, Lah JK, Meador K, et al. Optimum stimulus parameters for lateralized suppression of speech with magnetic brain stimulation. Neurology 1996;47:1590–1593.
9. Wassermann EM, Blaxton TA, Hoffman EA, et al. Repetitive transcranial magnetic stimulation of the dominant hemisphere can disrupt visual naming as well as speech in temporal lobe epilepsy patients [abstract]. Ann Neurol 1996;40:525.
10. Mottaghy FM, Hungs M, Brugmann M, et al. Facilitation of picture naming after repetitive transcranial magnetic stimulation. Neurology 1999;53:1806–1812.
11. Wassermann EM, Blaxton TA, Hoffman EA, et al. Repetitive transcranial magnetic stimulation of the dominant hemisphere can disrupt visual naming in temporal lobe epilepsy patients. Neuropsychologia 1999;37:537–544.
12. Stewart L, Meyer B, Frith U, et al. Left posterior BA37 is involved in object recognition: a TMS study. Neuropsychologia 2001;39:1–6.
13. Shapiro KA, Pascual-Leone A, Mottaghy FM, et al. Grammatical distinctions in the left frontal cortex. J Cogn Neurosci 2001;13:713–720.
14. Matthews PM, Adcock J, Chen Y, et al. Towards understanding language organisation in the brain using fMRI. Hum Brain Mapp 2003;18:239–247.
15. Sparing R, Mottaghy FM, Hungs M, et al. Repetitive transcranial magnetic stimulation

16. Topper R, Mottaghy F, Brugmann M, et al. Facilitation of picture naming by focal transcranial magnetic stimulation of Wernicke's area. Exp Brain Res 1998;121:371–378.
17. Cappa SF, Sandrini M, Rossini PM, et al. The role of the left frontal lobe in action naming rTMS evidence. Neurology 2002;59:720–723.
18. Sakai KL, Noguchi Y, Takeuchi T, et al. Selective priming of syntactic processing by event–related transcranial magnetic stimulation of Broca's area. Neuron 2002;35:1177–1182.
19. Claus D, Weis M, Treig T, et al. Influence of repetitive magnetic stimuli on verbal comprehension. J Neurol 1993;240:149–150.
20. Michelucci R, Valzania F, Passarelli D, et al. Rapid-rate transcranial magnetic stimulation and hemispheric language dominance: usefulness and safety in epilepsy. Neurology 1994;44:1697–1700.
21. Pascual-Leone A, Dhuna A, Gates JR. Identification of presumed Broca's area with rapid transcranial magnetic stimulation. J Clin Neurophysiol 1991;8:344.
22. Epstein CM, Meador K, Loring DR, et al. Localization and characterization of speech arrest during transcranial magnetic stimulation. Clin Neurophysiol 1999;110:1073–1079.
23. Stewart L, Walsh V, Frith U, et al. Transcranial magnetic stimulation produces speech arrest but not song arrest. Ann N Y Acad Sci 2001;930:433–435.
24. Stewart L, Walsh V, Frith U, et al. TMS produces two dissociable types of speech disruption. Neuroimage 2001;13:472–478.
25. Epstein CM, Woodard JL, Stringer AY, et al. Repetitive transcranial magnetic stimulation does not replicate the Wada test. Neurology 2000;55:1025–1027.
26. Khedr EM, Hamed E, Said A, et al. Handedness and language cerebral lateralization. Eur J Appl Physiol 2002;87:469–473.
27. Knecht S, Floel A, Drager B, et al. Degree of language lateralization determines susceptibility to unilateral brain lesions. Nat Neurosci 2002;5:695–699.
28. Pascual-Leone A, Houser CM, Reese K, et al. Safety of rapid-rate transcranial magnetic stimulation in normal volunteers. Electroencephalogr Clin Neurophysiol 1993;89:120–130.
29. Duzel E, Hufnagel A, Helmstaedter C, et al. Verbal working memory components can be selectively influenced by transcranial magnetic stimulation in patients with left temporal lobe epilepsy. Neuropsychologia 1996;34:775–783.
30. Ferbert A, Mussmann N, Menne A, et al. Short-term memory performance with magnetic stimulation of the motor cortex. Eur Arch Psychiatry Clin Neurosci 1991;241:135–138.
31. Hufnagel A, Claus D, Brunhoelzl C, et al. Short-term memory: no evidence of effect of rapid-repetitive transcranial magnetic stimulation in healthy individuals. J Neurol 1993;240:373–376.
32. Petrides M. Impairments on nonspatial self-ordered and externally ordered working memory tasks after lesions of the mid-dorsal part of the lateral frontal cortex in the monkey. J Neurosci 1995;15:359–375.
33. Goldman-Rakic PS. Circuitry of primate prefrontal cortex and regulation of behavior by representational memory. In Plum F, Mountcastle F (eds): Handbook of Physiology, vol 5. Washington, DC, The American Physiological Society, 1987:373–517.
34. Mottaghy FM, Krause BJ, Kemna LJ, et al. Modulation of the neuronal circuitry subserving working memory in healthy human subjects by repetitive transcranial magnetic stimulation. Neurosci Lett 2000;280:167–170.
35. Mottaghy FM, Doring T, Muller-Gartner HW, et al. Bilateral parieto-frontal network for verbal working memory: an interference approach using repetitive transcranial magnetic stimulation (rTMS). Eur J Neurosci 2002;16:1627–1632.
36. Mottaghy FM, Pascual-Leone A, Kemna LJ, et al. Modulation of a brain-behavior relationship in verbal working memory by rTMS. Brain Res Cogn Brain Res 2003;15:241–249.
37. Fitch WT. The evolution of speech: a comparative review. Trends Cogn Sci 2000;4:258–267.
38. Petrides M, Pandya DN. Dorsolateral prefrontal cortex: comparative cytoarchitectonic analysis in the human and the macaque brain and corticocortical connection patterns. Eur J Neurosci 1999;11:1011–1036.
39. Tokimura H, Tokimura Y, Oliviero A, et al. Speech-induced changes in corticospinal excitability. Ann Neurol 1996;40:628–634.
40. Fadiga L, Craighero L, Buccino G, et al. Speech listening specifically modulates the excitability of tongue muscles: a TMS study. Eur J Neurosci 2002;15:399–402.
41. Lin KL, Kobayashi M, Pascual-Leone A. Effects of musical training on speech-induced modulation in corticospinal excitability. Neuroreport 2002;13:899–902.
42. Meister IG, Boroojerdi B, Foltys H, et al. Motor cortex hand area and speech: implications for the development of language. Neuropsychologia 2003;41:401–406.
43. Seyal M, Mull B, Bhullar N, et al. Anticipation and execution of a simple reading task enhance corticospinal excitability. Clin Neurophysiol 1999;110:424–429.

19 Other Cognitive Functions

Marjan Jahanshahi

In its short lifespan, transcranial magnetic stimulation (TMS) has become a powerful technique for investigating brain function, including cognition. Application of TMS to cognitive function has provided the unique opportunity to create a transient functional "lesion" in normal brains to determine whether the contribution of a brain area to a given task is necessary, the timing of this contribution, and the functional connectivity and interaction between remote areas of the brain. TMS has been used to map the plasticity associated with learning and reorganization of cognitive function after brain injury or neurological illness and to assess the pharmacology of cognition. TMS has the potential for treating a range of cognitive deficits. These applications are rapidly making TMS an indispensable tool in cognitive neuroscience.

TMS is a descendent of the technique of mapping brain-behavior relations with intracranial electrical stimulation (IES) of the cortex during surgery. Compared with IES, TMS has several advantages. Most importantly, it is noninvasive; electrical stimuli that are delivered to the brain through the intact scalp. Second, although IES is only used in patients with brain pathology (commonly those with epilepsy undergoing surgery after years of chronic medication intake), TMS is most frequently applied to study brain-behavior relationships in healthy volunteers not taking any medication. Third, unlike IES, TMS provides the possibility of repeating the assessment over time or during performance of other tasks. These advantages have rapidly established TMS as a useful tool for investigation of brain-behavior relationships in neuroscience. TMS started life in the motor physiology laboratory,[1] but even in its infancy, it was applied to the study of aspects of cognitive function.[2] Application of TMS to the investigation of cognitive function, which as recently as 2000 was considered an emerging field,[3] has blossomed to such an extent that there are now a special edition of a journal,[4] a book,[5] and several reviews[3,6,7] specifically focusing on this topic.

In this chapter, I first outline some methodologic considerations of using TMS to study cognition. I then describe the range of applications of TMS to cognition and speculate about where this emerging field may be heading in the future. Although some of the pertinent literature is discussed or summarized in tables, an exhaustive review of the literature is outside the scope of the chapter. The preceding chapter covered the application of TMS to the study of language, and this chapter focuses on other facets of cognition.

A dictionary definition of *cognition* is any process whereby a person acquires knowledge. Cognition refers to higher mental processes, including perception, attention, memory, learning, language, thinking, and reasoning. It is debatable whether higher-order motor processes such as response selection, response set, timing, motor imagery, or action observation should also be considered at least partly cognitive. Although ordinarily I would argue that they are, I have not considered TMS studies concerned with these higher-order motor processes in this chapter primarily because they are covered elsewhere in this book. Cognition is the topic of interest in cognitive psychology and neuropsychology. The field of cognitive psychology deals with how we gain information of the world; how this information is represented, transformed, and stored; and how the resulting knowledge is used to direct our behavior. Neuropsychology is concerned with the behavioral expression of brain dysfunction and involves investigation of disorders of attention, perception, memory, language, thought, and reasoning in neurological patients or after injury to the brain. Modern cognitive neuroscience emerged from the bridging of neuropsychology and cognitive psychology with other disciplines such as physiological psychology and neurophysiology. TMS has become a powerful tool for the study of brain-behavior relationships by cognitive neuroscientists and has proved to be of value in the examination of cognitive psychological models and in demonstrating dissociations of function normally revealed by neuropsychological assessment of patients.

Multitechnique Approach to Cognitive Neuroscience

Churchland and Sejnowski[8] compared a number of techniques used in neuroscience in terms of temporal and spatial resolution (Fig. 19-1). Although positron-emission tomography (PET) and functional magnetic resonance imaging (fMRI) have excellent spatial resolution in the millimeter range but poorer temporal resolution in the seconds range, the reverse is true for high-density electroencephalography (EEG) and magnetoencephalography (MEG), with a temporal resolution in the millisecond range and a spatial resolution of about 0.5 cm. Compared with these techniques, TMS has intermediate spatial and temporal resolution, respectively, in the range of 0.5 to 1 cm on the scalp and of 50 to 100 ms. TMS has two major limitations. First, stimulation is not very focal and has a peak value at a defined position under the coil and then falls off with distance. Second, in contrast to PET and fMRI, which provide data from the whole brain, TMS can mainly be delivered to only one area of the brain close to the surface at any given time.

If PET and fMRI have provided neuroscientists with a sort of window through the skull for investigating which brain networks

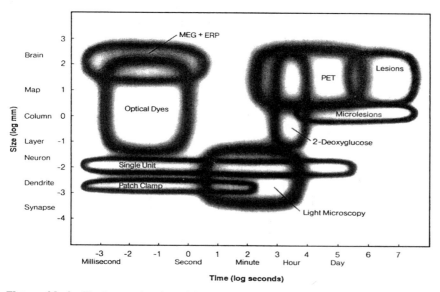

Figure 19-1 The temporal and spatial resolution of techniques used in neuroscience. (From Churchland PS, Sejnowski TJ. Perspectives on cognitive neuroscience. Science 1988;242:741-745.)

are active in a subject who is awake, alert, and engaged in task performance, TMS is the only technique that allows noninvasive intervention with these networks under the same circumstances. With PET, fMRI, or EEG, the effect of experimental manipulation of behavior on blood flow or electrical activity is recorded; with TMS, processing in the brain is experimentally "manipulated" and the effects of this intervention on behavior is measured. TMS has another unique feature. If task performance is disrupted by TMS, it shows that activity is necessary for, rather than just associated with, task performance.

■ Design and Procedural Considerations in the Application of Transcranial Magnetic Stimulation to Study Cognition

Choice of Sensitive Measures

There are similarities and differences between the application of TMS to investigate cognitive and motor function. The main and procedurally important difference is that for stimulation over the motor cortex, which is commonly the primary target site in the study of motor function, TMS produces an observable and measurable effect on muscle activity in the form of a muscle twitch or motor evoked potential (MEP). In contrast, with the exception of TMS over the occipital cortex, which can give rise to phosphenes,[9,10] TMS over other cortical sites that are commonly of interest in the study of cognition does not produce any observable effects and has to be quantified in terms of indices of task performance. Diverse measures have been used in studying the effects of TMS on cognition. Measures of the latency[11–13] or accuracy[14] of responses, the latency of the P300 component of cognitive event-related potentials,[13,15] the introduction or alteration of some specific response bias,[16,17] the interruption of ongoing behavior such as speech arrest,[18,19] and the production of a deficit[20,21] have all been used to assess the effects of TMS on specific cognitive processes. With the combined use of TMS and imaging,[22–26] blood flow or blood oxygen level–dependent (BOLD) measures of brain activation are also used to quantify the effect of TMS on cognition. The sensitivity of the measures selected is of key importance. For example, in studies of the effects of repetitive TMS (rTMS) on random number[16] or letter[17] generation, my colleagues and I have found that among numerous measures of randomness, those reflecting serial response bias [such as count score 1 that measures habitual counting by ones (e.g., 1, 2, 3, 4) or a digram repetition index that measures the repetition of stereotyped digrams (e.g., A, B, C)], which have been shown to be sensitive to capacity demands, were also the most sensitive to the effects of rTMS. In contrast, other measures of randomness quantifying the overall distribution of responses (e.g., chi-square test), repetition avoidance (e.g., repeated pairs), or cycling through the set of possible responses (e.g., median gap score) were not significantly affected by rTMS.

Experimental Design

TMS is relatively painless, but it is not devoid of sensation. A loud click accompanies the discharge of the stimulator through the coil, and the subject blinks and experiences a tactile sensation on the scalp and a muscle twitch at the site of stimulation. All of these sensory inputs may potentially interfere with performance of cognitive tasks. An important consideration for experimental design is to exclude any effects produced by these sensory phenomena and to show that any disruptive effects observed are caused by interference with brain function. Several types of experimental design can control for these sensory effects.[3] In a *control site* design, the effects of TMS at a target and a control site are compared, with TMS predicted to produce an effect over the target but not the control site. In a *control task* design, the effects of TMS on experimental and control tasks are compared, with TMS predicted to produce an effect on the target task incorporating the process of interest but not on the control task. In a third *control time* design, the effects of TMS at several points in time during task performance are compared, with TMS predicted to produce an effect at particular times in the course of task performance and not at other times. However, because sensory input at particular times could have the same effect, a control time design does not incorporate adequate control for the interfering effects of sensory phenomena unless combined with one of the other two designs. Two or all three of these designs can be employed in a single study.[27,28]

Another possibility is to compare real TMS with sham stimulation. Sham stimulation can be achieved in two ways: by tilting the coil away from the scalp or by using special sham coils. It has been suggested that sham stimulation through tilting the coil away from the head may still produce some effects on the underlying cortex. Although these sham stimulation techniques adequately control for the auditory stimulation associated with real TMS, they do not control as well or at all for the tactile sensations. This needs to be improved to increase the face validity of such sham techniques and to ensure that sham stimulation has similar credibility for subjects as real TMS.

Another design variation that allows for control of the nonspecific sensory effects of stimulation is to separate cognitive task performance and TMS in time. The demonstration that low-frequency or high-frequency rTMS applications have long-lasting depression and facilitation effects on a target cortical site[29,30] means that there are two main methods for using TMS in the study of cognition. The first method is to deliver TMS pulses during task performance. Most TMS studies of cognition have used this method and have employed one of the approaches delineated earlier to control for the nonspecific effects of TMS. The second method is to start with a period of 5 to 10 minutes during which TMS is delivered over the target site, after which the subject engages in the cognitive task of interest. This method has been successfully used in a number of TMS studies of cognition.[13,31–34] When using the second method, there is no need to control for the nonspecific sensory, attentional, or behavioral effects of stimulation because the TMS and cognitive task performance are separated in time.

Determination of Stimulation Parameters and Localization of the Stimulation Site

Several procedural considerations are common to all TMS studies. Decisions must be made about the size, shape, and orientation of the stimulating coil; the intensity of stimulation; and the number of trials for obtaining a reliable effect. Depending on the particular design adopted for the study, suitable control sites or control tasks or control times of stimulation have to be selected for comparison and to demonstrate the specificity of TMS effects on a target site, task, or time of stimulation. Choice of the site and time of stimulation in TMS studies is usually based on previous information, such as that from imaging or lesion or experimental studies. For TMS on cortical sites other than the motor cortex, the intensity and frequency of stimulation and coil orientation are often rather arbitrarily selected on the basis of information for the motor cortex. For rTMS, it is also necessary to decide on the number of pulses in a train and the intervals between them. Because rTMS does carry a risk of epileptic seizures, careful screening of subjects and adherence to safety guidelines[35,36] are crucial. All safety studies have been concerned with TMS over the motor cortex, and applications of TMS to other cortical sites to investigate cognitive function is based on the assumption that the ranges of intensity and frequency of stimulation that are safe over the motor cortex are also safe when applied to other cortical sites. A questionnaire for screening the suitability of subjects for participation in TMS studies is available.[37] Informed consent should be obtained from all subjects.

After the experimental design has been worked out and the most suitable parameters of stimulation have been chosen, the issue of the precise localization of the site of stimulation has to be considered. Several different approaches have been used to achieve localization of cortical sites other than the motor cortex. A common approach for localization in TMS studies[16,38,39] has been to use anatomical landmarks such as the nasion and inion together with standardized measurements set out in the 10–20 system employed for electrode placement in EEG and event-related potential (ERP) studies. This method has the problem of intersubject variability in the precise location of the target site in relation to the anatomical landmarks. A second approach has been to measure and mark the site of TMS on the head using the first method and then try to overcome intersubject variability by exploring the optimal stimulation site for each subject by moving the coil in 1-cm steps in a grid around the marked point.[40] Some investigators have used MRI as an aid to precise localization of the coil over the target area in individual subjects,[41,42] which also improves the reproducibility of coil position across stimulation blocks in the same subject.

Transcranial Magnetic Stimulation as an Investigatory Tool in Cognitive Neuroscience

The demonstration by Amassian and colleagues[2] that TMS over the occipital cortex interfered with visual perception of letter triads is probably the first application of the technique to disrupt a cognitive process. In the study of cognition, TMS has been used in a variety of ways: to establish functional specialization of target areas; to examine functional connectivity, functional interaction, and integration; to investigate the time course of cognitive processes; to demonstrate plasticity of function associated with learning or reorganization of function after brain injury or illness; and to assess the pharmacological substrates of cognition. Current and potential future applications of TMS to the study of cognition are considered in the following sections.

Establishing Functional Specialization

Traditionally, cognitive neuroscience has relied on information from animal lesion studies and investigation of patients with focal brain injury to identify the specific role played by key brain areas. The former method has the drawback of cross-species generalization of the results and the fact that some higher cognitive functions are unique to humans. The latter method has the problem that damage is rarely focal and what is being assessed is the functioning of the rest of the brain after long-term reorganization in the absence of the contribution from the damaged site. In light of this, the fact that TMS creates a transient and reversible functional "lesion" in the target area constitutes a powerful tool for establishing whether the contribution of the target area is necessary or essential for performance of the task under study and identifying the role played by it.

When reviewing the applications of TMS to cognition, Jahanshahi and Rothwell[3] observed that most investigations had used TMS to demonstrate that the target area was essential for task performance. This probably represents the widest application of TMS in the study of cognition. As summarized in Table 19–1, TMS over frontal, temporal, parietal, and occipital sites has been used to demonstrate functional specialization of target areas for aspects of attention, perception, memory, language, and vision. This type of application of TMS is often informed by previous information from imaging studies of animals or neuropsychological assessment of patients. In some of the studies summarized in Table 19–1, TMS revealed the functional relevance or role of the target area during task performance or allowed testing of specific models or hypotheses. An example of this is our own work on random number generation (RNG). With PET, my colleagues and I[70] found activation of the left dorsolateral prefrontal cortex (DLPFC) during RNG. To test the hypothesis that the role played by the left DLPFC in RNG is to suppress habitual counting to allow the strategic selection of numbers in a random fashion, we[16] compared rTMS over the left and right DLPFC and medial frontal cortex. Only TMS over the left DLPFC significantly increased measures of habitual counting (count score 1) and did not affect other measures of randomness such as repetition avoidance or cycling, confirming the proposed role for this area during RNG. Other investigators have used imaging and TMS in tandem to identify the functional relevance of the areas activated. Two examples of such work considered visuospatial imagery. Kosslyn and colleagues[32] provided convergent evidence from PET and rTMS that BA17 is involved in and essential for visual imagery. fMRI and 1-Hz rTMS during performance of a mental clock task were used by Sack and coworkers,[34] who found that although visuospatial imagery was associated with bilateral activation in the intraparietal sulcus region, only rTMS over the right posterior parietal

Table 19-1 STUDIES USING TRANSCRANIAL MAGNETIC STIMULATION TO INVESTIGATE FUNCTIONAL SPECIALIZATION OF SPECIFIC BRAIN AREAS IN PARTICULAR COGNITIVE TASKS

Study	Topic Investigated	TMS Parameters Used
TMS over the Prefrontal Cortex		
Pascual-Leone and Hallett (1994)[14]	Working memory in a delayed response task	rTMS, 5 Hz, 110% of MT
Grafman et al. (1994)[20]	Free recall of verbal material	rTMS, 20 Hz, 120% of MT
Sabatino et al. (1996)[43]	Controlled attention during visual search tasks with verbal or nonverbal stimuli	TMS started at 10% of maximum output (1.5 T) and increased in steps of 5% to 10%, TMS delivered continuously at intervals of 2 s (0.5 Hz)
Jahanshahi et al. (1998)[16]	Suppression of habitual counting during random number generation	rTMS, 20 Hz, active MT
Jahanshahi and Dirnberger (1999)[17]	Suppression of stereotyped responses during random letter generation	rTMS, 20 Hz, active, MT
Flitman et al. (1998)[27]	Linguistic processing in a picture-word verification	rTMS, 15 Hz, 20% of output of Cadwell High Speed stimulator
Hong et al. (2000)[44]	Visual working memory	rTMS, 5Hz at 80% of MT or 50% of maximal output of Magstim Rapid 2 T stimulator
D'Alfonso et al. (2000)[45]	Selective attention to perceived threat of angry faces	rTMS, 0.6 Hz for 15 minutes, 130% of MT
Evers et al. (2001)[13]	Impact of spTMS and rTMS on subsequent visual oddball P300 ERPs	20 Hz rTMS or spTMS, 95% of MT
Jing et al. (2001)[15]	Impact of rTMS on subsequent auditory oddball P300 paradigm	10 Hz rTMS, 2 trains, each with 30 pulses with inter-train interval of 5 min, 100% of MT
Rossi et al. (2001)[21]	Encoding and retrieval of visual information	rTMS trains, 20 Hz, 500 ms, 10% below MT
Boroojerdi et al. (2001)[12]	Analogical reasoning	rTMS 10% below MT, trains of 10-s duration, 5 Hz frequency
Mull and Seyal (2001)[46]	Working memory on a 3-back task	spTMS, 15% above MT
Oliveri et al. (2001)[47]	Visual-object and visual spatial working memory	spTMS, 130% of MT
Grosbras and Paus (2002)[41]	Shifting visual attention	spTMS, 5% above MT, average rate of one pulse every 5 s
Cappa et al. (2002)[48]	Action naming	rTMS, 500 ms, 20-Hz trains, 10% below resting MT
Mottaghy et al. (2002)[33]	Visual working memory	rTMS, 1 Hz at 90% of resting MT
Sandrini et al. (2003)[49]	Encoding and retrieval of words	rTMS, 500 ms, 20 Hz, 10% below MT
Rami et al. (2003)[50]	Episodic memory	rTMS, 5Hz, 10% below MT
Koch et al. (2003)[39]	Time perception in a time estimation reproduction task	rTMS, 1 Hz 10-min duration, 90% of MT
TMS over the Motor Cortex, SMA, and Premotor Cortex		
Muri et al. (1995)[51]	Performance of sequences of memory-guided saccades	spTMS, 80% to 90% of maximum output, TMS delivered during presentation, memorization or executive phase saccade sequences
Pascual-Leone et al. (1996)[11]	Implicit learning of motor sequences	rTMS, 5Hz, 115% of MT
Haggard and Magno (1999)[52]	Awareness of action	spTMS, 160% of resting MT
Ganis et al. (2000)[53]	Mental rotation	spTMS, 120% of MT
Harmer et al. (2001)[54]	Processing of angry versus happy facial expressions	rTMS, 300 ms duration, 110% of active MT
Rushworth et al. (2002)[55]	Task switching	rTMS, 10 Hz, 5% above MT

TMS over the Parietal Cortex

Study	Function/Task	TMS parameters
Pascual-Leone et al. (1994)[56]	Visual extinction of contralateral visual stimuli	rTMS, 25 Hz, 115% of MT
Seyal et al. (1995)[57]	Increased sensitivity to ipsilateral cutaneous stimulation	spTMS, intensities ranging from 85% to 100% of maximum output
Ashbridge et al. (1997)[40]	Pop-out versus conjunction search	spTMS, 80% of maximum output of Magstim 200
Oliveri et al. (2000)[58]	Discrimination of electrical stimuli	Paired-pulse TMS, conditioning and test stimuli at 70% and 130% of MT, inter-stimulus intervals of 1, 3, 5, 7, 10, 15 ms
Fierro et al. (2000)[59]	Contralateral neglect induced by right parietal rTMS	rTMS, 25 Hz 400 ms, 15% above MT
Hilgetag et al. (2001)[31]	Ipsilateral enhancement of visual attention	rTMS, 1 Hz 10 minutes, at 90% of MT
Oliveri et al. (2001)[47]	Visual spatial working memory	spTMS, 130% of MT
Lewald et al. (2002)[60]	Spatial hearing	rTMS, 1 Hz for 10 min, 60% of maximum output of Magstim 2.2 T
Sack et al. (2002)[34]	Visuospatial function	rTMS, 1 Hz, 110% of subject's MT, single train of 12 min
Bestmann et al. (2002)[61]	Visuomotor mental rotation	rTMS, 20 Hz at 120% of resting MT
Bjoertomt et al. (2002)[62]	Spatial neglect in near space	rTMS, 500 ms duration, 65% of maximal output of Magstim Super Rapid stimulator

TMS over the Temporal Cortex

Study	Function/Task	TMS parameters
Grafman et al. (1994)[20]	Free recall of verbal material	rTMS, 20 Hz, 120% of MT
Pascual-Leone et al. (1991)[18]	Arrest of speech with frontotemporal TMS	rTMS started with 60% maximal output intensity at 8 Hz, then 16 Hz, and finally 25 Hz
Jennum et al. (1994)[19]	Arrest of speech with frontotemporal TMS	rTMS, 30 Hz for 1 s, intensity increased until speech arrest occurred
Topper et al. (1998)[63]	Facilitation of picture naming	rTMS, 10% above MT, 35% to 55% of maximum output of Novametrix 1.5T; TMS preceded or followed picture presentation at different intervals
Stewart et al. (2001)[64]	Object recognition	rTMS, 10 Hz for 600 ms, at 75% of maximum output of 2 T Magstim 200
Oliveri et al. (2001)[47]	Visual-object working memory	spTMS, 130% of MT

TMS over the Occipital Cortex

Study	Function/Task	TMS parameters
Amassian et al. (1989)[2]	Letter identification	spTMS, TMS delivered at different intervals after presentation of letters
Kammer (1999)[9]	Induction of phosphenes and transient scotomas	spTMS, 70% of maximal output (0.8 T)
Beckers and Homberg (1991)[65]	Identification of letter trigrams and memory scanning	spTMS, 50%, 60%, 70%, 80%, 100% of maximum output of Cadwell 2 T TMS delivered at different intervals after presentation of stimuli
Beckers and Zeki (1995)[28]	Visual motion perception	spTMS, intensity of 1 or 1.4 T or more, TMS delivered at different intervals before or after presentation of visual stimuli
Zangaladze et al. (1999)[66]	Tactile discrimination of orientation	150% of relaxed MT, inter-stimulus interval of at least 1 s
Kosslyn et al. (1999)[32]	Visual imagery	1 Hz rTMS, 90% of MT, single train of 10 min
Cowey and Walsh (2000)[67]	Induction of phosphenes	spTMS or rTMS up to 10 Hz maximal output (2 T)
Rushworth et al. (2001)[68]	Orienting and attention	rTMS, 10 Hz 500 ms, at MT
Theoret et al. (2002)[69]	Perception and storage of motion aftereffect	4 Hz rTMS, 70% of maximum output or 110% of MT
Bjoertomt et al. (2002)[62]	Spatial neglect in far space	rTMS, 500 ms duration, 65% of maximal output of Magstim Super Rapid stimulator

ERPs, event-related potentials; MT, motor threshold; rTMS, repetitive transcranial magnetic stimulation; spTMS, single-pulse transcranial magnetic stimulation; T, Tesla.

cortex resulted in impaired performance. This indicates that the right parietal cortex can compensate for transient disruption of processing in the left parietal cortex. Such asymmetric parietal involvement is consistent with theories of hemineglect.

The main focus of neuropsychology is on revealing cognitive deficit in patients, and several approaches are adopted in achieving this. The first approach is establishing the nature of cognitive deficits in a particular group of patients with a brain lesion, injury, or neurological illness by comparing their performance with that of healthy controls. Second, the specificity of cognitive deficits is determined by showing that they are present in patient group 1 but not group 2. The third approach is identifying dissociations in function by demonstrating that a patient group is impaired on task A but not task B. Demonstrating double dissociations—that patient group 1 is impaired on task A but not task B, whereas patient group 2 shows the reverse pattern of impairment on task B but not task A—is another key neuropsychological approach. All these approaches are amenable for investigation of cognitive function with TMS.

In neuropsychology, dissociations (e.g., Broca's demonstration that speech-production systems could be impaired while speech comprehension was intact after damage to Broca's area) and double dissociations (e.g., Wernicke's demonstration of impaired comprehension with preserved production of speech with damage to Wernicke's area) are particularly powerful tools for establishing independence of cognitive functions. TMS is ideally suited for demonstrating such dissociations between cognitive functions in the intact brain. For example, the hemispheric encoding and retrieval asymmetry (HERA) model of prefrontal activation postulated by Tulving and colleagues[71] is supported by functional imaging studies. However, these show differential activation of the left and right prefrontal cortex during encoding and retrieval and do not establish whether these activations are respectively necessary for encoding versus retrieval of episodic information. Neuropsychological studies have provided crucial evidence about the brain structures involved in memory, but distinguishing encoding from retrieval deficits in patients is not easy, and from the neuropsychological evidence, the role of the frontal cortex in episodic memory is not immediately evident. It is possible to test the HERA hypothesis and establish the functional relevance of the frontal activation observed in imaging studies with rTMS, and Rossi and coworkers[21] have done just that. They applied trains of subthreshold rTMS (20 Hz, 500 ms) over the left or right dorsolateral prefrontal cortex in 16 subjects during encoding or retrieval of colored magazine pictures, resulting in four active TMS conditions: left DLPFC encoding, left DLPFC retrieval, right DLPFC encoding, and right DLPFC retrieval. A baseline, no TMS and a sham condition were also included. In addition to reaction times, measures of accuracy of recognition of the pictures were obtained. As predicted by the HERA model, rTMS over the left and right DLPFCs, respectively, reduced discrimination (d') of the "already seen" from "never seen" pictures only when applied in the encoding and retrieval phases, respectively. This interaction is shown in Figure 19–2. The results of this study are a good example of the way in which TMS can be employed to assess theoretical neuropsychological models, identify the functional relevance of brain areas activated during imaging studies, and confirm double dissociations of function. In a subsequent study with verbal material, Sandrini and associates[49] confirmed the role of the right DLPFC in retrieval but found that TMS over the right and the left DLPFCs during encoding disrupted episodic memory specifically for semantically unrelated but not related words. This suggests that for verbal material, the left and right DLPFCs play a role in encoding when novel information has to be remembered.

Another way in which TMS has been applied to demonstrate functional specialization is by the virtual lesion mimicking the effect of a true lesion. Imaging has established that a network consisting of the posterior parietal cortex, the dorsal premotor-prefrontal cortex, and the anterior cingulate is involved in visuospatial perception.[72,73] Fiero and colleagues[59] used a task that required subjects to judge whether a tachistoscopically

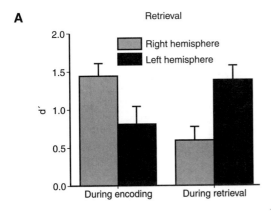

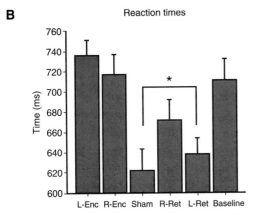

Figure 19–2 The effect of repetitive transcranial magnetic stimulation (rTMS) on retrieval (**A**) and reaction times (**B**). The hemispheric interaction of rTMS on the measure of retrieval, d' (i.e., discrimination, the ability to distinguish between "already seen" and "never seen" pictures) is shown in **A**. Reaction times are shortened by active rTMS and sham, regardless of the site. (From Rossi S, Cappa SF, Babiloni C, et al. Prefontal cortex in long-term memory: an "interference" approach using magnetic stimulation. Nat Neurosci 2001;4:948–952.)

presented horizontal line was transected by a short vertical bar centrally or to the right or left of center. They compared 25-Hz rTMS over the left or right parietal cortex with sham stimulation in 11 healthy volunteers. Without TMS, normal subjects showed pseudoneglect and overestimated the length of the left side of the line. With right parietal TMS, the pseudoneglect was reduced, and normal subjects underestimated the length of the left side of the line relative to results of the no stimulation trials. This demonstrates that with rTMS over the right parietal cortex, normal subjects performed similar to patients with right parietal lesions who show left spatial neglect. rTMS over the right parietal cortex induced a transient spatial neglect in healthy subjects. These results have been replicated by Bjoertomt and coworkers.[62]

Future applications of TMS may be of value in dissecting component processes of cognitive functions and skills and in testing various "box and arrow" cognitive psychological models of function. If more powerful stimulators are rendered safer at higher intensities and frequencies, future applications of TMS to study functional specialization are likely to be extended to hitherto unexplored sites, including deeper and large structures such as the anterior cingulate to unravel its contribution to cognition and affect. The cerebellum has been successfully stimulated with the coil over the inion,[74] and this may prove useful in exploring the role of this structure in cognition.

Examining Functional Connectivity, Functional Interaction, and Integration

The use of TMS to create a transient functional lesion has largely addressed questions of functional specialization in the brain. Understanding how different areas of the brain interact to produce integrated function is a necessary first step in appreciating how even the simplest cognitive task is performed. TMS can reveal such functional interactions. For example, Tokimura and colleagues[75] showed that for right-handed subjects, although spontaneous speech increased the size of EMG responses bilaterally, reading aloud increased EMG responses in the right hand only. This suggests that reading aloud increases the excitability of the motor hand area in the dominant hemisphere, implying functional interaction between the speech and motor output functions of the dominant hemisphere. Functional interactions between different cognitive systems, such as between attention and memory, can be explored with TMS.

Emotion can bias attention and affect memory.[76,77] Using TMS, an interaction between emotional and attentional processing of stimuli has been demonstrated by d'Alfonso and associates.[45] The right and left prefrontal cortex have been proposed to play differential roles in processing of emotions and goal-directed approach or withdrawal behaviors. More specifically, the right prefrontal cortex is thought to mediate negative emotions and withdrawal behaviors, whereas the left prefrontal cortex is thought to mediate positive emotions and approach behaviors.[78,79] To test this hypothesis, d'Alfonso and colleagues[45] examined the effect of 15 minutes of 0.6-Hz rTMS over the left and right DLPFCs at 130% of motor threshold on attention to the perceived threat of angry faces assessed with an emotional Stroop task in 10 young healthy female participants. Selective attention to angry faces was measured by comparing the latency of color naming for neutral and angry faces. rTMS did not induce any significant changes in self-rated mood. They found that rTMS over the left prefrontal cortex induced selective attention to angry faces, whereas with rTMS over right prefrontal cortex subjects attended away from the angry faces. The results support the differential roles of the right and left prefrontal cortex in behavioral withdrawal and approach respectively.

The neural basis of consciousness and awareness of our own actions, thoughts, and percepts remains a puzzle that continues to intrigue cognitive neuroscientists. The ability of TMS to induce movement or to interfere with visual perception makes it suitable for creating dissociations between performance of a task and conscious awareness of the stimuli on which the performance is based. This can provide valuable information about the neural substrates of conscious awareness. Cowey and Walsh[67] used TMS to investigate the neural substrates of visual awareness. They assessed a patient (G.Y.) with total destruction of V1 in the left hemisphere. With TMS of V5 over the hemisphere with the intact V1, the patient perceived moving phosphenes, similar to the normal subjects. In contrast, no motion perception was produced in the blind hemifield with V5 stimulation over the hemisphere with a damaged V1. In a second peripherally blind patient (P.S.), whose optic nerves had been destroyed but whose V1 was intact, TMS over V5 resulted in perception of moving phosphenes. These results highlight that interaction of V5 with V1 is necessary for visual motion perception.

For motor function, it has been shown that TMS can provide information about patterns of connectivity between various brain regions such as corticocortical[80] or transcallosal areas.[81,82] It was proposed to use the paired-pulse TMS technique, which examines the effect of a subthreshold conditioning TMS pulse over one site on the size of a response evoked by a suprathreshold test TMS pulse over a second site to examine corticocortical or transcallosal patterns of connectivity during performance of cognitive tasks.[3] Pascual-Leone and Walsh[83] used paired-pulse TMS to test the hypothesis that feedback from the secondary visual areas such as MT/V5 to V1 is necessary for visual awareness. When TMS was applied to V1 before V5 or with paired-pulse TMS over V5, there was no change in the perceived movement of the phosphenes. In contrast, when V5 was stimulated before V1 with an inter-stimulus interval (ISI) of 5 to 45 ms, five of the eight subjects reported absence of phosphenes at an ISI of 25 ms, and all subjects reported that the phosphenes were stationary rather than moving at an ISI of 45 ms. These results confirm the importance of the backprojections from V5 to V1 for perception and awareness of visual motion. The Cowey and Walsh[67] and Pascual-Leone and Walsh[83] studies are good examples of how TMS has demonstrated that functional interaction between components of a network necessary for visual awareness.

TMS has been successfully used concurrently with PET[22,84] and in an interleaved fashion with fMRI.[23,26] Paus and coworkers[22] found a significant correlation between cerebral blood flow and the number of TMS pulse trains at the stimulation site over the frontal eye fields and more distally in several areas of the visual cortex and the supplementary eye field. These results highlight the potential value of the combined techniques for exploring functional connectivity in the intact human brain in normal subjects and patients with neurological or psychiatric disorders. Because TMS provides the unique possibility of transient intervention with brain networks, its concurrent application with PET or fMRI is ideally suited for addressing questions of functional integration during performance of cognitive tasks.

What is the impact of altering activity at a target cortical site on the pattern of activation of the rest of the neural network engaged by the task? This is precisely the type of question that is starting to be addressed. Mottaghy and associates[85] used a 30-second train of rTMS at 4 Hz concurrently with PET and showed that stimulation over the left and right DLPFCs disrupted performance on an *n*-back working memory task. However, rTMS over the two sites resulted in different patterns of regional cerebral blood flow change in other regions of the distributed network. rTMS over the left DLPFC was associated with significant deactivations over this site, whereas rTMS over the right DLPFC deactivated this area, the bilateral parietal areas, and the anterior cingulate, with the latter change approaching significance. Performance on a control 0-back task was not altered by rTMS over any of the sites. The concurrent use of PET demonstrated that although the behavioral effects of rTMS over the left and right DLPFC were identical, the distributed networks involved in producing these effects were not.

The combined TMS and imaging technique may prove valuable in examining hypotheses about normal and abnormal patterns of functional connectivity. For example, in light of the proposal that word generation during first-letter word fluency depends on frontal-temporal connectivity,[86] what would the impact of TMS over frontal versus temporal sites be on word generation and on cerebral blood flow measures of connectivity? Psychiatric disorders such as schizophrenia are considered to be associated with disconnections between distributed brain areas, and PET has provided evidence for abnormal patterns of frontal-temporal connectivity in these patients.[87] It may be possible to mimic such abnormal patterns of functional connectivity in the normal brain with TMS and to examine the blood flow correlates with concurrent PET scanning. Another way in which TMS could be fruitfully combined with imaging would be in PET ligand studies to examine the impact of TMS over key cortical sites on receptor binding. As outlined in the section on "Pharmacology of Cognition," such studies have been already undertaken. [^{11}C]raclopride PET studies by Strafella and colleagues[88,89] established that rTMS over the frontal or motor cortex was associated with the release of dopamine in the striatum, consistent with the known frontostriatal connectivity.[90]

Besides combining TMS with imaging, concurrent use of TMS with high-resolution EEG may prove of value in unraveling functional connectivity and particularly the time course of this during cognitive task performance. The work of Illomniemi and coworkers[91] with high-resolution EEG has shown that the effects of TMS over the left sensorimotor hand area spread to adjacent premotor areas within 5 to 10 ms and to homologous areas in the contralateral hemisphere within 20 ms. EEG coherence analysis provides a measure of functional connectivity during task performance by examining the correlation between EEG power spectra at distant sites. With PET or fMRI, excitatory and inhibitory effects cannot be easily distinguished, whereas the increase and decrease in EEG coherence are respectively interpreted as indicative of greater interregional connectivity and relative functional disconnection.[92] TMS combined with measures of EEG coherence is ideally suited for examining dynamic changes in connectivity in specific networks, such as those that accompany perceptual or motor learning or encoding or consolidation in memory.

Revealing the Time Course of Cognitive Processes

TMS can be used to identify the time window during which the contribution of an area is necessary for performance or the relative timing of the contribution of two or more areas to task performance. This type of work can be reliably done only when single-pulse TMS rather than trains of stimuli are used. Amassian and colleagues[93,94] and Beckers and Zeki[28] have used TMS to identify the time course of transfer of information in the visual system, such as from striate to extrastriate cortices and from visual cortex to higher cortical areas such as the inferior temporal lobe. Amassian and associates[2] delivered TMS over the occipital cortex at 20-ms intervals from 0 to 200 ms after presentation of three letters on a monitor. When delivered at 80- to 100-ms delays, TMS abolished or reduced correct perception of the letters. When TMS was delivered at intervals of less than 60 ms (i.e., before the arrival of visual representation of the letters at the visual cortex) or at intervals longer than 120 ms (i.e., after transmission of the visual representation from the visual cortex to other cortical areas), the letters were correctly identified. These results demonstrated that the contribution of the visual cortex to letter recognition was crucial only in the time window of 80 to 100 ms after stimulus presentation.

Beckers and Zeki[28] employed TMS to identify the time course of transfer of visual motion information in the visual system. Over V5, TMS at intervals of −20 to +10 ms relative to the onset of visual stimulation abolished motion perception, whereas TMS over V1 was marginally effective and only at delays of 60 to 70 ms after visual stimulation. They concluded that visual motion signals reach V5 before V1 and estimated that it takes about 30 to 50 ms for signals to travel between the two cortical sites. The anti-saccade task requires subjects to move their eyes in the opposite direction to a target stimulus presented to the left or right of a central fixation point. This requires intentional suppression of reflexive saccades to the target to volitionally generate saccades in the opposite direction. In monkeys, performance on the anti-saccade task is associated with neuronal activity in the dorsolateral prefrontal cortex, which is considered to hold this information "on line" to suppress or produce a response.[95] Imaging studies have shown that the parietal and prefrontal cortices are activated during performance of the anti-saccade task.[96] Terao and colleagues[42] used TMS over parietal or frontal cortex at 80, 100, and 120 ms during performance of an anti-saccade task to map the temporal evolution of processing in the relevant cortical regions. TMS over posterior parietal cortex produced an effect earlier (80 ms) than TMS over the frontal eye fields (100 ms), indicating a posterior to anterior flow of information during the anti-saccade task.

Fierro and coworkers[59] demonstrated that rTMS over the right parietal cortex induced a transient spatial neglect in healthy subjects. In a subsequent study, Fierro and associates[97] examined the timing of right parietal and frontal activity in this type of visuospatial perception by using single-pulse TMS over these sites delivered 150, 225, or 300 ms after the onset of the visual stimulus. Transient bias was observed only with stimulation over the right parietal cortex 150 ms after stimulus presentation. The timing of this neglect-like spatial bias is consistent with the timing of transfer of information from occipital to parietal cortex.[2]

Evidence from functional imaging has implicated a network consisting of the prefrontal and parietal cortices and the anterior cingulate in working memory.[98] TMS has been used to establish which components of this network are essential for different types of working memory, verbal or visuospatial.[33,44,46,85] Mottaghy and colleagues[99] examined the time course of parietal and prefrontal activation during a verbal working memory tasks. They used single-pulse TMS in six healthy volunteers. Stimulation was delivered to the left or right parietal or prefrontal cortices at 10 time points with 40-ms intervals between them, from 140 to 500 ms into the delay of a 2-back task (Fig. 19–3). In this task, the subject has to decide whether the letter presented on the current trial is the same as the letter presented two trials earlier. The task requires holding information in working memory. Performance on the 2-back task was compared with a control choice reaction time task. TMS over the parietal cortex interfered with task accuracy earlier than prefrontal stimulation, and the interference effect occurred earlier for stimulation over the right than the left hemisphere. The results were interpreted as indicating propagation of processing from posterior to anterior cortical sites, with information flow being faster in the right than the left hemisphere and converging in the left prefrontal cortex.

Future applications of TMS with high-density EEG or event-related fMRI may provide investigatory tools with high temporal resolution, which can be used to examine the time course of functional integration and to determine the order or duration of activation of specific components of a neural network engaged in a cognitive task to build up a dynamic map of the sequence of activation of cortical areas.

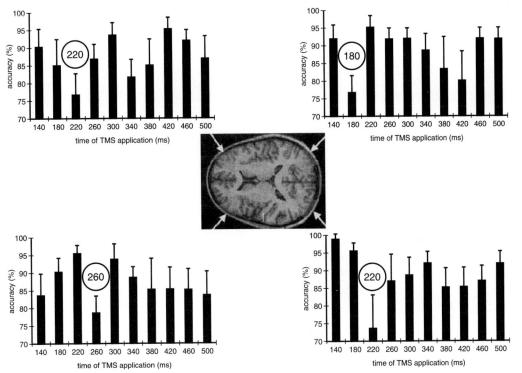

Figure 19–3 Accuracy in the 2-back working memory task as a function of time of transcranial magnetic stimulation (TMS) in the study of Mottaghy and colleagues.[99] The mean accuracy in the task was 87%. The TMS interference peaked at 180 ms at the right inferior parietal cortex, at 220 ms at the left inferior parietal cortex and right middle frontal gyrus, and at 260 ms at the left middle frontal gyrus.

Demonstrating Plasticity of Function

The noninvasive nature of TMS and the fact that it can be repeated across time has been valuable in the study of learning, plasticity, and reorganization of function after injury or in cases of neurological illness.

Change in Cortical Excitability with Learning. An important application of TMS has been to show the changes in excitability that occur in relevant brain areas with learning. Using motor cortical mapping with TMS, Pascual-Leone and colleagues[11] demonstrated that implicit learning during a serial reaction time task was associated with progressive expansion of the cortical output maps of the muscles involved in the task until explicit knowledge of the repeating sequence was developed, at which point the maps returned to their baseline topography (Fig. 19–4). This illustrated the rapid plasticity of cortical output in the course of learning. In a later study, Pascual-Leone and coworkers[100] showed that TMS over the contralateral DLPFC blocked implicit learning on a serial reaction time task, whereas TMS over the supplementary motor area did not interfere with learning, suggesting that the contribution of the former but not the latter area was essential for implicit learning. TMS also may produce differential effects on performance, depending on the degree of practice and training. Walsh and associates[101] examined the effect of TMS over the right parietal cortex during a visual search task before and after perceptual training on the task. After extensive training, the initially disruptive effect of TMS on performance disappeared but was reinstated when subjects were tested on a new visual search array. These studies indicate that TMS is a useful tool for investigating cortical plasticity and the changes that occur in function with learning.

294 Other Cognitive Functions

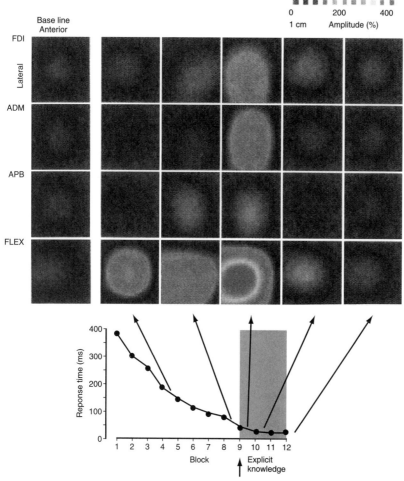

Figure 19–4 Cortical motor maps from all muscles tested and reaction times for a single representative subject during performance of the serial reaction time task. Motor output maps were obtained at baseline before the start of the serial reaction time (SRT) task and after blocks 4, 8, 9, 10, and 12 of the SRT task, with subjects at rest. Explicit knowledge was developed after block 9. (From Pascual-Leone A, Grafman J, Hallett M. Modulation of cortical motor output maps during development of implicit and explicit knowledge. Science 1994;263:1287–1289.)

Although the improved temporal resolution of event-related fMRI can help clarify the time course of engagement of cortical and subcortical areas during learning of cognitive or motor tasks, it is only with TMS that it is possible to establish which areas are *necessary* for learning or later skilled performance and when their contribution is essential. The noninvasive nature of TMS and the fact that it can be repeatedly administered to the same subject without any ill effects mean that the changes in cortical excitability and functional connectivity that occur with cognitive or motor learning in the course of development from childhood to adulthood to old age can be tracked through longitudinal follow-up studies.

Unraveling Reorganization of Function after Injury or Illness. Brain injury or neurological illness is often associated with reorganization of function to compensate for the loss of normal function resulting from the damage or illness. In a PET study of subjects blind from an early age, Sadato and colleagues[102] found activation of the visual cortex while discriminating tactile stimuli, including braille-type letters, whereas sighted subjects showed deactivation of this area. Cohen and coworkers[103] then used TMS to address the question of whether the increased activity in the visual cortex in the blind was contributing to tactile discrimination in someway or simply a feature of early damage unrelated to performance. Trains of TMS over the visual cortex increased errors during identification of braille or embossed roman letters by blind but not normal-sighted subjects, suggesting a functional reorganization of somatosensory processing in the blind individuals.

TMS has been used to examine the effect of neurological illnesses such as Parkinson's disease[104] or stroke[105] on patterns of cortical excitability and motor function. It should be possible to employ TMS to study the impact of neurological or psychiatric illness or brain injury on various aspects of cognitive function and the changes that may occur in these over time with progression of the illness or after surgery or rehabilitation. For example, TMS may help determine whether the restitution of activation patterns in networks involved in language processing that are associated with recovery of aphasia after stroke signify that the relevant areas have returned to normal functioning and to establish the nature and timing of any reorganization of function that may occur from the acute to more chronic phases. In seven previously right-handed individuals affected by left hemisphere stroke and presenting with aphasia, agraphia, and right hemiparesis (AARH), Papathansiou and coworkers[106] investigated the effect of a 40-minute session of rehabilitation therapy. This involved use of a writing prosthesis that enabled writing with the paralyzed hand by using proximal (shoulder) muscles. TMS revealed a significant increase in recruitment of ipsilateral cortical pathways after therapy, although only during writing and mainly in the distal muscles, which are normally involved in writing, rather than in the proximal muscles that actually performed the writing movements with the aid of the prosthesis during therapy. These results suggest that even in poorly recovered stroke patients with AARH, in the chronic phase, rehabilitation therapy aimed at increased use of the paretic hand, such as writing with a prosthesis, may induce recruitment of previously dormant ipsilateral corticoneuronal pathways.

Pharmacology of Cognition

Two PET studies have provided evidence for striatal dopamine release with rTMS over the prefrontal or primary motor cortex. In the first study, Strafella and associates[8] used [^{11}C]raclopride PET and showed that rTMS over the dorsolateral prefrontal cortex led to dopamine release in the ipsilateral caudate TMS. In a later study by the same group, 20-Hz rTMS over the left primary motor cortex at 90% of resting motor threshold resulted in a reduction of [^{11}C]raclopride binding in the left putamen compared with rTMS over the occipital cortex.[89] These studies have established that rTMS over the frontal cortex induces release of dopamine in the striatum, consistent with the known frontostriatal connectivity.[90] The same procedures could be fruitfully applied to study aspects of cognition, particularly attention and working memory, on which mesocortical dopamine is considered to exert a modulatory influence. For example, given the importance of dopamine to processing in the prefrontal cortex during working memory tasks,[107] what impact would rTMS over the DLPFC have on dopamine receptor binding while subjects engaged in a working memory task?

TMS has been used to investigate the effect of various pharmacological agents on motor cortex excitability.[108,109] In a similar fashion, TMS can be used to examine the modulatory influence of specific pharmacologic agents on cognition. The effects of TMS on performance of key tasks before and after administration of particular types of medication may clarify the modulatory impact of pharmacologic agents such as dopaminergic or cholinergic agonists or antagonists on cognitive processes of interest such as executive function and working memory mediated by the frontal cortices and memory processes mediated by the temporal cortices.

■ Transcranial Magnetic Stimulation as Therapy for Cognitive Deficit

There is evidence that rTMS alters the excitability of target cortical areas beyond the duration of the train so that its effects can be longer lasting.[110] This combined with the fact that TMS produces facilitatory effects on certain aspects of cognitive and motor function may open the way for future therapeutic applications of TMS. At particular intensities, rTMS over the motor cortex improved implicit learning in a serial reaction time task,[110] enhanced story recall, and produced faster reaction times in normal volunteers.[35] In a task requiring shifting visual attention, single-pulse TMS over the left frontal eye field facilitated responses to targets presented in the right hemifield.[41] rTMS over Wernicke's area reduced the latency of picture naming.[63] Similarly, high-frequency rTMS has been shown to result in an improvement in naming black and white drawings.[111] A 20-Hz subthreshold stimulation over the left prefrontal cortex enhanced action naming.[48] Application of 5-Hz rTMS over the left prefrontal cortex has also been reported to enhance analogic reasoning.[12] Significant reduction of scanning times on verbal and visuospatial selective attention tasks without any increase in errors has been documented with rTMS over the prefrontal cortex.[43]

Although 25-Hz rTMS over the parietal cortex has been shown to induce visual extinction of contralateral stimuli,[56] rTMS over the parietal cortex at parameters known to reduce cortical excitability (1 Hz for 10 minutes) resulted in ipsilateral enhancement of visuospatial attention.[35] The work of Olivieri and colleagues on the effect of TMS on detection of stimuli applied to the hands in normals[112] and patients with right hemisphere lesions[113] has revealed right hemisphere dominance for perception of such stimuli and established that spatial attention to the side opposite the lesion is enhanced and neglect is improved with transient disruption of the healthy left hemisphere. This suggests that rTMS may prove valuable in rehabilitation of neglect.

Evidence suggests that a large alpha power at rest and large event-related desynchronization (ERD) during task performance are associated with good cognitive performance.[114,115] In a study, specifically designed to investigate the possibility of enhancing cognitive performance with rTMS, Klimesh and colleagues[116] examined the effect of rTMS over the parietal or frontal cortex at individual upper alpha frequency (IAF)+1 Hz on subsequent cognitive performance during a mental rotation task. Two control conditions involved rTMS at lower IAFs: 3 Hz and 20 Hz (beta frequency). All conditions were compared with sham stimulation delivered with the coil rotated 90 degrees, which was applied in a randomized way with real rTMS. Only rTMS at upper IAF + 1 Hz over the frontal or parietal cortex resulted in significant improvement of accuracy of performance on the mental rotation task relative to sham stimulation. rTMS at IAF was associated with increased power during the reference interval and large ERD. It was interesting that most studies reporting facilitatory effects of TMS on performance used TMS at alpha frequency or its harmonics or subharmonics. They propose that these facilitatory effects may be partly mediated by the influence on alpha frequency.

Several different mechanisms may underlie the enhancement of cognitive performance by TMS. Some form of cross-modal facilitation associated with the acoustic and tactile components of TMS may contribute to the improvement of reaction times with stimulation. Changes in the excitability of the stimulated site may be a second mechanism. The enhanced performance on cognitive tasks with TMS may be related to the improved functioning of areas involved in task performance that are released from the inhibitory influence exerted by connected areas when the latter are targets of TMS.

It may be possible to exploit facilitatory effects of rTMS as part of cognitive rehabilitation programs, in the same way that rTMS over the DLPFC has been applied to treat chronic depression (reviewed by George and colleagues[117]). Such facilitatory effects, which have been mainly obtained in normal subjects, may have application in the rehabilitation of patients with memory or attentional deficits or those with unilateral neglect or certain types of aphasia.

High-frequency TMS above 5 Hz is considered to increase cortical excitability, and low-frequency TMS at 1 Hz is associated with decreased cortical excitability.[30,118] The increase in cortical excitability produced by high-frequency rTMS may prove useful in

Table 19-2 CURRENT AND FUTURE APPLICATIONS OF TRANSCRANIAL MAGNETIC STIMULATION IN COGNITIVE NEUROSCIENCE

TMS as an Investigatory Tool
- Functional specialization
    Whether the contribution of a target area is *necessary* for performance and what its specific role is
    Demonstration of specificity of a dysfunction to a particular patient group
    Demonstration of dissociations and double dissociations of function
- Functional connectivity, interaction, integration
    What is the impact of altering activity at a target cortical site on the pattern of activation of the rest of the network engaged by the task?
    For intracortical and transcallosal connectivity, what effect does subthreshold stimulation of one cortical area have on subsequent suprathreshold stimulation of another target cortical site?
- Timing of involvement of a target brain area
    The time window during which the contribution of a brain area is necessary for performance
    The relative timing of the contribution of two or more brain areas to task performance; the direction and timing of information flow between target sites
- Plasticity and reorganization of function
    What changes in excitability occur in target areas with learning?
    What changes in excitability occur in target cortical areas after brain damage, neurological, or psychiatric illness?
    What changes in excitability occur in target cortical areas after interventions such as surgery or cognitive rehabilitation programs?
- Pharmacology of cognition
    What changes in excitability occur in target cortical areas with different classes of medication and the time course of these changes?
    Effect of TMS on specific neurotransmitter systems involved in cognition examined with imaging

TMS as a Therapeutic Tool
- Increase speed of processing or responding
- Improve selective attention
- Enhance retention or accessibility of memories
- Improve picture and action naming
- Enhance implicit learning
- Improve attention to personal or extrapersonal space

TMS, transcranial magnetic stimulation.

overcoming "hypofrontality" and the associated cognitive deficits of working memory and executive function in cases of schizophrenia and Parkinson's disease. Similarly, psychomotor retardation in depression, poverty of action in schizophrenia, and akinesia in Parkinson's disease, which are all associated with prefrontal underactivation (reviewed by Jahanshahi and Frith[119]), may be modifiable with such high-frequency prefrontal rTMS. In the future, as we gain greater insight into the patterns and time courses of brain activations and inhibitions involved in specific aspects of cognition, the selective application of high- and low-frequency TMS to different cortical sites may allow fine tuning of patterns of cortical excitability to improve a specific aspect of cognitive performance in normal subjects or those with neurological illness or brain injury. Such fine tuning of cortical excitability with high- or low-frequency TMS may also prove beneficial as an adjunct therapy to enhance the effect of other therapeutic interventions such as medication or cognitive rehabilitation programs for a range of cognitive deficits.

Conclusion

Current and potential future applications of TMS to the study of cognition are summarized in Table 19–2. TMS is the latest of a number of powerful tools made available to cognitive neuroscientists in the last 2 decades. Future technical refinements of TMS are bound to improve its spatial and temporal specificity. We urgently need to gain a clearer understanding of the mechanisms of action of TMS over sites other than the motor cortex. With such improved understanding, we can fully exploit what TMS may be able to offer cognitive neuroscience as an investigatory or therapeutic tool and to ensure that future therapeutic applications of the technique are not surrounded by the same mist of ignorance and mystery that have engulfed electroconvulsive therapy.

Cognitive neuroscientists are increasingly likely to adopt a multitechnique approach to the study of any facet of cognitive function, because the convergence of evidence from the various techniques offers many advantages. TMS studies are guided by hypotheses generated from cognitive psychology and neuropsychology and the results of functional imaging. After brain areas involved in a particular cognitive task are identified by brain imaging and mapping methods, TMS allows many pertinent questions to be addressed. Is this activity necessary for task performance? Do different brain areas contribute to different component processes of the task? Are different brain areas necessary at different times during the task? What is the functional impact of altering activity at a target cortical site on the pattern of activation of the rest of the neural network engaged by the task? What changes in cortical excitability and functional connectivity occur with learning? What is the functional significance of altered patterns of brain activation after injury or illness? The potential of TMS as a therapeutic method or as an adjunct method to enhance the effect of other therapies in the rehabilitation of a range of cognitive deficits remains largely unexplored. To be able to address these questions in vivo using TMS is bound to make the future of cognitive neuroscience exciting and rich.

REFERENCES

1. Barker AT, Jalinous R, Freeston IL. Non-invasive magnetic stimulation of the human motor cortex. Lancet 1985;1:1106–1107.
2. Amassian VE, Cracco RO, Maccabee PJ, et al. Suppression of visual perception by magnetic coil stimulation of human occipital cortex. Electroencephalogr Clin Neurophysiol 1989;74:458–462.
3. Jahanshahi M, Rothwell JC. Transcranial magnetic stimulation studies of cognition: An emerging field. Exp Brain Res 2000;131:1–9.
4. Walsh V, Rushworth M. A primer of magnetic stimulation as a tool for neuropsychology. Neuropsychologia 1999;37:125–135.
5. Walsh V, Pascual-Leone A. Transcranial Magnetic Stimulation: A Neurochronometrics of Mind. Cambridge, MA, Bradford Books, MIT Press, 2003.
6. Pascual-Leone A, Walsh V, Rothwell J. Transcranial magnetic stimulation in cognitive neuroscience—virtual lesion, chronometry, and functional connectivity. Curr Opin Neurobiol 2000;10:232–237.
7. Cowey A, Walsh V. Tickling the brain: studying visual sensation, perception and cognition by transcranial magnetic stimulation. Prog Brain Res 2001;134:411–425.
8. Churchland PS, Sejnowski TJ. Perspectives on cognitive neuroscience. Science 1988;242:741–745.
9. Kammer T. Phosphenes and transient scotomas induced by magnetic stimulation of the occipital lobe: their topographic relationship. Neuropsychologia 1999;37:191–198.
10. Stewart LM, Walsh V, Rothwell JC. Motor and phosphene thresholds: a transcranial magnetic stimulation correlation study. Neuropsychologia 2001;39:415–419.
11. Pascual-Leone A, Grafman J, Hallett M. Modulation of cortical motor output maps during development of implicit and explicit knowledge. Science 1994;263:1287–1289.
12. Boroojerdi B, et al. Enhancing analogic reasoning with rTMS over the left prefrontal cortex. Neurology 2001;56:526.
13. Evers S, Bockermann I, Nyhuis PW. The impact of transcranial magnetic stimulation on cognitive processing: an event-related potential study. Neuroreport 2001;12:2915–2918.
14. Pasucal-Leone A, Hallett M. Induction of errors in a delayed response task by repetitive transcranial magnetic stimulation of the dorsolateral prefrontal cortex. Neuroreport 1994;5:2517–2520.
15. Jing H, Takigawa M, Okamura H, et al. Comparisons of event-related potentials after repetitive transcranial magnetic stimulation. J Neurol 2001;248:184–192.
16. Jahanshahi M, Profice P, Brown RG, et al. The effects of transcranial magnetic stimulation over the dorsolateral prefrontal cortex on suppression of habitual counting during random number generation. Brain 1998;121:1533–1544.

17. Jahanshahi M, Dirnberger G. The left dorsolateral prefrontal cortex and random generation of responses: Studies with transcranial magnetic stimulation. Neuropsychologia 1999;37:181–190.
18. Pascual-Leone A, Gates JR, Dhuna A. Induction of speech arrest and counting errors with rapid-rate transcranial magnetic stimulation. Neurology 1991;41:697–702.
19. Jennum P, Friberg L, Fuglsang-Frederiksen A, et al. Speech localization using repetitive transcranial magnetic stimulation. Neurology 1994;44:269–273.
20. Grafman J, Pascual-Leone A, Alway D, et al. Induction of a Recall Deficit by Rapid-Rate Transcranial Magnetic Stimulation. Neuroreport 1994;5:1157–1160.
21. Rossi S, Cappa, SF, Babiloni C, et al. Prefontal cortex in long-term memory: an "interference" approach using magnetic stimulation. Nat Neurosci 2001;4:948–952.
22. Paus T, Jech R, Thompson CJ, et al. Transcranial magnetic stimulation during positron emission tomography: a new method for studying connectivity of the human cerebral cortex. J Neurosci 1997;17:3178–3184.
23. Bohning DE, Shastri A, McConnell KA, et al. A combined TMS/fMRI study of intensity-dependent TMS over motor cortex. Biol Psychiatry 1999;45:385–394.
24. Mottaghy FM, et al. Repetitive TMS temporarily alters brain diffusion. Neurology 2003;60:1539.
25. Mottaghy FM, Krause BJ, Kemna LJ, et al. Modulation of the neuronal circuitry subserving working memory in healthy human subjects by repetitive transcranial magnetic stimulation. Neurosci Lett 2000;280:167–170.
26. Bestmann S, Baudewi J, Siebner HR, et al. Subthreshold high-frequency TMS of human primary motor cortex modulates interconnected frontal motor areas as detected by interleaved fMRI-TMS. Neuroimage 2003;20:1685–1696.
27. Flitman SS, Grafman J, Wassermann EM, et al. Linguistic processing during repetitive transcranial magnetic stimulation. Neurology 1998;50:175–181.
28. Beckers G, Zeki S. The consequences of inactivating areas V1 and V5 on visual-motion perception. Brain 1995;118:49–60.
29. Chen R, Classen J, Gerloff C, et al. Depression of motor cortex excitability by low-frequency transcranial magnetic stimulation. Neurology 1997;48:1398–1403.
30. Berardelli A, Inghilleri M, Rothwell JC, et al. Facilitation of muscle evoked responses after repetitive cortical stimulation in man. Exp Brain Res 1998;122:79–84.
31. Hilgetag CC, Theoret H, Pascual-Leone A. Enhanced visual spatial attention ipsilateral to rTMS-induced 'virtual lesions' of human parietal cortex. Nat Neurosci 2001;4:953–957.
32. Kosslyn SM, Pascual-Leone A, Felician O, et al. The role of area 17 in visual imagery: Convergent evidence from PET and rTMS. Science 1999;284:167–170.
33. Mottaghy FM, Gangitano M, Sparing R, et al. Segregation of areas related to visual working memory of the prefrontal cortex revealed by rTMS. Cereb Cortex 2002;12:369–375.
34. Sack AT, Sperling JM, Prvulovic D, et al. Tracking the mind's image in the brain. II. Transcranial magnetic stimulation reveals parietal asymmetry in visuospatial imagery. Neuron 2002;35:195–204.
35. Pascual-Leone A, Houser CM, Reese K, et al. Safety of rapid-rate transcranial magnetic stimulation in normal volunteers. Electroencephalogr Clin Neurophysiol 1993;89:120–130.
36. Wassermann EM. Risk and safety of repetitive transcranial magnetic stimulation: report and suggested guidelines from the International Workshop on the Safety of Repetitive Transcranial Magnetic Stimulation, June 5–7, 1996. Electroencephalogr Clin Neurophysiol 1998;198:1–16.
37. Keel JC, Smith MJ, Wasserman EM. A safety screening questionnaire for transcranial magnetic stimulation. Clin Neurophysiol 2000;112:720.
38. Fierro B, Brighina F, Oliveri M, et al. Contralateral neglect induced by right posterior parietal rTMS in healthy subjects. Neuroreport 2000;11:1519–1521.
39. Koch G, Oliveri M, Torriero S, et al. Underestimation of time perception after repetitive transcranial magnetic stimulation. Neurology 2003;60:1844–1846.
40. Ashbridge E, Walsh V, Cowey A. Temporal aspects of visual search studied by transcranial magnetic stimulation. Neuropsychologia 1977;35:1121–1131.
41. Grosbras MH, Paus T. Transcranial magnetic stimulation of the human frontal eye field: Effects on visual perception and attention. J Cogn Neurosci 2002;14:1109–1120.
42. Terao Y, Fukuda H, Ugawa Y, et al. Visualization of the information flow through human oculomotor cortical regions by transcranial magnetic stimulation. J Neurophysiol 1998;80:936–946.
43. Sabatino M, DiNuovo S, Sardo P, et al. Neuropsychology of selective attention and magnetic cortical stimulation. Int J Psychophysiol 1996;21:83–89.
44. Hong KS, Lee SK, Kim JY, et al. Visual working memory revealed by repetitive transcranial magnetic stimulation. J Neurol Sci 2000;181:50–55.

45. d'Alfonso AAL, van Honk J, Hermans E, et al. Laterality effects in selective attention to threat after repetitive transcranial magnetic stimulation at the prefrontal cortex in female subjects. Neurosci Lett 2000;280:195–198.
46. Mull BR, Seyal M. Transcranial magnetic stimulation of left prefrontal cortex impairs working memory. Clin Neurophysiol 2001;112:1672–1675.
47. Oliveri M, Turiziani P, Carlesimo GA, et al. Parieto-frontal interactions in visual-object and visual-spatial working memory: evidence from transcranial magnetic stimulation. Cereb Cortex 2001;11:606–618.
48. Cappa SF, Sandrini M, Rossini PM, et al. The role of the left frontal lobe in action naming—rTMS evidence. Neurology 2002;59:720–723.
49. Sandrini M, Cappa SF, Rossi S, et al. The role of prefrontal cortex in verbal episodic memory: rTMS evidence. J Cogn Neurosci 2003;15: 855–861.
50. Rami L, Gironell A, Kulisevsky J, et al. Effects of repetitive transcranial magnetic stimulation on memory subtypes: a controlled study. Neuropsychologia 2003;41:1877–1883.
51. Muri RM, Rivaud S, Vermersch AI, et al. Effects of transcranial magnetic stimulation over the region of the supplementary motor area during sequences of memory-guided saccades. Exp Brain Res 1995;104:163–166.
52. Haggard P, Magno E. Localising awareness of action with transcranial magnetic stimulation. Exp Brain Res 1999;127:102–107.
53. Ganis G, Keenan JP, Kosslyn SM, et al. Transcranial magnetic stimulation of primary motor cortex affects mental rotation. Cereb Cortex 2000;10:175–180.
54. Harmer CJ, Thilo KV, Rothwell JC, et al. Transcranial magnetic stimulation of medial-frontal cortex impairs the processing of angry facial expressions. Nat Neurosci 2001;4:17–18.
55. Rushworth MFS, Hadland KA, Paus T, et al. Role of the human medial frontal cortex in task switching: A combined fMRI and TMS study. J Neurophysiol 2002;87:2577–2592.
56. Pascual-Leone A, Gomes-Tortosa E, Grafman J, et al. Induction of visual extinction by rapid-rate transcranial magnetic stimulation of parietal lobe. Neurology 1994;44:494–498.
57. Seyal M, Ro T, Rafal R. Increased sensitivity to ipsilateral cutaneous stimuli following transcranial magnetic stimulation of the parietal lobe. Ann Neurol 1995;38:264–267.
58. Oliveri M, Caltagirone C, Filippi MM, et al. Paired transcranial magnetic stimulation protocols reveal a pattern of inhibition and facilitation in the human parietal cortex. J Physiol (Lond) 2000;529:461–468.
59. Fierro B, Piazza A, Brighina F, et al. Modulation of intracortical inhibition induced by low- and high-frequency repetitive transcranial magnetic stimulation. Exp Brain Res 2001;138:452–457.
60. Lewald J, Foltys H, Topper R. Role of the posterior parietal cortex in spatial hearing. J Neurosci 2002;22:RC207.
61. Bestmann D, Thilo KV, Sauner D, et al. Parietal magnetic stimulation delays visuomotor mental rotation at increased processing demands. Neuroimage 2002;17:1512–1520.
62. Bjoertomt O, Cowey A, Walsh V. Spatial neglect in near and far space investigated by repetitive transcranial magnetic stimulation. Brain 2002;25:2012–2022.
63. Topper R, Mottaghy FM, Brugmann M, et al. Facilitation of picture naming by focal transcranial magnetic stimulation. Exp Brain Res 1998;121:371–378.
64. Stewart L, Meyer BU, Frith U, et al. Left posterior BA37 is involved in object recognition: a TMS study. Neuropsychologia 2001;39:1–6.
65. Beckers G, Homberg V. Impairment of visual-perception and visual short-term-memory scanning by transcranial magnetic stimulation of occipital cortex. Exp Brain Res 1991;87:421–432.
66. Zangaladze A, Epstein CM, Grafton ST, et al. Involvement of visual cortex in tactile discrimination of orientation. Nature 1999;401:587–590.
67. Cowey A, Walsh V. Magnetically induced phosphenes in sighted, blind and blindsighted observers. Neuroreport 2000;11:3269–3273.
68. Rushworth MFS, Ellison A, Walsh V. Complementary localization and lateralization of orienting and motor attention. Nat Neurosci 2001;4:656–661.
69. Theoret H, Kobayashi M, Ganis G, et al. Repetitive transcranial magnetic stimulation of human area MT/V5 disrupts perception and storage of the motion aftereffect. Neuropsychologia 2002;40:2280–2287.
70. Jahanshahi M, Dirnberger G, Fuller R, et al. The role of the dorsolateral prefrontal cortex in random number generation: A study with positron emission tomography. Neuroimage 2000;12:713–725.
71. Tulving E, Kapur S, Craik FI, et al. Hemispheric encoding/retrieval asymmetry in episodic memory: Positron emission tomography findings. Proc Natl Acad Sci U S A 1994;91:2016–2020.
72. Corbetta M. Frontoparietal cortical networks for directing attention and the eye to visual locations: identical, independent, or overlapping neural systems? Proc Natl Acad Sci U S A 1998;95:831–838.
73. Gitelman DR, Nobre AC, Parrish TB, et al. A large-scale distributed network for covert

spatial attention: further anatomical delineation based on stringent behavioral and cognitive controls. Brain 1999;122(Pt 6):1093–106.
74. Ugawa Y, Day BL, Rothwell JC, et al. Modulation of motor cortical excitability by electrical stimulation over the cerebellum in man. J Physiol 1991;441:57–72.
75. Tokimura H, Tokimura Y, Oliviero A, et al. Speech-induced changes in corticospinal excitability. Ann Neurol 1996;40:32–37.
76. Erk S, Kiefer M, Grothe J, et al. Emotional context modulates subsequent memory effect. Neuroimage 2003;18:439–447.
77. Kensinger EA, Brierley B, Medford N, et al. Effects of normal aging and Alzheimer's disease on emotional memory. Emotion 2002;2:118–134.
78. Davidson RJ, Abercrombie H, Nitschke JB, et al. Regional brain function, emotion and disorders of emotion. Curr Opin Neurobiol 1999;9:228–234.
79. Davidson RJ. Anterior electrophysiological asymmetries, emotion, and depression: conceptual and methodological conundrums. Psychophysiology 1998;35:607–614.
80. Kujirai T, Caramia MD, Rothwell JC, et al. Corticortical inhibition in human motor cortex. J Physiol 1993;471:501–519.
81. Cracco RQ, Amassian VE, Maccabee PJ, et al. Comparison of human transcallosal responses evoked by magnetic coil and electrical stimulation. Electroencephalogr Clin Neurophysiol 1989;74:417–424.
82. Ferbert A, Priori A, Rothwell JC, et al. Interhemispheric inhibition of the human motor cortex. J Physiol (Lond) 1992;453: 525–546.
83. Pascual-Leone A, Walsh V. Fast backprojections from the motion to the primary visual area necessary for visual awareness. Science 2001;292:510–512.
84. Fox P, Ingham R, George MS, et al. Imaging human intra-cerebral connectivity by PET during TMS. Neuroreport 1997;8:2787–2791.
85. Mottaghy FM, et al. Modulation of the brain-behavior relationship in verbal working memory by repetitive transcranial magnetic stimulation. Brain Res Cogn Brain Res 2003;15:241–249.
86. Frith CD, Friston K, Liddle PF, et al. Willed action and the prefrontal cortex in man: a study with PET. Proc R Soc Lond B Biol Sci 1991;244:241–246.
87. Friston KJ, Frith CD. Schizophrenia: a disconnection syndrome? Clin. Neurosci 1995;3:89–97.
88. Strafella AP, Paus T, Barrett J, et al. Repetitive transcranial magnetic stimulation of the human prefrontal cortex induces dopamine release in the caudate nucleus. J Neurosci 2001;21:RC157.
89. Strafella AP, Paus T, Fraraccio M, et al. Striatal dopamine release induced by repetitive transcranial magnetic stimulation of the human motor cortex. Brain 2003;126:2609–2615.
90. Alexander GE, De Long M, Strick PL. Parallel organization of functionally segregated circuits linking basal ganglia and cortex. Annu Rev Neurosci 1986;9:357–381.
91. Ilmoniemi RJ, Virtanen J, Ruohonen J, et al. Neuronal responses to magnetic stimulation reveal cortical reactivity and connectivity. Neuroreport 1997;8:3537–3540.
92. Andrew C, Pfurtscheller G. Event-related coherence as a tool for studying dynamic interaction of brain regions. Electroencephalogr Clin Neurophysiol 1996;98:144–148.
93. Amassian VE, Cracco RQ, Maccabee PJ, et al. Unmasking human visual-perception with the magnetic coil and its relationship to hemispheric-asymmetry. Brain Res 1993;605:312–316.
94. Amassian VE, Maccabee PJ, Cracco RQ, et al. Measurement of information-processing delays in human visual-cortex with repetitive magnetic coil stimulation. Brain Res 1993;605:317–321.
95. Funahashi S, Chafee MV, Goldman-Rakic PS. Prefrontal neuronal activity in rhesus monkeys performing a delayed anti-saccade task. Nature 1993;365:753–756.
96. Connolly JD, Goodale MA, Desouza JF, et al. Comparison of frontoparietal fMRI activation during anti-saccades and anti-pointing. J Neurophysiol 2000;84:1645–1655.
97. Fierro B, Brighina F, Piazza A, et al. Timing of right parietal and frontal cortex activity in visuospatial perception: a TMS study in normal individuals. Neuroreport 2001;12:2605–2607.
98. Smith EE, Jonides J. Storage and executive processes in the frontal lobes. Science 1999;283:1657–1661.
99. Mottaghy FM, Gangitano M, Krause BJ, et al. Chronometry of parietal and prefrontal activations in verbal working memory revealed by transcranial magnetic stimulation. Neuroimage 2003;18:565–575.
100. Pascual-Leone A, Wassermann EM, Grafman J, et al. The role of the dorsolateral prefrontal cortex in implicit procedural learning. Exp Brain Res 1996;107:479–485.
101. Walsh V, Ashbridge E, Cowey A. Cortical plasticity in perceptual learning demonstrated

102. Sadato N, Pascual-Leone A, Grafman J, et al. Activation of the primary visual cortex by Braille reading in blind subjects. Nature 1996;380:526–528.
103. Cohen LG, Celnik P, Pascual-Leone A, et al. Functional relevance of cross-modal plasticity in blind humans. Nature 1997;389:180–183.
104. Ridding MC, Inzelberg R, Rothwell JC. Changes in excitability of motor cortical circuitry in patients with Parkinson's disease. Ann Neurol 1995;37:181–188.
105. Turton A, Wroe S, Trepte N, et al. Contralateral and ipsilateral EMG responses to transcranial magnetic stimulation during recovery of arm and hand function after stroke. Electroencephalogr Clin Neurophysiol 1996;101:316–28.
106. Papathanasiou I, Filipovic SR, Whurr R, et al. Plasticity in the motor system induced by writing rehabilitation therapy in aphasic and agraphic patients with severe right hemiparesis. Neurology 2003;61:977–980.
107. Sawaguchi T, Goldman-Rakic PS. D1 dopamine receptors in prefrontal cortex: involvement in working memory Science 1991;251:947–950.
108. Ziemann U, Bruns D, Paulus W. Enhancement of human motor cortex inhibition by the dopamine receptor agonist pergolide: evidence from transcranial magnetic stimulation. Neurosci Lett 1996;208:187–190.
109. Priori A, Berardelli A, Inghilleri M, et al. Motor cortical inhibition and the dopaminergic system. Pharmacological changes in the silent period after transcranial brain stimulation in normal subjects, patients with Parkinson's disease and drug-induced parkinsonism. Brain 1994;117:317–323.
110. Pascual-Leone A, Tarazona F, Keenan J, et al. Transcranial magnetic stimulation and neuroplasticity. Neuropsychologia 1999;37:207–217.
111. Mottaghy FM, Hungs M, Brugmann M, et al. Facilitation of picture naming after repetitive transcranial magnetic stimulation. Neurology 1999;53:1806–1812.
112. Oliveri M, Rossini PM, Pasqualetti P, et al. Interhemispheric asymmetries in the perception of unimanual and bimanual cutaneous stimuli. A study using transcranial magnetic stimulation. Brain 1999;122:1721–1729.
113. Oliveri M, Rossini PM, Traversa R, et al. Left frontal transcranial magnetic stimulation reduces contralesional extinction in patients with unilateral right brain damage. Brain 1999;122:1731–1739.
114. Pfurtscheller G, Aranibar A. Event-related cortical desynchronization detected by power measurements of scalp EEG. Electroencephalogr Clin Neurophysiol 1977;42:817–826.
115. Klimesch W. EEG alpha and theta oscillations reflect cognitive and memory performance: a review and analysis. Brain Res Brain Res Rev 1999;29:169–195.
116. Klimesch W, Sauseng P, Gerloff C. Enhancing cognitive performance with repetitive transcranial magnetic stimulation at human individual alpha frequency. Eur J Neurosci 2003;17:1129–1133.
117. George MS, Lisanby SH, Sackheim HA. Transcranial magnetic stimulation: applications in neuropsychiatry. Arch Gen Psychiatry 1999;56:300–311.
118. Maeda F, Keenan JP, Tormos JM, et al. Modulation of corticospinal excitability by repetitive transcranial magnetic stimulation. Clin Neurophysiol 2000;111:800–805.
119. Jahanshahi M, Frith CD. Willed action and its impairments. Cogn Neuropsychol 1998;15:483–534.

20 Individual Differences in the Response to Transcranial Magnetic Stimulation of the Motor Cortex

Eric M. Wassermann

One of the most striking aspects of transcranial magnetic stimulation (TMS) is the degree to which individuals differ in the magnitude of their muscle responses to stimulation of the motor cortex. Much of the TMS literature is devoted to the study of differences in response between groups of neurological patients and healthy individuals, and these can be dramatic. However, the range of variation within the healthy population is also interesting, and no study of any amplitude-related measure of the motor evoked potential (MEP) has ever shown a difference in response between two populations that was greater than the normal range.

Normal variation in the MEP has been generally ignored or treated as a nonsytematic and uninformative source of variance. Neurophysiologists in particular, who tend to perform intensive investigations on small numbers of human or animal subjects, tend to neglect the issue of individual variability entirely. To behavioral scientists, geneticists, and clinicians, however, differences among individuals hold considerable importance. Moreover, when such differences are physiologically meaningful, robust, consistent, and readily quantified and scalar, as many interindividual differences are, they present a unique opportunity for study. In this chapter, I discuss some of the factors known to contribute to the individual variation in the response to TMS of the motor cortex in populations of motorically normal individuals.

■ Variability of Motor Evoked Potential Threshold and Amplitude

The MEP threshold is a very widely used measure of the excitability of the corticospinal system to exogenous stimulation with TMS. It is not a true threshold, but a probabilistic index of corticospinal and spinal neuron responsiveness to stimuli of low intensity. Perhaps a better overall measure of aggregate excitability in the corticospinal system is the input-output or recruitment curve[1] relating stimulation intensity at a range of levels to MEP amplitude. Nevertheless, the resting MEP threshold is deeply embedded in the TMS literature and provides a useful standard measure for comparison across studies.

The MEP threshold is relatively stable across time within individuals, but varies widely across the population (i.e., from about 30% to more than the maximum output of conventional stimulators and coils[2]) (Fig. 20–1). Similar variability is present in the threshold during a mild voluntary contraction of the target muscle. Based on a sample of 151 healthy individuals that contained 19 whose thresholds were determined on three different occasions,[2] I estimated that experimental error (i.e., mistaken estimation of the true threshold) contributed approximately 6% and 11% to the population variance for the resting and active conditions, respectively. The validity of the resting MEP threshold as a measure of corticospinal system excitability has been

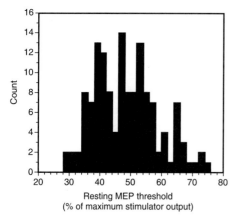

Figure 20–1 Histogram of the resting motor evoked potential thresholds in 151 healthy subjects. No counts are shown for individuals whose thresholds exceeded maximum stimulator output. (Adapted from Wassermann EM. Variation in the response to transcranial magnetic brain stimulation in the general population. Clin Neurophysiol 2002; 113:1165–1171.)

questioned because electromyographic silence does not imply the absence of excitatory traffic in the corticospinal tract. However, the MEP threshold measured during a standardized voluntary contraction is more variable than when measured at rest. Analogous dissociations between grip force and corticospinal cell activity have been described in primates.[3] The remaining 90% of the variance in our large group did not appear to be caused by measurement error or random variability and was likely to contain a substantial component related to stable biological differences between individuals and perhaps to experience, as in the increase in MEP amplitude found in individuals after motor learning.[4,5]

Scalp-to-Brain Distance and Age-Related Influences on Motor Evoked Potentials

One potential source of variation in the amplitude and threshold of the MEP is the distance of the coil from the stimulation target in the motor cortex. One might expect the MEP to be particularly sensitive to this distance, because the intensity of the induced magnetic field falls with the third power of distance from the source. In 17 healthy individuals 19 to 75 years old, McConnell and colleagues[6] found that the MEP threshold increased with the distance from scalp to cortex as determined from MRI scans. Earlier, the same group found that age and scalp to motor cortex distance were highly correlated in a sample of depressed patients.[7]

Genetic Factors

Some determinants of the MEP threshold may be genetic. In a study of 17 healthy sib pairs, aged 18 to 76 years,[2] we found a significant correlation between the MEP thresholds in the right (dominant) hand during both rest ($r^2 = 0.55$; $P < .001$) (Fig. 20–2) and voluntary activation ($r^2 = 0.30$; $P < .05$). There was no relation of age or sex to threshold in this sample. Although the similarity in threshold between siblings could easily have been caused by a gross anatomical factor, such as scalp to cortex distance, it was considerably weaker for the left hand, suggesting that the inherited factor might affect the organization of the hand representation in the dominant hemisphere.

Neurologically Normal Patients with Behavioral Disorders

It is not surprising that many neurological disorders, particularly those affecting movement, can alter the amplitude and threshold of

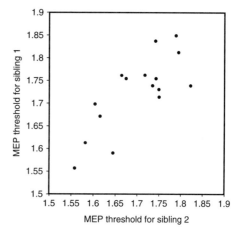

Figure 20–2 Plot showing the correlation of resting motor evoked potential thresholds between siblings. Axes are in logarithmic scale. (Adapted from Wassermann EM. Variation in the response to transcranial magnetic brain stimulation in the general population. Clin Neurophysiol 2002;113:1165–1171.)

the MEP. However, high-functioning psychiatric patients who have no clinically apparent neurological abnormalities can also differ from healthy individuals on various TMS measures of motor cortex function. For example, we studied a group of 16 patients with obsessive-compulsive disorder (OCD),[8] 11 of whom had no history or evidence of tics or any other neurological disorder and 7 of whom were on no medications. We found that, on average, these patients had MEP thresholds significantly lower than normal during rest and voluntary activation, and larger resting MEPs at a range of stimulation intensities (Fig. 20–3). The difference was apparent in the tic-free and the unmedicated patient subgroups when their data were analyzed separately and we have gotten similar results in subsequent studies.[9] The fact that the MEP threshold was reduced during both rest and overt voluntary contraction of the target muscle implies that the change was not due to subthreshold activation or disinhibition of the cortical output pathway, as might occur in anxious individuals. Rather, it implies a change located in series with the pathway conducting the stimulus from presynaptic cortical axon to muscle (i.e., located in a set of conducting synapses or neurons). The basis for this argument has been set forth elsewhere.[10,11]

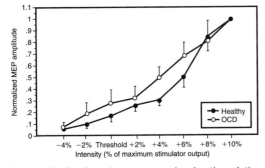

Figure 20–3 Recruitment curve showing the relation between stimulus intensity and motor evoked potential (MEP) amplitude in healthy individuals and patients with obsessive-compulsive disorder. The x axis is normalized to each individual's resting MEP threshold, and the units are the percent of maximum stimulator output. The y axis is normalized to each individuals largest MEP. Bars show the standard error.

Variability of Paired-Pulse Motor Evoked Potential Measures

The amplitude of the MEP in paired-pulse TMS studies of the type described by Kujirai and colleagues[12] and refined by Ziemann and coworkers[13] has become a popular measure of the relative degrees of intrinsic cortical inhibition and facilitation that are evoked by a subthreshold conditioning TMS pulse. It is important to note that this technique measures only the activities *evoked* by the conditioning stimulus and not the absolute or ongoing levels of inhibition and facilitation. It is therefore critical to know whether the MEP to a single TMS pulse is affected (as by tonic inhibition) before attempting to interpret paired-pulse data. Another important point is that factors (e.g., $GABA_A$ agonists) that increase inhibition also reduce facilitation, sometimes to a greater degree.[14] In many circumstances, it is not possible to distinguish between alterations in facilitation and inhibition unless there is a strong predictive hypothesis or a simultaneous change in the response to single stimuli.

Like the MEP to single pulses, paired-pulse measures vary substantially from measurement to measurement[15] and between ostensibly healthy subjects.[2] Although there is a well-known tendency to show inhibition at short intervals and facilitation at longer ones, there is an appreciable number of healthy subjects who show facilitation at short intervals and or inhibition at longer intervals even when large numbers of trials are obtained and experimental conditions are carefully controlled (Fig. 20–4A). A significant portion of this variation appears to result from stable individual differences rather than experimental error, because individuals seem to show similar tendencies across inter-stimulus intervals (see Fig. 20–4B).

Age-Related Differences

Two small studies, each comparing two groups with different mean ages have found disparate results. In one,[16] an elderly group showed significantly less paired-pulse inhibition than a group of young adults. In the other,[17] a middle-aged group showed more paired-pulse inhibition than young adults.

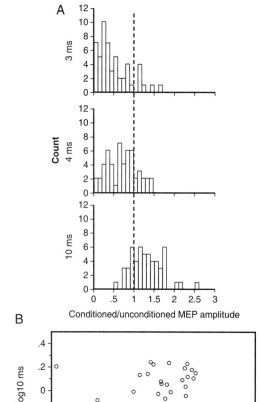

Figure 20–4 **A:** Histograms of the amplitude ratio of the mean conditioned to mean unconditioned motor evoked potential for 3, 4, and 10 ms from a paired-pulse transcranial magnetic stimulation study in 53 healthy subjects. **B:** Plot showing the correlation within individuals of the conditioned/unconditioned amplitude ratios at inter-stimulus intervals of 4 and 10 ms. (**A** adapted from Wassermann EM. Variation in the response to transcranial magnetic brain stimulation in the general population. Clin Neurophysiol 2002;113:1165–1171.)

Neither of these studies controlled for any differences between the groups other than age. In our own sample of 53 individuals, my colleagues and I found no age-related differences in the paired-pulse response.[2]

Sex Differences and Hormonal Effects

Steroid hormones are potent modulators of neuronal excitability. Estradiol itself facilitates of glutamatergic transmission.[18,19] Progesterone and cortisol are metabolized to neurosteroids that bind to a site on a subunit of the $GABA_A$ receptor, increasing its activity in a manner analogous to the action of the benzodiazepines. The effects of androgens on this system are less well known, but testosterone may decrease excitability.[20,21] Studies have shown that drugs shifting the balance of GABA and glutamate activity in the cortex in the direction of GABA activity increase TMS paired-pulse inhibition and decrease facilitation at moderate doses.[22–24] Therefore, one might expect that the same technique would be sensitive to the actions of neurosteroids in healthy individuals. In women immediately after menstruation, the circulating levels of estradiol and progesterone are low. Estradiol rises gradually throughout the follicular phase and progesterone begins to be secreted in the luteal phase during which estrogen remains high. In two studies,[25,26] we performed paired-pulse TMS experiments in groups of healthy, ovulating women across the menstrual cycle and found that intracortical inhibition decreased and facilitation increased late in the follicular phase when high estradiol levels were unopposed by progesterone. Then there was a drop in facilitation and increase in inhibition in the luteal phase when progesterone was present. The magnitudes of the increase in excitability from the early to the late follicular phase and the subsequent drop in the luteal phase were comparable to effects described for behaviorally significant doses of drugs.[22]

Not all neurologically normal women, however, show the expected decrease in excitability in the luteal phase. In a sample of women meeting rigorous behavioral criteria for premenstrual syndrome and premenstrual dysphoric disorder, inhibition actually *decreased* and facilitation *increased* in the luteal phase.[27] An aberrant brain response to a normal circulating level of progesterone is consistent with current theories regarding the pathogenesis of this disorder[28] and could be caused by an alteration in a component of the $GABA_A$ receptor complex or in the cerebral metabolism of progesterone.

Although the effects of cortisol-derived neurosteroids on human cortical excitability have not been measured with TMS, there is reason to believe that they act in a manner very similar to that of the neurosteroid metabolites of progesterone.[29] Cortisol levels vary in a circadian pattern that is lost in clinical depression, and the levels rise with exercise, illness, and other stressors. These factors should be taken into account in the composition of experimental groups and the timing of experiments and in explaining differences in the responses to paired-pulse TMS.

Neurologically Normal Patients with Behavioral Disorders

Tic-free and unmedicated patients with OCD have markedly reduced intracortical inhibition.[8] Similar findings exist in patients with attention deficit-hyperactivity disorder (ADHD),[30,31] and Gilbert and associates[32] found a strong inverse relationship between intracortical inhibition and impulsivity on scales for ADHD but not with the severity of tics or OCD in a group of children identified as having Tourette's syndrome. This is the first report of a direct correlation between a TMS measure and any index of disease. This reflection of behavioral abnormalities and traits in the motor cortex may not surprise behaviorally oriented clinicians. However, as discussed in the earlier section on MEP amplitude in OCD, the fact that the physiological disease phenotype can be expressed in the motor cortex, even in individuals without recognizable movement disorders, has potential importance for understanding the pathogenesis of neurobehavioral disorders and for the use of TMS as a measure of cortical function in healthy individuals and patients.

Personality and Paired-Pulse Cortical Excitability

In our study of OCD,[8] we found that the patients also had decreased intracortical inhibition relative to a sample of healthy individuals that had been screened with the Structured Clinical Interview for the DSM-IV[33] and interviewed by a psychiatrist. Individuals with high degrees of anxiety or a significant tendency for obsessions or compulsions, but who did not meet diagnostic criteria for OCD were excluded. By contrast, screening of healthy subjects in most neurophysiological and clinical studies involving motor cortex TMS consists at most of a brief medical history and physical examination. When we compared our OCD patients with the large general population sample mentioned earlier[34] who were screened only for psychiatric or neurological diagnoses, neurological abnormalities, and neuroactive medications, we found no such difference in excitability.

Nevertheless, this general population sample proved interesting: Because the difference in intracortical inhibition that we found between the psychiatrically screened normal subjects and the OCD patients could have been an artifact of the screening procedure, we looked at the unscreened general population sample for correlations between paired-pulse excitability and a range of scaleable differences between individuals, including measures of intelligence and temperament. The only correlation of any magnitude or statistical significance was between paired-pulse excitability (lower inhibition or higher facilitation) and the tendency to experience anxiety and other negative emotions (i.e., *neuroticism*, a dimension in the five-factor model of personality as tested with the NEO-PI-R inventory).[35] The association of cortical excitability and negative emotionality (common in OCD and related disorders) could have contributed to our paired-pulse findings in OCD. Interestingly, there was no effect of personality on MEP threshold, suggesting that the threshold change in OCD might be associated with actual pathology.

■ Conclusion

The fact that TMS of the motor cortex is sensitive to hidden but systematic differences among neurologically normal individuals has important implications for research using TMS. First, it should alert investigators to the importance of screening experimental subjects for individual factors known to influence cortical excitability (e.g., psychopathology) and balancing experimental groups for sex and demographic factors such as age and education that may produce unwanted differences. At the same time, this sensitivity to individual differences also opens new fields of study to motor neurophysiologists. For example, individual variation in the response to substances and the environment are increasingly acknowledged as major factors in the treatment and etiology of brain disease. Many of these are genetically determined. Physiological studies with TMS could prove particularly useful in identifying the physiologic phenotypes associated with genetic variations that affect behavior, the susceptibility to disease, and the response to chemical agents.

REFERENCES

1. Ridding MC, Rothwell JC. Stimulus/response curves as a method of measuring motor cortical excitability in man. Electroencephalogr Clin Neurophysiol 1997;105:340–344.
2. Wassermann EM. Variation in the response to transcranial magnetic brain stimulation in the general population. Clin Neurophysiol 2002;113:1165–1171.
3. Fetz EE, Cheney PD. Functional relations between primate motor cortex cells and muscles: fixed and flexible. Ciba Found Symp 1987;132:98–117.
4. Pascual-Leone A, Grafman J, Hallett M. Modulation of cortical motor output maps during the development of implicit and explicit knowledge. Science 1994;263:1287–1289.
5. Pascual-Leone A, Dang N, Cohen LG, et al. Modulation of muscle responses evoked by transcranial magnetic stimulation during the acquisition of new fine motor skills. J Neurophysiol 1995;74:1037–1045.
6. McConnell KA, Nahas Z, Shastri A, et al. The transcranial magnetic stimulation motor threshold depends on the distance from coil to underlying cortex: a replication in healthy adults comparing two methods of assessing the distance to cortex. Biol Psychiatry 2001;49:454–459.
7. Kozel FA, Nahas Z, deBrux C, et al. How coil-cortex distance relates to age, motor threshold, and antidepressant response to repetitive transcranial magnetic stimulation. J Neuropsychiatry Clin Neurosci 2000;12:376–384.
8. Greenberg BD, Ziemann U, Corá-Locatelli G, et al. Altered cortical excitability in obsessive-compulsive disorder. Neurology 2000;54:142–147.
9. Smith MJ, Jean-Mary J, Grafman J, et al. Abnormal cortical excitability in motor learning in OCD. Biol Psychiatry 2002;51(Suppl):81S.
10. Ridding MC, Rothwell JC. Reorganisation in human motor cortex. Can J Physiol Pharmacol 1995;73:218–222.
11. Touge T, Gerschlager W, Brown P, et al. Are the after-effects of low-frequency rTMS on motor cortex excitability due to changes in the efficacy of cortical synapses? Clin Neurophysiol 2001;112:2138–2145.
12. Kujirai T, Caramia MD, Rothwell JC, et al. Corticocortical inhibition in human motor cortex. J Physiol (Lond) 1993;471:501–519.
13. Ziemann U, Rothwell JC, Ridding MC. Interaction between intracortical inhibition and facilitation in human motor cortex. J Physiol (Lond) 1996;496:873–881.
14. Ziemann U, Lonnecker S, Steinhoff BJ, et al. The effect of lorazepam on the motor cortical excitability in man. Exp Brain Res 1996;109:127–135.
15. Boroojerdi B, Kopylev L, Battaglia F, et al. Reproducibility of intracortical inhibition and facilitation using the paired-pulse paradigm. Muscle Nerve 2000;23:1594–1597.
16. Peinemann A, Lehner C, Conrad B, et al. Age-related decrease in paired-pulse intracortical inhibition in the human primary motor cortex. Neurosci Lett 2001;313:33–36.
17. Kossev AR, Schrader C, Dauper J, et al. Increased intracortical inhibition in middle-aged humans; a study using paired-pulse transcranial magnetic stimulation. Neurosci Lett 2002;333:83–86.
18. Wong M, Thompson TL, Moss RL. Nongenomic actions of estrogen in the brain: physiological significance and cellular mechanisms. Crit Rev Neurobiol 1996;10:189–203.
19. Woolley CS. Electrophysiological and cellular effects of estrogen on neuronal function. Crit Rev Neurobiol 1999;13:1–20.
20. Beyenburg S, Stoffel-Wagner B, Bauer J, et al. Neuroactive steroids and seizure susceptibility. Epilepsy Res 2001;44:141–153.
21. Edwards HE, Burnham WM, MacLusky NJ. Testosterone and its metabolites affect afterdischarge thresholds and the development of amygdala kindled seizures. Brain Res 1999;838:151–157.
22. Ziemann U, Steinhoff BJ, Tergau F, et al. Transcranial magnetic stimulation: its current role in epilepsy research. Epilepsy Res 1998;30:11–30.
23. Ziemann U, Chen R, Cohen LG, et al. Dextromethorphan decreases the excitability of the human motor cortex. Neurology 1998;51:1320–1324.
24. Schwenkreis P, Witscher K, Janssen F, et al. Influence of the N-methyl-D-aspartate antagonist memantine on human motor cortex excitability. Neurosci Lett 1999;270:137–140.
25. Smith MJ, Keel JC, Greenberg BD, et al. Menstrual cycle effects on cortical excitability. Neurology 1999;53:2069–2072.
26. Smith MJ, Adams LF, Schmidt PJ, et al. Ovarian hormone effects on human cortical excitability. Ann Neurol 2002;51:599–603.

27. Smith MJ, Adams LF, Schmidt PJ, et al. Abnormal luteal phase excitability of the motor cortex in women with PMS. Biol Psychiatry 2003;54:757–762.
28. Schmidt PJ, Nieman LK, Danaceau MA, et al. Differential behavioral effects of gonadal steroids in women with and in those without premenstrual syndrome. N Engl J Med 1998;338:209–216.
29. Majewska MD. Neurosteroids: endogenous bimodal modulators of the $GABA_A$ receptor. Mechanism of action and physiological significance. Prog Neurobiol 1992;38:379–395.
30. Moll GH, Heinrich H, Rothenberger A. Methylphenidate and intracortical excitability: opposite effects in healthy subjects and attention-deficit hyperactivity disorder. Acta Psychiatr Scand 2003;107:69–72.
31. Moll GH, Heinrich H, Trott G, et al. Deficient intracortical inhibition in drug-naive children with attention-deficit hyperactivity disorder is enhanced by methylphenidate. Neurosci Lett 2000;284:121–125.
32. Gilbert DL, Bansal AS, Sethuraman G, et al. Association of cortical disinhibition with tic, ADHD, and OCD severity in Tourette syndrome. *Mov Disord* 2004;19:416–425.
33. First MB, Spitzer RL, Gibbon M, et al. Structured Clinical Interview for DSM–IV Axis I Disorders, Research Version, Non-patient Edition (SCID-I/NP). New York, Biometrics Research, New York State Psychiatric Institute, 1997.
34. Wassermann EM, Greenberg BD, Nguyen MB, et al. Motor cortex excitability correlates with an anxiety-related personality trait. Biol Psychiatry 2001;50:377–382.
35. Costa PT, McCrae RR. Revised NEO Personality Inventory (NEO-PI-R) and NEO Five-Factor Inventory (NEO-FFI) Professional Manual. Odessa, FL, Psychological Assessment Resources, 1992.

21 Potential Therapeutic Uses of Transcranial Magnetic Stimulation in Psychiatric Disorders

*Mark S. George, Ziad Nahas,
F. Andrew Kozel, Xingbao Li,
Kaori Yamanaka, Alexander Mishory,
Sarah Hill, and Daryl E. Bohning*

The potential therapeutic uses of transcranial magnetic stimulation (TMS) in psychiatry are simultaneously perhaps the most interesting and the most complex. Since its modern inception in 1985, TMS has been largely initially used as a research and clinical tool in clinical neurophysiology, as described in other chapters in this book. This chapter critically summarizes the studies using TMS as a potential therapeutic tool in psychiatry.

■ Limitations

Therapeutic uses of TMS in psychiatry have lagged behind the clinical neurophysiologic uses for several potential reasons. The first reason for a lag may be hearing about and becoming familiar with a new technology. TMS was developed in its modern form as a clinical neurophysiological tool, and information about the technology flowed out from these initial uses. It took several years for psychiatrists, as opposed to clinical neurophysiologists, to learn about TMS and begin to formulate how to use it therapeutically. More importantly however, the structural and functional neuroanatomy of several of the major psychiatric disorders is still inadequately understood, particularly compared with classic neurological disorders such as Parkinson's disease or amyotrophic lateral sclerosis. A major limitation in developing TMS as a psychiatric therapy has been to understand where to apply TMS for specific psychiatric disorders, given an inadequate understanding of the relevant functional anatomy. Although psychiatrists have used electroconvulsive therapy (ECT) for more than 60 years, most psychiatrists are not familiar with classic neurophysiological techniques, and there is a learning curve associated with using TMS for diagnosis, research, or therapy. With these caveats in mind, there is a large and rapidly growing literature on the therapeutic psychiatric uses of TMS.

Another major limitation of using TMS therapeutically is that there is inadequate understanding of what TMS is doing at a neurophysiological and neuropharmacological level, especially as a function of the use parameters. In an attempt to gain understanding in this area, psychiatric researchers have used TMS in animal models or combined TMS with functional imaging. Animal or imaging work has the promise of efficiently providing information about how TMS might work in a psychiatric condition, compared with the slow pace and high cost of a clinical trial.

Psychiatrically Relevant Animal Studies

Animal TMS studies offer many advantages over human clinical work. However, all studies are plagued by two major concerns. First, does the animal model validly reflect the human condition? Second, is TMS being applied in these animals in a way analogous to what is being done in humans? Although all animal models are vulnerable to the first question, the problem of the size of the TMS coil in small animals compared with humans is specific to TMS research and particularly worrisome. Even the smallest animal coils are several times larger relative to the brain size than human focal TMS coils. Nevertheless, initial rodent repetitive TMS (rTMS) studies reported significant antidepressant-like behavioral and neurochemical effects. In particular, rTMS enhances apomorphine-induced stereotypy and reduces immobility in the Porsolt swim test.[1] rTMS has been reported to induce electroconvulsive shock (ECS)-like changes in rodent brain monoamines, beta-adrenergic receptor binding, and immediate early gene induction.[2] The effects of rTMS on seizure threshold are variable and may depend on the parameters and chronicity of stimulation.[3] Pope and Keck have completed a series of studies using more focal TMS in rat models.[4] They have largely replicated earlier TMS animal studies using less-focal coils. However, even with the attempt at focal rat stimulation, the TMS-induced effects involve an entire hemisphere and cannot readily be extrapolated to what is happening in human TMS using focal coils.[5] In summary, TMS studies in animal models of stress, anxiety and depression have demonstrated antidepressant effects similar to those seen with ECS (analogous to human ECT) and other antidepressants. These studies, with one exception,[6] have not been very informative regarding the appropriate TMS use parameters for human clinical trials.

Several groups are considering performing analogous TMS animal studies using focal electrical stimulation, assuming that the induced electrical stimulation is actually what conveys the biological activity of TMS. However, creating electrodes that match the TMS field is a challenge, and the approach is subject to questions of comparative validity. A research group has designed TMS coils that weigh about 1 lb and are more focal than those currently produced (Epstein C, Davey K, Bohning D, personal communication, December 2002).[7] These may prove useful in animal TMS studies, with more focal stimulation than is currently done. At least one TMS manufacturer is advertising a small TMS coil for use with small animals such as mice or rats.

Combining Transcranial Magnetic Stimulation with Functional Imaging

Another method of evaluating TMS neurobiological effects for efficient therapeutic application is to combine TMS with functional neuroimaging. Combining imaging with TMS allows the physician to directly monitor TMS effects on the brain and to understand the varying effects of different TMS use parameters on brain function. Studies discussed elsewhere in this book suggest that TMS at different frequencies has divergent effects on brain activity.[8-12] Combining TMS with functional brain imaging promises to better delineate the behavioral neuropsychology of various psychiatric syndromes and some of the pathophysiologic circuits in the brain.

Several studies combining TMS with other neurophysiological and neuroimaging techniques have helped to elucidate how TMS achieves its effects in general and in psychiatric patients. Our group at MUSC developed and perfected the technique of interleaving TMS with blood oxygen level–dependent (BOLD) functional magnetic resonance imaging (fMRI), allowing for direct imaging of TMS effects with high spatial (1 to 2 mm) and temporal (2 to 3 seconds) resolution.[15-20] Another group in Germany has succeeded in interleaving TMS and fMRI in this manner, partially replicating the earlier work.[9,21] Work with this technology has demonstrated an intensity dose effect of TMS over the prefrontal cortex.[22,23] This had earlier been shown with TMS over motor cortex and is replicated each time a motor threshold is identified. An initial TMS/fMRI study found that prefrontal TMS at 80% motor threshold (MT) produced significantly less local and remote blood flow change than did 120% MT TMS.[24] Strafella and Paus used PET to show that prefrontal cortex TMS causes dopamine release in the caudate nucleus[25] and has reciprocal activity with the anterior cingulate gyrus.[26] Our group at MUSC[22,27] and groups in Scotland[28] and Australia[29] have all shown that lateral prefrontal TMS can cause changes in the anterior cingulate gyrus and other limbic regions in depressed patients. Imaging studies have consistently demonstrated that TMS delivered over the prefrontal cortex has immediate effects in important subcortical limbic regions. The exciting work in the field over the next decade is to determine whether the initial TMS effect on cortex and the secondary synaptic changes in other regions differs as a function of mood state, cortical excitability, and other factors that may change resting brain activity.

Paired-pulse TMS is a useful tool for evaluating cortical excitability, and there are several exciting findings regarding excitability in psychiatric disorders.[30] A fundamental limitation with paired-pulse TMS is that it measures excitability within the motor cortex, because it uses motor evoked potentials (MEPs) as the neurobiological end point. Psychiatric uses of TMS would be greatly enhanced if the clinician could directly assess the cortical excitability of other brain regions. Bohning and colleagues[31] have shown the feasibility of performing paired-pulse TMS within an fMRI scanner, using the regional BOLD response as the dependent variable. This paired-pulse TMS/fMRI technique involves a paired-pulse TMS setup as in a clinical neurophysiological laboratory, with a TMS positioner for holding the coil against the scalp within the scanner. The investigator also must interleave the paired-pulse TMS with fMRI scanner acquisition in a single event or block design. Pilot work has involved extracting the blood flow changes beneath the coil and comparing these TMS-induced changes with a theoretical model. Figure 21-1 demonstrates the initial results with this technique. These preliminary data demonstrate the feasibility of interleaved paired-pulse TMS/fMRI. They also demonstrate that it may be possible to use the modulation of the BOLD response by pairs of TMS pulses to test intracortical inhibition and facilitation and to investigate brain communication at time resolutions greater than that of the 3-second lag hemodynamic response.[9]

Extending the more theoretical imaging work described previously, several studies have used imaging to try and understand the therapeutic behavioral effects seen over time with TMS, particularly in depression. In contrast to imaging studies with ECT, which have found that ECT shuts off global and regional activity,[32,33] most studies using serial scans in depressed patients undergoing TMS have found increased activity in the cingulate and other limbic regions.[27,28] However, two studies have found divergent effects of TMS on regional activity in depressed patients, as determined by the frequency of stimulation and the baseline state of the patient.[29,34] For patients with global or focal hypometabolism, high-frequency prefrontal stimulation has been found to increase brain activity over time, with the opposite happening as well. Conversely, patients with focal hyperactivity have reduced activity over time after chronic daily low-frequency stimulation. However, these two small sample studies have numerous flaws. They simultaneously show the potential and the complexity surrounding the issue of how to use TMS to change activity in defined circuits. They also point out an obvious difference with ECT, for which the net effect of the ECT seizure is to decrease prefrontal and global activity.[32]

Combining TMS with functional imaging will likely continue to be an important method for understanding TMS psychiatric behavioral effects. Combination TMS and imaging will likely also evolve to be an important neuroscience tool for researching brain connectivity.[9,11,35–39]

■ Therapeutic Psychiatric Uses of Transcranial Magnetic Stimulation

Despite the lack of complete understanding of the pathophysiology of psychiatric disorders and with only limited knowledge of

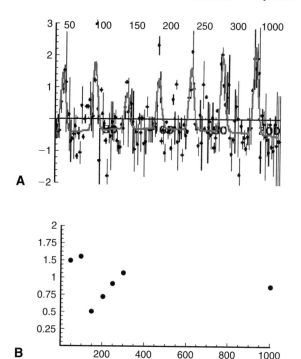

Figure 21-1 The paired-pulse transcranial magnetic stimulation/functional magnetic resonance imaging (ppTMS/fMRI) technique. These are blood oxygen level–dependent (BOLD) response data from directly underneath the TMS coil in motor cortex in two subjects. A mathematical model made of a hemodynamic response function multiplied by an exponential recovery function with independent amplitude scaling factors (relative to an inter-stimulus interval [ISI] = 1,000 amplitude) for the different ISI has been fit to the data and superimposed on the plots as a *thick line*. Further work is ongoing at shorter ISIs and using TMS over nonmotor regions. Although this is technically challenging and much work remains, ppTMS/fMRI could prove to be a useful tool in understanding the neurobiological effects of TMS and the pathophysiology of different neuropsychiatric disorders. The therapeutic uses of TMS in psychiatry have suffered because of inadequate understanding of disease pathophysiology and the neurobiological effects of TMS. Techniques such as this should improve future TMS therapeutic trials in psychiatry, making them more informed and focused in their hypotheses.

the translational neurobiologic effects of TMS, there has been much enthusiasm and controversy regarding TMS as a psychiatric treatment. The rapid launch of TMS in clinical trials was no doubt aided by the long history and widespread use of ECT. Although there has been much progress, no one understands how ECT works to treat depression,[40–44] although it remains clearly the most effective treatment for resistant depression. TMS was immediately adopted within psychiatry as a potentially more focal, nonconvulsive, and less-invasive method of ECT. With this background, it becomes clear why TMS was used initially in depression, despite there being other psychiatric disorders with a better defined and more regionally focused pathophysiology, such as obsessive-compulsive disorder (OCD).

Depression

Although there is controversy and much more work is needed, certain brain regions have consistently been implicated in the pathogenesis of depression and mood regulation.[45–52,53] These include the medial and dorsolateral prefrontal cortex, the cingulate gyrus, and other regions commonly referred to as limbic (e.g., amygdala, hippocampus, parahippocampus, septum, hypothalamus, limbic thalamus, insula) and paralimbic (e.g., anterior temporal pole, orbitofrontal cortex). A widely held theory during the past decade has been that depression results from a dysregulation of prefrontal cortical and limbic regions.[50,51,54,55]

In the modern era, the first few attempts to use TMS as an antidepressant were not influenced by this regional neuroanatomic literature, and stimulation was applied over the vertex.[56–58] However, working within the prefrontal cortical limbic dysregulation framework outlined previously and realizing that theories of ECT action emphasize the role of prefrontal cortex effects,[59] one of us (MSG) performed the first open trial of prefrontal TMS as an antidepressant in 1995,[60] followed immediately by a crossover double-blind study.[61] The theory behind this work was that chronic, frequent, subconvulsive stimulation of the prefrontal cortex over several weeks might initiate a therapeutic cascade of events in the prefrontal cortex and in connected limbic regions, thereby alleviating depression symptoms.[62] Functional imaging studies performed after these initial clinical trials suggest that this hunch was correct.[63] Prefrontal TMS sends direct information to important mood-regulating regions like the cingulate gyrus, orbitofrontal cortex, insula, and hippocampus. Beginning with these prefrontal studies, modern TMS was specifically designed as a focal, nonconvulsive, circuit-based approach to therapy. TMS was conceived of and launched to bridge from functional neuroimaging advances in circuit knowledge to the bedside as a focal, noninvasive treatment. Unfortunately, inadequate knowledge of the pathophysiologic circuit and of TMS effects has limited its use in this manner and much more work is needed in terms of the optimal anatomic location, TMS use parameters, and dosing regimen.

Since the initial studies, there has been continued interest in TMS as an antidepressant treatment. Multiple trials have been conducted from researchers around the world.[64,65] In general, there is not a large industry sponsoring or promoting TMS as an antidepressant (or therapy for other disorders), and the funding for these trials has largely come from foundations and governments. The sample sizes in these antidepressant trials are small (in all, less than 100 per trial) compared with industry-sponsored pharmaceutical trials of antidepressants. A thorough review of all of these trials is beyond the scope of this update. However, most of the more than 20 double-blind, randomized studies have found modest antidepressant effects that take several weeks to build. Not all TMS antidepressant treatment studies have been positive.[66]

Meta-analyses of Transcranial Magnetic Stimulation Antidepressant Effect

One way of succinctly reviewing the field of TMS as an antidepressant is to perform meta-analyses on the published trials. There have been five independent meta-analyses of the published or public TMS antidepressant literature, each varying slightly in the articles included and the statistics used.[67–71] Despite the differences in their methods, the results of all five meta-analyses are the same. Daily prefrontal TMS delivered over several weeks has antidepressant effects greater than sham treatment. For example, Burt and colleagues[67] examined 23 published comparisons for controlled TMS prefrontal antidepressant trials and found that TMS had a combined effect size of 0.67, indicating a moderate to large antidepressant effect. A subanalysis was done on the studies directly comparing TMS to ECT. The effect size for TMS in these studies was greater than in the studies comparing TMS to sham, perhaps reflecting subject selection bias. The investigators suggested that perhaps TMS works best in patients who are also clinical candidates for ECT. The meta-analysis conducted by Kozel and George[69] was confined to published double-blind studies with individual data using TMS over the left prefrontal cortex. The summary analysis using all 12 studies that met criteria revealed a cumulative effect size of 0.53 (Hedge's d; range, 0.24 to 0.82), and the total number of subjects studied was 230. Kozel and George[69] then used a funnel plot technique to assess whether there is a publication bias in the literature, and whether this bias might affect the results of the meta-analysis. [This technique assumes that with small sample studies, there is a large chance of both erroneous positive and negative results. As the sample size of studies increases, the effect sizes should begin to converge,

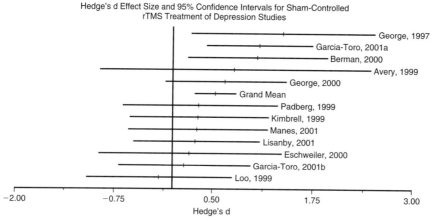

Figure 21-2 Size plot of the effect of prefrontal transcranial magnetic stimulation (TMS) as an antidepressant. This is a Forrest plot of the effect sizes of the sham-controlled studies of repeated (at least 2 weeks), daily left prefrontal TMS to treat depression. The effect sizes have varied widely, with all but one finding the effects significantly greater than the sham. The mean effect size is consistent with other antidepressant treatments. (From Kozel FA, George MS. Meta-analysis of left prefrontal repetitive transcranial magnetic stimulation [rTMS] to treat depression. J Psychiatr Pract 2002;8:270-275.)

resembling a funnel.] The funnel plot indicated that a publication bias is likely and that there are more positive small sample studies in the TMS antidepressant literature than should occur by chance (Fig. 21-2). These investigators then employed techniques to determine how large this publication bias would have to be to change the results of the meta-analysis. The fail-safe results indicated that there would have to be 56 nonsignificant unpublished studies of approximately the same average sample size as the published studies to change the cumulative meta-analysis effect to a nonsignificant result (i.e., 55 studies with Rosenthal's method and 20 with Orwin's method). The most critical meta-analysis of the TMS antidepressant field was conducted using the guidelines put forth in the Cochrane library.[70] However, even this stringent meta-analysis included 14 trials suitable for their analysis and found that left prefrontal TMS at 2 weeks produced significantly greater improvements in the Hamilton Rating Scale than did sham.[70] For reasons that are not clear, despite finding a statistically significant antidepressant effect at 2 weeks, they conclude that there is no strong evidence of a TMS effect.

In summary, all five meta-analyses of the TMS published literature concur that repeated daily prefrontal TMS for 2 weeks has antidepressant effects greater than sham.

Transcranial Magnetic Stimulation Compared with Electroconvulsive Therapy

Although there is consensus that TMS has statistically significant antidepressant effects, a more important question is whether these effects are clinically significant. The meta-analyses previously discussed concur on an effect size of Cohen's d of 0.65, which is a moderate effect, in the same range as the effects of antidepressant medications. For example, small to medium effect sizes (0.31 to 0.40) are common in randomized controlled trials of novel antidepressants.[72] With respect to whether or not TMS has clinical significance, an important clinical issue is whether TMS would be clinically effective in patients referred for ECT. This question has been addressed in a series of studies in which ECT referrals were randomized to receive ECT or rTMS. In an initial study, Grunhaus and colleagues[73] compared 40 patients who presented for ECT treatment and were randomized to receive ECT or TMS. ECT was superior to TMS in patients with psychotic depression, but the two treatments were not statistically different in patients without psychotic depression. The same group replicated this finding in a larger and independent cohort with an improved design.[74] Janicak and colleagues[75] reported a similar small series, finding near equivalence between TMS and ECT. The major differences between these studies and the rest of the controlled studies of TMS efficacy are the patient selection (suitable for ECT), the length of treatment (3 to 4 weeks), the lack of a blind, and the lack of a sham control.[76] Unfortunately, none of the studies explicitly measured differences in cognitive side effects, although presumably TMS has no measurable cognitive side effects, whereas ECT has several. In a similar but slightly modified design, Pridmore[77] reported a study comparing the antidepressant effects of standard ECT (three times per week), and one ECT per week followed by TMS on the other 4 weekdays. At 3 weeks, they found that both regimens produced similar antidepressant effects. Unfortunately, detailed neuropsychological testing was not performed, but one would assume that the TMS and ECT group had less cognitive side effects than the pure ECT group. The Israeli group found that relapse rates in the 6 months after ECT or rTMS were similar.[78] In sum, studies suggest that TMS clinical antidepressant effects are in the range of other antidepressants and persist as long as the clinical effects after ECT.

Transcranial Magnetic Stimulation and Sleep Deprivation Response

Psychiatrists have known for years that it is possible to transiently improve depression symptoms in depressed patients by keeping them awake all night. About one half of all depressed patients who are sleep deprived (SD) for one night report a substantial improvement in mood the following morning.[79] Unfortunately, more than one half of these subjects immediately relapse into depression when they next are allowed to sleep. This SD effect is remarkable for its speed of response (most other antidepressants, including ECT, take 2 to 3 weeks to improve mood) and because it is possible to predict who will respond to SD based on functional brain imaging assessing the resting activity in the cingulate gyrus.[80–82] Two studies have examined the relationship between sleep deprivation and TMS in depression. Padberg and colleagues[83] studied in an open trial whether the response to partial sleep deprivation might predict the clinical outcome of rTMS treatment. Thirty-three drug-free patients suffering from a major depressive episode underwent a partial sleep deprivation at least 5 days before rTMS and subsequently received 10 sessions of 10-Hz rTMS of the left prefrontal cortex. After rTMS, a significant overall improvement of 32% on the Hamilton Rating Scale for Depression was observed. Amelioration of depression after partial sleep deprivation was inversely correlated with improvement after rTMS, softly suggesting that those who responded to TMS were not those who responded to sleep deprivation. With this sample size, this was not significantly predictive. If these results were to replicate, an investigator could perhaps use a pre-TMS SD response to refine those getting TMS. Approaching this topic from a different perspective, another group used a controlled, balanced, parallel design to study whether rTMS, applied in the morning after partial sleep deprivation (PSD), is able to prevent depression relapse.[84] Twenty PSD responders were randomly assigned to receive active or sham stimulation for 4 days after sleep deprivation. Active stimulation prolonged significantly ($P < .001$) the antidepressant effect of PSD up to 4 days. This study suggests that rTMS may be an effective method to prevent relapse after depression improvement after partial SD or other antidepressant treatments[85] Further studies like these are needed, particularly given the role of pre-SD cingulate hyperactivity in predicting SD response and the multiple imaging studies showing that prefrontal TMS affects the cingulate acutely[22,24,26,86,87] and over the longer term.[88]

Transcranial Magnetic Stimulation as a Potential Adjunct to Other Antidepressant Therapies

Because of its noninvasiveness, some have wondered whether TMS might ultimately evolve into a treatment that would be given in addition to medications to speed their onset of action. This notion was particularly popular after an initial report of TMS antidepressant effects that occurred after just 1 week.[89]

These rapid responses have not been replicated despite many attempts, and it appears that TMS takes at least 2 weeks and probably longer to achieve maximum clinical effects.[76] There has never been a study showing faster onset of action of any two combined antidepressants, including ECT.[90] This is why it is not general clinical practice to start with combinations of antidepressant treatments. The sample sizes needed to demonstrate an additive or synergistic effect of TMS on top of a traditional antidepressant is unknown, but it would likely be more than 300. Nevertheless, at least two groups have used TMS as an adjunct treatment, with both failing to find an additive effect of TMS.[91,92] Both studies were underpowered to detect even a large and potentially clinically significant effect.

Other Thorny Issues

Although the literature suggests that prefrontal TMS has an antidepressant effect greater than sham and that the magnitude of this effect is at least as large as other antidepressants, many issues are not resolved. For example, it is unclear how best to deliver TMS to treat depression. Most, but not all,[65,93] studies have used focal coils positioned over the left prefrontal cortex. It is still not known whether TMS over one hemisphere is better than another or whether there are better methods for placing the coil. For the most part, the coil has been positioned using a rule-based algorithm to find the prefrontal cortex, which was adopted in the early studies.[60] However, this method was shown to be imprecise in the particular prefrontal regions stimulated directly underneath the coil, depending largely on the subject's head size.[94] An electroencephalographic montage, taking into account different skull sizes, would seem a better approach. However, it is unclear what scalp position to chose, resulting in the TMS coil being placed over which underlying cortical structure.

Most studies have stimulated with the intensity needed to cause movement in the thumb (i.e., MT). There is increasing recognition that higher intensities of stimulation are needed to reach the prefrontal cortex, especially in elderly patients, where prefrontal atrophy may outpace that of motor cortex, where the motor threshold is measured.[95–98] There are also emerging data that TMS therapeutic effects likely take several weeks to build. Consequently, many of the initial trials, which lasted only 1 to 2 weeks, were likely too brief to generate maximum clinical antidepressant effects. There are virtually no data on using TMS as a maintenance treatment in depression.[85,99]

Transcranial Magnetic Stimulation to Treat Mania

Grisaru and colleagues in Israel delivered right or left prefrontal TMS to a series of bipolar affective disorder (BPAD) patients admitted to their hospital for mania.[100] TMS was given daily in addition to the standard treatment for mania. After 2 weeks, the group receiving right-sided TMS was significantly more improved than the group that had received left-sided TMS. The investigators concluded that TMS might be useful as an antimanic agent. However, although subjects were assigned to the two groups at random, the left-sided group was more ill than the right-sided group on several measures. The investigators failed to replicate this antimanic effect in a follow-up study[101] and offered the suggestion that the antimanic effect of right prefrontal stimulation was a worsening (or antidepressant) effect of the group stimulated with left prefrontal stimulation.

■ Current State of Transcranial Magnetic Stimulation for Depression in Clinical Practice

TMS is a promising tool for treating depression acutely. It probably can also induce mania or hypomania in BPAD patients or susceptible patients. Its antimanic properties remain to be explored. Although it is approved in Canada and Israel as an antidepressant treatment, it is still considered investigational in the United States by the Food and Drug Administration (FDA). Despite the body of work showing antidepressant efficacy, prefrontal TMS is not an approved treatment from the standpoint of the FDA. The FDA treats the data from each TMS manufacturer separately, precluding consideration of the meta-analyses described earlier. A large-scale industry sponsored clinical trial designed for FDA approval is underway in the United States. The National Institutes of Mental Health (NIMH) is also funding a multisite trial. A small number of US, Canadian, and European psychiatrists are using TMS in clinical practice to treat depression under their general license to practice.

Review of Potential Antidepressant Mechanisms

How does TMS act to improve depression? The work done has provided evidence that prefrontal TMS produces immediate[24–26,102,103] and longer-term[27,104,105] changes in mood-regulating circuits. The original hypothesis about its antidepressant mechanism of action is still the most likely explanation.[22] What remains unclear is which specific prefrontal or other brain locations might be the best for treating depression and whether this can be determined with a group algorithm or requires individual imaging guidance. Much work remains to understand the optimum dosing strategy for the antidepressant effect of TMS. It is unlikely that the combinations of intensity, frequency, coil shape, scalp location, number of stimuli, or dosing strategy (e.g., daily, twice daily) used in the first decade of TMS as an antidepressant are the most effective for treating depression.[76] It is not understood how electrical stimulation of these circuits over time results in improvement of depression symptoms. The translational cascade of events remains undefined. Determining these answers using clinical trials alone would be a slow and expensive process. We hope that the work with TMS in animal models and functional imaging reviewed earlier will soon streamline this research area.

Some behavioral evidence from treatment trials is consistent with the functional imaging data showing repeated subtle changes in mood-regulating circuits. Szuba and colleagues[106,107] initially discovered that there is a subtle but statistically significant improvement in self-rated mood within each day over the 20 minutes of a daily TMS session (and that this is greater than with sham TMS). We found a nonstatistically significant trend confirming this in an independent study in bipolar depression.[108] A later clinical trial found this as well[109] and suggested that these subtle within-subject, within-session effects might predict eventual response. These three studies suggest that during each treatment session, the mood regulating circuit is being activated and slightly normalized. This gradual daily improvement then sums over several weeks when genuine clinical antidepressant effects emerge. Moreover, if they are important in eventual clinical response, one could consider dose-finding studies of different use parameters designed to find the parameters that maximally produced within-day changes. The parameters that produced the greatest within-day changes would be hypothesized to also be the most potent for eventual full treatment.

There are fewer data to suggest that TMS works to improve depression through activating normal anticonvulsant regulating systems—a widely held theory about the antidepressant mechanisms of action of ECT.[110] An appealing notion is that the brain "interprets" TMS-induced currents as potential seizures, with resultant activation of anticonvulsant cascades, which are tied to antidepressant efficacy. In support of this hypothesis, several animal studies have found that TMS has electroconvulsive shock (ECS)-like anticonvulsant effects.[111–113] However, there is only scant evidence to suggest that TMS has anticonvulsant effects in depressed patients. An initial open study found that the MT slightly increased during 2 weeks of TMS.[114] However, the MT does not always correlate with seizure threshold, and this was an open study with only small effects. Operator bias can influence MT determination, particularly with respect to coil location and angle. In a double-blind study, we examined for—and failed to find—a significant change in MT over the course of a TMS treatment trial.[108] Moreover, if TMS antidepressant efficacy were linked to its ability to initiate anticonvulsant cascades, the TMS use parameters closest to producing seizures would be predicted to be the most efficacious. However, there is no clear advantage of higher-frequency TMS,[115,116] even though it is clearly more likely to provoke seizures. Further work, perhaps using surrogate markers such as MR spectroscopy–measured γ-aminobutyric acid (GABA), are needed to explore this hypothesized antidepressant mechanism of action.

Transcranial Magnetic Stimulation as a Treatment for Other Psychiatric Conditions

TMS has also been investigated as a possible treatment for a variety of neuropsychiatric disorders. In general, the published literature about these conditions is much less extensive than for TMS as an antidepressant, and conclusions about the clinical significance of effects must remain tentative until large-sample studies are conducted.

Schizophrenia

Several studies have used TMS to investigate schizophrenia without consistent replications of early findings, which were compounded by medication issues.[117–120] The syndrome of schizophrenia involves many different symptoms ranging from auditory hallucinations to paranoid thoughts to blunted and restricted affect. The functional neuroanatomy of these different symptoms appears different. TMS treatment studies in schizophrenia have tended to enroll schizophrenia subjects with one particular symptom and then apply TMS to a brain region based on group functional imaging studies. The best example of this is the work of Hoffman and colleagues,[121] who have studied schizophrenia patients with auditory hallucinations and have stimulated them daily for several weeks over the temporal lobe at low frequencies, hypothesizing that they might inhibit a pathologically active region. An initial open study found that two of the three patients reported almost a total cessation of hallucinations for 2 weeks after their treatment.[121] A follow-up study by the same group using a sham condition found improvement of hallucinations after TMS over the left auditory cortex for 10 days.[122] Although an independent group has found similar results,[123] these exciting results need replication in larger studies before final acceptance. The idea of focally using TMS to modify symptoms is attractive.[124] TMS studies on the negative symptoms of schizophrenia have not been as positive. A 1-day prefrontal TMS challenge study by Nahas and colleagues[125] at MUSC failed to find significant effects on negative symptoms. Similarly, a well-conducted sham controlled clinical trial failed to find a treatment effect greater than sham.[126]

Anxiety Disorders

In a randomized trial of left and right prefrontal and mid-occipital 20-Hz stimulation in 12 patients with OCD, Greenberg and coworkers[127] found that a single session of right prefrontal rTMS decreased compulsive urges for 8 hours. Mood was also transiently improved, but there was no effect on anxiety or obsessions. Using TMS probes, the same group reported decreased intracortical inhibition in patients with OCD,[128] which has also been observed in patients with Tourette's disorder.[129] Somewhat surprisingly, OCD patients had a lowered MEP threshold in one study,[130] unrelated to intracortical inhibition, and which appears to replicate (Wassermann EM, personal communication, 2003). Only two other studies have examined possible therapeutic effects of rTMS in OCD. A double-blind study using right prefrontal slow (1-Hz) rTMS and a less-focal coil failed to find statistically significant effects greater than sham.[131] In contrast, an open study of a group of 12 OCD patients refractory to standard treatments who were randomly assigned to right or left prefrontal fast rTMS found clinically significant and sustained improvement in a third of patients.[132] Further work is warranted testing TMS as a potential treatment for OCD.

McCann and associates[133] reported that two patients with post-traumatic stress disorder (PTSD) improved during open treatment with 1-Hz rTMS over the right frontal cortex. Grisaru and associates[134] similarly stimulated 10 PTSD patients over motor cortex and found decreased anxiety. Grisaru and colleagues also reported a positive TMS study in PTSD patients (Grisaru N, personal communication, May 2001). Further work is needed.

Chronic Pain

The peripheral and central pathways for pain recognition and regulation are well understood compared with other psychiatric conditions. It is surprising that TMS has not been more widely investigated as a potential treatment for pain. An interesting study hints of potential therapeutic uses of TMS.[135] Lefaucher and colleagues[135] built on neurosurgical studies that found that electrical stimulation of the motor region could alleviate chronic facial pain. They applied a 20-minute session of sub motor threshold rTMS over the motor cortex at 10 Hz using a real or a sham coil in a series of 14 patients with intractable pain due to thalamic stroke or trigeminal neuropathy. They assessed pain levels using a 0 to 10 visual analog scale from day 1 to day 12 after the rTMS session. A significant pain decrease was observed up to 8 days after the "real" rTMS session. It is unclear why stimulation of the motor region would alter sensory pain. Further studies in this area are needed to see if this effect replicates.

Magnetic Seizure Therapy

Another interesting TMS development within psychiatry involves deliberately inducing an ECT-like seizure using TMS coils. The discussion throughout this chapter has focused on using TMS to change brain function without inadvertently causing a seizure. TMS at high frequencies and intensities can cause seizures. ECT produces a seizure through direct electrical stimulation, under anesthesia, of the scalp and skull. Although ECT is the most effective antidepressant, it has cognitive side effects and does not work in up to one half of treatment-resistant patients. If TMS was used to induce an ECT-like seizure, it might be able to focus on the point of origin of the seizure and spare some brain regions from unnecessary exposure to electrical currents and seizure spread. The direct application of electricity to the scalp with ECT loses focality and power due to the impedance of the overlying tissue. After a proof of concept demonstration in primates,[136] Lisanby and colleagues used an enhanced device with four times the usual number of charging modules to induce seizures in depressed patients referred for ECT.[137] Further clinical and preclinical work with this exciting technique, called magnetic seizure therapy (MST), has proceeded. An initial safety study found that MST seizures were briefer in duration than ECT seizures, that patients awoke from anesthesia much faster, and that their acute cognitive side effects were much less with MST.[138] Further work is underway to determine whether this technique has antidepressant effects. Because MST induces a seizure, it still requires repeated episodes of general anesthesia.

■ Conclusion

TMS is a powerful new brain stimulation tool, with extremely interesting research and one confirmed and several putative therapeutic psychiatric potentials. Although TMS clearly has the ability to engage subcortical-limbic circuits and to produce immediate, intermediate, and long-term effects, its use as a therapy in psychiatry has been limited by incomplete understanding of TMS neurobiological effects and of the underlying pathophysiology of the psychiatric disorders. Further understanding of the ways by which TMS changes neuronal function, especially as a function of its use parameters, will improve its ability both to answer neuroscience questions as well as to treat psychiatric diseases.

Acknowledgments

The authors' work with TMS has been supported in part by research grants from NARSAD, the Stanley Foundation, the Borderline Personality Disorders Research Foundation (BPDRF), the Dana Foundation (Bohning), NINDS grant RO1-AG40956, and the Defense Advanced Research Projects Agency (DARPA). Drs. George, Nahas, Kozel, and Bohning, alone or in combination, hold several TMS-related patents. These are not in the area of TMS therapeutics, but rather are for new TMS machine designs and combining TMS with MRI.

REFERENCES

1. Fleischmann A, Sternheim A, Etgen AM, et al. Transcranial magnetic stimulation downregulates beta-adrenoreceptors in rat cortex. J Neural Transm 1996;103:1361–1366.
2. Ben-Sachar D, Belmaker RH, Grisaru N, et al. Transcranial magnetic stimulation induces alterations in brain monoamines. J Neural Transm 1997;104:191–197.
3. Jennum P, Klitgaard H. Effect of acute and chronic stimulations on pentylenetetrazole-induced clonic seizures. Epilepsy Res 1996;23:115–122.
4. Pope A, Keck ME. TMS as a therapeutic tool in psychiatry: what do we know about neurobiological mechanisms? J Psychiatr Res 2001;35:193–215.
5. Weissman JD, Epstein CM, Davey KR. Magnetic brain stimulation and brain size: relevance to animal studies. Electroencephalogr Clin Neurol 1992;85:215–219.
6. Ben-shachar D, Gazawi H, Riboyad-Levin J, et al. Chronic repetitive transcranial magnetic stimulation alters beta-adrenergic and 5-HT2 receptor characteristics in rat brain. Brain Res 1999;816:78–83.
7. Davey KR, Epstein CM, George MS, et al. Modeling the effects of electrical conductivity of the head on the induced electrical field in the brain during magnetic stimulation. Clin Neurophysiol 2004;114:2204–2209.
8. Baumer T, Rothwell JC, Munchau A. Functional connectivity of the human premotor and motor cortex explored with TMS. Electroencephalogr Clin Neurophysiol 2003;56:160–169.
9. Bestmann S, Baudewig J, Siebner HR, et al. Subthreshold high-frequency TMS of human primary motor cortex modulates interconnected frontal motor areas as detected by interleaved fMRI-TMS. Neuroimage 2003;20:1685–1696.
10. Chouinard PA, Van Der Werf YD, Leonard G, et al. Modulating neural networks with transcranial magnetic stimulation applied over the dorsal premotor and primary motor cortices. J Neurophysiol 2003;90:1071–1083.
11. Lee L, Siebner HR, Rowe JB, et al. Acute remapping within the motor system induced by low-frequency rTMS. J Neuroscience 2003;23:5308–5318.
12. Tsuji T, Rothwell JC. Long lasting effects of rTMS and associated peripheral sensory input on MEPs, SEPs and transcortical reflex excitability in humans. J Physiol 2002;540:367–376.
13. Di Lazzaro V, Oliviero A, Berardelli A, et al. Direct demonstration of the effects of repetitive transcranial magnetic stimulation on the excitability of the human motor cortex. Exp Brain Res 2002;144:549–553.
14. Munchau A, Bloem R, Irlbacher K, et al. Functional connectivity of human premotor and motor cortex explored with repetitive transcranial magnetic stimulation. J Neurosci 2002;22:554–561.
15. Bohning DE, Shastri A, McConnell K, et al. A combined TMS/fMRI study of intensity-dependent TMS over motor cortex. Biol Psychiatry 1999;45:385–394.
16. Bohning DE, Shastri A, McGavin L, et al. Motor cortex brain activity induced by 1-Hz transcranial magnetic stimulation is similar in location and level to that for volitional movement. Invest Radiol 2000;35:676–683.
17. Bohning DE, Shastri A, Nahas Z, et al. Echoplanar BOLD fMRI of brain activation induced by concurrent transcranial magnetic stimulation (TMS). Invest Radiol 1998;33:336–340.
18. Bohning DE, Shastri A, Wassermann EM, et al. BOLD-fMRI response to single-pulse transcranial magnetic stimulation (TMS). J Magn Reson Imaging 2000;11:569–574.
19. Bohning DE, Shastri A, Wassermann EM, et al. BOLD-fMRI response to single-pulse transcranial magnetic stimulation (TMS). J Magn Reson Imaging 2000;11:569–574.
20. Shastri A, George MS, Bohning DE. Performance of a system for interleaving transcranial magnetic stimulation with steady state magnetic resonance imaging. Electroencephalogr Clin Neurophysiol Suppl 1999;51:55–64.
21. Baudewig J, Siebner HR, Bestmann S, et al. Functional MRI of cortical activations induced by transcranial magnetic stimulation (TMS). Neuroreport 2001;12:3543–3548.
22. Li XB, Nahas Z, Kozel FA, et al. Acute left prefrontal TMS in depressed patients is associated with immediately increased activity in prefrontal cortical as well as subcortical regions. Biol Psychiatry 2004;55:882–890.
23. Nahas Z, Lomarev M, Roberts DR, et al. Left prefrontal transcranial magnetic stimulation produces intensity dependent bilateral effects as measured with interleaved BOLD fMRI. Hum Brain Mapp 2000;11:520.
24. Nahas Z, Lomarev M, Roberts DR, et al. Unilateral left prefrontal transcranial magnetic stimulation (TMS) produces intensity-dependent bilateral effects as measured by interleaved BOLD fMRI. Biol Psychiatry 2001;50:712–720.
25. Strafella AP, Paus T, Fraraccio M, et al. Striatal dopamine release induced by repetitive transcranial magnetic stimulation of the human motor cortex. Brain 2003;126:2609–2615.
26. Paus T, Castro-Alamancos MA, Petrides M. Cortico-cortical connectivity of the human mid-dorsolateral frontal cortex and its modulation by repetitive transcranial magnetic

stimulation. Eur J Neurosci 2001;14:1405–1411.
27. Teneback CC, Nahas Z, Speer AM, et al. Two weeks of daily left prefrontal rTMS changes prefrontal cortex and paralimbic activity in depression. J Neuropsychiatry Clin Neurosci 1999;11:426–435.
28. Shajahan PM, Glabus MF, Steele JD, et al. Left dorso-lateral repetitive transcranial magnetic stimulation affects cortical excitability and functional connectivity, but does not impair cognition in major depression. Prog Neuropsychopharmacol Biol Psychiatry 2002;26:945–954.
29. Mitchel P. 15 Hz and 1 Hz TMS have different acute effects on cerebral blood flow in depressed patients. Int J Neuropsychopharmacol 2002;5:S7–S08.02.
30. Wassermann EM, Greenberg BD, Nguyen MB, et al. Motor cortex excitability correlates with an anxiety-related personality trait. Biol Psychiatry 2001;50:377–382.
31. Bohning DE, Walker JA, Mu Q, et al. Interleaved Paired Pulse TMS and BOLD fMRI [abstract]. Magn Res Med 2003.
32. Nobler MS, Oquendo MA, Kegeles LS, et al. Decreased regional brain metabolism after ECT. Am J Psychiatry 2001;158:305–308.
33. Sackeim HA. Functional brain circuits in major depression and remission. Arch Gen Psychiatry 2001; 58:649–650.
34. Speer AM, Kimbrell TA, Wasserman EM, et al. Opposite effects of high and low frequency rTMS on regional brain activity in depressed patients. Biol Psychiatry 2000;48:1133–1141.
35. Bestmann S, Baudewig J, Siebner HR, et al. Is functional magnetic resonance imaging capable of mapping transcranial magnetic cortex stimulation? Suppl Clin Neurophysiol 2003;56:55–62.
36. George MS, Bohning DE. Measuring brain connectivity with functional imaging and transcranial magnetic stimulation (TMS). In Desimone B (ed): Neuropsychopharmacology, Fifth Generation of Progress. New York, Lippincott, Williams and Wilkins, 2002:393–410.
37. Paus T, Jech R, Thompson CJ, et al. Transcranial magnetic stimulation during positron emission tomography: a new method for studying connectivity of the human cerebral cortex. J Neurosci 1997;17:3178–3184.
38. Siebner HR, Peller M, Lee L. Applications of combined TMS-PET studies in clinical and basic research. Suppl Clin Neurophysiol 2003;56:63–72.
39. Siebner HR, Peller M, Willoch F, et al. Lasting cortical activation after repetitive TMS of the motor cortex: a glucose metabolic study. Neurology 2000;54:956–963.

40. McCall WV, Reboussin DM, Weiner RD, et al. Titrated moderately suprathreshold vs fixed high-dose right unilateral electroconvulsive therapy: acute antidepressant and cognitive effects. Arch Gen Psychiatry 2000;57:438–444.
41. Sackeim HA. Memory and ECT: from polarization to reconciliation. J ECT 2000;16:87–96.
42. Sackeim HA, Devanand DP, Lisanby SH, et al. Treatment of the modal patient: does one size fit nearly all? J ECT 2001; 17:219–222.
43. Sackeim HA, Haskett RF, Mulsant BH, et al. Continuation pharmacotherapy in the prevention of relapse following electroconvulsive therapy: a randomized controlled trial. JAMA 2001;285:1299–1307.
44. Sackeim HA, Prudic J, Devanand DP, et al. A prospective, randomized, double-blind comparison of bilateral and right unilateral electroconvulsive therapy at different stimulus intensities. Arch Gen Psychiatry 2000;57:425–434.
45. George MS. An introduction to the emerging neuroanatomy of depression. Psychiatr Ann 1994;24:635–636.
46. George MS, Huggins T, McDermut W, et al. Abnormal facial emotion recognition in depression: serial testing in an ultra-rapid-cycling patient. Behav Modif 1998;22:192–204.
47. George MS, Ketter TA, Parekh PI, et al. Brain activity during transient sadness and happiness in healthy women. Am J Psychiatry 1995;152:341–351.
48. George MS, Ketter TA, Parekh PI, et al. Regional brain activity when selecting a response despite interference: an H2l5O PET study of the Stroop and an emotional Stroop. Hum Brain Mapp 1994;1:194–209.
49. George MS, Ketter TA, Parekh PI, et al. Blunted left cingulate activation in mood disorder subjects during a response interference task (the Stroop). J Neuropsychiatr Clin Neurol 1997;9:55–63.
50. George MS, Ketter TA, Post RM. Prefrontal cortex dysfunction in clinical depression. Depression 1994;2:59–72.
51. George MS, Ketter TA, Post RM. What functional imaging studies have revealed about the brain basis of mood and emotion. In Panksepp J (ed): Advances in Biological Psychiatry. Greenwich, CT, JAI Press, 1996:63–113.
52. Ketter TA, Andreason PJ, George MS, et al. Anterior paralimbic mediation of procaine-induced emotional and psychosensory experiences. Arch Gen Psychiatry 1996;53:59–69.
53. Kimbrell TA, Ketter TA, George MS, et al. Regional cerebral glucose utilization in patients with a range of severities of

unipolar depression. Biol Psychiatry 2002;51:237–252.
54. George MS, Post RM, Ketter TA, et al. Neural mechanisms of mood disorders. In Rush AJ (ed): Current Review of Mood Disorders. Philadelphia, Current Medicine, 1995:1.
55. Mayberg HS, Liotti M, Brannan SK, et al. Reciprocal limbic–cortical function and negative mood: converging PET findings in depression and normal sadness. Am J Psychiatry 1999;156:675–682.
56. Beer B. Uber das Auftretten einer objectiven Lichtempfindung in magnetischen Felde. Klin Wochenz 1902;15:108–109.
57. Grisaru N, Yarovslavsky U, Abarbanel J, et al. Transcranial magnetic stimulation in depression and schizophrenia. Eur Neuropsychopharmacol 1994;4:287–288.
58. Kolbinger HM, Hoflich G, Hufnagel A, et al. Transcranial magnetic stimulation (TMS) in the treatment of major depression—a pilot study. Hum Psychopharmacol 1995;10:305–310.
59. Nobler MS, Sackeim HA, Prohovnik I, et al. Regional cerebral blood flow in mood disorders, III. Treatment and clinical response. Arch Gen Psychiatry 1994;51:884–897.
60. George MS, Wassermann EM, Williams WA, et al. Daily repetitive transcranial magnetic stimulation (rTMS) improves mood in depression. Neuroreport 1995;6:1853–1856.
61. George MS, Wassermann EM, Williams WE, et al. Mood improvements following daily left prefrontal repetitive transcranial magnetic stimulation in patients with depression: a placebo-controlled crossover trial. Am J Psychiatry 1997;154:1752–1756.
62. George MS, Wassermann EM. Rapid-rate transcranial magnetic stimulation (rTMS) and ECT. Convuls Ther 1994;10:251–253.
63. Nahas Z, Teneback HC, Kozel A, et al. Brain effects of TMS delivered over prefrontal cortex in depressed adults: Role of stimulation frequency and coil-cortex distance. J Neuropsychiatry Clin Neurosci 2001;13:459–470.
64. Fitzgerald P. Is it time to introduce repetitive transcranial magnetic stimulation into standard clinical practice for the treatment of depressive disorders? Aust N Z J Psychiatry 2003;37:5–11.
65. Fitzgerald PB, Brown TL, Marston NAU, et al. Transcranial magnetic stimulation in the treatment of depression: a double-blind placebo controlled trial. Arch Gen Psychiatry 2003;60:1002–1008.
66. Loo C, Mitchell P, Sachdev P, et al. A double-blind controlled investigation of transcranial magnetic stimulation for the treatment of resistant major depression. Am J Psychiatry 1999;156:946–948.
67. Burt T, Lisanby SH, Sackeim HA. Neuropsychiatric applications of transcranial magnetic stimulation. Int J Neuropsychopharmacol 2002;5:73–103.
68. Holtzheimer PE, Russo J, Avery D. A meta-analysis of repetitive transcranial magnetic stimulation in the treatment of depression. Psychopharmacol Bull 2001;35:149–169.
69. Kozel FA, George MS. Meta-analysis of left prefrontal repetitive transcranial magnetic stimulation (rTMS) to treat depression. J Psychiatr Pract 2002;8:270–275.
70. Martin JLR, Barbanoj MJ, Schlaepfer TE, et al. Transcranial magnetic stimulation for treating depression [Cochrane review]. Oxford, The Cochrane Library, Update Software, 2002.
71. McNamara B, Ray JL, Arthurs OJ, et al. Transcranial magnetic stimulation for depression and other psychiatric disorders. Psychol Med 2001;31:1141–1146.
72. Thase ME. The need for clinically relevant research on treatment-resistant depression. J Clin Psychiatry 2001;62:221–224.
73. Grunhaus L, Dannon PN, Schreiber S, et al. Repetitive transcranial magnetic stimulation is as effective as electroconvulsive therapy in the treatment of nondelusional major depressive disorder: an open study. Biol Psychiatry 2000;4:314–324.
74. Grunhaus L, Schreiber S, Dolberg OT, et al. A randomized controlled comparison of ECT and rTMS in severe and resistant non-psychotic major depression. Biol Psychiatry 2002;53.
75. Janicak PG, Dowd SM, Martis B, et al. Repetitive transcranial magnetic stimulation versus electroconvulsive therapy for major depression: preliminary results of a randomized trial. Biol Psychiatry 2002;51:659–667.
76. Gershon AA, Dannon PN, Grunhaus L. Transcranial magnetic stimulation in the treatment of depression. Am J Psychiatry 2003;160:835–845.
77. Pridmore S. Substitution of rapid transcranial magnetic stimulation treatments for electroconvulsive therapy treatments in a course of electroconvulsive therapy. Depress Anxiety 2000;12:118–123.
78. Dannon PN, Dolberg OT, Schreiber S, et al. Three and six-month outcome following courses of either ECT or rTMS in a population of severely depressed individuals—preliminary report. Biol Psychiatry 2002;51:687–690.
79. Wu JC, Bunney WE. The biological basis of an antidepressant response to sleep deprivation and relapse: review and hypothesis. Am J Psychiatry 1990;147:14–21.
80. Ebert D, Feistel H, Barocka A. Effects of sleep deprivation on the limbic system and the frontal lobes in affective disorders: a study with

Tc-99m-HMPAO SPECT. Psychiatry Res 1991;40:247–251.
81. Ebert D, Feistel H, Barocka A, et al. Increased limbic flow and total sleep deprivation in major depression with melancholia. Psychiatry Res 1994;55:101–109.
82. Wu JC, Gillin JC, Buchsbaum MS, et al. Effect of sleep deprivation on brain metabolism of depressed patients. Am J Psychiatry 1992;149:538–543.
83. Padberg F, Schule C, Zwanzger P, et al. Relation between responses to repetitive transcranial magnetic stimulation and partial sleep deprivation in major depression. J Psychiatr Res 2002;36:131–135.
84. Eichhammer P, Kharraz A, Wiegand R, et al. Sleep deprivation in depression stabilizing antidepressant effects by repetitive transcranial magnetic stimulation. Life Sci 2002;70:1741–1749.
85. Li X, Nahas Z, Anderson B, et al. Can rTMS be used as a maintenance treatment for bipolar depression? Depress Anxiety 2004;20:98–100.
86. Nahas Z, Teneback CT, Kozel AF, et al. Brain effects of transcranial magnetic delivered over prefrontal cortex in depressed adults: the role of stimulation frequency and distance from coil to cortex. J Neuropsychiatry Clin Neurosci 2001;13:459–470.
87. Rushworth MF, Hadland KA, Paus T, et al. Role of the human medial frontal cortex in task switching: a combined fMRI and TMS study. J Neurophysiol 2002;87:2577–2592.
88. Teneback CC, Nahas Z, Speer AM, et al. Changes in prefrontal cortex and paralimbic activity in depression following two weeks of daily left prefrontal TMS. J Neuropsychiatry Clin Neurosci 1999;11:426–435.
89. Pascual-Leone A, Rubio B, Pallardo F, et al. Beneficial effect of rapid-rate transcranial magnetic stimulation of the left dorsolateral prefrontal cortex in drug-resistant depression. Lancet 1996;348:233–237.
90. George MS, Lydiard RB. Speed of onset of action of the newer antidepressants: fluoxetine and bupropion. Int Clin Psychopharmacol 1992;6:209–217.
91. Garcia-Toro M, Pascual-Leone A, Romera M, et al. Prefrontal repetitive transcranial magnetic stimulation as add-on treatment in depression. J Neurol Neurosurg Psychiatry 2001;71:546–548.
92. Lisanby SH, Pascual-Leone A, Sampson SM, et al. Augmentation of sertraline antidepressant treatment with transcranial magnetic stimulation. Biol Psychiatry 2001;49:81S.
93. Klein E, Kreinin I, Chistyakov A, et al. Therapeutic efficacy of right prefrontal slow repetitive transcranial magnetic stimulation in major depression: a double-blind controlled study. Arch Gen Psychiatry 1999;56:315–320.
94. Herwig U, Padberg F, Unger J, et al. Transcranial magnetic stimulation in therapy studies: examination of the reliability of "standard" coil positioning by neuronavigation. Biol Psychiatry 2001;50:58–61.
95. Kozel FA, Nahas Z, DeBrux C, et al. How the distance from coil to cortex relates to age, motor threshold and possibly the antidepressant response to repetitive transcranial magnetic stimulation. J Neuropsychiatry Clin Neurosci 2000;12:376–384.
96. McConnell KA, Nahas Z, Shastri A, et al. The transcranial magnetic stimulation motor threshold depends on the distance from coil to underlying cortex: a replication in healthy adults comparing two methods of assessing the distance to cortex. Biol Psychiatry 2001;49:454–459.
97. Mosimann UP, Marre SC, Werlen S, et al. Antidepressant effects of repetitive transcranial magnetic stimulation in the elderly: correlation between effect size and coil-cortex distance. Arch Gen Psychiatry 2002;59:560–561.
98. Padberg F, Zwanzger P, Keck ME, et al. Repetitive transcranial magnetic stimulation (rTMS) in major depression: relation between efficacy and stimulation intensity. Neuropsychopharmacology 2002;27:638–645.
99. Nahas Z, Oliver NC, Johnson M, et al. Feasibility and efficacy of left prefrontal rTMS as a maintenance antidepressant. Biol Psychiatry 2000;57.
100. Belmaker RH, Grisaru N. Anti-bipolar potential for TMS. Bipolar Disorders 1999;1:91–92.
101. Kaptsan A, Yaroslavsky Y, Applebaum J, et al. Right prefrontal TMS versus sham treatment of mania: a controlled study. Bipolar Disord 2003;5:36–39.
102. George MS, Stallings LE, Speer AM, et al. Prefrontal repetitive transcranial magnetic stimulation (rTMS) changes relative perfusion locally and remotely. Hum Psychopharmacol 1999;14:161–170.
103. Li X, Teneback CC, Nahas Z, et al. Interleaved transcranial magnetic stimulation/functional MRI confirms that lamotrigine inhibits cortical excitability in healthy young men. Neuropsychopharmacology 2004;29:1395–1407.
104. Levkovitz Y, Grisaru N, Segal M. Transcranial magnetic stimulation and antidepressive drugs share similar cellular effects in rat hippocampus. Neuropsychopharmacology 2001;24:608–616.
105. Levkovitz Y, Marx J, Grisaru N, et al. Long-term effects of transcranial magnetic stimulation on hippocampal reactivity to afferent stimulation. J Neurosci 1999;19:3198–3203.

106. Szuba MP, O'Reardon JP, Rai AS, et al. Acute mood and thyroid stimulating hormone effects of transcranial magnetic stimulation in major depression. Biol Psychiatry 2001;50:22–57.
107. Szuba MP, Rai A, Kastenberg J, et al. Rapid mood and endocrine effects of TMS in major depression. Proceedings of the American Psychiatric Association Annual Meeting, 1999:201.
108. Nahas Z, Kozel FA, Li X, et al. Left prefrontal transcranial magnetic stimulation (TMS) treatment of depression in bipolar affective disorder: a pilot study of acute safety and efficacy. J Bipolar Disord 2003;5:40–47.
109. Grunhaus L, Dolberg OT, Polak D, et al. Monitoring the response to rTMS in depression with visual analog scales. Hum Psychopharmacol 2002;17:349–352.
110. Sackeim HA, Decina P, Malitz S, et al. Anticonvulsant and antidepressant properties of electroconvulsive therapy: a proposed mechanism of action. Biol Psychiatry 1983;18:1301–1310.
111. Belmaker RH, Grisaru N. Magnetic stimulation of the brain in animal depression models responsive to ECS. J ECT 1998;14:194–205.
112. Ebert U, Ziemann U. Altered seizure susceptibility after high-frequency transcranial magnetic stimulation in rats. Neurosci Lett 1999;273:155–158.
113. Fujiki M, Steward O. High frequency transcranial magnetic stimulation mimics the effects of ECS in upregulating astroglial gene expression in the murine CNS. Mol Brain Res 1997;44:301–308.
114. Triggs WJ, McCoy KJ, Greer R, et al. Effects of left frontal transcranial magnetic stimulation on depressed mood, cognition, and corticomotor threshold. Biol Psychiatry 1999;45:1440–1446.
115. George MS, Nahas Z, Molloy M, et al. A controlled trial of daily transcranial magnetic stimulation (TMS) of the left prefrontal cortex for treating depression. Biol Psychiatry 2000;48:962–970.
116. Padberg F, Haag C, Zwanzger P, et al. Rapid and slow transcranial magnetic stimulation are equally effective in medication-resistant depression: a placebo-controlled study. CINP Abstracts 1998;21st Congress:103-st0306.
117. Feinsod M, Kreinin B, Chistyakov A, et al. Preliminary evidence for a beneficial effect of low-frequency, repetitive transcranial magnetic stimulation in patients with major depression and schizophrenia. Depress Anxiety 1998;7:65–68.
118. Davey NJ, Puri BK. On electromyographic responses to transcranial magnetic stimulation of the motor cortex in schizophrenia. J Neurol Neurosurg Psychiatry 1997;63:468–473.
119. Davey NJ, Puri BK, Lewis HS, et al. Effects of antipsychotic medication on electromyographic responses to transcranial magnetic stimulation of the motor cortex in schizophrenia. J Neurol Neurosurg Psychiatry 1997;63:468–473.
120. Puri BK, Davey NJ, Ellaway PH, et al. An investigation of motor function in schizophrenia using transcranial magnetic stimulation of the motor cortex. Br J Psychiatry 1996;169:690–695.
121. Hoffman RE, Boutros NN, Berman RM, et al. Transcranial magnetic stimulation of left temporoparietal cortex in three patients reporting hallucinated "voices." Biol Psychiatry 1999;46:130–132.
122. Hoffman RE, Boutros NN, Hu S, et al. Transcranial magnetic stimulation and auditory hallucinations in schizophrenia. Lancet 2000;355:1073–1075.
123. d'Alfonso AA, Aleman A, Kessels RP, et al. Transcranial magnetic stimulation of left auditory cortex in patients with schizophrenia: effects on hallucinations and neurocognition. J Neuropsychiatry Clin Neurosci 2002;14:77–79.
124. Hoffman RE, Cavus I. Slow transcranial magnetic stimulation, long-term depotentiation, and brain hyperexcitability disorders [review]. Am J Psychiatry 2002;159:1093–1102.
125. Nahas Z, McConnell K, Collins S, et al. Could left prefrontal rTMS modify negative symptoms and attention in schizophrenia? Biol Psychiatry 1999;45:37S#120.
126. Klein E, Kolsky Y, Puyerovsky M, et al. Right prefrontal slow repetitive transcranial magnetic stimulation in schizophrenia: a double-blind sham-controlled pilot study. Biol Psychiatry 1999;46:1451–1454.
127. Greenberg BD, George MS, Dearing J, et al. Effect of prefrontal repetitive transcranial magnetic stimulation (rTMS) in obsessive compulsive disorder: a preliminary study. Am J Psychiatry 1997;154:867–869.
128. Cora-Locatelli G, Greenberg BD, Harmon A, et al. Cortical excitability and augmentation strategies in OCD. Biol Psychiatry 1998;43:77s–#258.
129. Ziemann U, Paulus W, Rothenberger A. Decreased motor inhibition in Tourette's disorder: evidence from transcranial magnetic stimulation. Am J Psychiatry 1997;154:1277–1284.
130. Greenberg BD, Ziemann U, Cora-Locatelli G, et al. Altered cortical excitability in obsessive-compulsive disorder. Neurology 2000;54:142–147.
131. Alonso P, Pujol J, Cardoner N, et al. Right prefrontal TMS in OCD: a double-blind,

132. Sachdev PS, McBride R, Loo CK, et al. Right versus left prefrontal transcranial magnetic stimulation for obsessive-compulsive disorder: a preliminary investigation. J Clin Psychiatry 2001;62:981–984.
133. McCann UD, Kimbrell TA, Morgan CM, et al. Repetitive transcranial magnetic stimulation for posttraumatic stress disorder [letter]. Arch Gen Psychiatry 1998;55:276–279.
134. Grisaru N, Amir M, Cohen H, et al. Effect of transcranial magnetic stimulation in posttraumatic stress disorder: a preliminary study. Biol Psychiatry 1998;44:52–55.
135. Lefaucher JP, Drouot X, Nguyen JP. Interventional neurophysiology for pain control: duration of pain relief following repetitive transcranial magnetic stimulation of the motor cortex. Neurophysiol Clin 2001;31:247–252.
136. Lisanby SH, Luber B, Sackeim HA, et al. Deliberate seizure induction with repetitive transcranial magnetic stimulation in nonhuman primates [letter]. Arch Gen Psychiatry 2001;58:199–200.
137. Lisanby SH, Schlaepfer TE, Fisch HU, et al. Magnetic seizure therapy for major depression. Arch Gen Psychiatry 2001;58:303–305 (let).
138. Lisanby SH, Luber B, Barriolhet L, et al. Magnetic seizure therapy (MST): acute cognitive effects of MST compared with ECT. J ECT 2001;17:77.

(continued from previous: placebo-controlled study. Am J Psychiatry 2001;158:1143–1145.)

22 External Modulation of Visual Perception by Transcranial Magnetic and Direct Current Stimulation of the Visual Cortex

Andrea Antal, Michael A. Nitsche, and Walter Paulus

Most of our impressions about the world and our memories of it are based on sight. However, the mechanisms that underlie vision are not at all obvious to the perceiver. How do we see form, perceive a movement or motion, or distinguish colors? How can we guide body movements under visual control? One of the most challenging problems in neuroscience is to understand how the visual input is processed in primary and secondary visual areas, how it is transferred to higher cortical areas, and how visual perception is correlated with brain function.

The visual system has the most complex circuitry of all the sensory systems. Most of what we know about the functional organization of the visual system is derived from animal experiments in which parts of the visual cortex are experimentally lesioned or from clinical studies including patients who have visual dysfunctions related to trauma or dysfunctions of neurotransmitters. Nevertheless, in the past few years, with the aid of electrical stimulation or magnetic stimulation of the brain, a bridge between psychology and single-cell electrophysiology seems to be closer.

The first systematic studies of electrical stimulation of the visual cortex were carried out by intracranial electrical stimulation of the brains of neurological patients during neurosurgery.[1] Occipital stimulation produced light sensations called *phosphenes* whose position, color, and shape varied according to the position of the electrodes. Brindley and Lewin[2] stimulated the occipital cortex of a blind subject with implanted electrodes and described the properties of elicited phosphenes in detail; disappointed, however, they eventually left the field of visual neuroprothetics.

These early invasive approaches were followed by noninvasive stimulation methods. The first stimulation technique through the intact skull, transcranial electrical stimulation (TES), was developed by Merton and Morton in 1980.[3] Compared with invasive methods, the voltage had to be increased and discharge duration shortened. They showed that stimulation of the scalp over the occipital cortex could produce phosphenes. The main problem with this kind of stimulation is that the electric current flowing between the electrodes causes pain and contraction of the scalp muscles. In contrast, transcranial magnetic stimulation (TMS) provides a powerful, noninvasive, non-painful tool for activating neurons in the cortex. Single-pulse TMS can produce a brief excitation or inhibition in the brain for a few milliseconds. Repetitive TMS (rTMS) refers to regularly repeated stimuli to a single scalp site. With rTMS, longer lasting effects from hundreds of milliseconds to several seconds and minutes can be elicited, some of which outlast the period of stimulation. Mapping studies within the motor and visual cortex using single-pulse TMS showed a spatial resolution down to 0.5 to 1.0 cm at the scalp surface (reviewed by Walsh and Cowey,[4] Jahanshahi and Rothwell,[5] and Cowey and Walsh[6]). However, during prolonged rTMS, the effects along corticocortical connections may well produce less focal effects.

In the last few years, another noninvasive method has gained significant importance in studying human brain function. Transcranial direct current stimulation (tDCS) applied through the skull was shown to modulate directly the excitability of the motor[7–12] and visual[13–16] cortices in human subjects.

TMS and tDCS differ in basic aspects. Single-pulse TMS produces a brief (in the range of milliseconds), strong excitation or inhibition in the brain and in that way *stimulates or interrupts* ongoing cortical activity to a spatially and temporally restricted extent. In contrast, tDCS is usually applied for a much longer duration (seconds and longer), *modulates* cortical activity, and in this way induces functional changes in the human brain. In this, it shares certain similarities with rTMS. tDCS offers the possibility of inducing acute and persistent neuronal excitability changes without local discomfort, probably by shifting neuronal resting membrane potential.[17] Here, we summarize results derived from the use of TMS and tDCS on visual perception.

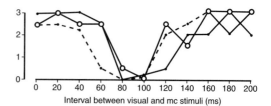

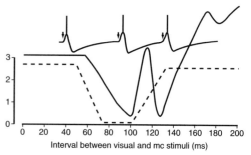

Figure 22–1 *Top:* The U-shaped curve represents the number of letters correctly identified in a composition of three briefly flashed dark letters on a light background, plotted as a function of the delay for the magnetic stimulus. When transcranial magnetic stimulation (TMS) was applied at 0 to 40 ms after the onset of the letters, there was no effect of TMS on recognition. TMS applied between 60 and 140 ms after the stimulus onset resulted in reduced performance and reduced recognition to zero between 80 and 100 ms. *Bottom:* Hypothetical mechanism of suppression by an inhibitory postsynaptic potential (IPS). The *solid line* is the visual evoked response, and the *dashed line* is the suppression. (From Amassian VE, Cracco RQ, Maccabee PJ, et al. Suppression of visual perception by magnetic coil stimulation of human occipital cortex. Electroencephalogr Clin Neurophysiol 1989;74:458–462.)

■ Transcranial Magnetic Stimulation

TMS is able to influence a variety of visual-perceptual aspects. Amassian and coworkers[19] were the first to demonstrate the inhibitory effect of TMS on the visual cortex systematically (Fig. 22–1). Single-pulse TMS was delivered over the occipital cortex of subjects while they were performing a letter-identification task. Performance was impaired when the TMS pulse was applied between 60 and 140 ms after the stimulus onset. They hypothesized that during and after magnetic stimulation of the visual cortex, an inhibitory postsynaptic potential component evoked by the magnetic pulse by means of polysynaptic circuits causes the suppression of visual perception. Since then, it has become obvious that this technique can transiently mimic the effect of neurological lesions on visual perception. Impairment of visual perception by TMS has been demonstrated by several investigators using different visual recognition tasks.[19–21] Perceptual impairment induced by TMS was also extended to extrafoveal positions.[22,23]

Phosphenes and Their Topography

Phosphenes are sensations of light that can be evoked by single-pulse or repetitive transcranial magnetic stimulation of the occipital cortex. They are commonly described as spots of light or stars that tend to persist for the duration of the stimulation and disappear with its cessation. Phosphenes are relatively variable in position or form, their induction also somewhat depending on the orientation and movement of the coil[22,24,25] and the direction of the induced current in the brain.[26] The stimulation over V1 results in stationary phosphenes, and stimulation over V5 produces moving light sensations.[27,28] Generally, the prevalence of phosphenes varies with varying stimulation characteristics.[24,28–32] Nevertheless, there are some general observations concerning the appearance of stationary phosphenes. Phosphenes are most easily perceived in the lower part of the visual field (due to the representation of the upper visual cortex closer to the skull convexity) and peripheral phosphenes are more common than central ones. Stationary phosphene thresholds were highly reproducible across test sessions held at least a week apart, did not correlate with the motor threshold,[32–24] and did not change with variations in ambient light.[35]

The pattern of moving phosphenes elicited by V5 stimulation corresponds to a strictly retinotopical organization. Stimulation of the left V5 produces moving phosphenes in the right hemifield, and moving the coil up and down changes the location of the moving phosphene conversely. The induction of moving phosphenes may provide the quickest and most reliable functional demonstration of V5 location.[27] Moving phosphene thresholds were also highly reproducible across test sessions held at least a week apart and did not correlate with the motor threshold.[34]

Where Do Phosphenes or Conscious Perception of Phosphenes Emerge? Studies investigating the cortical site of phosphene perception using different TMS coil positions have yet to determine the exact site in the visual system. Meyer and colleagues[24] suggested that the probable stimulation site for evoking phosphenes is the supracalcarine part of the occipital lobe. Others[22,25] argue that phosphenes are not evoked in the visual cortex itself, but rather within the underlying white matter by the activation of the optic radiation, in which adjacent fibers project to occipital cortical neurons. Marg and Rudiak[25] estimated the depth of stimulation in the visual cortex to be approximately 4 cm. Kammer and coworkers[33] suggested that unilateral phosphenes are generated in V2 and V3, whereas the stimulation of V1 yields phosphenes in both visual fields. Possibly, in the case of V2 or V3 stimulation, these extrastriate areas are first activated and then evoke activity in V1 by backprojection fibers, with the latter causing the perception of phosphenes. This may explain the observation that no phosphenes could be evoked in a patient with a lesion of V1.[36] A study conducted by Pascual-Leone and Walsh[28] also supports the V1 hypothesis, at least for the perception of moving phosphenes. A dual-pulse paradigm was used to study the role of V1 in the elaboration of the conscious illusory percept of a moving phosphene generated by stimulation of V5. Perception was disrupted by subthreshold stimulation applied to V1 5 to 40 ms after the suprathreshold stimulus to V5 (Fig. 22–2). No effect was observed when stimulation was applied to V1 before the V5 pulse. These results suggest that the functional integrity of V1 is necessary for the perception of moving phosphenes.

However, a study suggests that the excitability of extrastriate areas seems to be related to stationary phosphene perception, whereas the level of striate cortex excitability does not play a critical role.[37] It was shown by electrophysiology and functional magnetic resonance imaging (fMRI) that subjects reporting phosphenes activate specifically a larger extrastriate cortex network when exposed to a standard checkerboard stimulus pattern, compared with subjects lacking phosphene perception.

In summary, some divergent results concerning properties of phosphene perception need to be further characterized.

Monitoring Plastic Changes of the Visual Cortex by Phosphene Measurement: Relationship between Phosphene Thresholds and the Excitability of the Visual Cortex. Light deprivation, which is known to increase cortical excitability in the first 180 minutes, causes significant phosphene threshold reduction,[38] suggesting that phosphene threshold measurements provide information about excitability changes in the human

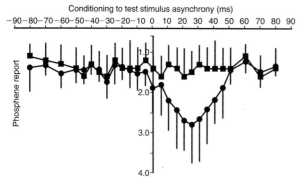

Figure 22–2 Mean responses of eight subjects to combined stimulation of V5 and V1 are shown by the *circles*. Results of a control experiment of five subjects with paired stimulation of V5 are shown by *squares*. The negative values indicate that V1 received transcranial magnetic stimulation (TMS) before V5; positive values indicate that V1 was stimulated after V5. The phosphenes elicited by V5 stimulation were present and moving (1.0), present but the subject was not able to judge whether it was moving or moving differently (2.0), present but stationary (3.0), or no phsophene was observed (4.0). TMS over V1 between 5 and 45 ms after TMS over V5 disrupted the perception of phosphenes. A conditioning stimulus to V5 did not change the effect of the V5 test stimulus, regardless of the interval *(squares)*. (From Pascual-Leone A, Walsh V. Fast backprojections from the motion to the primary visual area necessary for visual awareness. Science 2001;292:510–512.)

visual cortex. In migraine patients, whose visual cortex is thought to be hyperexcitable between and during migraine attacks, lower phosphene thresholds were found than in healthy subjects.[39]

Based on these results, several studies have applied TMS mapping in partially blind subjects, to study the neuroplastic changes of the visual cortex in patients. In a peripherally blind subject, stimulation of V1 resulted in easily elicited and reproducible phosphenes with normal thresholds, however, the spatial distribution was coarser compared with normal subjects.[36] The perception of moving phosphenes when V5 was stimulated was also normal. In cases of peripheral blindness, TMS studies do not seem to reveal abnormal function of the visual cortex. Gothe and associates[40] compared phosphene thresholds in subjects with some residual vision (measurable with Snellen test charts), subjects with very poor residual vision (i.e., with only light or movement perception but no measurable acuity), and subjects without any residual vision. In agreement with previous data, phosphene thresholds were normal in all of the subjects compared with healthy persons, but the number of effective stimulation sites was significantly reduced in subjects who had very poor or no residual vision. These results suggest that long-term visual deafferentation causes a reorganization of the visual system that reduces but does not eliminate the ability of blind subjects to perceive cortically elicited phosphenes.

Suppression of Color Perception

Colored phosphenes can be elicited by TMS,[23,25] but reports on the effect of TMS on color perception are rare. The reason probably is that it is difficult to selectively disturb color perception by external stimulation because V4, the area assumed responsible mainly for color vision, is located far from the skull surface. However, it is possible to separate the time course of achromatic and chromatic perception by TMS. Magnetic stimuli applied over the occipital cortex resulted in later suppression periods when chromatic stimuli (i.e., green letters against a red background) were used compared with those with achromatic stimuli.[41] Using threshold detection of chromatic and achromatic gaussian filtered dots and applying TMS to the primary visual cortex, Paulus and colleagues[42] have shown that the magnocellular (achromatic) system has a significantly higher susceptibility

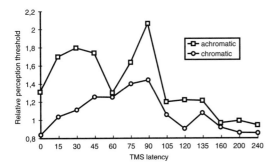

Figure 22-3 Differential inhibition of colored *(circles)* and achromatic *(squares)* gaussian dots by transcranial magnetic stimulation (TMS). At time zero, achromatic perception appears to be facilitated, whereas color perception appears to be inhibited by TMS. When TMS is applied 30 ms after presentation of the visual stimulus, a selective inhibition of black and white perception is apparent, whereas color perception is continuously more inhibited with a maximum at 75- and 90-ms inter-stimulus intervals. Overall, color perception is less susceptible to TMS than achromatic perception. (From Paulus W, Korinth S, Wischer S, et al. Differentiation of parvo- and magnocellular pathways by TMS at the occipital cortex. Electroencephalogr Clin Neurophysiol Suppl 1999;51:351-360.)

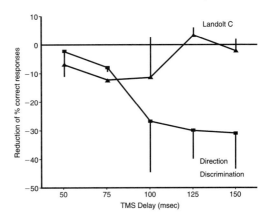

Figure 22-4 Plot of the average reduction of the percentage of correct responses produced by transcranial magnetic stimulation over the left V5 in the Landolt C spatial acuity and motion direction task. (From Hotson J, Brain D, Herzberg W, et al. Transcranial magnetic stimulation of extrastriate cortex degrades human motion direction discrimination. Vis Res 1994;34:2115-2123.)

to TMS than the parvocellular (chromatic) system, probably due to the larger axonal and neuronal size (Fig. 22-3). Using achromatic stimuli, TMS suppressed perception if given 30 to 45 ms after the onset of the stimuli and when applied with a delay of 90 ms. When chromatic stimuli were shown, the suppression was most effective when the magnetic pulse was applied between 60 and 120 ms after the stimulus presentation.

Suppression and Improvement of Motion Perception

Several studies have examined the effects of single-pulse TMS applied to V5 and found that motion perception could be impaired by TMS[4,43-47] (Fig. 22-4). Stimulation of this area can shorten motion after-effects and affect learning in a movement perception task.[27]

Beckers and Hömberg[43] showed that TMS-induced discrimination deficits were more marked for movements away from the fovea than toward it. These findings are consistent with previous monkey data showing that V5 includes more neurons tuned for motion away from the fovea than toward it.[48] In this study, TMS over V5 disrupted direction discrimination only when the stimulus appeared contralateral to the stimulated area. In another study, TMS elevated the discrimination threshold in both visual hemifields,[44] suggesting that unilateral stimulation could have bilateral physiological effects, probably by means of transcallosal connections.[49] The differences partly can be explained by methodological differences, different tasks, and the different stimulation parameters applied by the two studies.

TMS of V1 differentially affects speed and direction judgments, suggesting that sensory information processing which constrains speed discrimination is localized differently from the sensory information constraining direction discrimination.[50] In a motion-perception task, speed discrimination was significantly impaired, but direction discrimination remained intact when TMS was applied to this area.

TMS can improve performance in non-motion search tasks when motion serves as a distractor.[4] In a visual search task in which motion was present but irrelevant and attention to color or form was required, TMS applied over the left V5 improved performance, probably by disruption of motion perception.

rTMS also was used to modulate V5 function. In a visual motion priming task, rTMS using a 10-Hz frequency for a duration of 500 ms over V5, applied in the intertrial interval of a motion discrimination task, disrupted visual priming of motion, but motion perception was not affected when V1 or the posterior parietal cortex were stimulated.[51]

Stewart and coworkers[27] have shown that rTMS can affect the degree of learning in a visual motion task in a frequency-dependent manner. Subjects had to identify a moving stimulus, which was defined by the conjunction of shape and movement direction. Subjects who were stimulated with 3-Hz frequency over the left V5 learned significantly less during a 4-day session than the control group or the group receiving 10-Hz stimulation.[27] These frequency-dependent learning effects were also found during and after stimulation of the motor cortex, suggesting that rTMS effects on learning are generalized across modalities.

Transcranial Direct Current Stimulation

Although this method has been used from the 1960s, mainly in animals, human studies were rare up to 1998, when Priori and his coworkers[52] first used TMS to evaluate tDCS-induced changes of human cortex excitability. Later, it was found that by using a different electrode montage (i.e., motor cortex and contralateral orbit), an excitability enhancement could be elicited by anodal and a respective diminution by cathodal stimulation of the primary motor cortex, and that excitability changes outlasted the stimulation itself, if the stimulation duration was sufficiently long.[7]

Animal Studies

More detailed knowledge about the origin and mechanisms of cortical DC stimulation has been gained from early animal experiments. In rats and cats, it was shown that applying an anodal DC stimulus to the surface of the motor and the visual cortex increased cortical excitability, probably by depolarizing neuronal membranes at subthreshold level and increasing the spontaneous firing rate of the cells, whereas a cathodal current resulted in the reverse effect, probably because of hyperpolarizing neurons.[17,53–55] In cats, the stimulation effect on the visual cortex was less pronounced than on the motor cortex, probably because of the different structures of the cortices and different spatial orientations of the neurons.[53] The elicited effects were not restricted to the duration of stimulation itself but could outlast it for several hours if stimulation intensity and duration were high enough.[54]

Learning processes are accompanied by changes of neuronal activity and excitability, and these basic neurophysiological studies have raised the possibility that cortical excitability changes induced by weak direct current stimulation could modify higher-order cognitive processes. Early animal experiments demonstrated that DC stimulation affects behavior in an effective and reversible manner. Cathodal polarization of the striate cortex in the rabbit led to a large decrement in the performance of a conditioned response when light flashes were used as the conditioning stimuli.[56] The same effect was found in pattern or brightness discrimination tasks in rats.[57] Here, cathodal stimulation most probably compromised perception and impaired performance by decreasing the cortical firing rate. However, Ward and Weiskrantz[55] found that also anodal polarization applied to the surface of the striate cortex of monkeys resulted in impaired visual discrimination. This at a first glance surprising result can be explained by the specific structure of the task. In visual discrimination tasks, the correct decision must be met by comparing very similar stimuli, an overall cortical activity enhancement or increased cortical "noise" caused by anodal DC stimulation would make decision-making more difficult. The beneficial or worsening effect of DC stimulation may depend on task characteristics. As shown in rhesus monkeys, anodal stimulation enhances performance in a delayed reaction time task that includes no "noisy" aspects.[58] Apparently, an externally induced cortical excitability enhancement supports the activation of task-relevant engrams.

Human Applications

Most early human studies concentrated on possible therapeutic applications of tDCS and described some beneficial effects in psychiatric patients after stimulation.[59-61] However, these effects could not be replicated by all later studies, probably due to different patient subgroups and technical differences.[62,63] The main problem was to determine the psychophysiological and electrophysiological effects of tDCS objectively and to define optimal stimulation conditions. It was shown directly that transcranially applied DC can modulate excitability and activity of the motor cortex in healthy subjects, both during and after stimulation, as measured by TMS and fMRI.[7-12,52] The after-effects can last from minutes to hours beyond the end of the stimulation, depending on intensity of current and the duration of the stimulation.[7,8,64]

Transcranial Direct Current Stimulation Studies of Visual Perception

Concerning the visual modality, the perceptual effects of tDCS were found to be in accordance with its physiological effect. tDCS affects contrast sensitivity,[13] modulates the amplitude of the visual evoked potentials (VEP),[16] and modifies the perception of phosphenes[14,15] and motion-detection thresholds[65] in a polarity-specific way.

One study showed that cathodal stimulation decreases while anodal stimulation increases the amplitude of the N70 component of the VEP evoked by low-contrast luminance gratings[16] (Fig. 22-5). It was also shown that cathodal stimulation decreased while anodal stimulation slightly increased the normalized beta and gamma frequency powers, which are closely connected in time to the N70 peak. Because gamma activity is also related to higher-level information processing (e.g., different stages of perception and learning processes), tDCS may be a suitable method to induce alterations in higher-order cognitive processes.[66]

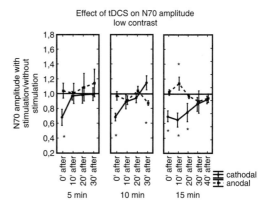

Figure 22-5 Effects of 5, 10, and 15 minutes of transcranial direct current stimulation on N70 amplitude of visual evoked potentials. The means of the N70 peak amplitude are shown for 50 low-contrast stimuli recorded at Oz position. A significant amplitude decrease was observed after cathodal stimulation. The increase of stimulation duration increased the duration of after-effects. Anodal stimulation yielded significant visual evoked potential amplitude changes only when stimulation was applied for 15 minutes. (From Antal A, Kincses ZT, Nitsche MA, et al. Excitability changes induced in the human primary visual cortex by transcranial direct current stimulation: direct electrophysiological evidence. Invest Ophthalmol Vis Sci 2004;45:702-707.)

One study evaluated the modulating effect of tDCS on the human visual cortex directly by measuring stationary phosphene thresholds (PTs).[14] Significantly reduced PTs were detected after the end of anodal stimulation, most likely due to increased cortical excitability, whereas cathodal stimulation resulted in an opposite effect, probably representing a decrement of the cortical excitability. The changes of PTs showed the same pattern independently of the magnetic waveform stimulation (monophasic or biphasic).

Also moving PTs can be modified by tDCS. Moderate shifts of excitability of V1 function influence moving phosphene perception in a polarity-specific way; reduced PTs were detected at the end of anodal tDCS over V1, and cathodal stimulation resulted in an opposite effect[15] (Fig. 22-6). This further supports the V1 hypothesis of moving phosphene perception[28] and suggests that not only a functional interruption of V1 but also

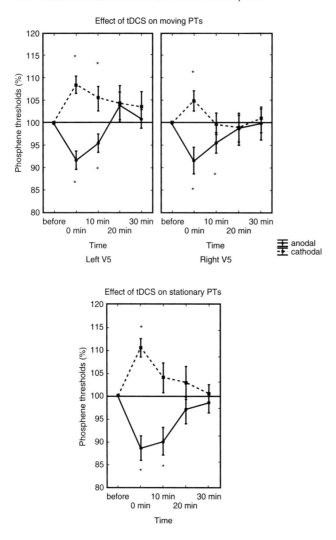

Figure 22–6 Normalized stationary and moving phosphene thresholds (PTs) before and after transcranial direct current stimulation. (From Antal A, Kincses ZT, Nitsche MA, et al. Modulation of moving phosphene thresholds by transcranial direct current stimulation of V1 in human. Neuropsychologia 2003;41:1802–1807.)

a shift in excitability of V1 influences moving phosphene perception.

tDCS can also impair contrast perception.[13] Significant contrast sensitivity loss (i.e., contrast perception threshold increase) was found after cathodal stimulation of V1, and anodal stimulation had no effect. The stimuli used in this study were black and white sinusoidal gratings with a spatial frequency of 4 cycles/degree, which is just at the top of the human contrast sensitivity curve. Thus the lack of an effect of anodal tDCS in this study is possibly caused by a ceiling effect. The stimulus with the optimal spatial frequency probably caused maximum excitation of the visual cortex, which cannot be increased more by anodal stimulation.

Functional Aspects: Modification of Visuomotor Behavior by Transcranial Direct Current Stimulation

It was demonstrated that tDCS of V5 is able to modify visuomotor performance in a visually guided hand-tracking task, suggesting an active role of this cortical area in visuomotor tasks.[66] An excitability decrement of V5 by cathodal tDCS improved the performance of subjects in an overlearned visuomotor paradigm, whereas anodal stimulation had no significant effect. The highly specific effect of reducing excitability in V5 that results in enhanced performance of this visually guided tracking task is most probably explained by the complexity of perceptual information processing needed for this task: high-resolution temporal-spatial analysis and comparison of motion velocity and direction of the target and the feedback cursor. During the execution of the task, the optimal and some suboptimal motion patterns are activated to a certain degree simultaneously. In this case, cathodal stimulation may focus the correct perception of these parameters by decreasing global excitation level and diminishing the amount of activation of concurrent patterns below threshold. This hypothesis was proved by additional experiments. Testing simple and complex motion perception in dot kinetograms, a diminution in excitability induced by cathodal stimulation improved the subject's perception of the direction of the coherent motion only if this was presented among random dots (i.e., complex motion perception) but worsened it if only one motion direction was presented (i.e., simple movement perception).[65]

Another study demonstrated that anodal stimulation of V5 and M1 during the early learning phase improves performance in a visuomotor coordination task.[66] According to these and the previously mentioned results, it seems that the effect of tDCS is phase specific. In the learning phase, probably due to the N-methyl-D-aspartate (NMDA)–receptor-pecific effects,[67] the performance improved by an excitability-enhancing anodal tDCS (cathodal stimulation has no effect in this case). Nevertheless, when the task is learned, plastic changes no longer occur, there are no changes in receptor strength, and cathodal stimulation over V5 is able to enhance visuomotor performance by decreasing gobal cortical excitability. Depending on task characteristics and accompanying activation states of a given cortical area, tDCS can result in different effects. These results raise the possibility of the usefulness of tDCS in rehabilitation strategies for neurological patients with visuomotor disorders.

■ Conclusion and Future Directions

One major aim of vision research is to find methods that are suitable to induce functional changes in the human brain in a controlled, safe way to explore visual perceptual and visuocognitive functions. Single-pulse TMS possesses a good temporal resolution and produces short-lasting effects. Mapping studies within the motor and visual cortex showed a spatial resolution down to 0.5 to 1.0 cm at the scalp surface (reviewed by Jahanshahi and Rothwell[5] and Cowey and Walsh[6]). The effect of tDCS is probably intracortical[8,12] and may be focally restricted,[68] but its temporal resolution is poor. Single-pulse TMS is ideally suited to deliver information about the global involvement of a given cortical area in the performance of a task and its time course. However, specific gradual changes in performance caused by tDCS-induced cortical excitability modulations may deliver additional information about the specific functional role of a given area and help to gain insight into details of cortical information processing. tDCS allows a manipulation of cortical network activity in the human and, in parallel, a psychophysical evaluation of correlated perceptual changes. It influences the brain's activity electrically and changes the organized cortical activity transiently and reversibly in a noninvasive, nonpainful way, similarly to TMS. However, in the mode of action, tDCS and TMS are at least partly complementary; whereas although single-pulse TMS induces externally triggered changes in the neuronal spiking pattern and *interferes* with cortical activity in a spatially and temporally restricted fashion, tDCS most probably modulates the spontaneous firing of neurons by changing resting membrane potential and does not disrupt but rather *modifies* ongoing neuronal activity. rTMS lies between the aforementioned methods, depending on the strength, frequency, and duration of stimulation.

Several studies have aimed to clarify the basic mechanisms of tDCS using the motor cortex as a model.[64,67] Pharmacological studies revealed that the induced after-effects may be NMDA-receptor dependent. There are no data available about the underlying cellular-molecular mechanisms of tDCS-induced effects over the visual areas, but we think that the after-effects are probably also NMDA-receptor dependent. However, given that the visual areas have rich cholinergic and dopaminergic innervations, it is possible that in the visual cortex other neurotransmitter or neuromodulator systems take part in the production of short- and long-term effects of tDCS.

According to our own results and those of others previously discussed, tDCS seems to be a promising method to induce acute and prolonged cortical excitability and activity modulations. It could evolve as a promising new tool in the field of neuroplasticity research. Significant efforts have been made to combine tDCS with other techniques, such as fMRI and electroencephalography.[10] The combination of these techniques seems to be a very promising approach to learn more about localization, time course and functional specifications of a given brain area involved in visual and visuocognitive tasks. Additional studies should be done to make this tool relevant for basic research purposes and for clinical application. To our knowledge, there are no studies in which the excitability of the occipital cortex was modified to gain clinical reimbursement. Nevertheless, in some cases of pathologically changed excitability of the occipital cortex (e.g., in photosensitive epilepsy, stroke, migraine with aura), the application of weak current could be possibly beneficial.

Acknowledgments

We thank Chris Crozier for the English corrections.

REFERENCES

1. Penfield W, Rasmussen W. The Cerebral Cortex of Man. New York, Macmillan, 1998:1–248.
2. Brindley GS, Lewin WS. The visual sensations produced by electrical stimulation of the medial occipital cortex. J Physiol 1968;194:54P–55P.
3. Merton PA, Morton HB. Stimulation of the cerebral cortex in the intact human subject. Nature 1980;285:227.
4. Walsh V, Cowey A. Magnetic stimulation studies of visual cognition. Trends Cogn Neurosci 1998;2:103–110.
5. Jahanshahi M, Rothwell J. Transcranial magnetic stimulation studies of cognition: an emerging field. Exp Brain Res 2000;131:1–9.
6. Cowey A, Walsh V. Tickling the brain: studying visual sensation, perception and cognition by transcranial magnetic stimulation. Prog Brain Res 2001;134, 411–25.
7. Nitsche MA, Paulus W. Excitability changes induced in the human motor cortex by weak transcranial direct current stimulation. J Physiol 2000;527:633–639.
8. Nitsche MA, Paulus W. Sustained excitability elevations induced by transcranial DC motor cortex stimulation in humans. Neurology 2001;57:1899–1901.
9. Rosenkranz K, Nitsche MA, Tergau F, et al. Diminution of training-induced transient motor cortex plasticity by weak transcranial current stimulation in the human. Neurosci Lett 2000;296:61–63.
10. Baudewig J, Nitsche MA, Paulus W, et al. Regional Modulation of BOLD MRI responses to human sensorimotor activation by transcranial direct current stimulation. Magn Res Med 2001;45:196–201.
11. Nitsche MA, Schauenburg A, Lang N, et al. Facilitation of implicit motor learning by weak transcranial direct current stimulation of the primary motor cortex in the human. J Cogn Neurosci 2003;15:619–626.
12. Nitsche MA, Nitsche MS, Klein CC, et al. Level of action of cathodal DC polarisation induced inhibition of the human motor cortex. Clin Neurophysiol 2003;114:600–604.
13. Antal A, Nitsche MA, Paulus W. External modulation of visual perception in humans. Neuroreport 2001;12:3553–3555.
14. Antal A, Kincses ZT, Nitsche MA, et al. Manipulation of phosphene thresholds by transcranial direct current stimulation in man. Exp Brain Res 2003;150:375–378.
15. Antal A, Kincses ZT, Nitsche MA, et al. Modulation of moving phosphene thresholds by transcranial direct current stimulation of V1 in human. Neuropsychologia 2003;41:1802–1807.
16. Antal A, Kincses ZT, Nitsche MA, et al. Excitability changes induced in the human primary visual cortex by transcranial direct current stimulation: direct electrophysiological evidence. Invest Ophthalmol Vis Sci 2004;45:702–707.

17. Purpura DP, McMurtry JG. Intracellular activities and evoked potential changes during polarization of the motor cortex. J Neurophysiol 1965;28:166–185.
18. Amassian VE, Cracco RQ, Maccabee PJ, et al. Suppression of visual perception by magnetic coil stimulation of human occipital cortex. Electroencephalogr Clin Neurophysiol 1989;74:458–462.
19. Beckers G, Hömberg V. Impairment of visual perception and visual short term memory scanning by transcranial magnetic stimulation of the occipital cortex. Exp Brain Res 1991;87:421–432.
20. Masur H, Papke K, Oberwittler C. Suppression of visual perception by transcranial magnetic stimulation—experimental finding in healthy subjects and patients with optic neuritis. Electroencephalogr Clin Neurophysiol 1993;86:259–267.
21. Miller MB, Fendrich R, Eliassen JC, et al. Transcranial magnetic stimulation: delays in visual suppression due to luminance changes. Neuroreport 1996;7:1740–1744.
22. Epstein CM, Verson R, Zangaladze A. Magnetic coil suppression of visual perception at an extracalcarine site. J Clin Neurophysiol 1996;13:247–252.
23. Kastner S, Demmer I, Ziemann U. Transient visual field defects induced by transcranial magnetic stimulation over human occipital pole. Exp Brain Res 1998;118:19–26.
24. Meyer BU, Diehl RR, Steinmetz H, et al. Magnetic stimuli applied over motor cortex and visual cortex: influence of coil position and field polarity on motor responses, phosphenes and eye movements. Electroencephalogr Clin Neurophysiol 1991;43:121–134.
25. Marg E, Rudiak D. Phosphenes induced by magnetic stimulation over the occipital brain: description and probable site of stimulation. Optom Vis Sci 1994;71:301–311.
26. Kammer T, Beck S, Puls K, et al. Motor and phosphene thresholds: consequences of cortical anisotropy. Suppl Clin Neurophysiol 2003;56:198–203.
27. Stewart L, Battelli L, Walsh V, et al. Motion perception and perceptual learning studied by magnetic stimulation. Electroencephalogr Clin Neurophysiol Suppl 1999;51:334–350.
28. Pascual-Leone A, Walsh V. Fast backprojections from the motion to the primary visual area necessary for visual awareness. Science 2001;292:510–512.
29. Kammer T, Nusseck HG. A recognition deficits following occipital lobe TMS explained by raised detection threshold? Neuropsychologia 1998;36:1161–1166.
30. Kammer T. Phosphenes and transient scotomas induced by magnetic stimulation of the occipital lobe: their topographic relationship. Neuropsychologia 1999;37:191–198.
31. Ray PG, Meador KJ, Epstein CM, et al. Magnetic stimulation of visual cortex: factors influencing the perception of phosphenes. J Clin Neurophysiol 1998;15:351–357.
32. Stewart LM, Walsh V, Rothwell JC. Motor and phosphene thresholds: a transcranial magnetic stimulation correlation study. Neurophsychologia 2001;39:415–419.
33. Kammer T, Beck S, Erb M, et al. The influence of current direction on phosphene thresholds evoked by transcranial magnetic stimulation. Clin Neuropyhsiol 2001;112:2015–2021.
34. Antal A, Kincses ZT, Nitsche MA, et al. No correlation between moving phosphene and motor thresholds: a transcranial magnetic stimulation study. Neuroreport 2004;15:297–302.
35. Kammer T, Beck S. Phosphene thresholds evoked by transcranial magnetic stimulation are insensitive to short-lasting variations in ambient light. Exp Brain Res 2002;145:407–410.
36. Cowey A, Walsh V. Magnetically induced phosphenes in sighted, blind and blind-sighted observers. Neuroreport 2000;11:3269–3273.
37. Meister IG, Weidemann J, Dambeck N, et al. Neural correlates of phosphene perception. Suppl Clin Neurophysiol 2003;56:305–311.
38. Boroojerdi B, Prager A, Muellbacher W, et al. Reduction of human visual cortex excitability using 1 Hz transcranial magnetic stimulation. Neurology 2000;54:1529–1531.
39. Aurora SK, Ahmad BK, Welch KM, et al. Transcranial magnetic stimulation confirms hyperexcitability of occipital cortex in migraine. Neurology 1998;50:1111–1114.
40. Gothe J, Brandt SA, Irlbacher K, et al. Changes in visual cortex excitability in blind subjects as demonstrated by transcranial magnetic stimulation. Brain 2002;125:479–490.
41. Maccabee PJ, Amassian VE, Cracco RQ, et al. Magnetic coil stimulation of human visual cortex: studies of perception. Electroencephalogr Clin Neurophysiol Suppl 1991;43:111–120.
42. Paulus W, Korinth S, Wischer S, et al. Differentiation of parvo- and magnocellular pathways by TMS at the occipital cortex. Electroencephalogr Clin Neurophysiol Suppl 1999;51:351–360.
43. Beckers G, Hömberg V. Cerebral visual motion blindness: transitory akinetopsia induced by transcranial magnetic stimulation of human visual area V5. Proc R Soc Lond Series B 1992;249:173–178.
44. Hotson J, Brain D, Herzberg W, et al. Transcranial magnetic stimulation of extrastriate cortex degrades human motion direction discrimination. Vis Res 1994;34:2115–2123.

45. Beckers G, Zeki S. The consequences of inactivating areas V1 and V5 on visual motion perception. Brain 1995;118:49–60.
46. Amassian VE, Cracco RQ, Maccabee PJ, et al. Transcranial magnetic stimulation in study of the visual pathway. J Clin Neurophysiol 1998;15:288–304.
47. Hotson JR, Anand S. The selectivity and timing of motion processing in human temporo-parieto-occipital cortex: a transcranial magnetic stimulation study. Neuropsychologia 1999;37:169–179.
48. Albright TD. Centrifugal directional bias in the middle temporal visual area (MT) of the macaque. Vis Neurosci 1989;2:177–188.
49. Paus T, Jech R, Thompson CJ, et al. Transcranial magnetic stimulation during positron emission tomography: a new method for studying connectivity of the human cerebral cortex. J Neurosci 1997;17:3178–3184.
50. Matthews N, Luber B, Qian N, et al. Transcranial magnetic stimulation differentially affects speed and direction judgments. Exp Brain Res 2001;140:397–406.
51. Campana G, Cowey A, Walsh V. Priming of motion direction and area V5/MT: a test of perceptual memory. Cereb Cortex 2002;12:663–669.
52. Priori A, Berardelli A, Rona S, et al. Polarization of the human motor cortex through the scalp. Neuroreport 1998;9:2257–2260.
53. Creutzfeldt OD, Fromm GH, Kapp H. Influence of transcortical dc-currents on cortical neuronal activity. Exp Neurol 1962;5:436–452.
54. Bindman LJ, Lippold OC, Redfearn JWT. The action of brief polarizing currents on the cerebral cortex of the rat (1) during current flow and (2) in the production of long-lasting after-effects. J Physiol 1964;172:369–382.
55. Ward R, Weiskrantz L. Impaired discrimination following polarization of the striate cortex. Exp Brain Res 1969;9:346–356.
56. Morell F, Naitoh P. Effect of cortical polarization on a conditioned avoidance response. Exp Neurol 1962;6:507–523.
57. Kupferman I. Effects of cortical polarization on visual discrimination. Exp Neurol 1965;12:179–189.
58. Rosen SC, Stamm JS. Cortical polarization: facilitation of delayed response performance by monkeys. Exp Neurol 1972;35:282–289.
59. Constain R, Redfearn JW, Lippold OC. A controlled trial of the therapeutic effect of polarisation of the brain depressive illness. Br J Psychiatry 1964;110:786–799.
60. Lippold OC, Redfearn JW. Mental changes resulting from the passage of small direct currents trough the human brain. Br J Psychiatry 1964;110:768–772.
61. Redfearn JW, Lippold OC, Constain R. A preliminary account of the clinical effects of polarizing the brain in certain psychiatric disorders. Br J Psychiatry 1964;110:773–785.
62. Lifshitz K, Harper P. A trial of transcranial polarization in chronic schizophrenics. Br J Psychiatry 1968;114:635–637.
63. Lolas F. Brain polarization: behavioral and therapeutic effects. Biol Psychiatry 1977;12:37–47.
64. Nitsche MA, Fricke K, Henschke U, et al. Pharmacological modulation of cortical excitability shifts induced by transcranial direct current stimulation in humans. J Physiol 2003;553:293–301.
65. Antal A, Nitsche MA, Kruse W, et al. Direct current stimulation over V5 enhances visuo-motor coordination by improving motion perception in humans. J Cogn Neurosci 2004;16:521–527.
66. Antal A, Nitsche MA, Kruse W, et al. Facilitation of visuo-motor learning by transcranial direct current stimulation of the motor and extrastriate visual areas in humans. Eur J Neurosci 2004;19:2888–2892.
67. Liebetanz D, Nitsche MA, Tergau F, et al. Pharmacological approach to the mechanisms of transcranial DC-stimulation-induced after-effects of human motor cortex excitability. Brain 2002;125:1–10.
68. Rush S, Driscoll DA. Current distribution in the brain from surface electrodes. Anesth Analg 1968;47:717–23.

23 Somatosensory System

Massimiliano Oliveri

Most applications of transcranial magnetic stimulation (TMS) for the study of the somatosensory system arise from its ability to transiently disrupt and modulate the neural activity in focal brain regions.[1–3] This kind of activity may allow us to look deeper into human somatosensory cortex physiology than it was possible with evoked potential measurements alone. In particular, the application of single-pulse and paired-TMS to the parietal cortex during the execution of tactile detection tasks provides interesting findings on the physiologic bases of conscious somatosensory perception.

In this chapter, I review a series of studies that analyzed the performance of normal subjects and patients with contralesional somatosensory deficits on tactile discrimination tasks. These studies exemplify three of the major potential contributions of TMS in this field: the transient disruption of focal cortical activity to establish the causal role and the timing of the contribution of a given cortical region in a behavior; the application of TMS to the study of functional brain connectivity; and the application of TMS to human patients to examine the compensatory cortical plasticity that occurs in response to a lesion.

■ Transcranial Magnetic Stimulation Modulation of Tactile Perception in Normal Subjects

The first group of studies used TMS as a "virtual lesion" method to transiently disrupt the cortical activity of focal brain areas in normal subjects mimicking the effects of specific neurologic lesion temporarily.

The work of Cohen and colleagues[4] provides the first example of this application of single-pulse TMS for the study of somatosensory system. In a series of experiments, TMS was delivered over the sensorimotor cortex or other control positions at various periods after the application of electrical stimuli to the contralateral or ipsilateral index fingers. Detection of somatosensory stimuli was blocked when TMS was applied simultaneously with and 20 ms after electrical stimuli, a period consistent with the arrival of the afferent volley at the primary sensory cortex; the recovery in detection of somatosensory stimuli was complete when TMS was delivered 200 ms after electrical stimulation. These effects were time and space specific, produced by TMS of restricted scalp positions contralateral to the finger stimulated.

Subsequent studies confirmed that the cortical processing required for the perception of cutaneous sensation from one hand can be transiently disrupted by TMS of the contralateral sensorimotor cortex. Seyal and coworkers[5] showed that TMS of the parietal cortex has a more profound effect on tactile localization than on simple detection of cutaneous stimuli delivered to the fingers of the contralateral hand. The region of the scalp over which TMS results in impaired cutaneous localization coincides with the region over which it suppresses simple perception.

An interesting issue addressed by these studies is the correlation of the TMS-induced tactile disruption with the modifications of SEPs generated by median nerve stimulation. Cohen and associates[4] found that TMS delivered immediately after a median nerve stimulus enhanced the early cortical components of the somatosensory evoked potential (SEP), but attenuated long latency components probably related to later cortical processing. These findings suggest that the site where detection block took place was probably cortical and that for conscious detection to occur, late cortical events in response to somatosensory stimulation are necessary.

Seyal and colleagues[6] showed that the period of maximum suppression of perception coincides with the period during which cortically generated SEPs are enhanced after TMS. The duration of suppression of perception, however, outlasts the duration of SEP enhancement. These findings can be interpreted by positing that the enhancement of the SEP is the result of synchronization of pyramidal cells in the sensorimotor cortex by TMS.

Another group of studies have looked at the role of interhemispheric interactions in the distribution of attention to tactile space. Seyal and coworkers[7] tested the effects of right parietal TMS on ipsilateral tactile perception, in a group of normal subjects carrying out a tactile detection task with the right thumb. Single-pulse TMS delivered 50 ms before the delivery of the electrical stimulus resulted in increased sensitivity to cutaneous stimulation compared with baseline or frontal TMS trials. These findings were interpreted as the result of a TMS-induced transient disruption of the ipsilateral parietal cortex, resulting in disinhibition of the contralateral (i.e., left) parietal cortex during the sensory discrimination task, probably mediated by transcallosal effects. This study provides an elegant example of how the effects of TMS are not limited to the target site but involve other anatomically correlated regions, even of the contralateral hemisphere.[8]

An example of how TMS can provide chronometric information about the timing of activation of parietal and frontal cortices in tactile attention tasks is provided by a study of my group.[9] We applied single-pulse TMS over left or right parietal and frontal scalp sites of normal subjects receiving unimanual or bimanual tactile stimulations. Anatomic localization of the stimulated scalp sites by means of MRI scans showed that the parietal site was located over the intraparietal sulcus and the frontal site over the inferior frontal sulcus.

The main results showed that TMS of the right parietal cortex interfered with the detection not only of contralateral, but also of ipsilateral stimuli. These effects were mainly evident during bimanual stimulation, reproducing a contralateral or ipsilateral tactile extinction. Regarding the timing of the effects, we found that the contribution of the right parietal cortex was critical between 20 and 40 ms after tactile stimulation, a finding reproducing the results obtained by previous investigators[4-6] with unimanual stimulation (Fig. 23–1).

This pattern of time-effects of TMS can also explain the different findings of our study and that of Seyal and colleagues[7] about ipsilateral sensory processing by the right parietal cortex. We showed that right parietal TMS disrupts ipsilateral tactile perception when delivered 20 to 40 ms after the electrical stimulus. However, Seyal coworkers documented an enhancement of ipsilateral perception when TMS was delivered 50 ms before the electrical stimulus. It is conceivable that the processing of ipsilateral sensory stimuli can be facilitated by transcallosal effects only provided that TMS is delivered before the sensory stimuli or that its effects last enough to produce contralateral hemisphere disinhibition. In contrast, when TMS is delivered after sensory stimulation, its effects could appear too late to allow expression of a disinhibition effect on the contralateral hemisphere; in this case, a mechanism of collision with the afferent volleys from ipsilateral stimulation could be prevalent.

All the cited studies used single-pulse TMS to disrupt function of the somatosensory cortex during somatosensory processing. Several different mechanisms probably contribute to the observed TMS-interference with tactile perception. It may be that volleys descending along the pyramidal tract collide with the ascending afferent volleys. The fact

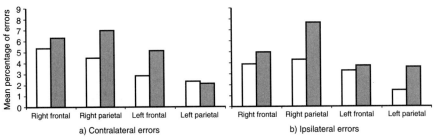

Figure 23–1 Mean percentage of errors of detection of contralateral (**A**) and ipsilateral (**B**) tactile stimuli as a function of the stimulated scalp site and of the tactile discrimination task. Gray histograms show the performance during bimanual stimulation; white histograms show the performance during unimanual stimulation. (Adapted from Oliveri M, Rossini PM, Pasqualetti P, et al. Interhemispheric asymmetries in the perception of unimanual and bimanual cutaneous stimuli: a study using transcranial magnetic stimulation. Brain 1999;122:1721–1729.)

that tactile detection is maximally impaired when TMS is delivered after the somatosensory stimulus suggests that motor activity before cortical arrival of the afferent volley is not a critical factor for the interference. However, it is well known that detection of a somatosensory stimulus may be attenuated when movement, even if triggered externally by TMS, starts after the cortical arrival of the afferent volley.[10] Another explanation, which has a general value for the TMS-interference effects with any sensory/cognitive activity, is that TMS induces a random noise activity in the context of the general activity of the somatosensory cortex, rendering it unresponsive to afferents from the periphery.[11]

Paired-TMS protocols can offer the way to modulate the excitability of a specific cortical region and even to facilitate task performance. Since its introduction, this technique enabled neurophysiologists to study specifically interneurons in human motor cortex. In the motor cortex of normal subjects, the test motor potential (MEP) evoked in the intrinsic hand muscles is inhibited by a preceding conditioning subthreshold stimulus (CS) at short interstimulus intervals (ISIs) of between 1 and 5 ms (i.e., intracortical inhibition [ICI]), whereas with longer ISIs of 8 to 15 ms, the test responses are facilitated (i.e., intracortical facilitation [ICF]).[12,13] The inhibition is likely to reflect the activity of GABAergic interneurons, whereas facilitation would depend on the activation of intracortical fibers by the CS, inducing local release of glutamate.

In one study,[14] my colleagues and I applied a paired-TMS protocol on the right parietal cortex of a group of normal subjects to investigate the presence of a pattern of excitatory and inhibitory interactions during tactile perceptual tasks. Fifteen healthy volunteers received threshold electrical stimuli to the left thumb. Electrical stimuli were followed after delays of 10 to 40 ms by paired-pulse TMS, with a CS (70% of motor threshold) followed at various ISIs (e.g., 1, 3, 5, 10, 15 ms) by a suprathreshold (130% of motor threshold) stimulus (TS). Baseline trials (without TMS) were intermingled with TMS conditions.

The main results showed that paired-pulse TMS of the parietal cortex can selectively modulate the perception of contralateral tactile stimuli, compared with single-pulse TMS, depending on the ISI between the CS and the TS. Paired-TMS disrupted tactile detection (as compared with single-pulse TMS alone) when delivered 1 ms apart; in contrast, at an ISI of 5 ms, paired-TMS induced a transient recovery of the baseline level in tactile stimulus detection rate, which overwhelmed the disrupting effects of the single-pulse TMS alone (Fig. 23–2).

These findings provide the first example of selective modulation of cortical excitability by paired-TMS in the somatosensory system. At this level, by analogy with the motor cortex, paired-TMS is likely to activate local inhibitory and excitatory interneurons depending on the ISI between the CS and the TS. Future

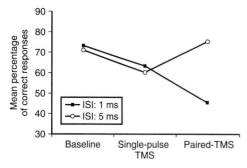

Figure 23–2 Mean percentage of correct responses in detection of contralateral tactile stimuli during baseline, single-pulse transcranial magnetic stimulation (TMS), and paired-TMS of the right parietal cortex with 1 and 5 ms inter-stimulus intervals. (Adapted from Oliveri M, Caltagirone C, Filippi MM, et al. Paired-transcranial magnetic stimulation protocols reveal a pattern of inhibition and facilitation in the human parietal cortex. J Physiol 2000;529:461–468.)

extinction can perceive a single contralesional stimulus but are unaware of the same stimulus when another is presented simultaneously on the ipsilesional side. Extinction, like neglect, is more frequent after right hemisphere damage,[16] and it has been interpreted as the result of a pathologic attentional bias toward the right space due to the disinhibition of the healthy (i.e., left) hemisphere after the release of the reciprocal inhibition from the affected one.[17,18]

The role of interhemispheric interactions in spatial attention can be explored by applying TMS over the unaffected hemisphere of patients with extinction, thereby analyzing the compensatory cortical plasticity that occurs in response to a lesion. An example of this approach is a study in which we applied single-pulse TMS to frontal, parietal, and prefrontal scalp sites of the unaffected hemisphere of right-brain–damaged (RBD) and left-brain–damaged (LBD) patients after a fixed interval of 40 ms from unimanual or bimanual tactile stimulation.[19]

TMS of the left frontal cortex at 110% of motor threshold intensity significantly reduced contralesional extinction rates during bimanual stimulation in RBD group, but it was ineffective in the LBD group and during unimanual stimulation (Fig. 23–3). The extinction improvement was not limited to patients with neglect or extinction but was also evident in those showing only contralateral somatosensory deficits.

This work represents the first example of the application of TMS in human patients to improve a sensory/cognitive function. The

studies could investigate the presence of specific alterations of these interactions in patients with somatosensory disorders. This aspect is analyzed in the following section.

Transcranial Magnetic Stimulation Modulation of Tactile Perception in Patients with Somatosensory Deficits

Among the multiple deficits of perception and exploratory behavior that constitute the neglect syndrome, extinction of contralesional sensory stimuli is a cardinal sign indicating an attentional disorder.[15] Patients with tactile

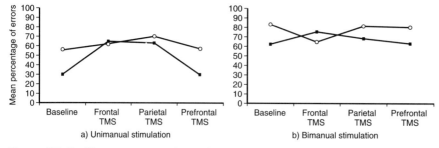

Figure 23–3 Mean percentage of contralesional errors according to tactile stimulation task—unimanual (**A**) and bimanual (**B**)—and transcranial magnetic stimulation scalp positions, in RBD (*circles*) and LBD (*squares*) patients. (Adapted from Oliveri M, Rossini PM, Traversa R, et al. Left frontal transcranial magnetic stimulation reduces contralesional extinction in patients with unilateral right brain damage. Brain 1999;122:1731–1739.)

results show that, as in the animal models of neglect,[20] transient disruption of one hemisphere restores the distribution of attention to the contralesional side of space, thereby improving neglect phenomena. These findings support the model of the hemispheric imbalance, with a disinhibition of the healthy hemisphere, as the physiologic basis of the right hyperattention in the neglect syndrome. The results confirm previous observations about hemispheric asymmetries in the processing of somatosensory stimuli. It has been assumed that in LBD patients, contralateral somatosensory deficits mainly reflect damage to the primary sensory processing of the stimulus in the anterior parietal lobe, without left or right asymmetries. In contrast, in RBD patients, regardless of the presence of neglect or extinction, contralateral somatosensory deficits would involve both sensory and attentional factors due to the right hemisphere dominance of conscious sensory perception.[21] This explains why a maneuver such as TMS, which, by disrupting the interhemispheric distribution of attentional information, would affect the more central levels of somatosensory processing, was effective only in RBD patients. It is also worth noting that in these patients only contralateral extinctions, primarily produced by deficits at the attentional level, were influenced by left hemisphere TMS, whereas contralateral unimanual stimuli were not influenced. This finding further confirms that the contralateral deficits ameliorated by TMS-disruption of the healthy hemisphere have an attentional component.

In another study[22] we tested the effects on contralesional extinction of single-pulse versus paired-pulse TMS with 1- and 10-ms ISIs, delivered over left parietal and frontal scalp sites of a group of eight RBD patients with tactile extinction. The aim of the study was to combine the interfering mechanism of action of single-pulse TMS with distinct facilitatory and inhibitory effects on task performance due to the modulatory actions of the preceding conditioning stimulus. According to the model postulating a relative disinhibition of the unaffected hemisphere in extinction patients, left paired-pulse TMS with ISIs that potentiate inhibitory cortical interneurons should transiently improve left tactile extinctions; however, paired-pulse TMS with ISIs that potentiate (or do not inhibit) cortical excitability should worsen or leave unchanged left extinctions. Another aim of the study was to analyze the chronometry of the spatial distribution of tactile attention in frontal and parietal regions. This question was addressed by tracing the effects of TMS of the parietal and frontal areas on left extinctions at various intervals (10, 20, 30, 40 ms) after tactile stimulation. The CS intensity was 70% of motor threshold, and the TS intensity was 130% of motor threshold.

The main results showed that paired-TMS had distinct effects on contralesional extinction depending on the ISI and on the time of application: at the ISI of 1 ms, there was an improvement in extinction rate greater than that induced by single-pulse TMS; at the ISI of 10 ms, instead, there was a worsening of extinction, with a complete reversal of the effects of single-pulse TMS (Fig. 23–4). We also found two distinct periods during which the left frontal and parietal cortices were sensitive to the facilitatory effects of TMS on extinction: the effect on parietal cortex

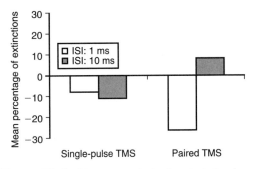

Figure 23–4 Mean percentage of contralesional extinctions during single-pulse and paired-pulse transcranial magnetic stimulation (TMS) of the left parietal cortex, with inter-stimulus intervals of 1 *(white histograms)* and 10 ms *(gray histograms)*. The baseline performance was subtracted from the performance during TMS. Negative values indicate an improvement, and positive values indicate a worsening of tactile extinctions. (Adapted from Oliveri M, Rossini PM, Filippi MM, et al. Time-dependent activation of parieto-frontal networks for directing attention to tactile space: a study with paired transcranial magnetic stimulation pulses in right-brain-damaged patients with extinction. Brain 2000;123:1939–1947.)

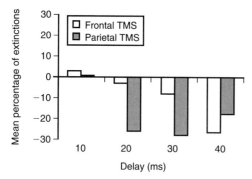

Figure 23–5 Mean percentage of contralesional extinctions after paired transcranial magnetic stimulation (TMS) of frontal *(white hystograms)* and parietal *(gray histograms)* scalp sites as a function of the delay between bimanual tactile stimulation and TMS. The baseline performance was subtracted from the performance during TMS. Negative values indicate an improvement, and positive values indicate a worsening of tactile extinctions. (Adapted from Oliveri M, Rossini PM, Filippi MM, et al. Time-dependent activation of parieto-frontal networks for directing attention to tactile space: a study with paired transcranial magnetic stimulation pulses in right-brain-damaged patients with extinction. Brain 2000;123:1939–1947.)

appeared significantly earlier (in a time-window of 20 to 30 ms after tactile stimulation) than the effect on the frontal cortex (reaching its maximum 40 ms after tactile stimulation) (Fig. 23–5).

These results provide an example of how TMS can be useful to test the chronometry of spatial attention, thanks to its time-resolution that is beyond the power of neuroimaging studies. The findings confirm that the ability to activate intracortical inhibitory and excitatory circuits extend outside of the motor cortex, suggesting the possibility to modulate the excitability disorders underlying sensory perception.

We tested the hypothesis that the pattern of inhibitory and excitatory interactions subserving right tactile perception is altered in the unaffected (left) parietal cortex of RBD extinction patients during bimanual stimulation due to the imbalanced inter-hemispheric competition for attention resources, which is selectively present in this stimulation condition.[23] We analyzed the effects of paired-pulse TMS, delivered at 1 ms ISI over the left parietal cortex, on right (i.e., ipsilesional) tactile detection, comparing the patients' performance with that of a control group of healthy subjects. The critical comparison was between the TMS-effects on right tactile detection during unimanual right (a nonextinction condition) versus bimanual (when contralesional extinction emerged) stimulation tasks.

The paradigm of paired-TMS consisted of a subthreshold conditioning stimulus (CS) followed by a suprathreshold test stimulus (TS) at 1 ms ISI. The CS intensity was 70% of motor threshold, and the TS intensity was 130% of motor threshold. The TS was delivered at a 30-ms delay after the onset of the tactile stimulus.

The main findings showed that paired-TMS with 1 ms ISI to the left parietal cortex affects detection of right finger stimuli during bimanual stimulation differently in RBD patients with tactile extinction than in normal subjects. In the control subjects, this form of paired-TMS disrupts right tactile perception more than single-pulse TMS, as previously demonstrated.[14] In RBD patients, this effect of paired-pulse TMS on right perception is abolished, and it is even reversed in sign. However, when patients are tested with unimanual right stimulation, both single-pulse and paired-pulse TMS to the left parietal cortex have the same disrupting effect as in normal subjects (see Fig. 23–5). The results confirm previous findings,[19,22] showing that single-pulse and paired-pulse TMS of the left (undamaged) hemisphere ameliorate left tactile extinctions of RBD patients, with the effect of paired-pulse TMS with a 1-ms ISI being significantly greater than that of single-pulse TMS.

The basic mechanism of disruption of contralateral tactile perception by single-pulse TMS seems to be intact in RBD patients during both unimanual and bimanual stimulation tasks, as proved by the fact that single-pulse TMS similarly disrupts right tactile perception in both patients and controls in the two stimulation conditions. What is altered in the left parietal cortex of RBD patients is the interaction between conditioning and test magnetic shocks, which, by analogy with the motor cortex, probably reflects the excitability of intrinsic interneuronal GABAergic and glutamatergic circuits.

Inhibition and facilitation at this level are generated by cortical elements, such as cortico-cortical connections, oriented parallel to the surface of the brain. The activation of such fibers by CS during paired-TMS induces release of glutamate, which activates both excitatory and inhibitory interneurons, leading to local increases in cerebral blood flow (CBF), as detected with neuroimaging recordings. Consistent with this hypothesis, a study using positron-emission tomography (PET) showed an increase of CBF with paired-TMS at inhibitory and excitatory ISIs between CS and TS.[24] By extension, a reduced intracortical inhibition may be likely to determine a reduced metabolic activation, as studied with PET or functional magnetic resonance imaging (fMRI) in the corresponding cerebral regions. These considerations suggest that the lack of inhibitory effects of paired-TMS in the left parietal cortex of RBD patients during bimanual stimulation could represent the physiologic counterpart of the reduced activity of the left hemisphere observed with fMRI or event related potentials during extinction, as compared with nonextinction, conditions.[25]

Conclusion

We reviewed a number of studies, using single-pulse and paired-TMS and different experimental paradigms, which show how this technique can be of fundamental importance in elucidating the mechanisms of somatosensory perception, as well as of its deficits (i.e., extinction and neglect). Some of these studies open the way to the possible application of TMS not only as a research method, but also as a tool for inducing long-lasting brain excitability changes, which could be useful for the treatment of various neurologic disorders.

REFERENCES

1. Walsh V, Cowey A. Magnetic stimulation studies of visual cognition. Trends Cogn Neurosci 1998;2:103–110.
2. Hallett M. Transcranial magnetic stimulation and the human brain. Nature 2000;40:147–150.
3. Pascual-Leone A, Walsh V, Rothwell J. Transcranial magnetic stimulation in cognitive neuroscience—virtual lesion, chronometry, and functional connectivity. Curr Opin Neurobiol 2000;10:232–237.
4. Cohen LG, Bandinelli S, Sato S, et al. Attenuation in detection of somatosensory stimuli by transcranial magnetic stimulation. Electroencephalogr Clin Neurophysiol 1991;81:366–376.
5. Seyal M, Siddiqui I, Hundal NS. Suppression of spatial localization of a cutaneous stimulus following transcranial magnetic pulse stimulation of the sensorimotor cortex. Electroencephalogr Clin Neurophysiol 1997;105:24–28.
6. Seyal M, Masuoka LK, Browne JK. Suppression of cutaneous perception by magnetic pulse stimulation of the human brain. Electroencephalogr Clin Neurophysiol 1992;85:397–401.
7. Seyal M, Ro T, Rafal R. Increased sensitivity to ipsilateral cutaneous stimuli following transcranial magnetic stimulation of the parietal lobe. Ann Neurol 1995;38:264–267.
8. Ilmoniemi RJ, Virtanen J, Ruohonen J, et al. Neuronal responses to magnetic stimulation reveal cortical reactivity and connectivity. Neuroreport 1997;8:3537–3540.
9. Oliveri M, Rossini PM, Pasqualetti P, et al. Interhemispheric asymmetries in the perception of unimanual and bimanual cutaneous stimuli. A study using transcranial magnetic stimulation. Brain 1999;122:1721–1729.
10. Starr A, Cohen LG. 'Gating' of somatosensory evoked potentials begins before the onset of voluntary movement in man. Brain Res 1985;348:183–186.
11. Pascual-Leone A, Tarazona F, Keenan J, et al. Transcranial magnetic stimulation and neuroplasticity. Neuropsychologia 1999;37:207–217.
12. Kujirai T, Caramia MD, Rothwell JC, et al. Corticocortical inhibition in human motor cortex. The J Physiol 1993;471:501–519.
13. Ziemann U, Rothwell JC, Ridding MC. Interaction between intracortical inhibition and facilitation in human motor cortex. J Physiol 1996;496:873–881.
14. Oliveri M, Caltagirone C, Filippi MM, et al. Paired-transcranial magnetic stimulation protocols reveal a pattern of inhibition and facilitation in the human parietal cortex. The J Physiol 2000;529:461–468.
15. Bisiach E, Vallar G. Unilateral neglect in humans. In Boller F, Grafman J (eds): Handbook of Neuropsychology. Amsterdam, Elsevier, 2000:459–502.
16. Vallar G, Rusconi ML, Bignamini L, et al. Anatomical correlates of visual and tactile extinction in humans: a clinical CT scan study. J Neurol Neurosurg Psychiatry 1994;57:464–470.
17. Kinsbourne M. Hemi-neglect and hemisphere rivalry. In Weinstein EA, Friedland RP (eds): Hemi-inattention and hemisphere

specialization. Advances in Neurology. New York, Raven Press, 1977:41–49.
18. Kinsbourne M. Mechanisms of neglect: implications for rehabilitation. Neuropsychol Rehabil 1994;4:151–153.
19. Oliveri M, Rossini PM, Traversa R, et al. Left frontal transcranial magnetic stimulation reduces contralesional extinction in patients with unilateral right brain damage. Brain 1999;122:1731–1739.
20. Lomber SG, Payne BR. Removal of two halves restores the whole: reversal of visual hemineglect during bilateral cortical or collicular inactivation in the cat. Vis Neurosci 1996;13:1143–1156.
21. Vallar G, Bottini G, Rusconi ML, et al. Exploring somatosensory hemineglect by vestibular stimulation. Brain 1993;116:71–86.
22. Oliveri M, Rossini PM, Filippi MM, et al. Time-dependent activation of parieto-frontal networks for directing attention to tactile space. A study with paired transcranial magnetic stimulation pulses in right-brain-damaged patients with extinction. Brain 2000;123:1939–1947.
23. Oliveri M, Rossini PM, Filippi MM, et al. Specific forms of neural activity associated with tactile space awareness. Neuroreport 2002;13:997–1001.
24. Strafella AP, Paus T. J Neurophysiol 2001;85:2624–2629.
25. Rees G, Wojciulik E, Clarke K, et al. Unconscious activation of visual cortex in the damaged right hemisphere of a parietal patient with extinction. Brain 2001;123:1624–1633.

24 Eye Movements

René M. Müri, Christian W. Hess, and Charles Pierrot-Deseilligny

Transcranial magnetic stimulation (TMS) had a slow start in oculomotor research, because it has not been capable of eliciting eye movements.[1-3] After it became apparent that TMS, in addition to exciting the motor cortex, might in an appropriate experimental setting also be used to specifically interfere with complex cognitive and executive cortical processing,[3-6] it was appreciated as a valuable tool in studying cortical control of eye movements. Depending on the paradigm used, TMS produces apart from activating effects transient inactivation of a cortical target region, thereby inducing a functional "lesion" and enabling to intervene with cortical processing with a well-defined timing. This opened new promising perspectives in oculomotor research, because many inhibitory and excitatory cortical regions are involved in eye movement control. By a thoughtfully planned manipulation of excitatory or inhibitory regions involved in eye movement control, their interactions and hierarchies can be investigated.

The first part of this chapter concisely reviews anatomy and physiology of the oculomotor system, which is required to understand the TMS results. In the second part of the chapter, experiments that attempted to elicit or to facilitate eye movements by TMS are discussed. The third part deals with studies that used TMS to interfere with saccade programming and execution or used TMS to map the temporal organization of saccade control. Mapping experiments of the frontal and parietal cortical regions are explained.

■ The Physiology and Anatomy of Saccadic Eye Movement Control

A network of brain structures extending from the cerebral cortex to the caudal medulla controls eye movements. Premotor areas such as the rostral interstitial nucleus of the medial longitudinal fasciculus (riMLF) and the paramedian pontine reticular formation (PPRF) are localized in the brainstem and are important for the generation of saccades (reviewed by Leight and Lee[7]). The vestibular nuclei are involved in the vestibulo-ocular reflex (VOR), pursuit movements, and optokinetic responses. However, these and other brainstem structures seem not to be influenced by TMS.

Saccadic eye movements may be classified according their behavioral components (Table 24–1). Saccade paradigms used in TMS studies are represented in Figure 24–1.

Such a classification makes sense because the control of the different aspects involves distinct cortical regions (reviewed by Pierrot-Deseilligny and colleagues[8]).

Three cortical regions of the cerebral cortex are involved in saccade triggering. First, the parietal eye field (PEF) in humans is located in

Table 24–1 BEHAVIORAL CLASSIFICATION OF SACCADES

Classification	Description

I. Reflexive saccades
  a. Visually guided
  b. Auditory guided
	Triggered by a sudden appearance of a visual target (a) or by a sudden noise (b) in the immediate environment

II. Internally triggered, voluntary saccades
  a. Intentional visually guided
  b. Predictive saccades
  c. Memory-guided saccades
  d. Antisaccades
	Saccades with a behavioral goal such as to catch a target on the fovea that has been visible on the retina for a time (a) or a target that is not yet (b) or no longer visible (c), or saccades made in the direction opposite to a suddenly appearing visual target (d)

III. Spontaneous saccades	Internally triggered saccades without obvious goals occurring during other motor activities

the intraparietal sulcus, near the angular and supramarginal gyrus (Fig. 24–2). The PEF triggers visually guided, reflexive saccades and plays a role in the disengagement of fixation. In contrast to the frontal eye fields (FEFs), the PEF is more involved in the *reflexive* exploration of the visual environment. However, the role of the FEFs and the PEF overlap somewhat in the control of visually guided saccades. Second, The FEFs, which in humans are located around the precentral sulcus, include the posterior margin of the middle frontal gyrus and the adjacent precentral sulcus (see Fig. 24–2), just anteriorly to the motor cortex near the hand area. The FEFs trigger mainly intentional saccades, which are encoded in retinotopic coordinates, with the motor vector of the saccade (i.e., the amplitude) equaling the retinal vector given by the position of the target image on the retina. Such saccades are intentional visually guided saccades, predictive saccades, anti-saccades, and memory-guided saccades with visual input. The FEFs are involved in the disengagement of fixation. The FEFs therefore play an important role in the visual exploration of the environment by intentional saccades. Third, the supplementary eye field (SEF) in humans lies in the posteromedial part of the superior frontal gyrus, anteriorly to the SMA and in the upper part of the paracentral sulcus. The SEF is mainly involved in saccade triggering if the amplitude calculation is performed in other than retinotopic coordinates (e.g., memory-guided saccades with a vestibular input or the amplitude of the second saccade in the double-step paradigm). The SEF seems also be important for the initiation of saccade sequences, and in the preparation of motor programs comprising saccades and other body movements. The pre-SEF, located anteriorly to the SEF, could be more involved in the learning of saccade sequences.[9]

Another important cortical region, although not directly able to trigger saccades, is the dorsolateral prefrontal cortex (DLPC). In humans, the DLPC is located anteriorly to the FEF, and is involved in the inhibition of reflexive saccades, such as in the anti-saccade paradigm for inhibiting unwanted reflexive saccades toward the visual target. This region is substantial for the spatial aspect of working memory, independent of the used coordinate system.

Studies Trying to Elicit or Facilitate Eye Movements by Using Transcranial Magnetic Stimulation

The first published studies about TMS effect on eye movements reported negative findings, because single-pulse TMS was not able to elicit eye movements.[1–3] Wessel and Kömpf[1] tried to elicit eye movements in two subjects by stimulating the frontal, precentral, posterior and inferior parietal, occipital, and temporal cortices. Magnetic stimuli were delivered with increasing intensity up to the maximal output of the stimulator. Stimulation

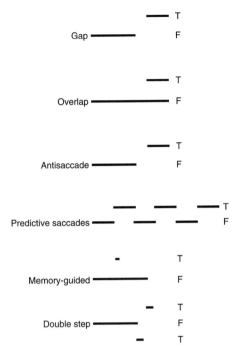

Figure 24–1 Some ocular motor paradigms used in TMS studies. Visually guided saccades with a gap: extinguishing of the central fixation (CF) point before the appearance of the lateral target (T). Such saccades are reflexive triggered by the sudden appearance of the lateral target. For the anti-saccade task, the presentation of the visual target is the same as for visually guided saccades, but the instruction given to the subject is different. Instead of looking immediately to the target (T), the subject has to look in the opposite direction (i.e., away from the visual target). An error is observed when a reflexive saccade is made toward the visual target first instead of the correct saccade, looking away from the target. For memory-guided saccades, the subject has the look at the central fixation point (CF). Then, a visual target (T) is presented for a short time, and after the extinguishing of the CF, the subject performs a saccade toward the remembered position of the target. For predictive saccades, if the change of the visual target (T) is predictive in time and direction, subjects begin very rapidly to start saccades with or just before the target changes. Such saccades are not visually guided; they are internally triggered based on an internally represented prediction of the change of the target. For double-step saccades, two consecutive targets (T1 and T2) are presented shortly within the latency time of the first saccade. Both saccades are performed without visual guidance, but the responses are based on different internal representation systems of the spatial location. The amplitude of the first saccade can be calculated in

was applied during rest and during reflexive saccades or smooth pursuit without modification of these eye movements. It was speculated that TMS was not able to stimulate ocular motoneurons specific enough to elicit eye movements and five possible explanations were given: insufficient stimulation strength, poor focalization of stimuli, inappropriate type of stimulation induced by TMS, inability to facilitate the oculomotor system in the same way as the skeletal muscles, and possible stimulation of inhibitory pathways. Meyer and colleagues[2] applied TMS over the region of the FEFs on both hemispheres and did not elicit any eye movements. With high stimulator output, eye blinks were observed in about 30% of the trials when the frontal region was stimulated. These blinks were associated with small vertical eye movements as part of the magnetically induced blink reflex. Clockwise or counterclockwise coil currents made no difference. The second region investigated without success in this study was the occipital cortex.

Müri and coworkers[3] stimulated the FEF using a figure-of-eight coil in three subjects during different gaze positions, or just before visually guided saccades (i.e., about 50 to 100 ms before a saccade). TMS did not elicit eye movements during these different eye positions, and was not able to influence saccade latency or amplitude. Another study performed by Elkington and coworkers,[10] stimulating the posterior parietal cortex before saccades, observed only an increased divergence of the eyes before saccades without affecting saccade triggering.

The main conclusion of these studies was that single-pulse TMS is not capable of eliciting saccades or smooth pursuit and therefore is not an appropriate method to elicit eye movements.

Li and coworkers[11] applied repetitive TMS (rTMS) over the premotor frontal cortex during double step saccades (see Fig. 24–1, saccade paradigms): rTMS was not capable of eliciting saccades but could, under certain

retinotopic coordinates, but for the second saccade, the brain must calculate the amplitude with a craniotopic coordinate system.

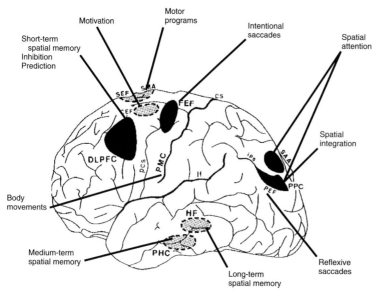

Figure 24–2 Cortical organization of saccade control. cs, central sulcus; pcs, precentral sulcus; DLPFC, dorsolateral prefrontal cortex; FEF, frontal eye field; HF, hippocampal formation; IPS, intraparietal sulcus; lf, lateral fissure; pcs, precentral sulcus; PEF, parietal eye field; PHC, parahippocampal cortex; PMC, primary motor cortex; PPC, posterior parietal cortex; SAA, spatial attentional area; SEF, supplementary eye field; SMA, supplementary motor area; CEF, cingulate eye field. (Modified from Pierrot-Deseilligny C, Ploner CJ, Müri RM, et al. Effects of cortical lesions on saccadic eye movements in humans. Ann N Y Acad Sci 2002;956:216–229.)

circumstances, facilitate saccade triggering. By using the first saccade to trigger the stimulus train, short-latency, multistep saccades were elicited in 3 of 9 subjects when rTMS was delivered over the premotor cortex (i.e., 5 to 7 cm lateral and 2 to 4 cm anterior to the vertex). The intervals between evoked saccades were proportional to the intervals between rTMS pulses delivered at different frequencies. The results of this study suggest that TMS has not invariably an inhibitory effect on structures involved in eye movement control but is also, under certain conditions, capable of facilitating saccadic eye movements.

Wipfli and colleagues[12] used double-pulse TMS (dTMS) to study triggering mechanisms of saccades in the FEF. Double pulses may have the advantage of intensifying the stimulus without loosing the temporal resolution to the extent as rTMS does. Depending on the stimulus parameter used, contrasting effects are also possible. These investigators stimulated the right FEF in twelve healthy subjects. By testing different inter-stimulus intervals (ISI) of 35, 50, 65, or 80 ms, only dTMS with an ISI of 50 ms reduced memory-guided saccade latency of contralateral saccades significantly (Fig. 24–3).

Stimulation over the occipital cortex had no significant effect. These results show that by using an appropriate ISI, dTMS is able to facilitate saccade triggering, and the investigators hypothesized that dTMS might

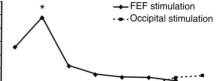

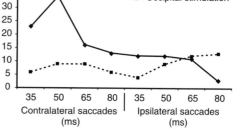

Figure 24–3 Effect of double-pulse transcranial magnetic stimulation (TMS) on latency of memory-guided saccades. A significant decrease for contralateral saccades was found when TMS was applied with an inter-stimulus interval of 50 ms.

specifically interfere with processing of presaccadic movement cells[13] or with an inhibition of suppression cells in the FEF.[14] It has been shown[15–17] that saccade latency is closely correlated with the increasing activity of these cells. An increase in this activity due to dTMS might explain the shortening effect of contralateral saccade latency observed in Wipfli's study.

Facilitation of saccade triggering may also be obtained by an inhibition of a region that is able to inhibit saccade triggering. The DLPC has been shown to be such a region. Patients with lesions of the frontal lobe including this region[18,19] present with an increased frequency of express saccades, especially in the gap task. Müri and coworkers[20] used single-pulse TMS to interfere with DLPC functioning by stimulating the right DLPC in five subjects with two different time intervals during the gap: a stimulation at 100 ms (i.e., half-time of the gap) and at 200 ms (i.e., end of the gap). In all subjects, a significant increase in the percentage of contralateral express saccades was found when TMS was applied at the end of the gap. Stimulation at 100 ms (i.e., during the gap) had no significant influence. Control stimulation over the posterior parietal cortex (PPC) with the same time intervals had no significant effect on the percentage of express saccades. The results of DLPC stimulation may be explained by a transient inhibition of the inhibitory function of the region, resulting consequently in a decreased inhibition of the superior colliculus, which is strongly connected with the prefrontal cortex.[21] Such a mechanism may then facilitate the triggering of express saccades. The described effect was only observed during stimulation at the end of the gap, which is consistent with neurophysiological data from monkey experiments,[22] showing that the occurrence of express saccades is correlated with the activity of buildup neurons of the superior colliculus at the end of the gap.

■ Interference With Processing of Eye Movements

Using single-pulse TMS during the anti-saccade paradigm (see Fig. 24–1 paradigms), Müri and coworkers[3] were the first in 1991 to report a positive effect of TMS interference with programming and performance of eye movements. The right FEF was stimulated in 11 subjects by a focal coil, and TMS significantly increased the latency of anti-saccades when applied some 60 to 100 ms after the visual target was presented. The vulnerable time window during which TMS had a significant effect on anti-saccade latency was found to vary from subject to subject lying between 60 and 100 ms. TMS did not increase the percentage of errors (i.e., the number of unwanted reflexive saccades toward the visual target, which was speaking against an inhibitory function of the FEF on reflexive saccades). To rule out unspecific effects, electrical stimulation of the median nerve or the facial nerve instead of TMS was performed at the same time interval during the anti-saccade task, and both control procedures had no significant effect on anti-saccade performance. Taken together, these results gave important arguments against the FEF being involved in the suppression of unwanted reflexive saccades.

Elkington and colleagues[10] disrupted saccade programming by stimulating the PPC in two subjects. They stimulated with a focal figure-of-eight coil 80 ms after the appearance of the visual target. The PPC was stimulated by placing the coil 3 cm posterior to the vertex and 3 cm laterally. The output was set at 60% of the maximal output (Magstim 200). TMS induced in this study an increase in saccade latency and a tendency to undershoot the visual target, which was more pronounced for contralateral saccades.

Priori and coworkers[23] examined specific and unspecific TMS effects on eye movement performance in detail. They applied single-pulse TMS during visually and acoustically guided saccades by means of a circular coil centered at the vertex. TMS was applied simultaneously with the appearance of the visual or acoustic target. They found an increase in saccade latency by 40 to 50 ms if TMS was given about 60 ms before the expected saccade onset. Moreover, there was a correlation between saccade latency increase and stimulus intensity, and the effect depended on the placement of the coil, because the effect diminished if the coil was moved away from the vertex. Saccade amplitude, however, was not affected by TMS. They studied unspecific effects on reaction time by applying the noise of the discharging apparatus only instead of TMS, or by applying electrical stimulation of the supraorbital nerve. The electrical stimulation of the supraorbital nerve induced a very small effect on saccade latency, and the noise of the discharging stimulator alone had no effect on saccade latency, suggesting that TMS interferes specifically with the function of cortical regions involved in the execution of saccades. However, because a nonfocal coil was used to stimulate over the vertex, identification of putatively involved regions was not possible.

Zangemeister and associates[24] stimulated during two paradigms of predictive saccades. Stimulation was performed with a nonfocal coil, placed with one segment tangentially on the scull, in an attempt to render the stimulation more focal and to reduce the number of blink artifacts on the eye movement infrared recording, infrared oculography being prone to pick up blink artifacts. The six stimulation positions were defined according to the 10–20 system: F3, F4, Fz, Cz, Pz, and Oz. An intensity of 70% of maximal output was used in the first experiment (Cadwell MES-10 stimulator) and of 90% of maximal output in the second experiment (Novametrix stimulator). They found in the first experiment that TMS could influence saccade trajectory and amplitude when applied 100 ms before saccade onset. This global effect was more marked over the parietal regions. In the second experiment, in which the level of predictability was lower, TMS immediately before (i.e., less than 50 ms before the saccade onset) affected saccade latency more significantly.

Ro and colleagues[25] examined the influence of frontal and parietal stimulation on reflexive and intentional saccades. The paradigm they used was a visually guided choice saccade task, where the subject had to perform a saccade in response to a central cue or to a peripheral stimulus. Ten subjects were tested by stimulating right and left hemispheres. TMS was applied in relation to the Go signal: 100 ms before, simultaneously with, and 100 ms after the Go signal. They found a significant increase in saccade latency for contralateral, endogenous saccades when the superior frontal cortex, including the FEF, was stimulated. Stimulation of the superior parietal cortex, however, had no significant influence of saccade latency. Ipsilateral saccades were not affected. In contrast to the endogenous saccades, exogenous (reflexive) saccades were not affected at all. These results confirm the important role of the FEF in the triggering of endogenous or intentional saccades.

Two studies[26,27] investigated how the brain updates visual space across saccades. To make successful saccades the brain must combine visual information about the retinal position of objects with extraretinal information about the actual eye position. The role of such extraretinal information processing may be studied by the double step saccade paradigm where two visual targets are successively presented within the latency of the first saccade (see Fig. 24–1). Humans can perform such double step saccades quite accurately. The brain can calculate the amplitude of both saccades from the retinal positions of the visual targets. To have a correct second saccade amplitude, the intrinsic variability of the first saccade amplitude requires, in addition to the retinotopic information, extraretinal information about the actual position after completion of the first saccade. This is illustrated in Figure 24–4.

If the brain calculates the amplitude of the second saccade in craniotopic coordinates, it will be able to compensate for any variability of the metrics of the first saccade (see Fig. 24–4, *upper left side*), and a linear regression calculated for the relationship between gain of the first and second saccade will result in a positive slope (see Fig. 24–4, *upper right side*). However, if the central nervous system calculates the amplitude of the second saccade only in retinotopic coordinates, the slope of the linear regression should be near zero (see Fig. 24–4, *lower part*).

The neural mechanisms underlying the craniotopic updating are poorly understood, and many cortical regions participate in the control.[27–32] Van Donkelaar and Müri[26] studied the role of the PPC in the control of double step saccades. They stimulated in three subjects the PPC just after (i.e., at

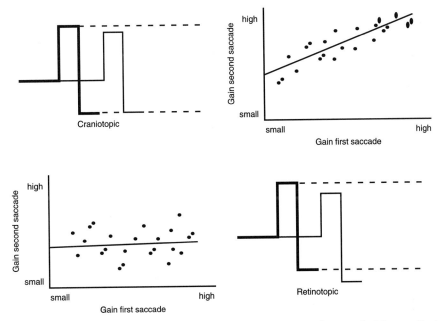

Figure 24–4 Coordinate systems. Craniotopic coding *(upper part)* means that the amplitude of the second saccade compensates for error of the first saccades and results in a positive slope. In a retinotopic coding, there is no compensation of the error by the second saccade, resulting in a slope of zero.

0 ms), 100 ms after, or 150 ms after the first saccade. They found that craniotopic updating becomes no longer possible when TMS was applied just before the second contralateral saccade (Fig. 24–5) by showing that TMS applied 150 ms after the first saccade reduced the slope near to retinotopic calculation.

In a second study, Tobler and Müri[27] examined the influence of single-pulse TMS on performance of double step saccades during FEF and SEF stimulation. In nine healthy subjects, the right FEF and the region of SMA were stimulated before executing the first saccade. FEF stimulation significantly increased the percentage of errors in amplitude for the contralateral second saccade as compared with no stimulation. There was no influence on the first contralateral saccade. Further analysis showed that this was due to an interference with retinotopic but not craniotopic gain calculation. Stimulation of the SEF region interfered with saccade ordering by increasing the errors in the sequence of the double step saccades. The FEF region might participate in the target memorization whereas the SEF is important for coding ordering information for sequential saccades,

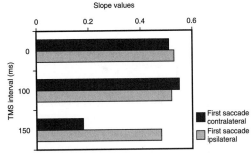

Figure 24–5 Effect of transcranial magnetic stimulation (TMS) on the slope values in the double-step saccades. TMS applied 150 ms after the first saccade reduced the slope near to retinotopic calculation. The slope value without stimulation is the same as for TMS applied coincident with or 100 ms after the first saccade.

even in the double step saccade paradigm. From these two studies, it may be concluded that parietal and frontal regions have complementary functions in updating visual space.

Because the cortical substrate of vergence movement control has been largely unknown in humans, a study investigated the role of the PPC in vergence control with single-pulse

TMS.[33] It is known from monkey studies[34] that vergence components of eye movements can be elicited by stimulating widely distributed cortical regions such as frontal, parietal, and occipital areas. Neurons in the lateral intraparietal area of the PPC are sensitive for visual disparity and depth. Kapoula and coworkers[33] studied in nine healthy subjects the effect of TMS on reflexive saccades with a gap and on vergence movements with a gap by applying TMS 80 ms after the appearance of the visual target. Right PPC stimulation increased saccade and vergence latency significantly in both directions, compared with latencies without stimulation. In the control experiment with TMS over the motor cortex, such an increase was not observed. The results confirm the view that the PPC is involved in reflexive saccade control and in vergence control. The similarity of the observed interference of saccade and vergence movements further supports the idea of a common circuitry controlling both types of eye movements.

Mapping Experiments

Topological Mapping of Cortical Regions Important in Eye Movement Control

By analogy to the mapping of the motor strip, several mapping experiments of cortical structures involved in eye movement control were performed. Oyachi and Ohtsuka[35] used single-pulse TMS to map the PPC during a memory-guided saccade task. They stimulated in four subjects with a figure-of-eight focal coil at various latencies (10, 50, 100, or 150 ms) after extinguishing of the central fixation point. Right hemisphere stimulation affected the accuracy of memory-guided saccades when applied 100 ms after the offset of the central target. Such an effect was found only at a precise position over the PPC, which was defined by three-dimensional magnetic resonance imaging (MRI). The stimulation point was located in the posterior portion of the intraparietal sulcus and the superior part of the angular gyrus in the right hemisphere. This location corresponds to the posterior eye fields, as confirmed by a functional MRI study.[36]

Thickbroom and associates[37] mapped the FEF by interfering with acoustically triggered saccades. They stimulated with a focal coil placed in relation to the vertex and the interaural line using 110% to 120% of the individual motor threshold. In five subjects, eye movements were recorded with infrared oculography, and from two subjects with minimal blink artifacts during TMS a spatial map of the FEF could be composed. TMS was applied 50 ms before mean reaction time of saccades without stimulation, and the latency of the saccades was determined. Typically, the greatest delay provoked by TMS was observed at stimulus sites on or 2 cm anterior to the interaural line, at a distance from the vertex of approximately 6 cm. By moving the coil away from this site, saccade latency declined to values without stimulation. They determined the anatomical relationship of cortical maps of the abductor pollicis brevis muscle and the orbicularis oculi muscle. They found that all three maps lie close together, with the center of each map separated only by about 5 to 10 mm. These results have practical consequences because the localization of the FEF in TMS studies may now be reliably located with respect to the abductor pollicis brevis area.

Ro and colleagues[38] mapped the FEF by single-pulse TMS in two subjects in relation to the hand area. They localized first the hand area and then mapped the region anterior to it during an endogenous saccade task. Single-pulse TMS was applied 50 ms before the onset of the Go signal. They defined a region where TMS produced delays in generating contralateral saccades. By combining these results with anatomical MRI of the subjects, the critical region was localized 2 cm anterior to the hand area on the middle frontal gyrus.

A further study by the same group[39] tested the variability of the FEF location in 10 subjects by TMS. They stimulated with 110% of the motor threshold a sequence of points over the prefrontal cortex until an effective site of the TMS was found that induced contralateral saccade delay. The location of this functionally defined FEF region across the subjects was approximately 1.5 cm anterior to the motor hand area. In three subjects, they were not able to find an effect with this stimulation strength.

Hill and coworkers[40] examined the influence of the orientation of the figure-of-eight coil and current direction on the performance of an eye movement task. It is known from earlier studies that the orientation of the coil influences motor parameters such as reaction times. They stimulated the prefrontal cortex during memory-guided saccades with eight different orientations of the coil and found that the most effective TMS-induced current orientation was antero-lateral, which is different from the optimal current orientation for the hand area of the motor cortex.

Temporal Mapping of Cortical Regions Important in Eye Movement Control

Another perhaps even more rewarding use of TMS in eye movement research is mapping in the temporal rather than topographical domain. Because TMS can be applied with a very precise timing during an experimental task, it is an excellent method to investigate the temporal organization of cortical processes. As such, TMS is complementary to functional imaging methods with their poor temporal resolution.

Two consecutive papers[41,42] on temporal mapping in a task of memorized sequences of saccades (Fig. 24–6) studied the temporal organization. In humans, it is known that patients with lesions including the SEF have difficulties to perform such saccade sequences in the correct order.[43,44] The paradigm can be divided into four subsequent functional phases: (1) the presentation phase of the visual targets, during which the subject has to learn the chronological order of the saccade sequence; (2) the memorization phase to hold this information on-line in the brain, (3) the preparation phase of the motor program, and (4) the execution phase of the saccade sequence (see Fig. 24–6).

In the first study with eight normal subjects,[41] TMS was applied by a nonfocal coil over the SMA region including SEF at different times during the preparation phase of the saccade sequence (i.e., around the Go signal), which corresponded to the extinguishing of the central fixation point. With TMS being moved closer to the Go signal, the number of errors in order increased (see Fig. 24–6, sequences); stimulation 80 ms

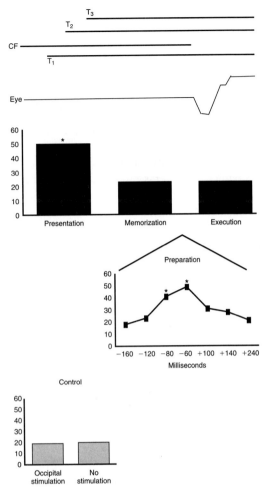

Figure 24–6 Summary of the results of temporal mapping of memory-guided saccade sequences.[41,42] Transcranial magnetic stimulation interfered significantly with sequence performance by increasing the percentage of errors in the sequence when applied during the presentation phase, and during the preparation phase.

before or 60 ms after the Go signal produced a significant increase of erroneous sequences of saccades. In the control experiment, the occipital cortex was stimulated with the same timing, and there was no influence on the number of erroneous sequences. Such an increase of errors found during SMA stimulation around the Go signal suggests that the movement preparation and possibly the initiation of the sequence are controlled by the SMA region. Single-cell studies of the SEF and the SMA[45–47] have shown that more than 50% of the neurons may be active during the premovement phase of internally guided movements, and sequence-specific neurons of the SMA are activated before a specific, learned sequence.[47] Such neurons normally cease their activity before the movement onset, which may explain why TMS over the SEF close to the execution of the sequence is no longer able to influence sequence performance.

TMS during other phases of the paradigm confirmed and expanded these results.[42] Stimulation over the SEF and pre-SEF region during the execution of the sequence did not affect the performance of memorized saccade sequences. TMS increased the percentage of errors only if applied during the presentation of the visual targets (i.e., during the learning phase of the chronological order rather than during the memorization delay). This indicates that the SEF region, including the pre-SEF, is important during the learning phase of the chronological order of saccade sequences. The fMRI studies have suggested that the pre-SEF may be specifically involved in learning of saccade sequences.[48,49]

The temporal organization of memory-guided saccade control was evaluated by TMS in two subsequent studies.[50,51] It is generally accepted that both the PPC and the DLPC are involved in the control of memory-guided saccades, and sustained neuronal activity is found in both regions during the memory delay. Patients with isolated lesions of either region have impaired performance of memory-guided saccades. However, neither from single-cell recordings nor from lesion studies could the examiner infer the precise time at which the PPC and the DLPC exert their influence during the paradigm.

In eight subjects, TMS was applied by a nonfocal coil over the PPC or the DLPC on either hemisphere. TMS was delivered in relation to the appearance of the visual target whose location had to be memorized during a two second memory delay. Stimulation was performed during three alternative phases: during an early phase, with stimulation at 160, 260, and 360 ms after the presentation of the flashed target; during a mid-memorization phase, with stimulation between 700 and 1,500 ms; and 2,100 ms after the presentation (i.e., 100 ms after the extinguishing of the central fixation point within the saccade latency).

We found that TMS significantly increased the percentage of error in amplitude (PEA) of contralateral memory-guided saccades at different time intervals (Fig. 24–7, only results for contralateral memory-guided saccades are shown), depending on the stimulated region. Right PPC stimulation increased the PEA of contralateral saccades when applied in the early phase (i.e., 260 ms after the appearance of the visual target), which was interpreted as interference of TMS with sensorimotor integration known to be performed in the PPC.[32] There was no such effect after left PPC

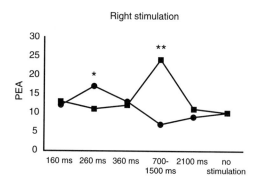

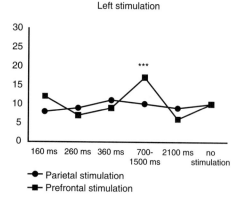

Figure 24–7 Temporal mapping of memory-guided saccades. There was a region and hemispheric specific effect of transcranial magnetic stimulation in time of memory-guided saccade performance. The percentage of errors in amplitude (PEA) for contralateral saccades is shown. Early stimulation (i.e., 260 ms after target presentation) over the right posterior parietal cortex (but not over the left) increased PEA. Stimulation during mid-memorization had a significant effect when applied over the prefrontal cortex.

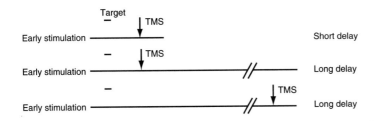

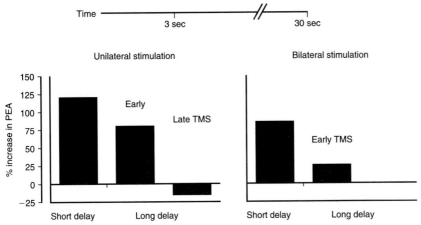

Figure 24–8 Role of the dorsolateral prefrontal cortex in memory-guided saccades. Transcranial magnetic stimulation was applied unilaterally or bilaterally in memory-guided saccades with a delay of 3 or 30 seconds. In both conditions, the percentage of increase in the percentage of errors in amplitude was significantly higher in the 3-second delay, suggesting an additional memory information transfer in parallel.

stimulation at the same time interval, indicating a hemispheric asymmetry of the PPC for eye movement control. Such a hemispheric asymmetry in eye movement control was also found in patients with PPC lesions.[52]

TMS of the DLPC showed a different pattern of interference. Stimulation at 260 ms after the visual target had influence on the PEA, but there was a significant increase in the PEA when TMS was applied during the mid-memorization period. This was found after right and left DLPC stimulation suggesting that the DLPC plays a major role only during the mid-memorization delay. These results are in accordance with experimental findings in monkeys on the organization of spatial working memory.[53,54]

In two TMS studies in healthy subjects[55,56] with stimulation of the DLPFC during a short delay (3 seconds) or a long delay (30 seconds), abnormalities in the amplitude of memory-guided saccades were found (Fig. 24–8). In the first study, double-pulse TMS was used. Early stimulation (i.e., 1 second after target presentation) significantly increased in both delays the percentage of error in amplitude (PEA) of contralateral memory-guided saccades compared with the control experiment without stimulation. TMS applied late in the delay over the DLPFC had no significant effect on PEA.

Moreover, the effect of early stimulation in the long-delay paradigm was smaller than in the short-delay paradigm. Late stimulation had no significant effect of the percentage of increase of PEA. These results revealed a functional dominance of DLPFC during early memorization and a serial information processing is suggested, because the DLPFC stimulation disordered the saccades irrespective of the duration of the preceding delay. However, there is strong evidence for an additional parallel information processing component: the degrading effect of TMS on the saccades was significantly greater with the short saccade delay, suggesting that during longer memorization periods, the MTL got access to additional spatial information from other structures enabling a partial rectification of the stimulation-induced saccade inaccuracy. Additional information to the medial temporal lobe (MTL) may come from the PPC or by the DLPFC of the other hemisphere; both types of anatomic connections exist between these regions. In the second study, single-pulse TMS was applied simultaneously over the left and right DLPFC 1 second after target presentation. Only by inhibiting both DLPFC simultaneously, the model of a serial organization may be corroborated or contradicted. If stimulus-induced inaccuracy of early bilateral stimulation is similar under both delay conditions, serial processing is likely with an information transfer through both DLPFCs. If the degrading effect of early bilateral DLPFC stimulation on saccade accuracy is significantly greater in short- than in long-delay saccades, an additional information transfer to the MTL, probably from the PPC, has to be assumed, indicating a parallel processing component for long memorization periods.

Terao and associates[57] studied the topography of anti-saccade control by focal TMS. TMS was applied at different time intervals (80, 100, and 120 ms) after the presentation of the target over frontal and parietal regions. They found a mild but significant delay of anti-saccade latency when the frontal (2 to 4 cm lateral and 2 to 4 cm anterior to the hand motor region) or parietal (6 to 8 cm posterior to the hand area and 0 to 4 cm lateral) regions were stimulated. There was no difference between right or left hemisphere stimulation. They also showed that posterior stimulation at the 80-ms interval induced a significant latency shift and that frontal stimulation at the 100-ms interval increased latency suggesting an information flow from posterior to anterior cortical regions during the saccade latency period.

Hashimoto and Ohtsuka[58] studied the effect of cerebellar stimulation on visually guided saccade performance. They stimulated with a focal coil over the cerebellum 0, 20, 40, or 60 ms after saccade onset. TMS with 80% of output had a significant effect on saccades when it was applied about 7 mm lateral and caudal to the inion. Stimulation at 0 ms (i.e., simultaneously with saccade onset) produced significant hypometria of contralateral saccades followed by corrective saccades. On ipsilateral saccades, stimulation with a latency of 0 to 40 ms induced saccade hypermetria, followed by a postsaccadic drift. The peak velocity of ipsilateral saccades was higher than that of saccades without TMS of the same amplitude. These results are consistent with neurophysiological data in the monkey showing that microstimulation of the oculomotor vermis induces hypometria of contralateral saccades and hypermetria of ipsilateral saccades.

▪ Repetitive Transcranial Magnetic Stimulation Studies

The use of rTMS in eye movement research is relatively recent. The first study was performed by Li and colleagues[10] is discussed previously. Brandt and coworkers[59] published a study dealing with rTMS and memory-guided saccades. They applied rTMS with a frequency of 20 Hz for 500 ms over the PPC and DLPC in 10 subjects. Their main result was that rTMS affected accuracy of memory-guided saccades when applied over both DLPC. The influence was more marked for contralateral saccades.

Stimulation of the right PPC had an effect on accuracy when stimulation started early (i.e., 50 ms after target offset) but not when started late (i.e., 500 ms after target offset). The stimulation sites were verified by three-dimensional MRI reconstruction, marking the scalp positions of TMS sites. There was a good alignment with the region of DLPC and PPC, and the results obtained in this study confirmed the findings found in single-pulse studies concerning temporal mapping of memory-guided saccades.[50]

Leff and associates[60] used rTMS in 14 subjects to study the influence of parietal and frontal stimulation on reading saccades. They stimulated the PPC and the FEF over both hemispheres. The nonspecific influence of rTMS on reading saccades was controlled by stimulation over the vertex. Left PPC stimulation delayed each new reading saccade by about 50 ms, irrespective of the position of a word in the array. No significant delay, however, was observed after right PPC stimulation or during control stimulation. The effects after FEF stimulation were different: rTMS over the right but not the left FEF increased the latency of the first saccade, but only when rTMS application started before the appearance of the visual target. They hypothesized that left PPC may maintain reading saccades along a line of text and the right FEF may by involved in the preparation of the motor plan for the scan path at the start of each new line of text.

REFERENCES

1. Wessel K, Kömpf D. Transcranial magnetic brain stimulation: lack of oculomotor response. Exp Brain Res 1991;86:216–218.
2. Meyer B, Diehl R, Steinmetz H, et al. Magnetic stimuli applied over motor and visual cortex. Influence of coil position and field polarity on motor responses, phosphenes, and eye movements. Electroencephalogr Clin Neurophysiol 1991;43(Suppl):121–134.
3. Müri RM, Hess CW, Meienberg O. Transcranial stimulation of the human frontal eye field by magnetic pulses. Exp Brain Res 1991;86:219–223.
4. Amassiam VE, Quirk GJ, Stewart M. A comparison of corticospinal activation by magnetic coil and electrical stimulation of monkey motor cortex. Electroencephalogr Clin Neurophysiol 1990;77:390–401.
5. Day B, Rothwell JC, Thompson PO, et al. Delay in the execution of voluntary movement by electrical or magnetic brain stimulation in intact man. Brain 1989;112:649–663.
6. Pascual-Leone A, Brasil-Neto J, Valls-Sole J, et al. Simple reaction time to focal transcranial magnetic stimulation. Comparison with reaction time to acoustic, visual and somatosensory stimuli. Brain 1992;115:109–122.
7. Leight RJ, Zee DS. The neurology of eye movements, 3rd ed. Oxford, UK, Oxford University Press, 1999.
8. Pierrot-Deseilligny C, Ploner CJ, Müri RM, et al. Effects of cortical lesions on saccadic eye movements in humans. Ann N Y Acad Sci 2002;956:216–229.
9. Pierrot-Deseilligny C, Müri RM, Ploner CJ, et al. Cortical control of ocular saccades in humans: a model for motricity. Prog Brain Res 2003;142:3–17.
10. Elkington P, Kerr G, Stein J. The effect of electromagnetic stimulation of the posterior parietal cortex on eye movements. Eye 1992;6:510–514.
11. Li J, Olson J, Anand S, et al. Rapid-rate transcranial magnetic stimulation of human frontal cortex can evoked saccades under facilitating conditions. Electroencephalogr Clin Neurophysiol 1997;105:246–254.
12. Wipfli M, Felblinger J, Mosimann UP, et al. Double-pulse transcranial magnetic stimulation over the frontal eye field facilitates triggering of memory-guided saccades. Eur J Neurosci 2001;14:571–575.
13. Bruce CJ, Goldberg M. Primate frontal eye fields: I. Single neurons discharge before saccades. J Neurophysiol 1985;53:603–635.
14. Burman DD, Bruce CJ. Suppression of task-related saccades by electrical stimulation in the primate's frontal eye field. J Neurophysiol 1997;77:2252–2267.
15. Hanes DP, Thompson KG, Schall JD. Relationship of presaccadic activity in frontal eye field and supplementary eye field to saccade initiation in macaque: poisson spike train analysis. Exp Brain Res 1995;103:85–96.
16. Hanes DP, Schall JD. Neural control of voluntary movement initiation. Science 1996;274:427–430.
17. Everling S, Munoz DP. Neuronal correlates for preparatory set associated with pro-saccades and antisaccades in the primate frontal eye field. J Neurosci 2000;20:387–400.
18. Guitton D, Buchtel HA, Douglas RM. Frontal lobe lesions in man cause difficulties in suppressing reflexive glances and in generating goal-directed saccades. Exp Brain Res 1985;58:455–472.
19. Pierrot-Deseilligny C, Rivaud S, Gaymard B, et al. Cortical control of reflexive visually-guided saccades. Brain 1991;114:1473–1485.
20. Müri RM, Rivaud S, Gaymard B, et al. Role of the prefrontal cortex in the control of express saccades. A transcranial magnetic stimulation study. Neuropsychologia 1999;37:199–206.
21. Leichnetz GR, Spencer RF, Hardy SGP, et al. The prefrontal corticotectal projection in the monkey: an anterograde and retrograde horseradish peroxidase study. Neuroscience 1981;6:1032–1041.
22. Dorris MC, Munoz DP. A neural correlate for the gap effect on saccadic reaction times in monkey. J Neurophysiol 1995;73:2558–2562.
23. Priori A, Bertolasi L, Rothwell JC, et al. Some saccadic eye movements can be delayed by transcranial magnetic stimulation of the cerebral cortex in man. Brain 1993;116:355–367.
24. Zangemeister WH, Canavan AG, Hoemberg V. Frontal and parietal transcranial magnetic stimulation (TMS) disturbs programming of

saccadic eye movements. J Neurol Sci 1995;133:42–52.
25. Ro T, Henik A, Machado L, et al. Transcranial magnetic stimulation of the prefrontal cortex delays contralateral endogenous saccades. J Cogn Neurosci 1997;9:433–440.
26. Van Donkelaar P, Müri RM. Craniotopic updating of visual space across saccades in the human posterior parietal cortex. Proc R Soc Lond B Biol Sci 2002;269:735–739.
27. Tobler PN, Müri RM. Role of human frontal and supplementary eye fields in double step saccades. Neuroreport 2002;13:253–255.
28. Rivaud S, Müri RM, Gaymard B, et al. Eye movement disorders after frontal eye field lesions in humans. Exp Brain Res 1994;102:110–120.
29. Duhamel JR, Goldberg ME, Fitzgibbon EJ, et al. Saccadic dysmetria in a patient with a right frontoparietal lesion. The importance of corollary discharge for accurate spatial behaviour. Brain 1992;115:1387–1402.
30. Heide W, Blankenburg M, Zimmermann E, et al. Cortical control of double step saccades: implications for spatial orientation. Ann Neurol 1995;38:739–748.
31. Sommer MA, Tehovnik EJ. Reversible inactivation of macaque dorsomedial frontal cortex: effects on saccades and fixations. Exp Brain Res 1999;124:429–446.
32. Andersen RA. Multimodal integration for the representation of space in the posterior parietal cortex. Phil Trans R Soc Lond B 1997;352:1421–1428.
33. Kapoula Z, Isolato E, Müri RM, et al. Effects of transcranial magnetic stimulation of the posterior parietal cortex on saccades and vergence. Neuroreport 2001;18:4041–4046.
34. Gnadt JW, Mays LE. Neurons in monkey parietal area LIP are tuned for eye-movement parameters in three-dimensional space. J Neurophysiol 1995;73:280–297.
35. Oyachi H, Ohtsuka K. Transcranial magnetic stimulation of the posterior parietal cortex degrades accuracy of memory-guided saccades in humans. Invest Ophthalmol Vis Sci 1995;36:1441–1449.
36. Müri RM, Iba-Zizen MT, Derosier CA, et al. Location of the human posterior eye field using functional magnetic resonance imaging. J Neurol Neurosurg Psychiatry 1996;60:445–448.
37. Thickbroom G, Stell R, Mastaglia F. Transcranial magnetic stimulation of the frontal eye field. J Neurol Sci 1996;144:114–118.
38. Ro T, Cheifet S, Ingle H, et al. Localization of the human frontal eye fields and motor hand area with transcranial magnetic stimulation and magnetic resonance imaging. Neuropsychologia 1999;27:225–231.
39. Ro T, Farne A, Chang E. Locating the human frontal eye fields with transcranial magnetic stimulation. J Clin Exp Neuropsychol 2002;24:930–940.
40. Hill AC, Davey NJ, Kennard C. Current orientation induced by magnetic stimulation influences a cognitive task. Neuroreport 2000;11:3257–3259.
41. Müri RM, Rösler KM, Hess CW. Influence of transcranial magnetic stimulation on the execution of memorised sequences of saccades in man. Exp Brain Res 1994;101:521–524.
42. Müri RM, Rivaud S, Vermersch AI, et al. Effects of transcranial magnetic stimulation over the region of the supplementary motor area during sequences of memory-guided saccades. Exp Brain Res 1995;104:163–166.
43. Gaymard B, Pierrot-Deseilligny C, Rivaud S. Impairment of sequences of memory-guided saccades after supplementary motor area lesions. Ann Neurol 1990;28:622–626.
44. Gaymard B, Rivaud S, Pierrot-Deseilligny C. Role of the left and right supplementary motor areas in memory-guided saccade sequences. Ann Neurol 1993;34:404–406.
45. Schlag J, Schlag-Rey M. Evidence of a supplementary eye field. J Neurophysiol 1987;57:179–200.
46. Schall JD. Neuronal activity related to visually guided saccadic eye movements in the supplementary motor area of rhesus monkeys. J Neurophysiol 1991;66:530–558.
47. Mushiake H, Inase M, Tanji J. Neuronal activity in the primate premotor, supplementary, and precentral motor cortex during visually guided and internally determined sequential movements. J Neurophysiol 1991;66:705–718.
48. Grosbras MH, Leonards U, Lobel E, et al. Human cortical networks for new and familiar sequences of saccades. Cereb Cortex 2001;11:936–945.
49. Heide W, Binkofski F, Seitz RJ, et al. Activation of frontoparietal cortices during memorized triple-step sequences of saccadic eye movements: an fMRI study. Eur J Neurosci 2001;13:1177–1189.
50. Müri RM, Vermersch AI, Rivaud S, et al. Effects of single-pulse transcranial magnetic stimulation over the prefrontal and posterior parietal cortices during memory-guided saccades in humans. J Neurophysiol 1996;76:2102–2106.
51. Müri RM, Gaymard B, Rivaud S, et al. Hemispheric asymmetry in cortical control of memory-guided saccades. A transcranial magnetic stimulation study. Neuropsychologia 2000;38:1105–1111.
52. Pierrot-Deseilligny C, Rivaud S, Gaymard B, et al. Cortical control of memory-guided saccades in man. Exp Brain Res 1991;83:607–617.
53. Goldman-Rakic PS. Circuitry of primate prefrontal cortex and regulation of behavior

by representational memory. In Plum F, Mountcastle VB (eds): Handbook of Physiology: The Nervous System, vol 5. Bethesda, The American Physiological Society, 1987:373–417.
54. Pierrot-Deseilligny C, Müri RM, Rivaud-Péchoux S, et al. Cortical control of spatial memory in humans: the visuo-motor model. Ann Neurol 2002;52:10–19.
55. Nyffeler T, Pierrot-Deseilligny C, Felblinger J, et al. Time-dependent hierarchical organization of spatial working memory: a transcranial magnetic stimulation study. Eur J Neurosci 2002;16:1823–1827.
56. Nyffeler T, Pierrot-Deseilligny C, Pflugshaupt T, et al. Information processing in long delay memory-guided saccades: further insights from TMS. Exp Brain Res 2004;154:109–112.
57. Terao Y, Fukuda H, Ugawa Y, et al. Visualization of the information flow through human oculomotor cortical regions by transcranial magnetic stimulation. J Neurophysiol 1998;80:936–946.
58. Hashimoto M, Ohtsuka K. Transcranial magnetic stimulation over the posterior cerebellum during visually guided saccades in man. Brain 1995;118:1185–1193.
59. Brandt S, Ploner C, Meyer B, et al. Effects of repetitive transcranial magnetic stimulation over dorsolateral prefrontal and posterior parietal cortex on memory-guided saccades. Exp Brain Res 1998;118:197–204.
60. Leff AP, Scott SK, Rothwell JC, et al. The planning and guiding of reading saccades: a repetitive transcranial magnetic stimulation study. Cereb Cortex 2001;11:918–923.

25 Intraoperative Monitoring of Corticospinal Function Using Transcranial Stimulation of the Motor Cortex

David Burke

Monitoring the function of major ascending and descending pathways in brain and spinal cord during neurosurgical and orthopedic operations is becoming the standard of care to avoid inadvertent neural injury during the procedure. The cost-effectiveness of many forms of intraoperative monitoring may not have been established conclusively. However, Western medicolegal systems are based on individual cases and create their own imperatives regardless of economic justification, and for an individual, it is appropriate that expensive measures be used to reduce the risk of a preventable catastrophe.

Corticospinal function can be monitored during operations on brain and spinal cord using transcranial stimulation of the motor cortex, and there are almost as many techniques as there are different authorities (Fig. 25–1). Corticospinal axons can be activated at the *cortical level* by transcranial stimulation through the intact scalp or by direct stimulation of the motor cortex during craniotomy, at the *decussation of the pyramidal tracts* by transpalatal or transmastoid stimulation, or in the *spinal cord* by direct spinal cord stimulation. The evoked activity can be recorded *from the spinal cord* using epidural or subarachnoid leads (Fig. 25–2), *from peripheral nerve*, or *from innervated muscles* as a compound muscle action potential (Fig. 25–3). However, it is likely that the peripheral nerve volley produced by spinal cord stimulation (the so-called neurogenic MEP) represents an antidromic volley in sensory axons, rather than or in addition to a discharge in a motor axons.[1–5]

This chapter focuses on the potentials recorded from spinal cord and muscle in response to transcranial stimulation. However, it is appropriate at this stage to enunciate a number of general principles that are equally applicable to other forms of intraoperative monitoring.

There is no one technique that is optimal for every operative procedure. The monitoring procedures to be employed need to be planned in advance in consultation with the surgeon. This is particularly so when the operations are neurosurgical because the presentation, site of lesion, extent of preexisting damage to the pathways to be monitored and the procedure to be undertaken will affect access by the neurophysiological team to neural structures and the ability to monitor corticospinal activity. It is not worth the effort to set up a monitoring program if the surgeon does not want the monitoring more than the neurophysiological team that supplies it. The surgical desire must be based on perceived need, not merely medicolegally driven "lip-service" to modern technology.

Whatever monitoring procedure is used, it may be impossible to measure waveforms or to interpret changes in them if the anesthetic conditions are not stable. The anesthetist or anesthesiologist must be committed to the monitoring and prepared to tailor the anesthetic regimen to suit the specific needs of the monitoring technique.

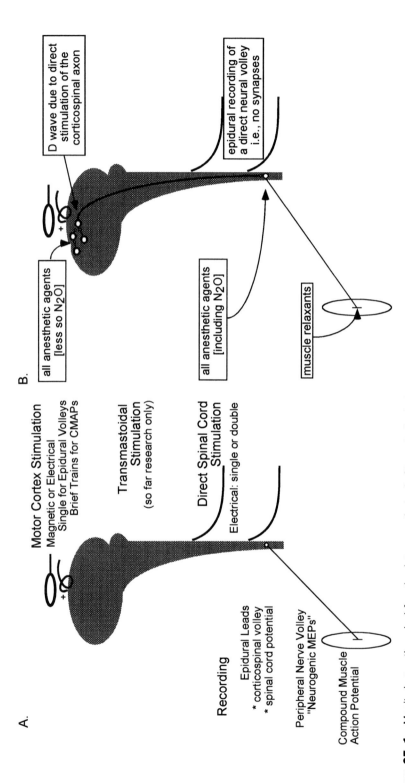

Figure 25–1 Monitoring corticospinal function intraoperatively. **A:** Sites of stimulation and recording. **B:** Sites of action of anesthetic agents that alter the evoked volley significantly. Notice that all depressant actions involve transmission across at least one synapse and that no synapses are involved in epidural recordings of the D wave to transcranial stimulation.

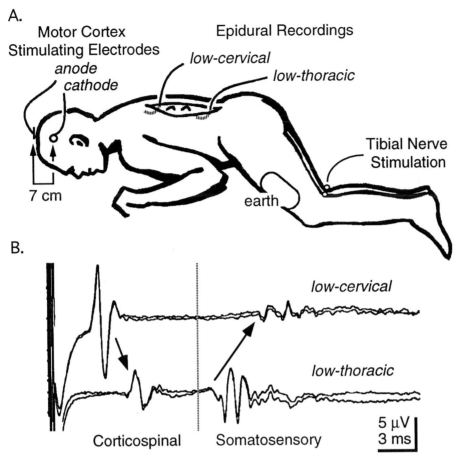

Figure 25-2 **A:** The technique used for recording simultaneously descending corticospinal volleys in response to anodal electrical stimulation of the motor cortex and ascending somatosensory volleys in response to stimulation of the tibial nerves in the popliteal fossae. The stimuli were delivered simultaneously, and the evoked volleys were recorded at two levels from the spinal cord *(lower traces)*. The descending corticospinal volley had a shorter latency and propagated down the spinal cord, whereas the ascending somatosensory volley had a longer latency and propagated up the spinal cord. **B:** The traces are duplicate averages of 10 sweeps. Negativity for the corticospinal volley is shown as an upward deflection, and negativity for the somatosensory volley is shown as a downward deflection, reflecting the fact that the volleys approach the bipolar recording electrodes from opposite directions.

If the deficit is severe in the modality being tested, it may not be possible to record any evoked neural or muscle potential. Many authorities undertake preoperative studies to document formally preexisting pathology, but this unit does not specifically endorse the practice because the situation is quite different in the anesthetized paralyzed patient. The assessment of corticospinal function preoperatively using transcranial magnetic stimulation may provide important information to guide the surgeon and the neurophysiologist, but it provides little insight into whether corticospinal activity can be monitored using transcranial electrical stimulation in the anesthetized patient. The neurophysiologist should commence such monitoring with a fall-back position that will allow a useful monitoring service to be provided if the original technique proves inappropriate in that patient. This is particularly so with neurosurgical procedures.

The ideal technique can assess more than one modality of function at more than one level of the neuraxis. There are false negatives and false positives with all monitoring procedures.

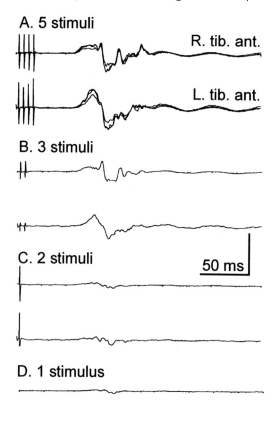

Figure 25-3 The effects of train duration on the evoked compound muscle action potential (CMAP). Recordings are from the right and left tibialis anterior, illustrating the responses to trains of stimuli containing 1, 2, 3 and 5 stimuli, each with a duration of 50 μs, inter-stimulus interval of 5 ms, and intensity of 500 V. Vertical calibration is 20 μV. Notice that the latency of the CMAP after the last stimulus in the train shortens as the number of stimuli increases. The major reason for this is probably that the corticospinal volley is complex and that motoneuron discharge involves temporal summation in the components of the volley, in addition to temporal summation between volleys. (From Bartley K, Woodforth IJ, Stephen JPH, et al. Corticospinal volleys and compound muscle action potentials produced by repetitive transcranial stimulation during spinal surgery. Clin Neurophysiol 2002;113:78–90.)

The latter are of little real importance because they do not reflect neural damage, although false positives may prolong the operation, and too many unsubstantiated interruptions will erode the surgeon's faith in the monitoring service. The former are critical because they reflect a failure of the technique to detect dysfunction resulting from inadvertent neural damage. Both can be reduced if the same modality is measured at two levels of the neuraxis and if two modalities are tested. This is the basis of the technique used in this unit for spinal cord monitoring by means of epidural recordings (see Fig. 25–2). When recording at two levels, it is wise to have one recording above the likely site of injury and one below; for example, compound muscle action potentials (CMAPs) from upper and lower limb muscles during spinal surgery or CMAPs from different lower limb muscles during nerve root surgery. However, when monitoring two different modalities of function, it is wise to remember that there is differential vulnerability of different spinal pathways. Although both corticospinal and somatosensory function commonly deteriorate together when there is spinal cord dysfunction, most monitoring units have seen cases when corticospinal function deteriorated before somatosensory function, perhaps even in isolation.

A monitoring service is relevant only when it assesses function that is potentially at risk. This principle is intuitively obvious, but it underlies the necessity for preoperative discussions between the surgeon and neurophysiologist, particularly when the patient has preexisting pathology or the operative procedure is not one for which monitoring has previously been requested. It is particularly important to assess transmission at or across the likely site of intraoperative dysfunction.

The identification of abnormal function should occur with sufficient time for the surgeon to do something about the dysfunction. The ideal technique would provide "real-time-feedback," but this situation cannot apply when the stimulus repetition rate is low, averaging is required to define the evoked volley, and the putative abnormality must be reproduced to show that it is not due to noise or chance variation in waveforms. In practice, however, it often takes less time to record duplicate averages than to convince the surgeon that something is amiss and that surgery needs to be interrupted, reversed, or even aborted.

Stimulation

Site of Stimulation in the Central Nervous System

Although this chapter focuses on transcranial stimulation, there are circumstances when stimulation at other sites may be valuable, such as when the operation is on the low spinal cord, cauda equina, or nerve roots. The stimuli are usually electrical, usually single when recording the evoked spinal cord volley or the "neurogenic MEP" from a peripheral nerve but usually multiple when recording the CMAP from appropriate lower limb muscles bilaterally. With spinal stimulation, many spinal tracts, both descending and ascending, may be activated, and there is evidence that the optimal stimulus parameters for pulse trains are not the same as with transcranial stimulation. For example, the 2-ms interval has been recommended by advocates of spinal stimulation, but longer intervals appear to be preferable for transcranial stimulation.[6,7] The recording conditions for, and advantages and disadvantages of the neurogenic MEP and CMAPs are discussed in the following sections.

Although the cathode is posterior to the cord, collision studies have demonstrated that the corticospinal pathways are activated by spinal cord stimulation at the threshold for the evoked spinal cord volley.[8] However, as the stimulus is increased, the volley will contain activity due to all accessible large fiber tracts in the spinal cord, and it is inappropriate to consider the evoked spinal cord potential a reflection of corticospinal activity. The major use of the evoked spinal cord volley is that it may allow a monitoring service to be offered when preexisting disease renders conventional monitoring techniques of little use.

Similarly, the corticospinal system can be accessed at the decussation of the pyramids (where there is a bend in the axons) by transpalatal or transmastoid stimulation.[8-11] The latter stimulus site does not appear to have been exploited for spinal cord monitoring, but may be appropriate for some operations. Collision studies have demonstrated that stimuli delivered at this site access corticospinal axons and may do so almost exclusively.[8,11]

Electrical or Magnetic Stimulation

Transcranial magnetic stimulation offers no advantages over transcranial electrical stimulation in the anesthetized or unconscious patient, except that similar studies can be undertaken preoperatively, intraoperatively, and postoperatively. However, this benefit is neutralized by the changes that occur in the response to magnetic stimulation when patients are anesthetized. Both forms of stimulation ultimately produce a current flow that depolarizes axons; they differ merely in the method of delivering that current flux to neural structures. The term *magneto-electrical* has been used instead of *magnetic* merely to reflect this fact. Nevertheless, some authorities have used this stimulus modality for intraoperative monitoring,[12-15] although it has been stated that the evoked CMAPs in the anesthetized patient are too variable to be useful.[14]

Compared with transcranial electrical stimulation, there are several disadvantages of TMS for intraoperative monitoring:

1. The conventional coil is cumbersome and prone to move and, even with the skullcap coil,[16] evoked EMG waveforms are quite variable, depending on cap fit.[17]
2. The stimulating coil intrudes more into the anesthetic field, and the activities of the anesthetist may displace it. With operations on the brain and brainstem, the coil intrudes more into the operative field.
3. The coil may heat up with repetitive use, and excessive heating may cause the machine to be disabled temporarily, interrupting monitoring, something that would be undesirable during hazardous surgery. When using strong stimuli in pulse trains, as is necessary to record reliable CMAPs, the energy in each pulse may be less in the later pulses in the train. These problems constituted important limitations on the use of older stimulators, but both have been addressed with current-generation stimulator design.
4. Strong magnetic stimulation produces D waves, but I waves are more prominent in the corticospinal volley produced by TMS, although this depends on coil orientation.[18-20] In the anesthetized patient, the

D wave may have a lower threshold than I waves,[21] but the magnetic D wave, if present, is usually smaller than the electrical D wave. The trial-to-trial variability of D and I waves depends on their size.[22] These considerations may not be important when monitoring CMAPs, because the morphology of the descending corticospinal volley and the size of individual waves are of little importance provided that the volley produces an adequate discharge of the target motoneuron pool. However, when recording corticospinal volleys using epidural electrodes, the magnetic volley will have greater variability than the electrical volley.

However, the only real limitation of transcranial electrical stimulation is pain experienced by conscious patients. It can be argued that there is no defined role for TMS in the operating theater, and accordingly, this chapter is restricted to the responses to electrical stimulation.

Site of Stimulating Electrodes

With transcranial electrical stimulation, the threshold for a corticospinal volley is lower for anodal stimulation than cathodal, much as has been shown to be the case with direct stimulation of the motor cortex in animals.[23-25] This is because current is believed to enter the long apical dendrite of the pyramidal neuron and to produce depolarization where it exits at the initial segment or the first few nodes of Ranvier.[24] This rationale probably explains the responses to transcranial stimulation over the hand area. However, it is uncertain whether it explains the greater efficacy of anodal stimulation when the pyramidal cell is orientated horizontally relative to the scalp, as would be the case with lower limb areas of motor cortex.

In the initial studies from this Unit, the anode was sited at the vertex and the cathode some 7 cm laterally, approximately over the hand area of motor cortex. The data in Figure 25–2 were recorded with this montage, which produces larger volleys for less stimulus artifact than a cathode at a more anterior site.[26,27] Theoretically, the current flow might have disadvantaged one motor cortex. However, with the stimulus intensities used for monitoring, this is more likely to have affected the ability to stimulate corticocortical axons, activity of which is responsible for I waves. Access to corticospinal axons was probably little affected by the location of the cathode, and it is likely that the D wave involved volleys from both motor cortices.

With CMAPs, all components of the complex corticospinal volley can be important in producing the motoneuron discharge. Since the adoption of CMAP recordings in response to repetitive transcranial stimulation as a worthwhile additional technique for monitoring corticospinal function, this unit has changed to a C3-C4 montage for the stimulating electrodes, even when recording epidural volleys. However, we have the impression that this has resulted in a greater stimulus artifact in the epidural recordings.

Stimulating Electrodes

Spiral subdermal needle electrodes or collodion-secured surface electrodes are equally satisfactory, the former chosen in this unit because of their convenience. Theoretically, the needle electrodes could result in a more focused current density in the scalp: in practice, there have been no untoward side effects with their use.

Single Stimuli or Trains of Stimuli

When recording epidural volleys, single stimuli are appropriate, the recommended repetition rate being approximately once every 3 seconds. However, single transcranial stimuli have no place when recording CMAPs because, even with the lightest of anesthetic regimens, only a few motoneurons in the target pool discharge in response to intense transcranial stimuli.[28] The temporal summation produced in the motoneuron pool by repetitive corticospinal volleys may be sufficient to compensate for the depressant effects of anesthesia,[29,30] allowing reliable CMAPs to be recorded,[6,7,31-33] provided that choice of anesthetic agent, depth of anesthesia, and muscle relaxation are appropriate.[34-39]

Stimulus Parameters

Stimulus Intensity

Transcranial electrical stimulation at and just above threshold activates corticospinal axons at nodes of Ranvier close to the neuronal cell body in motor cortex.[8,24,26,40] However, quite modest increases in stimulus intensity cause subcortical shifts in the site of activation of some axons in the volley,[26] and with strong stimuli (>750 V), all or most corticospinal axons can be activated at the decussation of the pyramidal tract, where there is a bend in the axons.[8] The variability of small components of the corticospinal volley, whether they be subcomponents of the D wave or I waves, is quite high.[22] Accordingly, when recording epidural volleys from the spinal cord, the stimulus should be modest, adjusted so that it produces a large simple D wave, without evidence of components arising from stimulation of corticospinal axons deep in the brain or brainstem. The necessary voltages are usually 250 to 350 V, although this varies with the choice of stimulus waveform and stimulus duration and with other variables such as the age of the patient and the thickness of the skull, the type of electrodes used, and their impedance.

Similarly, when monitoring epidural volleys, I waves add little to the clinical value of the procedure if the D wave at the low cervical region is more than 10 μV. If recording corticospinal and somatosensory volleys at two spinal levels corresponding to the upper and lower limb outflows, as in Figure 25–2, I waves in the corticospinal volley may be superimposed on the somatosensory potential at the low thoracic level. A modest stimulus intensity combined with judicious adjustment of the depth of anesthesia can effectively suppress the I waves, resulting in a corticospinal volley consisting of only a large simple D wave (see Fig. 25–2).

When monitoring CMAPs, however, the more intense the stimulus, the lower the level of anesthesia and the more complex the descending corticospinal volley, the greater will be the motoneuron discharge. In practice, square-wave stimuli of 500 V with a duration of 50 μs are usually adequate for reproducible CMAPs when trains of stimuli are used.

Stimulus Width

There is no consensus about stimulus width.[6,7,41] The strength-duration properties of axons dictate that the threshold current for a potential is smaller the longer the stimulus duration (Fig. 25–4A), but the energy in the stimulus (i.e., the stimulus charge, equivalent to current in mA multiplied by duration in ms) is lower by a factor of three or four (see Fig. 25–4B). The effect on the stimulus artifact is not predictable. Deletis and colleagues[6] suggest that with a pulse width of 0.5 ms, more I waves are recruited by the later stimuli when stimulus trains are used to produce CMAPs—a form of cortical warm-up phenomenon. In the author's experience, stimuli lasting 50 μs are satisfactory for epidural recordings of the descending corticospinal volley to single stimuli and for CMAPs to trains of stimuli.

Stimulus Trains

The number of stimuli in the pulse train has been between five and eight in different studies. In the author's experience, five stimuli are optimal (see Fig. 25–3), and little is gained by increasing the number of stimuli in the train further.[7] However, this experience is based on a 5-ms inter-stimulus interval in the train, and it is conceivable that longer trains may produce larger CMAPs when the interval is shorter.

The most frequently used inter-stimulus interval has been 2 ms,[31–33,35–39] but few studies have systematically compared different intervals. This choice is supported by studies of spinal cord stimulation in the cat.[42] However, the optimal intervals for spinal stimulation and transcranial stimulation are not necessarily the same, and what is optimal in the cat may not be so for human patients. At 2 ms, the second and subsequent responses would fall within the relatively refractory period after the previous discharge[7,40,41] (Fig. 25–5), and this could attenuate D waves (and also I waves).[7] At intervals of more than 4 ms, the second and subsequent discharge would fall within the supernormal period (see Fig. 25–5), and for these reasons, intervals of

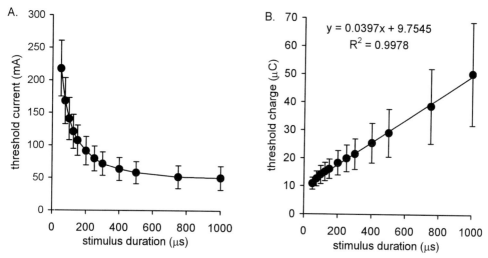

Figure 25-4 Strength-duration properties for the corticospinal D wave. Data are given as the mean ± SEM (n = 10). **A:** The strength-duration curve shows that the stimulus intensity necessary to produce a submaximal D wave of constant amplitude decreased dramatically (in a hyperbolic rather than exponential fashion) as the duration of the stimulus was increased. **B:** The corresponding charge-duration curve shows that the stimulus charge (i.e., stimulus current × stimulus duration) increased with stimulus duration. (From Bartley K, Woodforth IJ, Stephen JPH, et al. Corticospinal volleys and compound muscle action potentials produced by repetitive transcranial stimulation during spinal surgery. Clin Neurophysiol 2002;113:78-90.)

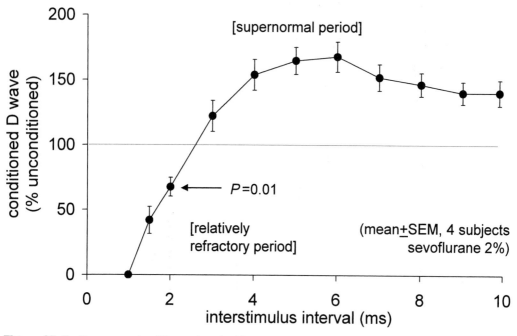

Figure 25-5 Recovery cycle of the corticospinal D wave. Data are given as the mean ± SEM for four subjects, showing the change in amplitude of a submaximal D wave at different inter-stimulus intervals after a near-maximal D wave produced by a strong transcranial stimulus (as in Fig. 25-2B). There was significant suppression of the D wave ($P = .01$) at the 2-ms conditioning-test interval, with supernormality at intervals longer than 3 ms. Supernormality peaked at 5 to 6 ms. (From Bartley K, Woodforth IJ, Stephen JPH, et al. Corticospinal volleys and compound muscle action potentials produced by repetitive transcranial stimulation during spinal surgery. Clin Neurophysiol 2002;113:78-90.)

5 ms have been recommended.[7,40,41] When studied systematically,[7] the CMAPs produced by trains of stimuli at 5-ms intervals proved to be larger in amplitude, longer in duration, and much more complex than those using 2-ms intervals in most patients (Fig. 25–6). There was no evidence that preexisting cerebral pathology favored either interval. It is therefore the practice of this unit to begin with stimuli of 50-μs duration at 5-ms intervals and to shorten the interval if satisfactory CMAPs cannot be recorded with an intensity of 500 V.

Recording

Spinal Cord

The descending corticospinal volley is generally recorded using epidural electrodes inserted by the surgeon and secured by sutures. The recording can be referential, against a remote electrode inserted into adjacent muscle. Bipolar electrodes distort the descending volley, affecting particularly poorly synchronized activity, but their use eliminates one extra lead in the operating field, and the recordings are less affected by artifact. Cardiac pacing leads with an interelectrode distance of 2 to 3 cm are quite satisfactory. It is convenient to insert these leads at two levels, so that they overlie the cervical and lumbar enlargements, but it is difficult to record any descending volley if the lower leads are too caudal, over cauda equina or conus medullaris.

In this unit, the signal is filtered using a bandpass of 500 Hz to 5 kHz. The high-pass filter of 500 Hz distorts the potential, reducing its amplitude and altering its latency, but these distortions are of no consequence for monitoring. The heavy filtering stabilizes the trace and minimizes stimulus artifact, allowing continuous recording, interrupted only by excessive operative artifact or the use of the diathermy.[26,27]

Under these conditions, the amplitude of the maximal D wave is commonly about 10 to 30 μV at the low cervical electrodes and about 6 to 10 μV at the low thoracic electrodes. These amplitudes are such that averaging is not necessary to define the potential, but this unit normally averages 10 to 20 sweeps because the D wave is smaller at the more caudal site and because ascending somatosensory volleys are recorded in the same sweeps (see Fig. 25–2). With a stimulus repetition rate of once every 3 seconds, it would take 1 to 1.5 minutes to identify and then confirm an abnormality, and this is probably an acceptable delay.

There are four advantages of epidural recordings of the corticospinal volley:

1. The recordings involve stimulation of a group of axons and recording from the same axons at a caudal site without any intervening interneurons. As a result, the potentials are relatively resistant to the depressant effects of anesthesia.
2. Complete muscle relaxation is possible. Indeed, it is preferable.
3. The amplitude of the D wave has a very low trial-to-trial variability, such that a decrease in amplitude of 20% at a low cervical electrode can be confidently considered abnormal, provided that anesthetic conditions and electrode placement are stable.
4. Corticospinal volleys can be recorded using epidural leads in neurologically normal subjects, almost without exception, and can often be obtained in patients with preexisting pathology.

There are five disadvantages of epidural recordings of the corticospinal volley:

1. Epidural volleys require the insertion of recording electrodes into the epidural space, and this is usually appropriate only when there is a posterior approach to the cord or posterior fossa.
2. Epidural leads cannot record postsynaptic activity in cauda equina or nerve roots, presumably because single transcranial stimuli produce little such activity in deeply anesthetized patients.
3. Epidural recordings do not allow identification of which corticospinal tract is dysfunctional.
4. The caudal epidural lead is often dislodged during spinal surgery and must then be replaced. (The insertion and reinsertion of epidural leads carry a risk of epidural hemorrhage, but this has not occurred

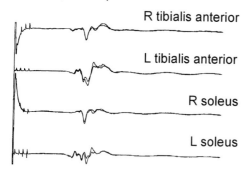
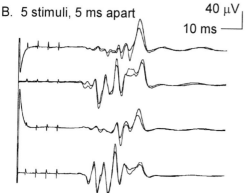
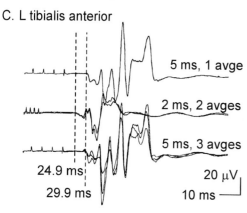

Figure 25–6 The effects of different inter-stimulus intervals. **A:** Recordings from the right and left tibialis anterior and soleus using five stimuli with a duration of 50 μs, inter-stimulus interval of 2 ms, and intensity of 500 V. **B:** The stimulus parameters were identical, except that the inter-stimulus interval was 5 ms. With the 5-ms interval, the evoked compound muscle action potential (CMAP) was larger in amplitude, more prolonged in duration, and much more complex. **C:** The recordings are from the left tibialis anterior and were made in the illustrated sequence, initially using the 5-ms interval, then two recordings using the 2-ms interval, and then three recordings using the 5-ms interval. The interrupted *vertical lines* indicate that the CMAP using the 5-ms interval had a latency approximately 5 ms longer than that using the 2-ms interval. Notice the apparently spontaneous variability of the averaged CMAPs in the lowest traces. (From Bartley K, Woodforth IJ, Stephen JPH, et al. Corticospinal volleys and compound muscle action potentials produced by repetitive transcranial stimulation during spinal surgery. Clin Neurophysiol 2002;113:78–90.)

in this unit in the 12 years that these techniques have been used routinely.)
5. Epidural leads are more expensive than surface EMG electrodes.

Nerve

Peripheral nerve volleys set up by single stimuli delivered to the spinal cord have been recorded from, usually, the tibial nerve in the popliteal fossa, and the resulting "neurogenic MEPs" have been proposed as a technique for monitoring corticospinal function peripherally.[43] An advantage of this technique is that full muscle relaxation can be used; in fact, the recordings would be difficult without full relaxation. There is considerable evidence that the recorded volley contains and may be dominated by antidromic activity in sensory axons.[1-5] This is consistent with the fact that a single shock, even to the spinal cord, does not cause many motoneurons to discharge.[30] The problem of low excitability of the motoneuron pool provided the impetus behind the introduction of repetitive stimulation to exploit the property of temporal summation, for both spinal cord stimulation[30,42] and cortical stimulation.[29]

Muscle

The recording of CMAPs requires attention to the anesthetic regimen and light muscle relaxation. The trial-to-trial variability of CMAPs to single transcranial stimuli is extremely high, even when no volatile or intravenous anesthetics are administered.[28] Single transcranial electrical stimuli, even if intense, occasionally result in intermittent failure to produce a CMAP, or the evoked CMAP will be small.[28] This problem can be reduced (but not eliminated) by delivering trains of transcranial stimuli, much as in Figure 25–3.

The muscles from which CMAPs are recorded vary with the operative indication, but bilateral recordings are generally advisable so that unilateral dysfunction can be detected. Recordings from an appropriate muscle in each of the four limbs allow dysfunction in the thoracolumbar region to be differentiated from a false-positive change. With caudal lesions or those of nerve roots, it may be appropriate to make a number of recordings from muscles innervated by different nerve roots for the appropriate limb, but a contralateral muscle is also recommended.

CMAP recording has three advantages:

1. CMAPs can be used for monitoring for almost any neurosurgical or orthopedic procedure and do not require the posterior approach to the cord normally necessary for the insertion of epidural leads.
2. CMAPs can be used for monitoring when the pathology (and operation) is caudal, involving conus medullaris, cauda equina, or nerve roots.
3. Recordings from muscles bilaterally can allow an abnormality to be lateralized.

CMAP recordings have four disadvantages:

1. The CMAP is much more variable in amplitude, even when produced by trains of stimuli and the responses are averaged.
2. CMAPs are very sensitive to the depth of anesthesia and the degree of muscle relaxation.
3. CMAPs have not been recordable in every normal subject in any of the series reported (presumably because of inadvertent undocumented anesthetic reasons). However, as with epidural volleys, they may be recordable in some patients with preexisting corticospinal pathology.
4. Sometimes CMAPs may be recordable but be too variable to form the basis for a reliable monitoring service.

■ Anesthetic Considerations

Anesthesia aside, merely being unconscious reduces the excitability of motor cortex and the excitability of spinal cord circuits. The former would depress the ability of transcranial stimulation to generate I waves, and the latter the ability to translate the descending corticospinal volley into a motoneuron discharge. Neither action would alter significantly the ability to produce the D waves required when epidural recordings are used for monitoring, but both would jeopardize the production of CMAPs.

All anesthetic agents, even nitrous oxide, can depress excitability further at these levels, but there are differences in the effects of different agents on CMAPs.[14,35–37,44] Intravenous and volatile anesthetics (specifically thiopentone and propofol, sevoflurane and isoflurane) seem to have similar depressant effects on I waves in the corticospinal volleys produced by transcranial electrical stimulation when given in equipotent anesthetic concentrations.[45,46] However, volatile anesthetics are generally eschewed in favor of propofol when recording CMAPs, though etomidate may be better than propofol.[14] Propofol may be given by a targeted infusion giving a steady plasma level of about 4 μg/mL, together with intermittent fentanyl.

It is commonly stated that anesthetics have no effect on the D wave. This is not so, at least for volatile anesthetics. Isoflurane and sevoflurane both reduce the size of the D wave, an effect that is particularly obvious for liminal D waves.[47] This occurs because they can increase the threshold for the D wave by about 35%, probably through an effect on Na^+ channels at nodes of Ranvier, affecting particularly those channels responsible for persistent Na^+ currents.[40] In practice, this does not constitute a limitation because merely increasing stimulus strength compensates for the depressant action. The depressant effect on I waves cannot be overcome as readily by increasing stimulus strength and, as a result, the corticospinal D wave can be considered relatively immune to the effects of anesthetics when compared with the CMAPs.

Muscle relaxation is desirable for epidural recordings; contraction of paraspinal muscles can produce a stimulus-locked artifact of slow onset and waveform in the epidural recording when the degree of muscle relaxation is too light. However, the use of any muscle relaxation will interfere with the recording of CMAPs. There is little overt twitching of the body in response to the transcranial stimulus trains when a muscle relaxant such as atracurium is administered by intravenous infusion so that the twitch ratio of a train of four in adductor pollicis to stimulation of the ulnar nerve at the wrist is approximately 10%.

For CMAP recordings, anesthesia must be light and supplemented by opiates (even though opiates may depress the CMAP recording[44]), and muscle relaxation must be incomplete. The commonest cause for attenuation of the CMAPs is a change in the anesthetic regimen. Any such change, such as bolus injections of agents, should be notified to the neurophysiologist before they occur, and should be avoided during hazardous stages of the operation.

Reproducibility of Potentials and Criteria for Abnormality

Large simple D waves, such as in Figure 25–2, are highly reproducible provided that anesthesia and anesthetic conditions (blood pressure, temperature) remain stable.[22] The coefficient of variation (standard deviation divided by mean amplitude) for large simple D waves of amplitude 10 to 30 mV is less than 0.08, but that of smaller waves of the corticospinal volley (I waves and subcortically originating components of the D wave) is more than 0.2.[22] The former finding implies that, under near-perfect conditions, a 20% decrease in amplitude in single trials would be outside 2.5 SD of the normal mean. These data apply to the low cervical recording in Figure 25–2. The D wave at more caudal sites is smaller and is presumably more variable. The protocol used in my unit measures ascending somatosensory potentials from the cord in the same sweeps as the descending corticospinal volley, as in Figure 25–2. It is recommended that 10 or 20 trials be averaged.

A 20% decrease in amplitude of the D wave can be detected by eye, and this is the criterion used in this unit to notify the surgeon that something could be amiss.

The variability of CMAPs is much higher, as seen in the averaged traces of Figures 25–3 and 25–6. This unit therefore recommends averaging about 10 trials, but even then, any decrease in amplitude should be more than 50% and preferably be accompanied by a change in waveform before a warning is issued.

Choice of Technique and Suggested Protocols

When setting up a monitoring procedure, it is prudent to begin with a technique that is simple and can be applied to most operative indications, and for these reasons the recording of CMAPs to repetitive transcranial electrical stimulation are recommended. However, the most reproducible recordings (and the ones less prone to false-positive changes) are those made directly from the spinal cord using epidural electrodes. For those conditions for which epidural recordings can be made, they are preferred because they allow a superior monitoring service. When the surgical and anesthetic teams have confidence in the service, it is recommended that the epidural recording technique be added to the monitoring armamentarium. Useful additional information can be obtained during neurosurgical procedures if both techniques are employed in the same operation.[6,41,48]

Compound Muscle Action Potentials to Repetitive Transcranial Electrical Stimulation

1. Insert spiral subdermal needle electrodes into the scalp at C3 and C4 for stimulation.
2. Trigger the recording EMG or EP machine from the transcranial stimulator.
3. Record surface EMG patterns from four appropriate muscles, set filters at 100 Hz and 2 kHz, amplify as appropriate (usually up to 20 to 50 μV/division), set the sweep duration to 100 ms, and disable the artifact reject facility (or the stimulus artifacts due to the later stimuli in the train can cause sweep rejection). Average 10 responses.
4. Maintain stable light anesthesia with nitrous oxide and propofol (4 μg/mL) or etomidate, supplemented by fentanyl.

5. Infuse muscle relaxants such that the twitch ratio with train-of-four stimulation is about 10%.
6. Use trains of five square-wave stimuli with a duration of 50 μs, inter-stimulus interval of 5 ms, and strength of 500 V. (This unit uses a commercial stimulator [Digitimer D185, UK] that satisfies regulatory authorities, including the U.S. Food and Drug Administration.) Stimulus trains are repeated not more than once every 3 seconds.
7. If satisfactory CMAPs cannot be recorded, check anesthetic level and the degree of muscle relaxation. If these are appropriate, change stimulus parameters (e.g., inter-stimulus interval to 2 ms).

Epidural Volleys to Single Transcranial Electrical Stimuli

1. Insert spiral subdermal needle electrodes into the scalp at C3 and C4 for stimulation, as for CMAPs. (There may be less stimulus artifact with the anode at the vertex and cathode 7 cm to one side, but C3 and C4 allow for a change to CMAP recordings.)
2. Trigger the recording EMG or EP machine from the transcranial stimulator.
3. Insert bipolar cardiac pacing electrodes at the high and low thoracic levels, thread them up the epidural space to overlie the cervical and lumbar enlargements, and then secure them by sutures.
4. Filter recorded activity at 500 Hz and 5 kHz, amplify to 10 μV/division, and use a sweep duration of 30 ms. Enable the artifact reject facility. Average 10 to 20 sweeps.
5. Use single 50-μs stimuli from the Digitimer D185 stimulator, repeated about once every 3 seconds, and increase stimulus intensity as necessary, usually to 250 to 350 V, to produce a large simple D wave. (Earlier studies used a capacitively coupled high-voltage stimulator [Digitimer D180 or D180A]. This stimulus waveform was developed to minimize pain when transcranial electrical stimulation was performed in conscious subjects.) The D wave should appear at the low cervical electrodes at about 3 to 4 ms and at the low thoracic electrodes after another 3 to 4 ms (see Fig. 25–2).
6. Consider delivering simultaneous stimuli to the motor cortex and using the stimulators on the recording machine to the tibial nerves in the popliteal fossae so that somatosensory function can be monitored in the same recordings. A large standing wave should appear in the low thoracic recording at 11 to 13 ms, and the smaller traveling wave should reach the low cervical level about 4 ms later (see Fig. 25–2).

REFERENCES

1. Su CF, Haghighi SS, Oro JJ, et al. "Backfiring" in spinal cord monitoring. High thoracic spinal cord stimulation evokes sciatic responses by antidromic sensory pathway conduction, not motor tract conduction. Spine 1992;17:504–508.
2. Haghighi SS, York DH, Gaines RW, et al. Monitoring of motor tracts with spinal cord stimulation. Spine 1994;19:1518–1524.
3. Leppanen R, Madigan R, Sears C, et al. Intraoperative collision studies demonstrate descending spinal cord stimulated evoked potentials and ascending somatosensory evoked potentials are mediated through common pathways. Clin Neurophysiol 1999;110:2265–2266.
4. Toleikis JR, Skelly JP, Carlvin AO, et al. Spinally elicited peripheral nerve responses are sensory rather than motor. Clin Neurophysiol 2000;111:736–742.
5. Minahan RE, Sepkuty JP, Lesser RP, et al. Anterior spinal cord injury with preserved neurogenic 'motor' evoked potentials. Clin Neurophysiol 2001;112:1442–1450.
6. Deletis V, Rodi Z, Amassian VE. Neurophysiological mechanisms underlying motor evoked potentials in anesthetized humans. Part 2. Relationship between epidurally and muscle recorded MEPs in man. Clin Neurophysiol 2001;112:445–452.
7. Bartley K, Woodforth IJ, Stephen JPH, et al. Corticospinal volleys and compound muscle action potentials produced by repetitive transcranial stimulation during spinal surgery. Clin Neurophysiol 2002;113:78–90.
8. Rothwell J, Burke D, Hicks R, et al. Transcranial electrical stimulation of the motor cortex in man: further evidence for the site of activation. J Physiol (Lond) 1994;481:243–50.
9. Levy WJ, York DH. Evoked potentials from the motor tracts in humans. Neurosurgery 1983;12:422–429.
10. Levy WJ, York DH, McCaffrey M, et al. Motor evoked potentials from transcranial stimulation of the motor cortex in humans. Neurosurgery 1984;15:287–302.

11. Ugawa Y, Rothwell JC, Day BL, et al. Percutaneous electrical stimulation of corticospinal pathways at the level of the pyramidal decussation in humans. Ann Neurol 1991;29:418–427.
12. Edmonds HL, Paloheimo MP, Backman MH, et al. Transcranial magnetic motor evoked potentials (tcMMEP) for functional monitoring of motor pathways during scoliosis surgery. Spine 1989;14:683–686.
13. Herdmann J, Lumenta CB, Huse KO. Magnetic stimulation for monitoring of motor pathways in spinal procedures. Spine 1993;18:551–559.
14. Taniguchi M, Nadstawek J, Langenbach U, et al. Effects of four intravenous anesthetic agents on motor evoked potentials elicited by magnetic transcranial stimulation. Neurosurgery 1993;33:407–415.
15. Glassman SD, Zhang YP, Shields CB, et al. Transcranial magnetic motor-evoked potentials in scoliosis surgery. Orthopedics 1995;18:1017–1023.
16. Kraus KH, Gugino LD, Levy WJ, et al. The use of a cap-shaped coil for transcranial magnetic stimulation of the motor cortex. J Clin Neurophysiol 1993;10:353–362.
17. Emerson RG, Adams DC, Nagle KJ. Monitoring of spinal cord function intraoperatively using motor and somatosensory evoked potentials. In Chiappa KH (ed): Evoked Potentials in Clinical Medicine, 3rd ed. Philadelphia, Lippincott-Raven, 1997:647–689.
18. Werhahn KJ, Fong JK, Meyer BU, et al. The effect of magnetic coil orientation on the latency of surface EMG and single motor unit responses in the first dorsal interosseous muscle. Electroencephalogr Clin Neurophysiol 1994;93:138–146.
19. Kaneko K, Kawai S, Fuchigami Y, et al. The effect of current direction induced by transcranial magnetic stimulation on the corticospinal excitability in human brain. Electroencephalogr Clin Neurophysiol 1996;101:478–482.
20. Wilson SA, Day BL, Thickbroom GW, et al. Spatial differences in the sites of direct and indirect activation of corticospinal neurones by magnetic stimulation. Electroencephalogr Clin Neurophysiol 1996;101:255–261.
21. Burke D, Hicks, R, Gandevia SC, et al. Direct comparison of corticospinal volleys in human subjects to transcranial magnetic and electrical stimulation. J Physiol (Lond) 1993;470:383–393.
22. Burke D, Hicks R, Stephen J, et al. Trial-to-trial variability of corticospinal volleys in human subjects. Electroencephalogr Clin Neurophysiol 1995;97:231–237.
23. Phillips CG, Porter R. Corticospinal neurones. Their role in movement. London, Academic Press, 1977.
24. Amassian VE, Stewart M, Quirk GJ, et al. Physiological basis of motor effects of a transient stimulus to cerebral cortex. Neurosurgery 1987;20:74–93.
25. Rothwell JC, Thompson PD, Day BL, et al. Stimulation of the human motor cortex through the scalp. Exp Physiol 1991;76:159–200.
26. Burke D, Hicks RG, Stephen JPH. Corticospinal volleys evoked by anodal and cathodal stimulation of the human motor cortex. J Physiol (Lond) 1990;425:283–299.
27. Burke D, Hicks R, Stephen J, et al. Assessment of corticospinal and somatosensory conduction simultaneously during scoliosis surgery. Electroencephalogr Clin Neurophysiol 1992;85:388–396.
28. Woodforth IJ, Hicks RG, Crawford MR, et al. Variability of motor evoked potentials recorded during nitrous oxide anesthesia from the tibialis anterior muscle after transcranial electrical stimulation. Anesth Analg 1996;82:744–749.
29. Taniguchi M, Cedzich C, Schramm J. Modification of cortical stimulation for motor evoked potentials under general anesthesia: technical description. Neurosurgery 1993;32:219–226.
30. Taylor BA, Fenelly ME, Taylor A, et al. Temporal summation—the key to motor evoked potential spinal cord monitoring in humans. J Neurol Neurosurg Psychiatry 1993;56;104–107.
31. Jones SJ, Harrison R, Koh KF, et al. Motor evoked potential monitoring during spinal surgery: responses of distal limb muscles to transcranial cortical stimulation with pulse trains. Electroencephalogr Clin Neurophysiol 1996;100:375–383.
32. Pechstein U, Cedzich C, Nadstawek J, et al. Transcranial high-frequency repetitive electrical stimulation for recording myogenic motor evoked potentials with the patient under general anesthesia. Neurosurgery 1996;39:335–344.
33. Calancie B, Harris W, Broton JG, et al. "Threshold-level" multipulse transcranial electrical stimulation of motor cortex for intraoperative monitoring of spinal motor tracts: description of method and comparison to somatosensory evoked potential monitoring. J Neurosurg 1998;88:457–470.
34. Kalkman CJ, Ubags LH, Been HD, et al. Improved amplitude of myogenic motor evoked responses after paired transcranial electrical stimulation during sufentanil/nitrous oxide anesthesia. Anesthesiology 1995;83:270–276.

35. Pechstein U, Nadstawek J, Zentner J, et al. Isoflurane plus nitrous oxide versus propofol for recording motor evoked potentials after high frequency repetitive electrical stimulation. Electroencephalogr Clin Neurophysiol 1998;108:175–181.
36. Ubags LH, Kalkman CJ, Been HD. Influence of isoflurane on myogenic motor evoked potentials to single and multiple transcranial stimuli during nitrous oxide/opioid anesthesia. Neurosurgery 1998;43:90–94.
37. Van Dongen EP, ter Beek HT, Schepens MA, et al. Effect of nitrous oxide on myogenic motor potentials evoked by a six pulse train of transcranial electrical stimuli: a possible monitor for aortic surgery. Br J Anaesth 1999;82:323–328.
38. Van Dongen EP, ter Beek HT, Schepens MA, et al. Within patient variability of lower extremity muscle responses to transcranial electrical stimulation with pulse trains in aortic surgery. Clin Neurophysiol 1999;110:1144–1148.
39. Van Dongen EP, ter Beek HT, Schepens MA, et al. Within-patient variability of myogenic motor-evoked potentials to multipulse transcranial electrical stimulation during two levels of partial neuromuscular blockade in aortic surgery. Anesth Analg 1999;88:22–27.
40. Burke D, Bartley K, Woodforth IJ, et al. The effects of a volatile anaesthetic on the excitability of human corticospinal axons. Brain 2000;123:992–1000.
41. Deletis V, Isgum V, Amassian VE. Neurophysiological mechanisms underlying motor evoked potentials in anesthetized humans. Part 1. Recovery time of corticospinal tract direct waves elicited by pairs of transcranial electrical stimuli. Clin Neurophysiol 2001;112:438–444.
42. Yamada H, Transfeldt EE, Tamaki T, et al. General anesthetic effects on compound muscle action potentials elicited by single or dual spinal cord stimulation. J Spinal Disord 1995;8:157–162.
43. Owen JH, Laschinger J, Bridwell K, et al. Sensitivity and specificity of somatosensory and neurogenic-motor evoked potentials in animals and humans. Spine 1988;19:1111–1117.
44. Thees C, Scheufler KM, Nadstawek J, et al. Influence of fentanyl, alfentanil, and sufentanil on motor evoked potentials. J Neurosurg Anesth 1999;11:112–118.
45. Hicks R, Burke D, Stephen J, et al. Corticospinal volleys evoked by electrical stimulation of human motor cortex after withdrawal of volatile anaesthetics. J Physiol (Lond) 1992;456:393–404.
46. Woodforth IJ, Hicks RG, Crawford MR, et al. Depression of I waves in corticospinal volleys by sevoflurane, thiopental and propofol. Anesth Analg 1999;89:1182–1187.
47. Hicks RG, Woodforth IJ, Crawford MR, et al. Some effects of isoflurane on I waves of the motor evoked potential. Br J Anaesth 1992;69:130–136.
48. Kothbauer K, Deletis V, Epstein FJ. Intraoperative spinal cord monitoring for intramedullary surgery: an essential adjunct. Pediatr Neurosurg 1997;26:247–254.

26 Magnetic Stimulation in the Assessment of the Respiratory Muscle Pump

Michael I. Polkey and John Moxham

Respiratory failure occurs when the load placed on the respiratory muscle pump exceeds its capacity, and if progressive and untreated, it leads to death. Although respiratory failure most commonly arises because of an excessive load (e.g., lung disease), in the context of neurological disease, it can also occur because of respiratory muscle dysfunction, and recent developments in magnetic stimulation techniques have allowed significant progress to be made in the evaluation of respiratory muscle function.

■ Function of the Respiratory Muscles

Broadly, the respiratory muscles may be considered inspiratory or expiratory. The inspiratory muscles are subdivided into the diaphragm and the extradiaphragmatic inspiratory muscles. The diaphragm is of particular interest because it accounts for about 70% of minute ventilation in normal humans[1] and because it is the only inspiratory muscle for which force output (as transdiaphragmatic pressure [Pdi])[2] can be quantified. Because of its origin in the neck and long course through the thorax, the phrenic nerve is susceptible to damage often after medical interventions.[3-6] The other inspiratory muscles include the sternomastoids, the scalenes and some intercostals. The force output from these muscles cannot be quantified though it is possible to record electromyographic data. In humans with normal lungs, respiratory failure occurs when inspiratory muscle strength is approximately 30% to 40% of predicted values or less.[7] The abdominal muscles are the most important muscles of expiration and are recruited at high levels of ventilation in normal subjects (as in exercise) and in patients with lung disease.[8,9] These muscles serve important functions in humans in relation to phonation, cough,[10] and as accessory muscles of respiration.[11]

■ Diaphragm and Phrenic Nerve Anatomy

The diaphragm consists of a single tendon that inserts into the muscular portion of each hemidiaphragm. Each hemidiaphragm has a costal component that arises from the costal cartilages of ribs 7 to 12 and a crural portion that arises from the vertebrae L1 to L3. The distinction is important because electromyographic (EMG) signals from skin and needle electrodes reflect costal diaphragm activity, whereas an esophageal electrode overlies the crural portion. However, functionally the costal and crural portions of the diaphragm are considered to act in tandem, although they have different mechanical actions.[12]

A phrenic nerve supplies each hemidiaphragm; each nerve arises from the fourth and, to a lesser extent, the third and fifth cervical roots.[13] The contribution from the fifth cervical root may descend into the thorax

Table 26–1 PROPERTIES OF DIFFERENT METHODS OF ELECTROMYOGRAPHIC RECORDING

Electrodes	Invasiveness	Side Effects	Contamination	Quantifiable	Available
Surface electrode	–	None	High risk	No	Yes
Needle electrode	+++	Pneumothorax, bleeding	Low risk	No	Yes
Esophageal electrode	+	Not significant	Low risk	Yes	Difficult

–, nil; +++, very; +, slight.

before joining the nerve trunk, and this is sometimes of practical relevance when considering whether a given field activates the entire nerve. The nerves then run down the mediastinum to the borders of the heart and then into the diaphragm muscle. Within the muscle, although the nerve trunks spread radially the sub-branches spread in a circumferential manner and proximal incisions (whether radial or circumferential) in the diaphragm carry a risk of diaphragm dysfunction.[14] The diaphragm and other respiratory muscles are subject to voluntary and automatic control. Voluntary control is necessary to allow the coordination of various muscle groups necessary during, for example, speech. Nonvolitional mechanisms are mediated from brainstem centers that maintain respiration during sleep and wakefulness. A rapidly conducting pathway exists between the motor cortex and phrenic motor neurons. This pathway is interrupted by capsular but not noncapsular ischemic stroke,[15] demonstrating that each hemidiaphragm has a unilateral contralateral hemispheric representation, although the two hemidiaphragms always act in tandem in vivo. Brainstem injury can produce well-described abnormal patterns of respiration,[16] but magnetic stimulation is of little value for investigating these syndromes because of the difficulty in stimulating the brainstem.

Peripheral Stimulation of the Phrenic Nerves

Before the advent of magnetic stimulation, the phrenic nerves could only be stimulated using electrical stimulation (ES). The motor point of the phrenic nerve has been known for almost 200 years[17,18]; ES can be useful for measurement of phrenic nerve conduction time,[19] but great care must be taken to avoid inadvertent brachial plexus coactivation.[20] Precise focusing is required, and the technique therefore is not suitable for patients with thick necks or indwelling venous catheters. Because of the difficulties in reliably stimulating the nerves and because of the problem of twitch potentiation,[21] it has not proved possible to use ES to make measurements of transdiaphragmatic pressure in patients which are sufficiently accurate to exclude any but the most profound diaphragm weakness.[22]

The options available for magnetic phrenic nerve stimulation are cervical magnetic stimulation (CMS), unilateral anterior phrenic nerve stimulation, bilateral anterior phrenic nerve stimulation, and anterior (presternal) phrenic nerve stimulation. The relative merits of these approaches are summarized in Table 26–2, and examples of coil stimulating positions are shown in Figure 26–1.

Cervical Magnetic Stimulation

Cervical magnetic stimulation (CMS) was the first method described for stimulation of the phrenic nerves in humans.[23] Essentially, a 90-mm circular coil is discharged over the cervical spines at a level approximating to C7. It was originally considered that this approach activated the phrenic nerve roots, but it has been also argued that the field passes through the mediastinum and activates the phrenic nerve trunks anteriorly.[24] The main strengths of this technique are that only one stimulator is required and that normal values for twitch transdiaphragmatic pressure (Tw Pdi) are described,[25] but there are some limitations. Because of the ease of use, this is the stimulation technique that has been used in most published clinical studies.[10,25–27]

Despite obvious advantages, there are some disadvantages. The most important limitation is that there is significant coactivation of the extradiaphragmatic muscles of the upper thorax,[28] which can influence both pressure and compound muscle action potential (CMAP) recordings. Contraction of the upper thoracic muscles has the effect of stiffening the chest wall[28-30] so that the esophageal component of the twitch transdiaphragmatic pressure, Tw Pdi (and therefore the Tw Pdi itself) is greater than that obtained with supramaximal ES. Because the twitch esophageal pressure (Tw Pes) continues to increase as stimulator output increases, it is in practical terms difficult to assess supramaximality when using CMS. From an electromyographic perspective, especially with surface electrodes, it is common to record potentials from extradiaphragmatic muscles, which can lead to the erroneous reporting of a falsely short phrenic nerve conduction time.[31]

A final disadvantage is that, as classically described, the subject must sit upright with the head in forward flexion. This limits the value of the technique in subjects and patients who are supine, such as those receiving ventilatory support.

Unilateral Anterior Magnetic Stimulation

This mode of stimulation entails the discharge of a more focused field through a 45-mm figure-of-eight coil positioned anteriorly over the phrenic nerve adjacent to the sternomastoid muscle at the level of the cricoid cartilage.[32]

Unilateral magnetic stimulation is only suitable for assessing the patency or function of a unilateral phrenic nerve and diaphragm unit (Fig. 26-3), but in this context, unilateral magnetic stimulation (UMS) has advantages compared with ES. First, because precise localization of the nerve is not required (as is the case with ES), it has proved possible to elicit a twitch response with UMS but not ES in some cases.[32] Second because close apposition to the skin is not mandatory, the technique is practical in patients in intensive care, and these patients are recognized to have a high incidence of phrenic nerve or diaphragm injury, such as after cardiac surgical procedures.[33]

UMS seems to be a relatively clean technique with regard to pressure measurements but it is acknowledged that there is crossover

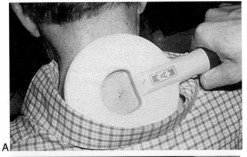
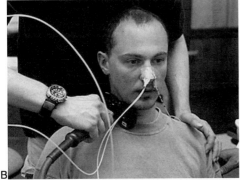
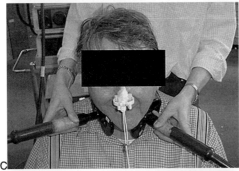
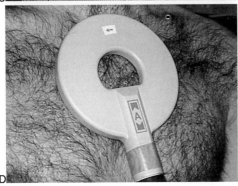

Figure 26-1 Photographs show the coil positions for cervical magnetic stimulation (**A**); unilateral anterior phrenic nerve stimulation (**B**); bilateral anterior phrenic nerve stimulation (**C**); and anterior (presternal) phrenic nerve stimulation (**D**).

stimulation of approximately 10% so that a small signal can be recorded from the contralateral diaphragm. The magnitude of contralateral EMG signals is minimized if diaphragm needle electrodes are used,[32] suggesting that they may arise in part because of contamination from contralateral upper thoracic muscle activation. Contamination of the ipsilateral surface EMG signal may occur,[34] although this can be minimized by the use of a high stimulating position.[35] Even when signals are recorded from an esophageal electrode the latency is marginally shorter (<1 ms) with UMS than ES.[34]

Bilateral Anterior Magnetic Phrenic Nerve Stimulation

Bilateral anterior magnetic phrenic nerve stimulation (BAMPS) entails the bilateral application of UMS[36]; hence, it requires two stimulators and two 45-mm coils. The main advantage of BAMPS is that it reliably produces a signal that is supramaximal and closely similar to optimal ES.

Two studies, as it happens both from our research group, have compared BAMPS with bilateral ES.[36,37] In both studies, we found that the magnitude of the Tw Pdi obtained with BAMPS was close to that obtained by ES, and when we have performed ramp studies, we have consistently found BAMPS to be demonstrably supramaximal.[36,38] More importantly, the esophageal and gastric contributions to the Tw Pdi are similar between ES and BAMPS, suggesting that the observed local coactivation of upper thoracic muscles is, in contrast to CMS, not of practical importance.

Because BAMPS can be applied in the supine position, it is our preferred option for the assessment of diaphragm function in the intensive care unit (ICU) (reviewed by Polkey and Moxham[39]) or in other situations when it is difficult for the patient to sit upright and for the assessment of the effects of anesthetic drugs on diaphragm function.[40] The main disadvantages of BAMPS are related to the expense and bulk of the necessary equipment. It is also acknowledged that when evaluating electromyographic outcomes BAMPS adds little to UMS performed sequentially on each side.

Anterior (Presternal) Magnetic Stimulation

Anterior (presternal) magnetic stimulation (aMS) is a method in which the stimulating coil (90-mm circular coil) is placed anteriorly to the sternum with the cranial rim at the level of the sternal notch.[24,37] As expected, this technique stimulates both phrenic nerves. Our study[37] found that at maximal stimulator output the Tw Pdi was close to that achieved by supramaximal ES but that a supramaximal condition could not be demonstrated. As predicted, phrenic nerve conduction time (PNCT) was reduced (6 ms for aMS, 7.2 ms for BAMPS, and 7.4 ms for ES). Small wave activity preceding the main action potential was sometimes observed with aMS confirming activation of extradiaphragmatic muscles. However, the technique was able to diagnose diaphragm weakness correctly both in ambulant patients undergoing respiratory muscle studies and in 10 patients studied on an intensive care unit. We believe this technique has a role for the clinical confirmation or exclusion of diaphragm weakness in supine patients when two stimulators are not available or access to the neck is not possible.

Central Stimulation Modes

In contrast to transcutaneous ES of the cortex, which is effective[41] but painful, transcutaneous magnetic cortical stimulation, with measurement of the motor evoked potential (MEP), offers a painless and straightforward method of assessing the cortico-spinal pathway, though it offers no information with regard to the brainstem.[42]

The conventional technique for stimulation of the motor cortex is to use a 90-mm circular coil positioned over the vertex, and this approach also is feasible in the diaphragm.[43–45] In these studies, TMS was applied during inspiration but it was subsequently demonstrated that it is possible to obtain an MEP in the unfacilitated condition,[46] although with this approach, the threshold is typically 80% or more and may be unachievable in some subjects. We have shown, however, that a double-cone coil may be used to demonstrate a stimulus response curve for the diaphragm motor area[47] (Fig. 26–4) and that the curves so elicited are sensitive to change induced by, for example, mechanical ventilation.[48] As with other muscles, facilitation (by performing an inspiratory effort) results in a larger amplitude signal with a shorter latency and a lower threshold.

Clinical applications of TMS for the diaphragm include elucidating the origin of breathlessness or hypoventilation particularly in wakeful subjects and the investigation of corticospinal pathways in patients being considered for the implantation of permanent phrenic nerve pacing systems.[49]

Double and Repetitive Stimulation Modes

Repetitive stimuli have been anecdotally applied in our laboratories (to us) in cervical and unilateral anterior stimulating positions in the hope that it would prove possible to stimulate the phrenic nerves supramaximally over a range of frequencies and construct a force-frequency curve of the diaphragm. The value of doing this is that the force-frequency curve of skeletal muscle is altered by the presence of fatigue[50] and length change.[51] Such stimulation is possible in the human diaphragm using ES,[52] but it is very painful, and it was hoped that repetitive MS might overcome this.

In practice, although high-energy peripheral repetitive MS (rMS) is tolerable in humans,[53,54] the problem with regard to the phrenic nerves is that the coactivation of other muscles causes extension of the neck with rCMS and head turning with rUMS. An additional problem with current stimulators is that they can only deliver full power to a maximal frequency of 25 Hz, but a minimum of 50 Hz is necessary for a force-frequency curve.

Alternative approaches include the use of paired stimuli and "train-of-four stimuli." Although other groups disagree, our view is that a train-of-four is unsatisfactory, being complex and expensive but still not tetanic. Paired stimuli are an alternative approach but complex software is need to subtract the force elicited by the first stimulus from that elicited by the total to leave that due just to the second, the T2[55]; from this can be constructed the T2 force-frequency curve. This approach can detect changes in length[38] and those resulting from fatigue[56] and is tolerable to patients[25] and to naïve elder subjects.[57] Paired stimuli also have a role when the intention is simply to generate a bigger stimulus, such as when simulating cough.[58]

Although it is possible to apply paired stimuli to the diaphragm motor area, this remains a research tool. Data suggest that the diaphragm motor area responds in a manner similar to other proximal skeletal muscle so that short inter-stimulus intervals reduce the size of the MEP and long (>5 ms) increase MEP amplitude.[59]

■ Measurements

Having achieved adequate phrenic nerve stimulation, nerve and muscle function can be assessed as an electromyographic or pressure response (Fig. 26–2). These approaches are not mutually exclusive, and each has advantages and disadvantages.

Electromyogram

Measurement of the compound muscle action potential (CMAP) elicited by a single stimulus provides information on both the conduction time and EMG amplitude. For the phrenic nerve, conduction time (PNCT) is more usually quoted than conduction velocity[24] because of the impossibility of accurately determining nerve length in vivo. Very broadly, demyelination results

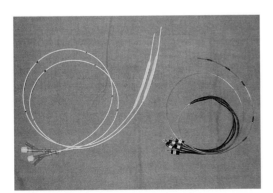

Figure 26–2 A pair of balloon catheters for measurement of esophageal and gastric pressure *(left)* and an esophageal electrode *(right)*. This example has five electrodes. (Electrode courtesy of YM Luo, M.D., Kings College Hospital, London, UK.)

Table 26–2 PROPERTIES OF DIFFERENT STIMULATOR MODES

Mode of Stimulation	No. of Stimulators	Hemidiaphragm	Other Muscles	Supine Possible	Painful	Focusing	Coil
Electrical stimulation	1 electrical	Yes	No	Yes	Yes	Difficult	
Cervical magnetic stimulation	1	No	Substantial	No	No	Easy	90 mm
Bilateral anterior magnetic stimulation	2	No	Minimal	Yes	No	Easy	45 mm (2)
Anterior magnetic stimulation	1	No	Minimal	Yes	No	Easy	45 mm (2)
Unilateral anterior magnetic stimulation	1	Yes	Minimal	Yes	No	Easy	90 mm

in a prolonged conduction time, and axonal loss results in reduced amplitude. The electromyographic approach has the advantage that signals are not influenced by the prior contractile history or lung volume.[60] PNCT and CMAP amplitude may be used as diagnostic evidence or to follow abnormalities when the diagnosis is known[61] but do not give insight into the functional properties of the diaphragm. Three types of EMG electrode are used; their properties are summarized in Table 26–1.

Surface electrodes have the advantage of being well tolerated by the patient and easy to apply. Standard electrodes are composed of silver discs 0.5 cm in diameter, which are coated with electrode paste and applied to the skin over the costal diaphragm. In our laboratory, we find, consistent with the data of Verin and colleagues,[62] that the best location to record signals is in an intercostal space level with the xiphoid sternum between the midline and the midclavicular line. The electrodes are placed close together with a ground electrode somewhat distant. We attempt to reduce noise by vigorous abrasion of the skin with an alcohol soaked swab. Other investigators have examined this and other electrode sites systematically.[63] The major disadvantage of surface electrodes is that they record potentials from other muscles, and this can result in signal contamination when using some magnetic stimulation techniques[31] or with ES if there is inadvertent brachial plexus coactivation.[20] Although surface electrodes are satisfactory (if a good quality signal is obtained) for the measurement of phrenic nerve conduction time, satisfactory normal values for amplitude do not exist.

Needle electrodes can be used to evaluate diaphragm function and have the advantages that they are free from contamination by other muscles and that they can be used to examine the spontaneous activity of the muscle. The main disadvantage is that they are inherently invasive in nature, and while enthusiasts report few side effects,[64,65] the risk of pneumothorax and bleeding is not insignificant in patients receiving mechanical ventilation and in those (e.g., with obstructive lung disease) in whom the diaphragm has a flatter orientation. An additional limitation is that, in contrast to surface and esophageal electrodes, a needle only samples the portion of the diaphragm into which it is inserted.

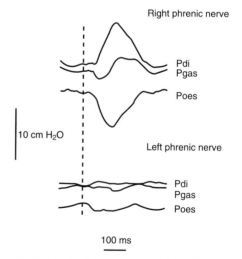

Figure 26–3 Twitch responses obtained in a patient with an elevated left hemidiaphragm. A satisfactory response is obtained on the right, but left-sided stimulation yields no twitch transdiaphragmatic pressure (Tw Pdi). Hemidiaphragm paralysis is diagnosed.

Esophageal electromyography has the advantage of being largely free of contamination from coactivated muscles and not having significant risks although the pernasal passage of the catheter (see Fig. 26–2) can occasionally be uncomfortable for some patients. The basic technology has been described for more than 40 years[66] but has not yet entered routine practice. In part, this may be because no simple device is commercially available, and this may be caused by differences in design. Early designs used a single pair of electrodes,[66] but alternative complex designs exist that use multiple pairs and require complex software to interpret the data.[67] Based on our experience,[31,34,68] the simplest design of an electrode for clinical use would contain three electrodes with a width of 1 cm separated by 3 cm. The purpose of having three electrodes is that when the central electrode is positioned over the electrical center of the diaphragm, the signal elicited from the two remaining electrodes referred to this one is equal and opposite in polarity and this aids positioning. A similar electrode has been produced by another group,[69] but it is not commercially available.

Pressure Measurements

The function of the inspiratory muscles is to produce negative intrathoracic pressure and this can be simply measured from a catheter placed in the esophagus. Usually, the gastric pressure is also measured, and the arithmetic difference between esophageal pressure (Pes) and gastric pressure (Pga) is the transdiaphragmatic pressure (Pdi); an example is shown in Figure 26–3. To measure pressure, we prefer air-filled catheters,[70] but fluid-filled and solid-state catheters are an acceptable alternative.[71]

It is possible to record mouth pressure (Pmo) as a noninvasive surrogate for esophageal pressure[68] provided a maneuver is performed to open the glottis. Although in the absence of lung disease Tw Pmo correlates well with Tw Pes, most laboratories capable of phrenic nerve stimulation currently prefer Tw Pdi to Tw Pmo. In clinical practice the Tw Pdi elicited by ES is insufficiently precise to diagnose any but the most profound diaphragm weakness,[22] but this is not so for

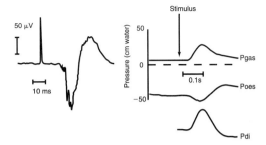

Figure 26–4 Examples of results produced by transcranial magnetic stimulation during a facilitated inspiration. The motor evoked potential *(left panel)* is shown, along with the lowest trace the transdiaphragmatic pressure (Pdi) *(lowest trace of right panel)* and the esophageal (Poes) and gastric (Pga) pressures *(upper panel)*.

magnetic stimulation. For example, a reduced Tw Pdi is a powerful predictor of daytime hypercapnia in patients with amyotrophic lateral sclerosis.[7] Particular applications of magnetic stimulation have already been discussed, but it is worth noting some aspects of pressure measurement.

The most important consideration is that of potentiation. For all skeletal muscle, preceding activity increases twitch tension; the magnitude of the increase may be 60% or more of the unpotentiated twitch.[21,72] This is less important when the aim is to investigate presence or absence of phrenic nerve or diaphragm activity but is very relevant where magnetic stimulation techniques are used in conjunction with pressure measurements to quantify diaphragm function sequentially. A second issue concerns lung volume. In brief, increasing lung volume causes a fall in Tw Pes irrespective of the mode of stimulation[28,38,70,71] and this is translated to a reduced Tw Pdi. The magnitude of the reduction is approximately 3.5 to 5 cm H_2O/L of lung volume increase. Because the lower limit of normal for Tw Pdi is, depending on the stimulation modality, approximately 20 cm H_2O, this is significant.

Interpretation of all pressure measurements depends on the respiratory system being isovolumic. In conscious subjects, this is easily achieved by asking the subject to relax with a closed mouth at end expiration (with a noseclip). However, in patients receiving tracheostomy or endotracheal ventilation, this is not feasible. To circumvent this problem, we use a custom-built occlusion valve[73] in which a balloon is transiently inflated (for about 500 ms) to cover the time of phrenic nerve stimulation and diaphragm contraction. Measurement of the tube pressure distal to the valve (i.e., patient's side) is then an approximation of esophageal pressure.[74]

Force Measurement in Other Muscle Groups

Magnetic stimulation of other nerves is also relevant to respiratory and critical care medicine. The expiratory muscles are important for the generation of cough[10] and as accessory muscles of inspiration. The thoracic nerve roots can be stimulated by a circular coil to determine conduction time[75] and, in conjunction with measurement of Pga, abdominal muscle strength.[10,76] Using this technique we have shown that, in amyotrophic lateral sclerosis, an effective cough (i.e., one in which there is transient supramaximal flow) depends on having a minimum Tw Pga of 7 cm H_2O.[10]

Critical care neuromyopathy may lead to a requirement for prolonged mechanical ventilation. Assessment of the respiratory muscle pump in patients in the ICU is beyond the scope of this book but has been covered elsewhere.[39] Patients with critical care neuromyopathy have a requirement for prolonged rehabilitation and it may be desirable to assess limb muscle function noninvasively to diagnose and monitor weakness. Magnetic stimulation techniques to measure force have been described for the quadriceps[77] and adductor pollicus[78] and are feasible in the ICU environment.

■ Indications and Contraindications to Magnetic Stimulation of the Phrenic Nerve

There are no absolute contraindications to magnetic stimulation except for the presence of implanted ferrous metal within the stimulation field; in the field of thoracic medicine, it is particularly important to check for the presence of cardiac pacemakers. The relative merits of different stimulation modes are summarized in Table 26–1.

The main indications for magnetic phrenic nerve stimulation are as follows:

1. Peripheral phrenic nerve stimulation
   a. Diagnosis of hemidiaphragm paralysis or paresis
   b. Diagnosis of bilateral diaphragm paralysis or paresis when voluntary tests are inconclusive
   c. Investigation of unexplained dyspnea
   d. Investigation of unexplained respiratory failure or weaning failure
   e. Investigation of respiratory muscle involvement in neurological diseases
2. Cortical stimulation
   a. Investigation of central pathways in patients with disease known to potentially affect these pathways
   b. Assessment of patients being considered for phrenic nerve pacing
3. Therapeutic stimulation (research based at present)
   a. Simulation of cough in patients with intact peripheral thoracic nerves

■ Normal Values

Because this is a relatively new field, it is probably appropriate for laboratories to derive their own normal values, but representative data from the literature are included in Table 26–3, and some general considerations are worth noting. Although magnetic stimulation is feasible and well tolerated in children with disease,[79,80] most investigators have not found it ethical to generate control data from older children. Some data concerning PNCT using ES exists in children,[81] and it is reasonable to assess magnetic stimulation derived values against this. Paradoxically, some data exist from a study in neonates in whom we found the most useful technique was to use a circular coil positioned so that one edge lies over the phrenic nerve in the neck.[82] In this studies supramaximal responses were obtained and the mean bilateral Tw Pdi was 8.7 cm H_2O.

Table 26-3 NORMAL VALUES FOR MAGNETIC STIMULATION OF THE RESPIRATORY SYSTEM

Test	Number Studied	Measured Outcome	Normal Range	References	Notes
Peripheral Nervous System					
Electrical stimulation	7	PNCT—SE	4.8–6.9 ms	Similowski et al., 1997[24]	Values quoted for left; right ~0.3 ms longer
Electrical stimulation	203	PNCT—SE	Age dependent	Ross Russell et al., 2001[81]	Studied children
Cervical magnetic stimulation	35	PNCT—SE	6.1–11.9 ms	Zifko et al., 1996[44]	
Cervical magnetic stimulation	35	Amplitude—SE	100–520 μV	Zifko et al., 1996[44]	
Cervical magnetic stimulation	10	PNCT—SE	4.2–6.7 ms	Luo et al., 1998[31]	Left 0.3 ms longer than right
Cervical magnetic stimulation	6	PNCT—EE	6.6–8.6 ms	Luo et al., 1998[31]	
Cervical magnetic stimulation	7	PNCT—SE	4.2–6.5 ms	Similowski et al., 1997[24]	Values quoted for left; right ~0.2 ms longer; PNCT falls as stimulator intensity increases
Cervical magnetic stimulation	23	Twitch Pdi	18–45 cm H_2O	Hamnegård et al., 1996[83]	Includes elderly subjects
Cervical magnetic stimulation	30	Twitch Pdi	15.1–52.3 cm H_2O	Polkey et al., 1997[56]	Left (43 mm coil)
Unilateral magnetic stimulation	5	Twitch Pdi	16 ± 3 cm H_2O	Mills et al., 1995[32]	
			12 ± 4 cm H_2O	Mills et al., 1995[32]	Right (43 mm coil)
Unilateral magnetic stimulation	8	PNCT—SE	4.6 ± 0.6 ms	Luo et al., 1999[34]	Values quoted for right nerve; left 0.4 ms longer
		PNCT—EE	6.9 ± 0.8 ms	Luo et al., 1999[34]	Values quoted for right nerve; left 0.8 ms longer
Bilateral anterior magnetic stimulation	6	Twitch Pdi	22.7–30.2 cm H_2O	Mills et al., 1996[36]	
Bilateral anterior magnetic stimulation	10	Twitch Pdi	24.2–33.9 cm H_2O	Polkey et al., 2000[37]	
Anterior magnetic stimulation	10	Twitch Pdi	16.4–30.5 cm H_2O	Polkey et al., 2000[37]	
Central Nervous System					
Cortical electrical stimulation	3	Latency—EE	11.4–12.3 ms	Gandevia et al., 1987[41]	Facilitated
Cortical magnetic stimulation	35	Latency—SE	10.9–16.5 ms	Zifko et al., 1996[44]	Facilitated
Cortical magnetic stimulation	10	Latency—SE	16.2 ± 0.33 ms	Lissens et al., 1994[45]	Facilitated
Cortical magnetic stimulation		Latency—SE	10.0–16.0 ms	Gea et al., 1993[43]	Facilitated; patients with respiratory disease
Cortical magnetic stimulation	7	Latency—EE	10.0–15.5 ms	Gea et al., 1993[43]	Facilitated; patients with respiratory disease
Cortical magnetic stimulation	9	Latency—SE	13.8–19.5 ms	Similowski et al., 1996[46]	Unfacilitated

EE, esophageal electrode; Pdi, transdiaphragmatic pressure; PNCT, phrenic nerve conduction time; SE, surface electrode.

Conclusion

The principal advantage offered by magnetic stimulation is the fact that the stimulation is less painful than the equivalent electrical stimulus and that less precise focusing is required. This means that the stimulus is acceptable for a much larger range of clinical applications technically and as defined by patient populations likely to accept stimulation. It is hoped that this can be translated into better care of patients with respiratory muscle pump failure.

REFERENCES

1. Mead J, Loring SH. Analysis of volume displacement and length changes of the diaphragm during breathing. J Appl Physiol 1982;53:750–55.
2. Agostini E, Rahn H. Abdominal and thoracic pressures at different lung volumes. J App Physiol 1960;15:1087–1092.
3. Depierraz B, Essinger A, Morin D, et al. Isolated phrenic nerve injury after apparently atraumatic puncture of the internal jugular vein. Intensive Care Med 1989;15:132–134.
4. De Vito EL, Quadrelli SA, Montiel GC, et al. Bilateral diaphragmatic paralysis after mediastinal radiotherapy. Respiration 1996;63:187–190.
5. McAlister VC, Grant DR, Roy A, et al. Right phrenic nerve injury in orthotopic liver transplantation. Transplantation 1993;55:826–830.
6. Rigg A, Hughes P, Lopez A, et al. Right phrenic nerve palsy as a complication of indwelling central venous catheters. Thorax 1997;52:831–833.
7. Lyall RA, Donaldson N, Polkey MI, et al. Respiratory muscle strength and ventilatory failure in amyotrophic lateral sclerosis. Brain 2001;124(Pt 10):2000–2013.
8. Estenne M, Derom E, De Troyer A. Neck and abdominal muscle activity in patients with severe thoracic scoliosis. Am J Respir Crit Care Med 1998;158:452–457.
9. Ninane V, Yernault JC, de Troyer A. Intrinsic PEEP in patients with chronic obstructive pulmonary disease. Role of expiratory muscles. Am Rev Respir Dis 1993;148(Pt 1):1037–1042.
10. Polkey MI, Lyall RA, Green M, et al. Expiratory muscle function in Amyotrophic Lateral Sclerosis. Am J Respir Crit Care Med 1998;158:734–741.
11. Ewig JM, Griscom NT, Wohl ME. The effect of the absence of abdominal muscles on pulmonary function and exercise. Am J Respir Crit Care Med 1996;153(Pt 1):1314–1321.
12. De Troyer A, Sampson M, Sigrist S, et al. The diaphragm: Two muscles. Science 1981;213:237–238.
13. Rajana MJ. Anatomical and surgical considerations of the phrenic and accessory nerves. J Int Coll Surg 1947;60:42–53.
14. Fell SC. Surgical anatomy of the diaphragm and phrenic nerve. Chest Surg Clin N Am 1998;8:281–294.
15. Similowski T, Catala M, Rancurel G, et al. Impairment of central motor conduction to the diaphragm in stroke. Am J Respir Crit Care Med 1996;154:436–41.
16. Plum F, Swanson AG. Central neurogenic hyperventilation in man. Arch Neurol Psychiatry 1959;81:535–549.
17. Ure A. An account of some experiments made on the body of a criminal immediately after execution, with physiological and practical observations. Q J Sci 1819;6:283–294.
18. Duchenne GB. Physiologie des Mouvements. Paris, Balliere, 1867.
19. Newsom-Davis J. Phrenic nerve conduction in man. J Neurol Neurosurg Psychiatry 1967;30:420–426.
20. Luo YM, Polkey MI, Lyall RA, et al. Effect of brachial plexus co-activation on phrenic nerve conduction time. Thorax 1999;54:765–770.
21. Wragg SD, Hamnegard C-H, Road J, et al. Potentiation of diaphragmatic twitch after voluntary contraction in normal subjects. Thorax 1994;49:1234–1237.
22. Mier A, Brophy C, Moxham J, et al. Twitch pressures in the assessment of diaphragm weakness. Thorax 1989;44:990–996.
23. Similowski T, Fleury B, Launois S, et al. Cervical magnetic stimulation: a new painless method for bilateral phrenic nerve stimulation in conscious humans. J Appl Physiol 1989;67:1311–1318.
24. Similowski T, Mehiri S, Duguet A, et al. Comparison of magnetic and electrical phrenic nerve stimulation in assessment of phrenic nerve conduction time. J Appl Physiol 1997;82:1190–1199.
25. Hughes PD, Polkey MI, Harrus ML, et al. Diaphragm strength in chronic heart failure. Am J Respir Crit Care Med 1999;160:529–534.
26. Attali V, Mehiri S, Straus C, et al. Influence of neck muscles on mouth pressure response to cervical magnetic stimulation. Am J Respir Crit Care Med 1997;156:509–514.
27. Polkey MI, Kyroussis D, Hamnegard C-H, et al. Diaphragm strength in chronic obstructive pulmonary disease. Am J Respir Crit Care Med 1996;154:1310–1317.
28. Laghi F, Harrison MJ, Tobin MJ. Comparison of magnetic and electrical phrenic nerve stimulation in assessment of diaphragmatic

contractility. J Appl Physiol 1996;80:1731–1742.
29. Wragg S, Aquilina R, Moran J, et al. Comparison of cervical magnetic stimulation and bilateral percutaneous electrical stimulation of the phrenic nerves in normal subjects. Eur Respir J 1994;7:1788–1792.
30. Mills GH, Kyroussis D, Hamnegard C-H, et al. Cervical magnetic stimulation of the phrenic nerves in bilateral diaphragm paralysis. Am J Respir Crit Care Med 1997;155:1565–1569.
31. Luo YM, Polkey MI, Johnson LC, et al. Diaphragm EMG measured by cervical magnetic and electrical phrenic nerve stimulation. J Appl Physiol 1998;85:2089–2099.
32. Mills GH, Kyroussis D, Hamnegard C-H, et al. Unilateral magnetic stimulation of the phrenic nerve. Thorax 1995;50:1162–1172.
33. Diehl J-L, Lofaso F, Deleuze P, et al. Clinically relevant diaphragmatic dysfunction after cardiac operations. J Thorac Cardiovasc Surg 1994;107:487–498.
34. Luo YM, Johnson LC, Polkey MI, et al. Diaphragm EMG measured with unilateral magnetic stimulation. Eur Respir J 1999;13:385–390.
35. Luo YM, Mustfa N, Lyall RA, et al. Diaphragm compound muscle action potential measured with magnetic stimulation and chest wall surface electrodes. Respir Physiol Neurobiol 2002;130:275–283.
36. Mills GH, Kyroussis D, Hamnegard C-H, et al. Bilateral magnetic stimulation of the phrenic nerves from an anterolateral approach. Am J Respir Crit Care Med 1996;154:1099–1105.
37. Polkey MI, Duguet A, Luo Y, et al. Anterior magnetic phrenic nerve stimulation: laboratory and clinical evaluation. Intensive Care Med 2000;26:1065–1075.
38. Polkey MI, Hamnegard C-H, Hughes PD, et al. Influence of acute lung volume change on contractile properties of the human diaphragm. J Appl Physiol 1998;85:1322–1328.
39. Polkey MI, Moxham J. Clinical aspects of respiratory muscle dysfunction in the critically ill. Chest 2001;119:926–939.
40. Fauroux B, Cordingley J, Hart N, et al. Depression of diaphragm contractility by nitrous oxide in humans. Anesth Analg 2002;94:340–345.
41. Gandevia SC, Rothwell JC. Activation of the human diaphragm from the motor cortex. J Physiol (Lond) 1987;384:109–118.
42. Corfield DR, Murphy K, Guz A. Does the motor cortical control of the diaphragm 'bypass' the brain stem respiratory centres in man? Respir Physiol 1998;114:109–117.
43. Gea J, Espadaler JM, Guiu R, et al. Diaphragmatic activity induced by cortical stimulation: surface versus esophageal electrodes. J Appl Physiol 1993;74:655–658.
44. Zifko U, Remtulla H, Power K, et al. Transcortical and cervical magnetic stimulation with recording of the diaphragm. Muscle Nerve 1996;19:614–620.
45. Lissens MA. Motor evoked potentials of the human diaphragm elicited through magnetic transcranial brain stimulation. J Neurol Sci 1994;124:204–207.
46. Similowski T, Straus C, Coic L, et al. Facilitation-independent response of the diaphragm to cortical magnetic stimulation. Am J Respir Crit Care Med 1996;154:1771–1777.
47. Sharshar T, Ross E, Hopkinson NS, et al. Effect of voluntary facilitation on the diaphragmatic response to transcranial magnetic stimulation. J Appl Physiol 2003;95:26–34.
48. Sharshar T, Ross E, Hopkinson NS, et al. Depression of diaphragm motor cortex excitability during mechanical ventilation. J Appl Physiol 2004;97:3–10.
49. Similowski T, Straus C, Attali V, et al. Assessment of the motor pathway to the diaphragm using cortical and cervical magnetic stimulation in the decision making process of phrenic pacing. Chest 1996;110:1551–1557.
50. Reid M, Grubwieser G, Stokic D, et al. Development and reversal of fatigue in human tibialis anterior. Muscle Nerve 1993;16:1239–1245.
51. Fitch S, McComas AJ. Influence of human muscle length on fatigue. J Physiol 1985;362:205–213.
52. Moxham J, Morris AJ, Spiro SG, et al. Contractile properties and fatigue of the diaphragm in man. Thorax 1981;36:164–168.
53. Polkey MI, Luo Y, Guleria R, et al. Functional magnetic stimulation of the abdominal muscles in humans. Am J Respir Crit Care Med 1999;160:513–522.
54. Lin VWH, Singh H, Chitkara RK, et al. Functional magnetic stimulation for restoring cough in patients with tetraplegia. Arch Phys Med Rehabil 1998;79:517–522.
55. Yan S, Gauthier AP, Similowski T, et al. Force-frequency relationships of in vivo human and in vitro rat diaphragm using paired stimuli. Eur Respir J 1993;6:211–218.
56. Polkey MI, Kyroussis D, Hamnegard C-H, et al. Paired phrenic nerve stimuli for the detection of diaphragm fatigue. Eur Respir J 1997;10:1859–1864.
57. Polkey MI, Harris ML, Hughes PD, et al. The contractile properties of the elderly human diaphragm. Am J Respir Crit Care Med 1997;155:1560–1564.
58. Kyroussis D, Polkey MI, Mills GH, et al. Simulation of cough in man by magnetic stimulation of the thoracic nerve roots. Am J Respir Crit Care Med 1997;156:1696–1699.

59. Demoule A, Verin E, Ross E, et al. Intracortical inhibition and facilitation of the response of the diaphragm to transcranial magnetic stimulation. J Clin Neurophysiol 2003;20:59–64.
60. Luo YM, Lyall RA, Harris ML, et al. Effect of lung volume on the oesophageal diaphragm EMG assessed by magnetic phrenic nerve stimulation. Eur Respir J 2000;15:1033–1038.
61. Zifko U, Chen R, Remtulla H, et al. Respiratory electrophysiological studies in Guillain-Barré syndrome. J Neurol Neurosurg Psychiatry 1996;60:191–194.
62. Verin E, Straus C, Demoule A, et al. Validation of improved recording site to measure phrenic conduction from surface electrodes in humans. J Appl Physiol 2002;92:967–974.
63. Markand ON, Kincaid JC, Pourmand RA, et al. Electrophysiologic evaluation of diaphragm by transcutaneous phrenic nerve stimulation. Neurology 1984;34:604–614.
64. Bolton CF. Clinical neurophysiology of the respiratory system. Muscle Nerve 1993;16:809–818.
65. Saadeh PB, Crisafulli CF, Sosner J, et al. Needle electromyography of the diaphragm: a new technique. Muscle Nerve 1993;16:15–20.
66. Agostini E, Sant'Ambrogio G, Carasso HDP. Electromyography of the diaphragm in man and transdiaphragmatic pressure. J Appl Physiol 1960;15:1093–1097.
67. Daubenspeck JA, Leiter JC, McGovern JF, et al. Diaphragmatic electromyography using a multiple electrode array. J Appl Physiol 1989;67:1525–1534.
68. Luo YM, Lyall RA, Harris ML, et al. Quantification of the esophageal diaphragm electromyogram with magnetic phrenic nerve stimulation. Am J Respir Crit Care Med 1999;160(Pt 1):1629–1634.
69. Sinderby C, Navalesi P, Beck J, et al. Neural control of mechanical ventilation in respiratory failure. Nat Med 1999;5:1433–1436.
70. Mead J, McIlroy MB, Selverstone NJ, et al. Measurement of intraesophageal pressure. J Appl Physiol 1955;7:491–495.
71. Stell IM, Tompkins S, Lovell AT, et al. An in vivo comparison of a catheter mounted pressure transducer system with conventional balloon catheters. Eur Respir J 1999;13:1158–1163.
72. Mador M, Magalang U, Kufel T. Twitch potentiation following voluntary diaphragmatic contraction. Am J Respir Crit Care Med 1994;149:739–453.
73. Spicer M, Hughes P, Green M. A non-invasive system to evaluate diaphragmatic strength in ventilated patients. Physiol Meas 1997;18:355–361.
74. Watson AC, Hughes PD, Harris ML, et al. Measurement of twitch transdiaphragmatic, esophageal, and endotracheal tube pressure with bilateral anterolateral magnetic phrenic nerve stimulation in patients in the intensive care unit. Crit Care Med 2001;29:1325–1331.
75. Chokroverty S, Deutsch A, Guha C, et al. Thoracic spinal nerve and root conduction: a magnetic stimulation study. Muscle Nerve 1995;18:987–991.
76. Kyroussis D, Mills GH, Polkey MI, et al. Abdominal muscle fatigue after maximal ventilation in humans. J Appl Physiol 1996;81:1477–1483.
77. Polkey MI, Kyroussis D, Hamnegard C-H, et al. Quadriceps strength and fatigue assessed by magnetic stimulation of the femoral nerve in man. Muscle Nerve 1996;19:549–555.
78. Harris ML, Luo YM, Watson AC, et al. Adductor pollicis twitch tension assessed by magnetic stimulation of the ulnar nerve. Am J Respir Crit Care Med 2000;162:240–245.
79. Rafferty GF, Greenough A, Dimitriou G, et al. Assessment of neonatal diaphragm paralysis using magnetic phrenic nerve stimulation. Pediatr Pulmon 1999;27:224–226.
80. Rafferty GF, Greenough A, Manczur T, et al. Magnetic phrenic nerve stimulation to assess diaphragm function in children following liver transplantation. Pediatr Crit Care Med 2001;2:122–126.
81. Ross Russell RI, Helps B-A, Helms PJ. Normal values for phrenic nerve latency in children. Muscle Nerve 2001;48:1548–1550.
82. Rafferty GF, Greenough A, Dimitriou G, et al. Assessment of neonatal diaphragm function using magnetic stimulation of the phrenic nerves. Am J Respir Crit Care Med 2000;162:2337–2340.
83. Hamnegård C-H, et al. Clinical assessment of diaphragm strength by cervical magnetic stimulation of the phrenic nerves. Thorax 1996;51:1239–1242.

27 Clinical Applications of Functional Magnetic Stimulation in Patients with Spinal Cord Injuries

Vernon Lin and Ian Hsiao

Magnetic stimulation is a technique that can activate nerves or muscles from a distance. Unlike electrical stimulation, magnetic stimulation does not rely on the passage of electric current through the tissue. The basic difference between electrical and magnetic stimulation is that the former injects current into the body by means of electrodes. In the case of magnetic stimulation, the magnetic field generated from the magnetic coil (MC) is able to pass through high-resistance structures such as skin, fat, or bone to stimulate underlying nerves, including brain tissues. With time-varying and sufficient amplitude, this magnetic field induces an electrical field, which causes ions to flow and results in depolarization of the nerve. The mechanism of stimulation at the neuronal level is thought to be the same for magnetic and electrical stimulation: Current passes across a nerve membrane and into the axon, resulting in depolarization and the initiation of an action potential that propagates by the normal method of nerve conduction.

The considerable advantage of magnetic stimulation is the remarkable lack of painful sensation compared with stimulation with skin-surface electrodes. With such electrodes, a high current density develops under the electrode, thereby favoring stimulation of skin receptors. With magnetic stimulation, there is no localized high current density at the skin surface under the MC. The ability of magnetic stimulation to induce electrical currents to flow within body tissues, especially in deep neural structures such as the motor cortex and spinal nerve roots, allows health care practitioners to monitor and influence many of these functions. As a result, this technique has become of great interest in studies of motor and sensory cortex functions and in psychiatry, for example, in the treatment of mood disorders. It also provides further aids to the clinician for the therapeutic treatment of muscle deconditioning in spinal cord injury (SCI) patients and spasticity in multiple sclerosis or stroke patients. In this chapter, we strive to provide an overview of the various clinical applications of functional magnetic stimulation (FMS) for treating patients with SCI, with an emphasis for stimulating the respiratory muscles. We define FMS as a technique that uses magnetic stimulation to produce useful bodily function.[1]

■ Clinical Applications of Functional Magnetic Stimulation for Patients with Spinal Cord Injury

Most FMS applications in individuals with SCI have aimed for noninvasive stimulation of the peripheral and central motor pathways. These applications cover the uses for diagnosis, prognosis, monitoring, and rehabilitation. Most therapeutic applications developed involve the activation of the spinal nerves and other peripheral nerves.[2–6] For example, if the MC is placed near C3 to C5, stimulation of the spinal nerves results in the activation of the phrenic nerves. Similarly, if the MC is located at the lower cervical region, the upper extremities are stimulated. Upper intercostal nerves are activated if the MC is placed at the upper thoracic region, and lower intercostal nerves are activated if MC is placed at lower thoracic region. The focus of magnetic stimulation of the spinal nerves has been an active area of research for many years. Using a cadaver thoracic spine model, Maccabee and colleagues[7] measured the voltages across the neural foramina during magnetic stimulation and determined that the highest voltages were measured within the foramina, suggesting that the most probable site of nerve activation was at the intervertebral foramen.

Functional Magnetic Stimulation of the Respiratory Muscles

FMS of the respiratory muscles for patients with SCI can be considered as one of the most important clinical applications of FMS. Major muscles of inspiration include the diaphragm (C3 to C5), parasternal muscles, and external intercostal muscles (T1 to T6). Able-bodied subjects are usually capable of generating approximately 100 cm H_2O of inspired pressure during a maximal effort, to which the diaphragm contributes 55% to 60% and intercostal muscles contribute 20% to 25%. Major muscles of expiration include the abdominal muscles (T7 to L1) and the internal intercostal muscles. Able-bodied subjects generate approximately 150 cm H_2O of maximal expiratory pressure. Respiratory dysfunction is a major cause of mortality and morbidity in subjects with SCI.[8–10] When the spinal cord is damaged, the respiratory muscles below the level of the injury become paralyzed and are devoid of supraspinal control. This interferes with the power and integration of the remaining muscles, reducing their ability to drive the chest wall efficiently. Those with paraplegia at and below level T12 have essentially no respiratory dysfunction. With levels of injury from T12 through T5, the progressive loss of abdominal motor function causes an impairment of forceful expiration and cough. With levels of injury from T5 through T1, the remainder of the intercostal volitional function is lost, with further impairment of inspiratory and expiratory effort. Patients with levels of tetraplegia from C8 through C4 have no intercostal or abdominal muscle activity and have a dramatically impaired cough. At the C4 level, the diaphragm might have lost some innervation, further compromising even quiet respiratory effort. At the C3 level, the diaphragm is typically weakened to the point at which ventilation cannot be self-sustained, resulting in ventilatory failure.

Respiratory dysfunction measured by pulmonary function tests in chronic SCI patients is characteristic of restrictive lung disease, defined as a breathing disorder resulting from impairment of the elastic properties of the lung and chest wall and marked by static or reduced lung volumes and capacities.[10,11] Patients with cervical cord injury have paralysis of the inspiratory and expiratory muscles and therefore have marked reduction of vital capacity (VC), little or no expiratory reserve volume, and an inspiratory capacity equivalent to VC.[12–14] The reduction of VC is most severe in tetraplegic patients. VCs in individuals with acute cervical lesions range from 24% to 31% of predicted VC.[14,15] The total lung capacity is reduced and the ratio of residual volume to total lung capacity is increased in subjects at all injury levels. The lung function of patients tested more than 12 months after injury is not significantly different from the function in those tested 6 to 12 months after injury.[16] Decreases of static-inspired and static-expired breathing pressure measured at the mouth indicate impairment of respiratory muscle functions.

Improvement of early surgical spine stabilization, pharmacological treatment, and chest physical therapy has improved survival of SCI patients.[17,18] Adequate alveolar ventilation can be maintained in most SCI patients during the acute rehabilitation period of SCI. Chest physical therapy techniques include bronchial hygiene, postural draining, chest percussion, quad cough, and high-frequency chest wall oscillation.[19–21] Application of some of these techniques in a timely manner has decreased mortality in SCI. In addition to chest physical therapy, functional exercise of respiratory and accessory muscles has been used effectively in the treatment of many SCI patients. Functional exercise includes the use of glossopharyngeal breathing and breathing against resistance.[22,23]

Functional electrical stimulation (FES) of the respiratory muscles has been an active area of research in the last several decades. Various types of electrical implants have been employed to generate inspired and expired pressures, by stimulating the phrenic nerves, ventral roots, intercostal nerves, or abdominal muscles.[24–29] Electric phrenic nerve stimulation can be applied as a mode of assisted ventilation to subjects with bilateral diaphragm paralysis, such as in cervical SCI above C6.[29–31] Upper intercostal muscles (T1 to T6) have been demonstrated to act as inspiratory agonists by lifting and expanding the rib cage.[32,33] FES of the intercostal nerve is achieved in human subjects by ventral root stimulation, and inspired volumes of 600 to 900 mL can be generated.[34] By placing a plate electrode along the ventral surface of the spinal cord near T9, significant expired pressure was also demonstrated in dogs.[35] FES of the abdominal muscles has been shown to generate significant expiratory pressures. By placing percutaneous electrodes on the abdominal wall motor points, significant expiratory flow and pressure were measured in patients with SCI.[28,29,36,37] These studies demonstrate efficient FES of inspiratory and expiratory agonists; FES in general requires surgery for electrode implants in most instances and can be rather painful for patients who have preserved sensation.

FMS of the respiratory muscles is effective and has many advantages when compared with the existing FES technology. FMS is easier to use and does not require surgery or electrode implants, preventing complications such as infection, bleeding, wire breakage, and implant failures. FMS is not as painful as FES and is well tolerated by conscious individuals. FMS can be applied over clothing, because it does not require skin contact or the use of electrode gel. There are also disadvantages of the FMS technology when compared with FES, one of which is the bulky size of the stimulator; another one is the nonfocal nature of magnetic stimulation, which may activate unwanted neuromuscular tissue.

Magnetic stimulation has been applied to stimulate the cerebral cortex to evaluate the nerve conduction velocity for the respiratory system. Using transcranial brain stimulation, diaphragm compound muscle action potential had a mean $\pm$ SD latency of 16.21 ± 0.33 ms, an amplitude of 3.52 ± 2.40 mV, and a central motor conduction time of 8.39 ± 0.41 ms.[38] Similarly, nerve conduction studies of the thoracic nerves using FMS[39] and transcranial stimulation for determining the central conduction time for upper intercostal muscles have been reported.[40] FMS of the phrenic nerve was first reported by Similowski and coworkers.[41] The MC was positioned above the spinous process of the C7 to stimulate both phrenic nerves with a single stimulus. They were able to obtain reproducible supramaximal compound muscle action potential in five of six subjects. Wragg and associates[42] compared FMS of the phrenic nerve with electric stimulation and showed that mean twitch transdiaphragmatic pressure obtained by FMS was significantly higher than electrical twitch transdiaphragmatic pressure. They speculated that FMS might have resulted in stimulation of extradiaphragmatic musculature. Mills and colleagues[43] also showed that cervical FMS in patients with known diaphragmatic paralysis could not generate transdiaphragmatic pressure. The stimulation of the extradiaphragmatic musculature most likely played a role in stiffening the upper thoracic cage to allow the diaphragm to act efficiently. Laghi and coworkers[44] also reached the conclusion that cervical FMS recruited extradiaphragmatic and diaphragmatic muscles, whereas FES had a more selective action on the diaphragm. They further concluded that FMS could be an equally effective method for detecting diaphragmatic fatigue.

FMS of the respiratory muscles was first performed in dogs,[27,45,46] with successful FMS of the phrenic nerves, upper intercostal nerves, and lower intercostal nerves. The mean inspired volume and pressure produced by FMS of the phrenic nerves in the carotid MC placement before phrenectomies were 373 ± 20.5 mL and -20 ± 2.0 cm H_2O, respectively. In postphrenectomized animals, the maximal inspired volume (219.0 ± 12.2 mL)

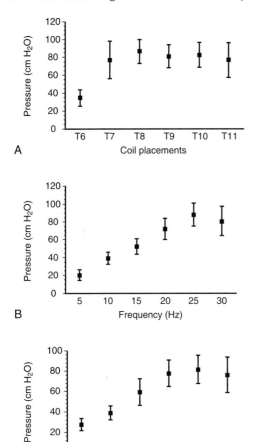

Figure 27–1 **A:** Optimal magnetic coil placement. Maximal expired pressure was obtained when the magnetic coil was placed on the T8 spinous process. **B:** Frequency profile, showing changes in expired pressure in response to changing frequencies while keeping the intensity and stimulation duration constant at 70% and 2 seconds, respectively. **C:** Intensity profile, showing changes in mean expired pressures in response to changing stimulation intensities while the frequency and stimulation duration were kept at 20 Hz and 2 seconds, respectively.

pressures, and they established the foundation for FMS as a potential technology for inspiratory and expiratory muscle training, ventilatory assistance, and cough production.

Expiratory functions produced in humans by FMS were first demonstrated by stimulating the lower thoracic spinal nerve root.[47] A 12.5-cm, round MC was placed in the lower thoracic region for stimulating the lower intercostal muscles and the abdominal muscles. Maximal expired pressure, volume, and flow rate generated by FMS were 83.6 ± 16.4 cm H_2O, 1.54 ± 0.18 L, and 4.75 ± 0.35 L/s, respectively. These values corresponded to 73%, 100%, and 90% of the maximal voluntary expired pressure, volume, and flow, respectively. The optimal coil placement was at T7, and the optimal stimulation parameters were with a frequency of 25 Hz and an intensity of 70% to 80% as shown in Figure 27–1. When similar stimulation protocol was applied to patients with SCI, FMS of the expiratory muscles also produced a substantial maximal expired pressure (68.2 ± 24.1 cm H_2O), volume (0.77 ± 0.48 L), and flow rate (5.27 ± 1.49 L/s).[2] Figure 27–2 illustrates the forced expiratory flow generated by one subject's maximal voluntary effort and by FMS in a flow volume loop. These values corresponded to 118%, 169%, and 110% of their voluntary maximums, respectively. The MC placement at the T10 to T11 spinous process and stimulation intensity at 80% produced the highest expiratory pressure and flow.

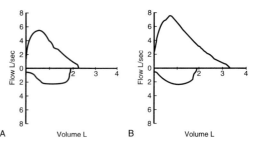

Figure 27–2 An example of a flow volume loop with **(B)** and without **(A)** functional magnetic stimulation (FMS). The forced expired flow generated by FMS was 7.65 L/s, which was 138% of forced expiratory flow produced at the end of tidal volume (V_T). The stimulation parameters were 70% intensity, 20-Hz frequency, and a 2-second burst length.

and pressure (-10.0 ± 1.0 cm H_2O) were observed in the C6 to C7 MC placement. Maximal mean expired volume and pressure occurred when the MC was placed at T9 to T10 (-199.0 ± 22.5 mL and 11.0 ± 2.3 cm H_2O, respectively). The stimulation profile was a 30-Hz frequency, 1-second burst length, and 70% intensity. These animal studies demonstrated the ability of FMS to generate inspired or expired volumes and

Like other respiratory muscle exercise training techniques such as diaphragmatic breathing, lateral costal breathing, and inspiratory resistive breathing, FMS can be used as a device for conditioning of the intercostal, diaphragm, and the abdominal muscles. A 4-week expiratory muscle conditioning study in patients with SCI resulted in significant improvement in voluntary maximal expired pressure (116%), volume (173%), and flow rate (123%) compared with their baseline data.[48] The FMS conditioning program was 20 minutes twice per day and 5 days per week.

Functional Magnetic Ventilation

Phrenic pacers have been used for patients with high cervical SCI for maintaining ventilatory support. Although the application of mechanical ventilation has reduced death rates of patients in the intensive care unit, mechanical ventilation accelerates disuse of respiratory muscles and leads to difficulty in weaning from mechanical ventilation. FES of the phrenic nerves involves placement of electrodes directly on the phrenic nerves in the neck or thorax[49–51] or placement of intramuscular electrodes in the diaphragm[52–55] using a laparoscopic procedure. Upper intercostal muscle pacing by electric ventral root stimulation facilitates inspiratory function in subjects with partial phrenic function.[56,57] FES was also applied to abdominal muscles as a method of enhancing ventilation in human subjects with SCI.[58] The previous FES methods are invasive, require surgery to implant electrodes, and expose patients to the risks associated with chronic implants, such as infection and hemorrhage.[59] Functional magnetic ventilation (FMV) is a newer mode of noninvasive, negative-pressure ventilation that employs a magnetic stimulator. This mode of ventilation has similarities to electrophrenic pacing, in that the phrenic nerves are electrically stimulated without surgical interventions.

FMS was used for ventilation in an animal study for a 2-hour period after complete C2 spinal cord transaction. This study reported that the average tidal volume (V_T) produced during FMV was 0.35 ± 0.05 L, with a maximum of 0.8 L and a minimum of 0.2 L. Figure 27–3 shows the V_T and tracheal pressure (P_{tr}) during spontaneous breathing of a dog and the pattern during FMV. The V_T and P_{tr} over time during FMV are shown in Figure 27–4. The mean V_T increased from 0.31 to 0.49 L at 15 minutes and fluctuated between 0.49 and 0.43 L until the termination

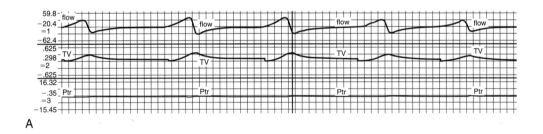

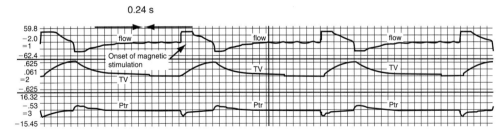

Figure 27–3 Data acquisition samples of expired ventilatory flow (L/min), tidal volume (V_T) in liters (L), and tracheal pressure P_{tr} (cm H_2O). **A:** Spontaneous breathing. **B:** Functional magnetic ventilation.

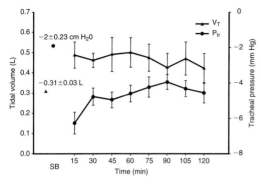

Figure 27–4 Changes of tidal volume and tracheal pressure versus time (minutes) during functional magnetic ventilation. SB, spontaneous breathing.

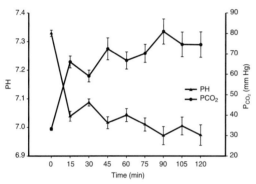

Figure 27–5 Changes of pH and P_{CO_2} (mm Hg) versus time during functional magnetic ventilation.

of FMV. At the end of FMV, the mean V_T (0.42 L) produced by FMV was higher than the mean spontaneous V_T of 0.31 L. The average P_{tr} was -6.30 ± 1.51 cm H_2O after 15 minutes of FMV, which is more negative than the spontaneous P_{tr} of -2.00 ± 0.23 cm H_2O. After the first 15 minutes, the P_{tr} ranged from -3.95 to -4.78 cm H_2O for the remaining 2 hours of FMV. The baseline mean arterial blood P_{CO_2} of the dogs was 33.20 ± 0.71 mm Hg. At 15 minutes of FMV, mean P_{CO_2} increased to 66.0 ± 2.9 mm Hg. For the rest of the 2-hour period of FMV, the mean P_{CO_2} fluctuated from a low of 59.1 ± 2.8 mm Hg at 30 minutes to a high of 80.9 ± 8.1 mm Hg at 90 minutes and decreased to 74.5 ± 6.2 mm Hg at the end of FMV (Fig. 27–5). The changes in mean pH, as expected, reflected the corresponding increases and decreases in arterial P_{CO_2}. The mean pH changed from a baseline level of 7.33 ± 0.01 to 6.99 ± 0.03 during 2 hours of FMV. The corresponding rise in the K^+ level probably occurred because of an extracellular shift of K^+ as the pH decreased. The modified tension-time index dropped from 0.17 to 0.14 ± 0.05 during the same period, suggesting that FMV did not induce inspiratory muscle fatigue. The only chemistry values that demonstrated significant changes over the 2-hour period were creatine kinase (CK) and K^+.

This study demonstrated a new mode of noninvasive negative pressure ventilation using FMS. It can also be used for inspiratory muscle conditioning for patients with SCI, similar to the expiratory muscle conditioning application.[48] FMV was achieved for 2 hours in dogs with C2 spinal cord transection. Sufficient V_Ts with elevated P_{CO_2} were produced during magnetic ventilation processes to maintain life. Although FMV is only in its infancy, with further technical improvements, this novel application may prove to be a practical clinical tool for patients needing ventilatory support.

Functional Magnetic Stimulation of the Gastrointestinal Tract

Neurogenic bowel is a common source of morbidity and mortality among patients with acute and chronic SCI. Signs and symptoms include fecal impaction,[60] constipation, abdominal distention,[61,62] prolonged bowel care,[61–63] and delayed colonic transit.[64–68] The prevalence of chronic gastrointestinal symptoms also appears to increase with time after a SCI.[62] Bowel management for patients with SCI includes a balanced diet, judicial use of oral and rectal medications, and digital stimulation.[69] Some SCI patients with long transit times find bowel care to be complicated, laborious, and difficult to accomplish. These patients often spend several hours each day emptying their colons, which decreases their quality of life and independence. For patients with very difficult bowel evacuation or severe gastrointestinal dysfunction, colostomies have simplified their care.[62,70]

Placing the MC at the lumbosacral region, FMS has been shown to activate the colon and improve colonic transit and gastric emptying in human[3] and animal studies.[71,72] In one study of patients with

SCI,[2] significant changes in rectal pressure and colonic transit time due to FMS were observed. Rectal pressures increased from 26.7 ± 7.4 to 48.0 ± 9.9 cm H_2O with lumbosacral stimulation and from 30.0 ± 6.4 to 42.7 ± 7.9 cm H_2O with transabdominal stimulation. With FMS, the mean colonic transit time decreased from 105.2 hours to 89.4 hours. The rectosigmoid transit times decreased from 50.4 ± 9.8 to 34.8 ± 9.4 hours. The important findings in this study are that FMS can activate the colon and improve colonic transit. FMS of the colon can be performed by placing the MC at the lower abdomen for transabdominal stimulation or by placing the MC at the lumbosacral region for sacral nerve stimulation. With transabdominal stimulation, tensing of the abdominal musculature was observed along with an increase in rectal pressure. With lumbosacral stimulation, a significant elevation of rectal pressure was observed.

When placing the MC at a higher level (T9), FMS enhanced gastric emptying in able-bodied subjects and SCI patients.[73] The gastric emptying half-time ($GE_{t1/2}$) was shortened more than 8% and 33% for able-bodied and SCI subjects, respectively. The $GE_{t1/2}$ of gastric emptying with FMS was accelerated by approximately 38 minutes compared with the baseline values. One possible mechanism is that FMS functions similarly to FES to accelerate gastric emptying through direct activation of the underlying gastric neuromuscular apparatus. Instead of stimulating at the stomach level, the thoracic spinal nerves are activated as they exit the neuroforamen.[6] FMS may be acting in a similar manner as gastric pacing, which uses electrodes implanted on the surface of the stomach muscle to cause rhythmic gastric contractions. Another postulate is related to the fact that rhythmic abdominal muscle contractions bring about improvement in gastric emptying. This is because the abdominal muscles (i.e., major expiratory agonists) were innervated by the lower thoracic spinal nerves.

Functional Magnetic Stimulation of the Bladder

The first clinical study investigating the efficacy of FMS of the bladder in SCI individuals was reported in 1994.[4] In that study, micturition by FMS was observed in 17 of 22 subjects. All subjects, except the one with a lower motoneuron lesion and areflexic bladder, responded well to FMS of the bladder. The mean changes in bladder pressure (P_{ves}) by sacral stimulation were 24.4 ± 4.9 cm H_2O. The sacral FMS produced a greater increase in P_{ves} than did suprapubic stimulation ($P < .05$). Higher intensities and frequencies of sacral stimulation were associated with greater changes in P_{ves}. Micturition by FMS was observed in 17 subjects. Induced micturition was reproducible by means of suprapubic or sacral FMS. Complete bladder emptying was observed in one subject using a water-cooled coil applied suprapubically. With 100 mL of water in the bladder and a sequence of intermittent stimulation that lasted 4.5 minutes, complete bladder emptying occurred. Other important findings included detrusor modulation by sacral FMS, external urethral sphincter muscle fatigue by suprapubic FMS, and the usefulness of an intermittent sacral FMS sequence to facilitate micturition. It was found that patients with reflex bladders were ideal candidates for FMS of the bladder and that patients with lower motoneuron lesions or flaccid bladders were not good candidates.

Micturition by FMS was observed in an earlier dog study.[5] The stimulation parameters were a 30-Hz frequency, 70% of maximal intensity, and 2-second burst length. The mean increase in ΔP_{ves} (change in bladder pressure) by suprapubic magnetic stimulation, 40.7 ± 8.1 cm H_2O, was significantly lower compared with lumbosacral stimulation, 68.0 ± 13.0 cm H_2O ($P < .05$). Voiding was achieved three in additional dogs in one protocol using the water-cooled coil. With the intermittent stimulation sequence applied to the lumbosacral spine, significant micturition could be observed repeatedly and reliably. This animal study demonstrated the effectiveness of FMS on micturition. Another study comparing sacral root FMS and FES was performed. FES was achieved at three locations along the sacral nerves or roots: proximal ventral sacral root using the Cortac platinum contact electrode near the L4 vertebra; distal right S2 root (intraforaminal) by using a bipolar hook electrode placed around the right S3 root at

L6 vertebra, 2 cm proximal to the foramen; and right S2 sacral nerve (extraforaminal), 2 cm distal to the neuroforamen. FMS was applied by placing the 9-cm MC at L3 to L5. The results showed that the latency for FMS when placing the MC between L3 and L5 was 1.8 ± 0.06 ms, which was exactly between the latencies obtained for extraforaminal and intraforaminal stimulation (1.70 ± 0.03 ms and 1.9 ± 0.23 ms) using hook electrodes placed along the right S2 sacral root or nerve. This corresponded with findings from earlier studies that predicted the foci of stimulation to be located at the foramen.[7] Moreover, the average amplitude obtained by FMS (1.4 ± 0.3 mV) was similar to that obtained from the contact electrode (1.5 ± 0.4 mV). These amplitudes were more than double those obtained by unilateral S2 stimulation. This illustrated that FMS could stimulate multiple sacral nerves simultaneously, producing similar compound muscle action potentials (CMAPs) as ventral sacral root stimulation. Similar levels of ΔP_{ves} and change in urethral pressure (ΔP_{ur}) produced by FMS and FES in this study confirmed this result.

Functional Magnetic Stimulation of the Lower Limb for Enhancing Fibrinolysis

FMS as a noninvasive and potentially more effective method of stimulating endogenous fibrinolysis was performed on 20 normal subjects.[74] This FMS study was designed to test changes in fibrinolysis in blood samples of patients taken before, 15 minutes after, and 60 minutes after FMS. The test for fibrinolysis was a modified dilute whole-blood clot lysis method and was performed on site. The subject was placed in the prone position, with the MC placed on the popliteal region, alternating from side to side every 30 seconds. The frequency of stimulation, burst length, and burst interval were 30 Hz, 4 seconds, and 26 seconds, respectively. Fibrinolysis was tested by the dilute whole-blood clot lysis time (WBCLT) and a modified semi-automated method by using a Sonoclot coagulation analyzer.

The following results illustrate the effect of the FMS protocol on fibrinolytic response for subjects. The mean WBCLT values decreased from 17.0 ± 1.3 hours at baseline to 12.0 ± 1.0 hours and 11.0 ± 0.8 hours at 10 minutes and 60 minutes, respectively, after FMS. The 10-minute and 60-minute post-FMS values were significantly different from the pre-FMS lysis time measurements ($P = .0001$). The degree of fibrinolytic enhancement at 10 and 60 minutes after FMS was not uniform in all individuals. This study demonstrated that FMS of the leg muscles could produce enhancement of systemic fibrinolysis that exceeded the increase in fibrinolytic activity accounted for by changes in posture and the influence of the circadian rhythm. This enhancement is apparently sustained for at least 60 minutes after cessation of the stimulation protocol. Using similar protocol on 22 patients with SCI, preliminary results show that the WBCLT values at 10 minutes after FMS (15.0 ± 1.6 hours) and 60 minutes after FMS (16.0 ± 1.5 hours) were significantly decreased from before FMS (19.9 ± 1.2 hours; $P < .001$). The values of WBCLT at 10 minutes and 60 minutes after FMS did not change significantly ($P > 0.45$). These results show that FMS provides a more rapid enhancement of systemic fibrinolysis than modalities such as FES and a more sustained enhancement than voluntary exercise. It could be an excellent tool for use when voluntary exercise cannot be performed, such as in patients with SCI, multiple sclerosis, stroke, central nervous system tumors, or spinal dysraphias.

Functional Magnetic Stimulation–Induced Analgesia

Analgesic medications can provide transient pain relief, but the chronic use of opioid and nonopioid analgesics is associated with a plethora of limiting side effects. Increasingly, patients are turning to "alternative" pain treatments, including nonpharmacologic analgesic modalities such as electrical stimulation. Electric stimulation–induced analgesia has been widely used in the management of chronic pain for more than 30 years. Transcutaneous electrical nerve stimulation (TENS), percutaneous electrical nerve stimulation (PENS), and electroacupuncture (EA) have been used for the treatment of chronic musculoskeletal and neuropathic pain. Spinal cord stimulation (SCS), a more invasive procedure requiring the permanent implantation of an epidural electrical stimulator, has been widely used for the treatment of patients with severe and intractable chronic pain. Clinical data suggest that TENS is the least effective form of electrical stimulation analgesia,[75–77] PENS is more effective,[78,79] and SCS is most effective.[80,81] The invasive electrical stimulation techniques such as PENS, EA, and SCS can stimulate deeper nerves or spinal cord neurons, but they require needle electrodes to overcome skin impedance or surgical implantation of stimulation systems, which are prone to break down.

An earlier study investigated the antinociceptive effect of FMS applied over the lumbosacral spine of rats.[82] The effects of FMS on acute mechanical and thermal hind paw withdrawal thresholds were determined and neuroablative and pharmacologic manipulations were used to further characterize the mechanism and anatomic site of FMS antinociceptive action. Immediately after 5 minutes of FMS, the mechanical withdrawal threshold increased from 66 ± 3 g to 160 ± 10 g, and the heat withdrawal threshold increased from $50.1 \pm 0.1°C$ to $51.7 \pm 0.2°C$. The frequency of stimulation, burst length, and burst interval were 20 Hz, 5 seconds, and 15 seconds, respectively. The FMS effect gradually diminished, lasting 30 minutes for mechanical antinociception and 40 minutes for heat antinociception (Fig. 27–6). Sham-treated rats had no significant changes over time for mechanical or heat withdrawal thresholds. When the nociceptive thresholds of the FMS treated rats were compared with the sham-treated rats, groups differed significantly over time ($P < .01$, for both mechanical and heat, two-way repeated measures ANOVA), which persisted 30 minutes after FMS for mechanical and heat antinociceptive effects.

To determine the possible role of endogenous neurotransmitters in mediating FMS antinociceptive effects, attempts to block FMS antinociception were made by using the opiate receptor antagonist naloxone (5 mg/kg) and the α_2-adrenoceptor antagonist atipamezole (5 mg/kg). The drugs were administered by the intraperitoneal route (1 mL/kg). Systemic naloxone completely blocked FMS evoked antinociceptive effects,

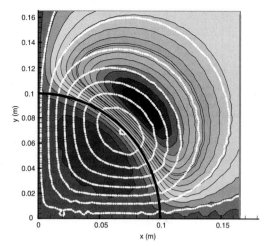

Figure 27–6 All antinociceptive effects are presented as the percentage of the maximum possible effect (%MPE), which is equal to (post-treatment threshold – pretreatment threshold)/(cutoff threshold – pretreatment threshold) × 100. The cutoff thresholds were 52°C for heat nociception and 170 g of force for mechanical nociception. Immediately after 5 minutes of lumbosacral magnetic stimulation, there was a robust antinociceptive effect for mechanical ($92 \pm 8\%$ MPE) and heat ($82 \pm 10\%$ MPE) stimuli applied to the hind paw (n = 10). This effect gradually diminished, persisting for 30 minutes for mechanical antinociception and 40 minutes for heat antinociception. $*P < .05$, $**P < .01$, $***P < .001$ versus baseline thresholds.

indicating that FMS activates an endogenous opioidergic analgesic mechanism to induce antinociception. When naloxone (5 mg/kg, given intraperitoneally) was injected 25 minutes before FMS, the FMS treatment had no effect on mechanical and heat thresholds (Fig. 27–7). Saline-injected rats had FMS antinociceptive responses similar to naive animals, with elevated withdrawal thresholds persisting for 30 minutes after stimulation. Control rats injected with naloxone (5 mg/kg, given intraperitoneally) and not treated with FMS had stable preinjection and postinjection withdrawal thresholds over a 60-minute period, indicating that this dose of naloxone did not affect nociceptive thresholds. Naloxone completely antagonized FMS antinociception, indicating that an endogenous opiate mechanism mediates this effect (Fig. 27–8).

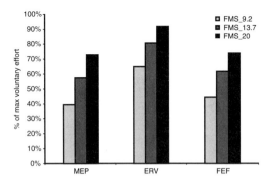

Figure 27–7 Comparison of coil performance (9.2, 13.7, and 20.0 cm) in maximum expiratory pressure, expiratory reserved volume, and forced expiratory flow. The average results produced by functional magnetic stimulation in normal subjects are expressed in percentage of their maximal voluntary efforts.

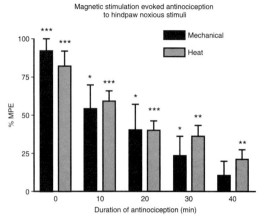

Figure 27–8 The opiate receptor antagonist naloxone (5 mg/kg, given intraperitoneally) completely blocked the antinociceptive effect of functional magnetic stimulation (n = 9).

Systemic atipamezole did not block the antinociceptive effect of FMS, suggesting that endogenous adrenergic analgesic mechanisms do not play a major role in FMS antinociception. When atipamezole (5 mg/kg, given intraperitoneally) was injected 25 minutes before FMS, the FMS treatment still evoked long-lasting mechanical (10 minutes) and heat (30 minutes) antinociception. There was no significant difference in the mechanical nociceptive thresholds between the atipamezole and saline treatment groups over time, but there was a significant difference between these treatment groups for heat stimuli ($P < .05$). The saline treated rats had a greater FMS-evoked heat antinociceptive effect than the atipamezole treated rats, and this difference persisted for 20 minutes after FMS. These data suggest that the FMS heat antinociceptive effect may be partially mediated by activation of the α_2-adrenoceptor. Control rats injected with atipamezole (5 mg/kg, given intraperitoneally) and not treated with FMS had stable preinjection and postinjection withdrawal thresholds over a 60-minute period, indicating that this dose of atipamezole did not affect nociceptive thresholds. The attempt to reverse FMS-evoked antinociception with the α_2-adrenoceptor antagonist atipamezole resulted only in a partial reversal of FMS antinociception for heat stimuli and no effect on FMS antinociception for mechanical stimuli. These results suggest that endogenous noradrenergic mechanisms do not play major roles in FMS antinociception and that FMS uses different analgesic mechanisms than those mediating stress analgesia.

This study demonstrated that 5 minutes of FMS over the lumbosacral spine at an intensity high enough to evoke a slight motor response in the tail could induce a long-lasting, robust antinociceptive effect that resembled the antinociception produced by EA, requiring endogenous opioidergic mediation. These results indicate that FMS can provide a transcutaneous method of evoking action potentials in the deeper nerves or spinal cord neurons, resulting in analgesia. The magnitude and duration of this analgesic effect is comparable to that reported in other studies measuring electroanalgesia to acute nociceptive stimuli in animals and humans.[83–88] Further animal and clinical investigations are required to determine the optimal treatment parameters for FMS evoked analgesia and to determine whether FMS is more effective than transcutaneous, percutaneous, or spinal electrical stimulation in analgesia trials.

Technical Considerations for Functional Magnetic Stimulation

Studies of the Respiratory Muscles

MC design is one of the most important aspects of FMS technology for clinical applications. In some applications, very focused stimulation is required,[89] whereas in other applications, stimulation that is more general is used to cover multiple nerves or a region of excitable neuromuscular tissue. It is imperative that a properly designed coil is used for any clinical application. The design approach is generally conducted in the following steps: computer modeling, experimental measurements, and animal or human trials. The following sections provide examples of coil design approaches that we have undertaken to meet the needs of some clinical applications.

Magnetic Coil Design Approach. Computer modeling is a common approach to begin the MC design process. Computer models can be built to evaluate the qualitative distributions of induced electric field strength (**E**) and the first spatial derivative of electric field, $-\delta E_y/\delta y$ (y is the orientation of nerve), generated by various coil designs; $-\delta E_y/\delta y$ is derived as a function of nerve activation for a long straight axon.[90] A proper computer model can effectively predict the coil's performance and therefore shortens the time for coil design. After a suitable coil has been determined by computer modeling, prototype coils can then be built. This is followed by the experimental evaluations of the prototype coils to validate the results of computer models. Animal or human clinical trials can be performed once improved coils are constructed.

Computer Modeling of Magnetic Coils. The magnetic field and the induced electric field generated from the MC are dictated by Maxwell's equations. To arrive at solutions for the electric field component E_y induced by magnetic stimulation, Maxwell's equations need to be simplified and solved. The details are described in Appendix 27.1.

Experimental E Measurements. The induced **E** in the proximity of the MC needs to be systematically mapped for all MCs. The detailed experimental setup and procedures are listed in Appendix 27.2.

Focal Magnetic Stimulation

The most common commercially available coils are circular units with diameters of 9 to 14 cm. The advantage of this geometry is that they can be easily constructed and can be applied in many anatomical regions. Because of the size and the nonfocal nature of this coil design, adjacent tissues may be inadvertently stimulated. An essential aspect of MC design is the identification of the area with respect to the coil where nerves are more likely to be depolarized. To improve the focality of magnetic stimulation, the planar double or butterfly coil was introduced and became the first major improvement over the conventional planar circular coil. Double coils consist of two windings usually placed side by side. Small-sized double coils are much more selective than conventional round coils.

Double coils were followed by the planar four-leaf coil.[91] The four-leaf coils further focalized the stimulus at the center of the coil and reduced the chances of peripheral stimulation. These coils often become too cumbersome to maneuver conveniently. A three-dimensional slinky coil design has shown even greater promise.[92–94] Let N*I be the total current in the central leg of the coil and let N be the number of loops. By unfolding N/k turns of this coil k times successively and spatially locating them at successive angles of i*180/(k − 1) degrees, where i = k − 1, k − 2, ..., 1, 0, we obtain the slinky-k coil. The center limb of the coil still carries the total current N*I, whereas each return limb now carries a current N*I/k. Under this definition, the circular or rectangular coil was considered a slinky-1 coil, and the butterfly (or figure-of-eight) coil was considered a slinky-2 coil. The work on the three-dimensional slinky MC design using computer simulations and experimental studies has demonstrated improved focality for the field distributions. The results showed that the slinky coil design produced more focalized stimulation when compared with the planar round coils. The primary-to-secondary peak ratios of the induced electric field from slinky-1 to -5 were 1.00, 2.20, 2.85, 2.62, and 3.54, with slinky-1 and -2 being equivalent to the round and double coils, respectively.[94]

Coil Design for Functional Magnetic Stimulation of Expiratory Muscles

The principal aim of our coil design was to optimize expiratory function produced by FMS. This was carried out by investigating different sizes and winding configurations of MCs. The size of the coil was a factor in determining the range of stimulation; larger coils could certainly cover a larger area of stimulation. The other factor, the winding configuration, governed the pattern of stimulation. Winding configuration refers to the way in which the copper cable is wound in the coils. Conventional round coils usually have wire tightly wound and placed in a narrow space, resulting in concentrated distributions of nerve activation function $(-\partial Ey/\partial y)$ along the outer edge of the coils[91,95] and very low activation function inside the coils as illustrated in Figure 27–9. To smooth the nerve activation function within the coil area, an option was spiral winding, which winds the wire gradually and evenly from the center to the circumference of the coils. Six coils of three sizes and two winding structures—round 9.2, 13.7, and 20.0 cm and spiral 9.2, 13.7, and 20.0 cm—were modeled. The current I in the coils was set randomly at a magnitude of 1,000 A and frequency of 4,000 Hz in all the computer modeling.

The computer modeling results showed that the $-\delta Ey/\delta y$ for the three spiral coils had a peak of −91.0, −95.5, and −90.4 V/m² at distances of 4.8, 6.7, and 9.5 cm from the center, respectively, with each coil along a 45-degree line of the x axis. The $-\delta Ey/\delta y$ peaks were −156.6, −174.6, and −177 V/m² at 6.7, 9.0, and 12.9 cm from the center of the same line for the round coils. Compared with the spiral coils, the round coils generated higher $-\delta Ey/\delta y$; however, they also decreased from their peaks much faster than the spiral coils (see Fig. 27–8). As a result, the $-\delta Ey/\delta y$ of the round coils, for the most part, were limited outside of the coils leaving a low $-\delta Ey/\delta y$ zone inside the coils; on the other hand, the $-\delta Ey/\delta y$ of spiral coils distributed more smoothly inside and outside of the coils. E measurements were plotted for three MCs: round 9.2-cm, spiral 13.7-cm, and spiral 20-cm

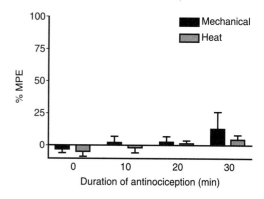

Figure 27–9 Contour plots of nerve activation function strength comparing a spiral 20-cm coil (dotted bars) and a round 20-cm coil (solid bars).

coils. All three coils had zero field intensity at the center of the coils (x = 0.0 cm), and the field intensity increased to a peak of 29.7, 50.1, and 65.3 V/m at x = 4.0, 4.0, and 7.0 cm, respectively. The locations of the peaks were inside the spiral coils rather than at the outer edges of the coils (6.8 and 10.0 cm, respectively), as observed with the round coils.

The mean maximal expired pressure FMS (MEP-FMS) produced by the round 9.2-, spiral 13.7-, and spiral 20-cm coils were 52.0 ± 17.2, 75.0 ± 15.8, and 96.0 ± 8.4 cm H₂O in five human subjects, which were 40%, 58%, and 73% of their maximal voluntary efforts (see Fig. 27–9). The mean maximal expired reserve volume FMS (ERV-FMS) were 0.90 ± 0.22, 1.10 ± 0.23, and 1.20 ± 0.27 L for the 9.2-, 13.7-, and 20-cm coils, respectively, which were 65%, 81%, and 92% of their maximal voluntary efforts. The mean maximal forced expiratory flow FMS (FEF-FMS) were 2.9 ± 0.5, 4.0 ± 0.3, and 4.8 ± 0.3 L/s for the 9.2-, 13.7-, and 20-cm coils, respectively, which were 44%, 62%, and 74% of their maximal voluntary efforts. Under the same stimulation parameters, the 20-cm spiral coil produced better expiratory function results than those of the 13.7- or 9.2-cm coils, as seen in the placement profile, for averages of 15% and 37% higher MEP, respectively.

Coil Design for Functional Magnetic Stimulation of Upper Intercostal Muscles

The goals of this design effort were to simultaneously stimulate the T1 to T6 spinal nerves that innervate the upper intercostal muscles and to minimally activate other nerves and muscles. A racetrack-shaped coil, with a longitudinal span covering T1 to T6, was designed to minimize the activations of unwanted nerves. The length of the racetrack coil was set at 16 cm, which is about the average length of T1 to T6 in adult humans. For the width, we used 8 cm for this racetrack coil design, with considerations that the averaged width between two neuroforamina at the same level in an adult human is 4 to 6 cm at the thoracic level.

Two coils were used for computer modeling, the racetrack coil (8×16 cm) and the spiral coil (12.5 cm). The computer simulation results showed that the 12.5-cm spiral coil mapped out a single phase with a smooth $-\delta E_y/\delta y$ distribution within and outside the perimeter of the coils. The peak of $-\delta E_y/\delta y$ was located at a distance of the coil's radius along the 45-degree line with the x axis, and $-\delta E_y/\delta y$ decreased gradually from its maximum to 0 toward the x and y axes. The wide $-\delta E_y/\delta y$ distribution profile demonstrated that this coil could potentially activate muscle groups. With a dimension of 8×16 cm, the $-\delta E_y/\delta y$ distribution profile of the racetrack coil was narrower, and this coil could stimulate all (T1 to T6) spinal nerves and avoid activating other muscles (e.g., shoulder muscles). The racetrack coil shape also allowed for more uniform stimulation across the covered area, resulting in better patient tolerance.

■ Conclusion

FMS as a technology has evolved from being a diagnostic tool to a therapeutic tool with many clinical applications. With the improvements in the magnetic stimulator technology and the MC designs, repetitive magnetic stimulation can result in useful bodily function, or what is defined as FMS. This chapter has focused on the potential clinical applications of FMS for patients with SCI, with an emphasis on the stimulation of the respiratory muscles. FMS was effective for restoring cough in patients with SCI and was able to provide adequate ventilatory support for longer than 2 hours in animals with high cervical injury. Improvements in MC designs were made for respiratory muscle stimulation. Additional clinical applications of FMS for SCI include FMS of the gastrointestinal tract, bladder, and the lower limbs. FMS was effective for enhancing gastric emptying and colonic transits in patients with SCI. Similarly, FMS enhanced bladder emptying in SCI, improved fibrinolysis in able-bodied and SCI subjects, and induced analgesia through an endogenous opiate mechanism.

We predict that the applications of FMS for patients with SCI and other neurologically impaired individuals will grow significantly in the future as the technology continues to advance. Portable FMS units will be made available for patients to be used at home or on the road. These advances will improve health and quality of life for patients with disabilities.

Acknowledgments

The authors would like to thank Stephen F. Figoni, PhD, RKT, for his editorial assistance in preparing the manuscript.

REFERENCES

1. Lin VWH, Hsiao I. Functional magnetic stimulation. In Lin V (ed): Spinal Cord Medicine: Principles and Practice. New York, Demos Medical Publishing, 2003:749–764.
2. Lin VWH, Singh R, Chitkara RK, et al. Functional magnetic Stimulation for restoring cough in patients with tetraplegia. Arch Phys Med Rehabil 1998;79:517–522.
3. Lin VWH, Nino-Murcia M, Frost F, et al. Functional magnetic stimulation of the colon in patients with spinal cord injuries. Arch Phys Med Rehabil 2001;82:167–173.
4. Lin VWH, Wolfe V, Perkash I. Micturition by functional magnetic stimulation. J Spinal Cord Med 1997;20:218–226.
5. Lin VWH, Hsiao I, Perkash I. Micturition by functional magnetic stimulation in dogs—a preliminary report. J Neurourol Urodyn 1997;16:305–314.
6. Singh H, Magruder M, Bushnik T, et al. Expiratory muscle activation by functional magnetic stimulation of thoracic and lumbar spinal nerves. Crit Care Med 1999;27:2201–2205.

7. Maccabee PJ, Amassian VE, Eberle L, et al. Measurement of the electric field induced into inhomogeneous volume conductors by magnetic coil: applications to human spinal geometry. Electroencephalogr Clin Neurophysiol 1991;81:224–237.
8. Lemons VR, Wagner FC Jr. Respiratory complications after cervical spinal cord injury. Spine 1994;19:2315–20.
9. Fishburn MJ, Marino RJ, Ditunno JF. Atelectasis and pneumonia in acute spinal cord injury. Arch Phys Med Rehabil 1990;71:197–200.
10. McMichan JC, Michel L, Westbrook PR. Pulmonary dysfunction following traumatic quadriplegia. JAMA 1980;243:528–531.
11. Roth EJ, Nussbaum SB, Berkowitz M, et al. Pulmonary function testing in spinal cord injury: correlation with vital capacity. Paraplegia 1995;33:454–457.
12. Hemingway A, Bors E, Hobby RP. An investigation of the pulmonary function in paraplegics. J Clin Invest 1958;37:773–782.
13. James WS, Minh V, Minteer A, et al. Cervical accessory respiratory muscles: function in a patient with a high cervical cord lesion. Chest 1977;71:59–64.
14. Gausch PA, Linder SH, Williams T, et al. A functional classification of respiratory compromise in spinal cord injury. Sci Nursing 1991;8:4–10.
15. Ledsome JR, Sharp JM. Pulmonary function in acute cervical cord injury. Am Rev Respir Dis 1981;124:41–44.
16. Anke A, Aksnes AK, Stanghelle JK, et al. Lung volumes in tetraplegic patients according to cervical spinal cord injury level. Scan J Rehabil Med 1993;25:73–77.
17. Hachen HJ. Idealized care of the acutely injured spinal cord in Switzerland. J Trauma 1977;17:931–936.
18. O'Donohue WJ Jr, Baker JP, Bell GM, et al. Respiratory failure in neuromuscular disease. Management in a respiratory intensive care unit. JAMA 1976;235:733–735.
19. Connors GL, Hilling L. Prevention, not just treatment [review]. Respir Care Clin North Am 1998;4:1–12.
20. Braun SR, Giovannoni R, O'Connor M. Improving the cough in patients with spinalcord injury. Am J Phys Med 1984;63:1–10.
21. Chiappetta A, Bakerman R. High-frequency chest-wall oscillation in spinal muscularatrophy. J Respir Care Pract 1995;8:112–114.
22. Dail CAW, Afield JE, Collier CR. Clinical aspects of glossopharyngeal breathing. JAMA 1955;158:445–469.
23. Gross D, Ladd HW, Riley EJ, et al. The effect of training on strength and endurance of the diaphragm in quadriplegia. Am J Med 1980;68:27–35.
24. Dimarco AF, Altose MD, Cropp A. Activation of the inspiratory intercostal muscles by electrical stimulation of the spinal cord. Am Rev Respir Dis 1987;136:1385–1390.
25. Dimarco AF, Budzinska K, Supinski G. Artificial ventilation by means of electrical activation of the intercostal/accessory muscles alone. Am Rev Respir Dis 1989;139:961–967.
26. Dimarco AF, Kovvuri S, Redtro J, et al. Intercostal muscle pacing in quadriplegic patients [abstract]. Am Rev Respir Dis 1991;143:A473.
27. Lin VH, Romaniuk JR, Dimarco A. Functional magnetic stimulation of the respiratory muscles in dogs. Muscle Nerve 1998;21:1048–1057.
28. Linder SH. Functional electrical stimulation to enhance cough in quadriplegia. Chest 1993;103:166–169.
29. Jaeger RJ, Langbein EW, Kralj AR. Augmenting cough by FES in tetraplegia: a comparison of results at three clinical centers. Basic Appl Myol 1994;4:195–200.
30. Glenn WW, Holcomb WG, Hogan J, et al. Diaphragm pacing by radiofrequency transmission in the treatment of chronic ventilatory insufficiency. Present status. J Thorac Cardiovasc Surg 1973;66:505–520.
31. Glenn WWL. The treatment of respiratory paralysis by diaphragm pacing. Ann Thorac Surg 1980;30:106–109.
32. Campbell EMJ. The role of scalene and sternomastoid muscles in breathing in normal subjects: An electromyographic study. J Anat 1955;89:378–386.
33. Taylor A. The contribution of the intercostal muscles to the effect of respiration in man. J Physiol 1960;151:390–402.
34. DiMarco AF, Supinski GS, Petro JA, et al. Evaluation of intercostal pacing to provide artificial ventilation in quadriplegics. Am J Respir Crit Care Med 1994;150:934–940.
35. Stani U, Kandare F, Jaeger R, et al. Functional electrical stimulation of abdominal muscles to augment tidal volume in spinal cord injury. IEEE Trans Rehabil Eng 2000;8:30–34.
36. DiMarco AF, Romaniuk JR, Supinski GS. Electrical activation of the expiratory muscles to restore cough. Am J Respir Crit Care Med 1995;151:1466–1471.
37. Jaeger RJ, Turba RM, Yarkony GM, et al. Cough in spinal cord injured patients: comparison of three methods to produce cough. Arch Phys Med Rehabil 1993;74:1358–1361.
38. Lissens MA. Motor evoked potentials of the human diaphragm elicited through magnetic

transcranial brain stimulation. J Neurol Sci 1994;124:204–207.
39. Chokroverty S, Deutsch A, Guha C, et al. Thoracic spinal nerve and root conduction: a magnetic stimulation study. Muscle Nerve 1995;18:987–991.
40. Lissens MA, Vanderstraeten GG. Motor evoked potentials of the human respiratory muscles. Muscle Nerve 1996;19:1204–1205.
41. Similowski T, Fleury B, Launois S, et al. Cervical magnetic stimulation: a new painless method for bilateral phrenic nerve stimulation in conscious humans. J Appl Physiol 1989;67:1311–1318.
42. Wragg S, Aquiline R, Moran J, et al. Comparison of cervical magnetic stimulation and bilateral percutaneous electrical stimulation of the phrenic nerves in normal subjects. Eur Respir 1994;7:1778–1792.
43. Mills GH, Kyroussis D, Hamnegard CH, et al. Bilateral magnetic stimulation of the phrenic nerves from an anterolateral approach. Am J Respir Crit Care Med 1996;154:1099–1105.
44. Laghi F, Harrison MJ, Tobin MJ. Comparison of magnetic and electrical phrenic nerve stimulation in assessment of diaphragmatic contractility. J Appl Physiol 1996;80:1731–1742.
45. Lin VWH, Romaniuk JR, Supinski GS, et al. Inspiratory volume production via direct intercostal muscle stimulation. Muscle Nerve 1992;15:1088.
46. Lin VWH, Romaniuk JR, Allen C, et al. High frequency magnetic stimulation of the inspiratory muscles. Muscle Nerve 1993;16:1088.
47. Lin VWH, Hsieh C, Hsiao IN, et al. Functional magnetic stimulation for cough [abstract]. Am J Respir Crit Care Med 1997;155:A917.
48. Lin VWH, Hsiao I, Zhu E. Functional magnetic stimulation for conditioning respiratory muscles. Arch Phys Med Rehabil 2001;82:162–166.
49. Glenn WW, Hogan JF, Phelps ML. Ventilatory support of the quadriplegic patient with respiratory paralysis by diaphragm pacing. Surg Clin North Am 1980;60:1055–1078.
50. Garrido-Garcia H, Martin-Escribano P, Palomera-Frade J, et al. Transdiaphragmatic pressure in quadriplegic individuals ventilated by diaphragmatic pacemaker. Thorax 1996;51:420–423.
51. Glenn WWL, Phelps ML, Elefteriades JA, et al. Twenty years of experience in phrenic nerve stimulation to pace the diaphragm. Pacing Clin Electrophysiol 1986;9:780–783.
52. Peterson DK, Nochomovitz M, DiMarco AF, et al. Intramuscular electrical activation of the phrenic nerve. IEEE Trans Biomed Eng 1986;33:342–351.
53. Nochomovitz ML, DiMarco AF, Mortimer JT, et al. Diaphragm activation with intramuscular stimulation in dogs. Am Rev Respir Dis 1983;127:325–329.
54. Peterson DK, Nochomovitz ML, Stellato TA, et al. Long-term intramuscular electrical activation of the phrenic nerve: efficacy as a ventilatory prosthesis. IEEE Trans Biomed Eng 1994;41:1127–1135.
55. DiMarco AF, Onders RP, Kowalski KE, et al. Phrenic nerve pacing in a tetraplegic patient via intramuscular diaphragm electrodes. Am J Respir Crit Care Med 2002;166:1604–1606.
56. DiMarco AF, Budzinska K, Supinski GS. Artificial ventilation by means of electrical activation of the intercostal/accessory muscles alone in anesthetized dogs. Am Rev Respir Dis 1989;139:961–967.
57. Dimarco AF, Kovvuri S, Pedtro J, et al. Intercostal muscle pacing in quadriplegic patients [abstract]. Am Rev Respir Dis 1991;143:A473.
58. Stanic U, Kandare F, Jaeger R, et al. Functional electrical stimulation of abdominal muscles to augment tidal volume in spinal cord injury. IEEE Trans Rehabil Eng 2000;8:30–34.
59. Glenn WW. Diaphragm pacing: present status. Pacing Clin Electrophysiol 1978;1:357–370.
60. Gore RM, Mintzer RA, Calenoff L. Gastrointestinal complications of spinal cord injury. Spine 1981;6:538–544.
61. Harari D, Sarkarati M, Gurwitz JH, et al. Constipation-related symptoms and bowel program concerning individuals with spinal cord injury. Spinal Cord 1997;35:394–401.
62. Stone JM, Nino-Murcia M, Wolfe VA, et al. Chronic gastrointestinal in spinal cord injury patients: a prospective analysis. Am J Gastroenterol 1990;85:114–119.
63. Glickman S, Kamm MA. Bowel dysfunction in spinal cord injury patients. Lancet 1996;347:1651–1653.
64. Menardo G, Bausano G, Corazziari E, et al. Large-bowel transit in patients with paraplegia. Dis Colon Rectum 1987;30:924–928.
65. Nino-Murcia M, Stone JM, Chang PJ, et al. Colonic transit in spinal cord-injured patients. Invest Radiol 1990;25:109–112.
66. Keshavarzian A, Barnes WE, Bruninga K, et al. Delayed colonic transit in spinal cord-injured patients measured by indium-111 Amberlite scintigraphy. Am J Gastroenterol 1995;90:1295–1300.
67. Beuret-Blanquart F, Weber J, Gouverneur JP, et al. Colonic transit time and anorectal manometric anomalies in 19 patients with complete transection of the spinal cord. J Auton Nerv Syst 1990;30:199–207.
68. Menardo G, Bausano G, Corazziari E, et al. Large-bowel transit in paraplegic patients. Dis Colon Rectum 1987;30:924–928.

69. Stein RB, Gordon T, Jefferson J, et al. Optimal stimulation of paralyzed muscle after human spinal cord injury. Am Physiol Soc 1992;72:1393–1400.
70. Kelly SR, Shashidharan M, Borwell, et al. The role of intestinal stoma in patients with spinal cord injury. Spinal Cord 1999;37:211–214.
71. Lin VWH, Hsiao I, Xu H, et al. Functional magnetic stimulation facilitates gastrointestinal transit in rats. Muscle Nerve 2000;23:919–1024.
72. Lin VWH, Hsiao I, Perkash I, et al. Functional magnetic stimulation facilitates colonic transit in rats. Arch Phys Med Rehabil 2001;82:969–972.
73. Lin VWH, Kim K, Hsiao I, et al. Functional magnetic stimulation facilitates gastric emptying. Arch Phys Med Rehabil 2002;83:806–810.
74. Lin VWH, Perkash A, Liu H, et al. Functional magnetic stimulation—a new modality for enhancing systemic fibrinolysis. Arch Phys Med Rehabil 1999;80:545–550.
75. Carroll C, Tramer M, McQuay H, et al. Transcutaneous electrical nerve stimulation in labour pain: a systematic review. Br J Obstet Gynaecol 1997;104:169–175.
76. Carroll D, Tramer M, McQuay H, et al. Randomization is important in studies with pain outcomes: systematic review of transcutaneous electrical nerve stimulation in acute postoperative pain. Br J Anaesth 1996;77:798–803.
77. Gadsby JG, Flowerdew MW. Transcutaneous electrical nerve stimulation and acupuncture-like transcutaneous electrical nerve stimulation for chronic low back pain. Cochrane Database Syst Rev 2000;2:D000210.
78. Ghoname EA, Craig WF, White PF, et al. Percutaneous electrical nerve stimulation for low back pain: a randomized crossover study. JAMA 1999;281:818–823.
79. Ghoname EA, White PF, Ahmed HE, et al. Percutaneous electrical nerve stimulation: an alternative to TENS in the management of sciatica. Pain 1999;83:193–199.
80. Long DM. The current status of electrical stimulation of the nervous system for the relief of chronic pain. Surg Neurol 1998;49:142–144.
81. Simpson BA. Spinal cord stimulation. Br J Neurosurg 1997;11:5–11.
82. Lin VWH, Kingery WS, Hsiao IN. Functional magnetic stimulation over the lumbosacral spine evokes a prolonged analgesic response in rats. Clin Neurophysiol 2002;113:1006–1012.
83. Danziger N, Rozenberg S, Bourgeois P, et al. Depressive effects of segmental and heterotopic application of transcutaneous electrical nerve stimulation and piezo-electric current on lower limb nociceptive flexion reflex in human subjects. Arch Phys Med Rehabil 1998;79:191–200.
84. Garrison DW, Foreman RD. Effects of transcutaneous electrical nerve stimulation (TENS) on spontaneous and noxiously evoked dorsal horn cell activity in cats with transected spinal cords. Neurosci Lett 1996;216:125–128.
85. McDowell BC, McCormack K, Walsh DM, et al. Comparative analgesic effects of H-wave therapy and transcutaneous electrical nerve stimulation on pain threshold in humans. Arch Phys Med Rehabil 1999;80:1001–1004.
86. Romita VV, Henry JL. Intense peripheral electrical stimulation differentially inhibits tail vs. limb withdrawal reflexes in the rat. Brain Res 1996;720:45–53.
87. Woolf CJ, Mitchell D, Barrett GD. Antinociceptive effect of peripheral segmental electrical stimulation in the rat. Pain 1980;8:237–252.
88. Ellis WV. Pain control using high-intensity pulsed magnetic stimulation. Bioelectromagnetics 1993;14:553–556.
89. Barker AT, Jalinous R, Freeston IL. Non-invasive magnetic stimulation of human motor cortex. Lancet 1985;1:1106–1107.
90. Roth BJ, Basser PJ. A model of the stimulation of a nerve fiber by electromagnetic induction. IEEE Trans Biomed Eng 1990;37:588–597.
91. Roth BJ, Maccabee PJ, Eberle LP, et al. In vitro evaluation of a 4-leaf coil design for magnetic stimulation of peripheral nerve. Electroencephalogr Clin Neurophysiol 1994;93:68–74.
92. Zimmermann KP, Simpson RK. Slinky coils for neuromagnetic stimulation. Electroencephalogr Clin Neurophysiol 1996;101:145–152.
93. Ren C, Tarjan PP, Popovic DB. A novel electric design for electromagnetic stimulation-the slinky coil. IEEE Trans Biomed Eng 1995;42:918–925.
94. Lin VWH, Hsiao IN, Dhaka V. Magnetic coil design considerations for functional magnetic stimulation. IEEE Trans Biomed Eng 2000;47:600–610.
95. Nilsson J, Panizza M, Roth BJ, et al. Determining the site of stimulation during magnetic stimulation of the peripheral nerve. Electroencephalogr Clin Neurophysiol 1992;85:253–264.

Appendix 27-1

To arrive at solutions for the electric field component E_y induced by magnetic stimulation, Maxwell's equations (Equations 1 a–d) were simplified in the following manner:

$$\nabla \times E = -j\omega\mu H \quad (1a)$$

$$\nabla \times H = J + j\omega\varepsilon E \quad (1b)$$

$$\nabla \cdot E = \rho/\varepsilon \quad (1c)$$

$$\nabla \cdot H = 0 \quad (1d)$$

Curling both sides of Equation 1a and combining with Equation 1b, if the medium is single and uniform, then we have

$$\nabla \times \nabla \times E = \nabla \times (-j\omega\mu H) = -j\omega\mu \nabla \times H$$
$$= -j\omega\mu(J + j\omega\varepsilon E) \quad (2)$$

$$\nabla(\nabla \cdot E) - \nabla^2 E = -j\omega\mu J + \omega^2\mu\varepsilon E \quad (3)$$

With $\rho = 0$, assuming no charge source, Equation 3 becomes

$$\nabla^2 E - j\omega\mu J + \omega^2\mu\varepsilon E = 0. \quad (4)$$

Separate the real part and imaginary part of E and J as $E = E_r + jE_i$ and $J = J_r + jJ_i$ or $\sigma E_r + \sigma j E_i$ and bring them into Equation 4:

$$\nabla^2(E_r + jE_i) - j\omega\mu\sigma(E_r + jE_i)$$
$$+ \omega^2\mu\varepsilon(E_r + jE_i) = 0 \quad (5)$$

Equate the real part and imaginary part of Equation 5 separately to 0:

$$\nabla^2 E_r + \omega\mu\sigma E_i + \omega^2\mu\varepsilon E_r = 0 \quad (6a)$$

$$\nabla^2 E_i - \omega\mu\sigma E_r + \omega^2\mu\varepsilon E_i = 0 \quad (6b)$$

Equations 6a and 6b can be simplified to a 2-dimensional problem in the case of a round coil.

$$\partial^2(E_{r,x})/\partial x^2 + \partial^2(E_{r,x})/\partial y^2 + \partial^2(E_{r,x})/\partial z^2$$
$$+ \omega\mu\sigma E_{i,x} + \omega^2\mu\varepsilon E_{r,x} = 0 \quad (7a)$$

$$\partial^2(E_{i,x})/\partial x^2 + \partial^2(E_{i,x})/\partial y^2 + \partial^2(E_{i,x})/\partial z^2$$
$$- \omega\mu\sigma E_{r,x} + \omega^2\mu\varepsilon E_{i,x} = 0 \quad (7b)$$

$$\partial^2(E_{r,y})/\partial x^2 + \partial^2(E_{r,y})/\partial y^2 + \partial^2(E_{r,y})/\partial z^2$$
$$+ \omega\mu\sigma E_{i,y} + \omega^2\mu\varepsilon E_{r,y} = 0 \quad (7c)$$

$$\partial^2(E_{i,y})/\partial x^2 + \partial^2(E_{i,y})/\partial y^2 + \partial^2(E_{i,y})/\partial z^2$$
$$- \omega\mu\sigma E_{r,y} + \omega^2\mu\varepsilon E_{i,y} = 0 \quad (7d)$$

For the purpose of computer modeling, the fourth term of Equations 7b and 7d can be rewritten as follows: $\omega\mu(\sigma E_{r,x} + J_{r,x})$ and $-\omega\mu(\sigma E_{r,y} + J_{r,y})$, respectively, to account for the operating current in the coils. $J_{r,x}$ and $J_{r,y}$ are the current density in the coil region in the computer model in perspective direction, and they are 0 everywhere else. A general solver for the solutions of systems of partial differential equations can be used to solve the above equations. Usually, this kind of solver incorporated a numerical technique based on the finite element method to solve unknowns such as induced electrical fields.

Appendix 27-2

The field probes used for this mapping are a bipolar probe and a gaussmeter probe (TBL STE92-0404). The magnetic coil (MC) is placed below a plastic container with dimensions of 30 × 30 × 25 cm. The container was filled with a resistive saline solution approximately equal to that of human tissue. The bipolar electric field probe was placed into the saline filled plastic container located directly above the MC. The electrodes of the probe were 2.5 mm apart and insulated everywhere except at their tips. The signal generated from this probe was displayed and digitized by an oscilloscope (HP 54602) and synchronized with the reading obtained from the gaussmeter (TBL 9200). The data were then stored in a personal computer for analysis. These data included on-screen visualization of the field distribution along a given plane.

The electric field measurements were made when the magnetic stimulation parameters were fixed at 20% of maximal intensity and 20-Hz frequency. To measure $\delta V/\delta y$ with respect to the MC, the two electrodes (1 and 2) of the electric field probe were placed parallel to the y axis. The probe was then moved along the x axis to measure $\mathbf{E}$. The voltage sensed at electrode 1, V_1, minus the voltage sensed at electrode 2, V_2, became the voltage difference, ΔV_{12}, between the two electrodes. This was monitored by inserting the two electrodes into channels 1 and 2 of the oscilloscope. If the distance, d, between points 1 and 2 was small, $\delta V/\delta y$ at the midpoint between the two electrodes could be approximated as $\Delta V_{12}/d$. Changes in $\delta V/\delta y$ along the z axis (i.e., vertical to the coil plane) were measured at the center of the coil.

28 Transcranial Magnetic Stimulation in Migraine

Alain Maertens de Noordhout, Anna Ambrosini, Peter S. Sándor, and Jean Schoenen

With the exception of rare monogenic forms such as familial hemiplegic migraine (FHM),[1] the common migraine phenotypes are heterogeneous and appear to be complex genetic disorders in which susceptibility genes and environmental factors interact.[2-5] Neuronal and vascular components are relevant in migraine pathophysiology and probably interrelated.[6-9] The neuronal structures involved are the cerebral cortex, the brainstem (e.g., periaqueductal gray matter, monoaminergic nuclei), and the peripheral and central components of the trigeminovascular system. The sequence of activation and the relative role of these structures are still controversial.

Several psychophysical studies of the visual system tend to support the hypothesis of neural dysfunction: more intense illusions to grating patterns,[10] faster low-level performance on psychophysical visual tests,[11] and the known clinical hypersensitivity to environmental light stimuli.[12] From these findings, it was proposed that there could be a generalized interictal hyperexcitability of the cerebral cortex in migraine that was more pronounced in visual areas because of its neuronal density[13] and that possibly favored the occurrence of spreading depression, which is thought to be responsible for the migrainous aura (reviewed by Lauritzen[14]). Other psychophysical tests of the visual system were thought to reflect hypoexcitability of the visual cortex.[15]

Clinical neurophysiology is particularly appropriate for the study of migraine pathophysiology because it is noninvasive and able to detect subtle functional abnormalities. During the past decade, numerous electrophysiological studies have yielded interesting but somewhat controversial results.

In this chapter, we review the results of transcranial magnetic stimulation (TMS) studies. To assess cortical excitability, TMS seems to be a tool of choice, because it can explore noninvasively excitatory and inhibitory neurons in visual and motor cortices. We review the available published data retrieved by a Medline search, present them comprehensively in tables, and discuss possible reasons for contradictions. Relevant studies presented at international meetings and published only as abstracts were included in the text but not in the tables. We examine possible neurobiological mechanisms of the reported abnormalities and propose that the predominant dysfunction of the central nervous in migraine is lack of habituation during sustained stimulation and not hyperexcitability that, in the strict physiological sense, implies lowering of stimulation threshold or increase of response amplitude to a supraliminal stimulus.

Transcranial magnetic stimulation has in theory the advantage of assessing directly excitability of the underlying cortex. It may be a tool of choice to search for cortical dysfunctions in migraine. TMS studies of the motor and the visual cortex have been performed. The former type has the advantage of relying on objective measurements (i.e., motor evoked potentials [MEPs]) in targets; the latter type most often uses the subjective perception of phosphenes.

Motor Cortex Studies

Data of published studies are summarized in Table 28–1. The first study of motor cortex TMS in migraine was published more than 10 years ago.[16] To overcome the problem of large interindividual motor threshold (MT) variability, migraine patients with symptoms always located on the same side were investigated, which allowed to use the other side as an additional control. In that pilot study conducted between attacks, MT was significantly increased on the affected cortical side of patients with migraine with auras (MA) compared with normal subjects or to the unaffected side. No MT differences were observed between normal subjects and patients with unilateral migraine without aura (MO) or between the normal and affected side of MO patients. Moreover, the maximal amplitude of MEPs, expressed as a ratio to maximal response to peripheral nerve stimulation (MEPmax/Mmax), was significantly reduced on the body side of auras in MA patients. Abnormally high MTs were reported later in menstrual migraine without aura[17] in the interictal phase and during attacks.

These results were not confirmed in a study by van der Kamp and colleagues,[18] who found increased MEP amplitudes and reduced MTs between attacks of patients with MA or MO. They also reported a positive correlation between MEP amplitudes and attack frequency, but they did not control for the occurrence of an attack in the days after the recordings. In a subsequent paper,[19] the investigators reported increased interictal MTs and reduced MEP amplitudes on the side of motor deficits in patients with FHM. These results were very similar to those obtained in our first study of patients with unilateral MA.

Afra and associates[20] studied a larger group of MA (n = 12) and MO patients (n = 19) with attacks occurring on either side and ensured that TMS was performed at least 3 days after the previous or before the next attack. Significantly higher mean MTs were observed during contraction in MA patients than in controls. Maximal MEPmax/Mmax values were normal in MA and in MO patients, whose attacks were not always located on the same side. Other parameters considered were the electromyographic silent period elicited by motor cortex stimulation and intracortical inhibition tested with paired TMS.[21,22] Both were normal in MA and MO patients.

In contrast, Aurora and coworkers[23] found that the cortical silent period was significantly shorter in MA patients than in controls. There was, however, no control in this study for the possible occurrence of a migraine attack in the days after the recordings. Others found no significant changes of MT to paired stimulations[24] or silent periods[25] in patients with MA or FHM.

Visual Cortex Studies

The conflicting results obtained in studies of phosphenes induced by occipital TMS are summarized in Table 28–2. Aurora and colleagues,[26] using TMS over the occipital lobe, reported an abnormally high prevalence of TMS-induced phosphenes in migraine with aura patients between attacks (11 of 11 MA compared with 3 of 11 normal subjects). Similar differences (100% of MA and 47% of controls) were reported by Brighina and coworkers.[24] Moreover, the threshold at which phosphenes appeared (PT), was much lower in MA patients than in controls.[26] The authors confirmed these results in a subsequent study of 15 migraineurs (14 MA) and 8 controls[27] Aguggia and associates[28] found a significant decrease of PT in MA (58.3%) compared with controls (83.7%) or tension-type headache (81.9%). In another study,[29] a reduced PT was reported in MA and in MO, but the proportion of patients who experienced TMS-induced phosphenes, although lower in migraineurs, was not significantly different from controls. The same group found

Table 28–1 TRANSCRANIAL MAGNETIC STIMULATION: MOTOR CORTEX

Study	Diagnosis (No. of Subjects)	Mean Age	Attack Control Before	Attack Control After	Methods Coil Shape	Max. Output	Motor Threshold (%) Rest	Motor Threshold (%) Active	Results MEP Amplitude	Central Motor Conduction Time (ms)	Cortical Silent Period (ms)	Intracortical Inhibition
Maertens de Noordhout et al., 1992[16]	MA (10)	36 ± 14	1 week	1 week	Circular 130 mm	1.5 T	55 ± 9 (↑)	41 ± 8 (↑)	↓	6.6 ± 0.7		
	MO (10)	39 ± 11					45 ± 6	33 ± 5	Normal	6.5 ± 0.7		
	HV (20)	40 ± 14					48 ± 6	33 ± 3	Normal	6.5 ± 0.6		
Bettucci et al., 1992[17]	MO (10)	333	No	No	Circular 130 mm	1.9 T	58 ± 5 (↑)			5.7 ± 0.3		
	HV (10)	31					48 ± 7			5.7 ± 1.3		
Van der Kamp et al., 1996[18]	MA (10)	35	3 days	No	Circular 130 mm	?	37 ± 4		↑	5.7 ± 1.2		
	MO (10)	50					38 ± 9		↑	6.6 ± 1.9		
	HV (10)	30					36 ± 5			5.7 ± 1.3		
Van der Kamp et al., 1997[19]	MA (10)	35	3 days	No	Circular 130 mm	?	37 ± 6		↑	5.7 ± 1.2		
	FHM (10)	30					44 ± 6 (↑)		Normal	6.8 ± 1.3		
	HV(6)	30					36 ± 7		Normal	5.7 ± 1.3		
Afra et al., 1998[20]	MA (25)	36 ± 15	3 days	3 days	Circular 130 mm	2.5 T	54 ± 18	43 ± 8 (↑)	Normal		101 ± 49	Normal
	MO (33)	36 ± 15					52 ± 12	41 ± 10			100 ± 49	Normal
	HV (27)	33 ± 10					47 ± 7	36 ± 6			101 ± 23	Normal
Aurora et al., 1999[23]	MA (10)	36 ± 7	1 week	None	Circular 95 mm	2 T	63 ± 14				63 ± 27	
	HV(10)	38 ± 6					58 ± 9				107 ± 20	
Werhahn et al., 2000[25]	MA (12)	38 ± 14	2 days	None	8-coil 90 mm	2.2 T	61 ± 12				183 ± 30	Normal
	FHM (9)	37 ± 13					60 ± 10				178 ± 5	Normal
	HV (17)	29 ± 6					55 ± 12				179 ± 30	Normal
Brighina et al., 2002[24]	MA (13)	39 ± 12	2 days	2 days	8-coil 45 mm	?	58 ± 5					
	HV (15)	32 ± 10					55 ± 7					

FHM, familial hemiplegic migraine; HV, healthy volunteer; ISI, interstimulus interval; MEP, motor evoked potential; MO, migraine without aura; MA, migraine with aura; TMS, transcranial magnetic stimulation; rTMS, repetitive transcranial magnetic stimulation.

Table 28–2 TRANSCRANIAL MAGNETIC STIMULATION: VISUAL CORTEX

Study	Diagnosis (No. of Subjects)	Mean Age	Attack Control Before	Attack Control After	Coil Shape	Max. output	Phosphene Prevalence	Phosphene Threshold	Others
Aurora et al., 1998[26]	MA (11)	37 ± 7	1 week	no	Circular 95 mm	2 T	100 (↑)	44.2 ± 8.6 (↓)	
	HV (11)	36 ± 7					27	68.7 ± 3.1	
Afra et al., 1998[20]	MA (25)	36 ± 15	3 days	3 days	Circular 130 mm	2.5 Tesla	56 (↓)	46	
	MO (33)	36 ± 15					82	50	
	HV (27)	33 ± 10					89	48	
Aurora et al., 1999[26]	MA (14) +MO (1)	40 ± 8	1 week	no	Circular 95 mm	2 Tesla	86.7 (↑)	45 (↓)	
	HV (8)	37 ± 6					25	81	
Mulleners et al., 2001[29]	MA (16)	43	1 day	no	Circular 130 mm	2 Tesla	75	47 ± 4.7 (↓)	
	MO (12)	46					83	46 ± 3.6 (↓)	
	HV (16)	43					94	66 ± 10.1	
Mulleners et al., 2001[30]	MA (7)	34 ± 12	No	no	Circular 130 mm	2 Tesla	no	no	↓ Visual extinction
	HV (7)	36 ± 13							
Battelli et al., 2002[32]	MA (16)	42 ± 14	2 weeks	no	Figure-of-eight 70 mm	2 Tesla	65 (left)	?80 (↓)	
	MO (9)	35 ± 15					67 (left)	?83 (↓)	
	HV (16)	40 ± 14					6 (left)	?110	
Brighina et al., 2002[24]	MA (13)	39 ± 12	2 days	2 days	Figure-of-eight 45 mm	?	100	56 ± 7	↓ by 1-Hz rTMS
	HV (15)	32 ± 10					47	57 ± 13	↑ by 1-Hz rTMS

HV, healthy volunteer; MA, migraine with aura; MO, migraine without aura; rTMS, repetitive transcranial magnetic stimulation.

that prophylactic treatment with valproate was able to increase PT in MA but not in MO[30] and that the ability of a TMS pulse over the occipital cortex to suppress visual perception was reduced in MA patients,[31] which was interpreted as reflecting reduced activity of inhibitory circuits in the occipital cortex of migraineurs. Significantly lower PTs for TMS over visual area V5 were found in migraine with or without aura compared with controls.[32] A study published in abstract form by Young and associates[33] also concluded that PTs for occipital TMS were lower in migraine with aura ($36 \pm 3\%$) or without aura ($40 \pm 6\%$) than in healthy subjects ($55 \pm 9\%$). Taken together, these studies favor the hypothesis of visual cortex hyperexcitability in migraine.

Afra and colleagues[20] obtained opposite results using a similar methodology and a circular coil: the prevalence of phosphenes was significantly lower in MA patients than in controls (10 of 18 versus 17 of 19), but no differences were found between controls and MO patients (18 of 22). Among subjects reporting phosphenes, mean PT were similar in all groups. These findings were replicated using more focal visual cortex stimulation with a figure-of-eight coil.[34] Others failed to find significant differences between migraineurs and healthy subjects in PTs for TMS over the primary visual cortex.[24]

Bohotin and coworkers[35] have used visual evoked potential (VEP) amplitude and habituation as a more objective measure than phosphenes to assess excitability changes induced in the visual cortex by repetitive TMS (rTMS) at two different frequencies. In normal subjects, 1-Hz rTMS, which is known to inhibit the underlying cortex, reduced amplitude in the first block of 100 averaged responses and induced lack of habituation over successive blocks. In MA and MO patients between attacks, 10-Hz rTMS, which most often activates the underlying cortex, increased first block VEP amplitude in migraineurs and transformed their lack of habituation into a normal habituation pattern. This study supports the concept that the visual cortex in migraineurs is hypoexcitable and not hyperexcitable between attacks. This was demonstrated in our laboratory by using repetitive transcranial magnetic stimulation. High-frequency rTMS over the occipital region, known to induce hyperexcitability of the underlying neural structures, was followed by a normalization of VEP habituation in migraineurs, whereas low-frequency rTMS, usually inhibitory, induced a deficit of VEP habituation in normal controls.[35]

■ Analysis

On the basis of the previously described studies, the use of TMS to assess motor and visual cortex excitability seems to have yielded conflicting results. Some discrepancies may reflect methodological differences that may be device and patient dependent. With two exceptions[25,32] in which a figure-of-eight coil was used, all published studies were performed with a circular coil. The size of the coil used was smaller in the studies by Aurora and colleagues[23,26] than in others. Types of coils differ substantially. A figure-of-eight coil produces a focal stimulation under the center of the coil, and a circular coil causes diffuse stimulation of the underlying cortical area.[36,37] It is likely that a larger cortical area was stimulated with large circular coils. The way human cortex is activated also differs depending on the direction of current flow in the coil. Other technical differences between the methods used for TMS may contribute to the discrepancy of results but are sometimes difficult to evaluate based on available information such as biphasic or monophasic magnetic pulses and maximum stimulator output.

With regard to patient selection, dramatic changes of evoked cortical responses and of cortical excitability occur 24 hours before and during the attack.[38,39] Although the occurrence of the last attack before the recording can be checked by history, occurrence of an attack within 24 hours after the recording has to be controlled by other means such as telephone calls. This was done in a restricted number of studies. Cortical excitability fluctuates with hormonal variations during the menstrual cycle.[40] We avoided such hormonal influences by recording all females at mid-cycle.

Nevertheless, some general lines can be drawn, particularly for *motor cortex* excitability. With the notable exception of one study,[18] the general findings were reduced interictal motor cortex excitability in several forms of migraine: unilateral or bilateral migraine with aura,[16,20] menstrual migraine without aura,[17] and FHM.[19] These findings do not favor the hypothesis[13] of a permanent cortical hyperexcitability in migraine. Excitability changes do not seem to result from dysfunction of cortical inhibitory interneurons, which were found to be normal in MA and MO patients.[20,25] Although such studies cannot be transposed to visual areas, these results do not favor the hypothesis of loss of GABAergic interneurons secondary to repetitive ischemic insults.[41] Moreover, in all studies except one, electromyographic silent periods to motor cortex stimulation were found normal. Although the origin of this silent period continues to be debated,[42] it also argues against abnormalities of inhibitory output pathways of the motor cortex. It seems that the increase of MT in several subtypes of migraine corresponds to decreased excitability of large pyramidal neurons, because changes of spinal motoneuron excitability is unlikely in migraine.

Published studies on *visual cortex* excitability give contradictory results. Although their exact generator remains a matter of debate, phosphenes elicited by brief, intense magnetic pulses directed to the occipital area of the brain are probably caused by activation of the primary visual cortex or subcortical sites such as the optic radiations adjacent to the posterior tip of the lateral ventricles.[28] One puzzling result in several studies[23,24,26,27,32] is the very low prevalence of phosphenes elicited in control groups (3 of 11, 2 of 8, 1 of 16, and 7 of 15), whereas all previous studies conducted in normal subjects report a much higher prevalence, usually between 60% and 80%.[43–46] Some migraineurs were included in "healthy volunteers group" in the latter studies. However, the normal phosphene prevalence rates found in our studies[20,34] in normal subjects, when personal and family histories of migraine were excluded, were similar at 89% and 64%.

In contrast with the findings of Afra and colleagues[20] and Bohotin and associates[34] of a significantly higher PT in both groups of migraineurs, Aurora and coworkers[23,26,27] and Mulleners and colleagues[29] reported a significantly lower PT. In a study in which TMS was applied laterally over visual area V5,[32] thresholds for magnetophosphenes were also lower in migraine patients. Results of the latter study are difficult to compare with all previous studies in which midline occipital TMS was used. In studies[23,26,27,29] reporting reduced PT, subjects who experienced no phosphenes were not taken into account for threshold calculations. If it is assumed that subjects without phosphenes have at least a 100% threshold, a recalculation of the figures in the paper of Mulleners and coworkers[29] would, for instance, increase the mean PT in the migraine without aura group from 46% to 55% and in the migraine with aura group from 47% to 60.2%, whereas in the healthy volunteer group, it would increase only from 66% to 68%. The differences between groups would lose statistical significance after such recalculations. However, some subjects may be resistant to induction of phosphenes by TMS and arbitrarily assigning them a PT of 100% may increase the risk for a type 2 error.

The fact that opposite results were obtained in different laboratories with similar methods suggests that phosphenes may be too subjective and variable to be used as measure of excitability of the visual cortex. Alternative, more objective and reliable methods should be developed. We have shown, for instance, that pattern-reversal VEP can be used to assess excitability changes of the visual cortex induced by rTMS.[35] Using this method and analyzing PR-VEP habituation, we have obtained evidence in favor of hypoexcitability, not hyperexcitability, of the visual cortex in migraine.

■ Conclusion

TMS studies disclose abnormalities of cortical information processing and excitability in migraine with and without auras between attacks. Although the results remain a matter of discussion, they underline at the least that migraine cannot be reduced to a "trigeminovascular disorder." Besides the studies of magnetophosphenes that have yielded contrasting results, chiefly because of lack of reliability of the method, several electrophysiological investigations of cortical activities in migraine favor lack of habituation and decreased cortical excitability as the predominant dysfunctions. Both abnormalities may be linked to each other by applying the "ceiling theory," postulating that a reduced preactivation excitability level of sensory cortices is responsible for a lack of habituation of cortical evoked responses during repeated stimulations. Whether the latter plays a role in migraine pathogenesis by increasing metabolic demands in the migrainous brain, which has a reduced mitochondrial energy reserve, must be proved. Additional studies are necessary to better define the precise nature of central nervous system dysfunction in migraine and to determine whether the cortical electrophysiological patterns allow other neurophysiological tests[47,48] to identify subgroups of migraineurs in whom correlations can be established with specific genotypes, responses to prophylactic agents, or interictal cognitive disturbances.

REFERENCES

1. Ophoff RA, Terwindt GM, Vergouwe MN, et al. Familial hemiplegic migraine and episodic ataxia type-2 are caused by mutations in the Ca^{2+} channel gene CACNL1A4. Cell 1996;87:543–552.
2. Tsounis S, Milonas J, Gilliam F. Hemi-field pattern reversal visual evoked potentials in migraine. Cephalalgia 1993;13:267–271.
3. Schoenen J, Wang W, Albert A, et al. Potentiation instead of habituation characterizes visual evoked potentials in migraine patients between attacks. Eur J Neurol 1995;2:115–122.
4. Tagliati M, Sabbadini M, Bernardi G, et al. Multichannel visual evoked potentials in migraine. Electroencephalogr Clin Neurophysiol 1995;96:1–5.
5. Rossi LN, Pastorino GC, Bellettini G, et al. Pattern reversal visual evoked potentials in children with migraine or tension-type headache. Cephalalgia 1996;16:104–106.
6. Afra J, Ambrosini A, Genicot R, et al. Influence of colors on habituation of visual evoked potentials in patients with migraine with aura and in healthy volunteers. Headache 2000;40:36–40.
7. Afra J, Proietti Cecchini A, et al. Comparison of visual and auditory evoked cortical potentials in migraine patients between attacks. Clin Neurophysiol 2000;111:1124–1129.
8. Khalil NM, Legg NJ, Anderson DJ. Long term decline of P100 amplitude in migraine with aura J Neurol Neurosurg Psychiatry 2000;69:507–511.
9. Yucesan C, Sener O, Mutluer N. Influence of disease duration on visual evoked potentials in migraineurs. Headache 2000;40:384–388.
10. Wilkins AJ, Nimmo-Smith I, Tait A, et al. A neurological basis for visual discomfort. Brain 1984;107:989–1017.
11. Wray SH, Mijovic-Prelec D, Kosslyn SM. Visual processing in migraineurs. Brain 1995;118:25–35.
12. Hay KM, Mortimer MJ, Barker DC, et al. Women with migraine: the effects of environmental stimuli. Headache 1994;34:166–168.
13. Welch KMA, D'Andrea G, Tepley N, et al. The concept of migraine as a state of central neuronal hyperexcitability. Neurol Clin 1990;8:817–828.
14. Lauritzen M. Spreading depression and migraine. Pathol Biol 1992;40:332–337.
15. Shepherd AJ. Increased visual after-effects following pattern adaptation in migraine: a lack of intracortical excitation? Brain 2001;124:2310–2318.
16. Maertens de Noordhout A, Pepin JL, Schoenen J, et al. Percutaneous magnetic stimulation of the motor cortex in migraine. Electroencephalogr Clin Neurophysiol 1992;85:110–115.
17. Bettucci D, Cantello M, Gianelli M, et al. Menstrual migraine without aura: cortical excitability to magnetic stimulation. Headache 1992;32:345–347.
18. Van der Kamp W, Maassen van den Brink, Ferrari M, et al. Interictal cortical hyperexcitability in migraine patients demonstrated with transcranial magnetic stimulation. J Neurol Sci 1996;139:106–110.
19. Van der Kamp W, Maassen van den Brink A, Ferrari M, et al. Interictal cortical excitability to magnetic stimulation in familial hemiplegic migraine. Neurology 1997;48:1462–1464.
20. Afra J, Mascia A, Gérard P, et al. Interictal cortical excitability in migraine: a study using transcranial magnetic stimulation of motor and visual cortices. Ann Neurol 1998;44:209–215.
21. Kujirai T, Caramia MD, Rothwell JC, et al. Corticocortical inhibition in human motor cortex. J Physiol (Lond) 1993;471:501–519.

22. Ziemann U, Lönnecker S, Steinhof B, et al. Effects of antiepileptic drugs on motor cortex excitability in humans: a transcranial magnetic stimulation study. Ann Neurol 1996;40:367–378.
23. Aurora SK, al-Sayeed F, Welch KM. The cortical silent period is shortened in migraine with aura. Cephalalgia 1999;19:708–712.
24. Brighina F, Piazza A, Daniele O, et al. Modulation of visual cortical excitability in migraine with aura: effects of 1 Hz repetitive transcranial magnetic stimulation. Exp Brain Res 2002;145:177–181.
25. Werhahn KJ, Wiseman K, Herzog J, et al. Motor cortex excitability in patients with migraine with aura and hemiplegic migraine. Cephalalgia 2000;20:45–50.
26. Aurora SK, Ahmad BK, Welch KMA, et al. Transcranial magnetic stimulation confirms hyperexcitability of visual cortex in migraine. Neurology 1998;50:1105–1110.
27. Aurora SK, Cao Y, Bowyer SM, et al. The occipital cortex is hyperexcitable in migraine: experimental evidence. Headache 1999;39:469–476.
28. Aguggia M, Zibetti M, Febbraro A, et al. Transcranial magnetic stimulation in migraine with aura: further evidence of occipital cortex hyperexcitability [abstract]. Cephalalgia 1999;19:465.
29. Mulleners WM, Chronicle EP, Palmer JE, et al. Visual cortex excitability in migraine with and without aura. Headache 2001;41:565–572.
30. Mulleners WM, Chronicle EP, Vredeveld JW, et al. Visual cortex excitability in migraine before and after valproate prophylaxis: a pilot study using TMS. Eur J Neurol 2002;9:35–40.
31. Mulleners WM, Chronicle EP, Palmer JE, et al. Suppression of perception in migraine: evidence for reduced inhibition in the visual cortex. Neurology 2001;56:178–183.
32. Battelli L, Black KR, Wray SH. Transcranial magnetic stimulation of visual area V5 in migraine. Neurology 2002;58:1066–1069.
33. Young WB, Oshinsky ML, Shechter AL, et al. Consecutive transcranial magnetic stimulation induced phosphene thresholds in migraineurs and controls [abstract]. Neurology 2001;56(Suppl 3):A142.
34. Bohotin V, Fumal A, Vandenheede M, et al. Excitability of visual V1-V2 and motor cortices to single transcranial magnetic stimuli in migraine: a reappraisal using a figure-of-eight coil. Cephalalgia 2003;23:264–270.
35. Bohotin V, Fumal A, Vandenheede M, et al. Effects of repetitive transcranial magnetic stimulation on visual evoked potentials in migraine. Brain 2002;125:1–11.
36. Cohen LG, Roth BJ, Nilsson J, et al. Effects of coil design on delivery of focal magnetic stimulation. Technical considerations. Electroencephalogr Clin Neurophysiol 1990;75:350–357.
37. Hallett M. Transcranial magnetic stimulation and the human brain. Nature 2000;406:147–150.
38. Kropp P, Gerber WD. Contingent negative variation during migraine attack and interval: evidence for normalization of slow cortical potentials during the attack. Cephalalgia 1995;15:123–128.
39. Judit A, Sandor P, Schoenen J. Habituation of visual and intensity dependence of auditory evoked cortical potentials tends to normalize just before and during the migraine attack. Cephalalgia 2000;20:714–719.
40. Smith MJ, Keel JC, Greenberg BD, et al. Menstrual cycle effects on cortical excitability. Neurology 1999;53:2069–2072.
41. Chronicle EP, Mulleners W. Might migraine damage the brain? Cephalalgia 1994;14:415–418.
42. Hallett M. Transcranial magnetic stimulation: a useful tool for clinical neurophysiology. Ann Neurol 1996;40:344–345.
43. Maccabee PJ, Amassian VE, Cracco JB, et al. Magnetic coil stimulation of human visual cortex: studies of perception. Electroencephalogr Clin Neurophysiol 1991;43:111–120.
44. Meyer BU, Diehl R, Steinmetz H, et al. Magnetic stimuli applied over motor and visual cortex: influence of coil position and field polarity on motor responses, phosphenes and eye movements. Electroencephalogr Clin Neurophysiol 1991;43:121–134.
45. Kastner S, Demmer I, Zieman U. Transient visual field defects induced by transcranial magnetic stimulation over human occipital lobe. Exp Brain Res 1998;118:19–26.
46. Stewart LM, Walsh V, Rothwell JC. Motor and phosphene thresholds: a transcranial magnetic stimulation correlation study. Neuropsychologia 2001;39:415–419.
47. Ambrosini A, Maertens de Noordhout A, Schoenen J. Neuromuscular transmission in migraine: a single fiber EMG study in clinical subgroups. Neurology 2001,56:1038–1043.
48. Sándor PS, Mascia A, Seidel L, et al. Subclinical cerebellar impairment in the common types of migraine: a 3-dimensional analysis of reaching movements. Ann Neurol 2001;49:668–672.

29 Transcranial Magnetic Stimulation in Sleep and Sleep-Related Disorders

Stefan Cohrs and Frithjof Tergau

■ Normal Physiology of Sleep Investigated by Transcranial Magnetic Stimulation

About one third of human life is spent in sleep. A simple behavioral definition of sleep is that it is a reversible behavioral state of perceptual disengagement from and unresponsiveness to the environment. Sleep is a complex amalgam of physiological and behavioral processes.[1]

Although sleep appears to be a passive state of being for the distant observer, it is a dynamic process incorporating two separate states that have been defined on the basis of a constellation of physiological parameters. After the identification of periodically occurring rapid eye movements (REMs) during sleep, it was divided into REM and non-REM sleep.[2] Non-REM sleep is subdivided into four sleep stages (i.e., 1, 2, 3, and 4) that roughly parallel a depth of sleep continuum.[1] Normal sleep usually is entered through light sleep (stage 1) characterized by slowing of the electroencephalographic (EEG) pattern, slow eye movements, and a slightly reduced muscle tone. Progression to sleep stage 2 is characterized by the occurrence of sleep spindles and K complexes. Stages 3 and 4, called slow wave sleep (SWS), are dominated by delta waves. REM sleep is a markedly different stage with a mixed-frequency EEG pattern, muscle atonia mediated by a powerful brainstem inhibitory mechanism, and episodic bursts of REMs. After about 90 minutes of non-REM sleep, the first REM episode is to be expected. During the night non-REM and REM sleep alternate every 60 to 120 minutes with the amount of slow wave sleep decreasing and REM sleep increasing progressively until the morning awakening.

Early in the course of transcranial magnetic stimulation (TMS) research, magnetic stimulation was used to investigate motor excitability during sleep in healthy subjects. During wakefulness, TMS is a noninvasive, well-tolerated neurophysiological tool. However, during sleep, the application of transcranial magnetic stimuli is often associated with arousal and disturbance of normal sleep processes. This must be taken into account interpreting the results of the following studies. Motor response amplitudes were shown to be depressed during SWS and enhanced or unchanged during REM sleep compared with wakefulness.[3] This was of particular interest because of the known inhibition of spinal motor centers during REM sleep mediating muscle atonia. The investigators interpreted that the increased motor response to TMS in REM sleep pointed to an enhanced excitability of the human motor cortex to stimulation during this sleep stage,[3] thereby overcoming the inhibitory input that converges to the same spinal motoneuron.

In a lager study, Fish and colleagues[4] found motor response amplitudes to be reduced in REM sleep compared with wakefulness in three healthy controls and in four patients

419

Table 29-1 MOTOR RESPONSE AMPLITUDE DURING DIFFERENT SLEEP STAGES AS A RESPONSE TO TRANSCRANIAL MAGNETIC STIMULATION

Study	SWS	REM	S2	Wakefulness
Hess et al., 1991[3]	↓	↑ / ↔	Not measured	Baseline
Fish et al., 1991[4]	Not measured	↓ and more variable	Not measured	Baseline
Stalder et al., 1995[6]	↓	↓ and highly variable	Not measured	Baseline
Grosse et al., 2002[7]	↓	↓	↓	Baseline

REM, rapid eye movement; S2, stage 2; SWS, slow wave sleep; ↑, increased; ↓, decreased; ↔, unchanged.

with torsion dystonia with a higher variability than during wakefulness. Prolonged latencies compared with wakefulness were observed in all subjects. The marked variability of response amplitudes to magnetic stimulation during REM sleep was hypothesized to reflect the rapid fluctuations in alpha motoneuron excitability that are known to occur in REM sleep.[4,5] The difference in results compared with Hess and coworkers[3] were attributed to not having stimulated during muscles twitches,[4] thereby possibly missing phases of increased excitability. Further evaluation of the influence of phasic versus tonic REM sleep on motor evoked potential (MEP) amplitudes of different muscles elicited by TMS demonstrated comparable results for these two substates within REM sleep. However, muscle responses are highly variable and independent of tonic or phasic episodes during REM sleep.[6] The investigators attributed the highly variable MEP responses to possible fluctuations in the cortical or spinal inhibition during REM sleep.[6] In a study examining additional TMS target parameters such as motor thresholds (MTs), stimulus-response curves, and latencies of MEPs in healthy subjects, Grosse and associates[7] showed an increased MT in stage 2, stage 4, and REM sleep (the sleep stages investigated) compared with wakefulness. Flattened response curves to stepwise increased stimulus intensities and prolonged latencies to the TMS stimulus were found and interpreted as a sign of reduced overall excitability of the corticospinal system in all sleep stages. Excitability appeared to be lower in sleep stage 2 than in stage 4, manifested by a higher MT in stage 2 sleep.[7] No direct statistical comparison of MTs between stage 4 and REM sleep is reported by Grosse and colleagues,[7] but their results point in the opposite direction compared with the data of Stadler and coworkers, who found smaller amplitudes in SWS than in REM sleep. An overview of TMS parameters in normal sleep is given in Table 29-1.

Even though the exact function of sleep remains to be elucidated, several aspects have been investigated by a multitude of neurophysiologic techniques. One way to explore the physiologic role of sleep in the homeostasis of life is to study the influence of sleep loss or total sleep deprivation on different target parameters. The effects of sleep deprivation on MT, MEP size and intracortical inhibition were first studied in a complex experiment examining mechanisms of deafferentation-induced plasticity in the human motor cortex.[8] One night of sleep deprivation was used as a second control condition to differentiate sedative drug effects from their direct effects on measured TMS parameters. Compared with the unaltered control condition, TMS parameters were not affected by sleep deprivation. However, because the purpose of this experiment was not to examine the effects of sleep deprivation, but deafferentation-induced plasticity, the other interventions (i.e., ischemic nerve block and slow repetitive TMS) might have masked possible effects of sleep deprivation on motor cortex excitability. The two studies directly examining the effects of sleep deprivation on motor cortex excitability report somewhat inconclusive results. Civardi and associates[9] found resting and active MTs and the silent period to be unchanged, whereas intracortical inhibition and facilitation measured by paired-pulse TMS were significantly reduced in a group of eight healthy volunteers after a minimum of 24 hours of sleep deprivation. These results were interpreted to reflect general hypoactivity of cortical area 4 interneurons. However, reduced inhibition and facilitation can be achieved by slight tonic contraction of the target muscle.[10] The previously reported results may instead reflect incomplete relaxation because of general distress caused by sleep deprivation. In contrast to the results of Civardi and colleagues,[9] Manganotti and coworkers[11] found an increased resting MT at 3:00 AM and 6:00 AM compared with baseline at 9:00 AM before sleep deprivation and report MT to return to baseline values at 9:00 AM after sleep deprivation. However, their data demonstrated a strong trend toward a still increased MT after 24 hours of sleep deprivation. Silent period measured with 130% MT showed a significant lengthening at 3:00 AM, normalizing in the morning. Intracortical inhibition showed an enhancement during the late night hours and statistically nonsignificant effects at 9:00 AM after sleep deprivation. Intracortical facilitation showed a trend toward a decrease in the late hours of the night. These alterations in cortical excitability were interpreted as a possible change in membrane excitability and as increased GABAergic mechanisms. However, data are difficult to interpret due to a lack of controlling for a possible influence of circadian rhythms in excitability independent of sleep deprivation. An overview of TMS parameters caused by sleep deprivation is given in Table 29–2.

The partially contradictory results for the effects on the MT, silent period, and intracortical inhibition remain to be clarified. Additional research is needed to explore the influence of sleep deprivation on motor cortex excitability to better understand its epileptogenic and antidepressive effects of this intervention.

Transcranial Magnetic Stimulation in Sleep Disorders

Restless Legs Syndrome and Periodic Limb Movements in Sleep

Most TMS work concerning sleep disorders has been carried out in patients suffering from restless legs syndrome (RLS) or periodic limb movement in sleep (PLMS) to evaluate excitability of the motor pathways of the central nervous system. RLS is a sensory-motor disorder characterized by unpleasant sensations in the limbs and an urge to move the limbs. The symptoms are worse or are exclusively present at rest, are most pronounced in the evening and at night, and can be at least partially relieved by moving the limbs.[12–14] Other features commonly seen in RLS include sleep disturbance, periodic limb movements in sleep and similar involuntary movements while awake, a normal neurological examination in the idiopathic form, and a tendency for the symptoms to be worse in middle to older age.[12] Periodic limb movements typically occur every 20 to 40 seconds for a duration of 0.5 to 5 seconds and may or may not be associated with an EEG arousal.[15] The relative frequency of periodic movements in sleep, their duration and their arousing effect decreases along the non-REM sleep stages, whereas the inter-movement interval increases.[16] During REM sleep the duration of movements is shortest and the inter-movement interval is longest. This suggests that the processes underlying PLM are most active at the transition from wakefulness to sleep and considerably attenuated during deep non-REM sleep and even more during REM sleep.[16] Using TMS, Smith and associates[17] found patients with periodic limb movement disorder (PLMD) to have motor

Table 29–2 EFFECT OF SLEEP DEPRIVATION ON TRANSCRANIAL MAGNETIC STIMULATION PARAMETERS

Study	Active Motor Threshold	Resting Motor Threshold	Silent Period	Intracortical Inhibition	Intracortical Facilitation
Civardi et al., 2001[9]	↔	↔	↔	↓	↓
Manganotti et al., 2001[11]	Not measured	↑	↑	↑	[↓]

↑, increased; ↓, decreased; ↔, unchanged.

conduction latencies comparable to those of normal subjects during wakefulness, non-REM sleep, and REM sleep. However, further TMS-parameters have not been determined, and possible differences might have been masked by the intake of clonazepam by 7 of 13 patients and the use of triazolam as a hypnotic by all subjects studied during sleep. Other neurophysiological techniques such as the blink reflex had demonstrated an increased excitability in patients with PLMS and located the underlying pathophysiology in subcortical structures possibly related to the dopaminergic striatopallidal system.[18] However, others found this reflex to be normal in patients with RLS.[19] Spinal cord excitability in patients with RLS and PLMS was studied by Bara-Jimenez and colleagues,[20] who evaluated the flexor reflexes by electrically stimulating the medial plantar nerve in 10 patients and compared them with matched controls. The patients showed significantly lower threshold and greater spatial spread, which was more prominent during sleep. They concluded that the PLMS in RLS might result from enhanced spinal cord excitability.[20] Supporting evidence for the subcortical origin of PLMS comes from the absence of any changes of cortical activity preceding periodic limb movements.[21,22]

Testing the hypothesis that the main site in the pathophysiology of PLMS is subcortical Tergau and coworkers[23] studied 18 patients with idiopathic RLS by single- and paired-pulse TMS. A number of parameters, including resting and active MT, MEP recruitment, cortical silent period, intracortical inhibition, and intracortical facilitation, were determined. Motor thresholds, MEP recruitment, and cortical silent periods showed comparable results in the patients and matched controls. However, intracortical inhibition was significantly reduced in the abductor digiti minimi muscle and the abductor hallucis muscle. Intracortical facilitation was increased in the abductor digiti minimi but reduced in the clinically affected abductor hallucis muscle. Comparing the patients symptomatic during the TMS procedure with those exhibiting the RLS symptoms at another time, the symptomatic patients showed larger changes. It was concluded that the reduced intracortical inhibition reflects disinhibited subcortical inputs to the cortex rather than a primary cortical phenomenon.

Tugnoli and associates,[24] studying 15 patients with primary RLS and 12 controls, came to the same conclusion that cortico-subcortical structures are involved in the pathogenesis of this disorder. In accordance with Tergau and colleagues,[23] they found unaltered MTs, MEP recruitment, cortical silent period, and central motor conduction time, whereas opposite results were reported for intracortical inhibition and facilitation. In contrast to the findings of Tergau and coworkers[23] and Tugnoli and associates,[24] two other groups reported decreased silent periods in hand and leg muscles[25] or only in leg muscles[26] in RLS. Whereas most groups reported unchanged MTs,[23,25] Stiasny and colleagues[26] found active MT to be unchanged only in the abductor digiti minimi muscle but increased in the anterior tibial muscle. Contrary to the observed circadian fluctuation in RLS symptoms no change in TMS parameters was observed between evening and morning measurements. Without describing details about TMS procedures, according to the given reference,[27] Provini and coworkers[21] reported normal MEP amplitudes and central and peripheral MEP latencies in a group of ten patients with idiopathic RLS.

In conclusion, different TMS parameters are altered in patients with primary RLS. However, some of the findings have not been replicated. The discrepancies in excitability measures in the different studies may reflect various study protocols, including TMS procedures, medication-free time before measurement, time of the day when measurements were done, and presence or absence of acute symptoms. RLS symptoms follow a circadian rhythm, with the highest prevalence and intensity of restlessness in the evening and early night.[28] Some excitability measures may vary in a similar way under certain circumstances. Tergau and associates[23] found more pronounced effects in the symptomatic subgroup in the late afternoon hours, pointing to the possibility that neurophysiologic findings are at least state influenced if not state dependent in RLS patients. Future studies should consider these aspects. The reported work shows that TMS is a valuable tool for investigating the pathophysiology of RLS and PLMS, demonstrating altered central nervous system excitability of structures somewhere between the spinal and the subcortical level. This is of special importance because TMS is the only neurophysiologic technique showing abnormalities in RLS. An overview of this work is provided in Table 29-3. Further research is needed to clarify the reported discrepancies.

Narcolepsy

Narcolepsy is a sleep disorder characterized by excessive daytime sleepiness, hypnagogic hallucinations, sleep paralysis, and cataplexy (i.e., attacks of reduced muscle tone during wakefulness as a reaction to strong emotions). The first TMS study of narcolepsy used magnetic brain stimulation in a patient with the narcolepsy-cataplexy syndrome during and after a cataplectic episode. Amplitudes and thresholds of responses in six muscles (i.e., diaphragm, lumbar erector spinae, trapezius, biceps, tibialis anterior, and abductor digiti V) remained unchanged during cataplexy compared with wakefulness without cataplexy. The investigators concluded that, similar to their earlier findings in REM sleep, an enhanced cortical excitability to magnetic brain stimulation could compensate for the postsynaptic spinal inhibition of muscle tone during cataplexy.[29] Another study examining three patients suffering from the narcolepsy-cataplexy syndrome determined the silent period and the effect of 2 seconds of 20-Hz rTMS at 110% MT on EMG activity while subjects performed a forceful grip with the index finger and the thumb.[30] The silent periods of the patients were found to be comparable to those of eight healthy control subjects. However, two of the patients off anticataplectic medication for at least 3 days showed a complete cessation of EMG activity lasting 0.6 to 3.5 seconds. One patient displayed an obvious reduction of EMG activity on the contralateral side. Stimulation of other cortical areas and stimulation of the peripheral nervous system did not induce muscle weakness episodes. It was concluded that the cessation of EMG activity represented an induction of cataplexy, and the researchers hypothesized that rTMS might transsynaptically activate brainstem nuclei involved in muscle atonia during REM sleep or inhibitory neurons at the spinal cord level in narcoleptic patients.[30]

The highlighted pathophysiological aspects are interesting, but only four narcoleptic patients, none of whom was drug naïve, have been studied. The collected data should build the base for further studies including drug naïve patients and determine other TMS parameters, including single- and paired-pulse studies. It would be helpful to determine the diagnostic usefulness of TMS in narcoleptic patients who do not yet suffer from cataplexy but fulfill the diagnostic criteria and to determine further anticataplectic drug effects in patients and in healthy controls.

Sleep Apnea

Very limited work has been reported in the area of sleep apnea. An uncontrolled investigation of seven patients suffering from sleep apnea came to the conclusion that TMS-induced MEP latencies of the abductor digiti minimi muscle were slightly increased compared with the literature.[31] Unfortunately, no explanation was presented for this finding, which remains to be clarified. Another TMS study investigated different causes of respiratory dysfunction, including sleep apnea, using TMS to determine the latencies and amplitudes of diaphragmatic compound action potentials.[32] Unfortunately, sleep apnea had not been classified as central or obstructive, and the data remain unclear. As in other areas for which TMS was used to study the pathophysiology of sleep disorders, interpretation is limited by the number of patients studied and the small number of possible TMS parameters determined.

■ Repetitive Transcranial Magnetic Stimulation and Sleep in Depression

Sleep disturbance belongs to the core symptoms of major depression experienced by

Table 29–3 TRANSCRANIAL MAGNETIC STIMULATION PARAMETERS IN RESTLESS LEGS SYNDROME AND PERIODIC LIMB MOVEMENTS IN SLEEP SYNDROME

Study	Resting Motor Threshold	Active Motor Threshold	Cortical Silent Period	Central Motor Conduction Time	MEP Recruitment	Intracortical Inhibition	Intracortical Facilitation
Entezari-Taher et al., 1999[25]	↔	Not measured	↓	Not measured	Not measured	Not measured	Not measured
Provini et al., 2001[21]*	Not reported whether measured	Not reported whether measured	Not measured	Not reported whether measured	Not reported whether measured	Not reported whether measured	Not reported whether measured
Smith et al., 1992[17]	Not measured	↑ in AT	↓ in AT	↔	Not measured	Not measured	Not measured
Stiasny et al., 2000[26]	Not reported whether measured	↔ in ADM	↔ in ADM	↔	Not reported whether measured	Not reported whether measured	Not reported whether measured
Tergau et al., 1999[23]	↔	↔	↔	Not measured	↔	↓ in ADM ↓ in AH	↑ in ADM ↓ in AH
Tugnoli et al., 2000[24]	↔	↔	↔	↔	↔	↑ in ADM and AT	↓ in ADM and AT

* Normal TMS findings but no details reported (according to given reference MEP amplitude and central and peripheral MEP latencies). ADM, abductor digiti minimi muscle; AH, abductor hallucis muscle; AT, anterior tibial muscle; MEP, motor evoked potential; ↓, reduced; ↑, increased; ↔, unchanged.

patients as problems initiating and maintaining sleep. Polysomnographically, sleep in major depression is characterized by an increased sleep latency, disturbance of sleep continuity, a reduction of SWS, and disinhibition of REM sleep, including a shortening of REM latency (i.e., time between sleep onset and the occurrence of the first REM period).[33,34] Several theories have been developed to explain sleep abnormalities of depression and tried to integrate the findings into pathophysiological models such as the two process-model of sleep[35] and the reciprocal interaction model of non-REM and REM sleep regulation.[36,37] Well-known treatment strategies of depression such as antidepressive drugs[38] and electroconvulsive therapy demonstrate early effects on non-REM and REM sleep regulation[39] that are predictive of a favorable treatment response. Several rTMS-studies have documented its antidepressive potential.[40–42] The observed effects of standard antidepressive treatment strategies on sleep documented the need to evaluate the effects of rTMS on polysomnographically monitored sleep.

In a study of healthy subjects, our group was able to demonstrate that suprathreshold rTMS is able to postpone REM sleep, prolonging REM latency and increasing the non-REM and REM cycle length.[43,44] The most pronounced effect was observed after left prefrontal rTMS. We concluded that the antidepressive action might be related to the influences of rTMS on circadian rhythms altered in depressed patients. Another group studying the influence of subthreshold rTMS on EEG and sleep parameters found only a small reduction of sleep stage 1 over the whole night and a small enhancement of sleep stage 4 during the first non-REM sleep episode. Other sleep variables, including REM latency, were not affected.[45] Subjective sleep quality in healthy subjects the night after slow rTMS appears to be unaltered.[46] It remains to be clarified whether stimulation parameters differentially influence sleep quality and sleep structure and whether the lack of an effect on REM latency after subthreshold rTMS[45] is related to the lower stimulus intensity.

The importance of stimulus intensity has been documented in other studies showing a stronger antidepressive response after higher rTMS stimulus intensity in depressed patients[47] and increased plasma thyroid-stimulating hormone (TSH) levels after left prefrontal suprathreshold compared with subthreshold stimulation in healthy subjects.[48] The possible importance of TSH increase for an antidepressive effect had been demonstrated in several treatment studies, including some assessing sleep deprivation and tricyclic antidepressants,[49–51] and has been documented for rTMS.[52]

In a study evaluating the predictive value of the response to partial sleep deprivation in a group of patients suffering from major depression, Padberg and colleagues[53] found that the amelioration of depression after partial sleep deprivation was inversely correlated with the improvement after 2 weeks of rTMS. However, there was no clinically applicable predictive value of the response to partial sleep deprivation for the outcome after rTMS. It was hypothesized that different subgroups of depressed patients responded to the two interventions. Another study investigating the possible use of rTMS and a possible relation to sleep in the treatment of depressed patients found that those responding to total sleep deprivation benefit from therapeutic rTMS the morning after the sleep deprivation. Usually, patients relapse rapidly, even after short naps, into a depressed mood. rTMS the morning after sleep deprivation appears to prolong this antidepressive effect up to 4 days after the initial sleep deprivation, despite intervening sleep.[54] The investigators attributed the positive treatment results of rTMS after sleep deprivation to a possible influence on circadian rhythms by activation of brain structures such as the suprachiasmatic nuclei. It was hypothesized that sleep deprivation might have changed the state of brain function, thereby influencing susceptibility to the antidepressive effects of rTMS.

Only a few rTMS treatment studies of major depression have specifically evaluated the effect on subjective sleep quality, and to our knowledge, no polysomnographic evaluation has taken place. Comparing the therapeutic effectiveness of rTMS and electroconvulsive therapy in major depression, the sleep-improving effect of the intervention appears to be more pronounced with electroconvulsive therapy than with rTMS in psychotic patients, as measured by a standard sleep questionnaire. No such difference was observed in nonpsychotic patients.[55] The results for nonpsychotic patients with major depression demonstrating comparable amelioration of depression and improvement of sleep complaints have been replicated.[56] In a study demonstrating successful treatment of major depression in medication-free patients, a subanalysis of the Hamilton Depression Scale items revealed a quicker response for the items of Depressed Mood and Insomnia, with a significant amelioration occurring after 1 week of rTMS treatment compared with other items that improved only after 2 weeks of treatment.[57]

Further studies evaluating the influence of rTMS on sleep in depressed patients, including the effects of varying stimulation parameters, are needed. Polysomnography and other parameters of circadian rhythms, such as hormones and body temperature, in depressed patients treated with rTMS should be used to better understand the mode of action of this promising new treatment strategy.

REFERENCES

1. Carskadon MA, Dement WC. Normal human sleep: an overview. In Kryger MH, Roth T, Dement WC (eds): Principles and practice of sleep medicine. Philadelphia, WB Saunders, 2000:15–42.
2. Aserinsky E, Kleitman N. Regularly occuring periods of eye motility, and concomitant phenomena, during sleep. Science 1953;118:273–274.
3. Hess CW, Mills KR, Murray NM, et al. Excitability of the human motor cortex is enhanced during REM sleep. Neurosci Lett 1987;82:47–52.
4. Fish DR, Sawyers D, Smith SJ, et al. Motor inhibition from the brainstem is normal in torsion dystonia during REM sleep. J Neurol Neurosurg Psychiatry 1991;54:140–144.
5. Chase MH, Morales FR. Subthreshold excitatory activity and motoneuron discharge during REM periods of active sleep. Science 1983;221:1195–1198.
6. Stalder S, Rosler KM, Nirkko AC, et al. Magnetic stimulation of the human brain during phasic and tonic REM sleep: recordings from distal and proximal muscles. J Sleep Res 1995;4:65–70.
7. Grosse P, Khatami R, Salih F, et al. Corticospinal excitability in human sleep as assessed by transcranial magnetic stimulation. Neurology 2002;59:1988–1991.
8. Ziemann U, Hallett M, Cohen LG. Mechanisms of deafferentation-induced plasticity in human motor cortex. J Neurosci 1998;18:7000–7007.
9. Civardi C, Boccagni C, Vicentini R, et al. Cortical excitability and sleep deprivation: a transcranial magnetic stimulation study. J Neurol Neurosurg Psychiatry 2001;71:809–812.
10. Ridding MC, Taylor JL, Rothwell JC. The effect of voluntary contraction on corticocortical inhibition in human motor cortex. J Physiol 1995;487:541–548.
11. Manganotti P, Palermo A, Patuzzo S, et al. Decrease in motor cortical excitability in human subjects after sleep deprivation. Neurosci Lett 2001;304:153–156.
12. Walters AS. Toward a better definition of the restless legs syndrome. The International Restless Legs Syndrome Study Group. Mov Disord 1995;10:634–642.
13. Allen RP, Picchietti D, Hening WA, et al. Restless legs syndrome: diagnostic criteria, special considerations, and epidemiology. A report from the restless legs syndrome diagnosis and epidemiology workshop at the National Institutes of Health. Sleep Med 2003;4:101–119.
14. Trenkwalder C, Walters AS, Hening W. Periodic limb movements and restless legs syndrome. Neurol Clin 1996;14:629–650.
15. Coleman RM. Periodic movements in sleep (nocturnal myoclonus) and restless legs syndrome. In Guilleminault C (ed): Sleeping and Waking Disorders: Indications and Techniques. Menlo Park, CA, Addison-Wesley, 1982:265–295.
16. Pollmächer T, Schulz H. Periodic leg movements (PLM): their relationship to sleep stages. Sleep 1993;16:572–577.
17. Smith RC, Gouin PR, Minkley P, et al. Periodic limb movement disorder is associated with normal motor conduction latencies when studied by central magnetic stimulation—successful use of a new technique. Sleep 1992;15:312–318.
18. Briellmann RS, Rosler KM, Hess CW. Blink reflex excitability is abnormal in patients with periodic leg movements in sleep. Mov Disord 1996;11:710–714.
19. Bucher SF, Trenkwalder C, Oertel WH. Reflex studies and MRI in the restless legs syndrome. Acta Neurol Scand 1996;94:145–150.
20. Bara-Jimenez W, Aksu M, Graham B, et al. Periodic limb movements in

sleep: state-dependent excitability of the spinal flexor reflex. Neurology 2000;54:1609–1616.
21. Provini F, Vetrugno R, Meletti S, et al. Motor pattern of periodic limb movements during sleep. Neurology 2001;57:300–304.
22. Trenkwalder C, Bucher SF, Oertel WH, et al. Bereitschaftspotential in idiopathic and symptomatic restless legs syndrome. Electroencephalogr Clin Neurophysiol 1993;89:95–103.
23. Tergau F, Wischer S, Paulus W. Motor system excitability in patients with restless legs syndrome. Neurology 1999;52:1060–1063.
24. Tugnoli V, Manconi M, Quartrale R, et al. Neurophysiological study of corticomotor pathway in patients with primary restless legs syndrome (RLS) [abstract]. Neurology Suppl 2000;54:A25.
25. Entezari-Taher M, Singleton JR, Jones CR, et al. Changes in excitability of motor cortical circuitry in primary restless legs syndrome. Neurology 1999;53:1201–1205.
26. Stiasny K, Haeske H, Mueller HH, et al. Impairment of cortical inhibition in restless legs syndrome-shortening of silent period induced by transcranial magnetic stimulation [abstract]. Sleep 2000;23:A131.
27. Ravnborg M, Liguori R, Christiansen P, et al. The diagnostic reliability of magnetically evoked motor potentials in multiple sclerosis. Neurology 1992;42:1296–1301.
28. Hening WA, Walters AS, Wagner M, et al. Circadian rhythm of motor restlessness and sensory symptoms in the idiopathic restless legs syndrome. Sleep 1999;22:901–912.
29. Rösler KM, Nirkko AC, Rihs F, et al. Motor-evoked responses to transcranial brain stimulation persist during cataplexy: a case report. Sleep 1994;17:168–171.
30. Hungs M, Mottaghy FM, Sparing R, et al. RTMS induces brief events of muscle atonia in patients with narcolepsy. Sleep 2000;23:1099–1104.
31. Cegla UH, Frode G. Transkranielle motorische Stimulation bei Patienten mit obstruktiver Schlafapnoe. Pneumologie 1993;47(Suppl 1):160–161.
32. Lu Z, Tang X, Huang X. Phrenic nerve conduction and diaphragmatic motor evoked potentials: evaluation of respiratory dysfunction. Chin Med J (Engl) 1998;111:496–499.
33. Riemann D, Berger M, Voderholzer U. Sleep and depression—results from psychobiological studies: an overview [review]. Biol Psychol 2001;57:67–103.
34. Benca RM, Obermeyer WH, Thisted RA, et al. Sleep and psychiatric disorders. A meta-analysis. Arch Gen Psychiatry 1992;49:651–668.
35. Borbely AA, Wirz Justice A. Sleep, sleep deprivation and depression. A hypothesis derived from a model of sleep regulation. Hum Neurobiol 1982;1:205–210.
36. Hobson JA, McCarley RW, Wyzynski PW. Sleep cycle oscillation: reciprocal discharge of two brainstem neuronal groups. Science 1975;189:55–58.
37. McCarley RW. REM sleep and depression: common neurobiological control mechanisms. Am J Psychiatry 1982;139:565–570.
38. Kupfer DJ, Spiker DG, Coble PA, et al. Sleep and treatment prediction in endogenous depression. Am J Psychiatry 1981;138:429–434.
39. Grunhaus L, Shipley JE, Eiser A, et al. Sleep-onset rapid eye movement after electroconvulsive therapy is more frequent in patients who respond less well to electroconvulsive therapy. Biol Psychiatry 1997;42:191–200.
40. Pascual Leone A, Rubio B, Pallardo F, et al. Rapid-rate transcranial magnetic stimulation of left dorsolateral prefrontal cortex in drug-resistant depression [see comments]. Lancet 1996;348:233–237.
41. George MS, Lisanby SH, Sackeim HA. Transcranial magnetic stimulation: applications in neuropsychiatry. Arch Gen Psychiatry 1999;56:300–311.
42. Lisanby SH, Kinnunen LH, Crupain MJ. Applications of TMS to therapy in psychiatry. J Clin Neurophysiol 2002;19:344–360.
43. Cohrs S, Tergau F, Riech S, et al. High-frequency repetitive transcranial magnetic stimulation delays rapid eye movement sleep. Neuroreport 1998;9:3439–3443.
44. Hajak G, Cohrs S, Tergau F, et al. Sleep and rTMS. Investigating the link between transcranial magnetic stimulation, sleep, and depression. Electroencephalogr Clin Neurophysiol Suppl 1999;51:315–321.
45. Graf T, Engeler J, Achermann P, et al. High frequency repetitive transcranial magnetic stimulation (rTMS) of the left dorsolateral cortex: EEG topography during waking and subsequent sleep. Psychiatry Res 2001;107:1–9.
46. Grisaru N, Bruno R, Pridmore S. Effect on the emotions of healthy individuals of slow repetitive transcranial magnetic stimulation applied to the prefrontal cortex. J ECT 2001;17:184–189.
47. Padberg F, Zwanzger P, Keck ME, et al. Repetitive transcranial magnetic stimulation (rTMS) in major depression: relation between efficacy and stimulation intensity. Neuropsychopharmacology 2002;27:638–645.
48. Cohrs S, Tergau F, Korn J, et al. Suprathreshold repetitive transcranial magnetic stimulation elevates thyroid-stimulating hormone in healthy male subjects. J Nerv Ment Dis 2001;189:393–397.
49. Baumgartner A, Graf KJ, Kurten I, et al. Neuroendocrinological investigations during sleep deprivation in depression. I. Early

morning levels of thyrotropin, TH, cortisol, prolactin, LH, FSH, estradiol, and testosterone. Biol Psychiatry 1990;28:556–568.
50. Parekh PI, Ketter TA, Altshuler L, et al. Relationships between thyroid hormone and antidepressant responses to total sleep deprivation in mood disorder patients. Biol Psychiatry 1998;43:392–394.
51. Prange AJ Jr, Wilson IC, Knox A, et al. Enhancement of imipramine by thyroid stimulating hormone: clinical and theoretical implications. Am J Psychiatry 1970;127:191–199.
52. Szuba MP, O'Reardon JP, Rai AS, et al. Acute mood and thyroid stimulating hormone effects of transcranial magnetic stimulation in major depression. Biol Psychiatry 2001;50:22–27.
53. Padberg F, Schule C, Zwanzger P, et al. Relation between responses to repetitive transcranial magnetic stimulation and partial sleep deprivation in major depression. J Psychiatr Res 2002;36:131–135.
54. Eichhammer P, Kharraz A, Wiegand R, et al. Sleep deprivation in depression stabilizing antidepressant effects by repetitive transcranial magnetic stimulation. Life Sci 2002;70:1741–1749.
55. Grunhaus L, Dannon PN, Schreiber S, et al. Repetitive transcranial magnetic stimulation is as effective as electroconvulsive therapy in the treatment of nondelusional major depressive disorder: an open study. Biol Psychiatry 2000;47:314–324.
56. Grunhaus L, Schreiber S, Dolberg OT, et al. A randomized controlled comparison of electroconvulsive therapy and repetitive transcranial magnetic stimulation in severe and resistant nonpsychotic major depression. Biol Psychiatry 2003;53:324–331.
57. George MS, Nahas Z, Molloy M, et al. A controlled trial of daily left prefrontal cortex TMS for treating depression. Biol Psychiatry 2000;48:962–970.

30 Transcranial Magnetic Stimulation Studies in Children

Marjorie A. Garvey

Single-pulse transcranial magnetic stimulation (TMS) provides a noninvasive, painless method of probing the motor system. The technique is of particular interest for studying maturation of the motor system and may provide insights into the developmental disabilities strongly associated with specific delays of motor development. This chapter reviews studies using single-pulse TMS in children, with particular reference to insights into neurodevelopment in children. It also briefly considers the potential of TMS as a diagnostic tool in neurological disorders. It does not address the use of repetitive TMS in children.

■ Safety and Acceptability

The main safety concerns of single-pulse TMS are psychological effects, the ability to provoke seizures, and effect on hearing. For the most part, safety of TMS in children has been extrapolated from adult studies, which are discussed in more detail elsewhere in this text. Since 1988, more than 75 TMS studies have been published including more than 2,000 children, most of whom were normal children. No adverse events in the healthy children were reported in any of these studies, and single-pulse TMS did not induce seizures in children with epilepsy.[1] These studies have reinforced the opinion that TMS is safe in children.

Only two studies have directly assessed safety in child populations. Transient changes in the quantitative electroencephalogram (i.e., autoregressive [AR] power analysis) have been induced by single-pulse TMS in children with cerebral palsy, but not in normal subjects.[2] These changes could not be seen on visual inspection of the electroencephalogram and were noticed only in the first-order elementary processes of the AR; the AR changes disappeared within 5 minutes. Hearing was tested before and after TMS in children with neurological disorders; no transient or permanent hearing loss was found in any subject.[3]

Acceptability of TMS in children was addressed in a study using a self-report questionnaire to assess children's subjective experience with TMS.[4] Thirty-eight subjects completed TMS, and 34 said they would repeat it. Descriptions of the sensation of the stimulus varied from a "pop" or "snap" to a "feeling like when you rub your feet on the carpet and then touch something metal." Children gave TMS an overall rating of 6.13 on 10-point scale (most disagreeable = 1; most enjoyable = 10) and ranked it fourth highest among common childhood events: higher than going to the dentist but lower than going to a birthday party. Two subjects (5%) discontinued TMS because they found it uncomfortable, suggesting that TMS does unsettle some children, although most tolerate it well.

Single-Pulse Transcranial Magnetic Stimulation

Most studies using TMS in children have concentrated on neurodevelopment. This is appropriate because an understanding of neurological disorders in children must be based on an in-depth understanding of normal development.

Motor Evoked Potential Thresholds

The motor evoked potential (MEP) threshold is higher in children than in adults, gradually decreasing to adult levels by mid-adolescence. As in adults, MEP threshold is lower when there is background muscle activation (i.e., active motor threshold [AMT]) than when the target muscle is at rest (i.e., resting motor threshold [RMT]). MEP responses can be elicited even from infants when there is background muscle activity by using 100% of stimulator output. However, even at maximal stimulator output, it may not be possible to elicit reliable MEP responses from children up to 6 years of age when muscles are at rest.[5–8]

Motor Evoked Potential Latency

An estimate of central motor conduction time (CMCT) can be obtained by measuring the latency of a TMS evoked MEP. There are two distinct developmental patterns of CMCT, depending on whether it is measured while muscles are at rest (i.e., resting CMCT) or with background muscle activation (i.e., active CMCT). Active CMCT reaches maturity in children by 2 years of age,[5,9] but resting CMCT does not reach maturity until early adolescence.[10] The difference between MEP latencies evoked at rest and those evoked during muscle activity (i.e., latency jump) is three to four times longer in preschool children than in adults and reaches maturity by mid-adolescence.[11] In adults, the latency jump is thought to reflect transsynaptic activation of cortical motoneurons by interneurons[12] and recruitment of faster pyramidal neurons at higher levels of muscle activation.[13] Mechanisms responsible for the development of this latency jump are still unclear but may include neuronal and synaptic maturation within the motor cortex, central myelinogenesis (which approaches maturity in early adolescence),[14,15] and developmental aspects of motoneuronal recruitment.[16,17]

Motor Evoked Potential Amplitude and Duration

The TMS-evoked MEP amplitude and duration have also been studied. MEP duration shows little change with age, although the MEP is polyphasic in children younger than 5 years of age compared with older children.[5,6] When stimulating at 10% above motor threshold or at maximal stimulator output, MEP amplitude shows a distinct maturational profile: little change in MEP amplitude before 10 years of age and gradual increase in MEP amplitude between 10 years of age and early adulthood.[6,18]

Cortical Maps

Cortical maps give an estimate of the somatotopic representations of muscles within the motor cortex. By stimulating at a number of different scalp positions, it is possible to assess the location of the optimal position for stimulation and the center of gravity, which defines the mean position of the map.[12,19] One study has examined cortical maps in healthy children as a comparison for children with cerebral palsy.[20] Cortical representation sites for the tibialis anterior, biceps brachialis, and abductor pollicis brevis muscle were identified between 1 and 4 cm, 4 and 6 cm, and 5 to 8 cm lateral to the cranial vertex, respectively. Although the investigators did not report the optimal stimulation site or the center of gravity, these data suggest that it is possible to study the somatotopic representations of muscles within the motor cortex in children. Future studies need to examine cortical maps more extensively throughout childhood to determine their developmental aspects.

Stimulus Response Curves

MEP amplitude increases with increasing stimulus intensity. Stimulus-response, or input-output, curves are used to assess this phenomenon. This measure assesses neurons that are intrinsically less excitable or spatially further from the center of activation. In adults, the shape of the curve is usually sigmoidal and its features are represented by threshold, steepness, and plateau level. No published studies have examined stimulus-response curves in children. Unpublished data from my laboratory for children 6 to 13 years of age show an age-related leftward shift in the threshold (reflecting decreasing MEP threshold with increasing age). A relative increase in slope was seen in children 11 to 13 years of age compared with younger children. None of the children in the age range studied reached a plateau level.

Ipsilateral Connections

Ipsilateral MEPs can be elicited in adult subjects. Corticofugal fibers mediating these ipsilateral MEPs appear to be dissociated from those mediating the contralateral MEPs because they are elicited at different coil orientations and they have discrete cortical maps.[21] When the dominant primary motor cortex is stimulated, MEPs are evoked simultaneously in the contralateral and ipsilateral target muscles in approximately 60% of normal children younger than 10 years. As in adults, the ipsilateral MEP occurs at a longer latency compared with the contralateral MEP.[22] After 10 years of age, these delayed ipsilateral MEPs become harder to elicit.[23] They may be mediated by ipsilateral oligosynaptic or polysynaptic projections, such as the corticoreticulospinal and corticopropriospinal pathways.[21,23] Large-amplitude ipsilateral MEPs have been found in a patient with complete agenesis of the corpus callosum[21] and in patients with cerebral palsy.[23] The functional purpose of these ipsilateral pathways is unclear. Their disappearance through normal development and persistence in patients with anomalous motor development suggest that they may be redundant pathways that remain persistently active only when the fast-conducting corticofugal pathways are malfunctioning.

Inhibitory Effects of Transcranial Magnetic Stimulation

The physiology of TMS-evoked silent periods has been extensively studied.[12,24,25] Inhibitory interneurons within the motor cortex are thought to be responsible for the ipsilateral (ISP) and contralateral (CSP) silent periods, and the corpus callosum is thought to mediate the ISP.[26] The ontogeny of these silent periods may reflect maturation of these inhibitory interneurons and myelogenesis of the corpus callosum.

The developmental trend of the CSP is not clear because of conflicting reports. Two studies have examined developmental trend of the CSP in children between 6 and 15 years of age. Duration of the CSP ranged between 3.5 and 207 ms using similar stimulation techniques.[7,8] Statistical analysis showed a significant age-related correlation in one study but no significant age-related differences in the other. The great variability of CSP duration in normal populations may explain the discrepancy and suggests that future studies will need to study much larger sample sizes to examine the developmental trajectory of CSP duration.

Transcallosal inhibition can be assessed by measuring the ISP.[27,28] The ISP is absent in children 4 to 6 years of age[29] and present in 10-year-old children.[30] Latency becomes shorter, and duration gets longer with increasing age. Both are close to maturity by 12 years of age.[7]

■ Paired-Pulse Transcranial Magnetic Stimulation

Intracortical excitability can be assessed by delivering two stimuli in a condition-test paradigm. Several different techniques have been described in adult subjects, but only one has been used in children. This is the short inter-stimulus intracortical inhibition (SICI) and intracortical facilitation (ICF) paradigm. The physiology of this technique has been discussed in detail elsewhere in this text. Only one study has examined SICI and ICF in children.[8] The investigators found no age-related changes in inhibition or facilitation in children between 8 and 13 years of age.

A two-coil conditioned-test paradigm has also been used to study intercortical inhibition.[31–33] Adults show inhibition when the conditioning pulse is given at any time between 5 and 15 ms before the test pulse (with maximal inhibition at 7 ms). In contrast, no clear developmental pattern was found for this in children 6 to 10 years of age, because they show facilitation and inhibition at all of the inter-stimulus intervals.[34]

Asymmetry of Cortical Function

The biological mechanisms underlying lateralization of the human nervous system are poorly understood but include structural asymmetries and genetic and environmental factors.[35] Human handedness is thought to arise from the anatomical and functional asymmetries of the primary motor cortex.[36] The most extensively studied phenomenon of cortical asymmetry is the MEP threshold. In adults, a mean difference of 5% (ranging between −10% and +15%) has been consistently found between the dominant (lower) and nondominant MEP threshold in adult subjects.[37-39] In adults, asymmetry has also been examined in cortical mapping (larger maps on the dominant hemisphere)[38,40,41]; the CSP (shorter duration when the dominant motor cortex was stimulated);[42] the SICI/ICF (conflicting reports)[43-45]; and the two-coil paired-pulse paradigm testing interhemispheric inhibition (greater inhibition after stimulation of the dominant hemisphere).[46]

Few TMS studies have examined cortical asymmetry in children. As in adults, MEP threshold is lower in the dominant motor cortex; this side-to-side difference is five times larger in 6- to 8-year-old children and is significantly larger in early adolescents compared with young adults.[7]

Asymmetry of inhibitory phenomena has been examined during development. Duration of the CSP evoked in the nondominant hemisphere is longer than that evoked in the dominant hemisphere.[7,47] Six-year-old children are more likely to have a dominant ISP than a nondominant ISP, and in these children, duration of the nondominant ISP is longer than the dominant ISP, but this asymmetry is not present in older children or in adults.

The mechanisms underlying the gradual decrease of cortical asymmetry during development are unclear. Increased use of the dominant hand may be a factor. When children are learning to write with their dominant hand, cortical asymmetry is marked; young adults, in contrast, are less likely to exercise their penmanship, writing instead using the computer. In these subjects, cortical asymmetry is present but smaller than that in children. Elderly, retired populations are more likely to be occupied with tasks that require the use of both hands equally. These subjects have no significant asymmetry of MEP thresholds.[48]

However, task-dependent plasticity (i.e., increasing use of the dominant hand leading to an increase in cortical excitability) cannot entirely explain cortical asymmetry of handedness. Hand dominance becomes apparent as early as 2 years of age and is firmly established by 4 years of age,[49] suggesting that that genetic and early (e.g., in utero hormonal) environmental factors may establish dominance and that continued use of the dominant hand adds stability. Future studies may shed light on these questions by examining the effects of sexual dimorphism, genetic loading, and environmental factors.

Transcranial Magnetic Stimulation Parameters and Neuromotor Development

Increasingly, investigators are examining the role of TMS parameters in normal motor function. It is possible that maturation of TMS-evoked excitatory and inhibitory phenomena reflects functional aspects of the neural substrate for neuromotor development. This is of particular interest in children, because abnormal motor function and delayed neuromotor development can be prominent features of developmental disabilities and of neurological and neuropsychiatric disorders. If abnormal TMS parameters are indicative of anomalous motor function, it may be possible to study these disorders using TMS.

Normal Neuromotor Function

Maturation of acoustic reaction times, tapping, and ballistic movements is similar to the developmental trajectory of resting CMCT[50] but not of active CMCT.[51] Speed of finger tapping is related to active MEP thresholds and latency but not to duration of the ISP, even when age is taken into account.[7] Duration of the CSP is not related to neuromotor function. However, similarity in age-related changes in two separate factors does not necessarily imply that these factors are related to each other. Statistical tests may provide a larger degree of confidence of the relationship but cannot provide certainty. Comparing healthy children with those who have developmental disabilities strongly associated with motor anomalies may provide more information.

Mirror Movements in Children

Mirror movements are normal in children up to 10 years of age. Adults also show mirror movements, but these can be elicited only during extremes of muscle effort, and they have minimal amplitude. Increased intensity or prolonged presence of significant mirror movements after 10 years of age is considered pathological. Pathological mirror movements have been described in children with congenital mirror movements (as part of Klippel-Feil syndrome and X-linked Kallmann's syndrome) and in children who have spastic hemiplegic cerebral palsy. Possible mechanisms for mirror movements include an inability of the active cortex to inhibit the corresponding motor cortex of the opposite hemisphere, immature or abnormally persistent ipsilateral motor pathways, and abnormal novel corticofugal pathways from the ipsilateral cortex that re-cross to the ipsilateral side.

TMS may be useful in distinguishing some of these mechanisms, especially when used in combination with other neurophysiologic tools or neuroimaging modalities. Combined cross-correlation electromyographic and TMS studies suggest that mirror movements in healthy children and adults arise from simultaneous activity of crossed corticospinal pathways originating from left and right motor cortices.[34] These findings concur with functional magnetic resonance imaging (fMRI) studies in adults that demonstrate inhibition of the ipsilateral motor cortex during unimanual phasic movements.

Evidence for novel corticofugal pathways from the ipsilateral cortex that re-cross to the ipsilateral side comes from studies in children with congenital mirror movements and those with congenital hemiplegia. Stimulation of the primary motor cortex of either hemisphere in congenital mirror movements and of the unaffected hemisphere in some children with congenital hemiplegia produces MEPs in contralateral and ipsilateral hands at exactly the same latency.[52–54]

Developmental Disabilities and Neurological Disorders

In neurological disorders, TMS may be able to detect the presence (or absence) of neurological involvement, provide information about disease progression, and monitor response to treatment. TMS may also have potential to reveal disease pathophysiology and mechanisms of action of pharmaceutical agents.

The only disorder in which TMS may play a diagnostic role is in younger patients with Rett's syndrome. These children show abnormally short resting CMCTs.[55–57] This has not been found in any other disorder and suggests abnormal synaptic organization within the motor cortex or abnormalities of the cortical or spinal motoneurons. The absence of abnormality, however, may also be useful when confirming a diagnosis. Children with Dopa-responsive dystonia can be misdiagnosed as having spastic diplegia but show no abnormalities of CMCT.[58]

Although TMS may not offer diagnostic insights in many disorders, it may be helpful in determining disease burden. Prolonged CMCT is present even in the early stages of Friedreich's ataxia,[59] and increasing severity of CMCT is related to disease progression.[60] Subclinical central nervous system abnormalities in the absence of overt neurological involvement have been found patients with mucolipidosis type III[61] and adrenoleukodystrophy.[62]

Treatment response has been observed in a patient with Wilson's disease,[63] and TMS has shown a dose-related increase in cortical excitability thresholds in children with well-controlled, benign childhood epilepsy with centrotemporal spikes who are treated with valproate, suggesting that the mechanism of action of this drug is to increase intracortical synaptic inhibition.[1]

TMS may shed light on the pathophysiology of certain diseases. The CMCT is close to normal in maple syrup urine disease, suggesting that functional abnormalities of dysmyelination are not a prominent feature, despite the injury to myelin composition that occurs in this disorder.[64] Malnutrition may also cause delayed CMCTs.[65]

TMS in other disorders appears to be less useful. No relationship could be found between disease severity and abnormality of the CMCT in Pelizaeus-Merzbacher disease.[66] In these patients, clinical improvement of neurological function during treatment with digitalis was not accompanied by changes in CMCT.[67]

Neuropsychiatric Disorders

Although abnormalities of TMS parameters have been found in neuropsychiatric disorders, they are nonspecific and, at most, give insights into possible neural substrates or have the potential for assessing treatment effect. Deficient intracortical inhibition using the paired-pulse paradigm occurs for children with attention deficit–hyperactivity disorder (ADHD); a single dose of methylphenidate reversed these abnormalities.[68] Decreased duration of the CSP occurs in patients with Tourette's syndrome, and patients with combined ADHD and Tourette's syndrome have decreased duration of the CSP and decreased intracortical inhibition.[69] Because decreased inhibition of intracortical inhibition and shorter duration of CSP have been found in many other disorders, these findings cannot be diagnostic, but they point to a possible neural substrate for these disorders and suggest that although ADHD and Tourette's syndrome are frequently comorbid, they are associated with distinct abnormalities of cortical inhibition.

■ Future Studies

TMS may prove to be a useful tool in the investigation of normal development, developmental disabilities, and neurodegenerative diseases in children. Although many investigators have published small case reports or case series on different disorders, few have gone on to fully characterize these disorders. The potential for assessing medication compliance and therapeutic effect in disorders treated with neuroactive drugs has not yet been explored.

REFERENCES

1. Nezu A, Kimura S, Ohtsuki N, et al. Transcranial magnetic stimulation in benign childhood epilepsy with centro-temporal spikes. Brain Dev 1997;19:134–137.
2. Yasuhara A, Niki T, Ochi A. Changes in EEG after transcranial magnetic stimulation in children with cerebral palsy. Electroencephalogr Clin Neurophysiol Suppl 1999;49:233–238.
3. Collado-Corona MA, Mora-Magana I, Cordero GL, et al. Transcranial magnetic stimulation and acoustic trauma or hearing loss in children. Neurol Res 2001;23:343–346.
4. Garvey MA, Kaczynski KJ, Becker DA, et al. Subjective reactions of children to single-pulse transcranial magnetic stimulation. J Child Neurol 2001;16:891–894.
5. Koh TH, Eyre JA. Maturation of corticospinal tracts assessed by electromagnetic stimulation of the motor cortex. Arch Dis Child 1988;63:1347–1352.
6. Nezu A, Kimura S, Uehara S, et al. Magnetic stimulation of motor cortex in children: maturity of corticospinal pathway and problem of clinical application. Brain Dev 1997;19:176–180.
7. Garvey MA, Ziemann U, Bartko JJ, et al. Cortical correlates of neuromotor development in healthy children. Clin Neurophysiol 2003;114:1662–1670.
8. Moll GH, Heinrich H, Wischer S, et al. Motor system excitability in healthy children: developmental aspects from transcranial magnetic stimulation. Electroencephalogr Clin Neurophysiol Suppl 1999;51:243–249.
9. Eyre JA, Miller S, Ramesh V. Constancy of central conduction delays during development in man: investigation of motor and somatosensory pathways. J Physiol 1991;434:441–452.
10. Müller K, Ebner B, Hömberg V. Maturation of fastest afferent and efferent central and peripheral pathways: no evidence for a constancy of central conduction delays. Neurosci Lett 1994;166:9–12.

11. Caramia MD, Desiato MT, Cicinelli P, et al. Latency jump of "relaxed" versus "contracted" motor evoked potentials as a marker of cortico-spinal maturation. Electroencephalogr Clin Neurophysiol 1993;89:61–66.
12. Abbruzzese G, Trompetto C. Clinical and research methods for evaluating cortical excitability. J Clin Neurophysiol 2002;19:307–321.
13. Rossini PM, Barker AT, Berardelli A, et al. Non-invasive electrical and magnetic stimulation of the brain, spinal cord and roots: basic principles and procedures for routine clinical application. Report of an IFCN committee. Electroencephalogr Clin Neurophysiol 1994;91:79–92.
14. Giedd JN, Blumenthal J, Jeffries NO, et al. Development of the human corpus callosum during childhood and adolescence: a longitudinal MRI study. Prog Neuropsychopharmacol Biol Psychiatry 1999;23:571–588.
15. Yakovlev PI, Lecours A-R. The myelogenetic cycles of regional maturation of the brain. In Minkowski A (ed): Regional Development of the Brain in Early Life. Oxford, Blackwell, 1967:3–70.
16. Paus T, Zijdenbos A, Worsley K, et al. Structural maturation of neural pathways in children and adolescents: in vivo study. Science 1999;283:1908–1911.
17. Huttenlocher PR, Dabholkar AS. Regional differences in synaptogenesis in human cerebral cortex. J Comp Neurol 1997;387:167–178.
18. Müller K, Hömberg V, Lenard HG. Magnetic stimulation of motor cortex and nerve roots in children. Maturation of cortico-motoneuronal projections. Electroencephalogr Clin Neurophysiol 1991;81:63–70.
19. Chen R. Studies of human motor physiology with transcranial magnetic stimulation. Muscle Nerve Suppl 2000;9:S26–S32.
20. Maegaki Y, Maeoka Y, Ishii S, et al. Central motor reorganization in cerebral palsy patients with bilateral cerebral lesions. Pediatr Res 1999;45(Pt 1):559–567.
21. Ziemann U, Ishii K, Borgheresi A, et al. Dissociation of the pathways mediating ipsilateral and contralateral motor-evoked potentials in human hand and arm muscles. J Physiol 1999;518:895–906.
22. Müller K, Kass-Iliyya F, Reitz M. Ontogeny of ipsilateral corticospinal projections: a developmental study with transcranial magnetic stimulation. Ann Neurol 1997;42:705–711.
23. Eyre JA, Taylor JP, Villagra F, et al. Evidence of activity-dependent withdrawal of corticospinal projections during human development. Neurology 2001;57:1543–1554.
24. Hallett M. Transcranial magnetic stimulation. Negative effects. Adv Neurol 1995;67:107–113.
25. Terao Y, Ugawa Y. Basic mechanisms of TMS. J Clin Neurophysiol 2002;19:322–343.
26. Meyer B-U, Röricht S, Gräfin von Einsiedel H, et al. Inhibitory and excitatory interhemispheric transfers between motor cortical areas in normal humans and patients with abnormalities of the corpus callosum. Brain 1995;118:429–440.
27. Wassermann EM, Fuhr P, Cohen LG, et al. Effects of transcranial magnetic stimulation on ipsilateral muscles. Neurology 1991;41:1795–1799.
28. Meyer B-U, Röricht S, Woiciechowsky C. Topography of fibers in the human corpus callosum mediating interhemispheric inhibition between the motor cortices. Ann Neurol 1998;43:360–369.
29. Heinen F, Glocker FX, Fietzek U, et al. Absence of transcallosal inhibition following focal magnetic stimulation in preschool children. Ann Neurol 1998;43:608–612.
30. Heinen F, Kirschner J, Fietzek U, et al. Absence of transcallosal inhibition in adolescents with diplegic cerebral palsy. Muscle Nerve 1999;22:255–257.
31. Ferbert A, Priori A, Rothwell JC, et al. Interhemispheric inhibition of the human motor cortex. J Physiol 1992;453:525–546.
32. Schnitzler A, Kessler KR, Benecke R. Transcallosally mediated inhibition of interneurons within human primary motor cortex. Exp Brain Res 1996;112:381–391.
33. Hanajima R, Ugawa Y, Machii K, et al. Interhemispheric facilitation of the hand motor area in humans. J Physiol 2001;531(Pt 3):849–859.
34. Mayston MJ, Harrison LM, Stephens JA. A neurophysiological study of mirror movements in adults and children. Ann Neurol 1999;45:583–594.
35. Geschwind N, Galaburda AM. Cerebral lateralization. Biological mechanisms, associations, and pathology. I. A hypothesis and a program for research. Arch Neurol 1985;42:428–459.
36. Hammond G. Correlates of human handedness in primary motor cortex: a review and hypothesis. Neurosci Biobehav Rev 2002;26:285–292.
37. Macdonell RA, Shapiro BE, Chiappa KH, et al. Hemispheric threshold differences for motor evoked potentials produced by magnetic coil stimulation. Neurology 1991;41:1441–1444.
38. Cicinelli P, Traversa R, Bassi A, et al. Interhemispheric differences of hand muscle representation in human motor cortex. Muscle Nerve 1997;20:535–542.
39. Triggs WJ, Calvanio R, Macdonell RA, et al. Physiological motor asymmetry in human handedness: evidence from transcranial

magnetic stimulation. Brain Res 1994;636:270–276.
40. Triggs WJ, Subramanium B, Rossi F. Hand preference and transcranial magnetic stimulation asymmetry of cortical motor representation. Brain Res 1999;835:324–329.
41. Volkmann J, Schnitzler A, Witte OW, et al. Handedness and asymmetry of hand representation in human motor cortex. J Neurophysiol 1998;79:2149–2154.
42. Priori A, Oliviero A, Donati E, et al. Human handedness and asymmetry of the motor cortical silent period. Exp Brain Res 1999;128:390–396.
43. Cicinelli P, Traversa R, Oliveri M, et al. Intracortical excitatory and inhibitory phenomena to paired transcranial magnetic stimulation in healthy human subjects: differences between the right and left hemisphere. Neurosci Lett 2000;288:171–174.
44. Civardi C, Cavalli A, Naldi P, et al. Hemispheric asymmetries of cortico-cortical connections in human hand motor areas. Clin Neurophysiol 2000;111:624–629.
45. Maeda F, Gangitano M, Thall M, et al. Inter- and intra-individual variability of paired-pulse curves with transcranial magnetic stimulation (TMS). Clin Neurophysiol 2002;113:376–382.
46. Netz J, Ziemann U, Hömberg V. Hemispheric asymmetry of transcallosal inhibition in man. Exp Brain Res 1995;104:527–533.
47. Masur H, Althoff S, Kurlemann G, et al. Inhibitory period and late muscular responses after transcranial magnetic stimulation in healthy children. Brain Dev 1995;17:149–152.
48. Matsunaga K, Uozumi T, Tsuji S, et al. Age-dependent changes in physiological threshold asymmetries for the motor evoked potential and silent period following transcranial magnetic stimulation. Electroencephalogr Clin Neurophysiol 1998;109:502–507.
49. Illingsworth RS. The Development of the Infant and Young Child: Normal and Abnormal, 8th ed. Edinburgh, Churchill Livingstone, 1983.
50. Müller K, Hömberg V. Development of speed of repetitive movements in children is determined by structural changes in corticospinal efferents. Neurosci Lett 1992;144:57–60.
51. Heinen F, Fietzek UM, Berweck S, et al. Fast corticospinal system and motor performance in children: conduction proceeds skill. Pediatr Neurol 1998;19:217–221.
52. Cincotta M, Borgheresi A, Liotta P, et al. Reorganization of the motor cortex in a patient with congenital hemiparesis and mirror movements. Neurology 2000;55:129–131.
53. Reitz M, Muller K. Differences between 'congenital mirror movements' and 'associated movements' in normal children: a neurophysiological case study. Neurosci Lett 1998;256:69–72.
54. Carr LJ, Harrison LM, Evans AL, et al. Patterns of central motor reorganization in hemiplegic cerebral palsy. Brain 1993;116:1223–1247.
55. Eyre JA, Kerr AM, Miller S, et al. Neurophysiological observations on corticospinal projections to the upper limb in subjects with Rett syndrome. J Neurol Neurosurg Psychiatry 1990;53:874–879.
56. Nezu A, Kimura S, Takeshita S, et al. Characteristic response to transcranial magnetic stimulation in Rett syndrome. Electroencephalogr Clin Neurophysiol 1998;109:100–103.
57. Heinen F, Petersen H, Fietzek U, et al. Transcranial magnetic stimulation in patients with Rett syndrome: preliminary results. Eur Child Adolesc Psychiatry 1997;6(Suppl 1):61–63.
58. Müller K, Hömberg V, Lenard HG. Motor control in childhood onset dopa-responsive dystonia (Segawa syndrome). Neuropediatrics 1989;20:185–191.
59. Cruz Martinez A, Anciones B. Central motor conduction to upper and lower limbs after magnetic stimulation of the brain and peripheral nerve abnormalities in 20 patients with Friedreich's ataxia. Acta Neurol Scand 1992;85:323–326.
60. Cruz-Martinez A, Palau F. Central motor conduction time by magnetic stimulation of the cortex and peripheral nerve conduction follow-up studies in Friedreich's ataxia. Electroencephalogr Clin Neurophysiol 1997;105:458–461.
61. Toscano E, Perretti A, Balbi P, et al. Detection of subclinical central nervous system abnormalities in two patients with mucolipidosis III by the use of motor and somatosensory evoked potentials. Neuropediatrics 1998;29:40–42.
62. Nezu A, Kimura S, Kobayashi T, et al. Transcranial magnetic stimulation in an adrenoleukodystrophy patient. Brain Dev 1996;18:327–329.
63. Meyer B-U, Britton TC, Benecke R. Wilson's disease: normalisation of cortically evoked motor responses with treatment. J Neurol 1991;238:327–330.
64. Müller K, Kahn T, Wendel U. Is demyelination a feature of maple syrup urine disease? Pediatr Neurol 1993;9:375–382.
65. Tamer SK, Misra S, Jaiswal S. Central motor conduction time in malnourished children. Arch Dis Child 1997;77:323–325.

66. Nezu A, Kimura S, Takeshita S, et al. Magnetic stimulation of the corticospinal tracts in Pelizaeus-Merzbacher disease. Electroencephalogr Clin Neurophysiol 1998;108:446–448.
67. Nezu A, Kimura S, Osaka H, et al. Effect of digitalis on conduction dysfunction in Pelizaeus-Merzbacher disease. J Neurol Sci 1996;141:49–53.
68. Moll GH, Heinrich H, Trott G, et al. Deficient intracortical inhibition in drug-naive children with attention-deficit-hyperactivity disorder is enhanced by methylphenidate. Neurosci Lett 2000;284:121–125.
69. Moll GH, Heinrich H, Trott GE, et al. Children with comorbid attention-deficit-hyperactivity disorder and tic disorder: evidence for additive inhibitory deficits within the motor system. Ann Neurol 2001;49:393–396.

Index

Page numbers followed by f indicate figures; those followed by t indicate tabular material.

A

Abductor digit minimi, motor evoked potentials of, 123f
Abductor hallucis, motor evoked potentials of, 115f, 117f, 124f
Action observation and action execution matching system, 175–176
Active motor thresholds
  antiepileptic drugs effect on motor excitability in epileptic patients, 261
  description of, 49, 52, 253
  in epilepsy, 255–258
Active motor training, 146
Adrenoleukodystrophy, 119
Amyotrophic lateral sclerosis
  central motor conduction in, 94–95
  central motor conduction time prolongation in, 94–95
  cortical silent period in, 95–96
  corticomotor threshold in, 95
  corticospinal tract hyperexcitability in, 162
  definition of, 155
  diagnosis of, 94
  electromyography of, 155
  excitatory postsynaptic potentials, 156–157
  fasciculations in, 160
  genetic findings, 155
  masseter motor evoked potentials in, 130
  motor evoked potentials in
    abnormalities of, 156–157
    description of, 95, 155
    latency of, 157
    low-amplitude, 160
    resting motor evoked potential threshold, 156–157
    surface electrodes for recording of, 155–156
    waveforms, 160–161
  pathophysiology of, 161–162
  peri-stimulus time histogram in, 160
  single motor unit recording in, 158–160
  superoxide dismutase 1, 155
  tongue motor evoked potentials in, 138–139
  transcranial magnetic stimulation in
    description of, 94–96
    fasciculations evoked by, 160
    motoneuron firing evoked by, 156
  triple-stimulation technique applied to, 95
  with upper motoneuron signs, 160
Analgesia
  antinociceptive effects, 401
  functional magnetic stimulation-induced, 400–402
  transcutaneous electrical nerve stimulation for, 401
Anisotropy, 13–14
Anterior (presternal) magnetic stimulation, 384
Antidepressants, transcranial magnetic stimulation use as
  description of, 315–318
  mechanisms, 319–320
Antiepileptic drugs, single- and paired-pulse transcranial magnetic stimulation studies of
  active motor thresholds, 261
  cortical silent period duration, 263
  description of, 261

Antiepileptic drugs, single- and paired-pulse
        transcranial magnetic stimulation
        studies of (*Continued*)
    intracortical facilitation, 263
    long-interval intracortical inhibition, 263
    motor evoked potential size, 262–263
    resting motor thresholds, 261
    short-interval intracortical inhibition, 263
Anxiety disorders, 320–321
Apomorphine, 183
Area under the curve, 85
Ataxia
    central motor conduction time prolongation, 190
    cerebellar
        description of, 205–208
        early-onset, 97
    cerebellar stimulation in, 204–208
    early-onset cerebellar, 97
    Friedreich's
        central motor conduction time prolongation in, 97
        characteristics of, 97
    hypothyroidism-related, 206
    intracortical inhibition of motor cortex of patients with, 204–208
    motor threshold in, 190
    paired-pulse studies in, 190
    repetitive transcranial magnetic stimulation of, 190
    silent period in, 190
    of unknown pathophysiology, 206
Atipamezole, 402
Attention deficit-hyperactivity disorder, 307, 434
Autosomal dominant cortical myoclonus epilepsy, 256
Awareness, 290

B

Barthel Index, 244
Benign myoclonus epilepsy, 189
Betz cells, 165
Biceps brachii
    central motor conduction of, 109
    motor evoked potentials of, 108f, 213
Bilateral anterior magnetic phrenic nerve stimulation, 384
Biot-Savart law, 7
Biphasic pulses, 62f
Biphasic stimulators, 50
Biphasic waveform, 32, 32f
Bladder, functional magnetic stimulation of, 399–400
Blood oxygen level-dependent functional magnetic resonance imaging, 283, 312–313
Brain-derived neurotrophic factor, 72
Brainstem infarction, 135
Broca's area, 69
Brodmann area, 44, 167

C

Cadwell coil, 35–36
Canalicular stimulation, 87
Cartesian coordinates, 3
Central disc protrusions, 122
Central fatigue, 212
Central lumbar canal stenosis, 116, 116f
Central motor conduction
    abnormalities of, 92t
    amyotrophic lateral sclerosis findings, 94–95
    to cranial muscles, 95
    description of, 83
    from hand muscles, 85
    in hereditary spastic paraplegia, 96
    multiple sclerosis findings, 91
    multiple system atrophy findings, 98
    normal values of, 88–90
    in primary lateral sclerosis, 96
    progressive supranuclear palsy findings, 98
    psychogenic weakness findings, 98
    stroke findings, 97
    suspended abnormality of, 111–112, 112f, 116
Central motor conduction time
    abnormalities of, 90
    calculation of, 106
    in children, 430
    compressive myelopathy findings, 96–97
    definition of, 86, 157
    in dystonia, 186
    to lower limbs, 88
    myelopathy evaluations, 106, 108
    normal values for, 88–90
    Parkinson's disease findings, 181–182
    prolongation of
        in amyotrophic lateral sclerosis, 94–95, 157
        in ataxia, 190
        description of, 90–91
        in Friedreich's ataxia, 97
        in Huntington's disease, 98, 188
        in multiple sclerosis, 91, 93
        in multiple system atrophy, 185
        in progressive supranuclear palsy, 185
        in Wilson's disease, 185
Central nervous system
    functional recovery of, 239
    plasticity of
        compensatory, 239
        cross-modal, 148–149
        description of, 143

evaluation of, 143–144
    muscarinic receptor function in, 144–145
    use-dependent, 144
Cerebellar ataxia
    description of, 205–208
    early-onset, 97
Cerebellar stimulation
    in ataxia patients, 204–208
    limitations of, 208–209
    in nonataxia patients, 204–205
    in normal subjects
        cerebellar-cerebral interactions, 201–204
        descending volleys, 199
        description of, 197
        facilitatory effect, 201, 202f
        inhibitory effect, 197–201
    studies of, 208
Cerebellocortical inhibition, 56–57
Cerebellomotor-cortical interaction, 201–204, 203f
Cerebellothalamocortical pathway, 170
Cerebral blood flow
    regional, 69, 234
    repetitive transcranial magnetic stimulation
        measurement of, 69
Cerebral cortex, magnetic stimulation of, 395
Cerebrum, repetitive transcranial magnetic
    stimulation effects on, 223
Cervical cord compression, 109
Cervical roots, spondylotic compression of, 122
Cervical spondylotic myelopathy
    causes of, 109
    description of, 108
    dorsal cord compression, 113–115
    high cervical cord lesion, 109–110
    lumbosacral cord lesions, 115
    motoneuron disease, 110–111
    motor evoked potentials studies, 108–118
    with pure spastic paraparesis, 109
    spinal cord involvement evaluations, 113, 114f
Cervicomedullary junction stimulation, 110
Children
    attention deficit-hyperactivity disorder in,
        307, 434
    central motor conduction time in, 430
    cortical function asymmetry in, 432
    cortical maps in, 430
    developmental disabilities in, 433–434
    mirror movements in, 432–433
    neurological disorders in, 433–434
    neuromotor function in, 432–433
    neuropsychiatric disorders in, 434
    with Rett's syndrome, 433
    Tourette's syndrome in, 434
    transcranial magnetic stimulation studies in
        acceptability of, 429
        description of, 429

developmental disabilities, 433–434
inhibitory effects, 431
intracortical excitability, 431
ipsilateral motor evoked potentials, 431
motor evoked potentials
    amplitude, 430
    latency, 430
    thresholds, 430
neurological disorders, 433–434
neuropsychiatric disorders, 434
paired-pulse, 431–432
parameters, 432–434
safety of, 429
single-pulse, 430–431
stimulus response curves, 430–431
Chronic fatigue syndromes, 218
Chronic pain, 321
Cognition
    deficits in, 295–297
    definition of, 281–282
    functional magnetic resonance imaging studies of,
        281–282
    multitechnique approach to, 282–293
    pharmacology of, 295
    positron emission tomography evaluations of,
        281–282
    repetitive transcranial magnetic stimulation
        applications to, 68–69, 74, 291
    time course of, 291–292
    transcranial magnetic stimulation applications to
        blood oxygen level-dependent measures of
            brain activation used with, 283
        cognitive deficits, 295–297
        control site design, 283
        control task design, 283
        control time design, 283
        experimental design, 283–284
        functional connectivity, 289–291
        functional interaction, 289–291
        functional specialization, 285–289
        overview of, 281, 297t
        plasticity, 293–295
        random number generation, 285
        sensitive measures used in, 283
        sham stimulation compared with, 284
        stimulation parameters, 284–285
        stimulation site, 284–285
        time course evaluations, 291–292
Coils
    circular, 23, 50, 83, 84f
    cone, 23
    consistent orientation of, 36f, 37
    construction of, 26
    cooling of, 25
    8-shaped, 23, 23f, 32f, 33, 50
    with ferromagnetic core, 23–24

Coils (*Continued*)
  forces that affect, 28
  illustration of, 33f
  inconsistent orientation of, 36, 36f
  inductance of, 27
  orientations of, 36f, 36–37, 62
  peak magnetic energy in, 27
  sham, 25, 25f
  slinky, 25–26
  tilting of, 24f
  warming of, 27–28
Color perception suppression, 332–333
Compensatory plasticity, 239
Compound muscle action potentials
  corticospinal function monitoring, 376
  description of, 48, 85, 368, 371, 374
Compressive myelopathy
  central motor conduction time prolongation in, 96–97
  motor evoked potentials evaluation of, 96–97
Conduction block, 91
Conductors, in electrostatic field, 5–6
Cone coils, 23
Consciousness, 290
Consistent orientation of coils, 36f, 37
Constraint-induced movement therapy, 146, 246
Cortical excitability
  cortisol-derived neurosteroids' effect on, 306–307
  description of, 253
  in epilepsy, 255–260
  high-frequency repetitive transcranial magnetic stimulation effects on, 296–297
  injury/illness effects on, 294–295
  language and, 276–278
  learning effects on, 293–294
  measures of, 253–255
  in migraine, 416
  in obsessive-compulsive disorder, 307
  paired-pulse transcranial magnetic stimulation of, 313
  Parkinson's disease effects on, 295
  personality effects on, 307
  pharmacological agents' effect on, 295
  selective modulation of, 343
  stroke effects on, 295
  transcranial direct current stimulation-induced modulations of, 337
Cortical infarcts, 241
Cortical inhibition
  paired-pulse transcranial magnetic stimulation measurement of, 66
  repetitive transcranial magnetic stimulation and, 65–66
Cortical maps, 430
Cortical myoclonus epilepsy, 189

Cortical plasticity, 67
Cortical silent period
  in amyotrophic lateral sclerosis, 95–96
  antiepileptic drugs' effect on, 263
  definition of, 254
  description of, 87–88
  in epilepsy, 258–259
  in stroke, 97–98
Corticobasal degeneration, 185
Corticofugal pathways of motor system
  contralateral responses, 168–169
  corticospinal projections targeting inhibitory spinal interneurons, 169
  monosynaptic corticomotoneural projections, 166–167
  oligosynaptic corticomotoneuronal projections, 167–168
Corticomotoneuronal projections
  monosynaptic, 166–167
  oligosynaptic, 167–168
Corticomotor excitability
  description of, 147
  gain function of, 232
  repetitive transcranial magnetic stimulation effects on
    in hand dystonia, 234–235
    in Parkinson's disease, 231–232
Corticomotor threshold
  in amyotrophic lateral sclerosis, 95
  description of, 88
Corticospinal axons, 365
Corticospinal excitability, 225
Corticospinal function, intraoperative monitoring of
  anesthesia, 375–376
  compound muscle action potentials, 368, 375–377
  cost-effectiveness of, 365
  description of, 365–368
  electrical stimulation for, 369–370
  epidural volleys, 373–374, 377
  ideal technique for, 367–368
  single transcranial electric stimuli, 377
  stimulation site, 369
  transcranial magnetic stimulation for
    corticospinal volley, 373–374
    electrical stimulation vs., 369–370
    recordings, 373–375
    reproducibility of potentials, 376
    stimulating electrodes, 370
    stimuli, 370
    stimulus parameters, 371–373
    stimulus trains, 371
    technique, 376–377
Corticospinal projections, 169
Corticospinal tract neurons, 201

Corticospinal volley, 373–374
Coulomb's law, 1–2
Cranial nerve(s)
   cortical control of, 129
   lateralization, 129
   motoneurons, 129–130
   V, 130–131
   VII, 131, 134–135
   IX, 133t
   X, 133t
   XI, 136–137
   XII, 133t, 137–139
Craniotopic updating, 353–354
Cricopharyngeal muscle, 136
Critical care neuromyopathy, 388
Cross-modal plasticity, 148–149
Current continuity, 6–7
Current density, induced, 26–27
Current pulse shape, 27

D

Deafferentation, 147
Demyelinating neuropathies, 122, 124
Dentate nucleus, in Wilson's disease, 207
Dentatothalamocortical pathway, 197
Depression
   brain regions involved in, 314–315
   electroconvulsive therapy for
     description of, 311, 314, 425
     transcranial magnetic stimulation compared with, 316–317
   fatigue in, 219
   repetitive transcranial magnetic stimulation for, 71, 219
   sleep deprivation and, 317, 423, 425
   transcranial magnetic stimulation for, 314–318
Descending corticospinal volleys
   description of, 46f, 48, 90
   epidural recording of, 373–374, 377
   multiple, 156
   recording of, 367f
Dextromethorphan, 144
Diaphragm
   anatomy of, 381–382
   automatic control of, 382
   electromyogram of, 385–387
   function of, 381
   voluntary control of, 382
Disynaptic reciprocal inhibition, 169
Dorsal cord compression
   central motor conduction abnormality in, 113
   intracranial demyelinating disorder vs., 114–115
   motor evoked potentials studies of, 113–114

Dorsolateral prefrontal cortex, 350, 359
Double-pulse transcranial magnetic stimulation, 352
Dual-pulse transcranial magnetic stimulation, 23
D waves, 43, 45, 373
Dysphagia, 135
Dystonia
   central motor conduction time in, 186
   hand, repetitive transcranial magnetic stimulation for
     after-effects of, 233–235
     description of, 232–233
     physiological rationale for, 233
   intracortical inhibition in, 187
   motor evoked potentials findings, 186
   motor threshold in, 186
   paired-pulse technique in, 187
   repetitive transcranial magnetic stimulation in, 187–188
   silent period in, 186–187

E

Early-onset cerebellar ataxia, 97
8-shaped coil, 23, 23f, 32f, 33, 50
Electrical brain stimulation
   experiments
     on animals, 43–45
     on humans, 45–46
   magnetic stimulation vs., 31–32
Electrical stimulation
   functional, of respiratory muscles, 395
   intracranial, 281
   magnetic stimulation vs., 31–32, 393
   phrenic nerve, 382
   transcranial. *See* Transcranial electrical stimulation
Electric charge, static, 1
Electric field
   bifocal stimulation generation of, 20f
   description of, 2–3
   induced
     definition of, 10, 26
     electrical conductivity boundaries on spatial distribution of, 12–14
     spatial distribution of, 12–14
   primary, 20
   secondary, 20
   strength of, 24f
   total, 12
   in transcranial electrical stimulation, 19–20
   transneuroforaminal, 38, 38f
Electric field intensity, 2
Electric fields, description of, 1
Electric flux, 4–5

Electric potential, 3–4
Electroconvulsive therapy
  depression treated with
    description of, 311, 314, 425
    transcranial magnetic stimulation vs., 316–317
  description of, 311
  seizures, 321
  transcranial magnetic stimulation compared with, for depression, 316–317
Electroencephalography, 70–71, 76
Electromagnetic induction, 1
Electromotive force, 6
Electromyography
  amyotrophic lateral sclerosis evaluations, 155
  diaphragm, 385–387
  electrodes used in, 386
  esophageal, 387
  properties of, 382t
  respiratory muscles, 385–387
Electrostatic field, conductors in, 5–6
Electrostatic force, 2f
Epilepsy
  antiepileptic drugs for. See Antiepileptic drugs
  autosomal dominant cortical myoclonus, 256
  description of, 189
  focal, 259
  motor excitability in, 255–260
  repetitive transcranial magnetic stimulation applications
    description of, 263
    high-frequency, 263–265
    low-frequency, 263–264
    safety of, 265
  single- and paired-pulse transcranial magnetic stimulation in
    active motor thresholds, 255–258
    cortical silent period duration, 258–259
    description of, 256t–257t
    intracortical facilitation, 259–260
    long-interval intracortical facilitation, 260
    long-interval intracortical inhibition, 260
    motor evoked potential size, 258
    resting motor thresholds, 255–258
    safety of, 260–261
    short-interval intracortical inhibition, 259
Epileptic focus
  description of, 260–261
  high-frequency repetitive transcranial magnetic stimulation activation of, 264–265
Erb's point, 86
Esophageal electromyography, 387
Esophageal pressure, 387
Event-related desynchronization, 296

Event-related functional magnetic resonance imaging, 292
Excitability
  buildup of, for motor behavior, 171–172
  cortical
    cortisol-derived neurosteroids' effect on, 306–307
    description of, 253
    in epilepsy, 255–260
    high-frequency repetitive transcranial magnetic stimulation effects on, 296–297
    injury/illness effects on, 294–295
    language and, 276–278
    learning effects on, 293–294
    measures of, 253–255
    in migraine, 416
    in obsessive-compulsive disorder, 307
    paired-pulse transcranial magnetic stimulation of, 313
    Parkinson's disease effects on, 295
    personality effects on, 307
    selective modulation of, 343
    stroke effects on, 295
    transcranial direct current stimulation-induced modulations of, 337
  corticomotor, 147
  repetitive transcranial magnetic stimulation conditioning effects on, 224
Excitatory postsynaptic potentials, in amyotrophic lateral sclerosis, 156
Eye movements
  cortical regions important to, topologic mapping of, 356–357
  craniotopic updating, 353–354
  description of, 349
  frontal eye fields, 350–352, 355–356, 358
  interference with processing of, 353–356
  mapping experiments, 356–360
  repetitive transcranial magnetic stimulation studies of, 360–361
  saccadic
    description of, 349–350, 353
    double-step, 355
    focal transcranial magnetic stimulation effects on, 360
    intentional, 354
    predictive, 354
    reflexive, 354
    visual space across, 354
  single-pulse transcranial magnetic stimulation effects on, 353
  transcranial magnetic stimulation effects on, 350–353
  triggering of, 353

F

Facial nerve
  motor evoked potentials, 134
  proximal portion of, 135
  transcranial magnetic stimulation of, 87, 135
Faraday's law of electromagnetic induction, 9–10, 61
Farad per meter, 2
Fasciculations, in amyotrophic lateral sclerosis, 160
Fatigue
  central, 212
  in chronic fatigue syndromes, 218
  curves for, 211, 211f
  definition of, 211
  in depression, 219
  historical perspective of, 211–212
  neurological disorders with
    description of, 217
    multiple sclerosis, 217–218
  in post-polio syndrome, 218–219
  transcranial magnetic stimulation studies of
    under active conditions after fatiguing exercise, 216–217
    in chronic fatigue syndromes, 218
    description of, 212
    maximal sustained contractions, 212–213
    in multiple sclerosis, 217–218
    in post-polio syndrome, 218–219
    reduced resting motor evoked potentials amplitudes, 215
    under resting conditions after fatiguing contractions, 214–216
    silent period duration, 213
    submaximal contractions, 213–214
Feedback inhibitory interneurons, 168
Fibrinolysis, 400
First dorsal interosseous muscle, 155, 212
Fisher's syndrome, 206
Flux
  of a vector quantity, 4–5
  electric, 4–5
Flux density, 7–9
Focal magnetic stimulation, 403
Friedreich's ataxia
  central motor conduction time prolongation in, 97
  characteristics of, 97
Frontal eye fields, 350–352, 355–356, 358
Functional electrical stimulation
  functional magnetic stimulation vs., 395
  of respiratory muscles, 395
Functional magnetic resonance imaging
  blood oxygen level-dependent, 283, 312–313
  cognition studies, 281–282
  cortical activity and connectivity evaluated with, 69–70
  event-related, 292
Functional magnetic stimulation
  analgesia induced by, 400–402
  antinociceptive effects of, 401
  of bladder, 399–400
  endogenous neurotransmitters' effect on, 401
  functional electrical stimulation vs., 395
  of gastrointestinal tract, 398–399
  of lower limb, 400
  micturition evaluations, 399–400
  spinal cord injury applications
    analgesia, 400–402
    bladder, 399–400
    focal magnetic stimulation, 403
    functional magnetic ventilation, 397–398
    gastrointestinal tract, 398–399
    lower limb, 400
    magnetic coils
      computer modeling of, 403
      design of, 403–404
      double, 403
      for expiratory muscles, 403
      placement of, 393–394
      for upper intercostal muscles, 403
    overview of, 393–394
    respiratory muscles, 394–397, 402–403
    technical considerations, 402–405
Functional magnetic ventilation, 397–398
F waves
  description of, 50
  peripheral conduction time estimations using, 86–87

G

Gastric pressure, 387
Gastrointestinal tract, 398–399
Gauss's law, 1, 5
Guillain-Barré syndrome, 112, 112f

H

Hand dystonia, repetitive transcranial magnetic stimulation for
  after-effects of, 233–235
  description of, 232–233
  physiological rationale for, 233
Hand muscles, central motor conduction studies
  description of, 85
  suspended abnormality, 112, 112f

Headache, transcranial magnetic stimulation-
    related, 75
Hearing, repetitive transcranial magnetic
    stimulation effects on, 75–76
Hemimasticatory spasm, 130
Hemispherectomy, 150
Hemispheric infarction, 135
Hemispheric lesions, 150
Hereditary spastic paraplegia
    central motor conduction in, 96
    motor evoked potentials evaluation of, 118
High-frequency repetitive transcranial magnetic
    stimulation
    cognition effects, 68
    cognition evaluations, 284, 296–297
    corticospinal excitability effects of, 225
    description of, 64t, 64–65
    epilepsy findings, 263–265
H reflexes, 34, 35f, 50
Huntington's disease
    central motor conduction time prolongation in,
        98, 188
    motor threshold in, 188
    paired-pulse studies in, 188–189
    silent period in, 188–189
Hypothalamic-pituitary-adrenocortical system,
    repetitive transcranial magnetic
    stimulation effects on, 72
Hypothyroidism-related ataxia, 206

I

Inconsistent orientation of coils, 36, 36f
Induced current density, 26–27
Induced electric field
    definition of, 10, 26
    electrical conductivity boundaries on spatial
        distribution of, 12–14
    spatial distribution of, 12–14
Inductance, 27
Inflammatory lumbosacral multiradiculopathy,
    124, 125f
Interhemispheric inhibition, 55, 258
Interstimulus intervals, 182
Intracarotid amobarbital test, 271, 275
Intracortical excitability, 431
Intracortical facilitation
    antiepileptic drugs' effect on, 263
    definition of, 54
    in epilepsy, 259–260
    long-interval, 254
    repetitive transcranial magnetic
        stimulation and, 65
    short-interval, 53–54, 54f
    after stroke, 245

threshold for, 54
Intracortical inhibition
    definition of, 182
    in dystonia, 187
    long-interval, 54–55, 254
        antiepileptic drugs' effect on, 263
    loss of, 245
    in multiple system atrophy, 185
    in obsessive-compulsive
        disorder, 189
    in Parkinson's disease, 182–183
    short-interval
        antiepileptic drugs' effect on, 263
        description of, 51–53, 53f, 254
        in epilepsy, 259
    after stroke, 245
Intracranial demyelinating disorder, dorsal cord
    compression vs., 114–115
Intracranial electrical stimulation, 281
Intraoperative monitoring of corticospinal function
    anesthesia, 375–376
    compound muscle action potentials, 368,
        375–377
    cost-effectiveness of, 365
    description of, 365–368
    electrical stimulation for, 369–370
    epidural volleys, 373–374, 377
    ideal technique for, 367–368
    single transcranial electric stimuli, 377
    stimulation site, 369
    transcranial magnetic stimulation for
        corticospinal volley, 373–374
        electrical stimulation vs., 369–370
        recordings, 373–375
        reproducibility of potentials, 376
        stimulating electrodes, 370
        stimuli, 370
        stimulus parameters, 371–373
        stimulus trains, 371
        technique, 376–377
Ipsilateral motor cortex, 172–173
Ipsilateral motor evoked potentials, 166–167,
    244–245, 431
Ipsilateral silent period, 88
Ischemic nerve block, 67, 147
Isoflurane, 375
I1 wave, 52
I-wave facilitation, 263
I waves, 43–45, 44f, 47

L

Language
    definition of, 271
    lateralization of, 274–276

motor cortex excitability effects, 276–278
processing of
  disruption of, 271–272
  facilitation of, 272–274
  transcranial magnetic stimulation effects on, 271–274
Laplace's equation, 5, 20
Laryngeal superior muscles, 133t, 135–136
Lateralization, 129
Learning
  cortical excitability changes with, 293–294
  motor, 144–146
  neuronal activity and excitability changes, 334
Lenz's law, 11
Long axons, 17–18
Long-interval intracortical facilitation
  description of, 254
  in epilepsy, 260
Long-interval intracortical inhibition
  antiepileptic drugs' effect on, 263
  description of, 54–55, 57–58, 254
  in epilepsy, 260
Long-term depression, 143
Long-term potentiation, 143
Lorazepam, 52–53, 144, 147
Low-frequency repetitive transcranial magnetic stimulation
  cognition evaluations, 284
  cortical activity affected by, 68
  description of, 64t
  epilepsy findings, 263–264
  visual cortex excitability effects, 70
Lumbar stenosis, 116, 116f, 122
Lumbosacral cord lesions, 115

M

Magnetic fields
  description of, 8–9
  time-varying, 9–10
Magnetic flux density, 7–9
Magnetic paravertebral stimulation, 121–122
Magnetic seizure therapy, 321
Magnetic stimulation
  advantages of, 393
  cerebellar
    in ataxia patients, 204–208
    limitations of, 208–209
    in nonataxia patients, 204–205
    in normal subjects
      cerebellar-cerebral interactions, 201–204
      descending volleys, 199
      description of, 197
      facilitatory effect, 202f, 201
      inhibitory effect, 197–201
      studies of, 208
  cerebral cortex applications, 395
  coils for, 32
  description of, 1, 31, 393
  electrical stimulation vs., 31–32, 393
  focal, 403
  motor evoked potentials application of, 31
  nerve root activation by, 38–39
  peripheral nerve activation by, 33–38
  at skull base, 39
  technical considerations of, 31–33
  transcranial. *See* Transcranial magnetic stimulation
  unilateral, 383–384
Magnetic vector potential, 7–9
Mania, 318
Masseter motor evoked potentials, 130–131, 132t
Maximal sustained contractions, 212–213
Memory
  emotion effects on, 289
  working, 276
Micturition, 399–400
Migraine
  familial hemiplegic, 411
  neural dysfunction associated with, 411
  transcranial magnetic stimulation studies of
    analysis, 415–416
    description of, 411–412
    motor cortex, 412, 413t
    visual cortex, 412, 414t, 415
Mirror movements, in children, 432–433
Monohemispheric lesions, 241–246
Monophasic pulses, 62f
Monophasic stimulators, 50
Monophasic waveforms, 32–33
Monosynaptic corticomotoneural projections, 166–167
Mood, repetitive transcranial magnetic stimulation effects on, 71–72
Motion perception, 333–334
Motoneuron disease
  high cervical cord lesions vs., 110–111
  suspended abnormalities in, 111
Motor attention, 174
Motor behavior
  complex movements, 173–174
  excitability buildup for, 171–172
  pointing movements, 173
  sequential movements, 173–174
  simple movements, 171–173
  translational movements, 173
Motor cortex
  abstract cognitive operations in, 176
  in ataxic patient, 204–208
  cerebellar suppression of, 205
  definition of, 165

Motor cortex (*Continued*)
  excitability
    description of, 253
    in epilepsy, 255–260
    injury/illness effects on, 294–295
    language and, 276–278
    learning effects on, 293–294
    measures of, 253–255
    in migraine, 416
    Parkinson's disease effects on, 295
    pharmacological agents' effect on, 295
    stroke effects on, 295
  functional magnetic resonance imaging of, 69–70
  hand area
    transcranial electrical stimulation of, 45–46
    transcranial magnetic stimulation of, 46–48
  ipsilateral, 172–173
  in migraine, 412, 413t
  plasticity of, 67
  premotor pathways of
    afferent pathways to primary motor cortex, 170–171
    cerebellothalamocortical pathway, 170
    description of, 169
    premotor-primary motor cortical pathway, 170
    transcallosal pathway from homologous motor cortex, 169–170
  primary
    afferent pathways to, 170–171
    contralateral, 172–173
    definition of, 165
    perception of movement by, 174
    transcallosal pathway from, 169–170
    transcranial magnetic stimulation of, 166–167
    for repetitive transcranial magnetic stimulation, 62
  short-interval afferent inhibition of, 55–56, 56f
  transcranial direct current stimulation effects on, 338
  transcranial magnetic stimulation of, 144, 412
Motor evoked potentials
  abductor digit minimi, 123f
  abductor hallucis, 115f, 117f, 124f
  abnormalities, 90–91
  absent response, 91
  during activation, 49
  adrenoleukodystrophy findings, 119
  age-related influences on, 304–306
  amplitudes
    in children, 430
    degradation of, 91
    description of, 49–50, 85–86
    in multiple sclerosis, 93
    reduction of, 90
  amyotrophic lateral sclerosis findings, 95
  androgens' effect on, 306
  cerebellar effects of, 56–57
  cerebellar findings, 208
  in children, 430
  in chronic fatigue syndromes, 218
  compound muscle action potentials vs., 48, 85
  compressive myelopathy findings, 96–97
  cortical excitability effects, 49–50
  cortical silent period and, 87–88, 254
  corticomotor threshold, 88
  definition of, 83
  demyelinating lesions' effect on, 91
  dystonia findings, 186
  epilepsy findings, 258
  estradiol effects on, 306
  exercise-related decreases in, 215
  facial nerve, 134
  genetic factors, 304
  hereditary spastic paraplegia findings, 118
  hormonal effects on, 306–307
  Huntington's disease findings, 188
  inherent variability of, 85
  intracranial demyelinating disorder findings, 114–115
  in intramedullary lesions, 97
  ipsilateral, 166–167, 244–245, 431
  laryngeal, 136f
  latency of
    in amyotrophic lateral sclerosis, 157
    body position effects on, 86
    in children, 430
    description of, 48, 49f, 85–87
  lower upper limb muscles, 86f
  masseter, 130–131, 132t
  masticatory muscles, 130–131
  multifocal motor neuropathy findings, 124, 124f
  in multiple sclerosis
    fatigue, 217–218
    findings, 91, 93
  myelitis evaluations, 119–120
  neck muscles, 136–137
  in neurologically normal patients with behavioral disorders, 304–305
  paired-pulse, 305–307
  paravertebral, 114–115
  Parkinson's disease findings, 181–182
  primary lateral sclerosis findings, 112–113
  psychogenic weakness findings, 98
  repetitive transcranial magnetic stimulation effects on, 63–65
  at rest, 49
  scalp-to-brain distance on, 304
  sex differences in, 306–307
  short-latency, 253–254

size of
  antiepileptic drugs' effect on, 262–263
  contralateral motor cortex stimulation effects on, 258
  description of, 253
  in epilepsy, 258
  peripheral nerve stimulation effects on, 258
spinal cord injury evaluations, 120–121
spinal cord vascular disorders evaluated using, 118
spinal excitability effects, 49–50
sternocleidomastoid, 137
stroke findings, 97, 241
surface electrodes for, 155–156
syringomyelia findings, 118, 119f
tongue muscles, 133t, 137–139
trapezius, 137
for upper limb muscles, 86f
variability of, 303–305
voluntary background contraction effects on, 83–84
Wilson's disease findings, 98
Motor intention, 174
Motor learning
  description of, 144–146
  somatosensory input for, 146–147
Motor root stimulation, 86f
Motor system
  action observation and action execution matching system, 175–176
  corticofugal pathways of
    contralateral responses, 168–169
    corticospinal projections targeting inhibitory spinal interneurons, 169
    monosynaptic corticomotoneural projections, 166–167
    oligosynaptic corticomotoneuronal projections, 167–168
  description of, 165–166
  higher cortical motor functions, 174–176
  pathways of, 166t
  premotor cortical pathways of
    afferent pathways to primary motor cortex, 170–171
    cerebellothalamocortical pathway, 170
    description of, 169
    premotor-primary motor cortical pathway, 170
    transcallosal pathway from homologous motor cortex, 169–170
Motor thresholds
  active
    antiepileptic drugs effect on motor excitability in epileptic patients, 261
    description of, 49, 52, 253
    in epilepsy, 255–258
    in ataxia, 190
    corticobasal degeneration findings, 185
    definition of, 253
    dystonia findings, 186
    in Huntington's disease, 188
    in Parkinson's disease, 181
  resting
    in amyotrophic lateral sclerosis, 156–157
    antiepileptic drugs effect on motor excitability in epileptic patients, 261
    definition of, 253
    description of, 49, 63
    in epilepsy, 255–258
    variability of, 303–304
  sleep deprivation effect on, 420
  variability of, 303–305
Motor training, 146
Motor volition, 175
Movement(s)
  complex, 173–174
  imagery used in, 240
  pointing, 173
  sequential, 173–174
  simple, 171–173
  translational, 173
Movement disorders
  hand dystonia. *See* Hand dystonia
  overview of, 223–224
  Parkinson's disease. *See* Parkinson's disease
  repetitive transcranial magnetic stimulation for, 235
M response, 108
Multifocal motor neuropathy, motor evoked potentials in, 124, 124f
Multiple sclerosis
  afferent evoked potentials studies in, 94
  central motor conduction abnormalities in, 91
  central motor conduction time in, 91, 93
  fatigue associated with, 217–218
  motor evoked potentials in
    description of, 91, 93
    sensitivity of, 94
    triple-stimulation, 94
  progressive
    central motor conduction time in, 94
    motor evoked potentials in, 93
    spinal motor conduction time in, 93
  relapsing-remitting, spinal motor conduction time in, 93
  visual evoked potentials for, 94
Multiple system atrophy
  central motor conduction in, 98
  central motor conduction time prolongation in, 185
  cerebellar signs associated with, 207
  intracortical inhibition in, 185

Myelitis
  motor evoked potentials evaluation of, 119–120
  spinal cord injury, 120–121
  subacute combined degeneration, 120
Myelopathy
  central motor conduction time evaluations, 106, 108
  cervical spondylotic
    causes of, 109
    description of, 108
    dorsal cord compression, 113–115
    high cervical cord lesion, 109–110
    lumbosacral cord lesions, 115
    motoneuron disease, 110–111
    motor evoked potentials studies, 108–118
    with pure spastic paraparesis, 109
    spinal cord involvement evaluations, 113, 114f
  compressive
    central motor conduction time prolongation in, 96–97
    motor evoked potentials evaluation of, 96–97
  definition of, 106
  incidence of, 105
  motor evoked potentials studies, 106, 107t
  neurophysiological evaluation for, 105–106
Myeloradicular lumbosacral lesion, 115–116
Myeloradiculitis, 119
Myoclonus, 189

N

Naloxone, 401
Narcolepsy, 423
Near-monophasic waveform, 32, 32f
Neck muscles, 136–137
Needle electrodes, for electromyography, 386
Nerve growth factor, 143
Nerve roots, activation of, 38–39
Neurogenic bowel, 398
Neuroticism, 307
Neurotoxicity, repetitive transcranial magnetic stimulation and, 74–75
Non-REM sleep, 419

O

Obsessive-compulsive disorder, 189, 307
Ohm's law, 1, 6–7
Oligosynaptic corticomotoneuronal projections, 167–168
Orbicularis oris, 131, 134
Orientations of coils, 36f, 36–37

P

Pain
  chronic, 321
  functional magnetic stimulation for, 400
Paired associative stimulation, 186
Paired-pulse transcranial magnetic stimulation
  antiepileptic drugs
    active motor thresholds, 261
    cortical silent period duration, 263
    description of, 261
    intracortical facilitation, 263
    long-interval intracortical inhibition, 263
    motor evoked potential size, 262–263
    resting motor thresholds, 261
    short-interval intracortical inhibition, 263
  ataxia findings, 190
  cerebellocortical inhibition, 56–57
  in children, 431–432
  contralesional extinction affected by, 345
  cortical excitability induced using, 313
  definition of, 254
  description of, 22–23
  dystonia evaluations, 187
  epilepsy findings
    active motor thresholds, 255–258
    cortical silent period duration, 258–259
    description of, 256t–257t
    intracortical facilitation, 259–260
    long-interval intracortical facilitation, 260
    long-interval intracortical inhibition, 260
    motor evoked potential size, 258
    resting motor thresholds, 255–258
    safety of, 260–261
    short-interval intracortical inhibition, 259
  Huntington's disease findings, 188–189
  interactions, 51–58
  interhemispheric inhibition, 55
  intracortical facilitation, 54
  long-interval intracortical inhibition, 54–55, 57–58
  motor evoked potential measures, 305–307
  of parietal cortex, 343
  Parkinson's disease evaluations, 182–183
  premotor-primary motor cortex inhibition, 56, 57f
  repetitive magnetic resonance stimulation effects on cortical inhibition measured using, 66
  short-interval afferent inhibition of the motor cortex, 55–56
  short-interval intracortical facilitation, 53–54, 54f
  short-interval intracortical inhibition, 51–53, 53f

Parietal cortex, 342
Parietal eye field, 349–350
Parkinsonian syndromes, 185
Parkinson's disease
  central motor conduction time in, 181–182
  cortical excitability effects of, 295
  corticospinal motor dysfunction in, 225–226
  deep brain stimulation, 184–185
  intracortical inhibition in, 182–183
  motor evoked potentials in, 181–182
  motor impairment in, 225–226
  motor threshold in, 181
  movement preparation and execution in, 183
  paired-pulse studies in, 182–183
  repetitive transcranial magnetic stimulation for
    after-effects of, on motor symptoms, 226–231
    conditioning effects, 228t–229t
    description of, 183–184
    physiological rationale for, 225–226
    short-term effects of, 231
  rigidity associated with, 182
  silent period in, 182, 185
Peak magnetic energy in the coil, 27
Periodic limb movements in sleep, 421–423, 424t
Peripheral conduction time, 86, 108
Peripheral nerves
  activation of, 33–38, 121–122
  lesions of, 121
  magnetic paravertebral stimulation of, 121–122
  motor axons of, 121
Pharyngeal muscles, 133t, 135–136
Phosphenes, 329–332, 412
Phosphene thresholds, 335
Phrenic nerve
  anatomy of, 381–382
  anterior (presternal) magnetic stimulation of, 384
  bilateral anterior magnetic stimulation of, 384
  central stimulation modes, 384–385
  double stimulation mode, 385
  electrical stimulation of, 382
  functional magnetic stimulation of, 395–396
  magnetic stimulation of
    cervical
      advantages, 382–383
      description of, 382
      disadvantages, 383
    contraindications, 388
    indications, 388
    normal values, 388, 389t
    peripheral, 382
    repetitive, 385
    unilateral, 383–384
    transcutaneous magnetic cortical stimulation of, 384
Phrenic nerve conduction time, 384–385
Placebo coils, 25
Plasticity
  central nervous system
    cross-modal, 148–149
    description of, 143
    evaluation of, 143–144
    muscarinic receptor function in, 144–145
    use-dependent, 144
  compensatory, 239
  cortical, 67
  motor cortex, 67
  neural, 240
Pointing movements, 173
Positron emission tomography, 281–282, 290
Post-polio syndrome, 218–219
Post-traumatic stress disorder, 320
Power consumption, 27
Premotor cortical pathways
  afferent pathways to primary motor cortex, 170–171
  cerebellothalamocortical pathway, 170
  description of, 169
  premotor-primary motor cortical pathway, 170
  transcallosal pathway from homologous motor cortex, 169–170
Premotor-primary motor cortex inhibition, 56, 57f
Premotor-primary motor cortical pathway, 170
Presynaptic inhibition, 169
Primary lateral sclerosis
  central motor conduction in, 96
  motor evoked potentials findings, 112–113
  spinal cord compression vs., 112–113
Primary motor cortex
  afferent pathways to, 170–171
  contralateral, 172–173
  definition of, 165
  perception of movement by, 174
  transcallosal pathway from, 169–170
  transcranial magnetic stimulation of, 166–167
Primary motor cortex isolation, 205
Primary peak, 159f, 159–160
Progressive supranuclear palsy
  central motor conduction in, 98
  central motor conduction time prolongation in, 185
Psychogenic weakness
  central motor conduction findings, 98
  motor evoked potentials in, 98
Pulmonary function tests, 394
Pure spastic paraparesis, 109
Purkinje cells, 203

Q

Quadruple-pulse devices, 22
Quasistatic approximation, 11–12

R

Radiculopathy
  cervical, 122
  description of, 121–122
  diagnosis of, 105
  inflammatory lumbosacral multiradiculopathy, 124, 125f
  neurophysiological evaluation for, 105–106
Random number generation, 285
Rapid eye movements, 419
Rapid-rate stimulation devices, 23
Rectus femoris, motor evoked potentials of, 117f
Regional cerebral blood flow, 69, 234, 276
Repetitive transcranial magnetic stimulation
  after-effects of
    description of, 224–225
    in hand dystonia, 233–235
    in Parkinson's disease
      corticomotor excitability, 231–232
      motor symptoms, 226–231
  antidepressive effects of, 71–72
  ataxia findings, 190
  barriers to use, 235
  brain-derived neurotrophic factor levels and, 72
  cerebellar studies, 208
  cerebral blood flow measurements using, 69
  cerebral dysfunction and, 223
  of cerebrum, 189
  cognition applications
    deficits, 295–297
    description of, 68–69, 74
  contraindications, 76
  contralateral cortex effects, 66–67
  cortical excitability effects, 224
  cortical inhibition effects, 65–66
  definition of, 263, 329
  depression applications of, 71, 219
  description of, 329–330
  dorsolateral prefrontal cortex effects, 68
  dystonia applications
    findings, 187–188
    hand dystonia, 232–235
  electroencephalography and, 70–71, 76
  epilepsy applications
    description of, 263
    high-frequency rTMS, 263–265
    low-frequency rTMS, 263–264
    safety, 265
  eye movement studies using, 360–361
  facilitatory effects of, 295–297
  focal, 224
  functional state effects on impact of, 225
  in hand dystonia
    after-effects of, 233–235
    description of, 232–233
    physiological rationale for, 233
  hearing effects of, 75–76
  high-frequency
    cognition effects, 68, 284, 296–297
    corticospinal excitability effects of, 225
    description of, 64t, 64–65
  history of, 61
  hypothalamic-pituitary-adrenocortical system response affected by, 72
  imaging studies and, 69–70
  intracortical facilitation and, 65
  language processing effects of, 271–272
  low-frequency
    cognition evaluations, 284
    cortical activity affected by, 68
    description of, 64t
    visual cortex excitability effects, 70
  magnetic resonance imaging and, 62
  mid-dorsolateral prefrontal cortex applications, 70
  monitoring during, 76
  mood effects of, 71–72
  motion perception effects, 334
  motor cortex
    contralateral, 66–67
    description of, 62
    excitability, 64t
    plasticity of, 67
  motor evoked potentials effect, 63–65
  motor physiology of, 63–67
  movement disorders application of, 235
  neuroendocrine axis affected by, 75
  neurotoxicity caused by, 74–75
  paired-pulse inhibition and facilitation effects, 66
  Parkinson's disease findings, 183–184
  principles of, 61–63
  saccadic eye movements affected by, 351–352
  safety of, 72–76, 265
  seizures induced by, 72–74, 265
  sensory system effects, 70
  silent period effects, 65, 66t
  sleep applications, 423, 425–426
  speech areas affected by, 68
  speech production interference caused by, 274
  stimulating devices for, 223
  tremor evaluations, 189
  virtual lesions induced by, 144
  visual cortex, 415

visual system effects, 70
working memory effects, 276
Respiratory muscles
    critical care neuromyopathy, 388
    electromyography for, 385–387
    expiratory functions, 396, 404
    facilitated inspiration, 387f
    force measurements, 388
    functional electrical stimulation of, 395
    functional magnetic stimulation of, 394–397, 402–403
    function of, 381
    pressure measurements, 387–388
    in spinal cord injury patients, 394–397
Resting motor thresholds
    in amyotrophic lateral sclerosis, 156–157
    antiepileptic drugs effect on motor excitability in epileptic patients, 261
    definition of, 253
    description of, 49, 63
    in epilepsy, 255–258
    variability of, 303–304
Restless legs syndrome, 421–423, 424t
Rett's syndrome, 433
Root motor conduction time, 108

S

Saccadic eye movements
    description of, 349–350, 353
    double-step, 355
    focal transcranial magnetic stimulation effects on, 360
    intentional, 354
    predictive, 354
    reflexive, 354
    visual space across, 354
Schizophrenia, 320
Seizures
    electroconvulsive therapy, 321
    magnetic seizure therapy, 321
    repetitive transcranial magnetic stimulation-induced, 72–74, 265
    transcranial magnetic stimulation and, 260, 321
Selective attention, 290
Sensory cortex, repetitive transcranial magnetic stimulation effects on, 70
Sequential movements, 173–174
Sevoflurane, 375
Sham coils, 25, 25f
Sham stimulation, 284
Short-interval afferent inhibition of the motor cortex, 55–56, 56f
Short-interval intracortical facilitation, 53–54, 54f

Short-interval intracortical inhibition
    antiepileptic drugs' effect on, 263
    description of, 51–53, 53f, 254
    in epilepsy, 259
Siemens per meter, 6
Silent period
    antiepileptic drugs' effect on, 263
    apomorphine effects on, 185
    in ataxia, 190
    cortical, 87–88
    definition of, 254
    description of, 65, 66t
    in dystonia, 186–187
    in epilepsy, 258–259
    in Huntington's disease, 188–189
    ipsilateral, 88
    in Parkinson's disease, 182, 185
Single-pulse transcranial magnetic stimulation
    antiepileptic drugs
        active motor thresholds, 261
        cortical silent period duration, 263
        description of, 261
        intracortical facilitation, 263
        long-interval intracortical inhibition, 263
        motor evoked potential size, 262–263
        resting motor thresholds, 261
        short-interval intracortical inhibition, 263
    in children, 430–431
    coils used in, 83, 84f
    description of, 22, 49
    epilepsy findings
        active motor thresholds, 255–258
        cortical silent period duration, 258–259
        description of, 256t–257t
        intracortical facilitation, 259–260
        long-interval intracortical facilitation, 260
        long-interval intracortical inhibition, 260
        motor evoked potential size, 258
        resting motor thresholds, 255–258
        safety of, 260–261
        short-interval intracortical inhibition, 259
    somatosensory cortex effects, 342
    uses of, 83, 84f
Skin depth, 11–12
Skull base, magnetic stimulation at, 39
Sleep
    motor response amplitude during, 420, 420t
    physiological role of, 420
    physiology of, 419–421
    rapid eye movements, 419
    repetitive transcranial magnetic stimulation in, 423, 425–426
    slow wave, 419
    transcranial magnetic stimulation of, 419–421
Sleep apnea, 423
Sleep deprivation response, 317

Sleep disorders
  in depression, 425
  narcolepsy, 423
  periodic limb movements in sleep, 421–423, 424t
  restless legs syndrome, 421–423, 424t
  sleep apnea, 423
Slinky coils, 25–26
Slow wave sleep, 419
Somatosensory evoked potentials, 70, 342
Somatosensory input, 146–148
Somatosensory system
  deficits of, 344–347
  description of, 341
  tactile perception
    in normal subjects, 341–344
    in somatosensory deficit patients, 344–347
  transcranial magnetic stimulation applications to, 341
Spatial attention, 344
Spatial neglect, 292
Speech production, 274
Spinal cord
  cervical spondylotic myelopathy findings, 113, 114f
  compression of, 112–113
  genetic disorders that affect, 118–119
Spinal cord injuries
  alveolar ventilation in, 394
  bladder abnormalities in, 399
  bowel management in, 398
  description of, 393
  functional magnetic stimulation applications
    analgesia, 400–402
    bladder, 399–400
    focal magnetic stimulation, 403
    functional magnetic ventilation, 397–398
    gastrointestinal tract, 398–399
    lower limb, 400
    magnetic coils
      computer modeling of, 403
      design of, 403–404
      double, 403
      for expiratory muscles, 403
      placement of, 393–394
      for upper intercostal muscles, 403
    overview of, 393–394
    respiratory muscles, 394–397, 402–403
    technical considerations, 402–405
  functional outcome evaluations, 121
  gastric emptying in, 399
  motor evoked potentials evaluation of, 120–121
  neurogenic bowel in, 398
  pulmonary function tests in, 394
  respiratory muscles in, 394–397
  vital capacity in, 394
Spinal cord stimulation, 400–401
Spinal cord vascular disorders, motor evoked potentials study of, 118
Spinocerebellar ataxia type 1, 97
Static electric charge, 1
Steady currents, 6–7
Stereotactic transcranial magnetic stimulation, 23–24
Sternocleidomastoid, 133t, 136–137
Strength-duration curve, 27
Stroke
  acute neuronal failure in, 241
  central motor conduction abnormalities in, 97
  constraint-induced movement therapy for, 246
  cortical excitability effects of, 295
  cortical silent period in, 97–98
  factors that affect, 241
  hyperactivation patterns after, 247
  intracortical inhibition and facilitation after, 245
  ipsilateral motor evoked potentials in, 245
  monohemispheric lesions and, 241–246
  motor evoked potentials findings, 97, 241
  motor output after, 241–243, 246
  recovery from, 247
  summary of, 246–247
Subacute combined degeneration, 120
Submaximal contractions, 213–214
Superoxide dismutase 1, 155
Supplementary eye field, 350
Supplementary motor area, 165, 204
Suprahyoid muscle, 132t
Supramaximal stimulation, 88
Swallowing, 135
Syringomyelia, motor evoked potentials evaluation of, 118, 119f

T

Tactile inhibition, 344
Tactile perception
  in normal subjects, 341–344
  in somatosensory deficit patients, 344–347
Thoracic spinal nerves, 125
Thyristors, 26
Tibialis anterior, motor evoked potentials of, 115f, 117f
Tics, 189–190
Time-varying magnetic fields, 9–10
Todd's paresis, 258
Tongue muscles, 133t, 137–139
Total electric field, 12
Tourette's syndrome, 189–190, 434
"Train-of-four" stimuli, 385

Transcallosal pathway from homologous motor cortex, 169–170
Transcranial direct current stimulation
  animal studies of, 334–335
  contrast perception impaired by, 336
  cortical excitability modulations induced by, 337
  description of, 235
  human applications of, 335
  motor cortex effects, 338
  transcranial magnetic stimulation vs., 330
  of V5, 336–337
  visual perception studies, 335–336
  visuomotor behavior modification by, 336–337
Transcranial electrical stimulation
  development of, 329
  D waves produced by, 47
  electric field in, 19–20
  fields, 21–22
  locus of activation in, 18–19
  of motor cortex hand area, 45–46
  motor evoked potential latencies, 50
  principles of, 18f, 19
  repetitive discharges from, 48
  spinal H reflexes combined with, 50
  three-dimensional focusing, 21
  transcranial magnetic stimulation vs., 20–22, 62
Transcranial magnetic stimulation
  advantages of, 281
  antidepressant uses of
    description of, 315–318
    mechanisms, 319–320
  anxiety disorders treated with, 320–321
  axon bends in, 19
  blood oxygen level-dependent functional magnetic resonance imaging and, 283, 312–313
  characteristics of, 21–22
  chronic pain treated with, 321
  cognitive function applications of, 281
  coils
    circular, 23, 50, 83, 84f
    cone, 23
    construction of, 26
    cooling of, 25
    8-shaped, 23, 23f, 32f, 33, 50
    with ferromagnetic core, 23–24
    forces that affect, 28
    inductance of, 27
    peak magnetic energy in, 27
    sham, 25, 25f
    slinky, 25–26
    tilting of, 24f
    warming of, 27–28
  definition of, 240
  depression treated with, 314–320
  descending corticospinal volleys produced by, 48
  description of, 281
  development of, 281, 329
  double-pulse, 352
  double-shock, 170
  dual-pulse, 23
  D waves produced by, 47
  electrical field in, 20
  electroconvulsive therapy compared with, for depression, 316–317
  electromyographic response to, 49
  electronics of, 26
  facial nerve stimulation by, 87
  fields, 21–22
  focality of, 28, 28f
  functional imaging and, 312–313
  headache secondary to, 75
  hemispheric lesions evaluated using, 150
  history of, 281
  intracranial electrical stimulation compared with, 281
  laryngeal muscles, 133t, 135–136
  of left frontal cortex, 344
  of leg area, 50–51
  locus of activation in, 18–19
  motor cortex hand area, 46–48
  motor evoked potential latencies, 50
  orbicularis oris, 131, 134
  orientation sensitivity of, 47
  paired-pulse
    cerebellocortical inhibition, 56–57
    definition of, 254
    description of, 22–23
    interactions, 51–58
    interhemispheric inhibition, 55
    intracortical facilitation, 54
    long-interval intracortical inhibition, 54–55, 57–58
    premotor-primary motor cortex inhibition, 56, 57f
    repetitive magnetic resonance stimulation effects on cortical inhibition measured using, 66
    short-interval afferent inhibition of the motor cortex, 55–56
    short-interval intracortical facilitation, 53–54, 54f
    short-interval intracortical inhibition, 51–53, 53f
  pharyngeal muscles, 133t, 135–136
  positron emission tomography and, 290
  post-traumatic stress disorder treated with, 320
  principles of, 18f, 20
  radial direction in, 20–21
  rapid-rate, 23

Transcranial magnetic stimulation (*Continued*)
  real, 230
  repetitive. *See* Repetitive transcranial magnetic stimulation
  repetitive discharges from, 48
  of right parietal cortex, 342
  schizophrenia treated with, 320
  seizures treated with, 321
  single-pulse
    antiepileptic drugs
      active motor thresholds, 261
      cortical silent period duration, 263
      description of, 261
      intracortical facilitation, 263
      long-interval intracortical inhibition, 263
      motor evoked potential size, 262–263
      resting motor thresholds, 261
      short-interval intracortical inhibition, 263
    in children, 430–431
    coils used in, 83, 84f
    description of, 22, 49
    epilepsy findings
      active motor thresholds, 255–258
      cortical silent period duration, 258–259
      description of, 256t–257t
      intracortical facilitation, 259–260
      long-interval intracortical facilitation, 260
      long-interval intracortical inhibition, 260
      motor evoked potential size, 258
      resting motor thresholds, 255–258
      safety of, 260–261
      short-interval intracortical inhibition, 259
    somatosensory cortex effects, 342
    uses of, 83, 84f
  sleep deprivation response and, 317
  sleep evaluations, 419–421
  sound during, 28
  speech production interference caused by, 274
  stereotactic, 23–24
  stimulators
    biphasic, 50
    circuitry of, 23f
    construction of, 26
    monophasic, 50
    types of, 22–24, 33
  stimulus pulse generation, 22
  therapeutic uses of
    animal studies, 312
    depression, 314–320
    description of, 313–314
    limitations, 311–313
    mania, 318
  three-dimensional focusing, 21
  time-effects of, 342
  transcranial direct current stimulation vs., 330
  transcranial electrical stimulation vs., 20–22, 62

Transcranial stimulation
  electrophysiologic basis of, 17
  long axons, 17–18
  strength-duration relationship, 18
Transcutaneous magnetic cortical stimulation, 384
Transient forearm deafferentation, 147
Translational movements, 173
Transneuroforaminal electric field, 38, 38f
Trapezius, 133t, 136–137
Tremor, 189
Triple-stimulation technique
  amyotrophic lateral sclerosis evaluations, 95
  description of, 86, 86f, 91
TSTtest, 86

U

Unified Parkinson's Disease Rating Scale, 184
Unilateral magnetic stimulation, 383–384
Unmasking, 143

V

V1, 331
V5, 336–337
Vector quantity, flux of, 4–5
Verbal working memory, 276
Visual cortex
  excitability of, 331–332
  in migraine, 412, 414t, 415
  plastic changes of, 331–332
  repetitive transcranial magnetic stimulation of, 415
  transcranial magnetic stimulation of, 412, 414t, 415
Visual evoked potentials
  multiple sclerosis evaluations, 94
  transcranial direct current stimulation effects, 335
Visual perception
  description of, 329–330
  phosphenes, 329–332
  transcranial direct current stimulation studies, 334–337
  transcranial magnetic stimulation effects on
    color perception suppression, 332–333
    description of, 329
    motion perception suppression and improvement, 333–334
    phosphenes, 329–332
Visuomotor behavior, 336–337
Vital capacity, 394
Vitamin B12 deficiency, 120

W

Waveforms
  biphasic, 32, 32f
  monophasic, 32–33
  near-monophasic, 32, 32f

Wilson's disease
  central motor conduction in, 98, 185–186
  dentate nucleus in, 207
  magnetic resonance imaging findings, 207f
  motor evoked potentials in, 98, 185–186
Working memory, 276